LAND RECLAMATION IN ECOLOGICAL FRAGILE AREAS

PROCEEDINGS OF THE 2ND INTERNATIONAL SYMPOSIUM ON LAND RECLAMATION AND ECOLOGICAL RESTORATION (LRER 2017), 20–23 OCTOBER 2017, XI'AN, P.R. CHINA

Land Reclamation in Ecological Fragile Areas

Chief Editor

Zhenqi Hu

College of Geoscience and Surveying Engineering, China University of Mining and Technology, Beijing, P.R. China

CRC Press is an imprint of the
Taylor & Francis Group, an **informa** business

A BALKEMA BOOK

CRC Press/Balkema is an imprint of the Taylor & Francis Group, an informa business

Typeset by V Publishing Solutions Pvt Ltd., Chennai, India
Printed and bound in Great Britain by CPI Group (UK) Ltd, Croydon, CR0 4 YY

Published by: CRC Press/Balkema
Schipholweg 107C, 2316 XC Leiden, The Netherlands
e-mail: Pub.NL@taylorandfrancis.com
www.crcpress.com – www.taylorandfrancis.com

ISBN: 978-1-138-05103-4 (Hbk)
ISBN: 978-1-315-16658-2 (eBook)

Land Reclamation in Ecological Fragile Areas – Hu (Ed.)
© 2017 Taylor & Francis Group, London, ISBN 978-1-138-05103-4

Table of contents

Land Reclamation in Ecological Fragile Areas – Hu (Ed.)
© 2017 Taylor & Francis Group, London, ISBN 978-1-138-05103-4

Preface

The combination of rapid urbanization and industrialization has boosts China's economy greatly, however, extensively human activities, including mining brings series negative effects on land and environment. Land reclamation is an effectively to mitigate and reverse those negative impacts. Most of China's northern and western regions were ecological fragile areas, where the climate is semi-arid and arid, and were eager to be influenced by human activities. This book is the proceeding of the 2nd International Symposium on Land Reclamation and Ecological Restoration (LRER 2017) which is a series international conference on land reclamation every three years organized by China Coal Society. The first one was Beijing International Symposium on Land Reclamation and Ecological Restoration (LRER 2014), which has promoted the communication of land reclamation and ecological restoration technology and experience internationally, advanced and renovated of reclamation and ecological restoration technology and practice, accelerating international collaboration in this field. And LRER 2017 is in order to promote the land reclamation and ecological restoration in mining areas to meet the requirements of "The Belt and Road Initiative".

The papers were presented at the symposium and addressed a range of land reclamation and restoration issues reflect the current work from around the world, covering land reclamation in ecological fragile areas. A total of 18 topics included in following themes: 1) Mining and reclamation policies, regulations and standard; 2) Monitoring, prediction and assessment of mining impact on land environment; 3) Surface mined land reclamation and ecological restoration; 4) Subsidence land reclamation and ecological restoration; 5) Solid wastes management, waste dump and tailings pond restoration; and 6) Contaminated land remediation. We received about 120 papers, some of them are selected to be published on International Journal of Coal Science & Technology, which is a special issue for the symposium. The contents of this proceeding will be interest to engineers, scientists, consultants, government officials and students in this area.

LRER 2017 is supported by many organizations, China Coal Society has contributed greatly to the conference as the host. All the organizers have taken great efforts, including China University of Mining and Technology (Beijing), Xi'an University of Science and Technology, Chinese Ecological Restoration Network. The support from the co-organizers is highly appreciated. To be noted are: International Affiliation of Land Reclamationists, American Society of Mining and Reclamation, International Journal of Mining, Reclamation and Environment, International Journal of Coal Science & Technology, China Industry Alliance of Mine Environment Restoration, National Natural Science Foundation of China.

In this book, we hope to provide some information about the latest progress of land reclamation technologies, policies, and management, etc.

Professor Zhenqi Hu
Chair of the Organizing Committee,
International Academic Committee
China University of Mining and
Technology (Beijing)

Land Reclamation in Ecological Fragile Areas – Hu (Ed.)
© 2017 Taylor & Francis Group, London, ISBN 978-1-138-05103-4

Organizers

CONSULTATIVE COMMITTEE OF THE CONFERENCE

CHAIR

Wang Xianzheng, *President of China Coal Industry Association*

ADVISORY COMMITTEE MEMBERS:

Peng Suping, *Academician of Chinese Academy of Engineering, Professor of China University of Mining and Technology (Beijing)*
Zhang Tiegang, *Chief Engineer of Pingdingshan Coal Group Co., Ltd.*
Yuan Liang, *Deputy General Manager of Huainan Mining Group Co., Ltd.*
Gu Dazhao, *General Manager of Science and Technology Development Department, Shenhua Group*
Wu Qiang, *Academician of Chinese Academy of Engineering, Professor of China University of Mining and Technology (Beijing)*
Tian Hui, *Vice Chairman of China National Coal Association*
Liu Feng, *Vice Chairman of China National Coal Association*
Yang Renshu, *President of China University of Mining and Technology (Beijing)*
Jiang Yaodong, *Vice President of China University of Mining and Technology (Beijing)*
Raj Singhal, *Mining and Mineral Industry Consultant, Canada*
W. Lee Daniels, *International Affiliation of Land Reclamationists, USA*

ORGANIZATION COMMITTEE

CHAIR: Liu Feng, Yang Gengshe, Zhenqi Hu
SECRETARY GENERAL: Yao Wanqiang, Wang Lei
DEPUTY SECRETARY-GENERAL: Tang Fuquan, Xiao Wu
MEMBERS:
Bao Yuying, *Inner Mongolia University*
Chen Qiuji, *Xi'an University of Science and Technology*
Fan Tianli, *Orient Landscape Industry Group CO., Ltd, Beijing*
He Xiao, *Inner Mongolia Agricultural University*
Lei Mei, *Institute of Geographic Sciences and Natural Resources Research, Chinese Academy of Sciences, China*
Li Bo, *Chinese Ecological Restoration Network (www.ER-CHINA.com)*
Li Xinju, *Shandong Agricultural University*
Liu Jingjing, *Beijing Construction Environment Remediation CO., Ltd.*
Xu Liangji, *Anhui University of Science & Technology*
Yang Jianjun, *Xinjiang University*
Zhang Hebing, *Henan Polytechnic University*
Zhang Yu, *Green Mining United Institute*

INTERNATIONAL ACADEMIC COMMITTEE

HONORARY CHAIR

Peng Suping, *China University of Mining and Technology (Beijing), China*

CHAIR

Zhenqi Hu, *China University of Mining and Technology (Beijing), China*

CO-CHAIR

Raj Singhal, *International Journal of Mining, Reclamation and Environment, Canada*
W. Lee Daniels, *International Affiliation of Land Reclamationists, USA*
Robert G. Darmody, *American Institute of mining and reclamation, USA*

COMMITTEE MEMBERS

A. Hoppenstedt, *Technical University of Berlin, Germany*
A.K. Waitkus, *Environmental Consulting, USA*
A. Başçetin, *Istanbul University, Turkey*
Bai Zhongke, *China University of Geosciences (Beijing), China*
Bi Yinli, *China University of Mining and Technology (Beijing), China*
Bian Zhengfu, *China University of Mining and Technology, China*
C. Drebenstedt, *Freiberg University of Mining and Technology, Germany*
Chang Jiang, *China University of Mining and Technology, China*
Chen Gang, *University of Alaska at Fairbanks, USA*
Chen Tongbin, *Institute of Geographic Sciences and Natural Resources Research, Chinese Academy of Sciences, China*
C. Tsadilas, *National Agricultural Research Foundation, Greece*
C.M. Neculita, *Research Institute on Mines and Environment (RIME), University of Québec in Abitibi-Témiscamingue, Canada*
D. Mulligan, *The University of Queensland, Australia*
E. Topal, *Nazarbayev University, Kazakhstan*
E. Yilmaz, *Department of Mining Engineering, Karadeniz Technical University, Trabzon, Turkey*
Feng Zhongke, *Beijing Forestry University, China*
Gao Jixi, *Nanjing Institute of Environment Science, China*
Huang Jinlou, *Research Center for Eco-Environmental Sciences, Chinese Academy of Sciences, China*
J. Skousen, *West Virginia University, USA*
K. Tursyn, *Kazakh National Research Technical University, Kazakhstan*
Li Shuzhi, *China Coal Technology & Engineering Group, Tangshan Research Institute, China*
Luo Ming, *China Land Consolidation and Rehabilitation Center, Ministry of Land and Resources, China*
M.A. Naeth, *University of Alberta, Canada*
M. Osanloo, *Amirkabir University of Technology, Iran*
N. Chaturvedi, *The Energy and Resources Institute, India*
P.D. Stahl, *University of Wyoming, USA*
P.J. Beckett, *Laurentian University, Canada*
P. Sklenicka, *Czech University of Life Sciences Prague, Czech Republic*
P.P. Manca, *DICAAR - Department of Civil and Environmental Engineering and Architecture—University of Cagliari, Via Marengo 2, Cagliari, Italy*
S.K. Maiti, *Indian Institute of Technology, India*
S. Mišák, *Technical University of Ostrava, Czech Republic*
S.L. Smith, *British Land Reclamation Society, UK*
V. Litvinenko, *State University of Mineral Resources, Russia*
Wang Yunjia, *China University of Mining and Technology, China*
Wang Shuangming, *Shaanxi Geological Survey, China*

W. Wende, *Leibniz Institute of Ecological Urban and Regional Development, Dresden, Germany*
Xu Youning, *Xi'an Geological Survey Center of China Geological Survey, China*
Y.P. Chugh, *Southern Illinois University at Carbondale, USA*
Zhao Tingning, *Beijing Forestry University, China*
Zhao Yanling, *China University of Mining and Technology (Beijing), China*
Zhou Lianbi, *Department of Environmental Engineering, Beijing General Research Institute of Mining & Metallurgy, China*

Land Reclamation in Ecological Fragile Areas – Hu (Ed.)
© 2017 Taylor & Francis Group, London, ISBN 978-1-138-05103-4

Members of editorial committee

Mining and reclamation policies, regulations and standard

Land Reclamation in Ecological Fragile Areas – Hu (Ed.)
© 2017 Taylor & Francis Group, London, ISBN 978-1-138-05103-4

Surface coal mine permit application for successful reclamation, semi-arid shortgrass prairie (Wyoming, USA)

A. Krzyszowska Waitkus
Environmental Consulting, Laramie, Wyoming, USA

ABSTRACT: Approximately 40% of U.S. coal originates in an ecologically sensitive area of semi-arid shortgrass prairie in Wyoming. Before a surface coal mine can begin operation in the U.S., it must secure a mining permit and comply with regulations and performance standards under the U.S. Surface Mining Control and Reclamation Act (SMCRA), other federal environmental acts, and state programs. The Wyoming Department of Environmental Quality (WDEQ)/Land Quality Division (LQD) administers Wyoming's coal regulatory program. Examples of Wyoming's regulations and environmental protection performance are presented. The permit application and bonding process for the largest surface coal mine permit in the U.S., North Antelope Rochelle Mine (NARM) located in short grassland prairie in the northeast Wyoming, is discussed. The permit application process begins with the collection of baseline environmental data that characterizes premining conditions of the permit area. The permit application includes adjudication information, baseline information, mine and operation plans, and reclamation plans. Fulfillment of permit commitments and requirements of rules and regulations are inspected monthly by the LQD's representative in the field. Before a mine permit is issued, the mine operator must submit a reclamation bond to secure the performance of reclamation obligations that is later revised annually. In Wyoming, four reclamation bond release phases (Area Bond, Phase 1, 2, and 3) indicate the completion of various stages of the reclamation process. NARM's specific bond release verification criteria, performance standards, and field verifications of bond release phases are discussed. The Bond Release Geodatabase (a GIS/GPS approach) was developed for this mine to monitor progress in meeting criteria and performance standards for incremental bond release. The Bond Release Geodatabase significantly reduces the time needed to track bond release progress, reach agreement between operator and regulator, and improve the state inspector's ability to assess reclamation adequacy and progress.

1 INTRODUCTION

The State of Wyoming produces approximately 40% of all mined coal in the U.S. (Carroll, 2011). The Land Quality Division (LQD) of the Wyoming Department of Environmental Quality (WDEQ) is in charge of Wyoming's coal regulatory program (State of Wyoming, 2016). The state's regulations and environmental protection performance standards are important in achieving successful reclamation results. According to WDEQ/LQD Coal Rules and Regulations (Wyoming State Rules & Regulations, 2017), the goal of surface coal mine reclamation is to restore the land to its pre-disturbance ecological function. To accomplish this goal, a thorough baseline inventory of environmental conditions must be performed prior to mining (Norton *et al.*, 2010). It is especially important in the semi-arid environment of Wyoming, where a native vegetation of the shortgrass prairie is sensitive to any disturbance due to the lack of the rainfall (Knight *et al.*, 2014). The baseline inventory includes surveys and assessments of: cultural and historical resources; soils; vegetation; wildlife, surface and groundwater hydrology; climatology; wetlands, and geological data (soil and rock structure, coal seam thickness). The company then develops mining and reclamation plans incorporating all of the environmental components of the baseline inventory.

Before a company can begin mining in the U.S., it must secure a mining permit and comply with regulations and performance standards of the USA Federal Surface Mining Control and Reclamation Act (SMCRA) of 1977 (Public Law 95–87, referred to as SMCRA), state programs and other federal environmental acts (U.S. Government Publishing Office, 2016). Verification of state rules and regulations, permit commitments, and performance standards are conducted through field inspections, annual reports, and permit revisions.

Few references discuss surface coal mine permit content and its role in achieving reclamation goals (U.S. Congress, Office of Technology Assessment,

1986). Specific studies conducted in the western U.S. states discuss the effects of mining and reclamation approaches on geomorphic processes (Toy & Black, 2000), soil properties, and management (Pinchak et al., 1985; Schroeder & Vinning, 1993; Buchanan et al., 2005; Reynolds & Reddy, 2012), and re-vegetation strategies and methods (Williams et al., 2002; Schuman et al., 2005; Strom et al., 2010).

Before a mine permit is issued, the mine operator must submit a reclamation bond to secure the performance of reclamation obligations (Bonogofsky *et al.*, 2015). The operator might reduce a reclamation bond by reclaiming disturbed areas simultaneously with mining coal. Reclamation progress in Wyoming occurs in four phases of bond release together with the verification of bond release criteria and performance standards (Krzyszowska Waitkus & Blake, 2011).

This paper discusses the permit content information and the bond release processes of the largest surface coal mine permit in the U.S., the North Antelope Rochelle Mine (NARM), located in the semi-arid (less than 300 mm rain/year) shortgrass prairie of northeast Wyoming. Information presented in this article were obtained from the Permit 569-T8 Peabody Powder River Mining LLC North Antelope Rochelle Mine stored in the Cheyenne office of the WDEQ/LQD (Permit 569-T8, 2014).

NARM is an active open pit mine that produced over 111 million tons of coal in 2015. The permit area covered 18590 ha in 2015. Since the initial operation began in 1982, approximately 7680 ha have been disturbed. Coal seam thickness averages approximately 20 m with an average overburden thickness approaching 90 m. The mine contains three major pits: East Pit (6 km long), North Pit (8 km long), and West Pit (6 km long). The mine uses three operating draglines and 11 shovels supported by numerous trucks and bulldozers.

2 RULES AND REGULATIONS

Since 1969 the state of Wyoming has created numerous environmental acts to help regulate the mining industry. They were revised after the primary federal law SMCRA was enacted in 1977, creating a basis for environmental laws, rules and regulations in various states (Squillace, 1990). SMCRA established minimum federal standards for the regulation of mining, introduced an idea of bonding, and provided penalties for violations of the laws, regulations, or permit conditions. A federal agency, the Office of Surface Mining, Reclamation and Enforcement (OSMRE), was created to oversee SMCRA and revise it on an ongoing basis. The Code of Federal Regulations lists the rules and maintains an up-to-date electronic version (US Government Publishing Office, 2016).

In addition to SMCRA, mining operations must abide by regulations put forth by a number of other federal and state environmental agencies. In Wyoming these include: U.S. Office of Surface Mining, Reclamation and Enforcement; U.S. Army Corps of Engineers; U.S. Environmental Protection Agency; U.S. Fish and Wildlife Service; Wyoming Game and Fish Department; and WDEQ.

SMCRA's major components were incorporated into the Wyoming Environmental Quality Act and Industrial Development Information and Siting Act (State of Wyoming, 2016) giving WDEQ/LQD the authority to establish standards for reclamation. These standards allowed for the creation of rules and regulations for the coal industry in Wyoming (Wyoming State Rules & Regulations, 2017), which became an official part of Wyoming's coal regulatory program after approval by the U.S. Secretary of the Interior. Rules quoted in this article are from the Wyoming Environmental Quality Act and Industrial Development Information and Siting Act (State of Wyoming, 2016) and from WDEQ/LQD Coal Rules and Regulations (Wyoming State Rules & Regulations, 2017) that are constantly being updated to reflect changes in SMCRA. All these rules, regulations, and guidelines can be found on the website of the WDEQ/LQD.

3 CONTENT OF THE PERMIT

Information presented in the permit follows federal and state rules and regulations and contains specific permit commitments. Surface coal mining permits are issued for a term not to exceed five years, after which the operator must request and submit a revised permit for a term renewal. A surface coal mining permit must be terminated if the permittee has not commenced surface coal mining operations covered by the permit within three years of permit issuance.

The coal mine permit application consists of four major portions: adjudication section, supporting data (baseline information), mine plan, and reclamation plan (Table 1). More than 60% of the information concerns premining baseline data.

The application, once approved, constitutes the enforcement contractual document (permit) with the WDEQ/LQD that is used during compliance inspections of mining operations. Any discrepancy from the permit commitment found during an inspection or through a revision of an annual report could result in a violation with associated penalties.

The permit must contain a time schedule to address the following: mining progression (topsoil

Table 1. Contents of the permit 569-T8, 2014.

Category	Number of volumes
Adjudication information	2
Baseline information including: land use, cultural resources, (history, archeological investigation, paleontology), climatology, geology, hydrology, soils, vegetation, wildlife, wetlands, alluvial valley floors	18
Mine plan including mining method, schedule and mining hydrology	7
Reclamation plan including reclamation schedule*	2

Note: *Describe processes of overburden and topsoil handling and provide a reclamation schedule in the form of numerous maps.

removal, overburden removal, coal removal), backfilling, grading, contouring, and seeding. The time schedule uses yearly blocks for five years followed by blocks of five year periods until the end of mining. Maps and cross-sections included in the permit must follow specific, consistent requirements related to scale, contours, legend, etc. All engineering maps, plans, and designs or reports including cross sections must be certified by a professional engineer registered under the laws of the State of Wyoming (State of Wyoming, 2016). All mining and reclamation maps are presented in the same scale of 1:1500.

3.1 *Adjudication information*

Adjudication files include forms showing the surface land to be affected by the mine, the license to mine coal, a certificate of public liability insurance, a list of surface and coal owners, documentation of surface owner consent and right of entry, the permit area boundary, and lands to be affected over the life of the mine. New permit applications must contain the reclamation performance bond estimate, covering all disturbances and construction anticipated during the first 12 months of the permitted operation (State of Wyoming, 2016). This bond amount is updated annually by the operator through annual reports and approved by the WDEQ/LQD according to WDEQ/LQD Guideline No. 12 (Wyoming Department of Environmental Quality, 2017).

A certificate of public liability insurance must be submitted with the permit according to requirements of the State of Wyoming (2016). An appendix of lands within and adjacent (within 0.8 km) to the permit area contains a list of surface owners and coal rights, and records of water rights owners. Surface ownership includes private, federal (e.g., U.S. Department of Agriculture), and state of Wyoming lands.

3.2 *Baseline information*

Pre-disturbance inventory is a critical part of reclamation planning and establishes a framework and schedule for successful restoration of severely disturbed land, monitoring, and evaluation (Norton *et al.*, 2010). According to rules and regulations the following information is required in the baseline information of the permit: land use, brief history of the area including description of historic places, archeological and paleontological resources, climatology, topography, geology and overburden assessment, hydrology (groundwater, surface water, water rights), soil assessment, vegetation inventory, and wildlife use (State of Wyoming, 2016; Wyoming State Rules & Regulations, 2017).

The premining land use summary is delineated on a map, statistically calculated, and presented in a table. For example, the categories of land use within the permit area include: surface coal mining, grazing, oil and gas development, wildlife habitat, and recreational uses. The permit contains the hectares of each land use category including a vegetation type.

Historical overview, cultural resource sites, and paleontological inventories are conducted on all permit areas according to requirements included in the Wyoming Environmental Quality Act. The individual inventory, evaluation, and mitigation reports are compiled in proprietary volumes to protect the confidentiality of cultural resources under the Archeological Resources Protection Act of 1979 (State of Wyoming, 2016). The characteristics of climatology presented in the permit include temperature, growing season, precipitation, humidity and evapotranspiration, and winds. Due to the potential for the air pollution from dust, the operator must present air quality monitoring plans including the locations of air quality monitoring stations.

The topography, geology, and overburden assessments describe the premining topography, geology, stratigraphy, quality and quantity of premining overburden (State of Wyoming, 2016). Geologic cross sections provide information on the lithology across the permit area including overburden, interburden, and coal thickness. Wyoming mines must follow the requirements of an Approximate Original Contour (AOC) type topography where surface configuration achieved by backfilling and grading of mined areas must closely resemble the general surface configuration of the land prior to mining and complement the drainage patterns of the surrounding terrain (Wyoming State Rules &

Regulations, 2017). The State of Wyoming and other states (Montana, Colorado, New Mexico, and North Dakota) require the establishment of AOC as per federal regulations unless specifically exempted due to excess or lack of overburden material.

During the mining and reclamation process two sampling programs are employed at NARM to define unsuitable overburden material ahead-of-mining and backfill material. In order to define unsuitable overburden material ahead-of-mining, overburden sampling procedures are established for material moved by a truck/shovel operation only. In NARM, at least one hole per 64 ha within the permit area is sampled for chemical analysis. Sample cuttings are collected on intervals at a maximum of 3 meters in specific lifts of 16.5 m to a depth of 148 m. Samples are analyzed for quality by comparing with parameters and concentrations described in WDEQ/LQD Guideline 1 (Wyoming Department of Environmental Quality, 2017). The location of unsuitable material collected from the baseline study is delineated on a map titled "Overburden Suitability" included in the permit and also is shown on the Geographic Position System (GPS) unit located on a shovel removing overburden. The map on the GPS of the shovel delineates places where these unsuitable materials could be stored (at least under 1.2 m of suitable material) and cannot be stored (such as in close proximity to a drainage channel or its flood plain).

Postmining backfill quality procedures include a description of the sampling methods, suitability classification, and selective placement. Composite samples of backfill material are collected at depths of 0–0.6 m and 0.6–1.2 m and analyzed for quality by comparing with parameters and concentrations described in WDEQ/LQD Guideline No. 1. Each sample site is located on 150 m centers (2 ha). If unsuitable material is found in the top 1.2 m of regraded spoil material, a mitigation plan is implemented to verify the quality of the initial backfill sample or to more accurately define the area of unsuitability within the initial sample area. The unsuitable backfill material is capped or removed and relocated deeper under at least 1.2 m of suitable material (excluding areas under floodplain channel bottom of reclaim drainages). Chemical properties of backfill material with a map showing the collection sites are submitted with the annual reports. Results are analyzed by the operator and reviewed by the LQD specialist. Best management practices concerning successful revegetation and chemical properties of surface material show a direct connection between the chemical results of the overburden baseline study and the practical implementation of burying unsuitable overburden and backfill material.

Premining hydrology and hydraulics for both surface and ground waters within the permit area and 0.8 km adjacent area around the permit are discussed in the baseline study (State of Wyoming, 2016; Wyoming State Rules & Regulations, 2017). Surface water includes rivers, creeks, lakes, and reservoirs within and adjacent to the permit area. Descriptions of premining stream characteristics (drainage density, channel sinuosity, stream channel morphology) are discussed, along with drainage basins and surface water runoff quantity and quality. At NARM, baseline water quality data for major streams are collected regularly and analyzed for specific parameters (e.g. ammonia, nitrogen, nitrate, bicarbonate, boron, carbonate, fluoride, sulfate, total dissolved solids, and dissolved arsenic, calcium, chloride, chromium, iron, molybdenum, potassium, sodium, and zinc). Information is presented on aquifers that may be affected, their areal extent and hydraulic properties, groundwater quality, and groundwater recharge and discharge areas. In 2016, NARM 93 groundwater monitoring sites and seven surface water stations were sampled regularly according to the permit commitments.

Soil at the mine is ranked as suitable or unsuitable for salvage according to chemical or physical criteria that affect revegetation success. These criteria are established by the state regulatory authority and reflect characteristics of climate and vegetation (U.S. Congress, Office of Technology Assessment, 1986). All suitable soils must be salvaged and replaced on the reclaimed area. Most prairie soils in Wyoming are thin with low organic carbon content. It was estimated that minimum concentration of 0.52% organic content or 0.89% organic matter is necessary to sustain an adequate level of nutrient cycling in Wyoming reclaimed soils (Ingram *et al.*, 2005).

WDEQ/LQD Guideline No. 1 A (Wyoming Department of Environmental Quality, 2017) includes criteria for suitability for soils or soil substitutes. Soil assessment requirements include a site specific soil inventory and suitability maps with tabulations of soil salvage depths for affected lands (State of Wyoming, 2016). At least one sample for each soil series in the affected area is collected and analyzed for specific elements (e.g. pH, conductivity, particle size, texture, sodium absorption ratio, selenium, boron, and organic matter). Results are compared with the criteria to determine soil suitability. The results of quantitative soil analyses are used to create a soil stripping depth map for affected lands. A table in NARM's permit lists the following for each soil mapping unit: acreage of areas where soil removal is planned, average depth of soil to be removed, and volumes of soil to be salvaged. NARM's permit includes a map showing various soil unit descriptions with soil salvage depths. This information assists in determining soil depth application during the reclamation process.

Vegetation inventory requirements include delineation, mapping and tabulation of major vegetation communities, cover sampling of vegetation communities, species lists, shrub density determinations, and tree counts. The vegetation inventory also includes the selection of a reference area located outside of a disturbed area. Data from a reference area is used to compare data from the reclaimed and revegetated areas. NARM's permit includes a map titled "Native Vegetation Types" delineating eight major vegetation communities.

The baseline study includes a description of potential and actual faunal distribution within the permit and areas adjacent to the permit with a list of indigenous vertebrate wildlife species. Wildlife studies are done in accordance with state and federal wildlife agency requirements. NARM's permit includes a map titled "Wildlife Features", showing prairie dog colonies, intact raptor nests, former raptor nests, nesting platforms, and active and inactive sage grouse leks.

3.3 *Mine plan*

According to State of Wyoming requirements, the mine plan must include a description of type and method of mining, life of the mine in years, affected acres, estimated annual production, mine facilities, blasting plan, equipment list, and mining hydrology. The permit application must contain a map showing the estimated order and progression of mining and reclamation on all proposed affected lands (Wyoming State Rules & Regulations, 2017). Mine sequence maps for soil, coal removal, and rough backfilling include identification of the area to be mined during the first five years in one-year blocks during the specific term and in five-year blocks for the remaining life of the mine. Soil salvage and handling techniques as well as erosion control plans are presented in the mine plan section. Tables of soil volumes that will be salvaged, material excavated and backfilled, and hectares of areas stripped of soil and graded with soil replaced and seeded are presented for each of the first five years, and in five-year blocks thereafter, until the end of mining. This information is included in NARM's permit table titled "Material Handling Summary." The information is verified during annual report reviews and inspections to ensure the operator is following permit commitments and reclamation is occurring according to the schedule. The operator demonstrates the ability to achieve AOC through the calculation of bulking factors (swell factors) and estimation of overburden material volume available for backfilling.

The mine plan must ensure the protection of the quantity and quality of, and rights to, surface water and groundwater both within and adjacent to the permit area (Wyoming State Rules & Regulations, 2017). Mining hydrology sections include a plan and timetable for control and treatment of surface and groundwater, discussion, and determination of probable hydrologic consequences within and outside the permit area for the duration of mining. The plan must estimate groundwater drawdown and provide a monitoring plan to include the collection, recording, and reporting of data on water levels and quality.

A hydrologic control plan for surface water includes a hydrologic control map, structure designs, and a maintenance plan. Six of the 29 volumes of NARM's permit application contain engineering plans of sedimentation reservoirs, highwall sumps, and permanent impoundments. Design requirements for each constructed feature of the surface water control plan follow WDEQ/LQD Guideline No. 8 (Wyoming Department of Environmental Quality, 2017) and Coal Rules and Regulations (Wyoming State Rules & Regulations, 2017).

3.4 *Reclamation plan*

The reclamation plan describes procedures and methods used to minimize adverse impacts during mining and to restore disturbed land to premining land use (Norton *et al.*, 2009). By definition, reclamation "shall restore the land to a condition equal to or greater than the highest previous use." The land, after reclamation, must be suitable for the previous use or must have a use which is of greater economic or social value than previous uses (Wyoming State Rules & Regulations, 2017). At NARM, reclamation must restore the land to livestock grazing and wildlife use functions. The reclamation plan coordinates with the mining plan and contains a time schedule to complete reclamation as quickly as possible.

NARM's reclamation plan includes information on soil removal, storage, protection, and replacement, along with a plan for backfilling, grading, and contouring of affected lands. Also included are methods of monitoring hydrologic restoration of disturbed land, revegetation plans with a means of measuring of success, and information on postmining land use.

Reclamation begins with the salvage and protection of soil from the disturbed land. Salvaged soil is either stockpiled and protected from erosion and degradation or placed directly on regraded areas (direct haul). According to NARM's experience, direct haul produced the best revegetation results. Direct haul can result in higher mycorrhizae levels, better physical characteristics, and a slightly greater potential for seed bank benefit as compared to stockpiled soil (Schuman, 2002; Anderson *et al.*, 2008). However, caution should be taken if soil contains infestations of weeds or other undesirable species. The soil replacement management plan in NARM's permit discusses erosion control and water conservation practices, soil replacement depth range, and soil characteristics for the restoration of critical and important habitat such as wetlands, floodplains etc.

Soil is usually applied at a uniform depth on regraded areas in Wyoming surface coal mines. According to the Wyoming State Rules & Regulations (2017), variable soil depths are also allowable: "Soil thickness may also be varied to the extent such variations help meet the specific revegetation goals identified in the permit." Some studies performed in the western U.S. showed that revegetation results can be affected by varying soil depths (Buchanan *et al.*, 2005; Bowen *et al.*, 2005; Williams *et al.*, 2002). To enhance the growth of hard-to-establish shrubs such as big sagebrush (*Artemisia tridentata*), NARM has developed two soil replacement depth ranges: typical (15 to 40 cm) and deep (40 to 120 cm).

Soil is applied on backfilled and graded areas as soon as possible following mining, according to performance standards (Chapter 4, Section 2(b)) of the Wyoming State Rules & Regulations (2017):

- "Rough backfilling and grading shall follow coal removal as contemporaneously as possible based upon the mining conditions. The operator shall include within the application for a permit to mine a proposed schedule for backfilling and grading with supporting analysis."
- "Backfilled materials shall be replaced in a manner which minimizes water pollution on and off the site and supports the approved postmining land use. Preparation of final graded surfaces shall be conducted in a manner that minimizes erosion and provides a surface for replacement of soil that will minimize slippage."
- "All affected lands shall be returned to their approximate original contour…"

The recontoured surface design has critical effects on postmining vegetation, land stability, and postmining land use (U.S. Congress, Office of Technology Assessment, 1986). In Wyoming's semi-arid environment, a complex (convex on top, concave at bottom) slope of less than 25% has been found to be least prone to erosion. To avoid long uniform slopes, constructed slope lengths do not exceed premining lengths and are shorter than 150 m long. No highwalls may be left unless they are approved by the Administrator of the WDEQ/LQD. During reclamation, pre- and postmining topographies are compared to determine if reclamation achieved requirements of AOC topography (Steward *et al.*, 2006). In NARM's permit, pre- and postmining slope frequency and distribution are illustrated graphically (showing e.g., average slope, median slope, and slope interval). Slope distribution analysis is performed using state-of-the-art terrain modeling and analysis software. Analysis of histograms and cumulative frequency curves indicate a strong correlation between pre- and postmining slopes at NARM.

A rebuilt hydrologic system must be erosionally stable to drain water and blend with surrounding drainage systems. For NARM, surface water quality is predicted to be similar to premining and meet premining class of use standards. Hydrologic restoration plans for surface water discuss postmining watershed characteristics, practices to be used for drainage reconstruction (including channel geometry), locations of permanent impoundments, and a postmining monitoring plan. The function of Porcupine Creek has been restored by rebuilding the creek structure, sinuous channel, and hydrology. The restoration established a maximum amount of wetland and riparian habitat and stabilized streambanks (Hansen & Murphree, 2003).

Structures (facilities and flood control reservoirs, sediment ponds, diversions, sediment traps, alternate sediment control measures) constructed in NARM's permit area control runoff and provide sediment control during mining. According to Wyoming State Rules and Regulations (2017), sediment can't leave the permit boundary and cause an off-site impact. Sediment ponds are depicted on the Mine and Reclamation Plan Map titled "Hydrologic Control Plan." Sixty one permanent impoundments as of 2014 in NARM's permit area will supply water for livestock and wildlife, and will be used as wetland mitigation. Permanent impoundments on reclaimed surfaces are delineated on the Mine and Reclamation Plan Map titled "Post-mining Contours with Hydrologic Restoration."

Hydrologic restoration plans for groundwater describe the source of recharge and provide a postmine monitoring plan with a map of permanent backfill well locations. A discussion of probable hydrologic consequences estimates final ground and surface water quantity and quality. The assessment of ground water quantity includes rate of recovery predictions for the affected and backfill aquifers. A map showing final anticipated potentiometric surfaces of the affected aquifer is provided in NARM's permit.

The major goal of the reclamation plan is to establish stable vegetation communities on chemically suitable soil and backfill material. According to the revegetation performance standard "The operator shall establish on all affected lands a diverse, permanent vegetative cover of the same seasonal variety native to the area or a mixture of species that will support the approved postmining land use in a manner consistent with the approved reclamation plan. This cover shall be self-renewing, and capable of stabilizing the soil" (Wyoming State Rules & Regulations, 2017). The revegetated area must support the approved postmining land use. NARM's revegetation plan provides a discussion on vegetation establishment methods (seed mixes, seeding techniques, mulching materials and rates, husbandry practices). Also included are interim vegetation monitoring and final revegetation success plans. Various seed mixes are developed based on baseline studies and on-site evaluation. Native plant species are preferred, with introduced species used only to provide utility in the mixes. The seed mixes correspond to the following vegetation types: grassland, riparian grassland, sagebrush grassland, playa grassland, shrub mosaic and sagebrush mosaics. All of these communities are native to the premining environment. High plant production on reclaimed soils in Wyoming, despite lack of fertilization, suggests that total nitrogen is not limiting production (Ingram *et al.*, 2007). It was found that reclaimed soils in Powder River Basin in Wyoming had recovered structurally towards a native soil condition after a period of 10–15 years (Wick *et al.*, 2009). Reclamation success is determined by a quantitative comparison of data from the reference area with data from the reclaimed area that include: cover (vegetative cover, total ground cover), production (total production, specific species production), density and distribution (shrubs, trees), species diversity, and species composition (Bilbrough & Howlin, 2012). This information is presented and discussed in the NARM's annual report. Also specific areas planted in the various seed mixes, along with seeding rates, are included in tables and maps.

Disturbed areas seeded with native shrub and grass species also provide habitat for wildlife. Rock piles are constructed to simulate natural features. To reduce the effects of mining on birds of prey, several mitigation measures have been incorporated into the reclamation plan. Wildlife mitigation and monitoring plans include information on all habitat restoration procedures.

The NARM permit includes following maps (scale 1:1500) and exhibits showing major steps in the mining/reclamation processes together with a time schedule: "Topsoil removal (disturbance) schedule"; "Hydrologic Control Plan"; "Mining schedule"; Reclamation schedule (topsoil replacement); Postmining contours with hydrologic restoration; "Postmining contours with surface slope distribution"; "Postmining vegetation"; and "Postmining environmental monitoring locations." These maps are reviewed in the field during inspections. Also reviews of annual reports are important tools in determining if the operator is on time with the planned schedule of mining/reclamation processes. In case of delay or change in the direction of mining and reclamation, the operator must submit a revision to the permit.

4 RECLAMATION BOND PERFORMANCE

Before a mine permit is issued the mine operator must submit a reclamation bond to secure the performance of reclamation obligations. The cost of reclaiming the disturbed areas of a coal mine is revised and updated annually through annual reports submitted to WDEQ/LQD. A reclamation performance bond or financial assurance must be large enough to cover the full costs of reclamation to ensure that the regulatory authority will have sufficient funds to reclaim the site if the permittee fails to complete the reclamation plan or hire a third party to complete the reclamation plan in the event of a bankruptcy of the coal mine company. The bond release can begin five years from the date of permanent seeding in the eastern U.S. and ten years after permanent seeding in the western U.S.

In Wyoming, there are four bond release phases (Area Bond, Phase 1, 2, and 3) for surface coal mines indicating the completion of various reclamation phases (Krzyszowska Waitkus & Blake, 2011) (Table 2). In general, performance standards related to the release of performance bonds for Area Bond and Phase 1 include the area being backfilled and regraded with a functional drainage system and applied soil. For Phase 2, vegetation cover must be established in stable areas without erosion. For Phase 3, the revegetation standards and goals must be completed (Wood & Buchanan,

Table 2. Verification of bond release criteria.

Bond release phase	Verification of bond release criteria
Area Bond	Backfilling, regrading completed according to approved plan
Phase 1	Stream channel reconstructed Soil applied
Phase 2	Vegetation established (ocular evaluation) Surficial stability established Permanent water impoundments construction/renovation designed and approved by the State Engineer's Office
Phase 3	Revegetation success standards established (statistical analysis) Mitigation of wetlands (approved by the US Army Corps of Engineers) Trees established

2000). Revegetation as a portion of the assessment of reclamation success created a lot of discussion between coal operators, regulators, and environmental specialists for long time in the state of Wyoming (Krzyszowska Waitkus *et al.*, 2000; Shuman, 2002).

The amount of the bond is revised annually through annual reports according to WDEQ/LQD Guideline 12 (Wyoming Department of Environmental Quality, 2017). Bond release submittals are not part of the permit, but the bond amount and bonding instruments (e.g. letter of credit, certificates of deposit, and treasury bills) are submitted with the permit.

As a result of cooperation between the federal agency, private industry, environmental consultants, and WDEQ/LQD representatives a set of criteria and performance standards (Table 2) was established that have to be fulfilled and verified before the reclamation bond can be released. Up-to-date information (WDEQ/LQD Guidelines 20, 21, 22, 23, and 25) regarding bond release verification and compliance standards is posted on the Wyoming DEQ website (Wyoming Department of Environmental Quality, 2017).

As reclamation phases are completed, the permittee may apply for incremental, phased bond release. Phased bonding encourages companies to begin reclamation as soon as possible (called contemporaneous reclamation). Initially, criteria and performance standards are verified in the field by the LQD representatives. Specific bond release requests are approved in the field after joint inspection by the permit operator, LQD, and OSMRE representatives. Additionally, the operator compares the data gathered during the baseline inventory with data gathered from the reclaimed land. After reviewing the submitted information and performing a field inspection, the LQD representative makes a decision to approve or deny the bond release request. If an Area Bond and Phase 1 request is approved, 60% of the bond can be released, for Phase 2, 15%, and for Phase 3, 25% of the bond can be released.

Various phases of bond release require verification of criteria and performance standards in the field which represents a challenge for the regulatory agency, especially for large mines. These challenges require a system to streamline field data collection which can be utilized in evaluating bond release and compliance requirements. As a result of these challenges, a Bond Release Geodatabase was developed for NARM in Wyoming. This geodatabase was the first spatially supported database developed in Wyoming and in the U.S. to process reclamation bond releases. The geodatabase was designed to support the tracking of areas which have achieved specific bond release criteria and have satisfied performance standards in support of incremental bond release. Compliance features (e.g., topsoil depth, erosion, or wildlife habitat features) were monitored and entered into the geodatabase by LQD personnel during inspections using a Global Position System (GPS) enabled data collector (Krzyszowska & Waitkus, 2008). Verified field data together with information submitted by the operator were compiled into one Bond Release Geodatabase. This geodatabase is a highly effective method which sufficiently reduces the time needed to track the bond release progress and reach agreement between the operator and the regulator. It also improves the state inspector's ability to assess reclamation adequacy and progress (Krzyszowska Waitkus & Blake, 2011).

5 SUMMARY

The primary governmental responsibility for the regulation of surface coal mining rests with the rules and regulations of states that follow the requirements of SMCRA, 1977. The coal mine permit is an agreement between the mine operator, state, and federal agencies. The permittee is required to meet performance standards and fulfill reclamation goals to restore disturbed land to premining land use.

Coal mine permit content follows the rules and regulations of state regulatory agencies. The permit presents data requirements and collection methods to assess the adequacy of baseline and monitoring data, predicts the impacts of mining, provides information on reclamation techniques, and presents methods evaluating the success of reclamation. Because mining and reclamation schedules included in the permit are constantly changing due to market demand, the permit must be constantly updated. The schedule, permit commitments, and compliance with rules and regulations are inspected monthly by the WDEQ/LQD's representative in the field. The permit is instrumental for the permittee and for the regulatory agencies to gather all information and present how all disturbed land can be restored to the premining land use for the western states of the U.S. The presented case of the permit shows how the regulations together with the operator's experience and their best management practices are advantageous in the reclamation of severely disturbed land.

Reclamation of large surface coal mines can be successful in a disturbance-sensitive, semi-arid environment. Data from the largest surface coal mine in Wyoming shows that a disturbed area can be sufficiently revegetated to fulfill SMCRA requirements over a period of 10 years. The set of criteria and performance standards established for the state of Wyoming accelerate the bond release application submittals and more reclaimed area should be released to the land owners in timely manner. A Bond Release Geodatabase developed by the WDEQ/LQD field inspector has streamlined the bond release process and helped with compliance inspections. The GIS/GPS approach was proven to be successful for the large mining/reclamation activities of severely disturbed areas.

REFERENCES

Anderson, J.D., Ingram, L.J. & Stahl, P.D. (2008). Influence of reclamation management practices on microbial biomass carbon and soil organic carbon accumulation in semiarid mined lands of Wyoming. *Appl. Soil Ecol.*, *40*, 387–397.

Bilbrough, C. & Howlin, S. (2012) *Handbook of Approved Sampling and Statistical Methods for Evaluation of Revegetation Success on Wyoming Coal Mines*. [Online] Wyoming Department of Environmental Quality, Land Quality Division. Available from: http://deq.wyoming.gov/lqd/coal/resources/handbook/ [Accessed 19th February 2017].

Bonogofsky, A., Jahshan, A., Yu, H., Cohn, D. & MacDonald, M. (2015) *Undermined Promise II*. [Online] National Wildlife Federation, Natural Resource Defense Council, & Western Organization of Resource Councils. Available from: http://www.underminedpromise.org/UnderminedPromiseII.pdf [Accessed 19th February 2017].

Bowen, C.K., Schuman, G.E. & Olson, R.A. (2005) Influence of topsoil depth on plant and soil attributes of 24-year old reclaimed mined lands. *Arid Land Res. Manag*. [Online] *19*, 267–284. Available from: doi: 10.1080/15324980590951441 [Accessed 19th February 2017].

Buchanan, B., Schuman, G.E. & Olson, R.A. (2005) Long-term plant community development in response to topsoil replacement depth on mined land in Wyoming. In: Barnhisel, R.I. (ed.) *Raising Reclamation to New Heights: Proceedings of a Joint Conference of American Society of Mining and Reclamation, 22 nd Annual National Conference, June 18–25, 2005, Breckenridge, Colorado, USA*. ASMR, 3134 Montavesta Rd. Lexington, KY 40502. pp.108–114.

Carroll, C. (2011) *Wyoming's Coal Resources*. Wyoming State Geological Survey. Summary Report.

Hansen, B.W. & Murphree, P. (2003) Reclamation of Porcupine Creek riparian and wetland habitats. In: Barnhisel, R.I. (ed.) *Working together for Innovative Reclamation: Proceedings of the Joint Conference of the 9th Billings Land Reclamation Symposium and the 20th Annual Meeting of the American Society of Mining and Reclamation and the 9th Billings Land Reclamation Symposium, June 3–6, 2003, Billings, Montana, USA*. ASMR, 3134 Montavesta Rd. Lexington, KY 40502. pp. 86–89.

Ingram, L.J., Schuman, G.E., Stahl, P.D. & Lowell, L.K. (2005) Microbial respiration and organic carbon indicate nutrient cycling recovery in reclaimed soils. *Soil Sci. Soc. Am. J.* [Online] *69*, 1737–1745. Available from: doi:10.2136/sssaj2004.0371 [Accessed 19th February 2017].

Ingram, L.J., Stahl, P.D. & Anderson, J.D. (2007) The influence of management practices on microbial and total soil nitrogen. In: Barnishel, R.I. (ed.) *30 Years of SMCRA and Beyond. Proceedings of the 24th National Meeting of the American Society of Mining and Reclamation 2007: 2–6 June, Gillette, Wyoming, USA*. ASMR, 3134 Montavesta Rd. Lexington, KY 40502. pp. 346–360.

Knight, D.H., Jones, G.P., Reiners, W.A. & Romme, W.H. (2014) *Mountains and Plains: The ecology of Wyoming*. Yale University Press.

Krzyszowska Waitkus, A., Smith, J. Cash, G. & Moxley, M. (2000) Successful reclamation techniques and bond release for coal mine in Wyoming. In: *Striving for Restoration, Fostering Technology and Policy for Reestablishing Ecological Function. Proceedings of the Billings Land Reclamation Symposium, March 20–24, 2000, Billings, Montana, USA*. Reclamation Research Unit Publication No. 00–01 Montana State University-Bozeman. pp. 33–44.

Krzyszowska Waitkus, A. & Calle, M. (2008) Mobile computing and geodatabase application for coal mine inspection. Paper presented at the 2nd National Meeting of SMCRA Geospatial Data Stewards, March 26, 2008, OSM Geospatial Conference "Incorporating Geospatial Technologies into SMCRA Business Processes", March 25–27, Atlanta, GA.

Krzyszowska Waitkus, A. & Blake, C. (2011) Tracking bond release at a large Wyoming coal mining operation. In: Barnishel R.I. (ed.) *Reclamation: Science Leading to Success: Proceedings of the National Meeting of the American Society of Mining and Reclamation, June 11–16, 2011, Bismarck, ND, USA*. ASMR, 3134 Montavesta Rd. Lexington, KY 40502. pp. 338–349.

Norton, J., Krzyszowska Waitkus, A. & Loubsky, T. (2009) *Successful restoration of severely disturbed lands: overview of critical components*. University of Wyoming. Cooperative Extension Service. Bulletin No. B–1202.

Norton, J., Krzyszowska Waitkus, A. & Strom, C. (2010) *Successful restoration of severely disturbed lands: predisturbance/baseline inventory*. University of Wyoming. Cooperative Extension Service. Bulletin No. B-1212.

Permit 569-T8. (2014) *Peabody Powder River Mining LLC, North Antelope Rochelle Mine*. Available: Wyoming Department of Environmental Quality, Land Quality Division, Cheyenne, Wyoming, USA.

Pinchak, B.A., Schuman, G.E. & Depuit, E.J. (1985) Topsoil and mulch effects on plant species and community responses of revegetated mined land. *J. Range Manage.*, *38*, 262–265.

Reynolds, B. & Reddy, K.J. (2012) Infiltration rates in reclaimed surface coal mines. *Water, Air, & Soil Pollution*, 223 (9), 5491–5958.

Schroeder, S.A. & Vinning, K.C. (1993) *Relation of compaction and soil physical parameters to productivity of reclaimed soils*. Final Report. North Dakota State University Land Reclamation Research Center.

Schuman, G.E. (2002) Mined land reclamation in the Northern Great Plains: have we been successful? In:

Barnishel, R.I. & Collins, M. (eds.) *Reclamation with a purpose. Proceedings of the 19th National Meeting of the American Society of Mining and Reclamation and the 6th International Affiliation of Land Reclamationists. June 9013, 2002, Lexington, Kentucky, USA*. 3134 Montavesta Rd. Lexington, KY 40502. pp. 842–864.

Schuman, G.E., Vicklund, L.E. & Belden, S.E. (2005) Establishing Artemisia tridentata ssp. wyomingensis on mined lands: Science and economic. *Arid Land Res. Manag.*, *19*, 353–362.

State of Wyoming. (2016) Wyoming Environmental Quality Act and Industrial Development Information and Siting Act. 2016 Edition, LexisNexis.

Steward, D.G.M., Burget, W., Carlson, R., Stowe, R.R. & Vicklund, L.E. (2006) *Handbook of Western Reclamation Techniques.* Second edition. [Online] University of Wyoming. Available from: https://www.osmre.gov/resources/library/ghm/WestReclHandbook2ndEd.pdf [Accessed 19th February 2017].

Strom, C., Norton, J. & Loubsky, T. (2010) *Successful restoration of severely disturbed lands: seeding essentials for reclaiming disturbed lands*. University of Wyoming. Cooperative Extension Service. Bulletin No.B-1204.

Squillace, M. (1990) *The Strip Mining Handbook*. [Online] Environmental Policy Institute. Available from: https://sites.google.com/site/stripmininghandbook/ [Accessed 20th January 2017].

Toy, T.J. & Black, J.P. (2000) Topographic reconstruction: the theory and practice. In: Barnishel, R.I., Darmody, R.G. & Daniels, W.L. (eds.) *Reclamation of drastically disturbed lands*. Agronomy Monography 41. ASA, CSSA, and SSSA, Madison, WI. pp. 41–77.

U.S. Congress, Office of Technology Assessment. (1986) Western Surface Mine Permitting and Reclamation. Washington, DC: U.S. Government Printing Office, June 1986, OTA-E-279. Available from: Library of Congress Catalog Card Number 86-600506.

U.S. Government Publishing Office. (2016) *Chapter VII-Office of Surface Mining Reclamation and Enforcement, Department of the Interior. Code of Federal Regulations. Title 30-Mineral Resources*. [Online] Available from: https://www.gpo.gov/fdsys/pkg/CFR-2016-title30-vol3/pdf/CFR-2016-title30-vol3-chapVII.pdf [Accessed 18th February 2017].

Wick, A.F., Ingram, L.J. & Stahl, P.D. (2009) Aggregate and organic matter dynamics in reclaimed soils as indicated by stable carbon isotopes. *Soil Biol. Biochem.*, *41,* 201–209.

Williams, M., Shuman, G.E., Hild, A. & Vicklund, L. (2002) Wyoming Big Sagebrush density: effects of seeding rates and grass competition. *Restor. Ecol.*, *10*, 385–391.

Wood, M.K. & Buchanan, B.A. (2000) Reclamation considerations for arid regions of the Southwest receiving less than twenty-five centimeters annual precipitation.. In: Barnishel, R.I., Darmody, R.G. & Daniels, W.L. (eds.) *Reclamation of drastically disturbed lands.* Agronomy Monography 41. ASA, CSSA, and SSSA, Madison, WI. pp. 303–322.

Wyoming Department of Environmental Quality. (2017) *Land Quality Division: Guidelines*. [Online] Available from: http://deq.wyoming.gov/lqd/resources/guidelines/ [Accessed 24th January 2017].

Wyoming State Rules & Regulations. (2017) *Environmental Quality, Department of Land Quality – Coal.* [Online] Available from: https://rules.wyo.gov/ [Accessed 24th January 2017].

Land Reclamation in Ecological Fragile Areas – Hu (Ed.)

40 years of SMCRA: What have we learned in the state of Wyoming

Brenda K. Schladweiler
BKS Environmental Associates, Inc., Gillette, WY, USA
Accord Resource Solutions, LLC, Gillette, WY, USA

ABSTRACT: The Surface Mining Control and Reclamation Act (SMCRA) was enacted in 1977 and was the overriding federal regulation governing mining and reclamation of surface coal mines in the United States of American (USA). Many of the newest surface mines in the USA at that time, were in the western portion of the nation. Wyoming surface coal mines numbered approximately 20 and were located throughout the state in the coal bearing regions, generally in the south/southwestern portion of the state and the northeastern corner.

The Office of Surface Mining Reclamation and Enforcement (OSMRE) is the federal agency tasked with oversight of the implementation of SMCRA. Individual states developed statewide programs that met or exceeded the requirements of SMCRA and, thus, obtained primacy over coal mine mining and reclamation within their boundaries as long as those regulatory conditions were maintained. OSMRE retained oversight on the Indigenous Nations programs within the USA.

Much information has been learned on the reclamation side of SMCRA, its regulations and state programs since its passage. This paper and presentation will present some of the basic changes made in reclamation programs over the last 40 years in the State of Wyoming.

1 INTRODUCTION

Many environmental issues, including those related to surface mining in the United States of America (USA), were being addressed by the federal government in the late 1960's and 1970's. However, reclamation was a word that was not new to the mining industry. Wyoming reclamation laws pre-dated the federal Surface Mining Control and Reclamation Act (SMCRA) of 1977. The Open Cut Land Reclamation Act of 1969 and the subsequent Environmental Quality Act of 1973 provided a regulatory framework for state oversight of surface mining, and, in particular, coal mining. Some researchers and agency personnel understood the vast coal resources in Wyoming and the need for such a framework (Thilenius and Glass, 1974).

The Office of Surface Mining Reclamation and Enforcement (OSMRE) is the federal agency tasked with oversight of the implementation of SMCRA. The federal government allows those states with surface coal mines to meet or exceed their implementation of the regulations and attain "primacy". States in the eastern USA have slightly different requirements than the western United States, e.g., length of the bonding period being 5 years instead of 10.

Learning curves abounded in the late 1970's and 1980's for industry, regulators, researchers and service providers. Much research occurred during the late 1980's and early 1990's in Wyoming under the Abandoned Coal Mine Land Research Program.

On the 40th anniversary of the passage of SMCRA, the Author reflects on the knowledge gained over this time. In addition, several people knowledgeable in past or current reclamation efforts were interviewed for insight into the extent of "what we have learned". Much of this information is from the western USA and, in particular, the State of Wyoming (where mines in 1977 were relatively new). The following general categories of knowledge were discussed.

1.1 *Influence of agriculture*

Many of the early reclamation-based regulations were influenced by agricultural practices in place at the time. Reclamation was considered a phase of mining and not necessarily a field of science in its own right. A simplistic way of looking at reclamation was "just getting something to grow and stabilize a site". In the eastern United States, that meant forested lands and subsequent land use. In the western United States, the dominant land uses were grazing lands for domestic livestock and wildlife habitat. Prime farmland was a part of the midwestern United States and cropland and hayland were dominant uses.

Eventual knowledge of seeding drastically disturbed lands did not necessarily mirror agricultural systems in the following ways:

1.1.1 *Seed mixes*

Seed mixes were rudimentary and much less diverse in 1977. The emphasis was on commercially available seed and quick-establishing plants with little understanding or regard to inter- and intraspecific species competition. Ecological diversity, longevity and wildlife habitat value were often secondary considerations. Species such as crested wheatgrass (*Agropyron cristatum*) and smooth brome (*Bromus inermis*) (often used for pasture grass) were commonly used in reclamation but are now considered invasive and undesirable. Commonly seeded legumes were also agriculturally-based, i.e., alfalfa (*Medicago sativa*) and yellow sweetclover (*Melilotus officinalis*).

Since 1977, the seed industry has essentially "caught up" to market demands. Natural Resource Conservation Service (NRCS) Plant Material Centers throughout the western USA developed hardy and prolific varieties of commonly seeded grasses, especially cool-season perennial grasses such as wheatgrasses. Many of these have the ability to outcompete other perennial grasses, forbs and shrubs; therefore, their overall use is now more limited. The seed for many warm season grasses such as blue grama (*Bouteloua gracilis*) was not commercially available and, until wildlife habitat for species of concern became an issue, the use of perennial forbs was limited. Commercial availability of perennial forb seed was also lacking.

In the late 1970's seed sources were limited, as were plant varieties. Markets for seed that could tolerate colder northern climates had not generally been developed prior to SMCRA. Often, seed sources were from more southern climates which, when planted in the north, caused many of the germinating seedlings to winter-kill or die. The push for adapted seed began in the mid-1980's. Hand collection of the less commercially available forb seed began about that time frame; however, such seed was costly and supplies limited and, in many ways, still is.

1.1.2 *Seeding equipment*

Many of the early seed drills, used for grains and other agricultural crops, were used for seeding reclaimed ground. Much of this ground was rocky with large clods of soil which caused problems for traditional agricultural equipment. Rangeland seeders have now been developed to handle the rockier soil.

In agriculture, a smooth firm seed bed is desirable. While that is still the case for flatter reclaimed areas, the likelihood of flat, smooth, clod-free seed beds in drastically disturbed areas is less likely. Rough is now considered better in some situations, especially steep slopes, especially those exposed to wind or water erosion. Therefore, pitters or imprinters can be helpful equipment in those instances, especially if followed with broadcast seeding/chaining to get good seed/soil contact.

The type of seed also made it difficult for passage through many traditional agricultural drills. Fluffy (trash) perennial grass seed (and some shrubs) containing awns and brackets did not feed adequately through seed chambers and tubing on drills. A new generation of drills was created by manufacturers to handle trash seed.

Broadcasting of shrub and forb seed was later found to improve chances of survival of those life forms. Seeding of shrubs and forbs, if drill seeded with perennial grasses, did not allow long-term survival of these species with the overly competitive cool-season grasses.

1.1.3 *Soil amendments*

Nitrogen application was once a commonly applied fertilizer to reclaimed systems in the western USA, especially if used in conjunction with surface hay or straw mulch. However, such applications were found to encourage annual weed growth and, if misapplied, could cause pollution of nearby waterways. Its use in many reclaimed areas is now discouraged. Phosphorus, on the other hand, can stimulate root growth, and, where limiting in the system, is a beneficial fertilizer.

In a semi-arid/arid environment such as Wyoming, topsoil is limiting. The dark, organic rich soils, such as those found in the Midwestern USA, is limited to the extreme eastern edge of Wyoming. Every effort is made to salvage all available material suitable for plant growth. Organic matter is considered a precious commodity. However, knowledge of how much organic carbon is needed to initiate nutrient cycling has changed since 1977; a much smaller percentage of organic carbon is necessary. This, however, does not negate the continued and judicial salvage of topsoil.

Understanding of the biotic component of soil was limited in 1977, especially as it related to mine land reclamation. The reestablishment of the soil biota and a functioning soil ecosystem is important to the nutrient cycling mentioned above. Organic fertilizers with low levels of nitrogen (to avoid the problems mentioned above) is an acceptable method of reestablishing this important community.

Mulching, in the form of native hay or straw, was once widely used but is now limited. Although native hay can be a desirable seed source, straw seed germination can have a deleterious effect from volunteer individuals during subsequent growing season. The high C:N ratios found in straw material degrade slowly over time, especially in semi-arid/arid environments such as Wyoming.

1.1.4 *Weed control*

As in most agricultural systems, weeds were often considered a negative aspect of plant growth and measures were taken to control them chemically (through herbicides) or mechanically (through mowing). In semi-arid/arid environments, weeds can play a vital successional role. However, it is important to distinguish the type of weed, i.e., non-noxious annuals or noxious perennials. The former can provide some beneficial aspects to reclamation such as shade for new seedlings with reduced surface temperatures, as well as reduced soil erosion and reduced grazing pressure. In addition, they can provide snow-catch in the winter time and improve soil moisture. Noxious perennials provide many more negatives than positives to desirable plant growth.

Since 1977, new invasive annual grasses such as cheatgrass (*Bromus tectorum*) have provided extreme challenges to reclamationists. More recently Ventenata grass (*Ventenata dubia*) and medusahead (*Taeniatherum caput-medusae*) have caused new concerns. Immediate control of small populations is often the best defense for such aggressive annual grasses. Once established, control can be problematic and often takes a multistep, multi-year approach to significantly reduce or control invasions of these annual grasses.

1.2 *Wildlife habitat*

Much has changed since the days in the 1970's when reestablishment of wildlife habitat was questioned. Doubt existed that large ungulates such as Pronghorn Antelope and Mule Deer would utilize and repopulate reclaimed areas in Wyoming. In addition, no one foresaw then that some species under the threat of listing as Threatened or Endangered by the U.S. Fish and Wildlife Service would lead much of the research being conducted today. Some considerations for wildlife habitat diversity include:

1.2.1 *Shrub establishment*

As part of reestablishing the land use of wildlife habitat, full shrubs such as Wyoming big sagebrush (*Artemisia tridentata wyomingensis*) had to be reestablished. Such reestablishment was difficult in the northeastern corner of the state which is transitional to the high plains and prairie ecosystems to the east and north. In the early years following the implementation of SMCRA, shrubs were assumed to invade, as was evident on many go-back agricultural hayland systems in northeastern Wyoming. Early seeding used fourwing saltbush (*Atriplex canescens*) as a substitute for big sagebrush. As mentioned earlier, southern sources of this seed did not allow for long-term establishment. Fall and spring seeding of shrub seed was conducted and often in the same drill rows as other plant lifeforms which often buried small seed and provided competition from perennial grasses. Winter broadcast seeding of big sagebrush and other shrub seeds is now considered a desirable method of shrub establishment.

Early on, many federal agencies were still trying to control big sagebrush through spraying, burning or chaining. It was not until much later that these practices transitioned to protecting big sagebrush stands. Since then, much has been learned about the need for such shrub patches in the landscape and associated research is still ongoing. Not all big sagebrush patches are considered equal in quality for wildlife habitat.

In 1996, the Wyoming Department of Environmental Quality-Land Quality Division (WDEQ-LQD), promulgated shrub density standards for shrub reestablishment. This was part of the regulatory performance standards a mine operator is required to meet in order for a reclaimed area to be deemed successful for release of the reclamation bond. The shrub density standard joined other revegetation performance standards for plant cover and plant production that had been in place for some time under Wyoming's primacy.

1.2.2 *Landscape diversity and topsoil*

In the early implementation of SMCRA, most reclaimed landscapes in Wyoming's surface coal regions were homogenous with less steep slopes as a result of coal removal and backfill sloping, as well as even redistribution of salvaged topsoil. Despite the fact that slope configurations of 5:1 were more desirable in high rainfall climates, these less steep slopes were established in semi-arid/arid areas. Remnant high walls or scarps could not generally be left, even if they were suitable habitat for many raptors. Currently, some (if well-planned in the mine permit and capable of demonstrating post mine stability) can be considered.

Recent changes in geomorphic reclamation has literally changed the landscape of the post-mining topography. Considerations for stabilizing drainage features allows the reclaimed area to mimic pre-disturbance stream channels.

Proper topsoil salvage and replacement was mandated by SMCRA. Direct haul of salvaged topsoil was an important consideration for the biological viability of the topsoil, as well as mine planning. Variable topsoil replacement depths are being used by some mines to create more plant diversity in the landscape and assist shrub growth.

Alleviating compaction by ripping of backfilled areas, travel paths for equipment, etc. is an important consideration. Contact between the overlying topsoil and underlying compacted layer enhances water movement and allows for plant roots to further break up those compacted zones.

Depending on the desired post-mining plant community, substrate replacement, both quantity and type, is critical. In order to get trees and some types of shrubs reestablished, it is necessary to use rockier substrate as replacement material on the reclaimed area. The uniform replacement of topsoil requirement of SMCRA did not foresee such need.

1.3 *Other considerations*

Some following does not fit into any earlier category:

1.3.1 *New technology*

No one could have foreseen how new technology has changed reclamation. The application of Geographical Information Systems (GIS) query capability in evaluating landscape problems through imagery is extremely valuable. Until recently, crude topographic and aerial maps that guided our thinking have given way to those produced with the use of Light Detection and Ranging (LIDAR), satellite, digital high altitude aerial imagery and drones or Unmanned Aerial Systems (UAS).

Global Positioning System (GPS) is used in heavy equipment to replace backfill and topsoil, as well as guiding topsoil salvage operations. Post-mining topography recreation is now at the dozer-level. In conjunction with the use of extensive databases, undesirable material can now be tracked in ways never understood in 1977.

2 WHAT HAVEN'T WE LEARNED YET

Regulations are often written with broad brush implications but implemented on a site-specific or regional basis. The one-size-fits-all mentality is often engrained in regulation simply as a result of the process. Complex problems deserve complex solutions and a regulatory framework that can adapt to finding those solutions. When SMCRA was devised, it was meant to address many of the problems in the eastern USA. Mines west of the 100th Meridian (or essentially the western USA) had similar categories of items to address such as topsoil salvage and stream protection but the environment was vastly different. Similarly, regions throughout the world have similar categories but their environments can be vastly different too. Fixing past problems or planning for new disturbance must take into account those regionally-specific conditions. Understanding the overall ecology of any site, its limitations and its challenges essential to achieving success.

Site specific considerations may include:

- Are there limited resources such as seed, water, topsoil etc.?
- How can those limited resources be used or preserved most effectively?
- How can organic matter best be incorporated into the system to increase soil fertility?
- Is stability a priority?
- Is another land use, such as wildlife habitat, a priority?

There is much continuing debate whether reclamation or restoration is the proper word for what SMCRA should accomplish. Ecosystem function and land uses can be restored. The restoration of late successional plant communities, however, is something only time can provide. What then is the best means for success comparison (for bond release purposes) with a young reclaimed area? Is it a similar undisturbed vegetation community, ecological site (as defined by the NRCS), or some type of technical standard? What is a reasonable expectation of trend given the five and ten year bonding time frames? That debate continues.

3 SUMMARY

Most interviewees agreed that technology has advanced sufficiently to provide the tools and resources needed to properly and cost-effectively reclaim a surface coal mine in Wyoming. These include, in part, available quality and quantity of seed, soil amendments, and the proper equipment. Potential variables impacting success include the preplanning needed to maximize the efficiency of effort, the money allocated to the necessary resources, the regulatory environment that allows flexibility in implementation, and the managerial and engineering commitments to consider reclamation a vital part of the mining process. As many western mines, such as in Wyoming, have gone from being "new" in 1977 to "mature" in 2017, how are the lessons learned over 40 years recognized and applied today?

As the nations of the world utilize their natural resources for the betterment of their people, how can impact be minimized while maintaining long-term sustainable use? Reclamation definitely has a role. Mining operators and regulators in Wyoming and the semi-arid/arid western USA have learned much in the 40 years since SMCRA was passed and implemented. According to Schroeder, et al., 2016, important positives during the course of a reclamation career include "dedication to the advancement of the sciences, professionalism, inspirational mentoring, and partnerships. Wyoming's reclamation practitioners have been a resilient group of people that have met many challenges over the 40 years since the passage of SMCRA. We likely have more to learn. Our charge is to pass on the knowledge we have gained.

ACKNOWLEDGMENT

Thank you to all the former and current industry personnel, as well as regulatory, that allowed the Author to interview them. Your contribution to this document was invaluable. The Author has chosen not to mention you by name to allow you to speak freely.

REFERENCES

Schroeder, S., T. Toy, and K. Vories. (2016) Challenges and Opportunities: Reflections on a Life in Mined Land Reclamation. Reclamation Matters, Spring 2016.

Thilenius, John F. and Gary B. Glass. (1974) Surface Coal Mining in Wyoming: Needs for Research and Management. Journal of Range Management 27 (5), 336–341.

Land Reclamation in Ecological Fragile Areas – Hu (Ed.)
 ISBN 978-1-138-05103-4

Risk-profit equilibrium of private capital in the Public-Private-Partnership (PPP) model of reclamation for coal mine subsidence

F. Li
College of Resources and Environment, Shandong Agricultural University, Taian, Shandong, China
College of Economics and Management, Shandong Agricultural University, Taian, Shandong, China

X.J. Li & X.Y. Min
College of Resources and Environment, Shandong Agricultural University, Taian, Shandong, China

Z.L. Wu
Jining Land Resources Bureau, Jining, Shandong, China

ABSTRACT: The reclamation for coal mine subsidence projects need a lot of capital, with the characteristics of huge investments, long project periods, and hard predicting risks. On the one hand, the government is facing enormous financial pressure. On the other hand, the government investment is inefficient due to the lack of expertise. Therefore, it is necessary to solve the problems using the PPP model to bring in private capital investments. The purpose of private capital investment is to obtain investment income. And earlier research indicated that a fair risk allocation among the stakeholders is essential for the success PPP projects. To improve negotiation efficiency, reduce financing costs, and ensure stakeholders' reasonable income level, a bargaining game-shapely model is built to solve the decision making problem of rational allocation of private capital returns based on full study of project risk. The paper provides a theoretical method, which contributes to improve the efficiency of financing negotiations, bring in private capital investment, and speed up the reclamation process.

Keywords: Coal Mine Subsidence; Reclamation; PPP; Risk; Profit; Game

1 INTRODUCTION

Coal dominates the energy consumption in China; a large number of well mining leads to ground subsidence and cause damage to land. By the end of 2013, there are 15,000 coal mines; the cumulative area of mining is 3600 million hectares, mined-out area is 700,000 hectares and coal mining collapse of the land is about 350,000 hectares (Shanggui Sun and Jinzhong Li, 2014). All of these are threatening the national food security. Therefore, the reclamation of coal mining subsidence, and the maximum protection of arable land is the necessary prerequisite to protect the national food production safety and the effective supply of mineral resources which has aroused the majority of our scholars and the government attaching great importance to the problems (Hu Zhenqi and Long Jinghua, 2016, Li Xinju Etc, 2014, Chen Longqian Etc, 2011, Li shuzhi, 2014, Bi Yinli Etc, 2014). This paper systematically studies the reclamation of coal mining subsidence area from the aspects of project willingness, reclamation technology, quality evaluation and benefit evaluation.

The starting point of private capital investment in reclamation is the mechanism of investment income, and the key to success lies in the reasonable sharing of risks. There are a lot of investigations from a qualitative point of view to study financing of reclamation (Zhang Qing-jun and Lu Jun-na, 2008). However, quantitative investigation was not found in literature. Therefore, based on the study of risk, this paper establishes a bargaining game—shapely model, carries out the rational allocation of private capital gains and provide a theoretical basis for financing negotiations which could improve the efficiency of negotiation, reduce the financing cost and guarantee the rationalized income level of government and private capital, and provide reference for application of PPP model in reclamation of coal mining subsidence area.

2 STUDY OF THE APPLICATION OF THE PPP MODEL IN THE PROJECT OF THE RECLAMATION IN COAL MINING SUBSIDENCE LAND

The management of coal mining subsidence area has a large scale of investment, a long cycle and difficult-forecasting risks. The reclamation rate of

China's land is still very low, the primary reason is that the reclamation funds is difficult to implement, especially the land that ware destroyed before the promulgation of "Regulations of Land Reclamation" are historical debts. The project of reclamation is huge and is difficult to ensure adequate financial funds. The second reason is that the project of reclamation is not well coordinated; there is a relationship between the relevant stakeholders and affecting the efficiency of reclamation (Wang Qiao-ni and Cheng Xin-sheng, 2008). The PPP model can solve the problem of shortage of reclamation funds and the coordination of relevant main-bodies in coal mining subsidence area, which is the only way to promote the process of reclamation in coal mining subsidence area in China.

According to the use of subsidence, it can be divided into two types: sub-governance can be transferred and still scattered operating by the farmers after treatment. If the land after governed can be transferred, the private capital investment entity will uniformly manage the land after reclamation by acquiring the franchise and putting operating income as part of the financial resources to repay the loan, and eventually recover the investment and obtain the return on investment which belongs to the quasi-business. The basic structure of the PPP project with the franchise for the governance of collapse zone is shown in the Figure 1. If there is no franchise, it is a non-business projects,

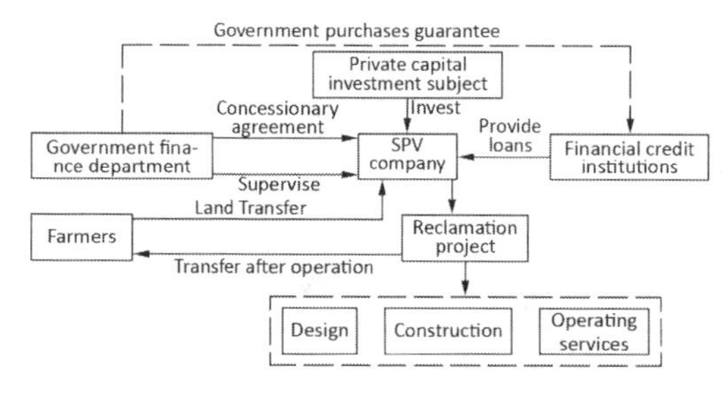

Figure 1. The basic structure of the PPP with franchise for subsidence governance.

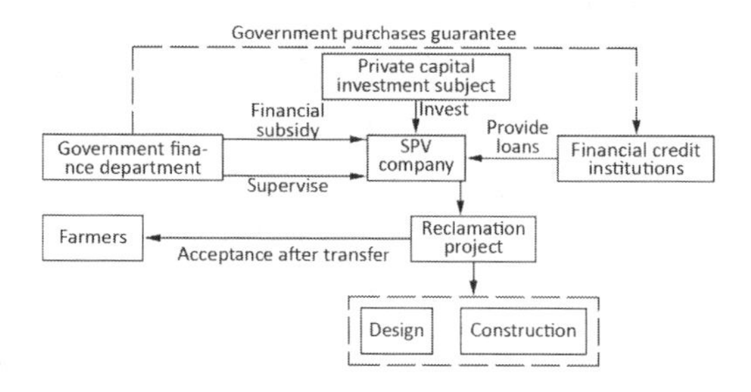

Figure 2. The basic structure of the PPP without franchise for governance subsidence.

relying entirely on government fees to achieve the project to repay the loan, recover investments and access to investment return. The basic structure of its PPP is shown in the Figure 2. The main sources of risk for PPP projects including: political risk, the risk of the project itself, financial and market risk, force majeure risk.

3 RESEARCH ON RISK BALANCE OF THE PPP PROJECT

Both the government and the private investment entities are able to anticipate the profit of unit time during the concession period. The focus is on the size of the government subsidy, the franchise of the operating cycle and the share of the risks. $K=(k_1,k_2... ...k_i)$ is the vector of the risky proportion which is undertaken by the government. The vector of the risky proportion for the private investment entities is $1-K$. The expectations of project disbursement, economic benefits, social benefits and other benefits are fixed, and can be predicted by both sides.

3.1 *Risk Sharing under asymmetric information*

Government departments G and private investment entity P are rational and they both hope to reduce their own risk and to obtain the maximum benefit when the cooperation is established. In the bargaining process, the government departments will threaten the private investment to transfer R $(R=(r_1,r_2... ...r_i); 0\le R\le K\le 1)$ of risk to them. Both sides need to consume a certain amount of strong strategy and non-strong strategy during the bargaining process. $\xi\,(1<\xi)$ represents a break factor in a turn of negotiation.

For an individual risk, we can draw conclusions as following: the proportion of the risk borne by the government and the private investor is the sub-game perfect Nash equilibrium expressed by the Equation $k^*=qr+[1/(1+\xi)]$ and $1-k^*=1-qr-[1/(1+\xi)]$. Therefore, sharing of the risk mainly depends on the coefficient ξ of contract impairment, qr is the proportion of risk that is transferred to the private investment by government departments with their strong position.

3.2 *Research about income distribution based on Shapley value model*

The number of the individuals that are involved in a project at the same time is N. $S(S\in N)$ is a composed alliance. $V(S)$ represents the union's income which is defined as a feature function on the union set. φ_i is the income of the direct party i who participates in the game. When the parties bear the same amount of investment and share of the risk, φ_i satisfies the Equation(1) according to the Shapley theorem (Li Bing, 2005).

$$\varphi_i = \sum_{S \in N} \frac{(|S|-1)!(n-|S|)!}{n!}[V(S)-V(S-i)] \qquad (1)$$

When the participants assume different investment quotas and the share of risk, it needs to add adjustment factors R_i. So that we can obtain the Equation $V_i = \varphi_i + (R_i - 1/n) \times V$. Among them $R = A \times B$, A is a normalized matrix that is the allocation of influential factors. B is the matrix that the extent to which each influencing factor affects the distribution of benefits.

3.3 *Case analysis*

A coal mining subsidence land reclamation project cooperated by the government and private investment through the implementation of the PPP model with a total investment of 50 million. Of which the government invested 5 million and the private investment fund invested 45 million. The two sides of the capital ratio is 0.1: 0.9. The total income is 100 million yuan. If the government departments and private investments do it separately, the two sides will receive earnings of 40 million and 60 million. After the consultation of experts and questionnaire survey, the matrix $B = [0.7, 0.3]$ that reflects the influential degree of the proportion of investment and risk commitment for profit distribution can be obtained. In addition, the parameter values listed in Table 1 can be known: q is the probability for each risk of government strong strategy; r is risk share; ξ is the factor of negotiation damage and ω is the importance of each risk factor.

Finally, it can be gotten that the government departments and private investments separately obtain the income of 26.9 million and 73.1 million.

Table 1. Risk parameter table for reclamation project of a coal mining subsidence.

	Risk parameters			
The type of risk	q	r	ξ	ω
Political risk	0.8	0.15	1.21	0.1
Projects' own risk	0.5	0.03	1.26	0.4
Financial and marketable risk	0.6	0.06	1.25	0.4
The risk of force majeure	0.6	0.08	1.13	0.1

4 CONCLUDING REMARKS

Based on the theory of bargaining game, this paper establishes a risk sharing model for the PPP financing model of coal mining subsidence under the condition of asymmetric information and sets the model parameters to calculate the plan of risk sharing. In this paper, the Shapley value method is used to further study the income distribution scheme of project, selecting the proportion of investment and the risk factors as the adjustment factors, and finally getting the proportion of the interests' distribution for the government and private investment. The project risk sharing is closely related to the distribution of income. Based on the reasonable risk sharing, the model establishes the calculation method of income distribution which makes the calculation result more realistic. It reduces the subjectivity and blindness of the two sides' consultations and speeds up the project process which is significant to promote the construction of reclamation projects in coal mining subsidence areas.

REFERENCES

Bi Yin-li, Wang Jin, Feng Yan-bo, Yu Hai-yang, Qin Ya-fei, Yu Miao. (2014) Effect of arbuscular mycorrhiza on root self-repairing action of Amorpha fruticosa L. In coal mining subsidence land in arid areas. *J. China Coal Society*, 08, 1758–1764.

Hu Zhenqi, Long Jinghua, Wang Xinjing. (2014) Natural restoration and artificial restoration of ecological environment for coal mining. *J. China Coal Society*, 08, 1751–1757.

Li Bing. (2005) The allocation of risk in PPP/PFI construction projects in UK. *International J. Journal of project management*, (23), 25–35.

Li Shu-zhi. (2014) Present Status and Outlook on Land Damage and Reclamation Technology of Mining Subsidence Area in China. *J. Coal Science and Technology*, 01, 93–97.

Li Wei, Chen Long-qian, Zhou Tian-jian, Zhang Kui, LI Long. (2011) Research Progress of Soil Quality in China Mining Subsidence Area. *J. Coal Science and Technology*, 05, 125–128.

Liu Ning, Li Xinju, Guo Bin, Min Xiangyu. (2016) Compaction Evaluation of Reclamation Soil in Coal Mining District Based on Modified Dielectric Constant Model. *J. Soil and Water Conservation*, 04, 318–322.

Shanggui Sun & Jinzhong Li. (2014) The Study of China Reclamation for Coal Mine Subsidence. *J. Contemporary Economics*, 21, 52–53.

Wang Qiao-ni, Cheng Xin-sheng, Zhang Zhi-guang. (2008) Analysis on the Status Quo, Problems and Reasons of the Reclamation for Coal Mine Subsidence in China. *J. Energy Environmental Protection*, 05, 49–53.

Land Reclamation in Ecological Fragile Areas – Hu (Ed.)

Analysis of influencing factors of land reclamation incentive policies and regulations implemented in a mining area—a case study of Wu'an city

L.J. Yin & M.C. Fu
School of Land Science and Technology, China University of Geosciences, Beijing, China

ABSTRACT: The exploitation of mineral resources occupied and destroyed a large amount of land and damaged the ecological system seriously. How to perfect the incentive policies and regulations in mining areas is an urgent institutional problem to be solved in the construction of ecological environment. This paper, taking the mining city Wu'an as an example, takes field investigation to interview the local government, mining area development enterprise, reclamation enterprise and local peasants, and conducts data analyzing by constructing the structural equation model with Amos 17.0 to clarify the dominant factors and obstacles that influence the land reclamation incentive policies and regulations enforcement. The results show that (1) different stakeholders have different cognitions to the factors that hinder the implementation of incentive policies and regulation; (2) the interactions between local governments and other stakeholders, especially local peasants, are obviously weak, and the government guaranteed the interests of local peasants insufficiently; (3) the government's supervision to the incentive policies and regulations enforcement is not enough, and the main responsibility between different stakeholders is not clear. This paper also proposes some reasonable suggestions to perfect the incentive policies and regulations of land reclamation in mining area.

Keywords: Land reclamation; Incentive policies and regulations; Structural equation model

1 INTRODUCTION

Land reclamation is a task with policy enforcement, widespread scope, and character of comprehensive. At present, the rate of mining area land reclamation in China is only 12%, far below the international 50% to 70%, lower than western 80% (Yang Huili & Fu Meichen, 2016). Mineral resources as an important resource of implementing sustainable development strategy in our country, for the social development and economic construction provides a strong guarantee (Zhang Haiyan, 2012). The late 1970s, the United States, Canada and other countries have developed a special land reclamation regulations. Canada *open pit and quarry control and reclamation law* on the land reclamation planning content, acceptance criteria, the source of reclamation funds, government departments' responsibility and reclamation technology made a detailed requirement (Jin Dan & Bian Zhengfu, 2009). In recent years, *land reclamation laws and regulations* continue to improve and perfect, China's scholars from different angles, studied land reclamation policies and regulations. Luo Ming (Luo Ming, 2011) compared the land reclamation system between China and the United States, put forward some suggestions on the direction of land reclamation in China. The current land reclamation policy in our country is not enough to pay attention to the incentive policy of land reclamation in mining area. It is a trend and policy requirement to improve the land reclamation policy and regulations. From the cognitive perspective of different stakeholders on policy, based on the structural equation model, this article will focus on the above research.

2 METHODS

2.1 *Research method*

Based on research ideas and purposes of this paper, choose the structural equation model as the mathematical method.

2.2 *Data sources*

2.2.1 *Survey of the region*

Wu'an is located in Handan City, belongs to Hebei Province, located in 113°45′~114°22′east longitude, and 36°28′~37°01′ north latitude. The terrain west high east low, surrounded by mountains, forming a small plain (Zhang Jianjun & Fu Meichen, 2007). Wu'an as a mining city in Hebei Province, is rich of mineral resources, has proven coal mine has five major categories, 23minerals, especially in coal, iron, limestone are the most abundant. In recent years, relying on resources, Its related industries also have been rapid development. Undoubtedly, reclamation is also in an extremely important position, which is closely related to the national policy system and the correct policy guidance.

2.2.2 *Questionnaire data source*

The questionnaire mainly for the local government, mining enterprises, mining reclamation enterprises and local farmers those four major stakeholders issue questionnaires. There are 196 valid questionnaires. The questionnaire include individual characteristics and investigation on the policies implementation in Mining area, including the local government's propaganda of policy, interaction between the local government and the stakeholders, etc, there are four aspects: "very important", "generally important", "important" and "not important" were recorded as 4,3,2 and 1 respectively. The individual characteristics of the respondents are shown in Table 1 and Structural equation model index selection is shown in Table 2.

3 MODEL STRUCTURE AND OPERATION

On the basis of the study hypothesis, draw a structural model, which was estimated by using the maximum likelihood, obtained the Unstandardized Regression Weights model diagram (Figure 1) and Standardized Regression weights model diagram (Figure 2),

Table 1. The individual characteristics of respondents.

Variable	Number	Percentage	Variable	Number	Percentage
Sex	196	100%	Cultural level	196	100%
Male	158	80.6%	Undergraduate and above	24	12.2%
Female	38	19.4%	Jounior college	52	26.5%
Age (year old)	196	100%	Senior high school	39	19.9%
18–30	38	19.4%	Middle school and below	81	41.4%
31–40	76	38.8%	Position	196	100%
41–50	54	27.5%	Local government	28	14.3%
51–60	28	14.3%	Mining development enterprises	25	12.7%
			Mining reclamation enterprises	29	14.8%
			Local farmer	114	58.2%

Table 2. Model symbol definition.

Potential variable	Observable variables	Symbol definition
Policy implementation	Local government's propaganda of policy	X1
	The degree of interaction between the local government and the various stakeholders	X2
	The degree of the local government to protect the legitimate rights and interests of the subjects	X3
	The local government to consider the depth of the interests of farmers	X4
	Sources of financing in the process of policy implementation	X5
	The supervision system and feedback mechanism in the process of policy implementation	X6
	Compensation Mode in the Process of Policy implementation	X7
	The way to deal with the ownership dispute in the process of policy implementation	X8
	Quality review mechanism in the process of policy implementation	X8
	Public participation in the process of policy implementation	X10
	Clear the rights and responsibilities of different stakeholders in the process of policy implementation	X11
	The amount of payment of the relevant enterprise deposit in the process of policy implementation	X12
	The degree of economic benefits after the implementation of the policy	X13
	The overall effect after the implementation of the policy	X14

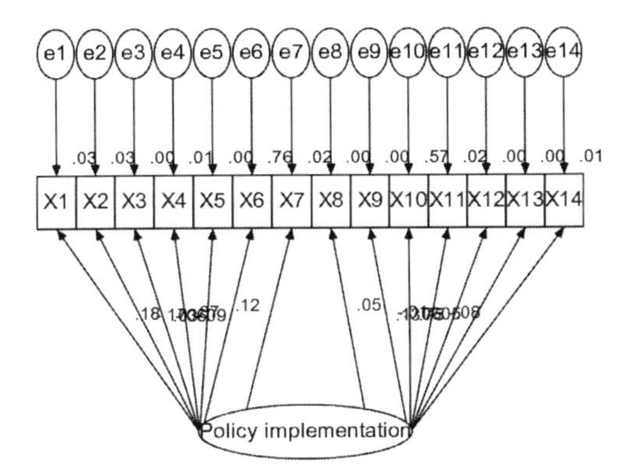

Figure 1. Unstandardized regression weights model diagram.

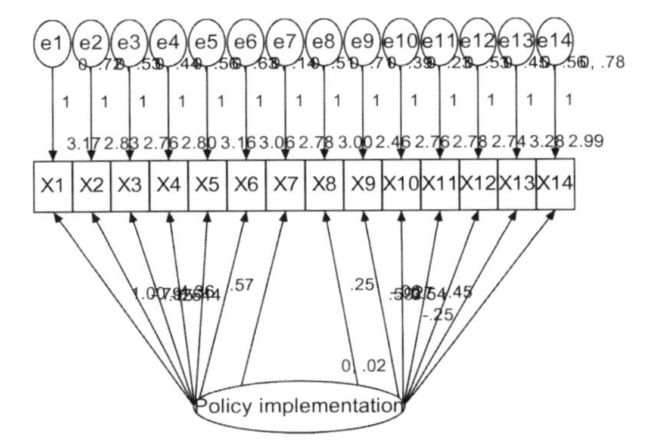

Figure 2. Standardized regression weights model diagram.

4 MODEL RESULTS AND ANALYSIS

All of indexes that the results of the model test indicators are meet the model fit standard. It is assumed that the model and the observed data are well fit, indicating that the model is ideal and can be accepted (Wu Minglong, 2010). Standardized regression weights of the model are shown in Table 3.

The above research results show that the implementation of the land reclamation incentive policy in Wu'an mining area is affected by many factors. The main influencing factors are, The supervision system and feedback mechanism in the process of policy implementation (X6) and Public participation in the process of policy implementation (X10), It's influence coefficient reached 0.874 and 0.752 respectively. Then, the more important influencing factors are Local government 's propaganda of policy (X1), The degree of interaction between the local government and the various stakeholders (X2), Clear the rights and responsibilities of different stakeholders in the process of policy

Table 3. Standardized regression weights of the model.

			Estimate
X1	<---	Policy implementation	.181
X2	<---	Policy implementation	.167
X3	<---	Policy implementation	–.059
X4	<---	Policy implementation	.092
X5	<---	Policy implementation	.029
X6	<---	Policy implementation	.874
X7	<---	Policy implementation	.124
X8	<---	Policy implementation	.047
X9	<---	Policy implementation	–.014
X10	<---	Policy implementation	.752
X11	<---	Policy implementation	.125
X12	<---	Policy implementation	.062
X13	<---	Policy implementation	–.052
X14	<---	Policy implementation	–.079

implementation (X11) and Compensation Mode in the Process of Policy Implementation (X7), the influencing coefficients were 0.181, 0.167, 0.125 and 0.124, respectively.

Mining land reclamation incentive policy implementation, can strengthen the management of land reclamation, improve the efficiency of land use, and provide an important guarantee for the development of local social economy. More than two thirds of the respondents believe that the implementation of the land reclamation incentive policy supervision process is extremely important. In fact, some of the operational measures of Wu'an City is not perfect, lack of specific implementation details, etc., seriously restricting the implementation of the policy effect. In the actual survey, it is found that the mining development enterprises and the mining reclamation enterprises generally believe that the interaction between the local government and the stakeholders is also the main influencing factors of the incentive policies implementation, while the local farmers are mainly concerned about compensation mode. To some extent compensation affects the local farmers' income, such as returning the cultivated land occupation tax, reclaiming the land and giving the economic subsidy. Undoubted, the compensation mode has become a topic of concern to farmers in wu'an mining areas.

5 CONCLUSIONS AND RECOMMENDATIONS

The main conclusions of this paper are, (1) The different stakeholders have different cognition to the factors that hinder the implementation of incentive policies and regulation; (2) the interactions between local governments and other stakeholders, especially local farmers, are obviously weak, and the government guaranteed the interests of local farmers insufficiently; (3) the government's supervision to the incentive policies and regulations enforcement is not enough, and the main responsibility between different stakeholders is not clear, while the local farmers are more concerned about corresponding compensation. There are four suggestions here, (1) It is suggested that Wu'an City should be build a systematic, practical land reclamation incentive policy supervision system with certain scope and renewal period. (2) Implement public participation mechanism, maximum limit protect the public's right to know. (3) wu'an local government should increase propaganda of policy, strengthen the degree of interaction with other stakeholders. The government can through effective way of teaching, so that the stakeholders can understand of the importance of land reclamation. (4) Clear the rights and responsibilities of different stakeholders in the process of policy implementation, and take full account of the interests of local farmers.

ACKNOWLEDGEMENTS

Finalization of the paper was supported by project Natural Science Foundation of China (41641008).

REFERENCES

Jin Dan & Bian Zhengfu (2009). Comparison and Reference of Land Reclamation Policies and Regulations at Home and Abroad. *J. China Land Science*, 23(10), 67–68.

Luo Ming. et al. (2011). Land reclamation legal system construction has a long way to go—Analysis from the perspective of the land reclamation system of Sino US comparison. *J. Land science*, (07), 44–46.

Wu Minglong. (2010). Structural equation model. AMOS Operation and Application, 2nd Edition. Chongqing University Press.

Yang Huili & Fu Meichen. (2016). Discussion on Reclamation Promotion Mechanism of Mining Land. *J. Theory and Discussion.*

Zhang Haiyan. (2012). Study on the Legal Mechanism of Ecological Compensation for Mineral Resources Development. *Xi'an: Chang'an University.*

Zhang Jianjun & Fu Meichen (2007). Study on Ecological Function Division of Wu'an City. *J. Resources and industry*, (03), 98–99.

Land Reclamation in Ecological Fragile Areas – Hu (Ed.)
© 2017 Taylor & Francis Group, London, ISBN 978-1-138-05103-4

Integrating green infrastructure strategies in post-mining areas

W. Wende, R.-U. Syrbe & P. Wirth
Leibniz Institute of Ecological Urban and Regional Development, Dresden, Germany

J. Chang
Chinese University of Mining and Technology, Xuzhou, China

ABSTRACT: Green Infrastructure (GI) is a new concept for Europe as contained in political documents from the EU Commission. GI comprises a strategically planned network of high-quality natural and semi-natural areas, not only in the countryside but also in cities. This paper aims to show initial approaches towards combining the concept of GI with individual strategies to develop post-mining landscapes in Chinese and German contexts. Using the German municipality of Bad Schlema as an example, it gives an impression of how pro-active post-mining development can overcome a desperate situation, using GI as a key concept. The restoration process was supported by government but also driven by the visions of local people. The main factor was an enhancement of the green potential for spa tourism and improved quality of life. The GI concept provided a structural backbone and guiding principle for the transformation of the city's appearance and identity.

1 INTRODUCTION

1.1 *Definition of Green Infrastructure related to the European Union*

Green Infrastructure (GI) is a new concept for Europe as contained in policy documents from the EU Commission (EU Communication: COM (2013) 249 final). However GI is also emerging as a new international concept in different countries of the world. GI comprises a strategically planned network of high-quality natural and semi-natural areas with other environmental features, designed and managed to deliver a wide range of Ecosystem Services (ESS) and protect biodiversity in urban settings. Thus, GI follows the target of improving urban ESS and human well-being, as well as promoting sustainable development in cities. The German government is currently launching a Federal GI Concept which also incorporates GI elements within post-mining areas (BfN 2017).

1.2 *GI in the general post-mining context of China and Germany*

This paper aims to develop initial ideas about combining the concept of GI with individual strategies to develop post-mining landscapes. Coal is the main source of energy in China, and remains an important energy resource in Germany. However, coal mining activities and rapid urbanisation have had a considerable impact on the GI structure in 'coal cities', particularly in China but also in Germany, resulting in the degradation and instability of overall ESS and damage to green infrastructure. Germany has accumulated some experience in both ecological restoration and green city construction using green infrastructure strategies. All these experiences could have strong implications for China. Thanks to the energy transition, Germany is currently facing a series of technical, methodological and planning adjustments of relevance to GI. This presentation initiates the idea of investigating coal mining cities in China and Germany as examples, with the aim of capturing temporal-spatial differences in scales and landscape structures, and revealing GI growth characteristics and laws. Based on these findings a full research project is to be developed and conducted in the future.

2 STATE OF THE ART: POST-MINING RESEARCH AND GI

Mining is a very dynamic process of landscape change that is accompanied by environmental impacts and fundamental social reorganisation, not only in the decades when resources continue to be mined but also when reserves are exhausted and the mines closed. Challenges in post-mining landscapes are similar in almost all mining regions (Wirth & Lintz 2007). Mining often causes considerable environmental degradation and severely affects the provision of ESS (Wende 2015, Chang & Yang 2015).

A main focus in research has been on transformation processes in former mining regions. Descriptive and explanatory studies of post-mining areas deal with political, cultural, social and economic change. German studies focus on the Ruhr district, once the largest old industrial area in Europe (Ache 2000; Wissen 2001). Rehabilitation has often become a task of national importance, such as the rehabilitation of lignite, copper and uranium mining areas in eastern Germany (Lintz & Wirth 2009, Wirth et al. 2012). It is therefore clear that the improvement of ecological and environmental planning tools as part of regional planning is a long-term task (Wende 2011, Wende 2015) and should focus on green infrastructure conception. It needs to be developed and conceptualised in detail before mining is initiated; this process must include an assessment of ESS (Syrbe 2015). However, there is a great lack of green infrastructure theory and assessment methods, particularly in the post-mining context (Feng et al. 2016; Syrbe & Chang 2017).

In China, mining development has resulted in a large number of post-mining landscapes, mainly situated in urban areas that grew strongly during the mining period. This has involved a number of negative impacts on the environment and living conditions (Chen et al. 2010, Luo & Chen 2011, Chang & Yang 2015, Ma & Hu 2015). Comprehensive land reclamation planning for mining areas and mine rehabilitation are still on an exploratory level and lack a theoretical foundation, while research on ecological restoration and rehabilitation is at a qualitative descriptive stage. Particularly, better restrictions and more mandatory safeguards under the relevant laws and regulations are necessary (Xia 2006). During the process of implementation, problems such as inadequate management, a lack of feedback and monitoring, and close interrelations with urban development have arisen (Wirth & Lintz 2007, Chang et al. 2011).

The damage to the landscape caused by mining should be repaired and compensated through a series of models, concepts, technical, landscape-structural and legal measures (Hennek & Unselt 2002, Wende et al. 2009, Wende 2015, Chang & Yang 2015). In Germany there are several examples of successful mining transformations and the re-use of post-mining landscapes, based on a combination of qualitative and quantitative methods (Ache 2000, Wissen 2001, Wirth et al. 2012, Harfst & Wirth 2011, Lockie et al. 2009). This experience is of great significance for China, which is undergoing rapid development. It could also provide ideas and opportunities for development of the corresponding geographical theory, ecological and environmental planning methodology and technology. German experience should thus not only be transferred to China but also used to enhance methodology and to improve implementation in both countries (Müller et al. 2005). The overall state of the art illustrates a significant research gap on the question of how to integrate the two concepts of post-mining and green infrastructure.

3 CASE STUDY BAD SCHLEMA (GERMANY)

3.1 *Description of the case study area*

The small community of Bad Schlema (4,848 inhabitants, 2015) is located in Saxony, a federal state in the eastern part of Germany. With an altitude of about 360 metres above sea level, it is embedded in the beautiful scenery of the Ore Mountains, a mountain range along the German-Czech border. Bad Schlema, and indeed the whole of the Ore Mountains, can look back on a long tradition of mining for tin, iron, silver, bismuth, copper, cobalt, nickel and other minerals since the 15th century (BMWi 2015, p. 42 ff.).

During mining explorations between 1909 and 1911 the world's most radioactive water at that time was found in Schlema (Wolkersdorfer 2008, p. 34). As the medicinal benefits of radioactive water were already established, the idea was born to develop a spa resort in Schlema. Consequently, a spa quarter was developed with a therapy centre, park, leisure facilities and hotels (Figure 1). The health spa business started in 1918 and the town quickly became one of Germany's most important radium spas.

3.2 *The destructive impacts of mining*

Spa development stopped as suddenly as it had started. In the aftermath of World War II, under the political conditions of the Cold War, the Soviet-German joint-stock company 'Wismut' began

Figure 1. The spa centre of Schlema in the 1930s (Source: Municipality of Bad Schlema).

mining the rich uranium deposits below Schlema for the Soviet nuclear power industry. The excavation started in the centre of the spa area. After a few years, up to 25,000 people were working in 54 underground pits. At the end of mining, the tunnels stretched over a total of 4,200 km (BMWi 2015, p. 42 ff.).

When uranium mining stopped with German reunification and the end of the Cold War in 1990, the environmental damage that had been caused was immense. Mining slag was piled up in 42 heaps, covering more than one-third of the town's area (311 ha; BMWi 2015, p. 42 ff.). Underground mine galleries increased the danger of surface collapse and subsidence. The whole townscape was defaced by slag-heaps, winding towers, the equipment and facilities of the SDAG Wismut company, mineworkers' estates and wastelands (Figure 2). At that time, the town was named a 'Valley of Death' in certain media because of the manifold sites with radioactive contamination (Lintz and Wirth 2009, pp 81/82). Against this dramatic background the situation of Schlema seemed fairly hopeless. Some experts proposed demolishing the remaining built-up areas and rebuilding elsewhere.

3.3 *From uranium mining to tourist destination – 'The miracle of Schlema'*

Finally, it was decided that it was not possible to establish a devastated and restricted area within an – all in all – densely populated region. In this situation it was of crucial importance that the German federal government launched an exclusive state rehabilitation programme for the former uranium mining areas in eastern Germany. To manage the tasks, the former mining company was restructured as Wismut GmbH, and remains in charge of all technical and organisational questions of rehabilitation to this day (BMWi 2015). The main focus of rehabilitation measures was and is on the stabilisation of the underground galleries and the mining dumps, as well as on dealing with radioactive materials.

Moreover, a group of visionaries converged in Schlema and together resolved to rebuild the spa resort. On the basis of a spa-development concept and state-supported remediation measures, significant investments in tourist infrastructure were made in the following years (for more detail see Lintz & Wirth 2009). A new spa area with medicinal spa, park, hotel and a huge spa garden area (Figure 3) was opened in 1998, and these days around 1000 visitors come to the health resort every day. The government of the federal state of Saxony, where Schlema is situated, awarded the spa title "Bad" to the city in 2004. And in 2018 the inhabitants will celebrate the hundredth anniversary of the spa resort. Because of the radical and unpredictable renaissance of the municipality, public media described it as 'The miracle of Schlema'.

3.4 *The role of GI*

Looking back at the spa development in Bad Schlema over the last 25 years, the 'miracle' can be interpreted as the result of combining a strong state rehabilitation programme with a local initiative to shape the post-mining landscape and apply a GI approach to Bad Schlema. The main factor was an enhancement of the green potential for future spa tourism and improved quality of life. This concept provided a structural backbone and guiding principle for the transformation of the appearance and identity of the town with the following GI components:

Figure 2. Schlema in the uranium mining period 1956 (Source: Municipality of Bad Schlema).

Figure 3. Spa gardens of Bad Schlema (2006) on the ground of the former mine (Source: Wirth).

- The spa garden of 18 hectares in the centre of the municipality on the site of the former mine facilities (see the foreground of Figure 3). In this area, underground mining galleries had to be refilled to ensure surface stability (Gemeinde Bad Schlema 2017).
- The mining dumps inside and outside the built-up area of Bad Schlema (see the right-hand background of Figure 3). All dumps had to be profiled and finally covered by a layer of non-contaminated soil (BMWi 2015, p. 10).
- The hiking and cycling trails connecting the municipality with a regional system of tourism trails. For this reason local infrastructure had to be upgraded (reconstruction of pathways and bridges).

Today both the inhabitants of the municipality and visitors appreciate the quality of the spa, which many perceive as a successful amalgamation of built and green infrastructure.

4 CONCLUSIONS

The findings of this paper illustrate the immediate necessity for basic research on the question of how to integrate the concept of post-mining landscapes with the concept of green infrastructure. The restoration of post-mining cities has to be planned at the time of mining development by the mining enterprises, otherwise the main expenses will be carried by the government, as was the case in Bad Schlema and other eastern German mining areas. The example shows how a restoration planning process of this kind should include the visions and ideas of local people for their own future and implement them in a new concept for an entire region after its mining era.

The restoration plan should upgrade subsidence areas and brownfields to new (secondary) habitats and green areas, both as compensation for the lost (primary) pre-mining habitats and as new potential for the successful post-mining development of the region. Thereby, a focus should be on the methodological development of GI assessment methods for innovative post-mining concepts. Individual case studies such as Bad Schlema, Germany, could serve as role models for this research and for the Chinese context. New assessment methods will lead the development of post-mining areas in a more sustainable direction, which will provide a good quality of life and new visions for local people.

REFERENCES

Ache, P. (2000): Cities in Old Industrial Regions Between Local Innovative Milieu and Urban Governance—Reflections on City Region Governance. European Planning Studies, 8 (6) 693–709.

BfN—German Federal Agency for Nature Conservation (2017): Bundeskonzept Grüne Infrastruktur [Federal concept green infrastructure]. Bundesamt für Naturschutz, Bonn, 68 p.

BMWi (2015): Wismut. Landscapes designed and preserved. German Federal Ministry for Economic Affairs and Energy [WWW Document]. URL http://www.wismut.de/www/webroot/de/download.php?download=3509 (accessed 3.28.17).

Chang, J.; Wende, W.; Luo, P.; Deng, Y. (2011): Re-use of the Mining Wasteland. Tongji University Press, Shanghai.

Chang, J.; Yang, Y. (2015): Suitability evaluation of abandoned mine lands supported by GIS: A case study of Yangzhuang mining area in Huaibei. In: Hu, Z.Q. (Ed.): Legislation, Technology and Practice of Mine Land Reclamation. CRC Press 117–125.

Chen, N.; Peng, W.; Xiao, D.; Lei, Q. (2010): Selective Solar Tunnel-Concept and Experiment. Acta Energiae Solaris Sinica. v31 (4) 442–446.

Feng S.; Chang J.; Hou W. (2016): A framework for setting restoration priorities for coal subsidence areas based on Green Infrastructure (GI). Acta Ecologica Sinica. 36(9), 2724–2731.

Gemeinde Bad Schlema (2017): Spa Gardens Bad Schlema. Homepage of the municipality [WWW Document]. URL http://www.kurort-schlema.de/en/nature/parks/spa-gardens/(accessed 3.28.17).

Harfst, J.; Wirth, P. (2011): Structural change in former mining regions: problems, potentials and capacities in multi-level-governance systems. In: Procedia – Social and Behavioral Sciences 14, 167–176.

Hennek, F.; Unselt, C. (2002): Sicherung von Naturschutzflächen in Bergbaufolgelandschaften. Landwirtschaftsverlag Münster.

Lintz, G.; Wirth, P. (2009): The Importance of Leitbilder for Structural Change in Small Towns, in: Strubelt, W. (Ed) Guiding Principles for Spatial Development in Germany, German Annual of Spatial Research and Policy. Springer, Berlin, pp. 75–95.

Lockie, S.; Franettovich, M.; Petkova-Timmer, V.; Rolfe, J.; Ivanova, G. (2009): Coal mining and the resource community cycle: A longitudinal assessment of the social impacts of the Coppabella coal mine. Environmental Impact Assessment Review, 29, 330–339.

Luo, P.J.; Chen, N. (2011): Abandoned coal mine tunnels: Future heating/power supply centers. Mining Science and Technology, v21 (5) 637–640.

Ma, B.G.; Hu, Z.Q. (2015): Bioremediation of acid-mine drainage contaminated with acid and heavy metals in coal mine by sulfate-reducing bacteria. In: Hu, Z.Q. (Ed.): Legislation, Technology and Practice of Mine Land Reclamation. CRC Press 423–427.

Müller, B.; Finka, M.; Lintz, G. (2005): Rise and Decline of Industry in Central and Eastern Europe. Springer, Berlin.

Syrbe, R.-U. (2015): Recultivation and sustainable development of post-mining landscapes. In: Hu, Z.Q. (Ed.): Legislation, Technology and Practice of Mine Land Reclamation. CRC Press 489–492.

Syrbe, R.-U.; Chang, J. (2017): Options and challenges for implementing green spaces in urban development.

In: Grunewald, K.; Li, J.; Xie, G.; Kümper-Schlake, L. (Eds.): Towards Green Cities - Urban Biodiversity and Ecosystem Services in China and Germany. Springer Briefs (in print).
Wende, W.; Marschall, I.; Heiland, S.; Lipp, T.; Reinke, M.; Schaal, P; Schmidt, C. (2009): Umsetzung von Maßnahmenvorschlägen örtlicher Landschaftspläne. Naturschutz und Landschaftsplanung, 41, (5) 145–149.
Wende, W. (2011): Ecological Potential and Environmental Planning of Post-Mining Landscapes—Germany as a Role Model for China. In: Chang, J.; Wende, W.; Luo, P.; Deng, Y. (Eds.): Re-use of the Mining Wasteland. Shanghai: Tongji University Press, 22–39.
Wende, W. (2015): Environmental planning of post-mining landscapes. In: Hu, Z. (Ed.): Legislation, technology and practice of mine land reclamation. Proceedings of the Beijing International Symposium Land Reclamation and Ecological Restoration, CRC Press, 471–474.
Wirth, P.; Cernic-Mali, B.; Fischer, W. (Eds.) (2012): Post-Mining Regions in Central Europe—Problems, Potentials, Possibilities. München: oekom Verlag.
Wirth, P.; Lintz, G. (2007): Strategies of Rehabilitation and Development in European Mining Regions. In: Good (Best) Practice Cases in Regional Development after Mining and Industry. Symposium, May 17–18, 2006, Institute of Geography and Regional Science, University of Graz, Austria. Graz: Institut f. Geographie u. Raumforschung d. Universität, (Grazer Schriften für Geographie und Raumforschung; 42) 75–85.
Wissen, M. (2001): Strukturpolitische Intervention und ungleiche Entwicklung. Zur Rolle des Staates im Strukturwandel. Geographische Revue 3 (1) 3–22.
Wolkersdorfer, C. (2008): Water Management at Abandoned Flooded Underground Mines. Fundamentals, Tracer Tests, Modelling, Water Treatment. Springer.
Xia, C. (2006): Regulating mine land reclamation in developing countries: The case of China. Land Use Policy, 24, 472–483.

Land Reclamation in Ecological Fragile Areas – Hu (Ed.)

Discussion on land reclamation and ecological restoration of mining wasteland based on a public recreation perspective

Q.W. Cui
Beijing Laboratory of Urban and Rural Ecological Environment, School of Landscape Architecture, Beijing Forestry University, Beijing, China

ABSTRACT: Besides reclamation of such normal lands as farmland, forest land, grassland, etc., urban parks, country recreation sites, outdoor theaters, theme parks and various exercise and fitness sites may be constructed on many mining wastelands so as to satisfy various outdoor recreation demands of the public. Therefore, how to eliminate environmental pollution, ensure geological security and recover natural ecology and simultaneously realize the potential public recreation value of mining wasteland reclamation to the maximum extent are key problems in mining wasteland reclamation works. In this paper, main types of domestic and foreign public recreation-guided landscape reconstructions on mining wastelands are summarized through case study. Then, existing disadvantages and main challenges in current mining land reclamation works in China are analyzed. Finally, some suggestions and prospects are proposed in allusion to more economical, high-efficient and sustainable mining wasteland reclamation and landscape reconstruction based on public recreation perspective.

1 INTRODUCTION

Mining wastelands' reclamation work in China has made great progress since the reform and opening up, and the work is transforming from simple farmland and forest land reclamation to further more reconstruction and recycle aspects like ecological restoration, estate development, etc. More and more restoration and reconstruction projects are now focusing on such requirements of the public as outdoor exercise, landscape, recreation, tour and sightseeing.

Public recreation oriented land reclamation and ecological restoration refers to the restoration, reconstruction and reutilization practice opened to the public and aiming at satisfying people's demands like outdoor leisure, landscape tour, sports, culture entertainment, science popularization education as well as tour and sightseeing. These projects have eliminated environmental pollution, ensured geological security, recovered natural ecology and realized potential public recreation value of mining wastelands to the maximum extent so as to create great social comprehensive benefits.

2 TYPES AND EXPERIENCE

2.1 *Reconstruction type*

There is no lack of public recreation-oriented reclamation and restoration practices in domestic and foreign mining wasteland reclamation works. In this paper, the practices are classified into four main types through case collection, i.e., urban parks, country recreation sites, cultural facilities and theme parks.

2.1.1 *Urban parks*

Park refers to municipal facility providing the urban residents with places for outdoor rest, view and admire, game, exercise and entertainment and operated by governments or public bodies [1]. Due to rapid urbanization in recent years and along with urban expansion and new town construction, many mining wastelands are incorporated as urban construction land and reconstructed into urban parks serving the surrounding citizens. As for most of such parks, mining pit topographic conditions are made full used of to construct landscape, and unique spatial structures and landscapes are thereby created. Parc des Buttes-Chaumont (Figure 1 L) in Paris, France opened in 1867 as an important case in modern urban park development history, is reconstructed by making use of a mining pit in limestone quarrying at northern suburb of the city at that time [2, 3]. There are many urban parks, like parc de la Creueta del Coll park (Figure 1 M), in Barcelona, Spain which are reconstructed by making use of abandoned mining pits. In China, Rizhao Yinhe Park, Tangshan South Lake Park, Xuzhou Longwanhu Park, Jiaozuo Fengshanzhen Park, Quarry Garden in

Shanghai Botanical Garden [4] (Figure 1 R) are all famous successful cases. Some mining wastelands are reconstructed into scenic spots like Shaoxing East Lake and Zhejiang Xinchang Giant Buddha beauty spots.

2.1.2 *Country recreation sites*

Country recreation sites refer to sites for leisure, exercise and recreation located in a place far away from city suburbs, need no precision management, need low-cost maintenance and with rustic charm. Generally, mine fields are far away from urban settlement, therefore they are much more likely to be reconstructed into different types of country recreation sites after being abandoned and become good places for people getting close to nature and relaxing. Some recreation sites are ecology conservation sites and wild animals and plants habitats formed through recovery, meanwhile, they have such function as science popularization exhibition and public recreation (Figure 2 L: American Quarry Park & Nature Preserve in Stearns County Parks). In addition, some of them make use of natural and wild scenery as well as quiet and comfortable atmosphere to provide the people with places for outing or picnic (Figure 2 R) or provide the outdoor enthusiasts with activity spaces for hike, rock climbing, boating, fishing and diving (e.g., Biville Quarry Park in France and Auchinstarry Quarry Park in Britain). For example, Lusatian Lake was once a lignite producing area in Germany, and after being mined for more than a century, many artificial lakes with large area has been formed there, and thus make it a paradise of surfing and canoeing enthusiasts, and regional revitalization is being carried out under the support of Internationale Bauausstellung (IBA).

2.1.3 *Cultural facilities*

Cultural facilities refer to public service facilities with the main purposes of satisfying the people's demands for such cultural activities as recreation and sports entertainment, science popularization exhibition, religious belief, etc. Outdoor theater is the most common reconstruction type of open pit (Figure 3 L: Dalhalla Theater in Sweden). There are also cases constructing recreational and athletic buildings like theater, gymnasium, etc. (Figure 3 M: Ópera de Arame Opera House in Curitiba, Brazil) which create unique show-watching experiences. Some mining wastelands, as productive cultural landscapes formed during nature remaking of human being, also possess industrial heritage even cultural relics value for its long history and significant status or become geologic research and science popularization sites for the exposed typical geological structures. In addition, there are also mining wastelands which are reconstructed into religious places due to factors of culture and location and they thereby possess more abundant culture connotations (Figure 3 R: Buddha mountain in Thailand).

2.1.4 *Theme park*

Theme park refers to a kind of commercially operated venue that satisfies the specific recreation and sightseeing requirements of the public with recreational facilities and the exhibition with a certain theme, including amusement parks that thrill the tourists mainly with roller coasters (see Figure 4 L: Six Flags Fiesta Texas Amusement Park, U.S.) and theme parks such as Disneyland, Sea World, and Garden of Eden (see Figure 4 M). Generally, the purpose of theme parks is to create a thrilling, novel, scary and exciting atmosphere for the public, for which it is generally a perfect match with the imposing, dangerous and unfamiliar topographic

Figure 1. Urban parks reconstruction cases.

Figure 2. Country recreation sites reconstruction cases.

Figure 3. Cultural facilities reconstruction cases.

Figure 4. Renovation cases of theme parks.

feature of mining area. Therefore, over the past few years, many local governments in China and the developers are trying to turn some discarded mining areas into theme parks, hoping to obtain greater economic return. For example, a park with the theme of Grimm's Fairy Tales is proposed to be constructed in the mining area of Mufu Mountain in Nanjing; in Changsha, the Snow World project is under construction in the tourist resort on Dawang Mountain with the aim of building the world's largest indoor ski resort and other ice-snow entertainment facilities on the base address of a 13-hectare dented limestone mining pit (see Figure 4 R).

2.2 *Experience reference*

In the above-mentioned projects, those mining wastelands with different conditions are adopted to create various recreation places for the public, "wastes" are turned into wealth and thus the comprehensive values of these wastelands are totally improved. The experience drawn from these successful cases are as follows:

2.2.1 *Select the renovation and reutilization type reasonably on the basis of different locations and site conditions*

The distance from the city and the natural landscape of the site have a combined influence on the possible orientation of land reclamation and landscape reconstruction for the mining wastelands, i.e., the mining wastelands with shorter distance to the city will more likely be used as construction land or public recreation sites; otherwise, they will more likely be subjected to ordinary ecological restoration and land reclamation according to their degree of uniqueness, or they may also be turned into outdoor activity space such as suburb recreational sites and theme parks.

2.2.2 *Effectively identify and adequately exploit the potential comprehensive values of the mining wastelands*

Whatever the orientation for restoration and reconstruction is, in the above-mentioned cases, the discarded mine fields have never been regarded as abandoned land, and attentions have been paid to the value and the revenue it may bring about in the aspects of meeting the requirements of ecosystem conservation and public recreation as well as promoting the development of local economy. It is based on such value identification that the landowners, the government management departments, the development and construction teams and planning and design personnel could creatively turn these wastelands into wealth and settle the problem of environment destruction while endow the land with new vitality.

2.2.3 *Adequately reserve and utilize the original mining landscape resources in the site*

Rather than erase and cover the traces of mineral extraction in the site, the above-mentioned renovation cases seek to fully reserve those traces as the historical information on the landscape evolution of the site and integrate them with the future function layout and landscape construction. For example, unique scenery and space for traveling and sightseeing may be created by arranging the site with the natural terrain formed by mining, such as pit body, water body, depression, and rock face, together with newly-developed routes. In this process, some mechanical constructions may also be deemed as mining relics and renovated as new functions.

3 PROBLEMS AND CHALLENGES

3.1 *Problems and disadvantages*

China has a large number of mine fields within its territory, and the area of discarded mine field is also enormous. Since the reform and opening up and especially from the year of 2000, the governments of all levels and the relevant management departments have been continuously strengthening the land reclamation and ecological restoration of mining wasteland with significant achievements. However, the author found that there are some problems and disadvantages in the generally adopted restoration and conservation practices for mining areas in China that takes terrain formation and vegetation restoration as the main working content; and these problems and disadvantages are mainly reflected in the following aspects:

3.1.1 *Take ecological restoration completely equivalent to green recovery project, and ignore habitat construction and biodiversity conservation*

The current ecological restoration work for the mining areas are usually large-area terrain leveling and earth sheltered planation, in an attempt to quickly restore the vegetation coverage. In order to improve the efficiency, people tend to choose a single species of arbor or bush, and plant them in rows and lines uniformly. Although such kind of artificial greening can quickly realize the vegetation restoration for the exposed ground surface, it is also easy to change the original complex terrain of the mining area, which may lead to excessively unitary habitat type and association of plants and animals; thus it is hardly conducive to the formation of healthy and stable natural ecosystems and let alone improving their biodiversity resources [5].

3.1.2 *Overemphasis on covering or removal of mining relics costs a lot and undermines the potential landscape resources*

Many of the restoration and conservation projects over exaggerate the environmental disturbances caused by discarded mines with excessive engineering measures taken, which not only cost a lot but also easily damage some of the naturally-aesthetic scenery resources of the mining area and even cause more serious destruction to the landscape. For example, in the case shown in Figure 5 (left), people tried to cover the rock face of a mining pit in Weihai suburbs by planting in the planters and board slots and even by spraying paint, which, however, made no difference in weakening the visual interference but to make it even less natural than the original rock texture. For another example, Figure 5 (right) shows a continuous rock wall of a mining pit in Jixian suburb of Tianjin. Spectacular and of highly landscape aesthetic value, the rock wall didn't get away with the effort of the local authorities who have tried to cover it by backfilling and piling up bags, which were so difficult to carry out that it only led to the collapse of the earth. Overdone is worse than undone!

3.1.3 *Emphasize only on the production and construction functions, and ignore exploitation in combination with public recreation and other comprehensive values*

In the face of complex and diverse mining wasteland landscape, the current land reclamation work is always a single choice to restore functions of agriculture, forestry, animal husbandry and fishery production or take the wasteland as construction land for real estate development, with rare consideration of combining them with the demand of scenic tours, public recreation and so on to carry out more careful restoration and reconstruction so as to create more comprehensive values and benefits. The result of dealing with the problem from a single perspective is to simplify the complex problem, which, although a unitary functional demand is achieved through a series of simple and rough engineering, denies and wastes more values yet to be exploited for the site.

Figure 5. More severe landscape damage resulting from inappropriate rock slope covering measures (Source: taken by Shen Xinshan).

3.2 *Challenges*

At present, China's land reclamation industry has formed a relatively fixed working mode and technical process, so we are faced with a good many of challenges to carry out more comprehensive and highly-efficient land reclamation and landscape reconstruction based on the public recreation.

3.2.1 *Value perception*

Currently, the lack of thorough understanding of mining wasteland and inadequate knowledge of its potential comprehensive value are the root causes of the above problems. In the minds of the people, mining wasteland is always full of ruins, disorders, racks and dangers, and the original cliffs, pits and stockpiles and other mining relics are all useless wastes and damaged sites. Such negative attitude directly leads to the current restoration treatment method of "erasure", and to improve such situation requires the establishment of a more comprehensive understanding of the mining wasteland landscape values of the management, design and construction personnel.

3.2.2 *Ecological concept*

Although the development of domestic ecological restoration industry is in full swing and "ecological" is widely used in all walks of life, but in fact people generally have a poor or even wrong understanding of ecological concept and ecological restoration. Land formation and afforestation are not conducive to achieving natural ecosystems and wild animals and plants habitats with abundant structural functions. Enrichment of biodiversity and improvement of ecosystem service functions should be the main purposes and evaluation criteria for ecological restoration engineering, which require the support of more in-depth researches and practices.

3.2.3 *Discipline cooperation*

At present, China's mining land reclamation and ecological restoration works are generally under the management of the land or forestry departments and most of them are completed mainly by the disciplines of geological disaster protection, soil and water conservation, environmental engineering, etc., with little participation of such disciplines as ecology, wildlife protection, urban and rural planning as well as landscape architecture, which inevitably leads to many discipline blind spots and cognitive misunderstandings during the process of practice. Therefore, a better system design is in need to promote mutual dialogue and cooperation of different departments and disciplines.

3.2.4 *Industry development*

The increasingly growing mining reclamation industry has formed a relatively mature industrial chain and a stable interest group, with a relatively fixed project operation mode established and engineering technical measures prepared. Therefore, with the ever solidified industry, how to carry out more conceptual and technological innovation, and guide the design and construction to transit from being unilateral to comprehensive and from extensive to delicate is one of the important challenges for future development.

4 SUGGESTIONS AND VISIONS

Finally, the following suggestions are put forward in this paper on land reclamation, ecological restoration and landscape reconstruction works of China's mining wasteland based on public recreation:

4.1 *Promote disciplinary exchanges and strengthen discipline cooperation to comprehensively improve the level of reclamation and restoration*

Strengthen communication and cooperation of different disciplines and facilitate people to understand the mining area reclamation and restoration in a more comprehensively way by collisions between different ideas. The reconstruction for recycle practice from the perspectives of public recreation and landscape construction calls for more participation of landscape architects, and the ecological restoration projects emphasizing ecological protection should give audience to the ecologists and biologists for opinions.

4.2 *Enhance status survey and comprehensive analysis and evaluation to rationally launch mining wasteland landscape reconstruction*

Multi-discipline teams should carry out more detailed and comprehensive site survey and evaluation, analyze the environmental hazards caused by mining, discover the landscape resources thus formed, and formulate reasonable landscape reconstruction objectives and landscape construction measures in combination with the orientation of reclamation and restoration. Regardless of the types of restoration and conservation, treat landscape reconstruction and scenery construction as one of the indispensable jobs, carefully carry out engineering intervention, advocate delicate design and construction, protect and make use of mining landscape resources.

4.3 *Carefully carry out engineering intervention, advocate delicate design and construction, protect and exploit mining landscape resources*

The mining wasteland is usually located remotely, by which the environmental interference caused is not that serious as imagined, so there is no need to adopt completely manually-controlled engineering restoration method for all the mine sites, whereas the concept of working by nature and time should be upheld, or moderate intervention and guidance may be conducted in the process of natural succession. Especially for the public recreation oriented reconstruction projects, simple and rough large-area terrain renovation and unnecessary damage to habitat diversity should be completely eradicated, and the rock slopes, pits, ponds and canyons should not be eliminated, covered or filled indiscriminately, while valuable landscape resources should be identified, protected and exploited purposefully so as to protect the site features at utmost, which requires cooperation of delicate site design and construction.

5 SUMMARY

Every coin has two sides. Although the exploitation of mineral resources has caused serious damage to the environment, but also formed unique mined industrial landscapes. The extremely complex terrain conditions of mined wastelands create a rich landscape and space experiences, and is easy to form more diverse wild biotopes. These characteristics endow the mined wastelands with potential values for nature conservation, landscape aesthetics and public recreation. The land reclamation measures of terrain leveling and vegetation planting being commonly used nowadays in China has achieved notable effects, but they always destroy the original useful landscape resources and lead to simplex ecosystem weakness because of excessive artificial intervention. Based on the perspective of public recreation, this paper emphasize that we should fully identify, protect and explore the potential values of the mining wastelands in terms of leisure & entertainment, exercise & fitness, cultural & natural exhibition as well as biodiversity conservation. It also advocates sufficient exchange and cooperation with the discipline of landscape architecture, so as to promote the scenery management and landscape reconstruction of mined wastelands in more kinds of reclaiming practices. These will promote the sustainable development of land reclamation and ecological restoration work toward a more comprehensive, rational and efficient practicing direction.

REFERENCES

Cui, Q. & Meng, F. (2013) From cavern pool to arcadia —Renaissance of the quarrying industry relics landscape of the mine park of Shanghai Chenshan Botanical Garden. *Landscape Design*, (01), 26–33.

Komara, A.E. (2004) Concrete and the engineered picturesque: The Parc des Buttes Chaumont, paris 1867. *Journal of Architectural Education*, 58 (1), 5–12.

Komara, A.E. (2009) Measure and map: Alphand's contours of construction at the Parc des Buttes Chaumont, paris 1867. *Landscape Journal*, 28 (1), 22–39.

Song, B. (2008) The studies on the ecological restoration in abandoned quarry in Xishan mountain in Beijing: Process, characters and mechanism of natural restoration. Jinan: Nanjing Forestry University.

Tang, X., Li, X. & Cao, L. (1997) *Landscape Design*. Beijing: China Forestry Press.

Land Reclamation in Ecological Fragile Areas – Hu (Ed.)
© 2017 Taylor & Francis Group, London, ISBN 978-1-138-05103-4

A simulation-based framework for estimating probable open-pit mine closure time and cost

Morteza Paricheh & Morteza Osanloo
Department of Mining and Metallurgical Engineering, Amirkabir University of Technology, Tehran, Iran

ABSTRACT: Today, the technical challenges for estimating closure time and cost are two significant parts of mine closure management. In this paper, a simulation-based framework is established to determine the probable open-pit mine closure times and mine closure costs. The model incorporates the price uncertainty into the probable mine closure time. To estimate more accurate mine closure costs the Hutchison's technique was modified using the predefined mine closure time. The Hutchison's technique uses Monte Carlo method and decision tree analysis to provide mine closure cost estimation in a probabilistic manner. The proposed framework is applied in a 2D hypothetical geological model. The results showed that the model works well and it can be used for real cases.

1 INTRODUCTION

Nowadays, Sustainable Development (SD) is one of the main critical and challenging tasks in most industries around the world. SD is the development that meets the needs of the present generations without compromising the future generations to meet their needs (Nathorial and Bardos, 2004; Galvic and Lukman, 2007; Laurence, 2011). The concept of SD gained its interest among the miners from the United Nation's Earth Summit Conference in Rio de Janeiro, Brazil in 1992. After that, many researchers have been discussing the subject and seeking the methods to incorporate the SD aspects into mining operations (Rahmanpour and Osanloo, 2017; Laurence, 2011; Adibi et al, 2015; Mirmohammadi et al, 2009; Folchi, 2003; Phillips, 2012). Now, the number of mining publications on this topic is increasing very rapidly (Moran, 2014).

Some researchers believe that mining is not part of SD because it is a temporary use of land and it has some negative impacts on environment, social and economic issues. As the response to these criticisms, minerals are clearly a part of SD because they are necessary if the quality of life, job creativity, and economic growth is to be improved, but at the same time, protecting the environment while supplying these minerals is essential (Villas Boas et al, 2005). In other words, mining operation would be sustainable when it is able to create a good balance among its three environmental, social and economic pillars (Botin, 2009). For this purpose, for a mine to contribute positively to SD, a comprehensive **reclamation and closure plan**, which is capable of considering social and environmental objectives and impacts (to avoid or at least minimize adverse impacts) in addition to the economical impacts is necessary from project inception (Laurence, 2011).

The closure plan (i.e. the reclamation and closure plan) defines a vision of the end results of the mining operation and sets the objectives to realize that vision. This forms an overall framework to channel all of the actions and decisions taken during the mine's life (Laurence, 2001, 2006 and 2011). Nowadays, in most of the countries, the mines with no closure plan are not permitted to start their operations. An appropriate closure plan guarantees a successful and on time closure without any negative environmental, social and economic impacts. Seeking a comprehensive closure plan, the closure time and cost are the most challenging issues which need accurate estimations.

Suddenly or premature mine closure is one hardly predictable undesired issue which overshadows all the closure objectives. The historical closure data show that around 75% of the mines which closed during the year 1981–2009 were unplanned or premature. Just 25% of the mines that closed were planned closures due to resource depletion. The main reasons for the premature mine closures involve economic, efficiency, community, environment and safety dimensions. The economic dimension is the most significant primary driver of premature mine closure (i.e. 44%) (Laurence, 2011). The point is that the mines are price taker, not price maker. Commodity price changes during the time and decreasing the price amplifies the premature closure. Mining is inherently variable and subject to a highly volatile commodity market. So, being at the lowest level of the costs helps mining companies to stay competitive in the condition of uncertainty.

The closure costs are often not given the recognition that they guarantee a part of the comprehensive mine plan success. Mine closure costs have the potential to affect the ultimate pit limit, production schedule, mine life and ultimately the profitability and net present value of an operation (Nehring and Cheng, 2016). So, incorporating true closure costs into financial feasibility studies for mines is an essential aspect of modern mining operation before the mine is even put into production. Closure goals and post-mining land uses prescribe the methods, the measures and the costs of mine closure (Soltanmohammadi et al, 2010). Then, to estimate the closure costs, the after-use option of land must be determined firstly. Estimating closure cost is not an easy task because of the finite number of mines which are closed safely according to the available standards. However, the topic of closure or reclamation cost estimation has scarcely been investigated.

Jones (2012) authenticated the Soltanmohammadi et al. (2010) and said cost estimates will vary depending on the purpose for which they are prepared, the organizations requiring or undertaking the estimate. He considered the different approaches by different organizations including operating companies, government authorities, and financial organizations and uses a hypothetical example based on practical experience to illustrate some of these differences in estimation methods. Gronhard and Scott (1979) divided the reclamation process into separate activities that must be performed, including preparation and planning, recontouring, topsoiling, and re-vegetation. The cost of reclamation was then estimated for a given set of mine conditions. An interdisciplinary approach employing hydrology, geographic information systems, and a recreation visitation function model is used by Mishra et al. (2012) to estimate the damages from upstream coal mining to lakes in Ohio. The estimated recreational damages to five of the coal-mining-impacted lakes, using dissolved sulfate as a coal-mining-impact indicator, amounted to $21 Million per year.

Leinart and Schumacher (2010) evaluated the role of cost estimating in mine planning and equipment selection and Sherpa was introduced as a cost estimator software which can be used in this regards. A combination of Decision Tree Analysis (DTA) and Monte Carlo Technique (MCT) was used to establish a cost probability curve that in turn provides closure costs at different levels of confidence by Hutchison in 2011. The main Hutchison's hypothesis was the pre-defined optimistic, pessimistic and most likely dimension of mine waste management and its probability and costs, which in turn are a function of future waste characteristics, the final configuration of pit, waste dump, tailings and water pile, future climate condition and future regulatory standards.

Pavloudakis et al. (2011) proposed a methodology for predicting the maximum budget that is required for financing environmental protection and land reclamation works throughout the entire life of the mine. The method was based on MCT and DTA as the same as Hutchison's technique. The method is able to calculate the maximum reclamation cost taking into account a minimum acceptable profitability index of the project by incorporating the uncertainty of parameters. Nehring and Cheng (2016) evaluated the effects of post-mining land use costs in optimal mine design and planning. It is mentioned that many countries have adopted a performance bond, which is included in the permitting procedure and is sufficient to cover the reclamation cost of the mining site in the event the mine owner does not complete all required land reclamation activities. Furthermore, according to Adibee et al. (2013), physical properties (slope, dip, erosion, the existing trees and plants and their characteristics) and chemical properties (acidity and alkalinity, heavy metal contamination, lack of nutrients, electrical conductivity, hardness) of rock and water in the site must be considered in order to identify the best suitable post-mining land use and its associated costs and benefits. The adverse effects of the waste dumps on the environment must be described and illustrated how determining the characteristics of the wastes can be used to predict their environmental impacts and reclamation costs across the site.

It is obvious that reclamation and closure costs are closely related to mine design for a specific mining project. In the case of an open-pit operation, the costs vary with the size of the ultimate pit, its average strip ratio, the mine life and also the surrounding mine's facilities. So, the reclamation cost estimator must take into account the pit, waste dump, tailing dam and other mine facilities and their sizes. In this paper, to tackle the suddenly mine closure the price volatility as the main affecting premature mine closure and mine expansion factor is studied. The probable pits concept was used to show the probable mine closure times. A simulation-based framework was established to provide probable mine closure times using the estimated cumulative distribution function of the commodity price. Then, as the other challenging area of research, it is trying to come up with the main disadvantage of the Hutchison technique using these probable pits. As it is said, the main shortcoming of the

Hutchison technique is the pre-defined optimistic, pessimistic and most likely dimension of mine waste management and its costs. In this paper, the term closure cost refers to the pit, waste dump and tailing dam areas.

2 PROPOSED FRAMEWORK

Figure 1 shows the proposed framework that is able to drive us to the before mentioned goals. As it is depicted in the picture, the historical commodity price data, geological block model, and post-mining land use are the items which are to be previously identified in the first step. Although the post-mining land use is somehow a function of closure cost itself, but it is assumed that it is identified previously. It can be distinguished through the application of the multi-criteria decision-making methods (Soltanmohammadi et al, 2010; Adibi et al, 2013). In the following, the suggested framework will be explained via a two-dimensional hypothetical geological block model of a copper deposit, step by step.

Figure 2 shows the geological block model containing 210 blocks. Block dimensions are assumed as 12.5 m × 25 m in height and width, respectively. The perpendicular to the plane expansion of the model is assumed as 25 m. The rock density (waste and ore) is assumed to be equal to 2.43 tons per cubic meter.

2.1 *Estimating the CDF of the price data*

Commodity price is one key contributing parameter in block economic value calculation. To model the price volatilities Monte Carlo Technique (MCT) was applied. The MCT creates the random numbers and links them with corresponding estimated variables using specified statistical distributions. Thus, the main step is to generate a Cumulative Distribution Function (CDF) from subjective or historical data for the variable. In the first step, the historical changes in mineral price are collected in order to simulate future possible price forecasts. The 316 monthly price data from Jan 1990 to April 2016 were collected. Table 1 shows the descriptive statistics of all price data.

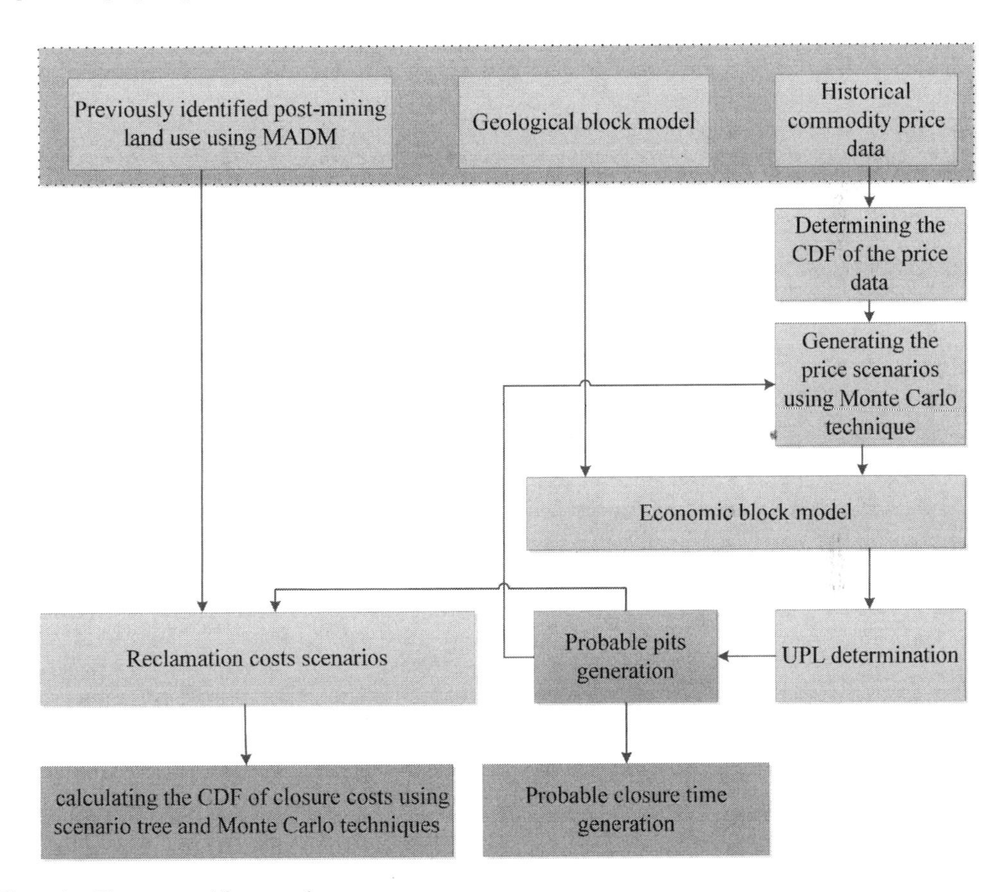

Figure 1. The proposed framework.

0	0	0	0	0	0	0	0	0	0.052	0.02	0.031	0	0	0	0	0	0	0	0	0
	0	0	0	0	0	0	0	0	1.87	2.68	2.64	0	0	0	0	0	0	0	0	
		0	0	0	0	0	0	0	1.56	3.53	1.67	0	0	0	0	0	0	0		
			0	0	0	0	0	0	2.19	3.78	1.69	0	0	0	0	0	0			
				0	0	0	0	0	1.01	2.94	0.89	0	0	0	0	0				
					0	0	0	0	1.29	2.01	1.09	0	0	0	0					
						0	0	0	0.733	1.5	0.59	0	0	0						
							0	0	0.36	0.27	0.824	0	0							
								0	1.26	0.32	0.33	0								
									0.29	1.4	0.35									

Figure 2. The 2D hypothetical geological block model (numbers are ore grade in percent).

Table 1. Descriptive statistics of the data ($/kg cu).

Mean	Maximum	Minimum	Std. deviation	Variance
4.08	9.87	1.37	2.52	6.39

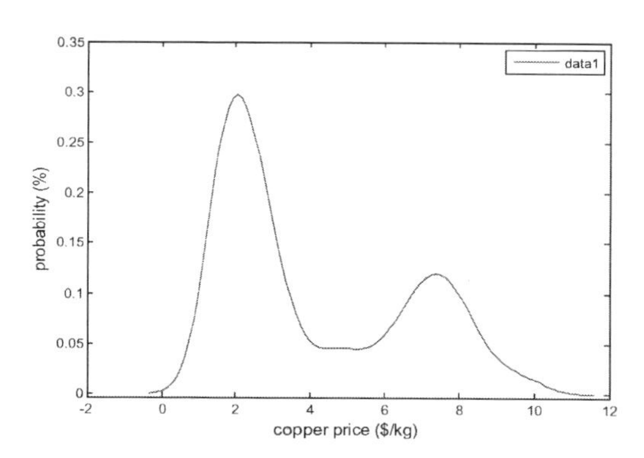

Figure 3. Probability distribution function of the price data.

As it is pictured in Figure 3, the price data have two different populations with averages of two and seven. These characteristics should be embedded into the CDF function appropriately. For this propose, the variation interval of the data (1.37–9.87 $/kg) was divided into twenty equal intervals and then the frequency of the data in each interval was calculated separately. After that, the best curve fitted to the cumulative frequencies of the data is selected as the finest CDF function for generating the price scenarios.

2.2 *Probable pits generation*

The Ultimate Pit Limit (UPL) defines what is economically mineable from a given deposit. It identifies which block should be left in the ground and which one should be elicited. Therefore, there is a break-even point for each deposit which yields the highest profit and each positive block beyond this threshold cannot pay for removal of overlying blocks. In another word, the optimum pit outline will guarantee maximum value gained from a given input geological block model and a given set of economic and geotechnical condition.

As it is shown in Figure 1, for each price scenario, the block economic model needs to be established. The block economic values that are resulted from these commodity price scenarios would be applied in the ultimate pit limit determination process. The block economic model is created from the geologic block model take into account production and processing costs and price scenario. The block economic values for ore and waste blocks are calculated using equations 1 and 2 respectively.

$$BEV_{ore} = M\left[10gy(p(r)-s)-c_0-c_w\right] \qquad g \geq g_c \quad (1)$$

$$BEV_{waste} = -c_w M \qquad g \leq g_c \quad (2)$$

where, g is average block grade in percent, M is block tonnage, y is recovery in percent, p(r) is commodity price scenario ($/kg cu), s is smelting and refining cost ($/ kg cu), c0 is processing cost, cw is mining cost ($/ ton rock), and gc demonstrates the break-even cut-off grade.

By identification of the economic block model, the next step is to identify the UPL subject to maximum profit. There are some methods for UPL selection and it can be found easily through the following equations. Equation 3 is the objective function, used to maximize the revenue. Equations 4 and 7 are the constraints of the model. Equations 4–6 define the slope requirements and the last one sets the binary condition for the variables.

$$\max \sum_i \sum_j c_{ij} x_{ij} \quad (3)$$

$$x_{ij} - x_{i-1,j-1} \leq 0 \quad (4)$$

$$x_{ij} - x_{i-1,j} \leq 0 \quad (5)$$

$$x_{ij} - x_{i-1,j+1} \leq 0 \quad (6)$$

$$x_{ij} = 0 \ or \ 1 \quad (7)$$

where c_{ij} is the block economic value and x_{ij} is a binary variable that gives 1 if the block located at row i and column j is extracted, otherwise, it is 0. The technical and economic data presented in Table 2 are used for calculating the blocks economic value.

The framework containing the before mentioned formulas is coded in MATLAB. Firstly, using the predicted CDF, the random price scenarios and then the UPLs are generated followed by economic block models. The simulation is done one thousand times. In every simulation, the price scenario is selected randomly from the price distribution using MCT. Since each run (replication) provides a single estimate of the UPL; several replications are required to assure a reliable estimate of the probable pits. Hence, the process is repeated and the new results are added to the previous ones. By this way, the probable pits are easily achievable. The results of the simulation are illustrated in Figure 4.

In this hypothetical 2D model, eight different pits are calculated using the proposed simulation-based method. The numbers inside the blocks demonstrate the likelihood of the pit occurrence. The number 1 means that the block is fallen down into the pit limit in all one thousand simulations. As the same way, the purple blocks are laid into the pit limit only sixty-six times out of one thousand times.

2.3 *Determination of probable closure times*

Now, one can follow up on the probable closure time easily using the probable pits. Suppose that only three blocks are planned to be extracted annually; consequently, nine blocks containing seven ore blocks and two waste blocks will be mined definitely (100%). In this case, the mine will survive three years and it will progress up to the depth 37.5 meters. Similarly, there is a 97.5% chance that the pit consists of ten ore blocks and six waste blocks becomes deeper up to fifty meters. Accordingly, the mine will be open for 5.3 years. The first nine columns of Table 3 show the characteristics of the probable pits and corresponding mine closure times. The three other columns will be explained later.

Table 2. The assumed technical and economic data.

Symbol	Description	Value	Unit
y	Recovery	95	%
s	Smelter and refinery cost	0.7	\$/kg cu
c0	Processing cost	7	\$/ton ore
cw	Mining cost	1.5	\$/ton rock
g_c	Cut-off grade	$\frac{8.5}{9.5(p(r)-0.7)}$	% cu

2.4 *Calculating closure cost scenarios*

In this paper, forestry is assumed as the best-adjusted post-mining land use for the region. Other than the after-use land option, the closure cost is highly depends on the square meter size of the disturbed land. It means closure cost depends on the greatness of open-pit mining operation, which is interrelated to the future potential of mine expansion. There are three regions associated with a mine that must be considered in planning, re-contouring and reclamation. These include the pit, waste dump and tailing dam areas. The pit areas are easily calculable for each of the defined probable pits. The tenth column of Table 3 shows these areas. As a rough estimation of the area of the other two parts, one can use the data presented by Johnson and Paone in 1982. Based on these data (Figure 5) surface mining is responsible for 85% of the total land utilized (consist of 69% for excavated area and 16% for waste dump area). Underground mining accounts for a comparatively small percentage (5%). Mineral processing waste engrosses the remaining (10%) of the land. So, ignoring the areas associated with underground mining, it would be a reliable premise that 70, 20 and 10 percent of the total lands are devoted to the pit, waste dump and tailing dam respectively. The two last columns of the Table 3 show these areas for the probable pits accordingly. Another method proposed by Xu et al (2014) could be applied in this regard that needs some more information related to how to design the waste dump and tailing dam area.

0.066	0.532	0.532	0.532	0.691	0.727	0.945	0.975	1	1	1	1	1	0.975	0.945	0.727	0.691	0.532	0.532	0.532	0.218
0	0.066	0.532	0.532	0.532	0.691	0.727	0.945	0.975	1	1	1	0.975	0.945	0.727	0.691	0.532	0.532	0.532	0.218	0
0	0	0.066	0.532	0.532	0.532	0.691	0.727	0.945	0.975	1	0.975	0.945	0.727	0.691	0.532	0.532	0.532	0.218	0	0
0	0	0	0.066	0.532	0.532	0.532	0.691	0.727	0.945	0.975	0.945	0.727	0.691	0.532	0.532	0.532	0.218	0	0	0
0	0	0	0	0.066	0.532	0.532	0.532	0.691	0.727	0.945	0.727	0.691	0.532	0.532	0.532	0.218	0	0	0	0
0	0	0	0	0	0.066	0.532	0.532	0.532	0.691	0.727	0.691	0.532	0.532	0.532	0.218	0	0	0	0	0
0	0	0	0	0	0	0.066	0.532	0.532	0.532	0.691	0.532	0.532	0.532	0.218	0	0	0	0	0	0
0	0	0	0	0	0	0	0.066	0.532	0.532	0.532	0.532	0.532	0.218	0	0	0	0	0	0	0
0	0	0	0	0	0	0	0	0.066	0.532	0.532	0.532	0.218	0	0	0	0	0	0	0	0
0	0	0	0	0	0	0	0	0	0.066	0.532	0.218	0	0	0	0	0	0	0	0	0

Figure 4. The probable pits (the numbers show the probability of pit occurrence).

Table 3. The characteristics of probable pits and probable mine closure times.

Pit	Probability (%)	No of block	Mine life (year)	Pit Depth (m)	No of ore block	No of waste block	Ore tonnage	Waste tonnage	Pit area (hectare)	Waste area (hectare)	Tailing area (hectare)
1	100	9.00	3.00	37.50	7.00	2.00	132890.63	37968.75	0.31	0.09	0.04
2	97.5	16.00	5.33	50.00	10.00	6.00	189843.75	113906.25	0.44	0.13	0.06
3	94.5	18.00	6.00	62.50	13.00	5.00	246796.88	94921.88	0.56	0.16	0.08
4	72.7	29.00	9.67	75.00	16.00	13.00	303750.00	246796.88	0.69	0.20	0.10
5	69.1	42.00	14.00	87.50	19.00	23.00	360703.13	436640.63	0.81	0.23	0.12
6	53.2	93.00	31.00	125.00	28.00	65.00	531562.50	1233984.38	1.19	0.34	0.17
7	21.8	103.00	34.33	125.00	29.00	74.00	550546.88	1404843.75	1.25	0.36	0.18
8	6.6	113.00	37.67	125.00	30.00	83.00	569531.25	1575703.13	1.31	0.38	0.19

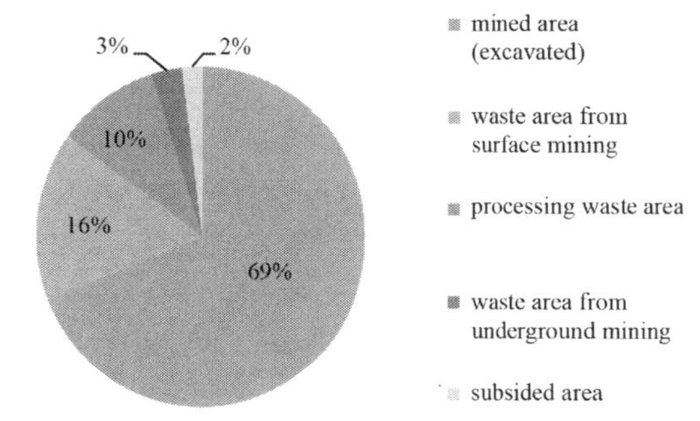

Figure 5. Percentage of land used by mining function, 1930–1980 (Johnson and Paone, 1982).

2.4.1 *Pit reclamation cost*

Given to the selected post-mining land use, it is to be evaluated that the slope of the ground is capable of bearing the defined land-use policy. The type of land use adopted is limited where the overall slope of the ground is more or equal to the angle of repose. The maximum slope of the ground, which in it the forestry option can be applied is 35 degrees (Osanloo, 2008). Therefore, this requirement must be checked to see either the overall pit slope is appropriate for the post land use option or not. Due to the 29-degrees slopes of probable pits, the pit slopes do not need any noticeable change. Then, it only needs to be smoothed properly. For this purpose, some drilling, blasting and re-countering activities need to be done. Consequently, topsoiling would be appropriately implemented. The uppermost layer in the soil profile often called topsoil (A horizon). It is a part of the soil in which organic matter is most abundant and where greatest leaching of soluble or suspended particles is inescapable. Topsoiling is the process whereby A horizon is removed, stored, and reapplied or put directly on re-contoured lands. At the end, the forestation activity would be finished. For this purpose, the economic and technical data presented in Table 4 are used.

2.4.2 *Waste dump reclamation cost*

As it is previously mentioned, the reclamation costs of the waste dumps are related to the physical and chemical characteristics of the waste rocks and surrounding waste waters. In this case, it is presumed that the waste dump is acidic and needs to be suitably buried using limestone and impermeable rocks. Hence, to prevent acid generation or mitigating the acidity of the wastes, a 30-cm layer of limestone was planned to be scattered before a 30-cm layer of impermeable clay. Finally, another 30-cm layer of topsoil will be sprinkled. The forestation is the final step, which needs to be completed suitably.

2.4.3 *Tailing dam reclamation cost*

Assuming that the tailing waste materials behind the dam are relatively alkaline, it does not need the neutralize material. Therefore, a 30-cm layer of the impermeable clay would be distributed to prevent the surface water penetration. Finally, the forestation would be implemented truly after the topsoiling process.

The reclamation costs associated to the three previously mentioned areas for each probable pit are calculated using the data presented in Table 4. Table 5 shows these costs.

2.5 *Prediction of reclamation cost using decision tree analysis*

Now, the probable scenarios for reclamation costs and their likelihoods are computed using the probable pit concept. This was the main shortcoming of the Hutchison's technique. The next step is to develop a decision tree and incorporate these scenarios and corresponding likelihoods. The combination of the DTA with MCA can improve the decision-making, incorporating the uncertainty of parameter using probability distributions. The effectiveness of such a simulation modeling depends on the reliability of these probability distributions.

Table 4. The proposed technical and economical data.

Technical parameters			Economic parameters		
Parameter	Value	Unit	Parameter	Value	Unit
Hole diameter	3	inch	Drilling cost	1	$ per meter
Burden	3	meter	ANFO cost	0.37	$ per kg
Spacing	3	meter	Booster price	0.64	$ per booster
Sub-drilling	30	% of drilling length	Detonating cord price	0.13	$ per meter
Stemming for each hole	2	meter	Detonator price	0.25	$ per unit
Hole length in section	14	meter	Forestation cost	1000	$ per hector
Sections in bench	8	meter	Transportation cost of rock or soil	6.3	$ per ton
Charge length in section	8	meter			
Specific charge	1.75	kg per meter			
Booster needed	1	per hole			
Detonating cord length in hole	1.65	meter			

Table 5. Reclamation cost of probable scenarios.

Pit	Probability (%)	Normalized probability (%)	Reclamation cost ($) Pit	Waste dump	Tailing dam	Total reclamation cost ($)	Reclamation cost ($/ton)
1	100	19.4	7399.7	5154.2	1732.9	14286.8	0.108
2	97.5	18.9	10281.0	7215.9	2426.1	19922.98	0.105
3	94.5	18.3	13162.3	9277.6	3119.3	25559.16	0.104
4	72.7	14.1	16043.6	11339.2	3812.5	31195.34	0.103
5	69.1	13.4	18924.9	13400.9	4505.7	36831.51	0.102
6	53.2	10.3	27568.9	19585.9	6585.2	53740.05	0.101
7	21.8	4.2	28813.2	20616.8	6931.8	56361.79	0.102
8	6.6	1.3	30057.5	21647.6	7278.4	58983.54	0.104

As shown in Table 5, there are eight different scenarios along with the relevant probability for reclamation cost of each part of the pit, waste dump and tailing dam. Accordingly, there would be 512 distinct scenarios. In order to reduce the numbers of scenarios, they are clustered based on the normalized probability of each scenario. The scenarios with more closed probabilities are placed in the same cluster. The weighted means of probabilities and costs of scenarios are considered as the probability and cost of the each cluster. In this way, the first three scenarios (1, 2 and 3) are located in the first cluster. Similarly, the two last ones are placed in the last cluster and the rests are in the second. Now, for each part of the closure (pit, dump and dam), there will be three separate scenarios with normalized probabilities of 55, 37 and 8 percent. The concluded decision tree containing 27 scenarios is illustrated in Table 6.

The Table indicates the likelihood and expected value of reclamation costs for each part (mine areas). Major factors affecting the cost of recontouring and topsoiling are the volume of rock and soil to be moved, haulage distance, and the operating cost of machinery. At the other hand, uncertainties on some other factor such as climate condition, acid generation potential may change the computed expected values for costs. So, a normal probability distribution function with 10% standard deviation was defined for every cost component.

Later, the simulation was run 1000 times using the Crystal Ball software. Random costs are selected by MCT for each cost component and finally, the total reclamation cost is calculated as the main output of the decision tree. Totally intended CDF of reclamation cost after 1000 simulation is depicted in Figure 6.

Table 6. Combined decision tree of the hypothetical example.

Pit		Waste dump		Tailing dam	
Probability (%)	Cost ($)	Probability (%)	Cost ($)	Probability (%)	Cost ($)
				0.55	2426.1
		0.55	7215.9	0.37	4967.8
				0.08	6931.8
				0.55	2426.1
0.55	10281.0	0.37	14775.4	0.37	4967.8
				0.08	6931.8
				0.55	2426.1
		0.08	21132.2	0.37	4967.8
				0.08	6931.8
				0.55	2426.1
		0.55	7215.9	0.37	4967.8
				0.08	6931.8
				0.55	2426.1
0.37	20845.8	0.37	14775.4	0.37	4967.8
				0.08	6931.8
				0.55	2426.1
		0.08	21132.2	0.37	4967.8
				0.08	6931.8
				0.55	2426.1
		0.55	7215.9	0.37	4967.8
				0.08	6931.8
				0.55	2426.1
0.08	29435.4	0.37	14775.4	0.37	4967.8
				0.08	6931.8
				0.55	2426.1
		0.08	21132.2	0.37	4967.8
				0.08	6931.8

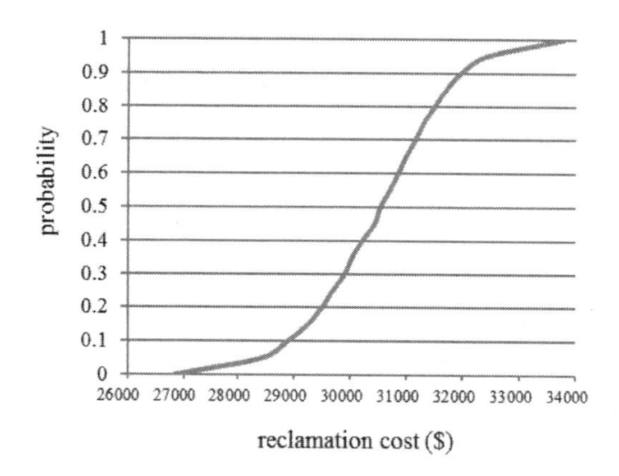

Figure 6. Cumulative probability distribution function of the closure cost.

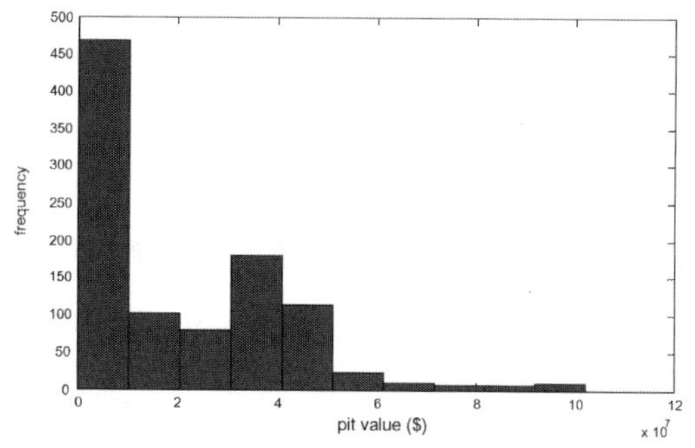

Figure 7. Histogram of probable UPLs values.

3 DISCUSSION

Historical commodity price data analysis showed that the PDF of the data has two different populations. Figure 7 shows that the histogram of UPLs values has the same behavior after 1000 simulations. This not only indicates that UPLs are varying following the price uncertainty, but also approves that the price uncertainty is considered very good.

The CDF of the reclamation cost is a useful tool for calculating the measure of Value at Risk (VaR) in the condition of uncertainty and risk. VaR is definable in both profit and loss probability distribution functions. In loss function, this measure is a level that loss would not be exceeded than with a specified confidence level. For instance, in the hypothetical proposed case study, the reclamation cost will not be greater than $32574 at a 95% confidence level. Similarly, the mine will not be open more than six years at the same confidence level (Table 3).

DTA can easily get complex when the number of variables increases. In this paper, three parts of the pit, waste dump, and tailing dam with eight discrete probabilities generate 512 possible outcomes. Making such a decision tree is very time-consuming and tedious. Hence, the clustering technique is necessary and should be applied more obsessively.

4 CONCLUSION

The closure is the final phase of the mining operation which involves decommissioning, reclamation and post-closure monitoring. Mine closure plan is required before a mining permit is granted, and it must demonstrate that in the future, the site will not pose a threat to the environment or the society. The plan has an inevitable expense that must be incurred once an operation commences; both during, and after the exploitation of an ore-body. In order to be feasible, mine closure costs must be accurately estimated ensuring that adequate funding is available when decommission time comes. Knowing the significant role of closure costs into the strategic mine planning process, mine planners still face the challenge of accurately incorporating the true cost of closure into the mine planning and cost estimation process. Choosing the appropriate method of estimation for a particular mine is vital since errors in closure cost generate great differences in the outcome of the environmental, social and economic performance. In doing so, this paper has sought to use the combination of both MCT and DTA, which is applied previously by Hutchison and modify that by probable pits concept. The probable pits are firstly used to identify the probable closure times under commodity price uncertainty and then to define a range of future potential mine closure cost scenarios. The proposed framework was tested by a hypothetical 2D geological block model. Finally, a cumulative distribution curve of closure cost is established that in turn provides for determining closure costs at different levels of confidence. It is worth nothing that, the physical, geological, and topographic conditions of mines vary greatly between and even within mine areas so that reclamation costs are site specific. Therefore, reclamation plan and cost are unique as the same as probable pits and these will need to be tailored to suit each operation.

REFERENCES

Adibee, N., Osanloo, M. & Rahmanpour, M. (2013). Adverse effects of coal mine waste dumps on the environment and their management. *Environmental earth sciences*, 70(4), 1581–1592.

Adibi, N., Ataee-pour, M. & Rahmanpour, M. (2015). Integration of sustainable development concepts in open pit mine design. *Journal of Cleaner Production*, 108, 1037–1049.

Botin, J.A. (2009). Sustainable Management of Mining Operations. Society for Mining, Metallurgy & Exploration (SME), p. 392.

Folchi, R. Environmental impact statement for mining with explosives: a quantitative method. *29th Annual Conference on Explosives and Blasting Technique. 2-5 February 2003, Nashville, Tennessee, U.S.A.*

Galvic, P. & Lukman, R. (2007). Review of sustainability terms and their definitions. *Journal of Cleaner Production*. 15 (18), 1875–1885.

Gronhard, D.E. & Scott, D.F. (1979). Reclamation costs of strip-mined land in western North Dakota. ND Farm Res. (United States), 37(1).

Hutchison, I., & Dettore, R. Statistical and probabilistic closure cost estimating. Proceedings Tailings and Mine Waste 2011, *6-9 November 2011, Vancouver, British Colombia*.

Johnson, W. & Paone, J. (1982). *Land Utilization and Reclamation in the Mining Industry, 1930–1980*. US Bureau of Mines, Washington, DC. Information Circular 8862.

Jones, H. Closure Cost Estimation – All Things to All Men?. International Mine Management Conference. *20–21 November 2012, Melbourne, Vic*, pp. 115–123.

Laurence, D. (2001). Closure risk factor-a simple model for optimizing mine closure outcomes. *Transactions-Society For Mining Metallurgy And Exploration, Inc*, 310, 125–133.

Laurence, D. (2006). Optimization of the mine closure process. *Journal of Cleaner Production*, 14(3), 285–298.

Laurence, D. (2011). Establishing a sustainable mining operation: an overview. *Journal of Cleaner Production*, 19 (2), 278–284.

Leinart, J.B. & Schumacher, O.L. The role of cost estimating in mine planning and equipment selection. *Mine Planning and Equipment Selection- MPES. 1-3 December 2010. Fremantle, Western Australia*, pp. 69–80.

Mirmohammadi, M., Gholamnejad, J., Fattahpour, V., Seyedsadri, P. & Ghorbani, Y. (2009). Designing of an environmental assessment algorithm for surface mining projects. *Journal of environmental management*, 90(8), 2422–2435.

Mishra, S.K., Hitzhusen, F.J., Sohngen, B.L. & Guldmann, J.M. (2012). Costs of abandoned coal mine reclamation and associated recreation benefits in Ohio. *Journal of environmental management*, 100, 52–58.

Moran, C.J., Lodhia, S., Kunz, N.C. & Huisingh, D. (2014). Sustainability in mining, minerals and energy: new processes, pathways and human interactions for a cautiously optimistic future. *Journal of Cleaner Production*, 84, 1–15.

Nathorial, C.P. & Bardos, R.P. (2004). Reclamation of Contaminated Lands. John Wiley & Sons Ltd, p. 212.

Nehring, M. & Cheng, X. (2016). An investigation into the impact of mine closure and its associated cost on life of mine planning and resource recovery. *Journal of Cleaner Production*, 127, 228–239.

Osanloo, M. (2008). Mine reclamation. Amirkabir University of Technology Publication, 2nd edition, 300 p (in Persian)

Pavloudakis, F., Roumpos, C. & Galetakis, M. (2012). Public acceptance of surface mining projects and the determination of the marginal environmental cost. *International Journal of Mining, Reclamation and Environment*, 26(4), 292–308.

Phillips, J. (2012). The level and nature of sustainability for clusters of abandoned limestone quarries in the southern Palestinian West Bank. *Applied Geography*, 32(2), 376–392.

Rahmanpour, M. & Osanloo, M. (2017). A decision support system for determination of a sustainable pit limit. *Journal of Cleaner Production*, 141, 1249–1258.

Soltanmohammadi, H., Osanloo, M. & Bazzazi, A.A. (2010). An analytical approach with a reliable logic and a ranking policy for post-mining land-use determination. *Land Use Policy*, 27(2), 364–372.

Villas Boas, C.,R., Shields, I.D., Solar, V.S., Anciaux, P. & Onal, G., (2005). A Review on Indicators of Sustainability for the Minerals Extraction Industries. CETEM/MCT/CNPq/CYTED/IMPC, Rio de Janeiro, Brazil. p. 230.

Land Reclamation in Ecological Fragile Areas – Hu (Ed.)
 ISBN 978-1-138-05103-4

The planning of land reclamation and ecological restoration in the abandoned land left over by history

Q.W. Zhu, J.W. Wang & T. Wang
School of Surveying Science and Technology, Xi'an University of Science and Technology, Xi'an, China
Key Laboratory of Western Mine and Hazard Prevention, Ministry of Education Xi'an University of Science and Technology, Xi'an, China

ABSTRACT: At present, the reuse of the abandoned land left over by history has been attracted the attention of the society, the reclamation in mining land and abandoned town village left over by history is abandoned land of industrial and mining and abandoned homestead reclamation, is restored the original production capacity of the land, is improved land utilization to meet the cultivated land, woodland and other ecological land. In this paper, through the engineering reclamation mode reclaim land abandoned, road reformed and irrigation repaired and drainage facilities for planning and research based on HuangLing County DianTou town and other 5 township (town) include 10 administrative villages (forest farm) land reclamation project. On the premise that repair the ecological environment, land reclamation can link to urban and rural construction land, make sure the total amount of construction sites keeping balance. After reclamation, we could increase the effective supply of land in HuangLing County in a certain extent, promote the improvement of the ecological environment and sustainable development. The agricultural land could improve the local ecological environment while playing against natural disasters in production process, it is also of great significance to guide land reclamation and ecological restoration in the abandoned land left over by history and implementing "the land reclamation ordinance".

1 INTRODUCTION

December 23, 2016, "the national land remediation plan (2016 ~ 2020)" (hereinafter referred to as "Plan") was approved by the State Council, issued and implemented by the Ministry of land and resources and the development and Reform Commission. "Plan" emphasize land regulation, soil remediation and ecological restoration to the importance of land layout in China, clear the "the main task of land reclamation" during the 13th Five-Year plan. This paper, through the full investigation and analysis as a basis for planning and use the feasible engineering measures and biological measures to secure the land reclamation quality reaches farming soil quality and productivity to meet the requirements of economic, social and ecological sustainable development.

2 THE BASIC SITUATION OF THE PROJECT AREA

In the project area, the backbone traffic is very convenient, but traffic is inconvenient in the individual area (the mining land) that located in the valley. In the project area, the main types of reclamation are abandoned homestead and mining plant, the terrain is relatively high, drainage ditch is relatively mature, power network of rural reformed completed—10 kV, 220 V arrive at every village, the forest coverage was high.

3 ENGINEERING DESIGN

3.1 *Data sources*

In this paper, combined with the specific situation of Huangling County Historical wasteland, the damaged land divided into rural roads, canals, villages and mining sites. By means of field survey and investigation, accurate measurement of the location, area and type of each damaged block (Table 1).

3.2 *Land leveling project*

In this project area, the abandoned land which are surface damage type. The ridge of field building and homestead reclamation should be used to dig deep shallow, terracing construction, direct use, truing and dressing and other engineering reclamation technology. According to the original terracing slope, and taking the road skeleton layout

Table 1. History of abandoned land information table.

Serial number	Ownership unit	Damaged area Hectare	Township owned
1	Mountain Fork Village	2.57	Long Fang Town
2	Small Garden Village	0.13	Long Fang Town
3	Cang Cun Xiang He Zhai Forest Farm	3.41	Long Fang Town
4	Xing Shu Zui village	0.41	A Dang Town
5	Car Village	6.7	Dian Tou Town
6	Huangling County Open Pit Coal Mine	0.59	Dian Tou Town
7	Dian Tou Forest Farm	3.59	Dian Tou Town
8	Ditch Village	1.44	Dian Tou Town
9	Shang Guan Chuan Village	1.91	Bridge Hill Street Office
10	Cao Wa Village	0.91	Tian Zhuang Town

ridge terrace during terrace construction. In the same area, it implement leveling that push high and pad low, with the muck sinotrans and covering with new soil.

3.3 *Field road project*

Try to retain the original field road, the newly built field road connected with the original roads and villages in the project area as far as possible in order to facilitate transportation and facilitate farmers travel during construction. As far as possible to reduce the area of covered, try to avoid the intersection of roads and ditches in order to reduce the number of canal buildings to optimize the cost of investment.

3.4 *Soil restoration project*

Soil restoration can be defined as (denotes refers to) the process of cultivating the reconstructed soil to satisfy the need (quality) of the filed during the process of land leveling or after the end. The cultivation of drought tolerant and fattening flora collaborated with the microbial technology make it possible to accelerate the formation of soil slope after the reclamation, the recovery of soil nutrient and the improvement of soil ecological value, in flat ground. Generally speaking, land leveling in the early stage of reclamation work, soil restoration may be carried out at various stages of reclamation work.

4 THE PROTECTION MEASURES AFTER RECLAMATION

4.1 *Adjustment for land ownership*

The adjustment for land ownership adhere to the principle that the total area of farmers contracted arable land basically unchanged, concentrated, convenient production and of life. It according to the area of farmland contracted by farmers, with the standard field as the basic unit to readjust the cultivated land after renovation and to sign letter of acceptance the land ownership limits and the right to land contractual management. The new cultivated land in reclamation projects managed and used by the village collective.

4.2 *Later management*

The management subject, management responsibility and management scope must be cleared after the completion of the reclamation work. In responsibility assignment of management subject, the department of land and resources management as the leader, the village committee as the dominator in their village, farmers and professional staffs as the participants. The scope of protection refers to the field facilities, field road, cultivated land quantity, farmland quality and so on.

5 THE ANALYSIS OF PROJECT BENEFIT

From the analysis of social benefit, the cultivated land area was increased after reclamation (Table 2), take land reclamation link to urban and rural construction land measure that not only can ensure that the total amount of arable land unchanged in the county but also priority arrangement development land for the people's livelihood, poverty alleviation, as well as the state's key industries and other infrastructure priorities for the development of land. Land reclamation increase the rate of arable land, improve the income of farmers, play an strong exemplary role for other towns, is conducive to promoting the development of land consolidation is good for increase the rational use of land and

Table 2. Table of land use structure.

Name	Before remediation	After remediation	Increment
	Hectare	Hectare	Hectare
Cultivated land, dry land	0	19.54	19.54
Transportation land, Rural road	0.78	0.78	0
Water and water conservancy facilities land, Ditch	0.1	0.1	0
Other land, Ridge of field	0	1.25	1.25
Town Village, Village	7.97	0	−7.97
Industrial land, Mining land	12.81	0	−12.81
Total	21.66	21.66	0

effectively protect the awareness of arable land. Based on the analysis of ecological benefit it is helpful to maintain soil and water conservation, reduce soil erosion, improve farmland productivity and reduce soil and water pollution. From the analysis of cultivated land quality the number and the quality of cultivated land in the project area are increased significantly after project implementation.

6 INCREASED SIGNIFICANTLY AFTER PROJECT IMPLEMENTATION. COMPREHENSIVE CONCLUSION

This project has great potential, comply with local master plan, feasible, significant benefit on society, ecology, economy, high enthusiasm for public participation, government leaders at all levels give great importance and support, the local departments cooperate well, and guaranteed funds. The plan through detailed fieldwork, data accurate and studied deeply, deemed that the project is feasible and can provide strong support for improving the local production and living conditions and accelerate the pace of new rural construction.

REFERENCES

Eger, P., Baysinger, C. & Cates, D., et al. (2014) Biochemical Reactors for Treating Mining Influenced Water. *Exploring New Frontiers in Reclamation Proceedings from the 31st National Meeting of the American Soci.*

Han, L.B., Shi, X.Y. & Yuan, Y. (2016) Method and application of reclamation land suitability zoning in open pit of Loess Plateau. *China mining*, 25(6): 66–71.

Li, X.L., Jiang, Y. & Wang, Y.F. (2011) The problems and countermeasures that land reclamation in damaged land left over by history of mining area. *China Science and technology information*, (23): 68–68.

Li, J., Liu, Y.P. & Hu, Z.Q. (2013) Research on the contents and methods of investigation of damaged land left over from history. *Geography and Geographic Information Science*, 29(3): 92–95.

Ma, R.H., Li, Q. & Li, X.B. (2002) Analysis of Partition method for agricultural land classification and evaluation unit at county level. *Geography and Territorial Research*, 18(2): 93–95.

Tang, X.M., Chen, B.M. & Zhang, L.M., et al. (2011) Discussion on the calculation method of national agricultural land consolidation potential under the new situation. China Land Science, 25(9): 67–71.

Wang, X.M. & Bian Z F. (2011) The consolidation potential of rural residential land in Ji Nan. *China Land Science*, 25(3): 52–57.

Yan, J.M., Xia, F.Z & Li, Q. (2012) Top strategy design of comprehensive land consolidation in China. *Transactions of the Chinese Society of Agricultural Engineering*, 28(14): 1–9.

Land Reclamation in Ecological Fragile Areas – Hu (Ed.)

 ISBN 978-1-138-05103-4

Study on renewal strategy of ecological restoration in an abandoned mining area based on the view of loess civilization

B. Wu
Xi'an University of Science and Technology, Xi'an, Shaanxi Province, China

ABSTRACT: Loess civilization belongs to the regional culture, but it has its own unique characteristics, special features and typicality, which makes it far beyond the limitations of its own region. This paper, from the perspective of loess civilization analysis, pointed out that focusing on the mining area is the foundation of achieving the harmony between nature and human beings. Only through the implementation of the harmonious development mechanism and advancement of the mining area construction via the perspective of ecological civilization, can the mining area achieve true harmony. Mining wasteland is the type of land type that causes tremendous damages to the environment, turning surrounding land resources into a waste. The purpose of this paper is to discuss various ways to rational reconstruction, to explore means to turn "wasteland" into 'treasure-land', and to perform in-depth research on the reasonable restoration of landscape and ecology on mining wasteland. Furthermore, this paper applies theories, methods and principles of landscape design and ecological restoration into concrete actionable tasks that help ecological restoration and landscape space design. These actions can meet the basic needs of ecological restoration and landscape reconstruction on wasteland, and the requirements of sustainable economic and social development.

Keywords: Ecological restoration in the mining area; Landscape Ecology; Landscape design

1 THE FUNDAMENTAL CHARACTERISTICS OF AN ABANDONED MINE AREA

As an important part of human civilization, mining has always been ubiquitous in the history of social evolution. With the rapid economic growth and infrastructure constructions, the needs for mineral resources have becoming more and more demanding. As a result, while the great economical values are achieved through mining activities, the ecological environment surrounding mining areas in China has been drastically damaged. This, in return, restricts the development of mining economy. This paper mainly discusses the influences of abandoned mining area having on ecological environment; analyze landscape patterns and explore various methods of mining area ecological restoration and reconstruction; and provide a scientific basis for the mining area landscape ecological recovery.

1.1 *Abandoned mines have an effect on the ecological environment*

1.1.1 *The destruction of the geomorphological landscape*

Mining of mineral resources include open pit and underground mining, both of which have various degrees of damage to the surface-to-surface table landscape. Strip mining is mainly by digging through mining land layer-by-layer, thus changes the geomorphological landscape; Underground mining utilize mineral extraction that leads to changes to the overburden.

1.1.2 *The destruction of land resources*

The mining of mineral resources is usually accounted for a large number of land. During the process of strip mining, stripped topsoil, post-mining wasted rocks, and mine tailings can destroy and the pressure of the mining area land. In the process of underground mining, the mining land will lead to ground subsidence and collapse, while accelerating soil and water loss and soil erosion.

1.1.3 *The damage to the water*

The mining of mineral resources has an impact on surface water resources and ground water resources. The various impacts on surface water resources include but not limited to the changes of hydrology and water quality pollution, etc. The major impacts if mining on groundwater resources are the decline in groundwater level and land subsidence, etc.

1.1.4 *The induction of geological disasters*

Mining of mineral resources and the related engineering constructions can cause disturbance to mine area geological structure. Underground mining, ground and slope excavation are both likely to trigger collapse, landslide, debris flow, ground subsidence, induced earthquake and other geological disasters.

1.1.5 *The destruction of biodiversity*

Mineral resources exploitation, in the process of vegetation destruction, such as slag heap, will damage the biodiversity of the mining area. At the same time, the ecosystem repair appears to be very slow due to the lower microbial activities in the waste land.

1.2 *Influence of abandoned mining area to the surrounding area*

Mufu refers to the mining land that lost its economic value during the mining process in open mining, due to heavy metals pollutions. Mufu is the kind of extreme bare land with sparse vegetation and serious soil erosion.

Mufu causes severe pollution to water, soil and trigger a series of economic, ecological and social problems. Foremost, the open coal mine with coal gangue pulverized coal, high sulfur coal gangue, deep rock, causes serious harm to surrounding areas. The pollution harm, with no urgent prevention, will incur irreversible consequences to the environment.

Mufu governance has become the key factor that restricts the development of the mining area economy. Ecological restoration on wasted lands not only is an important approach to the sustainable coordination between mining and agriculture and to the protection of mining area's social stability, but also a necessary step to protect the environment and to improve the ecological environment. Due to the disturbance of mining activities, mine land carries extremely harmful physical and chemical properties, such as high heavy metal content, extreme pH or salinization etc.

2 LANDSCAPE PATTERN ANALYSIS ON ABANDONED MINING AREA

2.1 *Mining area landscape spatial structure analysis*

The intertwine relationships among Natural environment, human factor and biological diversity will cause interference to landscape pattern. Human activity is the main factor of the mining area landscape pattern change. Under the interferences of such large excavation, strong changes have taken place in the mining area landscape structure. The mining of the native vegetation has been damaged; the physical and chemical properties of the soil has changed; what was suitable for animals, plants, soil microbes that provides balanced ecological system becomes unlivable, even generates poisons. What was originally advantages to the landscape design have become roadblocks now. The landscape heterogeneity structure leads to the formation of various sorts of plaque. Gradually, the plaque isolation degree increases and the space between the connection degree is reduced, eventually forms the fragmented landscape pattern.

2.2 *Mining area landscape internal energy cycle and the production capacity analysis*

The health of the mining area landscape plays an irreplaceable role sustainable development of regional economy, biodiversity and ecosystem landscape. In addition, the fragmentation of the mining area landscape structure indicates that the decline of mining area landscape functions and the landscape stability. To take Landscape features into the play, then green space should achieve the minimum standards in the landscape structure, meaning the entire green area should be relatively big and contains high level of connectivity. However, because of the topsoil stripping in mining areas, the waste rock piles, the accumulation of tailings, and acid and heavy metal pollution in non-ferrous metal mines, the surface of mining areas are becoming barren desert. Plus, water loss and soil erosion, rain erosion desertification intensifies the process of desert formation.

2.3 *Biological recovery method analysis of the mining area landscape*

Bioremediation of contaminated soil refers to the improvement of mining area through cultivating plant, grazing animals, bacteria, or microbes. During the process that these creatures grow on soil, hey play important physical and chemical roles to change the composition of the soil, therefore achieve the purpose of repairing the damaged soil.

Bioremediation improvement can be categorized into three major methods: the phytoremediation improvement method, animal repair improvement method and the modified bacteria and microbial remediation improvement method. Phytoremediation method improve the landscape by planting suitable plant. For example, in saline soil, planting alfalfa, clover plants to reduce salinity achieved fruitful results and remarkable effect.

In the process of mining area ecological environment restoration, biological method is very effective, especially in the restoration contaminated soil through microorganism.

3 ABANDONED MINING AREA ECOLOGICAL RESTORATION AND RENOVATION METHOD

3.1 *The vegetation of abandoned mine soil improvement*

Through the study of related study and analysis on the research of mining area and the area sampling, there is a variation in the effectiveness of vegetation improvement depends on the profile of the soil. The effective improvements are mainly concentrated on the top of the table, and the upper soil improvement tends to receive better treatment than lower level of soil.

3.2 *The change of landscape pattern*

The change of landscape pattern in mining areas are originally driven by resources exploitation, but dynamically change along with the spatial-temporal changes. With the rise of the modern mine survey and geographic information technologies, it becomes possible to monitor the dynamically changing landscape pattern in mining areas. The up-and-rising geographic information technologies also become indispensably important to research on regional land use. Ecological restoration and reconstruction of mufu, can be applied according to a predesigned path to a natural recovery, but also may be utilized with the aid of artificial support and induction. At present, the theory of landscape ecology has been applied to all aspects of the mine ecological restoration. Analyzing the changes in landscape patterns of the mining area before and after ecological reconstruction has also become important technical means and content to evaluate the efficiency and benefit ecological restoration.

ACKNOWLEDGMENT

Project funds: In 2015, special research project in shaanxi province department of education under the concept of the resources city transformation shanbei abandoned mining area landscape updates and ecological reclamation "(project number: 15Jk1481)".

REFERENCES

BoJie Fu, Lixiang Chen. The type of landscape diversity and significance. *Journal of geographical*, ploidy of 1996 (5): 454–464.

Junhua Zhang, Qingrui Chang, Keli Ja etc. Of the loess plateau vegetation restoration on soil fertility quality impact study [J]. *Journal of soil and water conservation Journal*, 2003 (4): 38–41.

Kongjian Yu, Dihua Li. Urban and regional planning of landscape ecological pattern. *Foreign urban planning*, 1997, (3): 27–4.

Longqian Chen, Dazhi Guo, Zhaoling Hu etc. Xuzhou mining area remote sensing monitoring of land use change and reclaimed-years reclamation study [J]. *Scientific progress*, 2004, 23 (2): 10–15.

Ranghui Wang, Zili Fan. Using remote sensing and Downstream of the tarim river Allah do five research areas of land desertification [J]. *Journal of remote sensing*, 998, 2 (2): 138–142.

RenCang Bu, Duning xiao, Xiuzhen Li. Ecological space theory and landscape heterogeneity. *Acta ecologica sinica*, 1997 (5): 453–7.

Xin-tong li. Sustainable agriculture landscape ecological planning and design. *Regional research and development*, 2000, 12 (3): 5–9.

Zhenqi Hu, Xiuhong Yang etc. Theory of the mining area ecological environment restoration [J]. *Journal of resources and environment*, 2005, 23 (1): 38–41.

Zhongke Bai, Jinlian Zhao, Yinni zhu. Try to talk about the mining area ecological reconstruction. L1. *Journal of natural resources*, 1999, 14 (1): 35–41.

Land Reclamation in Ecological Fragile Areas – Hu (Ed.)
 ISBN 978-1-138-05103-4

"The Belt and Road" strategy under the new mode of ecological restoration of mining areas

Q.W. Zhu & H.B. Yang
School of Surveying Science and Technology, Xi'an University of Science and Technology, Xi'an, China
Key Laboratory of Western Mine Exploitation and Hazard Prevention, Xi'an University of Science and Technology, Xi'an, China

ABSTRACT: China is the world's third largest mineral resources country. The mining is important cause of the mining area ecological damage. Our country put forward the strategic vision of "The Belt and Road" in September 2013, it is for the Chinese mining the Midwest brought foreign exchange window. This paper analyzes the "One Belt and Road" and strengthen the relationship between the mining area ecological restoration; Secondly, through the analysis of the effects of mining area ecological environment constitute, this paper analyzes the ecological restoration of mining area; Finally, the paper puts forward a new model of the mining area ecological environment restoration, this model is based on can be realized on the basis of their own advantages and disadvantages, and give full play to the advantages and avoid disadvantages, so as to achieve the mining area ecological optimization and the sustainable development of mining area of environmental protection.

1 INTRODUCTION

"Belt and Road Initiative" bring new opportunities for the economy along the line, but also take new challenges. Due to mining damage to the environment so that the fragile ecological environment along the silk road environment has become more serious, urgent mine ecological restoration. The national ecological restoration can be used to regulate and lead the "The Belt and Road" construction, and build a green silk road. To promote environmental protection with the ecological restoration, development escort "The Belt and Road".

2 "THE BELT AND ROAD" AND STRENGTHENING THE RELATIONSHIP BETWEEN THE ECOLOGICAL RESTORATION OF MINING AREAS

"The Belt and Road" initiatives put forward since 2013, and the 2014 planing, has entered the implementation stage now. We analyzed the Chinese mining cooperation prospects. On the basis of the study of "The Belt and Road" along the background, mineral resources and mineral resources distribution, mining investment environment. The regional mineral resources development and investment potential, but faces many risks. As everyone knows, "The Belt and Road" complex ecological environment; in the west along the line especially. "The Belt and Road" in promoting economic development along the line at the same time, bring great harm to the environment, ecological problems, of which the most important is the mining area. Mining caused serious ecological damage to the environment, reduce the carrying capacity of regional environment, and the implementation is not conducive to the sustainable development and ecological civilization. To build a "The Belt and Road", we must deal with the environmental challenges, to solve the ecological environment problem from the source; jointly promote the upgrading of the structure of "The Belt and Road"; construction of ecological security system, improve the bearing capacity of "The Belt and Road" construction environment. The implementation of "The Belt and Road" is not only to promote the development of economy along the road, but also to strengthen the development of the strategy of sustainable development and ecological civilization. So "The Belt and Road" is a great significance of strengthen the ecological restoration of mining areas.

3 ANALYSIS OF THE ECOLOGICAL ENVIRONMENT OF THE MINING AREA

China's mining land work started late in remediation and ecological restoration, from the beginning of the end of the 1950s, has now gone through 60 years of history,. But compared with some western countries, we still have a long way to go.

Table 1. Influence of ecological factors on the ecological environment of the mining area.

Factor type	Include content	On the ecological environment of mining area
Climatic factor	Light, temperature, moisture, air, etc	Through light, water, temperature and other factors affecting the growth of plants and animals and climate conditions
Soil factor	Soil structure, soil physical and chemical properties, soil fertility, etc.	Through the structure, fertility and other factors on the ecological restoration of the mining area, the situation of plants and animals have a direct impac
Terrain factor	The ups and downs, ground slope, sunny slope and shady slope etc.	The growth and distribution of plants indirectly affected by climate and soil
Biological factors	Various relationships among organisms	Through the relationship between the biological impact of the ecological restoration of mining areas, such as what kind of plant is conducive to the rapid restoration of the mining area
Artificial factor	Emphasize the particularity and importance of human function	The artificial factor is very important to the ecological restoration of the mining area

Table 2. Comparison of ecological restoration models in the mining area.

Repair strategies	Applicable conditions	Ecological factor	Repair and utilization	Repair technical measures	Input	benefit	cycle
Natural restoration	Human activities in mountain area, populated area	Terrain factor	Woodland and grassland	Ban, avoidance, aerial	Small	Small	long
Landscape development model	Adjacent urban areas, scenic spots, roads on both sides	Artificial factor	Park, tourism and leisure	Engineering measures or craft carving	large	large	short
Agricultural cycle model	Mountainous area, mining wasteland	Soil factor	Industrial land, farmland, etc	Flat filling, soil physical properties	large	large	long
Forestry Land Model	Traffic lines on both sides, urban planning area, scenic area	Artificial and soil factors	Orchard, garden land, forest land, green land	Ecological belt, hanging net spray sowing, planting, planting economic forest	large	large	short
Construction land model	Tailings, waste residue, waste heap, near the industrial zone	Artificial factor	Housing, factory	Leveling construction	large	large	short

Mining ecological restoration to local conditions, only to determine the factors, mainly in the ecological restoration process can be more effective with an antidote against the disease. There are many types of ecological factors, and the classification methods are not uniform. According to the nature of ecological factors, most scholars can be divided into the following five categories, as shown in the following Table 1:

Division of ecological factors are man-made, in the environment, and the role of various ecological factors are interrelated, mutual influence, therefore in the biological, ecological factor analysis, not only a one-sided attention to ecological factors. The interaction of various ecological factors also exist mutual compensation or enhancement.

4 THE NEW MODEL OF MINING ECOLOGICAL ENVIRONMENT RESTORATION

4.1 *The existing mode of ecological restoration in the mining area*

After years of development and improvement, the mining area can be divided into physical remediation, chemical modification, phytoremediation and so on. Based on the way of ecological restoration, and the effect of ecological factors, the ecological restoration of a series of models, such as natural restoration, construction land, forest pattern (fruit) industry recovery mode of land use, ecological agriculture circular economy mode, the development and utilization of landscape pattern.

4.2 *The new model of ecological restoration in the mining area*

With the development of science and technology, in the diversification of the information society, a model has been unable to meet the needs of mine repair. Only based on the fundamental, combining the ecological restoration in mining area, solve the city land demand and regional sustainable development, do the city, the population needs, economic and ecological planning, the implementation of the ecological factor is dominant, auxiliary in the mode of economic development, local multi-level and diversified development of the basic principle of planning, according to the ecological factors in the combination of "The Belt and Road" policy at the same time, according to local conditions and resources allocation, artificial and natural ecological system coordination, planning and planning, combined with the recent ecological and social economic benefit and win-win principle, the overall regulation of local control and synchronization principle. To determine the factors of ecological restoration of mining area, combined with the city planning and the needs of the people, according to local conditions to carry out the construction, the most suitable repair mode is local, not blindly imitate, but set hundred long in can realize their advantages and disadvantages on the basis, give full play to the advantages, avoid the disadvantages, so as to realize the sustainable the development and optimization of mining ecological environmental protection in mining areas.

5 SUMMARY

To build a "The Belt and Road", we need to deal with the mining area ecological environment challenges, improve the ecological construction "The Belt and Road" along, enhance the bearing capacity of the ecological environment, in order to better serve the "The Belt and Road" increased vitality and power. In the environmental problems produced in the process of mining is difficult to study in ecological restoration, but also hinder the sustainable development of society and economy, and influence factors of "The Belt and Road" construction. Mine ecological restoration must be detailed analysis of mine ecological factors, and seize the main contradiction and ecological restoration conditions, combined with economic development mode and the "The Belt and Road" strategy, improving the mining environment, and improving the quality of life of the people in maintaining social stability.

REFERENCES

Hu Angang, Ma Wei, Yan Yilong (2014) "Silk Road Economic Belt": the strategic connotation, and the realization of the path location. Journal of Xinjiang Normal University (*Philosophy and Social Sciences Edition*), 35 (2): 1–10.

Qi hisan (2015). The production situation and the security situation of coal production in developed countries of *J.* China coal, 41 (8), 140–143.

Wang Hairong, (2016) status quo and trend of land reclamation and ecological restoration in Zhengzhou mining area of Henan. *J. Green Science And Technology*, 06-0066-02.

Wang Lei, Song H.L., Xu Yanying, et al. (2012) Study on ecological restoration of abandoned coal mine. *J. Anhui Agricultural Science Bulletin,* 18 (5).

Zhang Shoujun (2013) Research Progress on ecological restoration of mining area, 6611, China. *J. Anhui agricultural sciences,* 0517-34-02-13362.

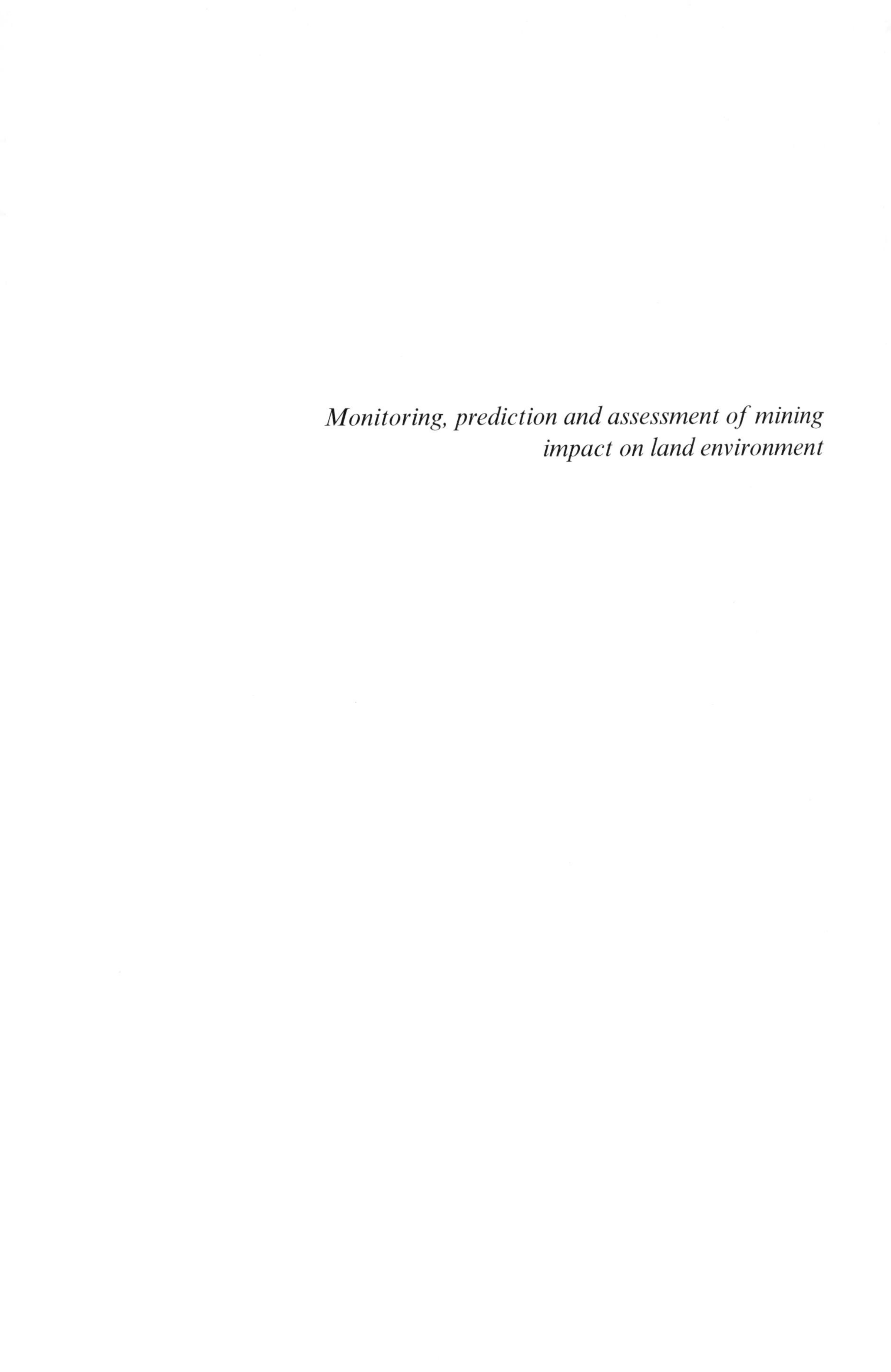

Monitoring, prediction and assessment of mining impact on land environment

Land Reclamation in Ecological Fragile Areas – Hu (Ed.)
© 2017 Taylor & Francis Group, London, ISBN 978-1-138-05103-4

Post-mining land-use selection by using a combination of PROMETHEE and SIR techniques

Sina Amirshenava & Morteza Osanloo
Department of Mining and Metallurgical Engineering, Amirkabir University of Technology, Tehran, Iran

ABSTRACT: Despite of positive impacts of mining activities on world economy, mining operations also caused unavoidably negative impacts on surrounding environment. Mine reclamation is one of the most important stages of the mining activities in line with the basic principles of sustainable development. Mine reclamation is a progressive activity, starts from the initial phase of mining and eventually leads to the preparation of all mined land areas such as pit, waste dump and tailing dam for Post-Mining Land-Use (PMLU). The PMLU can deeply affect on expenses and implementation of reclamation activities. In this study, different post-mining land-uses are evaluated in the Choghart iron ore mine which is located in Yazd province of Iran, by using a combination of PROMETHEE and A Superiority and Inferiority Ranking (SIR) techniques. According to the results, revegetaion with native species is the optimal PMLU solution for whole mined land of Choghart iron ore mine.

1 INTRODUCTION

Mining is one of the main activities for national economic development and provides the main materials for industrial production and soci-economic development (Bascetin, 2007; Yavuz and Altay, 2015). In contrast to these positive aspects, mining activities have some negative impacts that damages the environment. The most serious adverse environmental and social effects caused by surface mining activities are as follows (Dogan and Kahriman 2008; Yavuz and Altay, 2015):

- Nonrenewable resource depletion
- Destruction of natural landscape and habitats
- Health and safety problems
- Air and water pollution
- Contamination of soil
- Extinction of natural flora and fauna
- Abandoned mines and unreclaimed mine sites

Mine life is limited and it depends on the minable reserve and other factors such as economic, social and environmental factors that affect premature mine closure (Laurence, 2006; Maczkowiack et al, 2012). From one point of view, mining is in contradiction with the principles of sustainable development because of the negative impacts (Laurence, 2011). Mine reclamation is one of the most important stages of the mining activities in line with the basic principles of sustainable development and should be considered from the initial phase of mining (Narrei and Osanloo, 2011). Mine reclamation is a progressive activity, during which the pollutions left by mining activities are resolved as far as possible. Hence, the mine closure risk levels are reduced and it prepares the mined land such as pit, waste dump and tailing dam for PMLU. The PMLU is not necessarily the same as the pre-mining land-use and it is possible that a better land-use be introduced according to the regional potentials and the community needs (Osanloo, 2008; Mborah et al, 2016). The key objectives of mine reclamation given in Table 1 (Dimitrijevic et al, 2014, Shenavar and Osanloo, 2016).

The Bow-tie diagram of mine closure risks is depicted in Figure 1 where the negative impacts of mine closure are top event (overall risk) and mine reclamation plan is considered as a measure to reduce the risk level.

PMLU can deeply affect on expenses and implementation of reclamation activities. The optimal PMLU selection is the basic step in the successful implementation of the reclamation plan. There are

Table 1. The key objectives of mine reclamation.

No.	Mine reclamation objective
1	Protect the health and safety of mine workers and residents
2	Creating a profitable and sustainable land use
3	Improve the region's landscape
4	Reduce the negative impacts of mining activities
5	Social, economic and environmental stabilization

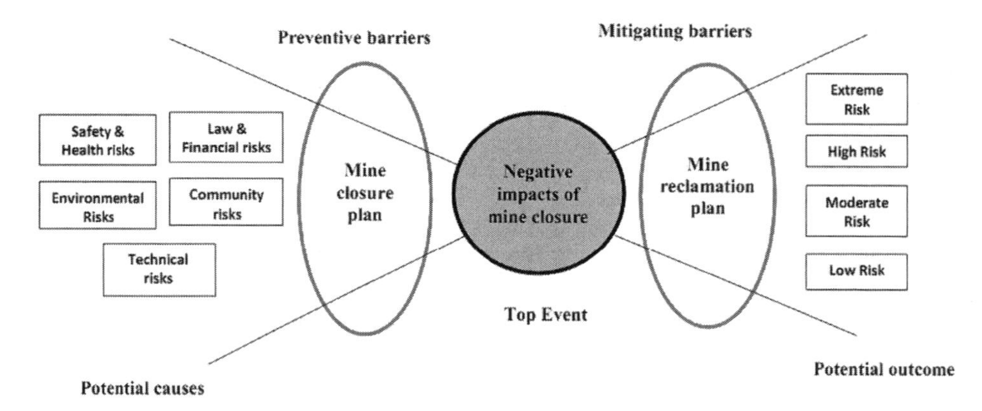

Figure 1. Bow-tie diagram of mine closure risks (Amirshenava et al, 2016).

three main viewpoints about reclamation of mined land as follows (Osanloo, 2008):

1. Agricultural land use anyway
2. Selection of an optimum land use option that is consistent with the existing conditions
3. Selection of an optimum land use option that improves the existing conditions

Considering the above perspectives, there are numerous land-uses for a mined land, but the most common PMLU include revegetaion with native species, agriculture (cropping, grazing and pasture), forestry, lake (for multiple purposes), recreation areas and wildlife habitat (Soltanmohammadi et al, 2008; Mborah et al, 2016).

PMLU selection is a multi-dimensional problem. MCDM methods are the best solution on this way (Zimmerman, 2016). Decision-making methods have widely been used in mining and have been applied by many researchers to PMLU selection. Bascetin (2007) used Analytical Hierarchy Process (AHP) method to select an optimal PMLU for open pit coal mines. Soltanmohammadi et al (2008 and 2009) developed a mined land suitability analysis framework based on the MCDM methods, namely AHP, ELECTREE and PROMETEHEE. This framework contains 50 attributes for evaluation of eight possible groups of PMLU options. Narrei and Osanloo (2011) proposed a combined MCDM approach to evaluate the PMLU methods. They utilized entropy, weighted least square method and AHP techniques to obtain the relative weights of attributes. SAW, TOPSIS, Compromise programming and three aggregate method used for ranking the alternatives. Betrie et al (2013) chooses the optimal remediation option for Acid Rock Drainage (ARD) from mine wastes using AHP and PROMETHEE methods. Dimitrijevic et al (2014) determines the PMLU in Klenovnik open pit mine in the Kostolac coal basin using MCDM methods, PROMETHEE and ELECTRE. They recommended to include at least two MCDM techniques in order to improve the reliability of ranking. Yavuz and Altay (2015) used two different fuzzy decision-making methods, Yager's and fuzzy AHP method to select the best-suited reclamation project. Zimerman (2016) describes an approach based on a GIS based decision-support systems to find the optimal reclamation method for abandoned mined-lands in the Appalachian region. Shenavar and Osanloo (2016) evaluates different reclamation alternatives using MCDM techniques, namely TOPSIS and ELECTRE for Sangan placer iron ore mine of Iran that operates by strip mining method.

In this study, different post-mining land-uses are evaluated in the Choghart iron ore mine, which is located in Yazd province of Iran, by using a combination of PROMETHEE and A Superiority and Inferiority Ranking (SIR) methods.

2 STUDY AREA

Choghart iron ore mine is located in Bafgh mining district of central Iran, 12 km northeast of Bafgh and 125 km southeast of Yazd. The annual production is approximately 3 Mton with an overall stripping ratio of 0.66:1. The total area of this region is 40 km^2. The region has a dry climate with very hot summers and the average annual rainfall is 53 mm. During the year, the temperature ranges between −7°c in winter to +47°c in summer (ICIOC, 2017).

3 METHODOLOGY

PMLU selection is a multi-criteria problem and different groups (Mine stakeholders, Local responsible persons) with various opinions are involved. A combination of MCDM methods is used to select the most appropriate PMLU in Choghart

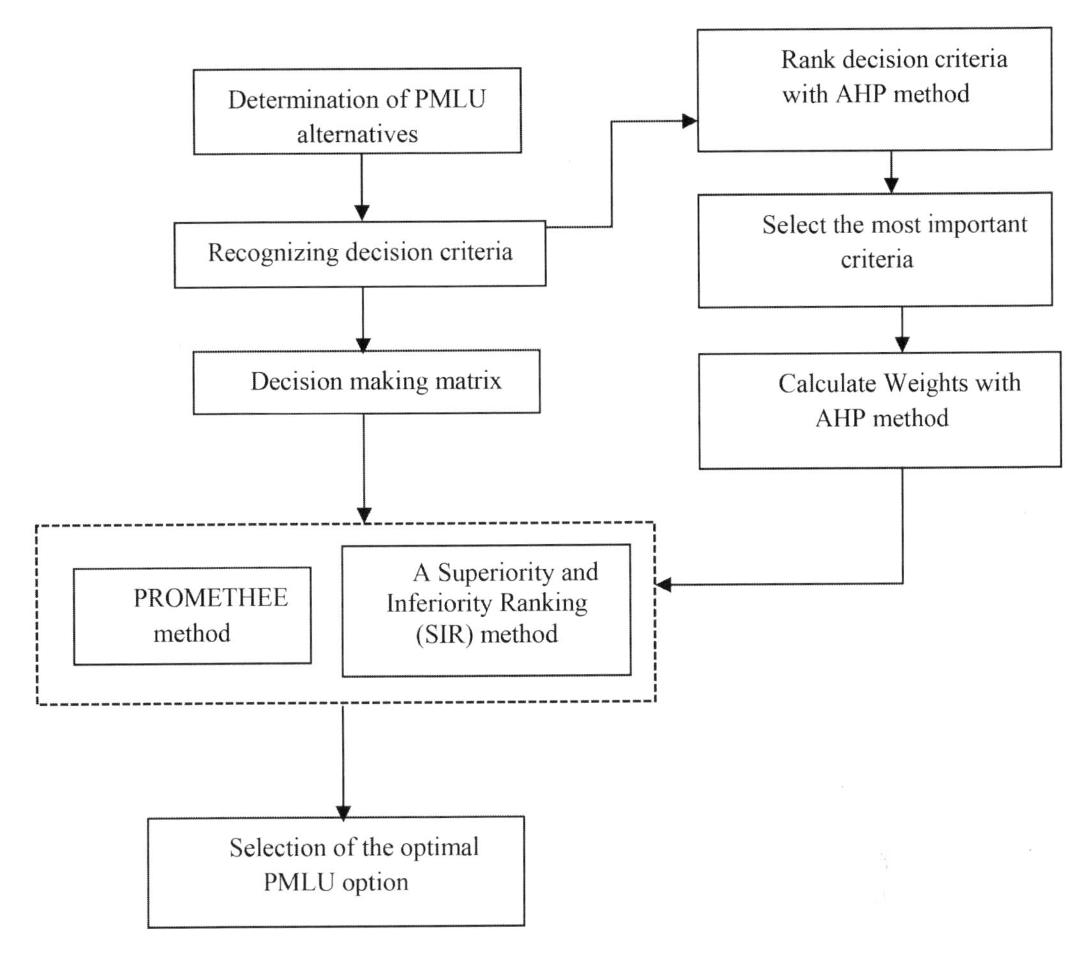

Figure 2. The framework of selecting the optimal PMLU alternative.

iron ore mine. The PMLU selection framework is demonstrated in Figure 2. The approach consists of the following stages:

1. Determine PMLU alternatives
2. Identify the decision making criteria
3. Rank the PMLU alternatives with PROMETHEE II method
4. Rank the PMLU alternatives with A Superiority and Inferiority Ranking (SIR) method
5. Select the optimal solution

3.1 *Determine the PMLU alternatives*

Determination of land-use alternatives is very important in mine reclamation. According to the key objectives of reclamation (Table 1) and considering regional conditions, six possible PMLU alternatives are selected for Choghart iron ore mine (Table 2).

Table 2. PMLU alternatives.

ID	PMLU alternative
A1	Solar power station
A2	Pit backfilling
A3	Educational centers
A4	Industrial construction
A5	Revegetaion (native and resistant species)
A6	Agriculture (Cereals, dates, pistachios)

3.2 *Identify decision making criteria*

PMLU selection is affected by five groups of factors including economic, social, landscape and environmental, technical and mine site factors (Table 3). The large numbers of criteria may cause confusing in the survey forms. So, at first the criteria (Table 3) are evaluated to select the most

Table 3. Decision making criteria in selecting the PMLU.

ID	Criteria	ID	Criteria
E	Economic factors	S	Social factors
E1	Capital cost	S1	Employment opportunities
E2	Operating cost	S2	Reducing of emigration from mining area
E3	Increasing income of local community	S3	life quality
E4	Changing real state value	S4	Government policy
E5	Maintenance cost	S5	Consistency with local lifestyle
E6	Payback period	S6	Concerns and needs of local people
T	Technical factors	S7	Land ownership
T1	Topography and size of the mined land	M	Mine site factors
T2	Distance to nearest water supply	M1	Soil's physical properties
T3	The complexity of the reclamation plan	M2	Soil's chemical properties
T4	Current land-use in surrounding area	M3	Exposure to sunshine
L	Landscape & environmental factors	M4	Precipitation
L1	Landscape quality	M5	Land slope
L2	Environmental acceptability	M6	Temperature
L3	Stop desertification		

Table 4. AHP Scale of pair-wise comparison.

Importance Scale	Compare factor of i and j
1	Equally important
3	Weakly important
5	Strongly important
7	Very strongly important
9	Extremely important
2,4,6,8	Intermediate values

important ones. To do that, AHP method is applied and the most important criteria in Choghart iron ore mine are determined.

3.2.1 *Rank the criteria by AHP method*

AHP method is applied for ranking the decision criteria. It is developed by Saaty and it is one of the methodological approaches applied for resolving highly complex decision-making problems by capturing both subjective and objective evaluation measures. It uses a pair-wise comparison matrix for finding the relative weight of the criteria. In this matrix, each criterion compares with others and their relative importance degree determines with a number according to Table 4 (Saaty, 1991). The pair-wise matrix is shown in Equation 1. The element a_{ij} indicates the relative importance of criteria "i" with respect to criteria "j" and determined according to Table 4.

$$D = \begin{bmatrix} a_{11} & a_{12} & \cdots & a_{1n} \\ a_{21} & a_{22} & \cdots & a_{2n} \\ \vdots & \vdots & \ddots & \vdots \\ a_{n1} & a_{n2} & \cdots & a_{nn} \end{bmatrix} \quad (1)$$

where: $a_{ii} = 1, a_{ij} = \frac{w_i}{w_j}, a_{ij} = \frac{1}{a_{ji}}, a_{ij} \neq 0$

To check for the matrix consistency, the following condition must be established:

$$a_{ik} \times a_{kj} = a_{ij} \quad (2)$$

To calculate the weight of criteria, if the pair-wise comparison matrix is consistent, then the weight of each criterion is determined as following equation:

$$w_i = \frac{a_{ij}}{\sum_{i=1}^{n} a_{kj}} \quad (3)$$

But this matrix is usually inconsistence, then the error or the difference should be minimized. The error is equal to:

$$e_{ij} = \left(a_{ij} \times w_j - w_i\right)^2 \quad (4)$$

So, w_i and w_j must be determined such that e_{ij} is minimized. To obtain the weight of the criteria, the mathematical model in Equation 5 should be solved.

$$MinZ = \sum_{i=1}^{n} \sum_{j=1}^{n} \left(a_{ij}.W_j - W_i\right)^2 \quad (5)$$

$$s.t: \sum_{i=1}^{n} W_i = 1$$

$$W_i \geq 0 \; i = 1,2,\ldots\ldots,n$$

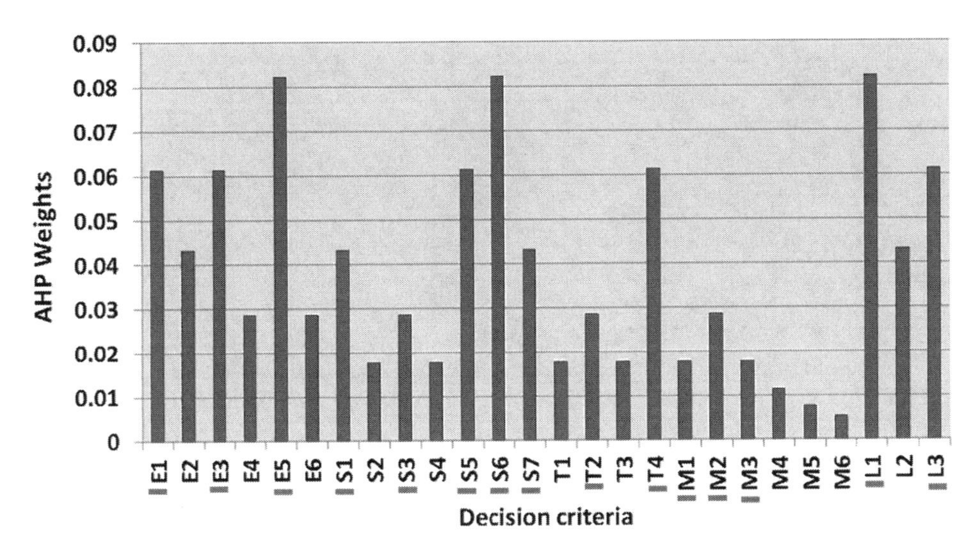

Figure 3. Relative weights of criteria calculated by AHP.

This model could be solved using Lagrange multipliers. So, with the following modification, the mathematical model in Equation 5 will be solved.

$$L = \sum_{i=1}^{n}\sum_{j=1}^{n}\left(a_{ij}.W_j - W_i\right)^2 + 2\lambda\left(\sum_{i=1}^{n} W_i - 1\right) \qquad (6)$$

With derivative of Equation 6 toward w_i:

$$\sum_{i=1}^{n}\left(a_{ij}.W_l - W_i\right)a_{il} - \sum_{i=1}^{n}\left(a_{ij}.W_l - W_i\right) + \lambda = 0 \qquad (7)$$

$i = 1, 2, \ldots\ldots, n$

From the above equations, (n+1) non-homogeneous linear equation and (n+1) variable is achieved. By solving these equations, the weight of each criterion will be determined.

This approach is conducted to rank the decision criteria in Choghart iron ore mine. Survey forms are sent to experts to identify the importance of each criterion on the reclamation process and reducing the mine closure risks level. The relative weights of criteria calculated by AHP are shown in Figure 3. During this step, the most important factors are selected from among the 26 criteria listed in Table 3. For this purpose, 15 criteria are selected and these 15 criteria should cover all 5 groups. So, they are selected with respect to the ratio of the number of criteria in each group. In each group the criteria with the highest relative weights are selected. The selected criteria are marked red in Figure 3. All criteria are positive except E1 and E5.

3.2.2 *Calculate the decision criteria relative weights*

The selected 15 criteria in previous step are not in the same power rank in the PMLU process selection. Therefore, to overcome to this situation, opinion of experts would be rating by AHP mechanism (Figure 4).

3.3 *Rank the PMLU alternatives*

In this stage, the post mining land uses are ranked by using the SIR and PROMETHEE methods. Regarding to the advantages and disadvantages of each method, a combination of them can guarantee reaching to optimal solution.

3.3.1 *Ranking with PROMTHEE method*

PROMETHEE (Preference Ranking Organization Method for Enrichment Evaluation) is an MCDM method that provided by Brans (1982) and further extended by Brans & Vincke (1985). Using this method, a finite number of alternatives can be ranked according to several, sometimes conflicting criteria (Brans et al, 1986). This method is one of the most frequently used methods of multi-criteria decisions (Tomić et al, 2011). The two kinds of PROMETHEE method builds a partial ranking (PROMETHEE I) and complete ranking (PROMETHEE II) (Betrie et al, 2013). The steps of the PROMETHEE II method are as follows (Brans et al, 1986; Albadvi et al, 2007; Behzadian et al, 2010; Tomić et al, 2011; Betrie et al, 2013):

Step 1. Construct a decision making matrix

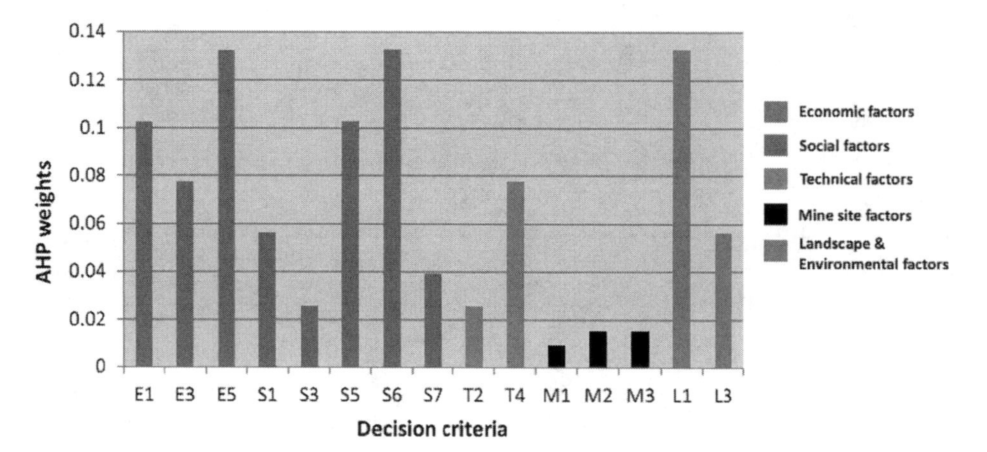

Figure 4. Final Relative weights of selected criteria calculated by AHP.

Each element ($C_i(A_i)$) of decision matrix (Equation 8) is a measure of a criterion over an alternative.

$$D = \begin{bmatrix} C_1(A_1) & \cdots & C_j(A_1) & \cdots & C_n(A_1) \\ \cdots & \cdots & \cdots & \cdots & \cdots \\ C_1(A_i) & \cdots & C_j(A_i) & \cdots & C_n(A_i) \\ \cdots & \cdots & \cdots & \cdots & \cdots \\ C_1(A_m) & \cdots & C_1(A_m) & \cdots & C_n(A_m) \end{bmatrix} \tag{8}$$

Step 2. Determination of deviations based on pair-wise comparisons

The difference of two evaluations (A_i and $A_{i'}$) on each criterion is determined using Equation 9. This Equation is inverted for the criterion with a negative nature.

$$d_j(A_i, A_{i'}) = C_j(A_i) - C_j(A_{i'}), \tag{9}$$

$$i = 1,2,\ldots,m \qquad i' = 1,2,\ldots,m$$

Step 3. Calculate the preference function value

The preference is expressed by a number ranging from 0 to 1 (0 for no preference or indifference to, 1 for strict preference). The preference of alternative A_i with regard to alternative $A_{i'}$ on each criterion as a function of $d_j(A_i, A_{i'})$ is determined using Equation 10.

$$P_j(A_i, A_{i'}) = f\left[d_j(A_i, A_{i'})\right] \tag{10}$$

There are six type of preference function namely: 1) usual criterion, 2) U-shape criterion, 3) V-shape criterion, 4) level criterion, 5) V-shape with indifference criterion and 6) Guassian criterion. The preference function is determined based on the nature of each criterion and decision maker opinion.

Step 4. Calculate the preference index

The preference index $\pi(A_i, A_{i'})$ is calculated as a weighted average of preference functions in accordance with Equation 11 and it expresses the preference of alternative A_i over $A_{i'}$ considering all criteria.

$$\pi(A_i, A_{i'}) = \sum_{i=1}^{k} \mathrm{Pj}(A_i, A_{i'}) w_j \tag{11}$$

w_j: the relative importance of each criterion

Step 5. Calculate the outranking flows

Positive outranking flow (leaving flow) and negative outranking flow (entering flow) are calculated with Equation (12) and (13) respectively.

$$\varnothing^{+}(A_i) = \frac{1}{m-1} \sum_{A_i \epsilon A} \pi(A_i, A_{i'}) \tag{12}$$

$$\varnothing^{-}(A_i) = \frac{1}{m-1} \sum_{A_i \epsilon A} \pi(A_i, A_{i'}) \tag{13}$$

Step 6. Calculate net outranking flow

In order to complete outranking in PROMETHEE II method, net outranking flow for each alternative is calculated by Equation 14. Each alternative that earns the most outranking flow is the optimal solution.

$$\varnothing(A_i) = \varnothing^{+}(A_i) - \varnothing^{-}(A_i) \tag{14}$$

3.3.2 *Ranking with SIR method*

The SIR (A Superiority and Inferiority Ranking) is an MCDM method that is developed by Xu (2001). This new and relatively complex method is a significant extension of the PROMETHEE method. The steps 1–3 of the SIR method are the same as PROMETHEE. The other steps of the SIR method are as follows (Xu, 2001):

Step 4. Calculate the superiority and inferiority index

For each alternative A_i, the superiority index $S_j(A_i)$ and inferiority index $I_j(A_i)$ with respect to the jth criterion are calculated by the Equations 15 and 16.

$$S_j(A_i) = \sum_{k=1}^{m} P_j(A_i, A_k) = \sum_{k=1}^{m} f_j\left(C_j(A_i) - C_j(A_k)\right) \quad (15)$$

$$I_j(A_i) = \sum_{k=1}^{m} P_j(A_k, A_i) = \sum_{k=1}^{m} f_j\left(C_j(A_k) - C_j(A_i)\right) \quad (16)$$

Step 5. Construct the superiority and inferiority matrix

When all the superiority and inferiority indexes are calculated, the superiority matrix (S-matrix) and the inferiority matrix (I-matrix) are as follows:

$$S = \begin{bmatrix} S_1(A_1) & \cdots & S_j(A_1) & \cdots & S_n(A_1) \\ \cdots & \cdots & \cdots & \cdots & \cdots \\ S_1(A_i) & \cdots & S_j(A_i) & \cdots & S_n(A_i) \\ \cdots & \cdots & \cdots & \cdots & \cdots \\ S_1(A_m) & \cdots & S_1(A_m) & \cdots & S_n(A_m) \end{bmatrix} \quad (17)$$

$$I = \begin{bmatrix} I_1(A_1) & \cdots & I_j(A_1) & \cdots & I_n(A_1) \\ \cdots & \cdots & \cdots & \cdots & \cdots \\ I_1(A_i) & \cdots & I_j(A_i) & \cdots & I_n(A_i) \\ \cdots & \cdots & \cdots & \cdots & \cdots \\ I_1(A_m) & \cdots & I_1(A_m) & \cdots & I_n(A_m) \end{bmatrix} \quad (18)$$

Step 6. Calculate the superiority flow and inferiority flow

The superiority flow ($\varphi^{>}(A_i)$) and inferiority flow ($\varphi^{<}(A_i)$) are calculated for each alternative using Equations 19 and 20.

$$\varphi^{>}(A_i) = \sum_{j=1}^{n} w_j \times S_j(A_i) \quad (19)$$

$$\varphi^{<}(A_i) = \sum_{j=1}^{n} w_j \times I_j(A_i) \quad (20)$$

Step 7. The SIR ranking

In this method, to rank the alternatives, at first the n–flow and r–flow should be calculated using Equations 21 and 22. Each alternative that earns the highest n–flow and highest r–flow is the best solution.

$$\text{n–flow} = \varphi^{>}(A_i) - \varphi^{<}(A_i) \quad (21)$$

$$\text{R} - \text{flow} = \frac{\varphi > (A_i)}{\varphi^{>}(A_i)\varphi^{<}(A_i)} \quad (22)$$

4 RESULT AND DISCUSSION

The presented MCDM approach (Figure 2) is conducted in Choghart iron ore mine to select the optimum PMLU. For this purpose, the decision matrix is constructed with regard to specified PMLU options and decision criteria based on the expert's opinions. The expert team including some locals, mine managers and the chief of the mine's HSE sector determine the score of alternatives for each criterion with a number from 1 to 9. The high score for the criteria with the positive nature represents a favorable situation and for the criteria with negative nature reflects the adverse conditions. The decision matrix can be seen in Table 5.

According to the matrix and following the stages of the proposed method, PMLU alterna-

Table 5. PMLU selection decision matrix in Choghart iron ore mine.

	E1	E3	E5	S1	S2	S6	S7	S8	T2	T4	M1	M2	M3	A1	A3
+/−	−	+	−	+	+	+	+	+	+	+	+	+	+	+	+
Specific weight	0.1	0.08	0.13	0.06	0.03	0.1	0.13	0.04	0.03	0.08	0.01	0.02	0.02	0.13	0.06
A1	8	6	8	6	6	4	6	8	5	4	4	3	9	5	3
A2	4	2	5	4	2	1	4	2	4	3	3	2	6	1	2
A3	7	7	6	8	7	5	8	8	2	4	4	2	7	3	3
A4	6	3	5	4	4	3	3	6	3	3	4	2	4	4	3
A5	3	3	2	4	7	7	6	5	5	7	7	6	6	7	8
A6	6	7	7	7	8	8	8	7	1	8	4	4	5	8	7

tives in Choghart iron ore mine are ranked by PROMETHEE II and SIR methods. In this study the V-shape criterion preference function is selected for all criteria in both methods. The final PROMETHEE calculation matrix is shown in Table 6. The values of leaving flow ($\varnothing^{+}$), entering flow ($\varnothing^{-}$), net outranking flow and final rate of alternatives are given in Table 7. Also, Table 8 presents the outcomes of SIR method.

According to the results (Table 7 and 8), the alternatives A1 to A4 have the same rank in both methods. In PROMETHEE II, alternatives A5 and A6 stay at top and second position. Also in SIR method, A5 and A6 gained the highest n–flow and r–flow respectively. In SIR method, the alternative that earns the highest n–flow and highest r–flow is the optimum one. But in case of Choghart iron ore mine of Iran, the optimum alternative is not clear because A5 has the highest n–flow and A6 has the highest r–flow. Thus, with sharing the results of these methods, the A5 (Revegetaion with native and resistant species) is the optimal PMLU in Choghart iron ore mine.

Table 6. Final PROMETHEE calculation matrix.

	A1	A2	A3	A4	A5	A6
A1		0.101	0.119	0.075	0.312	0.268
A2	0.261		0.292	0.119	0.409	0.473
A3	0.048	0.061		0.049	0.294	0.208
A4	0.166	0.049	0.210		0.344	0.373
A5	0.063	0	0.116	0.005		0.145
A6	0.025	0.070	0.036	0.039	0.150	

Table 7. Results of the PROMETHEE II method.

PMLU alternatives	$\varnothing^{+}$	$\varnothing^{-}$	$\varnothing$	Rating
A1	0.113	0.175	–0.062	4
A2	0.056	0.311	–0.255	6
A3	0.155	0.132	0.023	3
A4	0.058	0.228	–0.17	5
A5	0.302	0.066	0.236	1
A6	0.293	0.064	0.229	2

Table 8. Outcomes of the SIR method.

PMLU alternatives	$\varnothing^{>}$	$\varnothing^{<}$	n–flow	r–flow	Rating
A1	0.565	0.907	–0.342	0.384	4
A2	0.282	1.567	–1.285	0.152	6
A3	0.773	0.636	0.137	0.549	3
A4	0.288	1.135	–0.847	0.202	5
A5	1.5	0.362	1.137	0.806	1,2
A6	1.331	0.276	1.055	0.828	1,2

5 CONCLUSION

In this paper, a combination of MCDM approach, consist of AHP, SIR and PROMETHEE techniques, is applied to select the optimal PMLU. MCDM methods are the best tool for the PMLU selection problem. With regards to advantages and disadvantages of MCDM methods, and also considering the point that such methods leads to different results in the same problem and situation, therefore employment of a combination can guarantee of closing to optimal solution. The discussed approach is effective in the evaluation of the different PMLU options in a mining region. In this study, six PMLU alternatives were evaluated in Choghart iron ore mine according to the most important criteria and the result shows that the optimal PMLU is Revegetaion with native and resistant species.

REFERENCES

Albadvi, A., Chaharsooghi, S.k. & Esfahanipour, A. (2007) Decision making in stock trading: An application of Promethee. *European Journal of Operational Research*, 177, 673–683.

Amirshenava, S., Osanloo, M., Esfahanipour, A. & Nadimi, S. *Closure risk assessment in choghart iron ore mine using 3D risk model. SME Annual Meeting 2016*, 21–24 February, *Arizona, United States.*

Bascetin, A. (2007) A decision support system using analytical hierarchy process (AHP) for the optimal environmental reclamation of an open-pit mine. *Environ Geol*, 52, 663–672.

Behzadian, M., Kazemzadeh, R.B., Albadvi, A. & Aghdasi, M. (2010) PROMOTHEE: A comprehasive literature review on methodologies and applications. *European Journal of Operational Research*, 200, 198–215.

Betrie, G., Sadiq, R., Morin, K. & Tesfamariam, S. (2013) Selection of remedial alternatives for mine sites: A multicriteria decision analysis. *Journal of Environmental Management*, 119, 36–46.

Brans, J.P., Vincke, P. & Mareschal, B., (1986) How to select and how to rank projects: The PROMETHEE method. *European Journal of Operational Research*, 24, 228–238.

Dimitrijevic, B., Vujic, S., Matic, I., Majianac, S., Praštalo, J., Radosavljevic, M. & Čolakovic, V., (2014) Multi-Criterion Analysis of Land Reclamation Methods at Klenovnik Open Pit Mine, Kostolac Coal Basin. *Journal of Mining Science*, 50(2), 319–325.

Dogan, T. & Kahriman, A. (2008) Reclamation planning for coal mine in Istanbul, Agacil Region. *Environ Geol*, 56, 109–117.

ICIOC. (2017) *Iran Central Iron Ore Company*. [Online] Available from: http://www.icioc.ir/.

Laurence D. (2006) Optimization of mine closure process. *Journal of cleaner production*, 14, 285–298.

Laurence, D. (2011) Establishing a sustainable mining operation: an overview. *Journal of cleaner production*, 19(2–3), 278–284.

Maczkowiack, R.I., Smith, C.S., Slaughter, G.J., Mulligan, D.R., & Cameron, D.C. (2012) Grazing as a post-mining land use: A conceptual model of the risk factors. *Agricultural Systems*, 109, 76–89.

Mborah, C., Bansah, K.J. & Boateng, M.K. (2016) Evaluating Alternate Post-Mining Land-Uses: A Review. *Environment and Pollution*, 5(1), 14–22.

Narrei, S. & Osanloo, M. (2011) Post mining land-use optimum ranking, using multi attribute decision techniques with regard to sustainable resources management. *OIDA International Journal of Sustainable Development*, 11(2), 65–76.

Osanloo, M. (2008) *Mine reclamation*. Amirkabir University of Technology Publication, 2nd edition, 300 p (in Persian).

Saaty, T.L., (1991), *Multi criteria decision making—the Analytical Hierarchy Process*. Pittsburgh, RWS publications.

Shenavar, M. & Osanloo, M. *Land use selection and reclamation layout planning by MCDM – case study: Sangan placer iron ore mine of Iran, 16th International Symposium on Environmental Issues and Waste Management in Energy and Mineral Production (SWEMP)*, 5–7 October 2016, Turkey, Istanbul.

Soltanmohammadi, H., Osanloo, M., Rezaei, B. & Aghajani Bazzazi, A. (2008) Achieving to some outranking relationships between post mining land uses through mined land suitability analysis. *Int. J. Environ. Sci. Tech*, 5(4), 535–546.

Soltanmohammadi, H., Osanloo, M. & Aghajani Bazzazi, A. (2009) Deriving preference order of post-mining land-uses through MLSA framework: application of an outranking technique. *Eniviron Geol*, (58), 877–888.

Tomić, V., Marinković, Z. & Janošević, D. (2011) PROMETHEE method implementation with multi-criteria decisions. *Facta universitatis*, 9(2), 193–202.

Xu, X. (2001) The SIR method: A superiority and inferiority ranking method for multiple criteria decision making. *European journal of operational research*, (131), 587–602.

Yavuz, M., and Altay, B.L. (2015) Reclamation project selection using decision-making methods. *Environ Earth Sci*, 73(10), 6167–6197.

Zimmerman, M. (2016) Development of a Decision Support System for Post Mining Land Use on Abandoned Surface Coal Mines in Appalachia, *International Development, Community and Environment (IDCE)*, Paper 87.

Land Reclamation in Ecological Fragile Areas – Hu (Ed.)

Land ecological security evaluation in mining cities based on the PSR model: A case study of Wu'an

D.M. Fu & M.C. Fu
School of Land Science and Technology, China University of Geosciences, Beijing, China

ABSTRACT: China is a major mining and consuming country. Mining and utilization of mineral resources is bound to have a serious impact on regional land ecological security while promoting national economic development. In this paper, the mining city of Wu'an is used as an example, the "Pressure-State-Response" (PSR) model is adopted based on the land use data and the statistical data of national economy and social development, 23 evaluation factors are selected by using Analytic Hierarchy Process (AHP) from three aspects of land ecological security pressure, state and response. The evaluation index system of PSR model is established, the weight of each index is determined by the entropy method. The ecological security value and safety level of the study area are defined based on its background value. Land ecological security status in 2006, 2008, 2010, 2012 and 2014 of this area is evaluated, and the corresponding ecological security status is determined. The results show that the land ecological security value of Wu'an in 2006 is 0.3938, which is in a less secure state, and the land ecological security values of 2008, 2010, 2010 and 2014 are respectively 0.5096, 0.5716, 0.5834 and 0.5326, all in critical safety level.

1 INTRODUCTION

With the mining of mining, a number of mining cities appeared. Different scholars have different emphases in view of the related research of mining cities, someone focus on the industrial transformation of mining cities (Qiao Guotong et al., 2017), someone focus on its ecological environment (Ying Ning & FuMeichen, 2016), and someone focus on the its sustainable development (Martina S et al., 2016). This paper focuses on the research of land ecological security in mining cities, takes Wu'an which is a typical mining city as an example, structures the land ecological security evaluation index system by using PSR model and AHP, then calculates the value of ecological security index and divide the safety levels to evaluate the ecological security of the land.

2 METHODS AND MATERIAL

2.1 *Overview of Wu'an*

In the end of 2014, the total output value of Wu'an was 60.65 billion yuan, up by 7.5% over the previous year. The output value of the primary industry was 2.03 billion yuan, an increase of 12.8%; the output value of the secondary industry 39.11 billion yuan, an increase of 8.8%; the output value of the tertiary industry 19.51 billion yuan, an increase of 4.1%. The structure ratio of three industrial is 3.3: 64.5: 32.2, formed an industry-led, service industry as an auxiliary, agricultural-based industrial system. The total land area of Wu'an was 1818 km^2, among which the agricultural land area was 91562.57 hm^2, accounting for 50.36% of the total area, the construction land area was 23507.51 hm^2, accounting for 12.93% and Other land area was 66734.44 hm^2, accounting for 36.71%.

2.2 *Data and methods*

The original data are derived from the "Social and Economic Statistics of Wu'an" (2002–14), the "Land Use Planning for Wu'an" (2010–2020), and the data used in the evaluation process is calculated based on the original data.

This study adopts the PSR model, uses AHP, refers to relevant research results (Wang Lanxia et al., 2016; Pei Tingting et al., 2014) to construct the land ecological security evaluation index system of Wu'an (Table 1) based on its ecological environment characteristics.

Using the existing research results (Wang Lanxia et al., 2016) in the calculation method to calculate the land ecological security index and divide the safety levels (Table 2).

Table 1. Wu'an ecological security evaluation index system.

Target layer	Criterion layer	Index layer	Weight	Safety trends
Land ecological security	Pressure	Arable land reduction rate (%)	0.0581	–
		Forest land reduction rate (%)	0.0379	–
		Fertilizer usage (t)	0.0381	–
		Pesticide use (t)	0.0423	–
		Plastic film usage (t)	0.0435	–
		Urbanization rate (%)	0.0434	+
		Urban expansion strength (%)	0.0402	–
		Industrial SO_2 Emissions (t)	0.0372	–
		Urban registered unemployment rate (%)	0.0499	–
		Population density (cap/km^2)	0.0430	–
		The natural population growth rate (‰)	0.0494	–
		Per capita arable land (ha/cap)	0.0438	+
	State	Economic density (yuan/km^2)	0.0432	+
		The proportion of the secondary industry (%)	0.0446	–
		The proportion of the tertiary industry (%)	0.0444	+
		Forest cover rate (%)	0.0475	+
	Response	Energy consumption per unit of GDP (t/yuan)	0.0383	–
		Power consumption per unit of GDP (kW·h/yuan)	0.0497	–
		Built area greening rate (%)	0.0403	+
		Environmental investment ratio of total investment (%)	0.0529	+
		Industrial wastewater discharge compliance rate (%)	0.0347	+
		Industrial soot emissions compliance rate (%)	0.0404	+
		Urban sewage treatment rate (%)	0.0372	+

Table 2. Land ecological security level standard of Wu'an.

Value	Level	Features
$S \geq 0.8$	Safety	The structure of the land ecosystem is complete and the ecological function is basically perfect, which can effectively meet the needs of local development. It is less disturbed by human activities and has strong ability to restore it.
$0.7 \leq S < 0.8$	Relative safety	The structure of the land ecosystem is relatively complete and the ecological function is relatively perfect, which can meet the basic needs of local development in a certain period of time. It is subject to a certain degree of human activity interference, has a certain ability to restore itself.
$0.4 \leq S < 0.6$	Critical safety	The structure of the land ecosystem has deteriorated, and its basic function is barely maintained, which cannot be stable to meet the needs of local development in a certain period of time. It is subject to a relatively serious degree of human activity interference, has a weak ability to restore itself.
$0.2 \leq S < 0.4$	Relative unsafety	The structure of the land ecosystem is seriously deteriorated and the function is degraded, which cannot meet the demand of local development fully. it is more difficult to recover after being destroyed.
$S < 0.2$	Unsafety	The structure of the land ecosystem is incomplete and the function is very low, which cannot meet the needs of local development. It cannot restore relying on itself after been destructed.

3 RESULTS AND ANALYSIS

It could be seen from Figure 1 that during the study period the comprehensive index rose by 0.1387, whose average annually rose of 0.0173. The growth rate was slow, due to the unstably changes of pressure, state and response index. 2006, the comprehensive index was 0.3938 that the land ecology was in a less secure state; 2006–2008, the comprehensive index was 0.5096, which rose by 0.1158, and the safety level of the land system was improved to critical safety from relative safety,

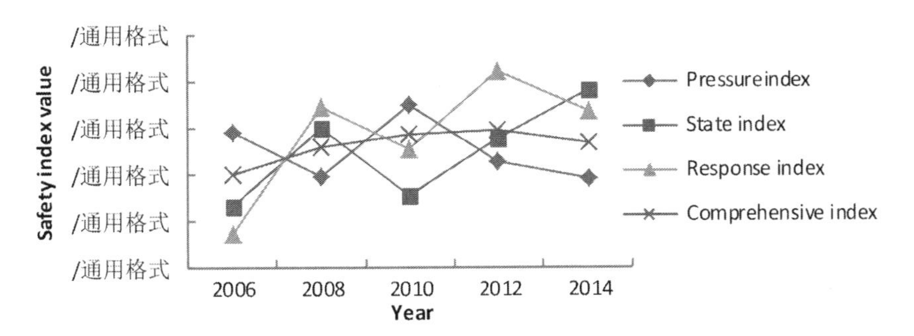

Figure 1. Land ecological security index value of Wu'an.

because the growth rate of the state and response index was over the pressure index; 2008–2010, the growth rate of the pressure index was over the reduction rate of the state and response index that made the comprehensive index slightly increase of 0.0620 to 0.5716 and the land system state still was in a critical safety level; 2010–2012, the growth rate of the response index was over the reduction rate of the pressure and state index, which made the comprehensive index slightly increase of 0.0118 to its highest value of 0.5834, and the land system state was still in a critical safety level; 2012–2014, the pressure and the response index had a larger reduction, which resulted the composite index decline of 0.0509 to 0.5326 although the land ecology security level is not downgraded.

4 CONCLUSION

This paper adopted the PSR model, used the AHP method to construct the evaluation index system, and used the entropy weight method to determine the weight of each index based on the relevant statistical data of Wu'an, then evaluated the land ecological security of it by calculating the comprehensive index. The evaluation results show that the land ecosystem security states of Wu'an was improving slowly With the actual situation, Wu'an should continue to promote industrial restructuring, develop ecological agriculture to reduce the usage of chemical fertilizers, pesticides and plastic film, introduce desulphurization technology to reduce the SO_2 emissions, control the urban expansion and population growth, increase investment in environmental protection, carry out actively land consolidation activities to improve the land ecological security situation and maintain the sustainable use of land resources.

ACKNOWLEDGEMENTS

Finalization of the paper was supported by project Natural Science Foundation of China (41641008).

REFERENCES

Martinat S, Dvorak P, Frantal B, et al. (2016) Sustainable urban development in a city affected by heavy industry and mining? Case study of brownfields in Karvina, Czech Republic. Journal of Cleaner Production, 118, 78–87.

Pei Tingting, Chen Ying, Zhao Yanan, et al. (2014) Evaluation on the Ecological Security of Baiyin City Based on P-S-R Model. Chinese Agricultural Science Bulletin, 30 (2), 215–221.

Qiao Guotong, He Gang, Zhu Yannan, et al. (2017) simulation of industrial structure adjustment and optimal policy for mining cities based on system dynamics. China Mining Magazine, 26 (2), 33–38.

Wang Lanxia, Qin Dahai, Meng Xiangmin, et al. (2016) Land Ecological Security Evaluation of Jixi City in Heilongjiang Province. Journal of Anhui Agricultural Sciences, 44 (30), 179–182.

Yang Ning, Fu Meichen. (2016) Ecological sustainability in a mining city from entropic perspective. China Mining Magazine, 25 (4), 68–71.

Land Reclamation in Ecological Fragile Areas – Hu (Ed.)

Experimental research on slope protection collaborated with vetiver and micropiles

Y.S. Deng
School of Civil Engineering and Architecture, Xi'an University of Science and Technology, Xi'an, China

P. Wu & H. Wang
School of Civil Engineering and Architecture, Hubei University of Technology, Wuhan, China

ABSTRACT: Slope protection with vetiver or micropiles were widely used in slope protection and consolidation engineering, but it was seldom researched in the slope protection collaborated with vetiver and micropiles. Through establishing a large scale test model similar to the actual project, the strain gauges and micro soil pressure cells were used to measure data, the effect on vegetation and micropiles on slope protection were researched by model tests through the top surface loading. The results showed that the vetiver and the micropiles would effectively improve the integrity of the slope and improve the load capacity of the slope, which provides a reference for the design of ecological and engineering cooperative slope protection.

Keywords: Vetiver, Micropiles, Collaborative slope protection, Safety factor, Model test

1 INTRODUCTION

With the continuous construction of China's infrastructure, a lot of artificial slope were produced which brings a series of safety problems of slope (Schiechtl, 1996); therefore, the slope protection problem has become an unavoidable research topic. The tensile strength and shear strength of the root is far greater than the soil and the root can effectively increase the shear strength of root-soil composite under the action of root-soil lateral frictional resistance, so as to achieve the slope supporting effect. Therefore, it is reasonable and effective to take the vegetation as the first choice of slope ecological protection (Xie, 1990; Wang, 1997; Nilaweera, 1999; Cannon, 1995). Compared with other herbs, vetiver has a short growth period (Figure 1–Figure 4, it takes only 1 months from seedling to grow lawns), strong resistance, strong penetration, high tensile strength and other characteristics, which can grow in the temperature of −10~50°C and annual rainfall 300~6000 mm, so it is widely used in protection project of highway slope, embankment, reservoir bank. CHENG planted vetiver in the opening of the hinge type concrete block paving system, which made the vetiver and foundation soil overlapped to a composite materials, and made a detail analysis based on the reinforcement mechanism of root slice method and the basic principle (Cheng, 2005). YE found that vetiver has strong adaptability by analysing the highway landslide soil basic properties of the Nanjing section, which can form a dense hedge in the short period and green the slope, and it also can effectively control soil erosion and reinforce the slope (Ye, 2004). WANG found that compared with the bare land, it can effectively reduce the amount of soil erosion by 90.45% through planting vetiver in the reservoir slope and abandoned quarry slope, and vetiver has significant effect on the soil and water conservation in slope protection (Wang, 2006). ZHANG found that the vetiver can improve the soil

Figure 1. The growth process of vetiver in the first week.

Figure 2. The growth process of vetiver in the second week.

Figure 3. The growth process of vetiver in the third week.

Figure 4. The growth process of vetiver in the fourth week.

cohesion and internal friction angle by mechanical test and analysis, so as to improve the shear strength of soil, thus increasing soil stability, but the shear strength can be improved until its root system reached a certain density, so as to play the role of slope protection (Zhang, 2006). But the hydrological effect and mechanical effect of vegetation will produce adverse effect on slope stability under certain conditions (Zhang, 2000), compared to the anti sliding pile, the tensile strength and shear strength of vegetation is far less than the anti slide pile, it often can't reach the expected effect if only use the slope planted, therefore, it is necessary to take other engineering protective measures to improve the stability of the slope.

The micropile is bored pile with less than 300 mm in diameter, it can withstand greater lateral force compared with the vegetation and it has the characteristics of small construction machinery, soil adaptability, flexible pile layout compared with the traditional anti slide pile, which is widely used in slope and landslide prevention (Ding, 2004; Gong, 2004; Hu, 2012). SUN discussed mechanism of micropiles group under the action of landslide thrust as well as the feasibility of replacing ordinary anti slide pile with micro pile group based on the comparative analysis of the anti sliding property of the micro pile group and the ordinary anti slide pile (Sun, 2009). WU, taking into account the uplift capacity of micro pile, combined the micropiles and concrete injection layer to a composite structure similar to soil nailing wall, and applied it to reinforce the slope engineering of high and steep road (Wu, 2005). ZHU realized that the deflection of steel pipe pile is smaller than reinforced pile, and the stiffness is larger than reinforced pile (Zhu, 2006). Therefore, the new technology of micropiles with steel pipe grouting is put forward to reinforce the slope and have good results.

The above studies have described the mechanism of vetiver and micropiles reinforced slope respectively, but the test analysis has not been carried out on slope protection collaborated with vetiver and micropiles. This paper has made some studies on vetiver and micropiles reinforcement mechanism of slope coordination combined with model test, which provide some reference for the similar slope reinforcement engineering.

2 TEST DESIGN

2.1 *Experimental material*

Planting vetiver grass in model slope and measuring the soil parameters. The soil is silty clay, and its density is 1.779 g/cm^3, the internal friction angle is 22.385°, the cohesion is 18.134 kPa.

According to the test conditions and operability, the micropiless are 6061 Aluminum tubes with 20 mm diameter and 5 mm thickness, the pile spacing and row spacing are 800 mm. The pile's bending stiffness value of the model is obtained based on the inverse beam under concentrated load of quarter point flexibility, the test was divided into two groups, each diameter and length of model pile was measured before the test and the stiffness of foundation pile was tested after the past of strain gauge in the end (Figure 5), the elastic modulus is 6.89E7 kPa, the poisson ratio is 0.33.

2.2 *Test device and model*

The site used in the test is a large scale model box (Figure 6), the height of model slope is 1000 mm, the long is 1000 mm, the width is 3200 mm and the angle is 45 degrees. The loading device of the test is composed by the weights, and each weight is 20 kg.

The micro strain soil pressure cells (Figure 7) were embedded to monitor changes of soil pressure between different rows of micro piles; there are 12 measuring points which were arranged in the row of soil. The internal force changes of micro pile was monitored by resistance strain gauge, three strain measurement points were laid on the bearing surface of each aluminum pipe, the data was collected by resistance strain gauge (Figure 8).

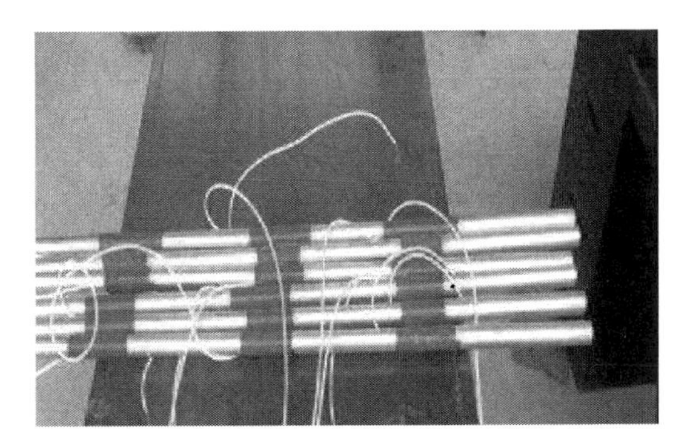

Figure 5. 6061 aluminum piles.

Figure 6. Test model diagram.

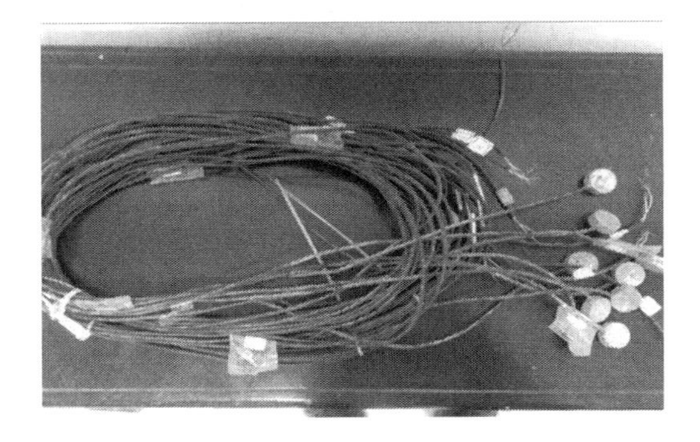

Figure 7. The micro strain soil pressure cell.

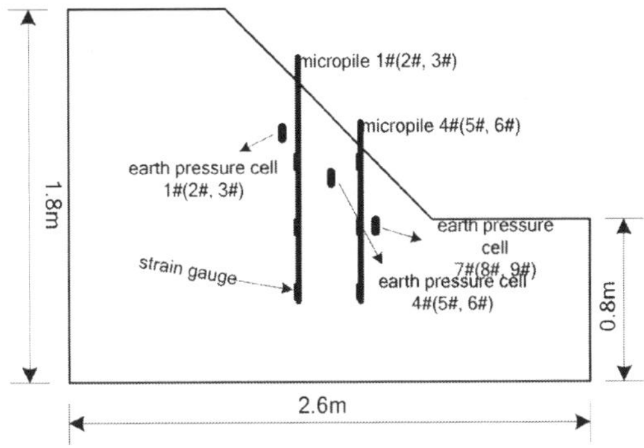

Figure 8. Layout of the test device.

3 MODEL TEST RESULTS AND ANALYSIS

3.1 *Model test*

The experiment was carried out three times, each stage was loaded with 11.6 kPa and each loading time was 10 hours, the strain gauges collect data once every two hours.

3.2 *Model test results*

The data of soil pressure box and strain gauge were collected by using resistance strain gauge.

3.3 *Model test analysis*

Through the numerical analysis of the micropile strain and the pile-side soil pressure:

1. The stress characteristics of same row of micro pile are roughly the same; there is little difference between the values of the stress, which the pressure of the middle micropiles is relatively large than both sides of the micropiles; The force characteristics of same column of micropiles are also roughly the same, but the values of the upper pile is obviously lower pile force.

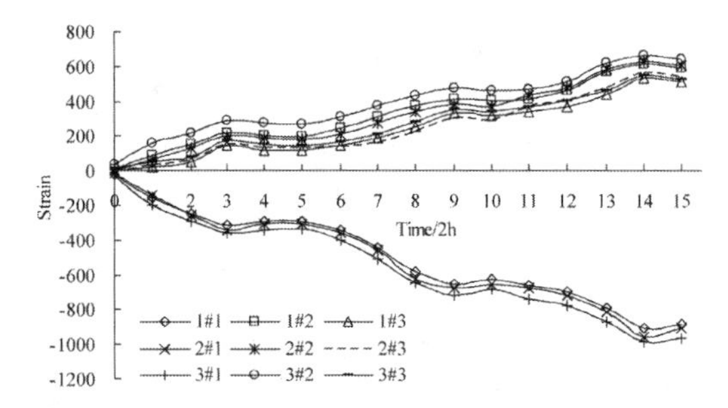

Figure 9. Strain value of strain gauge (1#~3#).

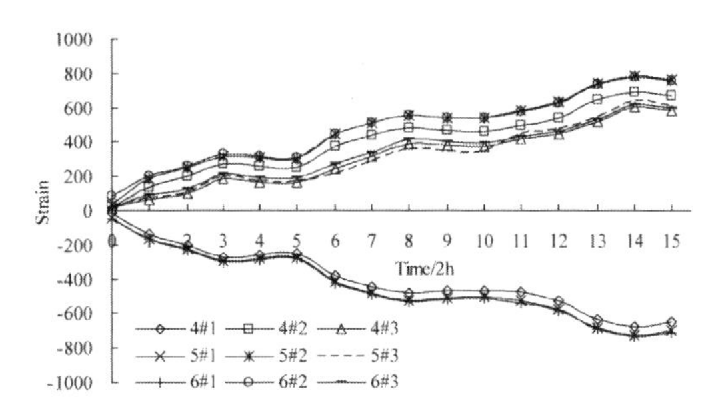

Figure 10. Strain value of strain gauge (4#~6#).

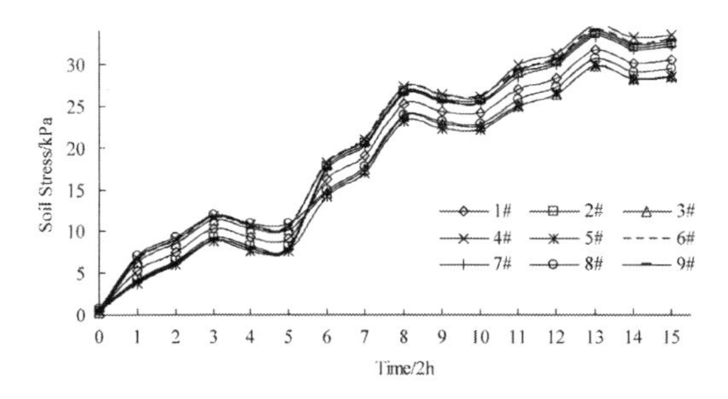

Figure 11. Soil pressure of micropiles.

2. The stress characteristics of each pile are almost the same, but the stress characteristics of the upper, middle and lower parts of same pile are significantly different. The upper part of the pile has the largest strain, the middle strain is the second, the lower part is the minimum strain, and the upper part of the pile is negative. Therefore, it is deduced that the pile is bent from the bottom to the upside; the bending moment distribution of pile is "S" shape. This is because the sliding body and the slide bed produce relative displacement in the sliding surface position, but the pile can rotate freely, so the bending moment of the slip surface is usually zero. The pile ends far away from the sliding surface are restricted by the soil on both sides and can not rotate freely to form the lateral earth pressure to balance the shear stress generated by the relative displacement of the slide body at the sliding surface, thus forming the "S" shape of the bending moment distribution.
3. The soil pressure cell has the same characteristics as the micropiles, and the soil pressure values of the same row of piles are not much different. The soil pressure values of the same pile are very different, the soil pressure value of the front part of pile is obviously higher than the back, it can be seen from the soil pressure value that when the top of the slope is under the load, the load is transferred downward from the top of the slope, showing a decreasing trend.
4. Whether the soil pressure cell or the micropiles, each changes of the value increase basically in equal amount when the increase mass load of three times to the same, and through the corresponding data curve, the value of soil and micropiles will suddenly increase and reach the maximum value at a certain time and then gradually stabilize.

4 CONCLUSION

We can draw conclusions from the model test and finite element analysis:

1. The strain of the upper, middle and lower parts of the pile is different, the upper has the largest strain, the middle is the second, the lower is the minimum, and the upper part is negative, which can be deduced that the moment direction is opposite from bottom to top, the middle is the reverse bend point.
2. The soil pressure values of the same pile are very different, the soil pressure value of the front part of pile is obviously higher than the back, and the difference decreases with the increase of the load, which indicated that when the top of the slope is under the load, the load is transferred downward from the top of the slope, showing a decreasing trend.

ACKNOWLEDGEMENTS

This project is supported by the National Natural Science Foundation of China (No. 51378182).

REFERENCES

Cannon SH, Powers PS, Phil RA, et al. (1995). Preliminary evaluation of the fire-related debris flows on Storm King Mountain. Colorado: Glenwood Springs, 95–508.

Cheng LF, Sun SL, Pei HJ. (2005). Analysis of mechanism of Vetiveria zizanioides-hinge jointed concrete bank protection system. Chinese Journal of Geotechnical Engineering, 27(5):562–566.

Ding WG, Wang X. (2004). Application of micropiling compound structure in a landslide treatment engineering. Geotechnical Engineering Technique, 18(1), 47–50.

Gong J, Chen RP, Chen YM, et al. (2004). Prototype testing study on micropiles under lateral loading. Chinese Journal of Rock Mechanics and Engineering, 23(20):3541–3546.

Hu YF, Wang TY, Ma L. (2012). Research on anti-sliding characteristics of single double-row and composite anti-slide micropiles. Chinese Journal of Rock Mechanics and Engineering, 31(7):1499–1505.

Nilaweera NS, Nutalaya P. (1999). Role of tree roots in slope stabilisation. Bulletin of Engineering Geology & the Environment, 57(4):337–342.

Schiechtl HM, Stern R. (1996). Ground bioengineering techniques for slope protection and erosion control. Embankments.

Sun SW, Zhu BZ, Ma HM, et al. (2009). Model tests on anti-sliding mechanism of micropile groups and anti-sliding piles. Chinese Journal of Geotechnical Engineering, (10):1564–1570.

Wang F, Jiang ZR, Li XJ. (2006). Effects for water and soil conservation of Veliveria zizanioides on the slopes of reservoir and stone pit. Journal of Gansu Agricultural University, 41(3):96–99.

Wang YJ, Xie MS. (1997). Effects of forests on slope stability in the granite mountainous area of the three gorges reservoir area. Journal of Beijing Forestry University, (4): 7–11.

Wu SC, Gao YT, Jin AB. (2005). Study on reinforcement of micro-pile and rockbolt for an unstable high-steep road cut slope. Chinese Journal of Rock Mechanics and Engineering, 24(21):3954–3958.

Xie MS. (1990). A study on determining the effective limits of the tree and bush roots strength and the best distribution of roots for stable slopes. Acta Conservations Soliet Aquae Sinica, (1): 17–24.

Ye Z, Tang JY, Xu LY. (2004). Test on vetiver system for embankment protection of ning-he expressway. Environmental Protection in Transportation, 25(6):23–26.

Zhang DL, Wang BL. (2006). Mechanics research on vetiver grass in railway side slope protection. Bulletin of Soil and Water Conservation, 26(6):94–96.

Zhang XG. (2000). Engineering geology in china. Beijing: Science Press.

Zhu BL, Hu HT, Zhang YF, et al. (2006). Application of Steel-Tube Bored Grouting Anti-Sliding Retaining Wall to Treatment of Landslide K108 in Beijing-Zhuhai Expressway. Chinese Journal of Rock Mechanics and Engineering, 25(2):399–407.

Land Reclamation in Ecological Fragile Areas – Hu (Ed.)
© 2017 Taylor & Francis Group, London, ISBN 978-1-138-05103-4

The design of hydraulic barriers: Integration between rational and observational methods

M. Cigagna, V. Dentoni & B. Grosso
DICAAR—Department of Civil and Environmental Engineering and Architecture, University of Cagliari, Cagliari, Italy

ABSTRACT: Hydraulic barriers can be used to control and prevent the migration of pollutants from contaminated sites to the surrounding area. Their barrier design requires a deep knowledge of the geological, geotechnical and hydrogeological characteristics of the site. Data collection and hydrogeological modeling can be very expensive, especially in complex geological conditions. Furthermore, despite the significant amount of data to be collected and the economic resources engaged, the resulting conceptual model of the site might not result sufficiently detailed as to allow a reliable design of the barrier. An alternative approach is represented by the implementation of subsequent observational phases alternated to the actual design and construction phases, so that the barrier design results the optimum for the specific site under consideration. This integrated design methodology (rational and observational) is discussed in the article with reference to a major case study located in Sardinia (Italy).

1 INTRODUCTION

In the remediation of contaminated sites, hydraulic systems are used for the reclamation of groundwater (*aquifer restoration*) and for the interception of the pollutant plume (*plume containment*), as well as for controlling the source of contaminant (*source control*) (Pearlman L., 1999). In the case of aquifer restoration, the planimetric setting of the extraction wells is such as to minimize the time required to remove the contaminants introduced into groundwater: the wells are aligned along the plume, in one or more lines, in the direction of the aquifer motion. This way, the maximum length of the contaminant path before extraction is equal to half the distance between two successive wells: the greater the number of wells, the shorter the pollutant path and the extraction time (Cohen R et al., 1997). When the objective of the remediation works is the plume containment or the source control, the wells are instead arranged in lines perpendicular to the direction of the aquifer motion, in order to capture all the polluted flow and prevent downstream propagation. In these cases, the hydraulic system works as a barrier (hydraulic barrier) against the propagation of the contaminated flow in the downstream direction. Hydraulic barriers are used to contain the plume or to control the contaminant source in the case of both *Emergency Securing Systems* and *Operative Securing Systems*.

1.1 *Emergency Securing Systems*

The *Emergency Securing Systems* are designed to limit downstream migration of aquifer contaminants caused by individual accidents (i.e.: breakage of pipes or reservoirs; spills from tankers; etc.). Often these types of systems operate for the time necessary to enable the source removal (or source isolation) and the remediation of the adulterated aquifer volume. This is the case of individual aquifer contamination events in which the source is limited and geometrically identifiable (abandoned mine dumps, mineral basins, unauthorized waste repositories, etc.), so that the source itself can be removed in a short time or permanently secured by means of physical isolation (U.S. EPA, 1998). Typical schemes of *Emergency Securing Systems* are reported in Figure 1, where it is possible to distinguish the following actions: permanent source insulation and aquifer restoration; source removal, plume control and aquifer restoration.

As mentioned above, those systems operate only for the time needed for the removal (or isolation) of the contaminant source and the restoration of the aquifer. Therefore, the variables that favour the implementation of such systems in order of priority are:

- The activation velocity: time for construction and start up;
- The effectiveness and action velocity: time needed for plume control and aquifer restoration.

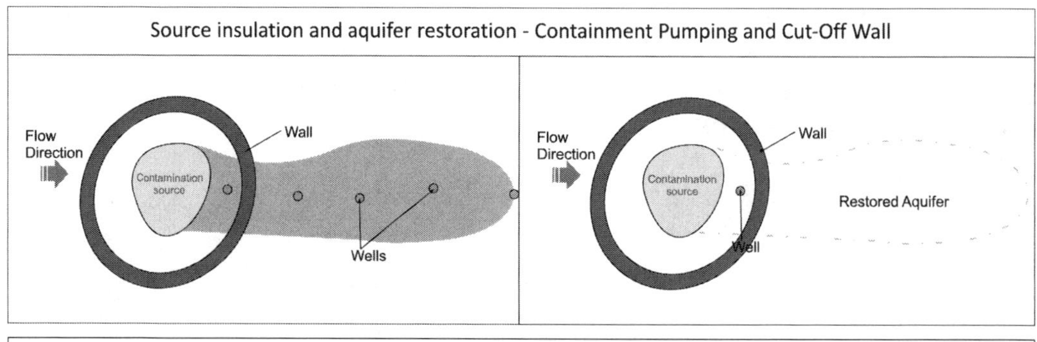

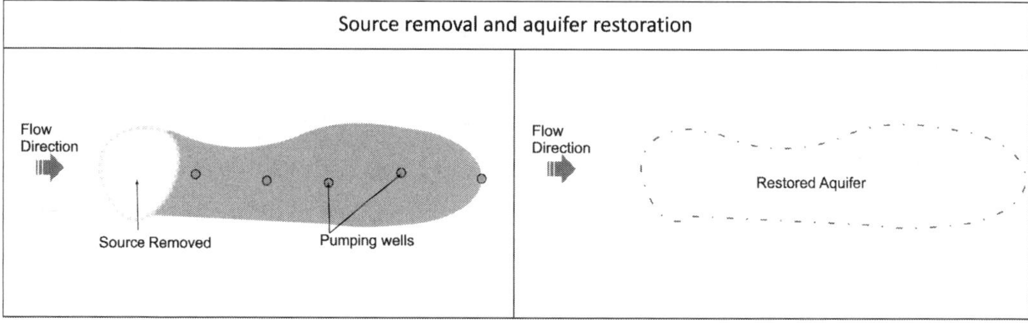

Figure 1. Typical schemes of Emergency Securing Systems.

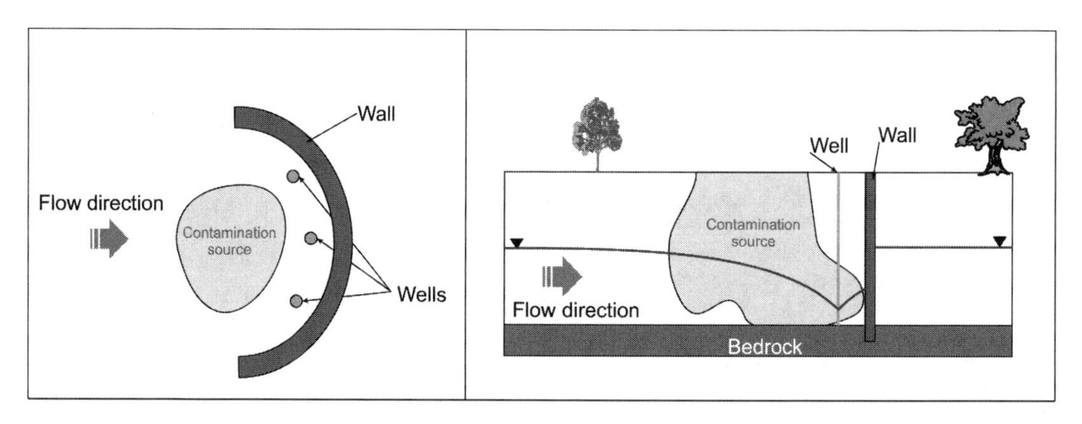

Figure 2. Operative Securing Systems - Scheme A.

The design of these works is necessarily developed in short time and cannot be preceded by an accurate phase of investigation; therefore, it is based on an approximate knowledge of the site and on a conceptual model somewhat detailed and accurate. On the other hand, due to the relatively short technical life of these systems, long enough to set remedy to single contamination events, oversized solutions become acceptable. They are in fact characterized by flow rate high enough to guarantee the effectiveness and the rapidity of the remediation action, even at low efficiency (high pumping and treatment costs).

1.2 *Operative Securing Systems*

The *Operative Securing Systems* are meant to prevent the impact on downstream sensitive targets when the aquifer contamination comes from permanent sources. This is typically the case of hydraulic barriers located at the downstream boundaries of industrial sites, where the internal sources of contamination can be removed only at the end of the plant's activity, after the decommissioning. These securing systems are meant to operate for a very long time, at least the life of the industrial activity. The design main objective is the reduction of the hydraulic barrier

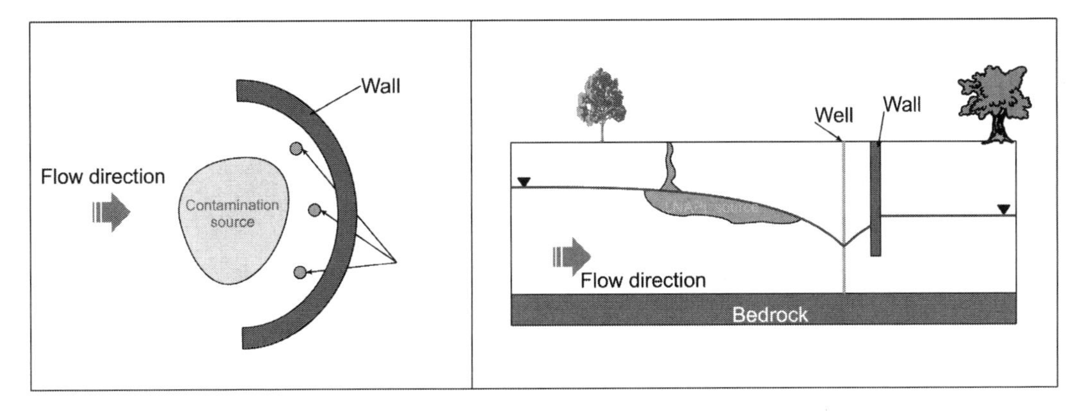

Figure 3. Operative Securing Systems - Scheme B.

operating costs, hence the containment of the extraction flow rate by which the pumping and water purification costs depend. The restraint of the investment cost and the construction duration represent secondary goals. The project is developed on the base of an in-depth environmental and hydrogeological characterization and often includes solutions that, although characterized by high construction costs, lead to consistent reduction of the operating costs. A typical example of this strategy is the combination of physical and hydraulic barriers, where the role of the physical barrier is to reduce the volume of groundwater flowing through the subsurface of the contaminated site (Beretta G. P., 2003). Typical schemes of such securing systems are in Figure 2 (Scheme A) and Figure 3 (Scheme A).

2 HYDRAULIC BARRIERS

In their most widespread configuration, hydraulic barriers are made of aligned wells by which the underground water is extracted. Their operation is based on the deviation of the flow lines exerted by the piezometric depression generated within the wells by the extraction of water: Figure 4 illustrates the concept. The effect of the piezometric depression on the flow lines outspreads planimetrically around the well, both in the direction of the natural outflow and perpendicularly, at distances that are proportional to the extracted flow rate (U.S. EPA. 2002, U.S. EPA, 2004). The same effect also extends in the vertical plane, underneath the well's bottom. When the hydraulic barrier alignment is orthogonal to the direction of the contaminant plume propagation, it results fully effective if all the flux lines that pass through the contaminant plume terminate in the extraction wells (i.e. the influence areas of the wells are mutually tangent or overlap). This condition is expressed by the effectiveness index of the barrier, which is the ratio between the sum of capture fronts and the plume width (Beretta G. P., 2003).

An effectiveness index less than one means that a portion of the polluted stream crosses the barrier. An index higher than the unit means the whole plume is captured with overlapping of the capture fronts around the wells (Figure 5). In this latter case, the barrier causes an increase in the flow rate proportional to the increase of the effectiveness index over the unit value. A unitary effectiveness index means that the entire flow is captured without overlapping of the single capture fronts. Clearly the ideal operating condition is represented by an index slightly higher than one (typically 1.2–1.3), as it ensures a safety coefficient with respect to the uncertainties of the hydrogeological model. The minimum value of the flow corresponds to a unitary effectiveness index. Higher flow rates determine an increase of the flow velocity and the extension of the capture solid, underneath the barrier and at its sides; at lower flow rates the contaminants cross the barrier and spread downstream.

2.1 *Design of hydraulic barriers*

The aim of a hydraulic barrier conceptual design is to define the location of the capture lines, the mutual distance between the wells, their depth and the pumping flow rate. The essential cognitive elements for the project development are the width, length and depth of the plume, the type of contaminants, the hydrogeological characteristics of the aquifer (type of aquifer, hydraulic conductivity, storage coefficient, effective porosity, etc.). A first indication of the wells mutual distance is given by Equation (1), valid for homogeneous and isotropic medium.

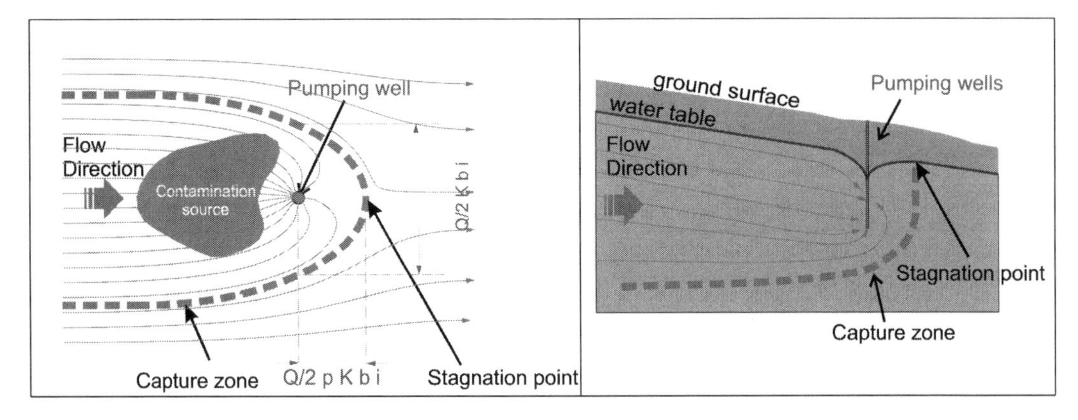

Figure 4. Deviation of the flow lines exerted by the piezometric depression.

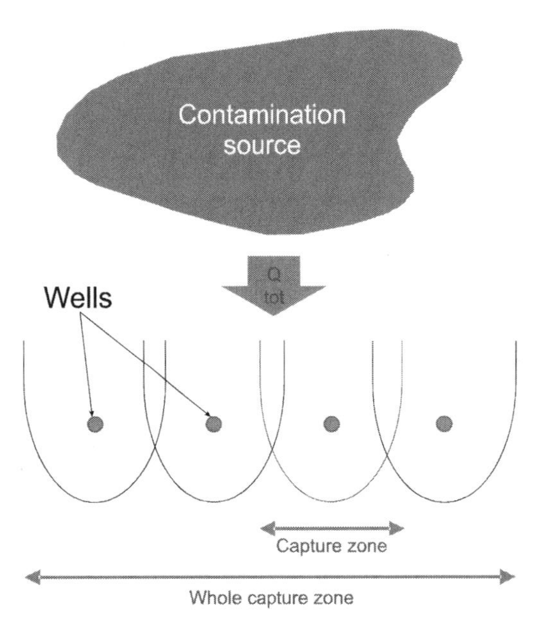

Figure 5. Overlapping of capture fronts around the wells.

$$y_{\max} = \frac{Q}{2Kbi} \tag{1}$$

Evaluations that are more accurate are attainable with numerical simulations carried out with the three-dimensional model of the aquifer and the wells system. This approach, certainly more correct, requires a detailed knowledge of the stratigraphic and hydrogeological characteristics of the aquifer. However, despite the apparent accuracy of the numerical simulation, given to the inherent complexity of the subsoil and the way groundwater motion occurs, it must be kept in mind that the perfect knowledge of the aquifer characteristics and, even more, the reproduction of its behaviour are not realistically attainable goals, even at extremely high investigation costs.

The alternative to the two methods described above is the observational approach, which seems more rational, economically more efficient and capable of generating projects that are more consistent with reality. The observational method is essentially based on the integration of the following elements:

- A construction plan of the barrier that alternates construction and experimental phases (i.e.: the experimental phases are carried out on the parts of barrier already built);
- A good knowledge of the aquifer key characteristics, to be integrated with the results of the tests performed on single or groups of wells;
- A good interpretive model (numerical or not) of the tests, to be refined on the base of the tests results.

This design approach permits to accomplish the construction of the barrier having optimized its characteristics on the base of the experimental results elaborated during the building process. This approach is more efficient, specifically in terms of adequacy of the final construction, than the oversizing approach due to a non-perfect knowledge of the aquifer characteristics, typical of traditional design schemes.

2.2 *Coastal hydraulic barriers*

Many hydraulic barriers are designed to protect the marine environment in front of industrial sites. The fresh water transported by the aquifer blends with marine waters in an interface zone that takes the form of a wedge, with its tip directed downwards, towards the hydraulic upstream. The inclination of the marine wedge depends on the hydraulic gradient of the aquifer. The Ghyben-Herzberg equation (2) describes the line of separation between fresh and saline water within the porous medium.

$$z = \frac{\rho_f}{\rho_s - \rho_{sf}} \cdot h \tag{2}$$

where ρ_s and ρ_f are respectively the salt water and the fresh water density, z is the depth of the salt/fresh water interface under the sea level and h is the level of the fresh water above the sea level (Figure 6).

The fresh water from upstream arrives at the marine wedge and in part runs above it, in part mixes with the saltwater in the interface zone. Due respectively to the marine wedge morphology and to the progressive mixing of fresh and marine water on the interface surface, the thickness of the fresh water stream and its flow rate gradually decrease (U.S. EPA, 1999).

The design of hydraulic barriers located near the coastline may benefit by the presence of the marine wedge. In fact, if the source of the contamination is located relatively close to the coastline and the contaminant plume has affected only the superficial part of the aquifer, the hydraulic barrier may be located in the area where the fresh water depth is less, in order to withdraw only the flow rate corresponding to the reduced aquifer section. Construction and operating costs consequently decrease, the first due to the reduction of the wells depth, the second because of the reduction of the flow rate. On the contrary, when the contamination source is located at considerable distance from the coastline and the pollution affects the whole aquifer depth, the hydraulic barrier must be located upstream of the mixing zone, in order to capture the entire plume. In fact, if the hydraulic barrier were localised downstream, where the fresh water reside only in the upper subsurface, it would not capture those contaminants that entered the volume of marine water through the mixing zone.

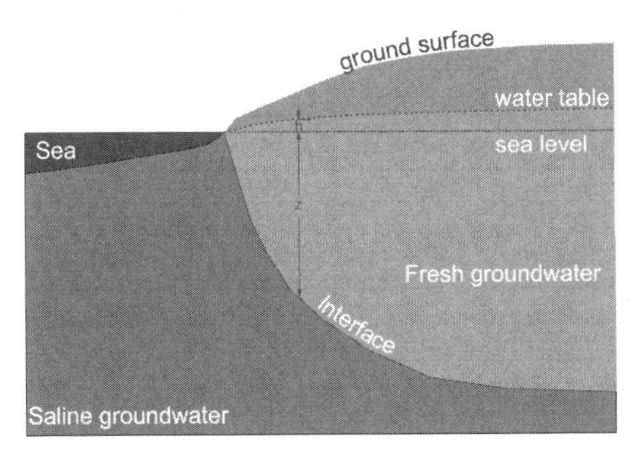

Figure 6. Formation of the saltwater wedge beneath the coastline.

3 THE CASE STUDY

According to the observational approach discussed above, a hydraulic barrier was designed to work as Operative Securing System of a large waste deposit located nearby the south west coast of Sardinia (Italy). The project included a first phase of hydrogeological survey aimed at:

- the definition of the site geology and stratigraphy and, in particular, the classification of the aquifer constituent soils and the description of the impermeable basement (depth and morphology);
- the identification of the aquifers and their hydrogeological characterization;
- the reconstruction of the piezometric surface and the evaluation of the hydrogeological parameters;
- the definition of the contamination extent and its chemical characteristics.

After a preliminary literature research, which ended up with the definition of the general geological and hydrogeological characters of the area, core drillings were extended to the basement. During the drilling, Lefranc tests were performed to measure the permeability of crossed the geological horizons. The resulting boreholes were equipped with piezometers which have been used to outline the horizontal and vertical extent of the contamination plume and the piezometric surface. Some wells were built along the hydraulic barrier alignment, which was previously defined on the basis of the preliminary information recorded on the site (extent of contamination and landfill boundaries); the log of the water physical parameters carried out in these piezometers revealed the location of the fresh/salt water interface and its morphology.

The reference conceptual model resulting from the survey above described is schematized in Figure 7, where the following elements are distinguishable:

- geological formation hosting the aquifer: medium-fine sands with silt-clay lenses
- type of aquifer: phreatic;
- maximum aquifer deepness: 80 m;
- horizontal hydraulic conductivity $Kx = 10^{-5}$ m/s;
- vertical hydraulic conductivity $kz = 10^{-6}$ m/s;
- storage coefficient $s = 10^{-3}$
- hydraulic gradient: 0.5%;
- aquifer contamination: heavy metals in solution and alteration of the pH.

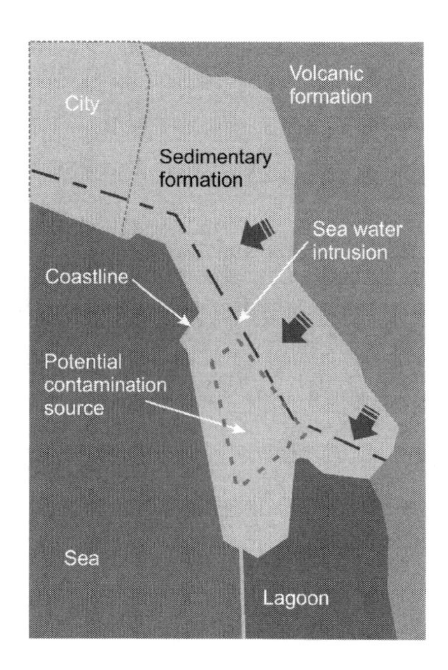

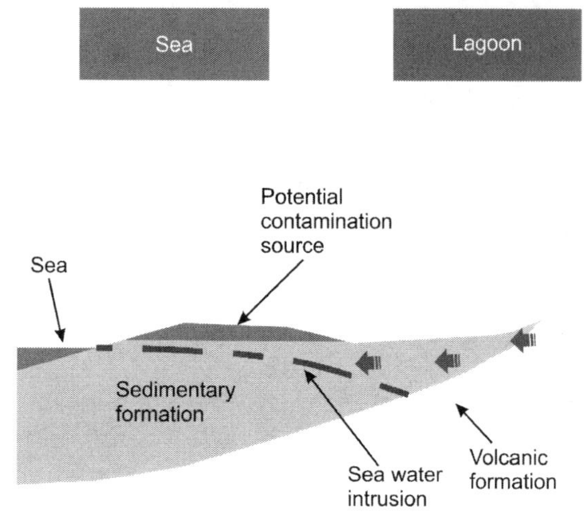

Figure 7. Conceptual model.

In the second phase (simulation and preliminary design), a preliminary draft of the barrier was defined, which included 20 wells deepened for about 5 meters into the fresh groundwater. The distance between the wells was between 50 and 100 m. The flow rates, the distance and the depth of the wells were determined with a numerical model defined on the basis of the survey data previously collected, which simulates the groundwater motion, the presence of the salt wedge and the effect of the barrier. The wells were located so that the piezometers set up during the first phase resulted in the midpoint of two successive wells.

In the third phase (realization of pilot wells) 5 more wells were drilled as schematized in Figure 8, the depth of which had been calibrated on the basis of the local thickness of the fresh water aquifer. In these wells, using the existing piezometers, long-term pumping tests were performed to determine the values of the hydrodynamic parameters referable to significant volumes of the aquifer, such as, in particular, the hydraulic conductivity and the storage coefficient.

In the fourth phase (hydrogeological model update and first project revision) the hydrogeological numerical model has been updated, using the hydrogeological parameters measured during the pumping tests. The revised model allowed the redefinition of the wells location, depth and flow rates, so as to permit the capture of the contaminated stream as a whole.

In the fifth phase (final construction) 15 more wells were built, together with the central piezometers for the control of the barrier effectiveness and the piping system to convey the contaminated water from the wells to the treatment plant (Figure 9).

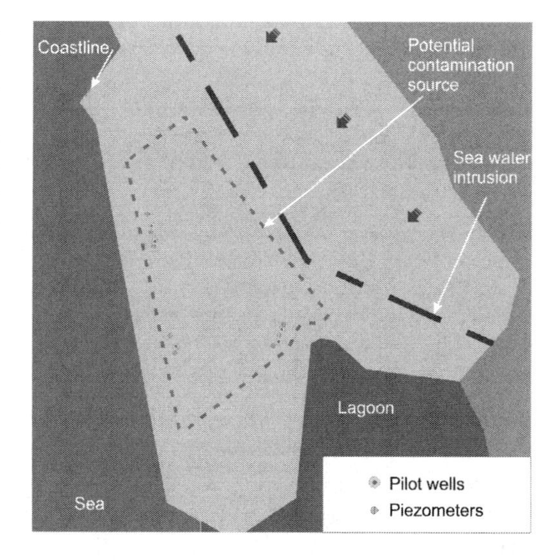

Figure 8. Pilot wells realization.

In the sixth phase (barrier calibration) the pumping tests were carried out on each of the 16 wells drilled as latest, in order to measure the local values of the hydrodynamic parameters and to check the interference between the wells.

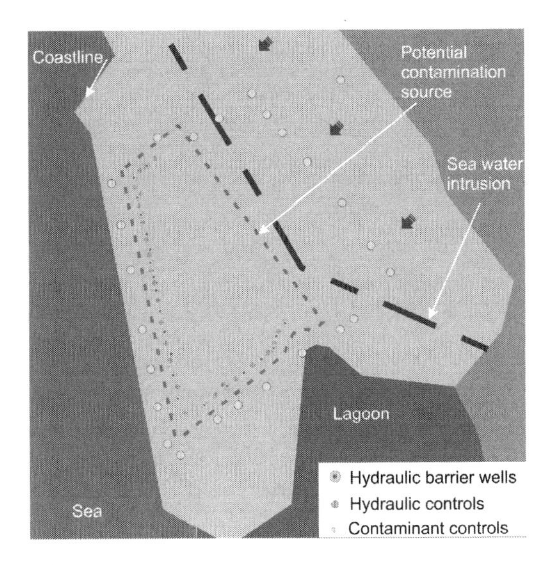

Figure 9. Final construction.

The latter activity consists in verifying if each pair of consecutive wells is able to determine the piezometric depression in the middle point, where the control piezometer is placed. This way the flow rate that guarantees the efficiency of the barrier is determined for each of the barrier wells. Evidently, the obtained flow rates guarantee the barrier effectiveness when the state of the aquifer corresponds to that being during the tests. In the seventh and final stage, the numerical model is further updated, so that it can describe, with good reliability, the dynamics of the aquifer/barrier system and thus can be adopted as a barrier management tool, in particular to face groundwater seasonal variations resulting from particularly rainy years.

4 CONCLUSIONS

The design of hydraulic barriers intended to work as environmental securing systems of contaminated industrial sites is made particularly difficult by two main aspects:

- The inherent complexity of the subsoil and the water flow that take place within it, in particular near the coastline, due to the interaction between fresh and marine water;
- The need to define a securing system that operates for long-term periods and thus must be characterized by high hydrodynamic efficiency and reduced management costs.

The implementation of an observational approach that integrates the survey and the construction phases results particularly suited to address the complexity above described. Its application to the design of a hydraulic barrier intended to secure a major waste deposit in the southwest coast of Sardinia led to following significant results:

- Limited investigation costs and a relatively short construction time for a barrier consisting of 21 wells whose depth, mutual distance and flow rate were optimized as to capture the whole contaminated stream.
- The development of a numerical model, which describes the behavior of the barrier/aquifer system in the presence of the fresh/marine water interface.
- The implementation of the same numerical model as a management tool to define the barrier operating modes depending on the variation of the aquifer regime.

ACKNOWLEDGEMENTS

Research carried out in the framework of projects conducted by CINIGeo (National Inter-university Consortium for Georesources Engineering, Rome, Italy).

REFERENCES

Beretta G.P., Bianchi M., Pellegrini R. (2003): *Linee guida per la verifica ed il collaudo delle barriere impermeabili per la messa in sicurezza di siti contaminate.* Provincia di Milano; Università degli Studi di Milano Dipartimento di Scienze della Terra 'A. Desio.

Beretta G.P., Pellegrini R. (2003), Linee guida per il monitoraggio delle barrier idrauliche. Provincia di Milano; Università degli Studi di Milano Dipartimento di Scienze della Terra 'A. Desio.

Cohen R.M., Mercer J.W., Greenwald R.M., and Beljin M.S. (1997): Design Guidelines for Conventional Pump-and-Treat Systems. U.S. EPA Groundwater Issue.

Pearlman L. (1999). Subsurface Containment and Monitoring Systems: Barriers and Beyond (Overview Report). National Network of Environmental Management Studies, Fellow For U.S. EPA, Office of Solid Waste and Emergency Response Technology Innovation Office.

U.S. EPA - Office of Solid Waste and Emergency Response (1998): Evaluation of Subsurface Engineered Barriers at Waste Sites.

U.S. EPA Contract No.: 68-S7-3002 (1-December 23, 2004) Hydraulic Barrier Intermediate Basis of Design Report Revision.

U.S. EPA National Service Center for Environmental Publication (December 2002): Elements for Effective Management of Operating Pump and Treat Systems.

U.S. EPA, Office of Ground Water and Drinking Water (4601), EPA/816-R-99-014t (September 1999): Salt Water Intrusion Barrier Wells.

Land Reclamation in Ecological Fragile Areas – Hu (Ed.)
 ISBN 978-1-138-05103-4

Study on calculating projection deformation of measured length using an amended particle swarm optimization

X.L. Wu, Y. Gong & F.Q. Tang
School of Geomatics, Xi'an University of Science and Technology, Xi'an, Shaanxi, China

ABSTRACT: In order to best control the projection deformations, this paper studied an amended Particle Swarm Optimization (PSO) algorithm, which could synchronous search multi-dimensional solution spaces those with large differences in magnitudes. To estimate the advantages of this new method, a contrast calculation between PSO algorithm and traditional optimization scheme was made subsequently. When using this modified PSO algorithm presented in this paper, the maximum deformation is 2.04 cm/km. This result indicate that both the location of the central meridian and the mean geodetic height within a same region affect the projection deformations. It could decrease affections of a certain central meridian longitude on projection deformations and expand the permitted differences of geodetic heights of a surveying area.

1 INTRODUCTION

China always specifies a unified 3°-zone Gaussian Gauss-Krueger plane rectangular coordinate system as the basic rectangular plane coordinate systems in engineering constructions. But in many cases, the locations of surveying area in China is far away from the official specified central meridian longitudes. Only if the official specified central meridian longitudes are close enough to the surveying areas with a relatively flat topography, the projection deformation of measured lengths could fit the permitted tolerance.

In order to ensure the accuracy of engineering surveying, it requires the deformation that generated when a measured length was projected onto the Gaussian plane meet a permitted tolerance 2.5 cm/km. Presently, the most widely used algorithms to control the projection deformation are on the basis of control variate method, which regard one of the unknown parameter as a known quantity and then calculate another parameter (Peng, 2012a. Wang, 2004a. Zhao, 2007a.). The limitation of these algorithms is obviously that the solution space is small and just useful in the case with one mean geodetic-height plane of a same area. Besides, the width of the projection zones is very limited. This paper studied an amended PSO algorithm to search the optimal solutions within a multi-dimensional solution space which has large differences in magnitudes with each other.

2 THEORETICAL MODEL

2.1 *Particle Swarm Optimization (PSO) algorithm*

Particle Swarm Optimization (PSO) algorithm simulate the process of the flock foraging in nature. In this model, the bird flock is treated as particle swarm and each bird is regard as an independent particle. Assume that there are D unknown parameters, the solution space should be initialized as a D-dimensional space. Define a set of particles in the solution space, the location of each particle in solution space could be expressed as a D-dimensional vector. The Process of a particle swarm optimization algorithm is to search a D-dimensional vector that best fit the goal function through iterative calculation (Zhuo, 2011).

During each iteration, the best solutions of particles are expressed as $P_i=(p_i^1,p_i^2,p_i^3,\ldots,p_i^D)$, and the best solutions of the whole particle swarm are $P_g=(p_g^1,p_g^2,p_g^3,\ldots,p_g^D)$. The location of each particle in the solution space can be updated according to the formula

$$\begin{cases} v_i^d = wv_i^d + c_1r_1\left(p_i^d - x_i^d\right) + c_2r_2\left(p_g^d - x_i^d\right) \\ x_i^d = x_i^d + \alpha v_i^d \end{cases} \tag{1}$$

where w = non-negative inertia factor; c1,c2 = acceleration constants; r1,r2 = non-negative random variables; α = weighting factor.

This paper made a further study on parameter w using control variable method. The result shows that larger w is, the convergence of the algorithm is weaker and the time of iteration takes longer (Figure 1). It turns that PSO algorithm could be useful in the 1-d solution space and when dealing with the case of multiple parameters, the normal PSO algorithm has its deficiency. According to formula (1), the searching intensity of a particle swarm is depend on the particles' speeds. If there is a large difference on the magnitudes among values of solution space, the PSO algorithm will failed of searching the optimal parameter in each dimension of solution space synchronously.

According to this study above, we could conclude that the magnitudes' differences between each dimension in the solution space need varies parameter w. So we transforms the parameter w to diagonal matrix (As shown in formula (2)).

$$W = \begin{bmatrix} w_1 & \cdots & 0 \\ \vdots & \ddots & 0 \\ 0 & \cdots & w_k \end{bmatrix}_{k \times k} \tag{2}$$

2.2 *Measured length projection deformation model*

The projection deformation generated by Gaussian projection could be divided into two part (Tian, 2004): ①The influence generated by the mean geodetic height of a measured length, which is according to formula (3)). ②The influence generated by central meridian longitude, according to formula (4).

$$\Delta S_1 = -\frac{H_m}{R_m} \tag{3}$$

where H_m = mean geodetic height of measured length, S = measured length, R_m = mean radius of curvature on ellipsoid.

$$\Delta S_2 = \frac{Y_m^2}{2R_m^2} S_1 \tag{4}$$

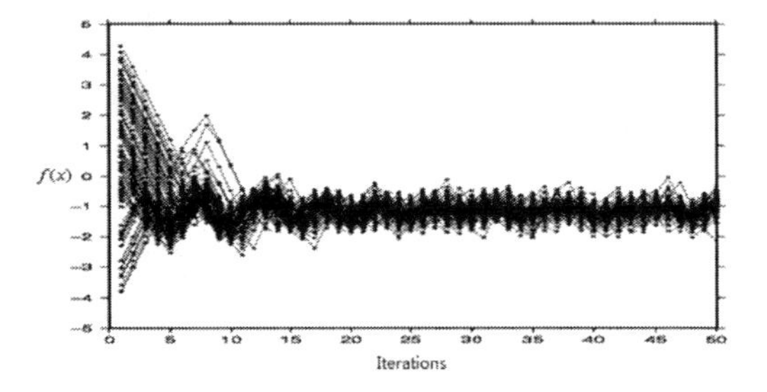

Figure 1. Comparison on momentum factor.

where y_m = average abscissa, S_1 = length of geodetic line on ellipsoid

$$\Delta S = \Delta S_1 + \Delta S_2 \tag{5}$$

$$\Delta S = \left(\frac{y_m^2}{2R_m^2} - \frac{H_m y_m^2}{2R_m^3} - \frac{H_m}{R_m} \right) \tag{6}$$

According to formula (6), we could calculate the per-kilometer measured length projection deformation, which changes with H_m and y_m.

3 ANALYSIS OF EXAMPLE

Assume that a surveying area of mine engineering is situated on the 38°N latitude, and the ranges of longitude is 109°13′45‴~109°32′09‴, the geodetic height is 1130 m~1190 m on the ground, 750 m~990 m underground. The available central meridian longitudes of our national Gaussian coordinate systems are 108°E and 111°E.

Projection deformations could be best controlled under the tolerance by choosing a proper compensation projection plane (Fan, 2000a, 2003a). Based on this viewpoint, an optimization algorithm is given in formulas (7) ~ (8).

$$H_m = \frac{y_{min}^2 + y_{max}^2}{4R_m} \tag{7}$$

$$H_m = \frac{\max\left\{ y_{min}^2, y_{max}^2 \right\}}{4R_m} \tag{8}$$

where formula (7) represents the case that the surveying area situated on one side of central meridian. Formula (8) represents the case that the surveying area stretch over the central meridian.

Thus, we could calculate the projection deformations according to this prioritization scheme and the results are as follows. ① When central meridian is set on the longitude 108°E and mean geodetic height of surveying area takes the mean ground geodetic height. The ground and underground measured length projection deformation are 0.23 cm/km, 17.51 cm/km. The underground part beyond tolerance. ② When central meridian is set on the longitude 111°E. The ground and underground measured length projection deformation are 25.19 *cm/km*, 7.46 *cm/km*. Both of them beyond the permitted tolerance. The results above indicate that the traditional optimization model do not meet the case with more than one mean geodetic height and the case that the surveying area located far away from the chosen central meridian. It highly depended on the location and topography of a surveying area.

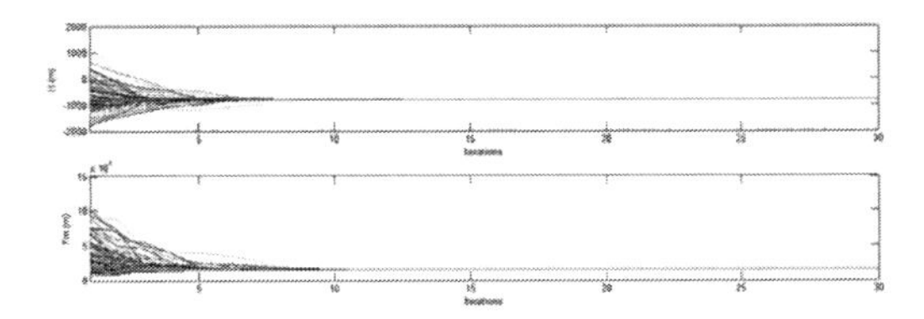

Figure 2. Process of the iterations and convergence.

When using the PSO algorithm with $\max\{|\Delta Si|\} = \min$ as its goal function. The max iterations were set to 50, number of particles were 100. Initialize the parameter w as a diagonal matrix. Set the goal function of this part with two undetermined parameters, one is the geodetic height of compensation projection plane and another is a compensation of the central meridian. Set the 2-d solution space to [−2000, 2000], [1,1 × 10^6] separately (unit: meters).. It only takes about 10 iterations before searching out the optimal solutions in each of it solution space (Figure 2). The results are H = −805.3381 *m*, Y_m = 150848.118 m. Based on the correction of this algorithm, the projection deformation on the ground is 2.04 cm/km, and the underground is 1.73 cm/km.

4 CONCLUSION

This paper presents a method of using Particles Swarm Optimization (PSO) algorithm to calculate the optimized parameters of controlling the measured lengths projection deformations. In part 3, the optimized projection deformation is 2.04 cm/km on the ground, and the underground part is 1.73 cm/km. Using stochastic search algorithm to solve the problem that searching for optimized parameters of controling measured length projection deformations have three advantages in this paper. First of all, it could better control the measured length projection deformations within surveying areas. Secondly, this method is suitable for the cases with more than one working elevation platforms. It enlarges the permitted elevation difference when established a regional Gaussian coordinate systems. The last, this method could deal with multivariable problems, which according to the given solution space and searching for optimized parameters.

REFERENCES

Fan, Y.Z., Wang, J.G., Zhao, L.H., (2000). On the Optimal Selection of the Compensating Projecting Plane. J. *Bulletin of Surveying and Mapping, 8,* pp. 20–21.

Fan, Y.Z., Wang, J.G., Zhao, L.H., (2003). On Optimal Selection of the Engineering Surveying Projection Plane. J. *Bulletin of Surveying and Mapping, 2,* pp. 46–47.

Peng, P.C., Li, Q.H., Zhu, D., (2012). Establishment of an independent coordinate system of the railway and coordinate conversion. J. *Science of Surveying and Mapping, 37(1)*, pp. 20–22.

Tian, Q.W., Liu, W.L., (2004). *Control surveying,* pp.160–164.

Wang, H.N., (2004). Theoritical Derivation of Optimal Compensation Projecting Plane. J. *Bulletin of Surveying and Mapping, 10*, pp.18–29.

Zhuo, J.W., (2011). Applications of MATLAB in Mathematical Modeling, pp. 60–67.

Land Reclamation in Ecological Fragile Areas – Hu (Ed.)

Study on the acquisition technology for information on damaged land in coal-mining areas

H.B. Zhang, X.C. Wang, S.D. Wang & W.Z. Hu
School of Surveying and Land Information Engineering, Henan Polytechnic University, Jiaozuo, Henan Province, China

ABSTRACT: The ecosystem service value is an important concept and an index that reflects the quality of the regional ecological environment status and the measurement of the total ecological benefits. In this study, based on the characteristics of land use and the types of ecosystem in the studied area, the ecological indicators including the net primary productivity of the vegetation and the vegetation cover were selected. Landsat TM remote sensing image, ground-based observation, meteorological data and statistical data etc. were applied to establish the remote sensing-based assessment criteria and assessment model for quantitative estimation of the ecosystem service value. The established assessment criteria and model were applied to conduct the quantitative calculation on the value of the single ecosystem service and the ecosystem service value per unit area. The results indicated that during the decade of 2000–2010, with the inter-conversion of different types of land use, the total value of the ecological service in the studied area displayed a gradually decreasing trend. Among which, the values of ecological service of the cultivation land, wetland and the total ecological service were reduced by 11.92, 11.75 and 5.74%, respectively. While the values of ecological services of forest land, waters and intertidal zone were increased to certain extend. However, these increased values did not change the continuously decreasing trend of the total value of ecosystem service in the studied area. Based on these assessments, the spatially- and temporally-changing patterns of the values of ecosystem service in the studied area were analysed to reveal their intrinsic relationship between the land use and the changes in values of ecosystem service.

1 INTRODUCTION

A large number of studies have been conducted on acquirement of information about the damaged land in both China and other countries. Many methods for the acquirement of information about the damaged land have been proposed, such as the extraction method from the existing data, image interpretation, GIS spatial analysis, field measurement on site, expert consultation, and survey/questioning method etc. have been widely applied in the survey of the damaged land. However, each of these methods has its own advantages and limitations. The extraction method from the existing data is extraction of the content and information from the survey of damaged land based on the existing data (Li *et al.*, 2013). Expert experience consultation and survey/questioning method are usually applied in sociological studies. During the early stage of survey, application of expert's experience consultation and the local knowledge of the local people can be helpful to gain the sketch information about the damaged land and its operation is easier but the data are somehow subjective. Thus, it can be used only as the essential assisted mean in the early stage but is not suitable for independent application (Zhang *et al.*, 2016). The survey on site and measurement method is widely applied and the data obtained are objective and accurate but they have the limitations of time-consumption and the requirement of larger manpower and material resources (Liu *et al.*, 2015).

In summary, currently, the methods for acquirement of information about the damaged land are mainly the traditional technologies, which are not effectively integrated with new technologies. The efficiency of information acquirement is lower and the technological studies are mainly focused on a single technology. The complete technological system is lacking. Few studies have been conducted on the effective fusion and integrations of information acquirement.

2 STUDIES ON THE TECHNICAL SYSTEM FOR ACQUIREMENT OF THE INFORMATION ON DAMAGED LAND IN COAL-MINING REGIONS

2.1 *Technologies for acquirement of the spatial information on land damage based on CORS-RTK and electronic tachometer field acquisition*

In this study, we proposed to use the cooperation and combination of CORS-RTK and electronic tachometer as the method to acquire the information about the damaged land. Firstly, we set the base stations based on the existing GPS sites. Then, we directly used RTK to collect the coordinates of the characteristic points of the spatial factors in the open clearance region. For the shelter and non-clearance region, we firstly used RTK to measure some mapping control points in its surrounding and then set electronic tachometer in these mapping control points, collected the coordinates of these characteristic points of these sheltered factors. After having finished the collection of coordinates, the collected coordinates were down loaded to the computer, officially edited, treated, and generated into the corresponding village and township factors, and expressed as symbols and output according to the figure-drawing requirement.

2.2 *The technologies for acquirement of information on damaged land based on the combination of multiple sources and multiple scales of vector data, thematic data and remote sensing images*

In this study, through combining GIS technology, we studied the key technologies including integration of coordinate translation and image registration of multiple sources of land damage data, the technology of semantic translation and integration of multiple sources of land damage data, the judgment of the information about the damaged land and up-grading technology etc. and established an information conversion model for damaged land and the technology system of information fusion.

1. Technology of Coordinate conversion and image registration based on multiple sources of land damage data.

There are many models for the conversion from one coordinate system to another coordinate system. The workloads for using different conversion models to carry out coordinate conversion are different and the precisions of the results obtained are also different. Currently, the commonly used models included the inter-spatial and inter-coordinate seven-parameter model, strict functional analytical model between space rectangular coordinates and earth coordinates strict functional analytical model, positive and negative gaussian projection operator between earth coordinates and plane coordinate system, and four-parameter model among plane coordinate system etc. Among them, all the conversion parameters for the 7-parameter model and four-parameter model are the unknown parameters. When coordinate conversion was conducted, the coordinates of several common points (characteristic points) within two different and already known coordinate systems were substituted into the conversion model. The conversion parameters between two coordinate systems were reversed with adjustment method. The obtained conversion parameters were then re-substituted into the model and coordinate conversion of the other part of characteristic points was conducted (Kefyalew *et al.*, 2016).

2. Semantic translation of multiple-sources of land damage data and integration technology.

Compared with the common difference in data formatting and the difference in coordinate system, shield or digestion of the semantic differences in multiple-sources of the spatial data is more difficult, its unfavourable influence on the integration results is stronger. Because of this situation, we conducted studies with the major focus on the technology of integrating multi-sources of vector spatial data under environment of semantic differences (abbreviation of semantic integration) (Vincent *et al.*, 2013) based on the systematic analysis and the expound of various differences among multiple sources of spatial data and the digestion strategy.

First, we designed the data bank of the integration results according to the mapping requirement and used this bank to store the integrated results. Then, we conducted formatting match between the data sources and the data bank of results to formulate the mapping rulers, which guided data extraction and format conversion. Later, according to the format mapping ruler, the key information was extracted from the data sources and the loaded to result data bank after being treated with corresponding conversion. Finally, we checked whether or not the complex and redundant elements were present in the integrated results, cleaned and combined the redundant elements.

3. Judgment of the information about the damaged land and up-dating technology.

For achieving the rapid acquirement of the information about the damaged land, the judgment of the information about the damaged land and up-date of technology based on remote sensing technology can be utilized. According to different types of damaged lands, we ensured that the remote sensing image data sources could be used to extract the information about the damaged land based on the analysis of the image resolution and the mapping scale relation and studied the judgment of the information about damaged land based on remote sensing image and technology up-dating and acquired the information about areas and distribution of digging damaged land, subsided land, and covered land etc.

2.3 *Integration of technologies for acquirement of information on land damage*

In this study, we used CORS-RTK in combination with electronic tachometer field corporation collection, vector dada, thematic data, remote sensing data fusion extraction as one system with multiple sources and multiple scales technical system for acquirement of information about the quantitative status and infrastructure status. We used the integration of ground investigation techniques and the hyperspectral remote sensing technology to acquire the information about the status of the quality of the damaged land. Finally, we conducted integration of the technologies mentioned above and the information obtained to form technical system for acquirement of the information fusion about the damaged land with multiple sources and multiple scales and the data bank of the information about damaged land.

3 CASE OF THE APPLICATION OF A TECHNICAL SYSTEM FOR ACQUIREMENT OF INFORMATION ON LAND DAMAGE IN COAL-MINING REGIONS

3.1 *An overview of the studied area*

Changcun coal-mining of Yima Coal-Mining Group of Henan province is located in the west section of Changcun Load, Yima City, Henan province. The mining area was 13.63 km^2 and is one of the key mining fields of Yima Coal-mining group. The designed annual production capacity was 1.8 million tons.

3.2 *Application analysis*

In September 2014, the technical system developed in this study for acquirement of the information about land damage in coal-mining region was applied to acquire the information about the damaged land in Changchun coal-mining. Firstly, we applied the existing data extraction method to acquire the basic information about the damaged land, such as the types of damaged land and the approximate range, so that we have the studied targets. Then, we applied the image fusion and the treatment method developed in this study to achieve the recognition of the subsided region and to increase the precision of recognition. We applied CORS-RTK in combination with technology of electronic tachometer field-acquirement of the spatial information about the land damage to further measure the detailed information about the damaged land. The precision and speed of data acquirement were increased. According to the survey on the pollution information about the polluted land, we applied the soil physicochemical properties hyperspectral remote sensing-based inversion method developed in this study, which greatly shortened the time for soil sample analysis and greatly increased the working efficiency. Finally, by using data fusion acquirement technical system based on vector data, vector dada, thematic data, remote sensing images with multiple sources and multiple scales, we treated and analyzed the data obtained with various methods mentioned above and united the coordinate systems of various sets of data, data formatting and semantic information, which was convenient for data application. The final application indicated that this new technical system developed in this study could reduce the workload of field survey by more than 25% and could significantly reduce the work cost and obviously increased the information-acquirement efficiency.

4 SUMMARY

To address the problems, such as low efficiency of information acquirement and the lack of complete technical system, existing in the traditional methods currently applied for acquirement of the information about damage land, in this study, we developed a new technical system that integrated CORS-RTK and electronic tachometer field acquisition, vector data, thematic data and the remote sensing image data fusion extraction as one system with multiple sources and multiple scales for acquirement of information about the damaged land. Its application indicated that this technical system greatly reduced the workload for field survey and the work cost, and obviously increased the working efficiency and thus, can provide the methodological support for rapid and precision acquirement of the information about the damaged land in coal-mining areas.

ACKNOWLEDGEMENTS

This research was supported by National Natural Science Foundation of China (No. 41541014, 41301617), China Postdoctoral Science Foundation (No. 2016M590679, 2015M580629), Key

Scientific Research Project of Henan Higher Education (No. 17A420001,16A420003), China Coal Industry Association Guidance Program (No. MTKJ-2015-284), Dr. Fund of Henan Polytechnic University (B2014-016). We also want to express our respect and thanks to the anonymous reviewers and the editors for their helpful comments in improving the quality of this paper.

REFERENCES

Kefyalew Sahle Kibret & Carsten Marohn. (2016) Georg Cadisch. Assessment of land use and land cover change in South Central Ethiopia during four decades based on integrated analysis of multi-temporal images and geospatial vector data. *Remote*, 3, 1–19.

Li Jing, Liu Yan-ping, Hu Zhen-qi, etc. (2013) Investigation Content and Survey Method on Abandoned Damaged Land. *Geograghy and Geo-Information Science*, 29 (3), 92–95.

Liu Li-feng, Du Fang, Yan Ma-ning, etc. (2015) Calculation on Soil Erosion Modulus Based on Sedimentation Investigation of Check Dam in First Subdivision of Loess Hilly-gully Region. *Bulletion of Soil and Water Conservation*, 358 (6), 124–129.

Vincent de Paul Obade & Rattan Lal. (2013) Assessing land cover and soil quality by remote sensing and geographical information systems (GIS). *CATENA*, 104, 77–92.

Zhang, X R, Bai Z K, Cao Y G, Zhao Z Q, etc. (2016) Ecosystcm evolution and ecological storage in outsize open-pit mining area. *Acta Ecologica Sinica*, 36 (16), 1–11.

Land Reclamation in Ecological Fragile Areas – Hu (Ed.)
© 2017 Taylor & Francis Group, London, ISBN 978-1-138-05103-4

Deformation mechanism and support measures of the soft rock roadway in Ronghua coal mine

F.Y. Ma
College of Geoscience and Surveying Engineering, China University of Mining and Technology, Beijing, China
College of Mining Engineering, Heilongjiang University of Science and Technology, Harbin, China

ABSTRACT: Using scanning electron microscopy and x-ray diffraction instrument to analyze the physics and chemistry of the rock samples. To solve the high stress soft rock roadway with large deformation problem at Ronghua coal mine, The experimental results show that the rock is mainly composed of quartz, feldspar and clay minerals, including clay minerals I/S mixed layer than 46%~96%, belonging to strong swelling soft rock roadway. Formulate different surrounding rock control schemes with finite difference method, and choose a better support scheme through three-dimensional numerical simulation. According to the site conditions, we design a steel truss reverse floor heave structure, combined with finite element simulation and field application. The monthly deformation of the roadway is less than 20 mm, provide a reliable guarantee for the actual production.

1 INTRODUCTION

Ronghua coal mine in Jixi City of Heilongjiang Province, the clay minerals are widely existed in the sedimentary rocks such as shale, mudstone and sandstone. Under the action of groundwater, the water absorption of clay minerals in these strata is expanded, which shows strong hydrophilic property, resulting in a great expansion pressure, which is a failure of the formation of the original structure. The east haulage roadway in order to open up a new project, buried depth −648 m, roadway design total length of 500 m, slope +4‰, lithology in sandstone, tuffaceous shale, mudstone, coal is given priority to, most of the rock fracture, serious fragmentations, more smooth surface, rock hardness 1~6.5 on the Richter scale, rock towards the 70°~150°,160°~240°, dip Angle, on average, about 20°. Structure of surrounding rocks is complex, fault is more, roadway construction, some sandstone fissure water and near fault fault of the water out, the deformation of surrounding rock of roadway is huge, which affected the normal production.

2 COMPOSITION ANALYSIS OF SURROUNDING ROCK IN ROADWAY

The surrounding rock of the main transportation roadway is composed of various kinds of mineral rocks. Ingredients include quartz, feldspar, calcite, feldspar, Illite (I), Montmorillonite (M) and a number of I/S mixed layer. Sampling the different areas in the East main roadway. Field sampling in different regions of the east main transport roadway. X ray diffraction analysis of the samples was carried out. Results show that the clay mineral content accounted for 32.1%~55.6% mineral aggregates, strong expansive minerals I/S (Illite/Smectite) accounted for the total content of clay minerals 46%~96% (Table 1).

Experimental results show that the surrounding rocks with high content of clay and strong swelling soft rock, and such rocks can easily lead to softening, disintegration and expansion. Therefore, the east main transportation roadway surrounding rock is the high expansive soft rock roadway. In order to further understand the internal structure of the whole rock, Scanning Electron Microscopy (SEM) analysis was carried out. Display in 5μm scanning electron microscope, the rock grain between the existence of a large number of filamentous I/S mixed layer flake. A large numbers of authigenic quartz crystal is coated with I/S mixed filamentous tablets. The I/S is distributed

Table 1. Mineral composition and content.

Sample number	Quartz	Potassium feldspar	Plagioclase	Calcite	Ling iron ore	I/S
1	36.9	0.9	1.4	1.4	3.8	96
2	43	—	1.3	1.3	5.4	97
3	42.4	1.0	3.9	1.1	4.3	91
4	31.5	10.4	23.4	—	2.6	46
5	42.6	—	1.7	1.8	5	94
6	25.6	5.0	12.9	—	1.1	90

widely, and the density is significantly different. The quartz crystal distribution is dense, and the surrounding is filled with I/S mixed layer. The samples are a lot of I/S mixed layer clay minerals, and widely distributed. Under the action of water, causing severe swelling and deformation. Therefore, the overall lithology is a softening and expansion characteristics.

3 CLAY MINERAL SWELLING MECHANISM

The hydration expansion of clay is the cation exchange between the crystals. Crystal surface adsorbed water molecules to form a water film. Increase the spacing of the lattice, form the phenomenon of expansion and dispersion. In dry or dry conditions, the distance between the layers is small, about 1 μm or so. In the case of water, the interlayer can adsorb and fill a large amount of water, and the distance between the layers can be increased to 2~3 μm. And under the action of clay, water molecules are dissociated into H^+ and OH^- ions, respectively, on the crystal plane and the end surface of the clay. Then the clay becomes "Hydration of clay". Compared with the dry clay, the clay has the colloidal properties. Including expansion, rheological properties, dynamic properties, dispersion or flocculation, etc.

When clay minerals are in water, the clay minerals are charged, and the exchange of cations and the hydrogen bonds between the water molecules into the clay minerals. Under the influence of the repulsion of the electric double layer, the clay particles or the crystal layer is further pushed away. This expansion phase is accompanied by a large volume change. Under the influence of the formation of the cross—linking, the formation of the gel is often caused by the formation of the particles. The osmotic pressure causes the water molecules to move to the high concentration of one side of the low concentration to reach the equilibrium of the infiltration. The ion concentration in the formation of clay is in equilibrium with the formation water. When the water enters into the formation of the formation. Because the ion concentration of the clay surface is higher than other water bodies, the water is sucked to the clay surface. A directional water film is formed in the clay minerals, and the repulsion force of the double layer is increased. Because the surface of clay particle electric double layer repulsion between the clay surface are open, so that the clay volume continues to expand. As shown in Figure 1.

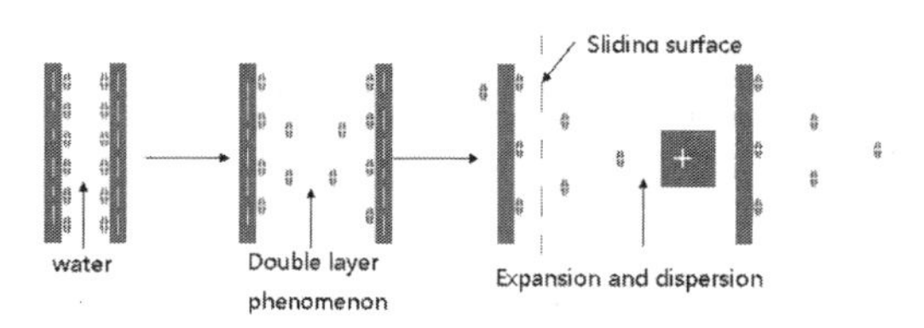

Figure 1. The formation principle of the diffusion double layer.

4 SUPPORT SCHEME OPTIMIZATION

The east haulage roadway belongs to high stress soft rock tunnel excavation of roadway rock strength after hydration rapidly reduced. So, first of all to ensure the airtight sex of the surrounding rock, spray processing, re-use of anchor cable "suspension" and bolt "composite", improve the overall strength of surrounding rock. Using the anchor net spray support, anchor rod specifications Φ18 mm × 2000 mm, anchor cable specifications Φ15 mm × 6000 mm. The east haulage roadway field situation analysis, in combination with other location support experience, respectively 7 kinds of support scheme design, through the computer three dimensional numerical simulation of the scheme optimization. Model size is 400 m × 36 m × 72 m, the total number of units of 168092, strata dip angle is 20°, the total length of 400 m of roadway, every 5 m excavation support simulation.

According to the simulation results of the program, finally we use the support scheme of whole section 15 roots anchor cable, implementation of smooth blasting in tunnels should immediately after spraying a layer of concrete, thickness of approximately 50 mm, then the layout of metal mesh, "W" steel band, anchor rope, and then spray a layer of concrete, thickness of about 100 mm. Floor drum up bottom cover with the bottom arch structure, serious area can use steel truss arch structure. Steel strand anchor cable used 15 mm, 6000 mm in length, anchor cable row spacing is 800 mm × 800 mm, and the section as shown in Figure 2.

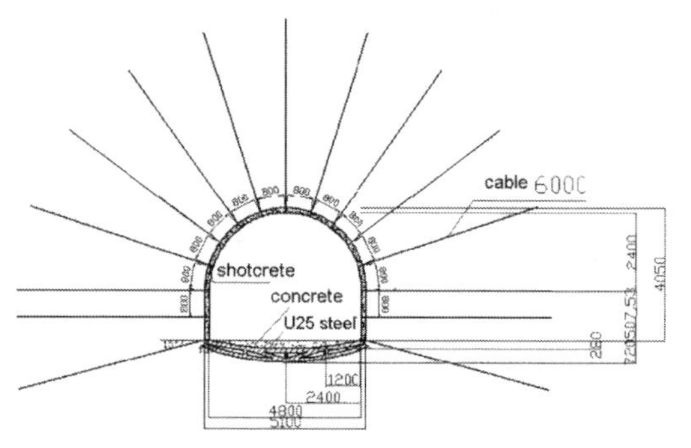

Figure 2. Roadway support design.

5 SUMMARY

Deformation monitoring results show that, The east haulage roadway deformation is bigger than others, stabilizing period is longer, more broken belt, water inflow is larger, the area of the large deformation, creep of soft rock characteristics, such as need to waterproof reinforcement. Clay minerals content is large, accounting for 96% of the total mineral, mineral of strong expansion get 30% of the clay minerals content. Such rocks are under the influence of water softening, collapse and expand, by methods such as spray grouting in tunnel, cut off from rocks and water molecules in the air, and improve the overall strength of surrounding rock, the expansion of the effective control of the soft rock deformation. The scene after the improved supporting plan, the overall deformation decreases obviously, The east haulage roadway section or a bottom drum up phenomenon, in the design the U25 the bottom support, steel truss effect is remarkable. In addition, west haulage roadway is using whole section anchor cable, the effect of significantly, underground work back to normal.

REFERENCES

Jing, H.H., Qi, Z.Y. & Zhao, L. (2010) Hydrophilic characteristics of rock at depth of Ronghua mine. *Journal of Heilongjiang Institute of Science and Technology*, 20 (6), 435–438.

Li, L.B., Lai, X.P., Li, Y.M. & Lin, H.M. (2010) Monitoring of the deformation and surrounding stress of a roadway under broken rock mass environment. *Journal of Xi'an University of Science and Technology*, 30 (1), 24–28.

Li, J., Feng, J.C., Zhang, S.K. & Liu, H.Y. (2015) Study on long and short bolt coordinate support technology of mine soft rock roadway. *Coal Science and Technology*, 43 (3), 17–21.

Li, Y.C., Zhang, S.L., Lu, Z., Wang, C. & Liu, H.Q. (2015) Anslysis of deformation and failure mechanism of roadway in the mud weakly consolidated soft rock strata. *Journal of Hunan University of Science & Technology (Natural Science Edition)*, 30 (1), 1–7.

Long, J.K. (2015) Study and application of synergetic bolting for deep heavily stressed soft–rock roadway. *Coal Engineering*, 47 (4), 41–43.

Niu, J.C., Gao, J.P., Zhang, S.J., Ding, Y.H. & Zhu, S.Y. (2015) Comprehensive analysis on surrounding rock stability of deep soft rock roadway and supporting practice. *Mining Safety & Environmental Protection*, 42 (2), 68–71.

Wang, L.G., Dai, G., Zhao, G.C. & Zhang, P. (2015) Stability and the feedback characteristics process of the soft rock. *Journal of Liaoning Technical University (Natural Science)*, 34 (9), 993–998.

Zhao, W.S., Han, L.J., Zhang, Y.D., Zhao, Z.N. & Wang, G.F. (2015) Study on the influence of principal stress on the stability of surrounding rock in deep soft rock roadway. *Journal of Mining and Safety Engineering*, 32 (3), 504–510.

Land Reclamation in Ecological Fragile Areas – Hu (Ed.)
© 2017 Taylor & Francis Group, London, ISBN 978-1-138-05103-4

Evaluation of the ecological environment in mining areas based on fuzzy comprehensive evaluation method

Q.W. Zhu, T. Wang, J.W. Wang, Y.J. Sun & W.M. Liu
College of Surveying Science and Technology, Xi'an University of Science and Technology, Xi'an, China
Key Laboratory of Western Mine Exploitation and Hazard Prevention, Xi'an University of Science and Technology, Xi'an, China

ABSTRACT: China is a country with coal as the main energy resources, but the problems of the environmental caused by coal mining is a huge challenge in China and in the world. Based on the example of coal mining, through the analysis of a series of environmental problems caused by coal mining, by using fuzzy comprehensive evaluation method, the comprehensive evaluation of the ecological environment in mining area. The results show that the method is effective, the ecological environmental protection of mining area more objective, so as to realize the sustainable development of mining area.

1 INTRODUCTION

Long term coal mining has a serious impact on the ecological environment of mining area, so it is of great significance to study the impact of coal mining on the environment and to evaluate the ecological environment effectively. The ecological environment of mining area is the result of multi-factor interaction, and the factors are uncertain and difficult to quantify. This paper uses fuzzy comprehensive evaluation method, combined with the Yulin city of Shaanxi province mining field survey and the State Environmental Protection Department March 13, 2015 approved the Technical Criterion for Ecosystem Status Evaluation (HJ192-2015) (hereinafter referred to as the standard), to evaluate the ecological environment of the mining area.

2 RESEARCH METHODOLOGY

Fuzzy comprehensive evaluation method is a comprehensive evaluation method based on fuzzy mathematics. This method has the characteristics of clear result and strong system, can better solve the fuzzy, difficult to quantify the problem. This paper chooses the data of the ecological environment index of the mine in 2015.

2.1 *The establishment of evaluation index system*

2.1.1 *The principle of selecting index*

The correct selection of evaluation index is the basis for the true reflection of the ecological environment quality of the mining area, the selection of evaluation index must be based on the standard and the certain principles as follow:

1. Representative and scientific principles, the selected indicators to be able to effectively represent the evaluation of the ecological environment of the region;
2. Comprehensive principle, select the evaluation index which has a great impact on the environment quality, to constitute a complete evaluation index system;
3. The principle of maneuverability, the selected indicators to be able to quantify the calculation, to clear the calculation method, easy to digital expression of indicators.

2.1.2 *Establish evaluation index system*

Using the AHP method, the evaluation index system of the ecological environment of mining area can be divided into three levels: the target layer, criterion layer and index layer. Combined with the characteristics of the ecological environment of the mining area, select the evaluation index, and the evaluation index of the ecological environment of the mining area is divided into two layers: three criteria layer index, corresponding to the nine evaluation indicators, and to select the corresponding indicators of the attributes of the nine to complete the system, as shown in Table 1.

According to the "standard", the ecological environment of the mining area is divided into I, II, III, IV 4 grades, the establishment of environmental assessment evaluation set V = {excellent, good, medium, poor}, and the basic characteristics of the various level of expression, see Table 2.

Table 1. Evaluation index system of ecological environment in mining area.

Target layer	Criterion layer	Index layer	Index attribute
Ecological environment evaluation	Resource development status	Water resources	Per capita water resources
		Land use	Land use structure ratio
	Ecological environment status	Vegetation	Vegetation coverage
		Coal gangue	Coal gangue accumulation area ratio
		Soil	The proportion of desertified land in mining area
		Collapse	Subsidence area ratio
	Economic development status	Population	Mining population density
		Quality of life	Per capita income
		Employment	Average employment rate of mining area

Table 2. Classification of ecological environment in mining area.

Grade	Corresponding state	Expression
I	Excellent	High vegetation coverage, abundant per capita resources, and stable ecosystems
II	Good	Vegetation coverage is high, per capita resources are abundant, suitable for human life
III	Medium	Vegetation coverage is medium and there are constraints that are not suitable for human life
IV	Poor	Vegetation coverage is poor, there are factors that clearly limit human life

2.2 *Determine the weight of evaluation index*

And then use the analytic hierarchy process to determine the weight of each index according to the established evaluation index system, respectively, to determine the relative importance of factors relative to the superior index, so as to obtain the judgment matrix A. Calculated the maximum eigenvalue $\lambda_{\max}$ of the judgment matrix A; In order to carry out the matrix consistency test, introduced the matrix random consistency ratio CR, only when CR is less than 0.1, to meet the consistency test, the calculation formula is as follows:

$$CR = \frac{CI}{RI} \tag{1}$$

$$CI = \frac{\lambda_{max} - n}{n-1} \tag{2}$$

Table 3. The weight coefficient of each index of environmental evaluation.

Criterion layer and weight coefficient	Index layer	Weight coefficient
Resource development status (0.218)	Per capita water resources	0.573
	Land use	0.427
Ecological environment status (0.545)	Vegetation	0.452
	Coal gangue	0.217
	Soil	0.173
	Collapse	0.158
Economic development status (0.237)	Population	0.201
	Quality of life	0.543
	Employment	0.256

where "n" represents the matrix order, CI represents the consistency index of the matrix, and RI represents the average random consistency index of the matrix. After calculating, CR passed the test, and then to determine the weight coefficient of each evaluation index, as shown in Table 3.

2.3 *Fuzzy comprehensive evaluation*

According to the membership function, the membership degree of each index can be obtained, so as to establish the single factor evaluation matrix:

$$R_1 = \begin{pmatrix} 0 & 0.21 & 0.79 & 0 \\ 0 & 0 & 0.45 & 0.55 \end{pmatrix} \tag{3}$$

$$R_2 = \begin{pmatrix} 0 & 0 & 0.63 & 0.37 \\ 0 & 0.67 & 0.33 & 0 \\ 0 & 0.32 & 0.68 & 0 \\ 0 & 0.49 & 0.51 & 0 \end{pmatrix} \tag{4}$$

$$R_3 = \begin{pmatrix} 0 & 0 & 0.19 & 0.81 \\ 0 & 0.55 & 0.45 & 0 \\ 0.2 & 0.8 & 0 & 0 \end{pmatrix} \tag{5}$$

The weight coefficients of each single factor can be obtained from Table 3: $W_1 = (0.573, 0.427)$, $W_2 = (0.452, 0.217, 0.173, 0.158)$, $W_3 = (0.201, 0.543, 0.256)$. The fuzzy relation matrix of environmental assessment of mining area can be obtained by multiplying the single factor weight coefficient and the corresponding single factor evaluation matrix:

$$R = \begin{pmatrix} 0 & 0.12 & 0.645 & 0.235 \\ 0 & 0.278 & 0.555 & 0.167 \\ 0.051 & 0.503 & 0.283 & 0.163 \end{pmatrix} \tag{6}$$

The elements in equation (6) represent the degree of membership of the three criteria of the criterion layer for the evaluation level. It shows that the development of resources is poor; second is the current situation of ecological environment; the current situation of the development of economy for the good membership reached 0.503, shows that the area of economic benefit is good. According to the principle of maximum degree of membership, the ecological environment of the mining area is classified as grade III (middle), the results of evaluation is basically consistent with the results of field survey.

3 CONCLUSION

The present situation of ecological environment in mining area is the result of many factors, this paper uses fuzzy comprehensive evaluation method, AHP, through the analysis of the present situation of the ecological environment evaluation index selection, reasonable classification, evaluation results objectively reflect the actual situation of the mine, with the field investigation results, which proves the rationality and scientificity of the evaluation the method, with reference to the role of the ecological environment in the mining area of the late treatment.

REFERENCES

Lin S.S. (2015) The application of fuzzy comprehensive evaluation in the evaluation of students' teaching. *Journal of Chongqing Electronic Engineering Career Academy,* 24:12–13.

State Environmental Protection Department (2015) (HJ192–2015). *Technical Criterion for Ecosystem Status Evaluation.* China.

Yang H., Zhang G.Z., Yang X.N., Wu F.P. & Li H.L. (2016) Comprehensive evaluation on water environment quality of the Tao River based on fuzzy comprehensive method. *Environmental Science & Technology,* 39(S1):382–384.

Zhang H.Y., Peng P., Xiao W. & Tang S.L. (2008) Evaluation of mine ecological environment based on AHP and GIS. *Coal science and technology*, 36 (9): 102–105.

Zhang Y. (2014) research on ecological environment impact assessment of mine. Master Thesis of Xi'an University of Science And Technology.

Land Reclamation in Ecological Fragile Areas – Hu (Ed.)
 ISBN 978-1-138-05103-4

Evaluation for land damage in mining areas with a high underground water level based on RS

Q.J. Chen
College of Geomatics, Xi'an University of Science and Technology, Xi'an, Shaanxi, China

ABSTRACT: Quickly and comprehensively acquiring the information of land damage in mining area is very important for land reclamation. Multi-period RS data were used in this study to obtain the land information of subsidence area with the method of Tasseled-Cap Transformation, then an evaluation model for land damage was set up based on a series of indexes containing the temporal in-formation of land condition. Through comparative analysis, the change patterns of indexes for different land use were found, which can help to tell apart the damage types and the evaluation results were conformed to the reality. This method can promote the monitoring level of the ecological environment in the mining area, and provide technical support for ecological restoration.

1 INTRODUCTION

China is the biggest coal production nation in the world, and the coal industry has provided the energy for the economic development, on the other hand, it also brought a lot of environmental problems, especially in the area with a high level of underground water. So land reclamation for damaged land resources has become an urgent task for the local government and the mining enterprises (Xiao et al., 2014; Hu 2014). Due to some reason of history or others, data related to land damaged information were not enough to meet the requirement of land reclamation design, which has affected the land reclamation progress. The appearance and development of RS (Remote Sensing) technologies provide an efficient tool for getting the information of ecological and environmental elements related to land reclamation (Nuray et al., 2011; Lu et al., 2007). In the mining area with a high underground water level, land damage situations in subsidence area were varied in different season. When evaluation the land damage, the temporal information shall been taken into consideration. In this study, multi-period RS images were used to get the land condition in different season, then an evaluation model including the temporal information of land condition was put forth to analyze the land damage.

2 STUDY AREA AND DATA COLLECTION

2.1 *Study area*

The study area is located in the north of Xinqiao Mine, Yongcheng City of Henan Provience (Fig. 1). Yongcheng City is rich in mineral resources, and the landform is flat. It also is the main area of crop production.

The bury depth of the underground water level is about 2 meters. Xinqiao Mine began to mine on December, 2007 with production capacity of 1.2 million tons/year. It extracted the coal layer with the thickness about 4.5 m. The subsidence coefficient was about 0.9 in this area. Plenty of land resources have been damaged due to mine subsidence. The subsidence area in this paper was formed in 2010 (Wang, 2005). Before taking actions to restore the land, the land damage information shall be obtained.

Figure 1. The location of the research area in Yongcheng.

2.2 *Data source and treatment*

The image of Landsat OLI was chosen as the data resource which was download from United States Geological Survey (USGS). The data type was L1T, path number 122 and row number 036. According to the image quality, four periods of image in 2013 were selected to conduct the study, and the acquired times for those images respectively were May 21, July 24, October 28 and December 31, 2013, which represented the season of spring, summer, autumn and winter. The boundary of district and Coal Mine were acquired from the map of mineral resources planning (2008–2015), which were registered in the RS image base on the characteristic points.

2.3 *Study method*

2.3.1 *Tasseled-Cap Transformation (TCT)*

In order to acquire the land condition from RS image, Tasseled-Cap Transformation was used to treat the Landsat data. Through TCT, some new bands were formed that were useful for vegetation mapping. For Landsat image, the first band was used to describe the overall brightness of the image. The second band represented greenness and was typically used as an index of vegetation. The third band was interpreted as an index of wetness. Other bands were the image's noise (Baig et al., 2014).

2.3.2 *Land damage evaluation model for subsidence area*

When ecosystems were disturbed, its structure and function would deviate from its original stable state. The degree of disturbance or damage can be measured through the difference of the states between subsidence and referenced land. The model was set up as below in Equation 1.

$$V = \sum_{i=1}^{n} \left| \frac{B(x_i) - A(x_i)}{A(x_i)} \right| \bullet W_i \tag{1}$$

where, V was the index to explain the state deviation of land in subsidence compared with referenced one; x_i was the index of land condition, i = 1,2,…n; B was the subsidence land and A was the referred land; W_i represented the weight of index i (i = 1,2,…n), they were calculated with the method of Analytic Hierarchy Process (AHP).

3 RESULTS ANALYSIS

3.1 *Band characteristics of TCT for different land use*

According to the difference of land damage in this study area, four types (farmland, abandon land, perennial water and seasonal water) were selected to analyze the characteristics of these new bands of TCT. Through compared analysis the curve of bands of different land use, it was showed that the shape formed by band 2, 3 and 4 was greatly different (Fig. 2). If the land was covered by water, the shape liked inverse V, because the wetness value was higher; if the land was covered by crop or vegetation, the shape near to a slope line. When the land is bare soil, the wetness further decrease, the shape changed its direction, became a flat V shape.

3.2 *Change of principal bands of TCT with time*

It was known that after Tasseled-Cap Transformation, most of the information was focused on the three principal bands. Through compared change of the foremost bands for different land use in different months, the following rules can be obtained. (1) Band 1: It represented the brightness, and can refer to the information of bare soil. It can be seen found that water body (perennial water) has the lowest value in a whole year. The value of seasonal water and perennial water change greatly from rainy season to dry season, because they became bare soil or thin water body. The value of abandoned land and farmland changed little in different months. (2) Band 2: It represented the greenness, and can be used as measure of vegetation. It can be found that there were two peaks for farmland, which reflected the growing crop habits in the area. From July to September, the land was planted with beans or corn, and then from October to June of next year, it was used for growing wheat. On October, it was the time for sowing seed, so the value was low. For abandon land, there was a great difference compared with farmland, on December, the value of it was very lower than farmland. As for perennial water or seasonal water, values were in the state of low level in the entire year. (3) Band 3: It represented wetness, and can be used to explain the soil and canopy moisture. For perennial water, except December, its value was the

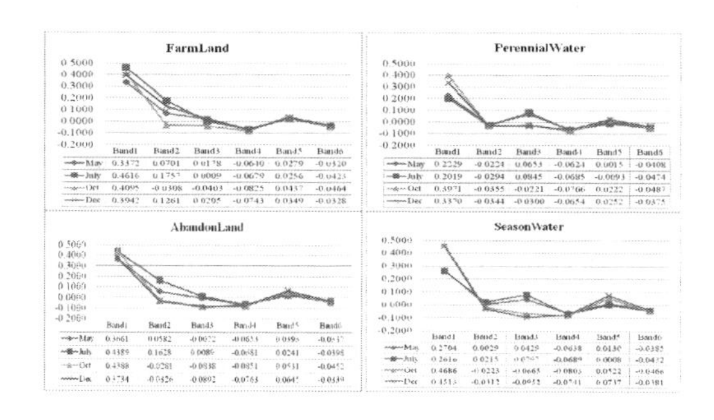

Figure 2. Band characteristics of TCT for different land use.

highest compared with other land types, but its value changed greatly within one year, which was explained with fluctuation of water abundance. In general, the value of abandon land was the lowest compared with other land types. Detail please see Figure 3.

3.3 *Land damage evaluation*

Based on the above analyses, it was found that it was not easy to discriminate farmland and abandon land in October, so, the data in this month was given up. Based on the information of other months, following indexes were set up to describe the land state.

1. Soil Information
These indexes can be got through band 1 of images after TCT as following:

$$\begin{aligned} B_m &= |(B_{m1} - B^f_{m1})| / (B_{m\max}); & B_{m\max} &= \max(|B_{m1} - B^f_{m1}|) \\ B_j &= |(B_{j1} - B^f_{j1})| / (B_{j\max}); & B_{j\max} &= \max(|B_{j1} - B^f_{j1}|) \\ B_d &= |(B_{d1} - B^f_{d1})| / (B_{j\max}); & B_{d\max} &= \max(|B_{d1} - B^f_{d1}|) \end{aligned} \tag{2}$$

where, B meant the brightness of land; m, j and d represented the months, that was May, July and December. B^f_{m1} represented the value of band 1 for farmland in May. B_{m1} represented the value of band 1 for the land in subsidence area in May.

2. Vegetation Information
The vegetation information can be measured through band 2 of images after TCT as following:

$$\begin{aligned} G_m &= |(G_{m2} - G^f_{m2})| / (G_{m\max}); & G_{m\max} &= \max(|G_{m2} - G^f_{m2}|) \\ G_j &= |(G_{j2} - G^f_{j2})| / (G_{j\max}); & G_{j\max} &= \max(|G_{j2} - G^f_{j2}|) \\ G_d &= |(G_{d2} - G^f_{d2})| / (G_{j\max}); & G_{d\max} &= \max(|G_{d2} - G^f_{d2}|) \end{aligned} \tag{3}$$

where, G meant the greenness of land, other variables are similar as above.

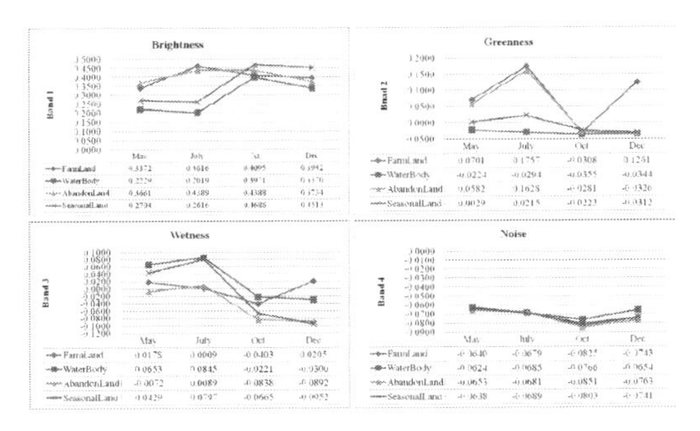

Figure 3. Change of principal bands of TCT with time for different land use.

3. Water Content Information
They can be measured through band 3 of images after TCT as following:

$$\begin{aligned} W_m &= |(W_{m3} - W^f_{m3})| / (W_{m\max}); \\ &\quad W_{m\max} = \max(|W_{m3} - W^f_{m3}|) \\ W_j &= |(W_{j3} - W^f_{j3})| / (W_{j\max}); \\ &\quad W_{j\max} = \max(|W_{j3} - W^f_{j3}|) \\ W_d &= |(W_{d3} - W^f_{d3})| / (W_{j\max}); \\ &\quad W_{d\max} = \max(|W_{d3} - W^f_{d3}|) \end{aligned} \tag{4}$$

where, W meant the wetness of land, other variables are similar as above.

4. Structure Information
The phenomenon was found that the shape of curve band 2, 3 and 4 of images after TCT was different for different land use. This information can be used to construct a new variable as following:

$$\begin{aligned} S_m &= |2 * b_{m3} - (b_{m2} + b_{m4})| / S_{m\max}; \\ &\quad S_{m\max} = \max(|2 * b_{m3} - (b_{m2} + b_{m4})|) \\ S_j &= |2 * b_{j3} - (b_{j2} + b_{j4})| / S_{j\max}; \\ &\quad S_{j\max} = \max(|2 * b_{j3} - (b_{j2} + b_{j4})|) \\ S_d &= |2 * b_{d3} - (b_{d2} + b_{d4})| / S_{d\max}; \\ &\quad S_{d\max} = \max(|2 * b_{d3} - (b_{d2} + b_{d4})|) \end{aligned} \tag{5}$$

where, S represented the structure relationship among the band 2, 3 and 4 of image after TCT; m, j and d represented the months; other variables are similar as above.

The weight of different index for land damage evaluation was calculated based on the method of AHP with help of excels software. The results were listed in the Table 1. Consistency checking result ($CR = 0.025 < 0.1$) indicated that the weighted coefficient was reasonable and efficient.

Based on the above model, the value of land damage in the study area was calculated, the result was showed in Figure 4. According to the value distribution, the land damage was classified into four types with equal interval division, which were light damage (0–0.1461), medium damage (0.1461–0.2707), serious damage (0.2707–0.3953) and heavy damage (0.3953–0.5199). The statistics of areas about different damage was listed in the Table 2.

Table 1. Index and weight for land damage.

Index	May	July	December	Weight
Brightness	0.035	0.064	0.02	0.119
Greenness	0.124	0.295	0.039	0.458
Wetness	0.029	0.057	0.115	0.201
Structure	0.063	0.127	0.032	0.222
Weight	0.251	0.544	0.205	1.000

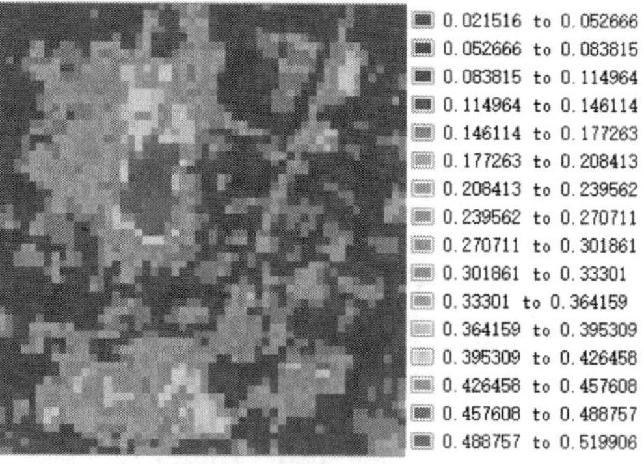

(a) Comprehensive value of land damage

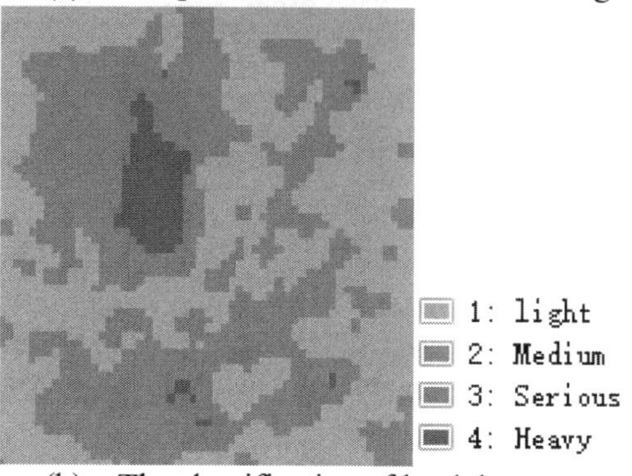

(b) The classification of land damage

Figure 4. The result of land damage distribution.

Table 2. Statistics of areas with different damage.

Type	Light	Medium	Serious	Heavy	Sum
Area (Ha)	142.38	105.48	21.87	12.15	281.88
Percent (%)	50.51	37.42	7.76	4.31	100.00

4 CONCLUSIONS

Temporal information about land condition in subsidence area can contribute to making systematic land reclamation. In this study, multi-period RS data were utilized to obtain the change rule of land state, and then evaluate the land damage. The following results were achieved:

1. Change pattern of index with time for different land damage was different which can facility the classification;
2. An evaluation model for land damage was established with consideration of temporal information of land condition;
3. Time series of land condition indexes were abstracted through RS images with the method TCT;
4. Evaluation result can reflect the degree of land damage in the subsidence area.

However, due to the restriction of RS resource, the time resolution for this study was not very accurate. In the future, we shall take multi-resource RS to work around this problem.

REFERENCES

Baig, M.H.A., Zhang, L., Tong, S., & Tong, Q. (2014). Derivation of a tasselled cap transformation based on landsat 8 at-satellite reflectance. *Remote Sensing Letters*, 5(5), 423–431.

Hu, Zhenqi. (2014) *Legislation, technology and practice of mine land reclamation*. CRC Press.

Lu Xia, Hu Zhenqi, Liu Weijie, Huang Xiaoyan. (2007) Vegetation Growth Monitoring Under Coal Exploitation Stress by Remote Sensing in the Bulianta Coal Mining Area. *Journal of China University of Mining & Technology*, 17(4):479–483.

Nuray Demirel, Şebnem Düzgün & Mustafa Kemal Emil. (2011) Landuse change detection in a surface coal mine area using multi-temporal high-resolution satellite images. *International Journal of Mining Reclamation & Environment*, 25(4):342–349.

Wang Dongpan. (2005) *Research on Character of Surface Movement Caused by Mining under Extra Thick Alluvium and High Diving Area*. Master Thesis of Henan Polytechnic University.

Xiao W, Hu Z, Fu Y. (2014) Zoning of land reclamation in coal mining area and new progresses for the past 10 years. *International Journal of Coal Science & Technology*, 1(2):177–183.

Land Reclamation in Ecological Fragile Areas – Hu (Ed.)
 ISBN 978-1-138-05103-4

Assessing ecological security based on fishnet: A case study of Guiyang City

B. Guo & F.Y. Xiao
College of Geomatics, Xi'an University of Science and Technology, Xi'an, Shaanxi, China

ABSTRACT: In this paper, with the support of GIS and RS, taking Guiyang City as an example, using Landsat TM images to extract information of land use and normalized difference vegetation index, combined with the river, the residents, soil erosion and DEM data, developed an indicator framework of ecological security assessment based on P-S-R mod-el, at the same time, used fishnet model in ArcGIS10.0 to fuse multiple heterogeneous data and produced a ecological security evaluation model to evaluate ecological security at fishnet scale.

1 INTRODUCTION

Taking Guiyang City as an example, extracting land use information and normalized difference vegetation index from Landsat TM images, combined with the river, resident data, soil erosion and DEM data, taking fishnet as the evaluation unit, to achieve effective integration of multi-source heterogeneous data, so as to carry out the assessment of ecological security and reveal the ecological security situation of Guiyang City. It can not only provide decision-making basis for the management and utilization of land resources in Guiyang City, but also provide reference for similar research in other regions of china.

2 DATA SOURCES AND PROCESSING

Data sources and processing methods of this paper are shown in Table 1. In this paper, the land use data, which is acquired from remote sensing image by the method of visual interpretation, was divided into farmland, forest, grassland, water area, residential and industrial land.

3 STUDY ON ECOLOGICAL SECURITY EVALUATION

3.1 *The measure indexes of ecological security*

This paper sets the comprehensive Ecological Security Index (ESI) of Guiyang City as the target level, pressure, state and response indicator as the criterion level and the index level is formed by the index which can be measured or calculated directly. The results are shown in Table 2.

3.2 *Process of the dimensionless values*

Different ES measure indicators are different in dimensions. To compare them with each other, their dimensions should be eliminated. So, in this study, the standard value method is adopted to unify the dimension of each index, and set each index in the range of 0–100, as shown below.

Table 1. Data sources and processing.

Data	Sources	Processing
Remote sensing image data	Earth science data FTP server of University of Maryland	format conversion (form TIFF to IMG), Band combination, Mosaic, Mask, coordinate transformation, Visual interpretation etc.
DEM	The GLS2005 Library in LANDSAT database	
Administrative divisions, soil erosion and population data of Guiyang	National Earth System Science Data Sharing Infrastructure	The definition and transformation of projection (Datum: Krasovsky, projection: Albers)

Table 2. Indicator framework of ecological security assessment.

Target level	Criterion level	Index level	
Comprehensive index of ecological security	Pressure	Human disturbance indices	
	State	Structural index	Landscape fragmentation index
		Elastic index	Vegetation coverage index
		Vigor index	Net primary productivity index
	Response	Soil erosion index	

Table 3. The weights of different indicators.

Indicator	Weight
Landscape fragmentation index (W_{jg})	0.15
Vegetation cover index (W_{fg})	0.15
Net primary productivity index (W_{jc})	0.25
Human disturbance index (W_{rl})	0.25
Soil erosion index (W_{tr})	0.2

$$Score = \frac{Xi - X\min}{X\max - X\min} \tag{1}$$

where Score is the assignment of different index, Xi is the true value of a index, Xmin is the minimum value of the index, Xmin is the maximum value of the index.

3.3 *Weighting*

In this study, the Delphi approach is used to determine the weights of different indicators, and the results are shown in Table 3.

3.4 *Calculation and grading for the degree of ecological security*

Based on the calculation results of each index, the method of weighted superposition is used to calculate the comprehensive index of ecological security in the study area. The formula is as follows:

$$ESI = E_{jc} \times W_{jc} \times E_{jg} \times W_{jg} \times E_{fg} \times W_{fg} \times E_{rl} \times W_{rl} \times E_{tr} \times W_{tr} \tag{2}$$

where ESI is the comprehensive index of ecological security evaluation, Wjc, Wjg, Wfg, Wrl and Wtr are the weights of different index.

4 RESULTS

Through the statistics of the ecological security situation of different security levels, we find that the area of Good is the most, its area is 727.29 km^2, accounting for 30% of the study area, and mainly distributed in the northeast and southern, and dispersedly in central city. On the contrary, the area of Very Poor is least, its area is 267.27 km^2, accounting for about 11% of the study area and mainly distributing in central area of the study area, especially in Baiyun District and Xiaohe Town. In addition, the area of Excellent is 292.75 km^2, accounting for 12% of the study area and mainly distributing in the northeast. The area of Common is 637.90 km^2, accounting for 27% of the study area and dispersedly distributing throughout the study area. The area of Poor is 475.22 km^2, accounting for 20% of the study area and its distribution is relatively dispersed in the middle of the study area.

5 CONCLUSION AND DISCUSSION

The results of this study indicate that the proportion of the Excellent (12%) is match that of the Very Poor (11%) of ecological status in Guiyang City. The difference is that there is obvious spatial distribution of these two security states, the Excellent areas are mainly distributed in the northeast of Guiyang City, while the Very Poor areas are mainly distributed in the Center District of Guiyang City, and south of Guiyang is in the Good. Therefore, in the future, Guiyang should continue to maintain the ecological status of the northeast and vigorously improve the ecological environment of the central region, in addition, the ecological environment in the southern region should be slightly improved, so as to realize the overall optimization of regional ecological environment.

This paper adopts the fishnet model based on the fusion of multiple heterogeneous data, which can reveal the ecological security situation of the study area in the fishnet scale and break through the original localization of evaluation scale of basin and administrative units. But the size of fishnet is subjective, so we will try to establish the mathematic model to determine the most suitable size of fishnet in the future. Besides, the ecosystem itself is a complex system together with many factors, it includes not only the natural factors but also human factors, and this research chooses 5 ecological security indexes containing the natural and human factors, but they are still less, this may cause the results of the study to have some contingency. Therefore, more indexes in similar research should be selected in the future, thus making the evaluation results more convincing.

ACKNOWLEDGEMENTS

This article is supported by the special project of Shaanxi Province Education Department (16JK1495).

REFERENCES

Fu B.J. (2010) Trends and priority areas in ecosystem research of China. *Geographical Research*, 29(3):383–396.

Pei H., Wei Y., Wang X.Y., Tan Z.H. & Hou C.L. (2014) Method of cultivated land landscape ecological security evaluation and its application. *Transactions of the Chinese Society of Agricultural Engineering (Transactions of the CSAE)*, 30(9):212–219.

Qin X.N., Lu X.L. & Wu C.Y. (2014) The knowledge mapping of domestic ecological security research: bibliometric analysis based on citespace. *Acta Ecologica Sinica*, 34(13):3693–3703.

Shi D.D., Dong X.F. & Liu X.G. (2015) Lanzhou ecological security assessment based on PSR model. *Resource Development & Market*, 31(8): 956–958.

Wang C.H., Tian D.X. & Liu Y.H. (2008) The contrary reseaches on the Chinese and foreign ecological security assessment. *Forward Forum*, 07:44–49.

Wei B., Yang X.S., Wu M., & Xiao J.H. (2009) Research review on assessment methodology of ecological security. *Journal of Hunan Agricultural University (Natural Sciences)*, 35(5): 572–579.

Wei W., Lei L., Zhou J.J., Shi W., Xie B.B. & Liu X. (2015) Evaluation of the ecological security in Shiyang River Basin based on Grid GIS and PSR model. *Chinese Journal of Soil Science*, 46(4): 789–795.

Xie X.F., Wu T., Xiao C., Jiao G.J., Bian H.J., Ma Y. & Chen J.H. (2014) Ecological security assessment of the Dongyang River Watershed using PSR modeling. *Resources Science*, 36(8): 1702–1711.

Yan C., Zhang A.M., Shi R.R., Guo H.H., Guo D.L., Wu S.H. & Song H.F. (2016) Spatiotemporal change analysis of Qianjiang's land ecological security evaluation in Chongqing City. *Bulletin of Soil and Water Conservation*, 36(4): 262–268.

Yao J.S. & Tian J.Y. (2007) Study progress and application of ecological security. *Journal of EMCC*, 17(2):47–50.

Yi W.Y., Su W.C. & Luo S.Q. (2014) Evaluation of land ecological safrty in Guiyang during 2002–2012. *Guizhou Agricultural Sciences*, 42(3):170–173.

Yu G.M., Zhang S., Yu Q.W., Fan Y., Zeng Q., Wu L., Zhou R.R., Nan N. & Zhao P.J. (2014) Assessing ecological security at the watershed scale based on RS/GIS: a case study from the Hanjiang River Basin. *Stochastic Environmental Research & Risk Assessment*, 28(2):307–318.

Zhang J.Q., Yi K.P., Hiroshi T., Wang X.F., Tong Z.J. & Liu X.P. (2011) Ecological security assessment of Baishan City in Jilin Province based on DPSIR. *Chinese Journal of Applied Ecology*, 22(1):189–195.

Zhao Y.Y., Long R.J., Lin H.L. & Ren J.Z. (2008) Study on pastoral ecosystem security and its assessment. *Acta Prataculturae Sinica*, 17(2): 143–150.

Zhou Y.X., Li Y.X., Sun Y.Y., Li J.T., Gao G.F. & Su C. (2016) Evaluation of land ecological security in Shandong Province based on PSR-TOPSIS model. *Journal of Agricultural Resources and Environment*, 33(4):320–326.

Zhu W.H., Miao C.Y., Zheng X.J., Cao G.L. & Wang F.F. (2014) Study on ecological safety evaluation and warning of wetlands in Tumen River watershed based on 3S technology. *Acta Ecologica Sinica*, 34(6): 1379–1390.

Zou C.X. & Shen W.S. (2003) Progress of the ecological security research. *Rural Eco-Environment*, 19(1), 56–59.

Land Reclamation in Ecological Fragile Areas – Hu (Ed.)

The application of classified assignment and power-law models for wound space in quantitative assessments of degraded land

L.L. Zhang, P. Zhao, L.H. Gou & Z.H. Luo
Beijing Orient Earth Environmental S&T Co. Ltd., Beijing, China

ABSTRACT: Land degradation occurs when in agriculture, forestry, mining, and the construction of varied facilities include highway, railway, hydropower and others. Evaluation factors singularity in wound space of degraded land include soil types, weathering stages, elements content, particle size etc. The spatial status of these non-linear processes have in common can be modeled by power-law. Establishing a basic classification system of evaluation factors, assign values to this system or use original value of each objects. The establishment of this data framework enhances the quantitative assessments by power-law models for wound space. The method described in this paper not only provides a new idea for mine-site remediation and ecological rehabilitation, but also facilities a near-natural landscape reinstatement planning. A case study of an abandoned quarry from area next to the Shiyan, Hubei Province, China is introduced.

Keywords: Wound space; Degraded land; Evaluation factors; Power-law models; Quantitative assessments

1 INTRODUCTION

Landscape damage, substrate exposure and vegetation deterioration are inevitable during the operation of massive earthwork in mining, railway construction and other civil engineering work. Taking the approach from ecological restoration perspective, this paper categorize influential factors before assign a value and make comprehensive assessment with historical data according to the various features of damaged land surface. The result could be developed into a disturbed land quantitively evaluation system based on power-law model. It may answer the following questions: a) how could it be possible to implement differential remediation? b) how could it be possible to evaluate degraded land relevant factors comprehensively and identify their correlation? c) whether the disturbed land surface historical data (e.g. geochemical data) is useful in the evaluation of its current status? d) whether it is practical to derive theoretical foundation for landscape reinstate and ecological restoration from big data in the practice of degraded land remediation?

2 QUANTITATIVE ASSESSMENTS OF DEGRADED LAND

2.1 *Value assignment*

There are many factors could change the pattern of land degradation. Depending on different interest of remediation target, the weight and types of impact factors are varied. From ecological restoration perspective, the texture and gradient of damaged land are of major concern. Land texture include the variety of rock, mineral composition, porosity, degree of weathering, etc.

When remediate contaminated land, then have to consider contamination level (heavy metal and organic pollutant concentration) and hydrological condition. All these factors contribute to characterize degraded land are called relevant indicators of degraded land. They can be expressed as the following formula (Zhang et al., 2017):

$$F = \{F_1, F_2 \cdots\cdots F_i\} \tag{1}$$

The establishment of value assignment system depend on reasonable classification of relevant indicators. Currently, there are some national standards (GB50218-1994; GB50500-2003; GB/T50145-2007) made primary categorization on texture, degree of rock weathering and mineral composition. With the precondition of plants survival and vegetation establishment, these primary categories can be sorted to establish systematic ecological restoration guidelines. Furthermore, each category can assign an integer value and add decimal digits according to observation in field. Contamination level can use element concentration directly. Each relevant indicator F_i has to be evaluated individually and the value of F can be calculated when the weight of each indicator is known.

Table 1. Ecological big data quantitatively analyzing structure.

Level of architecture	Name of system		Type of data
Ecological landscape (big data collection system)	Ecological land; Degraded land restoration	Classification system assignment data	Texture; Slope; Fabric; Weathering
		Historical data	Remote sensing image; Geochemistry, Geophysical; Basic geology
	Urban spatial suitability evaluation	River system; Wetland; Topographies	
Mathematical model	Power-law model; Weight; Interpolation		
Accumulation of case study	Differential remediation area division; Types of treatment measures; Near natural restoration		

2.2 *Power-law model*

Power-law model represent a non-linear process which is common in nature. Therefore, power-law model is widely employed in scientific research in quantitively assessment of subject's self-similarity. In mineral exploration, fractal-multifractal model is widely applied in deposition estimation and mineralization characteristics differentiation, such as the density-area (volume) model developed for anomaly analysis of mineralization (Cheng, 2006).

Assign a value $\gamma(V_{F_i})$ to disturbed land impact factor F_i when it vary within certain dimension spatially. With the increase of V, $\gamma(V_{F_i})$ shall decrease. The interaction between $\gamma(V_{F_i})$ and V can be expressed as:

$$\gamma(V_{F_i}) \propto aV^{\beta/3} \tag{2}$$

here, $\gamma(V_{F_i})$ represents the mean of V when it varies within certain dimension. In formula (2), a is a constant and β is independent. The discussion about the interval of β is very important. When $\beta = 3$ then assigned value is $\gamma(V_{F_i})$ in volume V. When $\beta \neq 3$ is abnormal could have two situation. If $\beta < 3$, then quadrat dimension is greater than evaluation indicator gradient. With the increasing of β, $\gamma(V_{F_i})$ will decrease but approach stable. When $\beta > 3$ means quadrat dimension smaller than evaluation indicator gradient. With the increment of β, $\gamma(V_{F_i})$ increase with greater scale which indicates accelerated increment of evaluation indicator. Therefore, $\gamma(V_{F_i})$ is a sensitive indicator with relative high abnormality.

2.3 *Quantative assessment of ecological data*

The general objectives of ecological restoration in degraded land include a) reinstate ecological environment; b) retain historical landscape and achieve aesthetic view. Therefore, it is necessary to differentiate remediation measures according to quantitatively analyzed relevant impact factors and aggregated weighted factors. By systematic classification, reasonable value assignment, historical data (e.g. geochemistry and fault) of land degradation relevant impact factors and rasterization analysis, a mathematical model which can differentiate area with different substrate, regional specification and contamination level to identify proper vegetation pattern and remediation measures. This quantitative ecological big data analyzing platform not only specify remediation measures for varied area, also offered a solution of landscape naturalization.

3 CASE STUDY

In Shiyan, where a dolerite steep slope has many fractures and the fractionation and weathering vary in great scale. In quadrat (5 m dimension) survey, rock structure fracture density and weathering level value assignment evaluation was carried out (Figure 1). A positive correlation between rock fractures and weathering level was found. According to different intensity of these two factors in combination, disturbed land can be divided into three categories: A is the area with complete rock structure and hard rock. Due to humid climate and relative high precipitation, moss can be used to naturalize local rock surfaces; B is the area with highly fractures rock structure and varied joints. Can use external soil to spray native herbaceous species; C is the area where rock structure almost all fractured and rock surface weathered totally. Herbaceous species and shrubs are recommended and woody species can be used in some designated place for landscape aesthetic. Transition zones between different type of area shall be arranged per local substrate to reach a naturalized landscape which would reduce maintenance cost significantly.

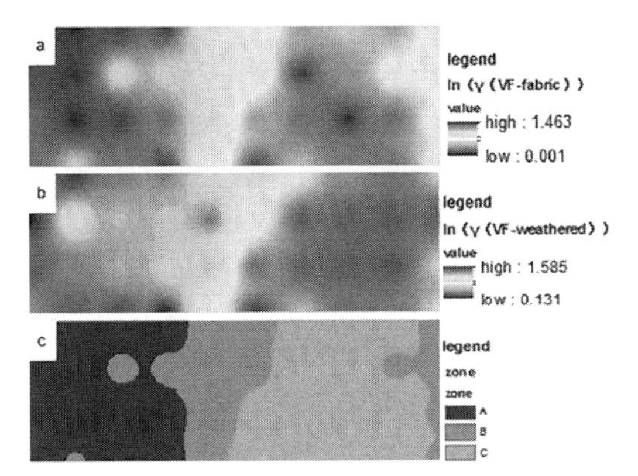

Figure 1. Doleritic slope disturbed surface relevant impact factor intensity area classification (Zhang et al., 2017) (a. ln (Valuation of fabric degree), b. ln (Valuation of weathering degree), c. differential remediation area division, Evaluation area: 50 m × 20 m, size of Window: 5 m × 5 m).

4 CONCLUSION

The restoration of disturbed land is a complicated topic which demand the expertise of geology, chemistry, hydrology and landscape planning. By establishing a value assignment system of disturbed land impact factors and analyzing them in mathematical model (power-law model), specified remediation suggestions per area can be produced which will give reliable support to naturalize disturbed land ecological restoration.

REFERENCES

1994, G. (1994). Engineering rock mass classification standard.

2003, G. (2003). People's Republic of China State Administration of quality supervision, inspection and Quarantine of construction engineering quantity list valuation specification.

2007, G.T. (2008). Soil engineering classification standard.

Cheng, Q. (2007). Mapping singularities with stream sediment geochemical data for prediction of undiscovered mineral deposits in gejiu, yunnan province, china. Ore Geology Reviews, 32(1–2), 314–324.

Longlong Zhang, Ping Zhao, Lihui Gou et al. (2017). Construction of digitized classification system of damaged land. Mine Surveying, 45(1), 95–99.

Land Reclamation in Ecological Fragile Areas – Hu (Ed.)
© 2017 Taylor & Francis Group, London, ISBN 978-1-138-05103-4

Preliminary analysis of the geological hazards prediction in the process of coal mining

X.Y. Sun & H.X. Fu
Country College of Geology and Environment, Xi'an University of Science and Technology, Xi'an, Shaanxi, China

D. Zhang
Rizhao Anzhuang Town Junior High School, Rizhao, Shangdong, China

Z.Q. Liu & X.H. An
Country College of Geology and Environment, Xi'an University of Science and Technology, Xi'an, Shaanxi, China

ABSTRACT: The coal mining will cause a series of damage to the overlying strata and surface soil, through the study of rock and soil of the earth surface movement and deformation mechanism, we can forecast range, three rock band distribution and ground surface movement and deformation degree, which has important significance for prevention of geological disasters, geological environmental protection and sustainable development in mining area. By means of numerical simulation and similar material simulation, this paper forecasts the change of overlying strata and surface soil layers, and provides a reference for mine environmental protection.

1 INTRODUCTION

This paper mainly studies the geological disasters of surface settlement, the research involves the mine field in Yulin of Northern Shaanxi province and the south of Ordos which is an important base of energy and chemical industry. In the process of coal mining or after mining will cause overlying strata destruction and surface settlement[2] or other geological disasters, destruction of the existing land resources, bring direct or potential harm to all kinds of buildings. Therefore a reasonable forecast of settlement range and the maximum settlement to reduce the secondary geological disaster, recovery of land utilization and protection of buildings and so on is of great significance[3].

After coal seam are mined, original mechanical equilibrium were destroyed in rock mass and will make overlying strata and ground surface deformation and movement. This originated in the stope nearby movement and destruction, with the development of mining will extend to surface, this paper referred to as surface subsidece[4]. Overing strata of coal seam into layered superimposed body output, which is structural "medium"[5] that conduct tectonic stress causing surface subsidence. It's primary goal for the research of it's change to study surface subsidence. This article embarks from coal mine actual strata, studies coal seam change law in three zones of overlying strata by using experimental method from inital stage of production to fully ofter mining, especially the distribution status of bending zone[1], detection of rock and soil layers changes and predict changes of coal overlying strata top soil layer by the field investigation of the surface soil deformation state compared with the experimental results.

2 GEOLOGICAL STRUCTURE AND COAL SEAM MINING IN THE STUDY AREA

This artical selected mine field that region geological structure is simple, no large fault and fold and obvious trace of magmatic activity, local development wide and gentle undulation, surface is covered by loose alluvium of Quanternary in mine field, the main stratum: Wucheng Formation of lower Pleistocene series Quanternary (Q1w), Lishi Formation of middle Pleistocene (Q1l), Salawusu Formation of upper Pleistocene (Q31s), Holocene series alluvial-pluvial layer (Q41al + pl, Q42al) and aeolian layer (Q42eol). Drill hole exposed stratum: Zhiluo Formation and Yanan Formation series, Fuxian Formation of lower Jurassic series, Wayaobao Formation of upper Triassic series. In addition to the contact relation between Fuxian Formation and Wayaobao Formation is micro angular unconformity, others are conformable contact. Angle of coal seam is 5°, mining with slicing large mining height fully mechanized method. It's of great representative significance to select the mine field to research it's law of surface subsidence in Yushen mining area.

3 EXPERIMENT RESEARCH

3.1 *Similar material experiment*

Similar material experiment is a kind of physics experiment, using the actual drilling data revealed the lithology of overling strata, integrated multiple drilling data, experiments were carried out in accordance with the premise of three similarity theory under the restricted conditions of physical experiment, the simulation experiment mainly observed changes in overlying strata, the overlying soil is replaced by heavy weigh. Considering the stability condition of overlying strata and the stability of the model of high requirements, determine the simulated strata height, size and the relevant parameters, leaving border protective coal pillar. According to time similarity radio excavate step by step, using omnidirectional positioning apparatus without prism observe data. This experiment model and the distribution of monitoring points is shown in picture1 (mm). Figure 1 Similar material experimental model and monitoring point distribution.

3.2 *Numerical simulation experiments*

Numerical simulation experiment is a numerical simulation method developed with the computer, The experimental software used in this paper is FLAC 3D, The profiles used were the same as those of similar materials, Access relevant literature[6–8] to adjust the numerical simulation parameters of overlying strata, Run the program and get the simulation results, Finally extract the curve of contact surface of bedrock and surface soil. Figure 2 numerical simulation experimental model and excavation results.

3.3 *Experimental results analysis*

Similar curves were obtained at the observation point of the bedrock and soil boundary as shown in Figure 3 Material simulation of bedrock

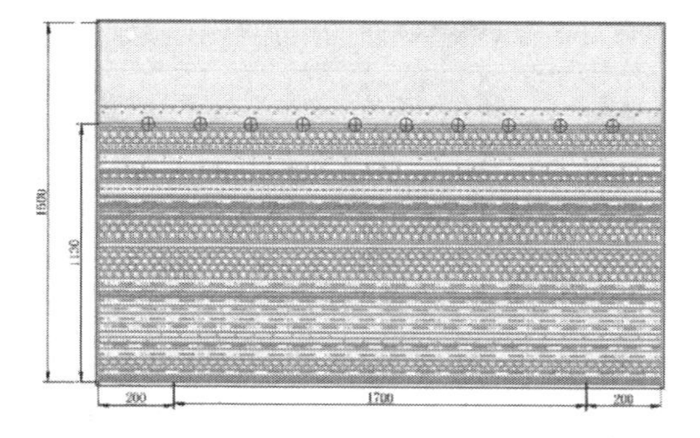

Figure 1.

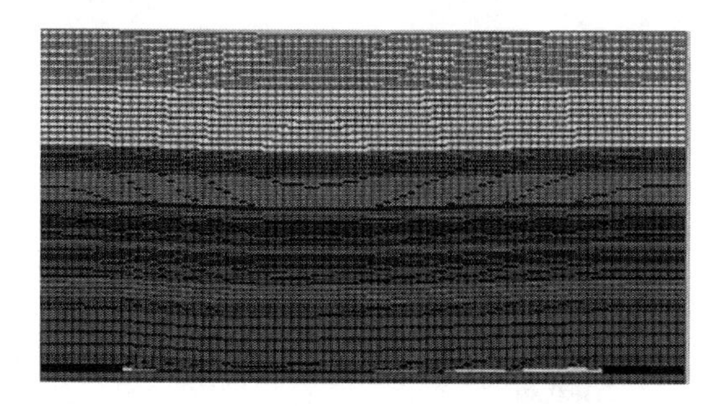

Figure 2.

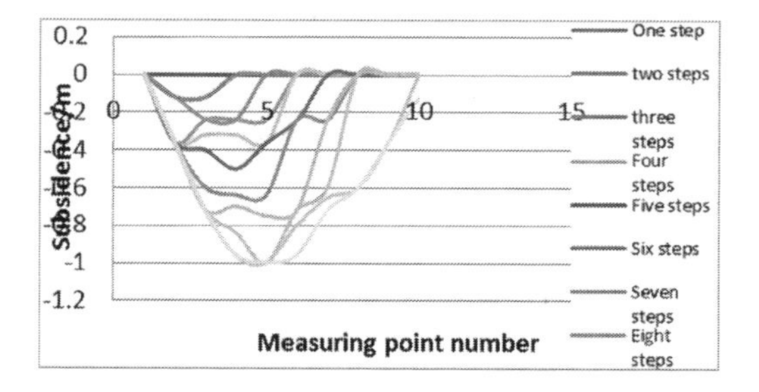

Figure 3.

soil boundary observation point of subsidence curve. The mid-point sinking curve is obtained by numerical simulation experiment as shown in Figure 4 Numerical simulation of the middle of the soil monitoring point of the sinking curve. The middle part of the experimental model sinks the curve as shown in Figure 5. Material simulation model in the middle of the monitoring point of the sinking curve.

1. The experimental field is medium thick coal seam, Depth shallow bedrock thickness of about 240 m, Moderate hardness, The experimental results show that the boundary between bedrock and soil is located at the bottom of curved subsidence, Closer to the fissure zone, There may be micro-fissure development, So that the amount of surface subsidence increases.
2. Through the experimental results, it is concluded that there is no subsidence at the boundary between bedrock and soil, From the seventh step that is excavated to 105 m out of the amount of subsidence. From the beginning of the fourteen steps to excavation to 210 m after the amount of sinking no longer change, And consistent with the location of the surface survey of the amount of subsidence. Similar materials experiment and numerical simulation experiments to get the same curve of the subsidence curve of bedrock and soil interface, The amount of subsidence is roughly the same,

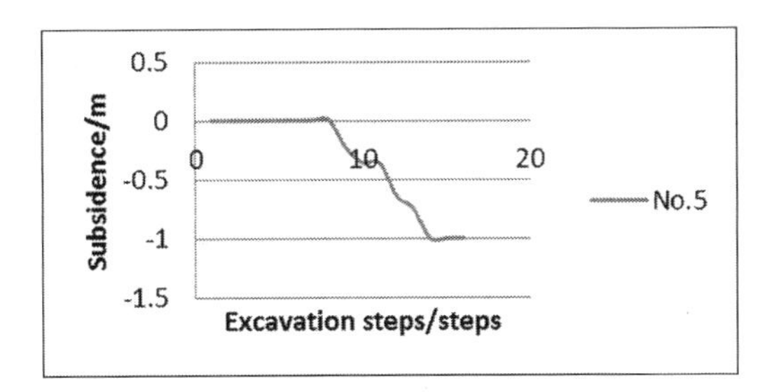

Figure 4.

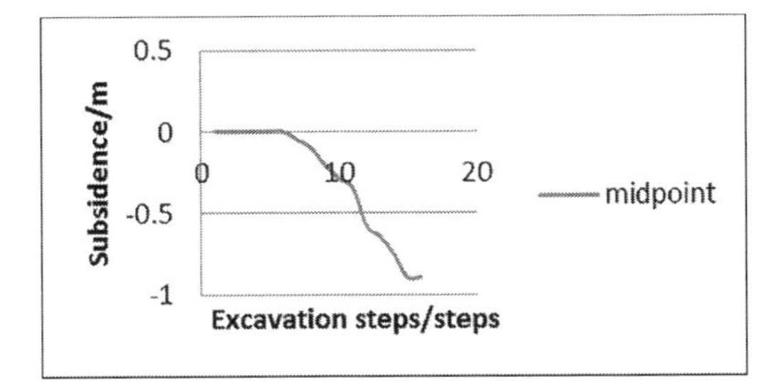

Figure 5.

It can be seen from Figure 3 that as the mining subsidence continues to grow, It no longer change after reaching a certain value, After fully mining the maximum settlement value of about 1.1 m or so, Accounting for about 37% of the thick. According to the law to analyze the surface subsidence of about 1 m, Consistent with the actual survey.

3. Whether it is in the mining process or fully mining, The sinking curve presents left and right asymmetric distribution, The absolute value of the mining slope is greater than the absolute value of the mining slope, The basement of the bedrock is located at the subsidence center on mined area in the rear, Leading to the ground subsidence basin at the same time in the mined area in the rear.
4. After fully mining, The fracture angle of the overburden is about 89°, The movement angle is about 70°, The boundary angle is about 64°, Three angles of the left side of the goaf, Slightly larger than the goaf of Mined area front, The numerical simulation results show that the topsoil angle is about 50°, Therefore, the subsidence area of the surface soil is larger than that of the subsurface basin in the interface between the bedrock and the soil, The boundary increases by about 84% of the soil thickness. Assuming that the soil thickness is d_1, Overburden thickness d_2, The subsidence basin boundary is larger than the mining boundary about $84\%d_1 + 49\%d_2$

4 CONCLUSION

1. The location of the surface subsidence caused by shallow coal seam in the experimental field is about 120 m in the coal seam, The subsidence of the subsidence basin is stable at 210 m, The boundary between bedrock and soil is obvious. Surface subsidence amount of about 1 m, Accounting for 37% of the total.
2. by the boundary angle and surface soil moving angle, The edge of the subsidence basin is $84\%d_1 + 49\%d_2$ higher than the mining boundary. The center of the subsidence basin is located in the center of the mined-out area.

REFERENCES

[1] Guo, S.S. (2017) Numerical Simulation and Aanlysis on Overlying Strata Movement of 3207 Working Face with Large Mining Height in Huoerxinhe Mine. *J. Mining Safety and Environmental Protection*. (01):1–4.

[2] Li, S.Z. (2014) Present Status and Outlook on Land Damage and Reclamation Technology of Mining Subsidence Area in China. *J. Coal Science and Technology*. (01):93–97.

[3] Xu, J. & Qian, M.G. (2001) Study and application of dominant stratum theory for control of strata movement. *China Mining*. (06):56–58.

[4] Song, S.J. & Zhao, X.G. & Zhang, Y. & Nie, W.J. (2016) Numerical Simulation Analysis of Slope Morphology and Erosion in Mining. Area. *J. Safety and Environment*. (04):368–373.

[5] Sun, X.Y. (2009) Influence of inherent anti - disturbance ability of geological environment on ground subsidence in coal mine. Shaanxi Coal Institute. The Proceedings of the Academic Annual Meeting of Shaanxi Coal Science Society- Geological Safeguard Technology for High Yield and High Efficiency Coal Mine Construction. Shaanxi Coal Institute. 2009:6.

[6] Sun, X.Y. & Xia, Y.C.(2010) Control Effect of Tectonic Stress and Joint Coupling on Mining Subsidence. *J. Mining and Safety Engineering*. (01):126–130.

[7] Yang, Y. & Liu, W.S. & Miao, X.X. & Feng, G.C. (2007) The research status and view of the mining subsidence and it's control technology in china. *J. China Mining*. (07):43–46.

[8] Zhao, C.H. & Jin, D.W. (2017) Numerical Simulation of Coal Mining on Groundwater System Disturbance in shallow Buried Coal Seam of Western Mining Area. *J. Coal mine safety*. (01):168–171.

Land Reclamation in Ecological Fragile Areas – Hu (Ed.)
© 2017 Taylor & Francis Group, London, ISBN 978-1-138-05103-4

The application of mobile GIS in mine land reclamation monitoring

R.M. Hu, W.M. Liu, S. Wang, X.Z. Zhang & Y. Li
College of Geomatics, Xi'an University of Science and Technology, Xi'an, China

ABSTRACT: Nowadays, there are many kinds of monitoring methods of land reclamation, such as remote sensing monitoring and plot monitoring. According to the current situation of mine land reclamation monitoring in China, and the inconvenience of field monitoring data collection, the author proposes a mobile GIS monitoring data acquisition system based on intelligent mobile terminal (smart phone, tablet). The system is easy to install and carry. It can collect the basic information of mining area such as the spatial location, images, vegetative cover extent, soil conditions and so on. What's more, the data collected in the field can be uploaded to the server and processed by people in real time, so the reclamation of the mining area can be monitored at any moment. And based on the data which is collected in the field, it is an important basis for the adjustment of reclamation targets, standards, measures and planning arrangements.

1 INTRODUCTION

In recent years, with the rapid development of Chinese economy, resources provides a strong guarantee for Chinese economical. However, due to the large-scale exploitation of mineral resources, there are a series of serious environmental problems in mining area, For example, land resources have been seriously wasted and the vegetation in the mining area has been seriously damaged. According to the relevant information, the land reclamation rate is generally 70%–80% in foreign countries, however, in China it's just about 1%. Therefore, It is of great significance for China to strengthen the land reclamation work, which is of great importance to effectively alleviate thc contradiction between people and land, Improve the quality ecological environment of the damaged area, and promote social stability and unity.

Therefore, the establishment of a reasonable land reclamation monitoring system, management and real-time monitoring of mine land reclamation work, can effectively supervise the mine land reclamation work on attention, and it is the guarantee of mine reclamation on time, quality and quantity increase quality and insure schedule of mine reclamation.

2 GIS SYSTEM INTRODUCTION

2.1 *Introduction of the system*

The GIS system is mainly used to collect the field data of land reclamation monitoring, which includes mine information, spatial location, reclamation area, vegetation information, soil information, field photos and other basic information. The specific use case of the software is as follows:

The software is based on intelligent mobile terminal development of mobile GIS system, which has great advantages compare with the previous handheld GPS. It's easy to install on almost all smartphones and tablets, and is easy to carry, the most important feature is that the data collected in the field can be sent to the server in real time when the network is in good condition.

2.2 *Introduction to development platform*

Android is a free and open source operating system based on Linux, which is mainly used in mobile devices, such as smart phones and tablet PCs, and

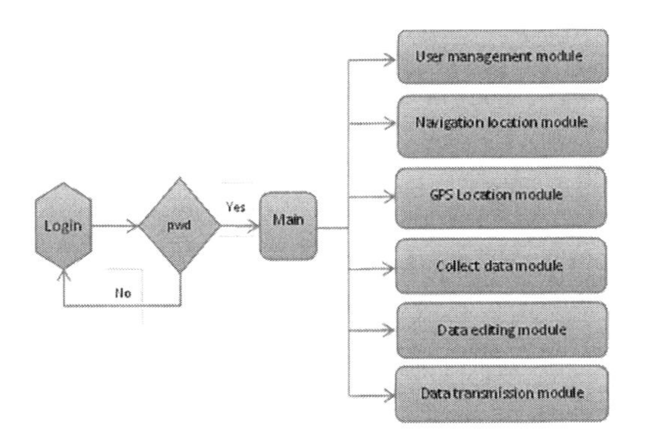

Figure 1. System structure diagram.

led by Google and Open Handset Alliance. The framework of Android system is composed four layers: the application layer, the application framework layer, the system runtime layer and the Linux kernel layer.

3 SYSTEM ANALYSIS

3.1 *The content of mine land reclamation monitoring*

3.1.1 *Primitive landform*

The contents of the original landform generally include: the basic topographic information, land use rights, soil conditions, vegetation coverage and other information.

3.1.2 *The damaged condition of land*

The damage of land in mining area exists during the step of mineral resources development. Different development periods and operation mode can cause different land damage mode and level. The damage condition of land is mainly to investigate clearly in the mining land loss, land occupation, land subsidence, land pollution and so on.

3.1.3 *The reclamation effect of land*

The most direct effect of land reclamation in mining area is the quality of soil and vegetation in reclamation area. The monitoring of land reclamation effect is mainly from four aspects: the progress of land reclamation, the implementation of funds, the reclamation facilities, soil quality, vegetation coverage.

3.2 *System functional requirements analysis*

The system mainly achieved the data collection work of land reclamation monitoring, The data can be collected to upload, save, edit, as well as analyzing requirements of the system according to the data needed.

The main part of the field work is the investigation and collection of data Mining land reclamation should be involved in the operation of the stage and the final inspection and acceptance phase. According to the working mode of data collection, the staff can use the portable mobile devices to complete the information collection and data collection.

1. Login and registration. Access to the system the registration must be finished before you access to the system and relevant functions.
2. Navigation. The system uses the map from Baidu maps, through the two secondary, you can achieve the line query and navigation, and will automatically track the function of the track.
3. GPS real time positioning. Today's mobile devices have GPS positioning function, in the open field, smart phones GPS positioning accuracy can reach the general handheld GPS accuracy level, you can use the mobile GIS system to measure the area of reclaimed land.
4. Collect data. The data acquisition is divided into two kinds: survey data and measurement data. The survey data mainly include the basic information of the mining area, soil information, landform, etc.; The survey data mainly include the measurement of the reclamation area of the mining area, the measurement of auxiliary facilities such as roads and water conservancy projects.
5. Editing and saving of data. The editing of data is the basic function that can query, modify and delete the data that has been recorded. The preservation of data can be stored in the Sqlit database of the Android phone.
6. The transmission of data. Under the condition of good network, the data stored in the database can be sent back to the server through the network in real time. The data stored in SD card can be collected when the network signal is not good. Until the end of the filed work through the WIFI transmission or direct export data.

4 DESIGN OF THE SYSTEM

4.1 *The analysis of system flow*

First of all, users need to log on to verify the identity of the system, After login, users will enter the main interface and manage the project, create new project or collect information using with previously saved items. After the project created, you can use the navigation interface to navigate precisely the route of the mine to be monitored.

4.2 *The analysis of system architecture*

The architecture of the system is divided into three layers: presentation layer, logic layer and database service layer.

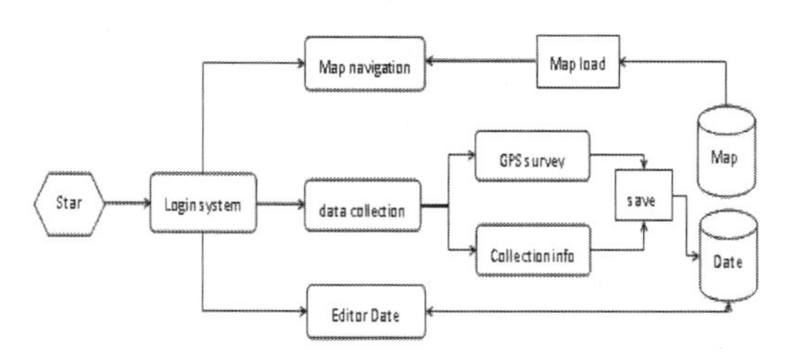

Figure 2. System flow chart.

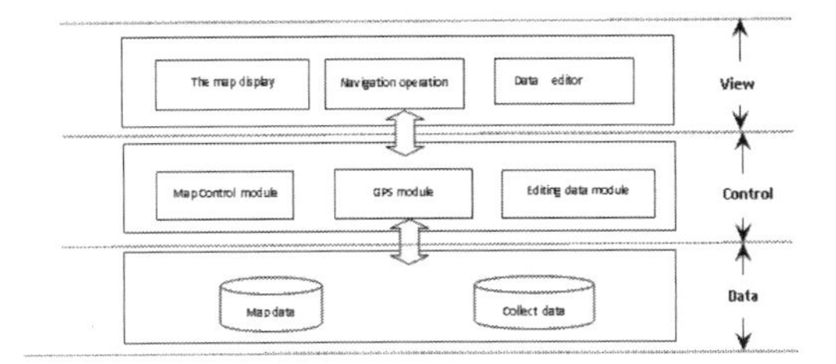

Figure 3. System architecture diagram.

Presentation layer provides interface for user interaction. Available services include: map display, map browsing operations such as zoom and translation, as well as the investigation of the form and the data collection interface display.

The logic layer implements the business procedure and specifies some business rules. It includes the module of spatial data editing, the module of reclamation area data acquisition and the modulre of positioning;

A database service layer mainly realizes the management of data storage, storage including survey information and spatial data storage, survey information is the basic information of mining area, topography, soil and other information, which is saved to the specified database.

4.3 *The design of system module*

In the process of implementation of GIS mobile field acquisition system in mine land reclamation monitoring, according to the demand analyzing function, this system can further analyses the management module of the user, line navigation module, GPS positioning module, data acquisition module, data storage and editing.

4.3.1 *The module of user management*

The module of user management achieves the user's registration, login, and personal data management. The user's registration is mainly for the first time the use of the system staff, and in order to use the corresponding function. The login checks the user's account and password. Personal information management is mainly for the user's basic information management, such as image, email and other basic information.

4.3.2 *The module of navigation*

Navigation module is mainly based on Baidu map which is the secondary development, users only need to enter start position and end point position, will enter into the ground coordinates of the corresponding field, and then the analysis and path planning, finally choose an optimal path.

4.3.3 *The module of GPS*

GPS positioning module is one of the most important modules of the system. At this stage, almost all of the intelligent mobile devices have GPS positioning function, which can accept the signal of GPS transmission in real time. In the field of data collection process, the staff generally in a static or slow state when the GPS satellite data signal is relatively stable.

Generally we can receive information from the GPS, which is the form of latitude and longitude (BL), the need for the longitude and latitude by Gauss positive (formula 4.1) into the corresponding geodetic coordinates. In the process of calculation, we must know the parameters of the ellipsoid to be converted. This procedure by default WGS-84 ellipsoid parameters for geodetic coordinates.

$$
\begin{aligned}
x &= X + \frac{N}{2} t \cos^2 BI^2 + \frac{N}{24} t\left(5 - t^2 + 9\eta^2 + 4\eta^2\right) \\
&\quad \cos^4 BI^4 + \frac{N}{720} t\left(61 - 58t^2 + t^4\right) \cos^6 BI^6 \\
y &= N \cos BI + \frac{N}{6}\left(1 - t^2 + \eta^2\right) \\
&\quad \cos^3 BI^3 + \frac{N}{120}\left(5 - 18t^2 + t^4 + 14\eta^2 - 58\eta^2 t^2\right) \cos^5 BI^5
\end{aligned} \tag{1}
$$

4.3.4 *The module of data collection*

The data acquisition module mainly consists of two parts, one part is the collection of the basic data of reclamation, including the collection of basic information and land reclamation information collection, with the help of mining staff. Another part is the realization of ancillary facilities and reclamation area measured by GPS function. To achieved a field survey.

The collection of basic data includes the reclamation areas of topography, land ownership, the degree of pollution of soil, vegetation types and coverage. The damaged condition, land occupation, digging, collapse, pollution and other information, the information effect of land reclamation, land reclamation project, the implementation of funds, reclamation facilities and the improvement of soil quality, vegetation status.

To achieve the necessary terrain spatial data acquisition based on GPS function, measurement, such as reclamation area of new roads and water conservancy facilities, power facilities and reclamation area measurement. These data will be to geodetic coordinates (XYZ) stored to the corresponding database.

4.3.5 *The module of edit and save data*

The data editing and saving module is divided into two parts: edit and save, data editing is the management of the collected information, to add or delete data which has been saved, then save the data again.

Data storage can be divided into two types: automatic save and manual save. The movement of the operator is automatically recorded every 5 seconds to the database, the information will be manually saved to the database.

4.3.6 *The module of data transmission*

The data transmission module realizes collection of the data to the server real-time, which is stored in the database through the server's processing and analysis.

The development of Android application often submits data to the server and get data from the server, this system mainly uses the HTTP protocol to submit data to the server by HttpClient, Submission of data is divided into two ways to get and post This system uses the post mode to transmit data to the background, the data encapsulate the JSON string in the form of transfer data to the backend server, JSON (JavaScript Object Notation) is a lightweight data interchange format, easy to read and write, but also easy to machine parsing and generation, interaction is very suitable for the server and the client.

4.4 *The design of database*

This system uses the embedded database Sqlite database with Android system, Sqlite is a lightweight database, its design goal is embedded, it occupied resources is very low, in the embedded devices, may only need a few hundred K of memory. However, the data can be stored 2TB, to meet the daily needs of the external data storage industry, the same as MySQL, SqlServer and other databases, Sqlite processing faster than they are.

The database consists of 6 tables, the basic information table of the mining area, the basic information table of the soil, the user table, the vegetation questionnaire, the spatial data collection table, the movement track record table, the basic structure of each form is shown below:

basic_info	
id	varchar(0) binary<pk>
name	varchar(0) binary
area	varchar(0) binary
ownership	varchar(0) binary
createTime	datetime
user_id	varchar(0) binary
x	decimal
y	decimal
z	decimal
landscape	varchar(0) binary
terrain	varchar(0) binary

soil_info	
user_id	varchar(0) binary
soilarea	varchar(0) binary
pollution	varbinary(0)
extend	varchar(0) binary

vegetation	
name	varbinary(0)
type	varbinary(0)
area	varbinary(0)
time	date
user_id	varbinary(0)

User	
user_id	varchar(0) binary<pk>
user_name	varchar(0) binary
user_email	varchar(0) binary
user_sex	boolean
user_age	int
user_dept	varchar(0) binary

Collect_data	
point_id	varbinary(0)
point_name	varbinary(0)
point_code	varbinary(0)
point_x	decimal
point_y	decimal
point_z	decimal

Figure 4. Database table structure.

5 SYSTEM IMPLEMENTATION AND EFFECT OF DISPLAY

5.1 *The implementation of user login module*

After opening the system, will enter a login screen, must input correct account number and password to enter the main interface, as shown in the Figure:

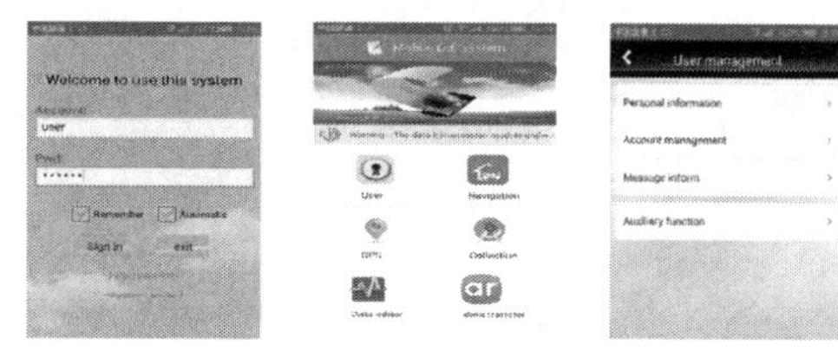

Figure 5. Login-interface.
Figure 6. Main-interface.
Figure 7. User manage.

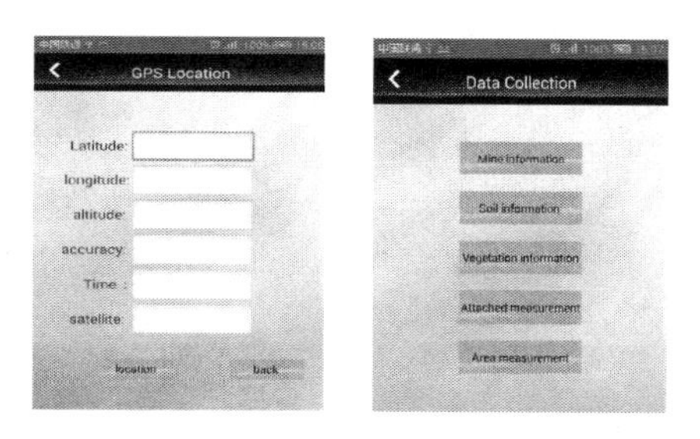

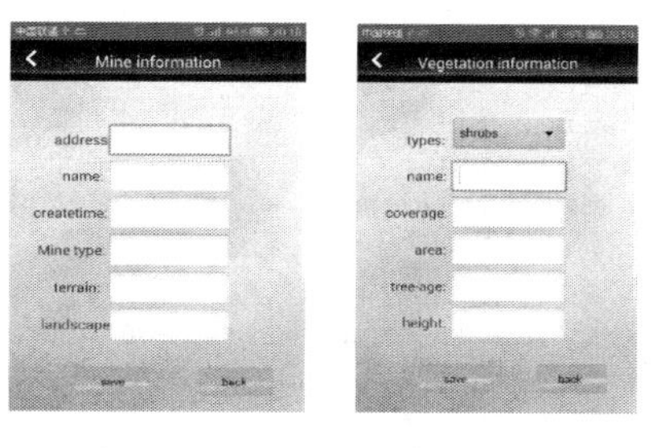

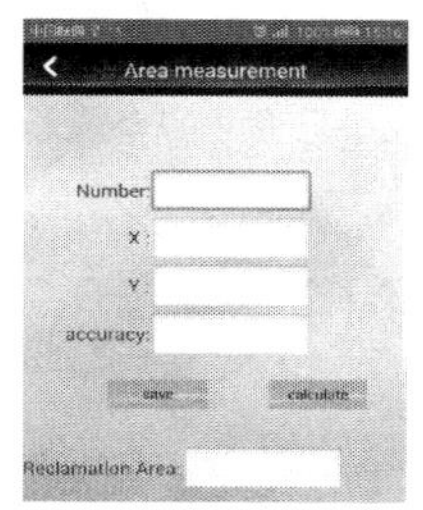

Figure 8. GPS location.
Figure 9. Acquisition.
Figure 10. Mine information.
Figure 11. Vegetation info.
Figure 12. Area survey.
Figure 13. Data upload.

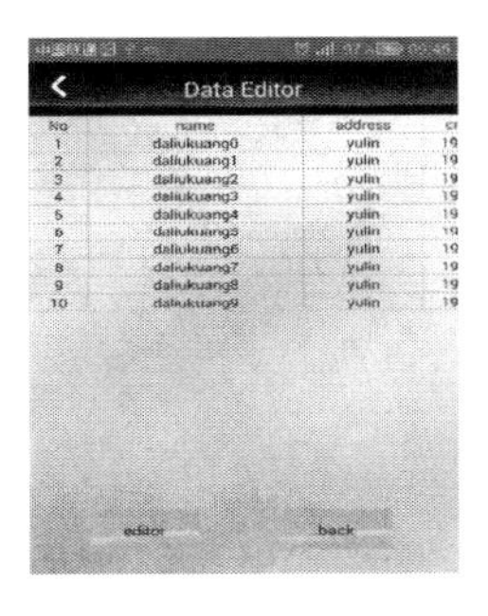

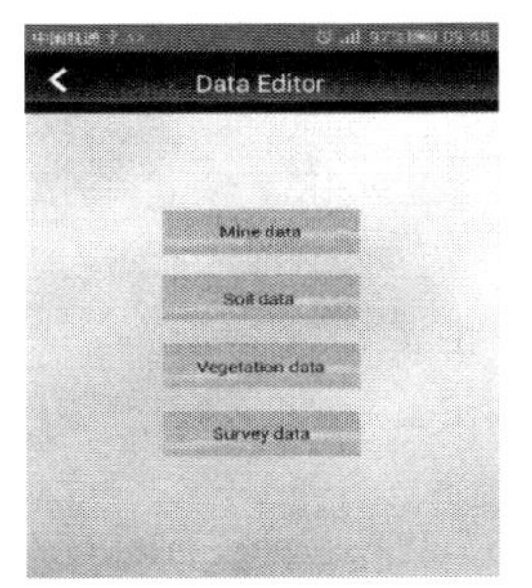

Figure 14. Data editing.
Figure 15. Details data.

5.2 *Implementation of GPS function*

The measurement of area and facilities is the main measurement in the module, click on the GPS function will first determine whether the device is open GPS function, if you do not open, the user is achieved to open GPS, enter the interface and select the button to measure the measurement work. As shown in Figure 8 to 12.

5.3 *The implementation of data editing*

The interface is mainly to collect information for editing, can only collect some information on the basic editing, the coordinates of the measurement can not be modified, can only be deleted, as shown in Figures 13 to 14:

6 CONCLUSION

In the monitoring of mine land reclamation, traditional information collection is the main mode of recording information in the form of book records and sketches, not only convenient and easy to remember. Mixed by mobile GIS acquisition system based on intelligent terminal can effectively reduce the field workload, only need to input the data corresponding to the intelligent terminal, will be upload to the server in real time, a lot of the process of eliminating the intermediate. The most important is that the system can take the time to send photos in the background, let the industry personnel on the site has a more intuitive understanding.

This system also has some shortcomings, such as in some higher requirements on the accuracy of the mining area, must use RTK aided measurement, because the system accuracy can only achieve the handheld GPS accuracy. The general intelligent terminal standby still need to carry spare batteries or mobile power equipment, in order to prevent the electricity shortage effect of operation.

REFERENCES

Bian Zhengfu. Study on land reclamation and ecological reconstruction in coal mining area of China. *Resources Industry,* 2005,7 (2): 18–24.

Feng Yuping, Wang Shuguang. Android based mobile positioning software. *Computer and Modernization,* 2015 (2): 11–13.

Hu Jinshan, Kang Jianrong. Design of land reclamation information system in mining area based on GIS. *Metal mine,* 2010, V39 (12): 113–116.

Huang Minghong, Luo Yongming. Land reclamation and ecological restoration in mining area. *Journal of soil science,* 2003, 40 (2): 161–169.

Liu Shujuan. Monitoring system of land reclamation in mining area [D]. *China University of Geosciences (Beijing),* 2012.

Liu Shujuan. Study on monitoring system of land reclamation in mining area [D]. *China University of Geosciences (Beijing),* 2012.

Wang Li. 3S technology in land reclamation in coal mining area [D]. *Taiyuan University of Technology,* 2013.

Wang Shiyun. Study on the monitoring and control of reclaimed land reclamation in open pit coal mine in Loess Plateau [D]. *China University of Mining and Technology (Beijing),* 2014.

Zhang Jie. Design and verification system of returning farmland to forest based on mobile GIS [D]. *Beijing Forestry University,* 2013.

Zhang Peng, Wang Ning, Yang Zhao. Study on land damage assessment of mining subsidence area based on GIS. *Coal Technology,* 2015 (11): 310–312.

Zhou Yan. The whole life cycle of mine land reclamation supervision system and information technology research [D]. *China University of Geosciences (Beijing),* 2014.

Land Reclamation in Ecological Fragile Areas – Hu (Ed.)
 ISBN 978-1-138-05103-4

Object-based change monitoring in mining areas—taking Pingshuo as an example

M.L. Zhang, W. Zhou & T. Yuan
School of Land Science and Technology, China University of Geosciences, Beijing, China

Y.H. Xie
Center for Urban and Environmental Change, Department of Earth and Environmental Systems, Indiana State University, Terre Haute, IN, USA

Y.F. Li
School of Environment, Beijing Normal University, Beijing, China

ABSTRACT: In this paper, RapidEye Remote Sensing Image (Map 2012) and SPOT7 Remote Sensing Image (Map 2015) in Pingshuo Mining Area are selected as the data. Combined with object-based classification and change vector analysis method, we studied the feasibility of high resolution remote sensing image for mining land classification and the accuracy of monitoring. The results show that the classification of reclaimed mining land has higher precision, the overall accuracy and kappa coefficient of the classification of Map2012 were 0.89 and 0.87, and the change region map were 0.87 and 0.84. It's obvious that object-based classification and change vector analysis which has a great significance to improve the monitoring accuracy can be used to monitor mining land, especially reclaiming mining land.

Keywords: land use/cover monitoring; object-based classification; change vector analysis; open-pit coal mine; land reclamation

1 INTRODUCTION

Mineral resources are not only the main source of energy, industrial raw materials, but also the basis for sustainable economic and social development. Coal as one of the major mineral sources plays an important role in the national socio-economic development (Bai et al. 1999). Scientific monitoring of coal mining areas, especially opencast coal mines, is of great theoretical and practical value for the new normal of China (Zhou et al. 2012).

The traditional method for mine area monitoring which is to take the dynamic monitoring for land resource, requires a lot of manpower and material resources to conduct field investigation of mining areas. And it is no longer applicable to real-time monitoring and data updates for large-scale open-pit mine. Compared with the traditional method, the use of remote sensing in the mining area, especially reclamation land, can build a complete database to facilitate data collation, reduce costs and improve efficiency. Since 1970s, change monitoring method has been applied abroad in low-resolution remote sensing image both in mining area by using pixel-based classification (Boerner et al., 1996). In China, these study began from 1980s, Hu et al. (2005) analyze the land use change in mining area using pixel-based method in Landsat TM and ETM data of Tangshan. Remote sensing change monitoring was used to monitor large-scale areas by using low-resolution remote sensing images in the past, which has lower resolution, less information, and cannot conduct depth analysis of certain areas. So object-based backdating approach was put forward, which achieved higher accuracy for change monitoring (Yu et al., 2016).

Now, object-based technology, especially multi-temporal, has been successfully applied in many studies. Al Khudhairy (2005) did the damage monitoring by using the pan-sharpened IKONOS data, while Chen and Hutchinson (2007) using the panchromatic QuickBird imagery. Bouziani et al. (2010) applied the change monitoring method to map updating by using IKONOS and QuickBird images. The results of many experts and scholars have shown that object-based classification can improve the classification accuracy so as to improve the accuracy of land use change analysis results and provide more reliable technical and data support for mining area monitoring (Leichtle et al., 2016).

In this paper, we combined with object-based classification and change vector analysis method, to study the feasibility of high-resolution remote sensing images for mining land classification and the accuracy of monitoring.

2 MATERIAL

2.1 *Study area*

Pingshuo open-pit mine is located in Shuozhou City of Shanxi Province, which between 112°10′E~113°30′E and 39°23′N~39°37′N. The study area with a total area of 30.43 km², which located in the west of Pingshuo mining area. It is sited in a fragile ecosystem in Loess Plateau, for the strongly eroded and less vegetation and has been reclaimed since 1980s (Bai & Xun, 2008). And the environmental problems here is very serious, which badly needs scientific monitoring and protection.

2.2 *Data sets*

This article uses RapidEye which is employed 5-band multi-spectral (blue, green, red, Near Infrared (NIR), red edge) image with resolution of 4.27 m collected in September 2012, referred to as t0 and SPOT7 which provides both image products at 4 m spatial resolution comprising multi-spectral bands of blue, green, red, and NIR and 1 m panchromatic image collected in August 2015, treated as t0-1. Both optical data sets are available as standard image products, i.e. radiometrically corrected, sensor corrected, and projected to the Earth's surface. In addition, height information is available by means of a Digital Elevation Model (DEM) with resolution of 2 meters in the study area, in order to reduce the effect of terrain on the classification accuracy.

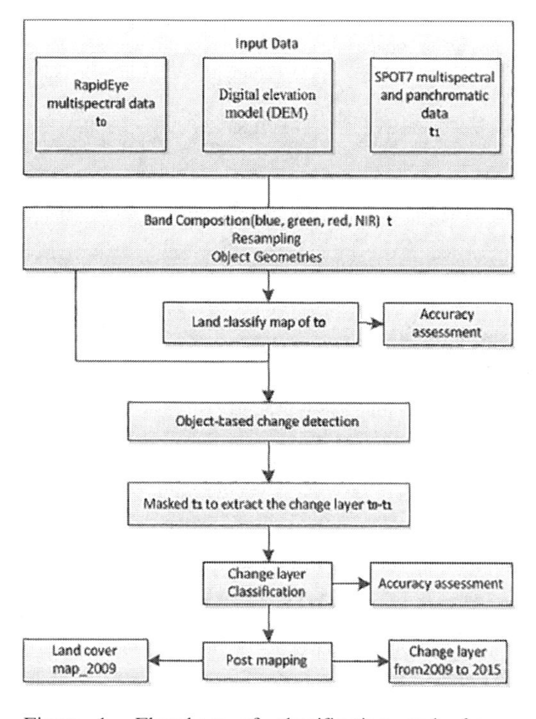

Figure 1. Flowchart of classification and change analysis.

3 METHODOLOGY

After got classification of 2012 images, we used Change Vector Analysis (CVA) to identify areas with changes from 2012 to 2015, and classified these areas with changes. We did not use any ancillary data to aid in classification, and not conduct manual editing.

And before classification, we need to perform band synthesis and image segmentation. The RapidEye and SPOT7 images are roughly consistent in the blue, green and NIR band so that their combination can be benefit to classification. So we choose the above bands of these images to build the data set in order to segment image objects. And we segmented the composition image into objects, and used the multi-resolution segmentation algorithm which was embedded in eCognition software (Baatz & Schäpe, 2000).

There are three values, scale, color and shape which are set to determine spatially adjacent segments, relative weighting of reflectance and shape in defining segments were determined by testing different parameter values and visually interpreting the image segmentation results. Therefore, we set the weights as 100, 0.9 and 0.1, respectively.

Then, we used the random sampling method to select the training sample properly, and classified RapidEye image into 7 classes, bare land, arable land, build-up land, industrial and mining land, forest, water by using Support Vector Machine (SVM) in object-based method.

As we know, CVA has been widely used for land cover change detection (Johnson & Kasischke, 1998) which was applied in this paper based on the classification map of 2012 (Fig. 2 (A)). And the first step of the CVA is to calculate the Change Vector (CV) based on normalized two dates images (Eq. 1), while change magnitude is calculated by Eq. (2).

$$\Delta V = R - S = \begin{bmatrix} r_1 - s_1 \\ r_2 - s_2 \\ \vdots \\ r_n - s_n \end{bmatrix} \quad (1)$$

$$|\Delta V| = \left[(r_1 - s_1)^2 + (r_2 - s_2)^2 + \cdots (r_n - s_n)^2 \right] \quad (2)$$

where R and S are the vectors of RapidEye and SPOT7; n is the number of bands.

The most critical step in CVA is to determine the threshold for land cover change (Chen et al., 2003). In this paper, we use multi-threshold to extract change area, that is, different land use types using different thresholds for extraction.

$$C_i = \begin{cases} |\Delta V_i(x,y)| \geq |\Delta \bar{V}_I| + a_i\sigma_i & \text{change} \\ |\Delta V_i(x,y)| < |\Delta \bar{V}_I| + a_i\sigma_i & \text{no change} \end{cases} \quad (3)$$

where i is a certain land use type; $|\Delta \bar{V}_I|$ is the absolute distance of the change vector of the land use type I; σ is the adjustable coefficient, which general value is 0.0 to 1.5 (Morisette & Khorram, 2000), and the value of this study is 1.5. On behalf of the standard deviation, $|\Delta \bar{V}_I|$ need to use classification map of 2012 to identify the specific type of each image object, then to determine the value σ of each class.

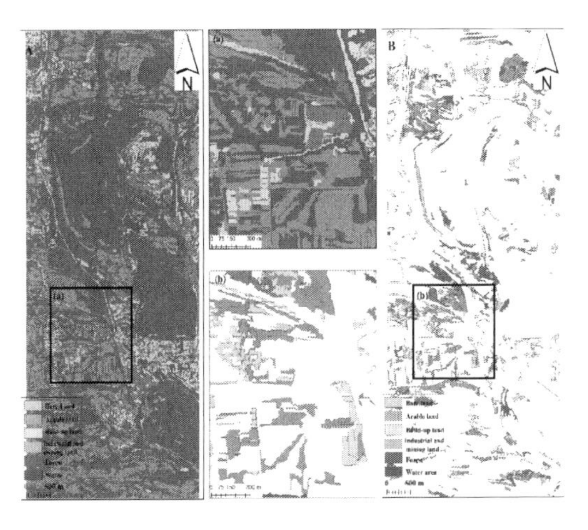

Figure 2 panel A, land cover classification map of 2012; panel B, land use/cover change map from 2012 to 2015.

CVA can extract change regions of all classes from 2012 to 2015 and we can obtain resultant layer of change analysis. Then SPOT7 image of 2015 (Map 2015) was masked by the resultant layer to get change region layer. Then land use/cover change map (Fig. 2B) was classified from change region layer by the same classification method above. With change analysis we can not only obtain land use/cover change map, but also the class hierarchy.

For all the layers, we used a stratified random sampling method to obtain training samples. We generated a total number of 280 sample points for classification map of 2012, with at least 40 samples for each class (Fuller, Smith, & Devereux, 2003). As the land cover change map also had less area, 120 sample points were selected for the accuracy assessment. Furthermore we used reference data created from visual interpretation of high spatial resolution image data which are RapidEye and SPOT7. Error matrices include the overall accuracies, user's and producer's accuracy, and the Kappa statistics.

4 RESULTS AND SUMMARY

The results (Table 1) show that the classification of mining land has higher precision, and the overall classification accuracy and kappa coefficient of Map2012 were 86.67% and 84.00%, and the land cover change map from 2012 to 2015 were 0.87 and 0.84.

Therefore Object-based classification and change vector analysis in high-resolution remote sensing image can not only be applied in urban change detection, but also be used to monitor land reclamation in mining land, and there is no doubt that it improves the monitoring accuracy. As we can see, the environmental disturbance of mining land always has a high degree, even more serious is the fact that mining completely destroys the native ecosystem and reclamation needs to reconstruct a new one to be adapted to local condition. Within the

Table 1. Error matrix for both the land cover classification of Map2012 (p1) and change area (p2) (%).

Land-use type	Accuracies Producer's (p1/p2)		User's (p1/p2)		Overall (p1/p2)		Overall Kappa statistics (p1/p2)	
Bare land	96.67	90.00	87.88	100.00	—	—	—	—
Arable	93.33	95.00	77.78	95.00	—	—	—	—
Build-up	83.33	100.00	100.00	60.61	—	—	—	—
Mining	90.00	75.00	93.10	100.00	—	—	—	—
Forest	83.33	95.00	83.33	100.00	—	—	—	—
Water area	90.00	65.00	100.00	86.67	—	—	—	—
	—	—	—	—	86.67	89.44	84.00	87.33

uncertainty of the adaption of the new ecosystem, it is of great importance that high monitoring accuracy in real time for mining land, especially reclaiming mining land.

REFERENCES

Alkhudhairy, D.H.A., Caravaggi, I., & Giada, S. (2005). Structural Damage Assessments from Ikonos Data Using Change Detection, Object-Oriented Segmentation, and Classification Techniques. *J. Photogrammetric Engineering & Remote Sensing*, 71(7), págs. 825–838.

Baatz, M., & Schäpe, A. (2000). An optimization approach for high quality multi-scale image segmentation. Paper presented at the Beiträge zum AGIT-Symposium.

Bai, Z.K., Xun, W.J. (2008). A case study on Pingshuo Mining Area: Land rehabilititation and reutilization in mining distericts. *J. Resources & Industries*, 10(05), 32–37. (in Chinese).

Bai, Z.K., Zhao, J.K., Li, J.C., et.al. (1999). Ecosystem damage in a large opencast coal mine: A case study on Pingshuo Surface Coal Mine, China. *J. Acta Ecologica Sinica*, 19(6), 870–875. (in Chinese).

Boerner, R.E.J., DeMers, M.N., Simpson, J.W., Artigas, F.J., Silva, A., & Berns, L.A. (1996). Markov Models of Inertia and Dynamism on Two Contiguous Ohio Landscapes. *J. Geographical Analysis*, 28(1), 56–66.

Bouziani, M., Goïta, K., & He, D.C. (2010). Automatic change detection of buildings in urban environment from very high spatial resolution images using existing geodatabase and prior knowledge. *J. Isprs Journal of Photogrammetry & Remote Sensing*, 65(1), 143–153.

Chen, J., Gong, P., He, C., Pu, R., & Shi, P. (2003). Land-use/land-cover change detection using improved change-vector analysis. *J. Photogrammetric Engineering & Remote Sensing*, 69(4), 369–379.

Chen, Z.Q., & Hutchinson, T.C. (2007). Urban damage estimation using statistical processing of satellite images: 2003 Bam, Iran earthquake. *J. Journal of Computing in Civil Engineering*, 21(3), 187–199.

Fuller, R.M., Smith, G.M., & Devereux, B.J. (2003). The characterisation and measurement of land cover change through remote sensing: problems in operational applications? *J. International Journal of Applied Earth Observation & Geoinformation,* 4(3), 243–253.

Hu, Z.Q., Xie, H.Q. (2005). Study on land use/cover change of coal mining area based on remote sensing images. *J. Journal of China Coal Society,* 30(1), 44–48. (in Chinese).

Johnson, R.D., & Kasischke, E.S. (1998). Change vector analysis: A technique for the multispectral monitoring of land cover and condition. *J. International Journal of Remote Sensing*, 19(3), 411–426.

Leichtle, T., Geiß, C., Wurm, M., Lakes, T., & Taubenböck, H. (2016). Unsupervised change detection in VHR remote sensing imagery—an object-based clustering approach in a dynamic urban environment. *J. International Journal of Applied Earth Observation and Geoinformation*, 54, 15–27.

Morisette, J.T., & Khorram, S. (2000). Accuracy assessment curves for satellite-based change detection. *J. Photogrammetric Engineering and Remote Sensing*, 66(7), 875–880.

Yu, W., Zhou, W., Qian, Y., & Yan, J. (2016). A new approach for land cover classification and change analysis: Integrating backdating and an object-based method. *J. Remote Sensing of Environment*, 177, 37–47.

Zhou, W., Cao, Y.G., Bai, Z.K., & Wang, J.M. (2012). Indicators for Monitoring Land Reclamation in Coal Mining Area. *J. China Land Sciences*, 26(11), 68–73. (in Chinese).

Land Reclamation in Ecological Fragile Areas – Hu (Ed.)
© 2017 Taylor & Francis Group, London, ISBN 978-1-138-05103-4

Fast surveying and large scale mapping in the mining area of the delta wing

K.N. Zhang, Z.Q. Yang & J.F. Wu
Academy of Surveying and Spatial Information, Chang'an University, Xi'an, Shaanxi, China

ABSTRACT: Land subsidence and ecological deterioration will be caused by underground coal mining, and a series of geological and environmental would occur frequently. The main mining areas in northwest of China are located in the arid loess mountain area, itself belongs to susceptible area of natural geologic hazard, and often happens mining subsidence. Fast large scale mapping of mine is crucial for disaster relief when the mine geological environment and disasters. The traditional photogrammetry cannot be responded the condition of real-time for high cost and long cycle. In recent years, low altitude remote sensing system develops rapidly based on sensor of UAV, trike, multi rotor aircraft by their low cost, miniaturization, customization, speed and other advantages. This paper used delta wing to survey the Ningtiaota Coal Mine in northern shaanxi. The research got DSM and DTM of mine by image control point and aerotriangulation, which comply with the requirements of the measurement accuracy, In this study, a fast mapping method of large area and large scale topographic map was developed to improve the monitoring, prediction and evaluation of land and environmental damage in mining area.

Keywords: Delta wing; Mining area; Fast surveying and mapping; Image matching

1 INTRODUCTION

Most mining areas of Northwest China are located in the semiarid loess gully area, its terrain and geomorphology are complex. It caused a series of geological and environmental disasters because itself belonged to prone area of natural disasters, and there were large area collapse caused by coal mining. The traditional satellite remote sensing and photogrammetry technology are limited by data collection capacity and poor current, cannot gain map of nature disasters timely. Low altitude photogrammetry System takes advantages of the flexible, efficient and low cost for obtain mining area image timely, it has been widely used in various engineering fields (Schmid C *et al.*, 2008). LeicaADS80 was used to surveying and mapping for Gansu Province (Wang J, 2012); UAV was used to monitor the environment of Zhangjiakou's important tailing (Hong Y F *et al.*, 2013); low altitude aerial technology was served to make basic scale topographic map of cities with large area (Li L *et al.*, 2012). However, there was little research that used low altitude photogrammetry system to survey large scale map. This paper surveyed map of Ningtiaota mine by delta wing in Shaanxi province, used SSK digital photogrammetric workstation for implementing aerial triangulation to gain DSM and DTM, and provided a mapping method for large scale topographic map of mine.

2 EQUIPMENT AND DATA

2.1 *Delta wing*

Delta wing aircraft is a kind of ultra light aircraft which has been developed rapidly in recent years. It is made of wing, fuselage, propulsion device, flight control devices and other components, the wing is lifting surface and Control surface. In this paper, the model of the power triangle wing XT912 is used, and its main technical parameters are shown in Table 1.

Delta wing equipped with PHASE IQ 180CCD camera, lens focal length of 55 mm and pixel 80 million, its shooting area was about 10 km, flying height was 500 m. The flight course was

Table 1. Technical parameters of delta wing.

Engine model	ROTAX912UL2	Trolley length	2.745 m
Wing model	STREAK2	Aircraft altitude	3.65 m
Wingspan	9.97 m	Tread	1.7 m
Wing Weight	49 kg	Maneuvering speed	80 knot/h
Wing area	15 m^2	Trim Speed	50 knot/h
Angle	130°	Maximum wind resistance	20 knot/h

designed 65% of course overlap and 55% lateral overlap, flied 6 trips, and gained 282 images.

2.2 *Image control points*

Ningtiaota Coal Mine is located in the central area of Shenmu county in Shaanxi Province, its width is about 9.9 km and length is 19.5, area is about 136 km^2. Due to the complex terrain in the survey area, multiple feature objects were selected as control points in the rage of each preseted control point, and many points were reserved, in order to avoid outdoor workers cannot reach and ensure specific position. In this paper, a total of 46 ground mark points were set up before the flight, and GPS were used to measure the spatial coordinates of the control points.

2.3 *Camera calibration*

In order to satisfy the data processing of the camera geometric calibration, a plurality of control points were selected in every test ground (Lower DG, 2004). For the view camera, 35 control points were often chose, and every control point was measured with point group (three points or more). Spatial model was established after points selection, and every point was measured. The paper used GPS network to measure by near trigonometrical station and checked the result. Finally, the camera was calibrated by the way of online camera calibration.

3 AEROTRIANGULATION

3.1 *Relative orientation and image matching*

This paper adopts the way of image matching to improve the accuracy. Delta wing is affected by wind, weather, shock and camera motion when it is flying, would cause a serious of problems, and the traditional image matching method cannot process data well (Zhang YJ *et al.*, 2011). Therefore, this paper proposes an improved SIFT matching algorithm to ensure precision. Harris corner is detected in the scale space firstly, then the key point is located precisely by Forstner which has accuracy of sub-pixel. Finally, 128 dimensional feature vector descriptor of traditional SIFT is simplified to 64. Then BBF-KD first matching method and second precise matching are used in the phase of feature points matching. Matlab2010 is used to extract and match points of SIFT and the presented method respectively from images of delta wing in Ningtiaota mine. As the experimental results, the green lines represent correct matching points, and the red lines represent error (Figure 1 & Figure 2). The more features are extracted by the proposed method, matching accuracy and speed are compared with the traditional SIFT algorithm have obvious advantages (Table 2).

3.2 *Absolute orientation*

This paper presents a method of absolute orientation regarding to large rotation angle and large scale, uses the information of data to orient free network. Firstly, the point set is processed of bar centralization and scale normalization, two scale parameter are kept the same; then four parameters and translation parameters are calculated by least squares iterative; seven parameters are obtained for absolute orientation finally.

Figure 1. Match results of the sift.

Figure 2. Match results of this method.

Table 2. Comparison the experimental results of SIFT and this method.

Operator	Left IP	Right IP	MP	Error MP	Correct MP rate	T/s
SIFT	390	329	168	18	89.29%	21.368
This method	1055	971	388	16	95.88%	19.273

IP: images points; MP: matching points; T: time.

4 STEREO MAPPING AND PRECISION ANALYSIS

4.1 *DSM & DTM*

Based on the results of aerotriangulation, the KKS mapping workstation is used to carry out the stereo mapping of the study area. Considering the data of six air routes, there may be the difference of model edge, then stereo map is surveyed by different model boundaries, and checked with points of outdoor annotation to build a stereo mapping model of Ningtiaota mine: DSM & DTM as shown in Figure 3.

4.2 *Accuracy analysis*

In this research, 165 planar points and 155 elevation points are selected to analyze the accuracy of the mapping (Table 3). The horizontal accuracy is superior to the elevation accuracy, this reason is mainly caused by non-metric camera of delta wing (De la Escalera A et al., 2010). The camera has been obtained the distortion parameters after checking, although camera is impacted during flying and transport to cause error. The delta wing images of small picture and short baseline, they are more smaller when be affected by wind, and influence and on elevation accuracy. Mean error of plane shall not be greater than 1.2 m, and error of elevation can not be more than 0.5 m, according to the national standard, accuracy for plane and vertical is consisted with requirements, to achieve a high precision digital mapping of mine.

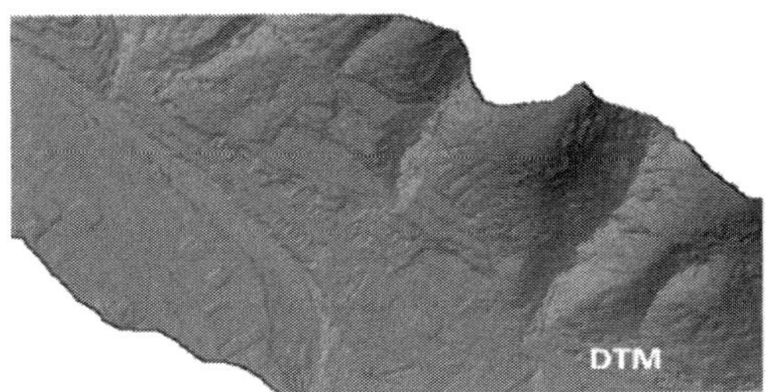

Figure 3. DTM & DSM (local enlargement).

Table 3. Accuracy of map (m).

	Mean error	Maximum error
x	± 0.072	0.210
y	± 0.093	0.206
Plane	± 0.115	0.308
Elevation	± 0.355	0.575

5 SUMMARY

In this study, a fast mapping method of large area and large scale topographic map was developed to improve the monitoring, prediction and evaluation of land and environmental damage in mining area. We used delta wing to survey Ningtiaota mine in Northern Shaanxi. SSK All Digital Photogrammetry Workstation was used for aerotriangulation; the improved SIFT algorithm was designed for relative orientation; big rotation angle and large scale was considered to absolute orientation. This paper got the DSM and DTM of mine, which accuracy was consisted with requirements, achieved a high precision digital mapping.

REFERENCES

De la Escalera A & Maria Armingol M. (2010) Automatic chessboard detection for intrinsic and extrinsic camera parameter calibration. *J. Sensors,* 10 (3), 2027–2044.

Hong Y F, Ysng Y P & Yang H J. (2013) Application of DP Grid-LAT in the processing of UAV remote sensing image for Tailings Pond. *J. Bulletin of Surveying and Mapping,* (11), 71–73.

Li L, Zhang H T & Li B. (2012) Application of UAV image in large area 1 - 2000 topographic map. *J. Bulletin of Surveying and Mapping,* (Z1), 457–461.

Lowe. D.G. (2004) Distinctive image features from scale-invariant keypoints. *J. International Journal on Computer Vision,* 60(2), 91–110.

Schmid C, Mohr R & Bauckhage C. (2008) Evaluation of interesting point detectors. *J. International Journal of Computer Vision,* 37(2), 151–172.

Wang J. (2012) Application of ADS80 aerial photogrammetry system in basic surveying and mapping of Gansu. *J. Bulletin of Surveying and Mapping,* (S1), 433–435.

Zhang Y J, Xiong J X & Hao L J. (2011) Photogrammetric Processing of Low-altitude Images Acquired by Unpiloted Aerial Vehicles. *J. The Photogrammetric Record,* 26 (134), 190–211.

Land Reclamation in Ecological Fragile Areas – Hu (Ed.)

Study on the limit angle of mining subsidence basin based on the field survey data in the mining area

Z.Z. Lian, R. He, H.B. Zhang & C.Y. Li
School of Surveying and Landing Information Engineering, Henan Polytechnic University, Jiaozuo, Henan, China

ABSTRACT: Study on the limit angle of mining subsidence basin can reflect the influence on environment due to the underground coal excavation. In this paper, a large number of observation station data of ground movement in mining filed had been collected in order to study the relationship between the limit angle of mining subsidence basin and unconsolidated layer thickness and bedrock thickness. Based on these field survey data, the regression relationships between the trend limit angle, the downward limit angle, the upward limit angle and the ratio of unconsolidated layer thickness and bedrock thickness have been established. The conclusions showed that with the increase of unconsolidated layer thickness and bedrock thickness, the limit angle decrease. It provides a new way to determine the limit angles in mining areas.

1 INTRODUCTION

China is the largest producer and consumer of coal in the world and is the largest user of coal-derived electricity, generating an estimated 73% of domestic electricity production in 2014 from coal. So coal plays a very important role in China. It is known that in the process of underground coal mining, the original stress balance status of the mining work face area was broken when the underground useful minerals excavated, and then the stress redistributes and reaches a new equilibrium. In this process, the underground strata and ground surface are forced to produce a continuous move and deformation, and a non-continuous damage. After the mining influence spread to the surface, the surface affected by mining subsidence begins to move down, resulting in a mining subsidence basin which the area is much larger than the mining work face area. The regulation of coal mining under buildings, railways and water-bodies specify the boundary points with 10 mm displacement value in vertical direction and consider the different rock types to determine the limit angle of mining subsidence basin which is from 50° to 65°. The limit angle of mining subsidence basin is a most important parameter for the underground mining subsidence prediction. However, this method is not suit to the mining area with thick unconsolidated layer and thin bedrock. It can reflect not only the influence degree of underground mining on the surface, but also the influence on environment due to the underground coal excavation. And the thickness of the thickness of soil layers and bedrock directly affects the limit angle size. So, it is necessary to explore the impact of unconsolidated layer thickness and bedrock thickness on the limit angle of mining subsidence basin.

2 INFLUENCE FACTOR OF LIMIT ANGLE

a. Character of overlying lithology. It plays a very significant role for the overlying strata movement and deformation law and the size of limit angle. According to the field mining data and statistics, the surface movement and deformation is closely related to the overlying lithology, which more hard overlying rock, higher strength and limit angle value. Generally, the stiffness of overlying strata can be express by the mean value of coefficient of bedrock. The calculation formula is:

$$f = \frac{\sum_{i=1}^{n} m_i R_i}{10\sum_{i=1}^{n} m_i} \tag{1}$$

where m_i is the normal thickness of ith overlying strata, m; R_i is the uniaxial compressive strength of i-th overlying strata, MPa.

b. Unconsolidated layer thickness. The deformation and movement laws in unconsolidated layer thickness and bedrocks are different. Compare with the bedrock, the unconsolidated layer plays an important role for the size of limit angle. The deformation value in unconsolidated layer is bigger; the limit angle size is smaller. So the ground movement range is large due to the underground mining.

c. Underground coal mining depth and thickness. Mining depth and coal thickness are the main factors to influence the surface movement range. The research results of Soviet Union showed that mining depth is deeper, surface movement and deformation values are greater, and the movement course is more dramatic. So the range of ground movement becomes large and the limit angle of surface subsidence basin turns down.
d. Other factors. Except the influence factors for the size of limit angle that discussed above, there are some other influence factors also can impact the size of the limit angle, such as angle of coal seam, roof manage methods, mining procedure and time, extent of mining and mining coefficient.

3 RELATIONSHIP BETWEEN UNCONSOLIDATED LAYER THICKNESS AND LIMIT ANGLE

3.1 *Mining area*

The mining area located in the Huaibei Plain West, most of the area is flat terrain except the few parts of low mountains and monadnocks which the altitude is from +180.0 m to +350.0 m, and the natural ground elevation is generally between +20.0 m ~ +40.0 m. This mining area span Huaibei coal field and Huainan coal field, and all of the bedrock layers are covered by Cenozoic unconsolidated layers which followed the order from old to new for the Cambrian, Ordovician and Carboniferous, Permian and Cenozoic. Coalfield is a network of patterns that composed of a series of nearly East-West and North faults, which the nearly East-West fault contains Subei fault in the center part and Banqiao fault in the south part, and the North-East fault contains Guzhenchangfeng fault in the Eastern, Fengwo fault in the middle part and Xiayigushi fault in the West part. In order to study the law of ground movement due to mining, a number of 10 observation lines had been established in this mining area, and a complete ground movement data had been collected.

3.2 *Feature analysis between trend limit angle and the ratio of unconsolidated layer thickness and bedrock thickness*

The trend limit angle is one of the main parameters to describe the ranges of surface movement in the mining area. According to the analysis of field mining measured data, one regression formula had been established based on the relationship among the trend limit angle, unconsolidated layer thickness and bedrock thickness. The regression formula is as follows:

$$\delta_0 = 3.7001 \cdot x^2 - 15.379 \cdot x + 73.556 \quad (2)$$

where regression coefficient squared (R2) is 0.9034, x is the ratio of unconsolidated layer thickness and bedrock thickness;

From the results showed in the above chart, it can be known that the curve similar to a parabolic shape, and it also show that maximum difference between measured and fitted values is –2.05, regression coefficient squared is 0.9034, relative error up to 3.59 and the absolute error is 2.05. Meanwhile, it proves that the impact of the ratio of unconsolidated layer thickness and bedrock thickness on the size of trend limit angle is big and should not be ignored.

3.3 *Feature analysis between downward limit angle and the ratio of unconsolidated layer thickness and bedrock thickness*

The downward limit angle is a factor to reflect the surface subsidence scope, and it influences the calculation results of subsidence range. Based on the data that collected from the mining area, the regression among downward limit angle and the ratio of

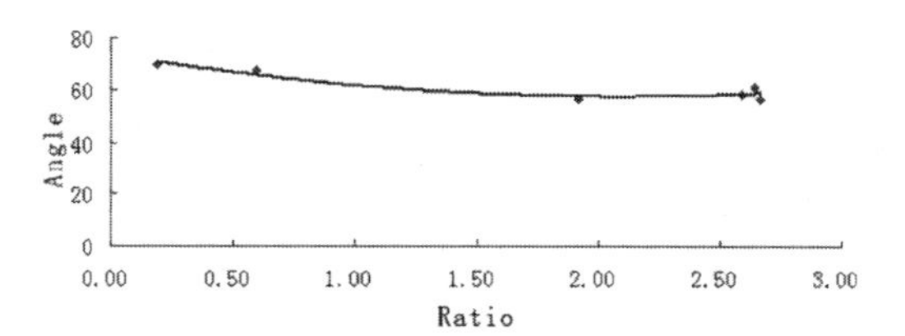

Figure 1. The relation graph of the trend limit angle and the ratio between surface soil thickness and bedrock thickness.

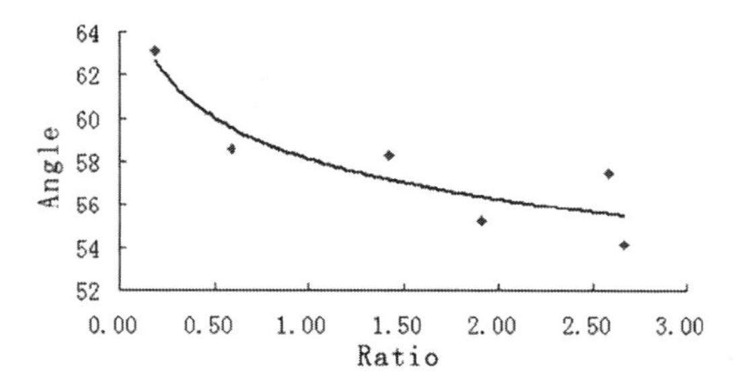

Figure 2. Relation graph of the downward limit angle and the ratio between surface soil thickness and bedrock thickness.

unconsolidated layer thickness and bedrock thickness had been established. It is as follows:

$$\beta_0 = -2.7294 \cdot \ln(x) + 58.137 \tag{3}$$

where regression coefficient squared (R2) is 0.8105, x is the ratio of unconsolidated layer thickness and bedrock thickness;

By the above chart, it can be got that the relationship between downward limit angle and the ratio of unconsolidated layer thickness and bedrock thickness is approximate to a logarithmic function. And it also shows that maximum difference between measured and fitted values is −1.96, regression coefficient squared is 0.8105, relative error up to 3.41 and the absolute error is 1.96.

3.4 *Feature analysis between upward limit angle and the ratio of unconsolidated layer thickness and bedrock thickness*

The upward limit angle is a parameter that can reflect the ground mining subsidence range above uphill direction of the mining work face. According to the analysis of the mining information, the regression among upward limit angle and the ratio of unconsolidated layer thickness and bedrock thickness had been established. It showed as follows:

$$\gamma_0 = -7.6905 \cdot x + 74.378 \tag{4}$$

where regression coefficient squared (R2) is 0.8433, x is the ratio of unconsolidated layer thickness and bedrock thickness;

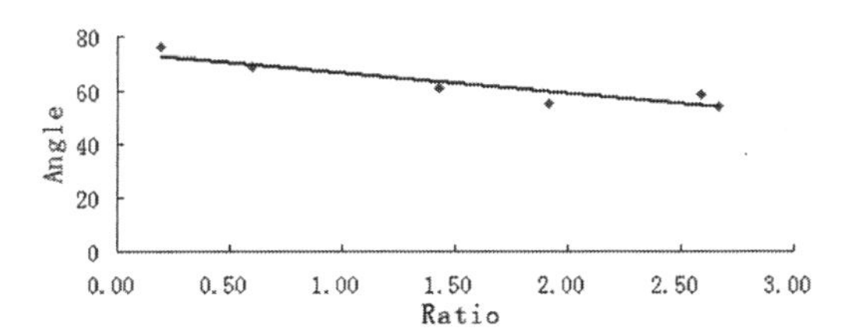

Figure 3. Relation graph of the upward limit angle and the ratio between surface soil thickness and bedrock thickness.

From the chart, it can be seen that the relationship between upward limit angle and the ratio of unconsolidated layer thickness and bedrock thickness is approximate to a line function. And it also shows that maximum difference between measured and fitted values is 4.37, regression coefficient squared is 0.8433, relative error up to 7.33 and the maximum absolute error is 4.37.

4 CONCLUSION

According to the large amount of relevant mining area's field observation station data of ground movement, the regression relationships between the trend limit angle, the downward limit angle, the upward limit angle and the ratio of unconsolidated layer thickness and bedrock thickness have been established. It is useful for the underground mining subsidence prediction and the mining safety. At the same time, it proves that with the increase of unconsolidated layer thickness and bedrock thickness, the limit angle decrease. The results confirmed assumptions about the influence of unconsolidated layer thickness and bedrock thickness for the size of the limit angle.

ACKNOWLEDGMENT

The authors would like to thank the Doctoral Scientific Fund Project of Henan Polytechnic University (B2017-10) and NSFC (NO. 41671507) for the finance support.

REFERENCES

He Guoqing, Yang Lun (1994). *Mining Subsidence*, Xuzhou: China Univercity of Mining and Technology Press.

Huang Le-ting (2003). Research and development of mining subsidence mechanism. *J. Coal Science and Technology*, 31(2): 54–56.

Peng S S (1992). Surfaee subsidence engineering soeiety for mining. Colorado, Metallurgy and Exploration Ine.

State Coal Industry Bureau (2000). Buildings, water, rail and major pillar for mine coal mining regulations and the pressure. Beijing: Coal Industry Press.

Land Reclamation in Ecological Fragile Areas – Hu (Ed.)
© 2017 Taylor & Francis Group, London, ISBN 978-1-138-05103-4

Formation mechanisms and characteristics of ground collapse during shallow coal seam mining

E.K. Hou
College of Geology and the Environment, Xi'an University of Science and Technology, Xi'an, Shaanxi, China

Q.H. Zhang
Qinghai Bureau of Coal Geological Exploration, Xining, Qinghai, China

Y.N. Xu
Xi'an Center of China Geological Survey, Xi'an, Shaanxi, China

X.Y. Che, Z.G. Shou & Z.C. Ji
College of Geology and the Environment, Xi'an University of Science and Technology, Xi'an, Shaanxi, China

ABSTRACT: Shallow seam mining is the main way that resources are extracted from Shenfu mining area. As a result, ground collapse induced by coal mining is very serious in this area, leading to serious damage to water resources, the eco-environment, buildings, and casualties. Taking the Shenfu coal mining area as an example, this article classifies and summarizes the characteristics of different kinds of ground collapse and their basic characteristics. According to the morphological characteristics, ground collapses can be subdivided into six types, namely collapse crater, collapse slot, crack, collapse basin, loess collapse, and loess landslide. Besides, three types of collapse caused by coal mining, fully-mechanized coal mining subsidence, room-and-pillar ground collapse, and other kinds of mining subsidence, are divided based on the relationship between ground collapse and the method of coal mining. What's more, the mechanisms of ground collapse formation in the case of fully-mechanized longwall and room-and-pillar mining are discussed.

Keywords: The Shenfu mining area; Ground collapse; characteristics; Formation mechanisms; Shallow seam

1 INTRODUCTION

Ground collapse induced by coal mining is a critical kind of geological disaster affecting mining areas, especially when exploiting shallow seams. This issue will be discussed in the context of the Jurassic coalfield in northern Shaanxi Province, an area with famous globally for its high-quality coal and favorable mining conditions.

The Shenfu mining area is located in the north of the Jurassic coalfield, itself in northern Shaanxi Province. Mines in this area were first developed in the late 1980s, the earliest shallow coal seams to be exploited within the northern Shaanxi Jurassic coal field. At present, ground collapse within this mining area is serious, induced by coal exploration, and has led to serious damage to water resources, the eco-environment, and buildings. Although previous studies have been carried out on the mechanisms and destructive effects of the movement of overlying strata, the local geological environment and hazards, as well as ecological restoration subsequent to coal mining and the retention of water, limited research has been carried out to date on the mechanisms of formation and characteristics of ground collapse (Fan LM et al. 2015, Shi PW & Hou ZJ 1996, Li WP et al. 2000, Xue DJ 2015, among others). Thus, the aim of this paper is to discuss these issues in relations to the ground collapse of shallow coal seams.

2 GROUND COLLAPSE CHARACTERISTICS

The eastern and western sides of the Shenfu mining area are marked by the Shenfu mine as well as Gushanchuan and Tuweihe, while the southern and northern margins are comprised of Shuimohe and the boundaries of Shaanxi Province. Thus, this total mining area encompasses 3,678 km^2. The coal-bearing strata within Shenfu mine comprise eight layers, numbered from one to eight from the top down. Recoverable coal has been found within layers three to seven, which consist of Middle and Lower Jurassic rocks of the Yanan group. Within these units, coal seams are buried at depths between 20 m and 180 m, numbered 1^{-2}, 2^{-2}, 3^{-1}, 4^{-2}, and 5^{-2}, and have average thicknesses of 10.28 m. Taking into account the burial depths of these coal seams as well as the present exploitation situation, to the south of Daliuta and to the north of Sunjiacha the main minable seams are 1^{-2} and 2^{-2}, respectively, at depths between 30 m and 170 m, while to the south of Sunjiacha and Majiata the main minable seams are 2^{-2} and 3^{-1}, respectively, at mining depths between 30 m and 120 m. Similarly, within the Dianta and Yongxing regions, the main minable seams are 3^{-1} and 5^{-2}, respectively, at mining depths between 40 m and 110 m. The Dachanghan-Laogaochuan-Sandaogou mine in Shenmu county comprises two main minable seams, 2^{-2} and 3^{-1}, at mining depths between 20 m and 110 m, while the main seams within the towns of Miaogoumen and Xinmin are 3^{-1} and 5^{-2}, respectively, at mining depths between 30 m and 150 m. There are numerous large-to-medium-sized mines within the Shenfu mining area that have both high output and are strongly susceptible to ground collapse. These mines are characterized as being within a severely affected zone of the geological environment (Fan LM et al. 2016).

The morphological characteristics of ground collapse are subdivided into six types, including collapse crater, collapse slot, crack, collapse basin, loess collapse and loess landslide.

1. To form a collapse crater, the ground above a mine appears to form an inverted conical funnel and an oval collapse crater. These craters are seen in the Shenfu mining area in places where shallow coal seams are covered with thin bedrock. These inverted conical funnels generally take the form of circles with reduced inverted cross-sections or elliptic collapse craters with ground diameter of 10 m or more, and can reach from 12 cm to more than 10 m in depth. One example is seen in the vicinity of Shendong Daliuta mine; in working face 1203 of this mine, an inverted conical funnel is present in loose strata. The ground diameter of an oval collapse crater is generally from 12 m to more than 100 m, attaining depths between 12 cm and more than 10 m, and with a nearly vertical pit wall. A structure matching this description formed in the summer of 1995 when the ground above the Shuijingqu mine collapsed (Fan LM & Yang HK 2000).
2. Collapse slots form in ground above shallow coal seams covered with thin bedrock within the Shenfu mining area. The formations of these slots are structurally similar to the formation of graben, as loose strata crack on both sides and the middle sinks. These structures can be 1 m up to greater than 10 m wide, while their sink depths range between greater than 10 cm to more than several meters. Such collapse slots can also range between 1 m to more than 10 m wide, while sink depths are generally more than 10 cm to more than to several meters. In the Shendong Daliuta mine, for example, first weighting and periodic weighting of working face 1203 first weighting and the coal wall corresponded with the ground and rear loose strata occurred in the collapse slot. This meant that a collapse slot occurred in Qianliang village, within the town of Dianta, in Shenmu county.
3. Cracks are common ground collapse features induced by coal within the Shenfu mining area. In general, the length a crack can range between a few meters and a few hundred meters, while their widths can range from a few centimeters to tens of centimeters. Based on the scattered characteristics of both sides of cracks, these features can be further divided into three, normal stepped, negative stepped, and those with no obvious caving. Of these, the tendencies of normal stepped cracks are consistent with slope, while the tendency of negative stepped cracks contradict slope, and have a roughly parallel arrangement. The height of these steps is commonly between few centimeters up to 1 m. However, in areas where the terrain is relatively flat, such as in the middle of a working face, the tendency of a normal stepped crack will contrast with the direction of the working face. Thus, almost on both sides of the working face, roadway cracks will confirm to a more inclined column stepped distribution, that tends to step along the working face, while on both sides of the working face far from the lateral side of the roadway, cracks will develop parallel to the working face and there will be stepped cracks on both sides. These cracks have a tendency that follows the working face. This type of crack formation was seen, for example, following exploitation of the Shendong Daliuta mine working face 12609. However, when the terrain of a more strongly sloped zone is cut, both normal stepped and negative stepped cracks can be seen, such as in the Fuyu mine when ground collapse was induced by coal. This took place within the loess loose strata sloped zone and negative stepped cracks developed in the vicinity of the town of Xinmin in Fugu county. In this area, the town of Fugu No. 2 mine experienced ground collapse induced by coal in the loess loose strata slope zone. As a result, normal stepped cracks developed in the town of Xinmin in Fugu county.
4. Collapse basins can form as the result of the style of coal cutting, for example when a shallow coal seam covered with thin bedrock was exploited in the Shenfu mine. In this case, the scope of the mine was larger than working face. Indeed, when terrain is relatively flat, the central basin generally has a flat bottom, and the surrounding working face presents a gentle

slope pattern caused by collapse. The depth of collapse is usually between 1 m and 3 m. One example of an elongated collapse basin was seen in the Shendong Daliuta mine where working face 1209 experienced ground collapse to a depth of 2.9 m.

5. Loess collapses often occur when coal is cut within a strong gully slope zone above thin bedrock. In these situations, coal is mined from a covered by loess terrain, and the horizontal sliding distance is slightly larger than vertical distance. This can lead to block slippage leaving the original structure of the loess, while in other situations the original structure can be completely lost, such as in the collapse area of the Xichagou mine. In this example, a gully edge developed into a loess landslide within the town of Xinmin town in Fugu county. In this case, the posterior step border attained a height between 4 m and 5 m, a width between 30 m and 50 m, and a length between 80 m and 90 m.
6. Loess landslides take place when cutting takes place in strong gully steep slope zones above thin bedrock. In these areas, coal is mined from an area covered by loess terrain. Collapsed materials in these examples comprise large blocks and the original loess structure is completely lost. Examples of these landslides are seen in the loess gully regions of old coal mining areas and have occurred many times around the town of Xinmin in Fugu county.

Based on the relationship between ground collapse and the time required for coal mining, collapses within the Shenfu mining area can be classified into two distinct types.

1. The first type of collapses is those that take place over short time intervals after mining is completed. In this case, ground collapse occurs at the mining working face during, and shortly after, exploitation. Following enlargement of the scope of the working face as it is mined out, periodic damage to overlying strata during the mining process, and the short time period, strata are damaged for a few minutes to a few days after the overlying loose strata are affected. This causes ground collapse, and forms collapse funnels, slots, cracks, collapse basins, loess collapses, and loess landslides. Thus, this kind of collapse will be quick in areas with thin bedrock, and slow in areas where bedrock is thick. These events generally tend to occur shortly after mining out is complete; examples include all of the fully-mechanized longwall mines belonging to the Shendong company that have collapsed. Indeed, small mines within the Shenfu mining area that utilize the room-and-pillar technique also frequently experience ground collapse induced by coal. For example, on November 16th, 2005, at about 12:00, a roof fall accident took place in the Fuyu mine within the town of Xinmin in Fugu county. In this case, a ground area of 110,000 m^2 collapsed, while the next day, on November 17th, 2005, at 21:50, a large roof fall accident again took place, expanding the area of ground collapse to 260,000 m^2.
2. The second type of collapses are those that take place over long periods of time after mining out is completed. These ground collapses occur after intervals that range between a few months to several years after mining out is complete. One example of such a collapse took place at 12:00 on October 14th, 2004, when a mined-out area of a mountain above the jointly-organized mine at Gaoshiya experienced ground collapse one year after mining was complete.

Thus, according to the relationship between ground collapse and mining methods, ground collapse events within the Shenfu mining area can be divided into three types.

1. The first of these collapses occur as the result of fully-mechanized coal mining subsidence. These are common in the Shenfu mining area as most oversize mines adopt a comprehensive mechanized longwall mining approach based on the overall height of roof caving. The key technical parameter that characterizes a fully-mechanized working face is that it has a continuously advancing length between 1,000 m and 6,000 m and a working face length of between 120 m and 350 m. At the same time, mining height can be between 4 m and 6 m. Thus, in cases when the mining coal seam is covered with thin bedrock, the ground forms a cutting collapse slot, while the internal part and the edge of the basin a normal stepped crack often develops, along with small amounts of inverted conical collapse funnels and collapse slots. In these situations, collapse basin sinkage can be up to 3 m; based on the relationship between collapse and goaf, such events mainly take place over short time intervals after mining out is complete.
2. The second of these ground collapses occur as the result of room-and-pillar mining. These are also common within the Shenfu mining area as most small mines adopt the room-and-pillar mining technique. In these cases, haulage and ventilation roadway spacings are generally between 80 m and 120 m, while all along the seam floor, the area for mining is accessed via roadway and the roof management method is used for coal pillar supports fixed with wooden bolts. These kinds of collapses also take place when shallow seams covered with thin bedrock are mined; in these cases, collapse areas

develop normal stepped and negative stepped cracks, as well as collapse slots, loess collapse, and loess landslides. These collapses take place suddenly after long time intervals when mining out is competed, depending on the relationship between the cause and the timing of goaf. The main hidden dangers of these long-time interval sudden collapses following mining out that large numbers of room-and-pillar spaces have often been mined out across a given area.

3. The third kind of ground collapses refer to other mechanisms of mining subsidence. These occur in oversize mines that do not utilize a fully-mechanized longwall mining area, but which use a continuous shear short wall mining method. Examples of these smaller mines, such as Yanghuopan and Nanliang mine, have been adapted to a longwall and double-longwall layout, and carry out coal mining by intermission. This means that all of the kinds of collapse discussed above can take place once mining out is completed, as the working face can be as long as 200 m and the average mining height is 2.2 m. The mined-out area caused by those mining methods are mainly subject to short time interval collapses, encompassing relatively small scale ground collapse basins, cracks, loess collapse, and loess landslides.

3 FORMATION MECHANISMS OF GROUND COLLAPSE

3.1 *Ground collapse in fully mechanized mines in thin-to-medium-thick bedrock*

3.1.1 *Examples of ground collapse*

Three common bedrock thicknesses are encountered within the Shenfu mining area, including special thin bedrock (less than, or equal to, 30 m), thin bedrock (between 30 m and 70 m), and medium-thick bedrock (between 70 m and greater than 100 m). All of these situations mean that the thickness of bedrock is relatively small above the main mining coal seam; generally, this thickness ranges between ten meters and tens of meters, while, at the same time, the overlying thickness of loose strata encompasses a range of different sizes.

An example of a mine with especially thin bedrock is fully-mechanized face 1203 of the Daliuta mine. This operation is mining coal seam No. 1^{-2} that has an average thickness of 6.3 m, is nearly horizontal, and is buried to depths between 50 m and 60 m. At the same time, the roof bedrock thickness is between 15 m and 40 m while the loose strata on top of this roof bedrock is between 15 m and 35 m thick. The roof management method in this mine relies on all collapse, while the length of the working face is 150 m and the mining height is 4 m. The ratio between bedrock thickness and mining height (i.e., referred to as the base-mining ratio) is between 3.75 and 8.75. In this mine, a spindle collapse crater appeared in the ground during the process of mining the coal seam, along with an inverted conical funnel and a stepped crack. The largest sink that formed in this area had a depth of 2.591 m, while the cracks in the ground exhibited ratios between the overlying strata (including loose strata) and mining height (referred to as the ground crack-mining ratio) between 7.5 and 18.75.

An example of a mine with thin bedrock is fully-mechanized face 12609 of the Daliuta mine. This operation is mining coal seam No.2^{-2} that has an average thickness of 5.08 m, is nearly horizontal, and is buried to depths between 80 m and 130 m. In this mine, the thickness of the roof bedrock is between 38.84 m and 65 m, while the loose strata on top of the roof bedrock is between 10 m and 87.93 m thick. The working face of this mine is 240 m in length, while the advance length is 5,658 m, the mining height is 4.72 m, and the base-mining ratio is between 8.23 and 13.77. Stepped cracks mainly appeared around this mine during exploitation of the coal seam at spacings between 5 m and 6 m, while the step fall is between 10 cm and 1 m. The ground crack-mining ratio of this mine is between 10.35 and 32.40.

An example of a mine with medium-thick bedrock is working face 32202 of the Bulianta mine. This mine has a fully-mechanized face and is exploiting coal seam No. 2^{-2}, which has a thickness between 5.99 m and 7.0 m, is nearly horizontal, and has a roof bedrock thickness between 70 m and 100 m. The loose strata on top of this bedrock is about 50 m thick, while the working face has a length of about 240 m and the mining height is 5.3 m. The base-mining ratio of this mine is between 13.2 and 18.87. Subsequent to mining out, 27 relatively large-scale cracks appeared in cut holes of the working face and extended to a length of 235 m. At the same time, lots of small cracks less than 15 m in length also formed; the width of these cracks was generally less than 40 mm, while the step fall was usually less than 100 mm. The ground crack-mining ratio of this mine is between 22.60 and 28.30.

On the basis of these real-life examples from the Shenfu mining area, it is clear that discontinuous deformation can occur in the ground above mines, while cracks can form when the ground crack-mining ratio is less 32.40.

3.1.2 *Mechanisms of ground collapse*

Ground collapse induced by coal mining is caused by the movement and destruction of overlying bedrock, and is influenced by numerous factors. As noted above, (section 3.1.1) in real-life examples of ground collapse, the key factors that influence the

Shenfu mining area types as well as the characteristics of fully-mechanized ground collapse are the bedrock thickness and mining height, specifically the base-mining ratio.

In the example of especially thin bedrock, the main weight on the roof of the mine, a cutting hole in working face 1203 of the Daliuta mine advanced 27 m. The main characteristic of this process was that the roof was cut down along the coal wall in the middle of the working face by 91 m, forming a sink step. Thus, due to the weight of bedrock, a large volume of phreatic water flooded into the working face through roof cracks at a maximum rate of up to 250 m^3/h. This, in turn, led to an inrush of sand into the tail of the working face after three days of inflowing water. The side step in working face subsidence was caused by bedrock cutting through the full thickness of the roof, while a graben formed in the ground to a height of about 20 cm when weight was first applied to the working face. This process demonstrates that the mine surface was controlled by this overburden of bedrock. The periodic loading interval of this process was between 9.4 m and 15.0 m (average: 12 m), while the period of loading was just one day. The pressure behavior of the mine working face was then alleviated following this first loading, while roof bedrock cutting tended to take place behind the shelf during periodic loading. The corresponding section of the ground then formed a stepped sinking crack when the working face was periodically loading, which shows that the overlying strata also underwent cutting across their full thickness. Based on a material simulation of similar behavior, the roof of the working face with a thickness of about 6 m fell after being weakened during the mining process. Indeed, this simulation showed that the strata above the roof are key as they directly determine what will happen to the bedrock surface as crack connectivity (pull apart) leads to fall cutting (shear) if enough land is available. This then also leads to initial and periodic loading, as loose strata are subsided and instant impulsive loads in the working face form. As shown in Fig. 1, the working face then exhibits characteristics of bedrock step cutting across its entire thickness (Hou ZJ 1996, Zhang XM & Hou ZJ 2007, Zhang JY et al. 1998).

Cutting the full thickness of bedrock is different from traditional caving, as the rock is not broken into smaller pieces but into a large rock mass. Thus, the development position is referred to as the cutting and caving zone. Slide in the bedrock is related to the thickness of overlying loose strata; when the load of overlying bedrock and the thickness of loose strata is large, this will promote step cutting across the full bedrock thickness. However, because the load bearing capacity of loose strata is small, this sinks as bedrock is cut and corresponding cracks form. These are probably connected to bedrock crack connectivity and ground stepped cracks form because loose strata follow the bedrock slide.

Figure 1. Schematic diagram to show stepped cutting across the entire bedrock thickness.

In the thin bedrock example, the fully-mechanized 20601 face of the Daliuta mine exploits coal seam No. 2^{-2}. This seam has an average thickness of 4.28 m, and is nearly horizontal, while the thickness of the roof bedrock is between 35 m and 55 m. The thickness of overlying loose strata is between 23 m and 55 m, the working face has a length of 220 m, an advance length of 2,660 m, and a mining height of 4 m. The first weighting interval of the working face is 35.4 m, while the average periodic weighting interval is 11.7 m. This means that sink step and along the coal wall of the step cutting with the full thickness cannot occur. In addition, this is not absolute as unconventionally drastic rock strata along the coal wall are cut with full thickness, the side step of the working face subsides, and the main roof is loaded. The difference between working faces 20601 and 1203 is that the bedrock thickness, rated operational resistance, and advance speed are all larger in the former. On the basis of simulation analysis, working face 20601 comprises hypogynous roof bedrock along with the advance of working face collapse; thus, epigynous rock breakage takes the form of a series of diamond rock pillars, before a caving zone appears above the goaf alongside a rock pillar fault zone (Zhang JY et al. 1998) (Fig. 2).

This rock pillar fault zone has attributes of both caving and fracture zones, as well as good integrity. Fracture zone characteristics also form among these rock pillars via crack connection, because of large dislocation among the rock pillars as well as the features of a caving zone. As a result, cracks are unable to penetrate through to the surface of the bedrock and enable integral connectivity; the dislocation of numerous rock pillars as distinct from working face 1203 enable cutting of the full thickness. Thus, loose strata subside and cracks occur alongside dislocation of the rock pillar, as load bearing capacity is small because of the regulation of this dislocation. This means that loose strata and the ground form parallel stepped cracks, and that the step fall is smaller than cutting across the entire thickness.

In the medium-thick bedrock example, working face 32202 of the Bulianta mine operates with working prop resistance values that are smaller than those rated for operation. This means that roof falls take place due to mining after weakening due to this process and sink step phenomena cannot form. In addition, ground collapse is related to the loading of the working face; sinking depth steps, however, are small, and ground collapse is caused by many loading events (Zhang JY et al. 1998, Huang QX 2002). On the basis of a simulation analysis, the roof bedrock of working face 32202 comprises hypogynous rock along with the advance of working face collapse, and the formation of a caving zone (Zhang J & Hou ZJ 2004). Rock strata above this caving zone developed fractures that extended through the roof of the bedrock to form a normal fracture zone (Zhang J & Hou ZJ 2004, Xue DJ 2015, Wu Q et al. 2005). Thus, because the load bearing capacity of loose strata is small, subsidence occurs with the bending of the strata fracture zone, and means that loose strata generate basically identical stepped cracks with overlying rock caving and the subsidence bend caused by periodic loading (Fig. 3).

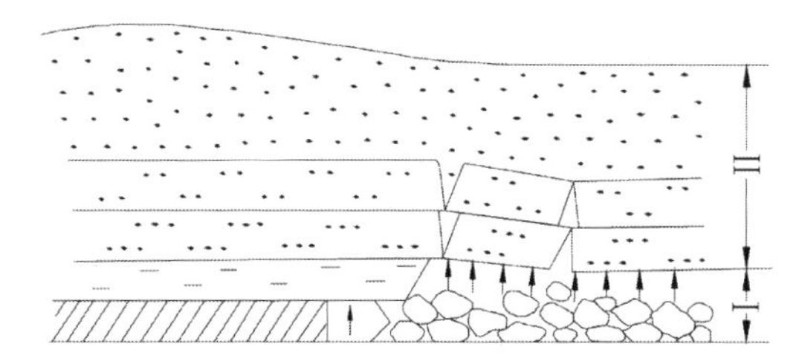

Figure 2. Schematic diagram to show a rock pillar fault.

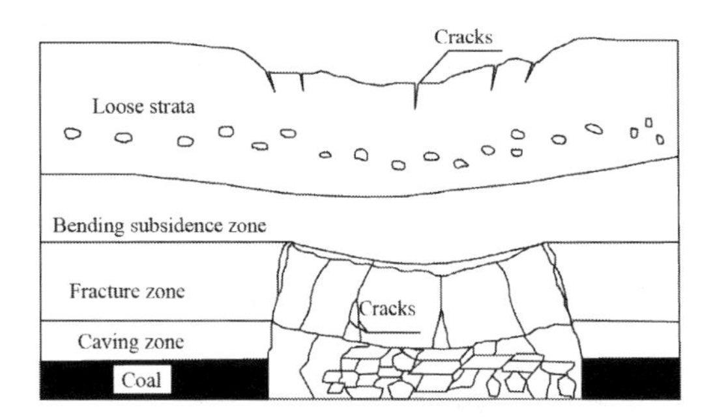

Figure 3. Schematic diagram to show overburden failure within medium-thick bedrock.

3.2 *Mechanisms of ground collapse in room-and-pillar mining*

3.2.1 *Practical examples of ground collapse*

The Xichagou mine and its neighbor, Fuguzhen No. 2, are both small mines within the town of Xinmin in Fugu county. In this region, mining employs the room-and-pillar method, while the management of mine roofs is based on support from coal pillars. These mine primarily exploit coal seam No. 3^{-1} that has an average thickness of 6 m, and a mining height of 3.5 m. The top of this coal is close to 2 m, while the bottom is 0.5 m, and the thickness of the roof bedrock is about 50 m, overlying a thickness of Quaternary loess of between 20 m and 80 m. A sudden ground collapse occurred to the west of Xichagou mine and to the southeast of the neighboring Fuguzhen No.2 goaf. This collapse took place a short time after mining out was completed, killing a miner and extending to a depth of 80,000 m^2. As discussed above, ground collapses can be divided into normal stepped ground fissures, negative stepped cracks, collapse slots, loess collapse, and loess landslides.

3.2.2 *Mechanisms of ground collapse*

Because the room-and-pillar mining method necessitates a small mined out area surrounded by pillar supports, old roof falls do not occur until after long periods of time. This method of mining also does not tend to cause ground collapse. However, when collapse does occur, such as in small mine shafts in the Shenfu mining area where room-and-pillar mining has been used, ground collapse disasters tend to take place in one mined out area after another. In these small mines, collapses tend to take place after short time periods during the process of mining, in addition to times when there is enough external force (such as an earthquake) to cause collapse. In these situations, the support of coal pillars is not large enough to attenuate intensity and instability over a short period, as they are under normal conditions when the majority of such shocks are not caused by the spontaneous combustion of coal. There is no doubt that the strength of supporting coal pillars is insufficient to present instability of mined-out areas; thus, under conditions of room-and-pillar mining over short time intervals, ground collapse only takes place when the actual size of coal pillars is insufficient or they are entirely absent. In contrast, over longer time periods before ground collapse, over half a year or more, the reasons underlying these events cannot include the external force of earthquake, but maybe due to insufficient coal pillar sizes, as well as the effects of weathering flake-and-creep, which leads to thinned coal pillars and reduces support intensity. Other factors can include instability caused by the spontaneous combustion of coal pillar. One example of this kind of event is the Gaoshiya jointly-organized mine where ground collapse occurred in a mined-out area of the mountains in Fugu county after mining had ceased for one year. This collapse was caused by the insufficient size of coal pillars as well as a decrease in their strength due to the effects of weathering flake-and-creep. In addition, while production stopped in 1995 at Shuijingqu mine, a ground collapse took place on the roadway goaf comprising four collapse craters and numerous ground fissures. This collapse was the result of the spontaneous combustion of a coal pillar in July 1995. Similarly, Lijiaban mine ceased production in early 1996 and a ground collapse took place later that spring, forming a 35-m diameter collapse crater between 5.5 m and 7 m in depth, also due to the spontaneous combustion of a coal pillar (Fan LM & Yang HK 2000). Because there are many small coal mines within the Shenfu mining area that have led to a large number of mined-out areas, there is a widespread problem due to the insufficiency of coal pillars and their likelihood of spontaneous combustion. Ground collapse is a very serious issue due to hidden underground spaces, the result of room-and-pillar mining.

4 CONCLUSIONS

1. Ground collapses within the Shenfu mining area can be divided into six types based on their morphological characteristics, collapse craters, collapse slots, cracks, collapse basins, loess collapse, and loess landslides. These types can be further classified into two groups based on their timescales, short time interval collapses and long time-interval sudden collapses, bearing in mind the relationship between ground collapse and the time needed for coal mining. Further, these collapse types can be further classified into fully-mechanized coal mining subsidence, room-and-pillar ground collapse, and other kinds of mining subsidence.
2. The ground crack-mining ratio within the Shenfu mining area falls to below 32.40. The use of fully-mechanized longwall mining, and the nature of special thin bedrock, thin bedrock, and medium-thick bedrock, all mean that the intensity and type of ground collapses are different within the mining area.
3. Fully-mechanized longwall mining, as well as coal seam mining from especially thin bedrock lead to ground collapses caused by the overlying bedrock step cutting and sinking. In these cases, cracks develop which connect to the bedrock, while thinner bedrock coal seam mining causes ground collapses due to rock pillar faulting. In contrast, coal mining in medium-thick bedrock leads to ground collapses caused by periodic bend-and-sink of bedrock. As the result of room-and-pillar mining, ground collapse can be caused by the insufficient size of coal pillars as well as their spontaneous combustion.

ACKNOWLEDGEMENTS

The project was supported by the National Natural Science Foundation Program of China (Ref: 41472234) and the Program of China Geological Survey (Ref: 12120101 1000150022).

REFERENCES

Fan LM, Li C, Chen JP, Ning JM (2016) Mineral resources geological hazard and prevention and control technology in the high strength exploration area. Beijing: Science Press.

Fan LM, Yang HK (2000) The study about the causes and situation of ground subsidence in Shenfu mining. *J. Shaanxi Coal Technology* (01): 7–9.

Fan LM, Zhang XT, Xiang MX, Zhang HQ, Shen T, Lin PX (2015) Characteristics of ground fissure development in high intensity mining area of shallow seam in Yushenfu coal field. *J. Journal of China Coal Society* 40(06): 1442–1447.

Huang QX (2002) Ground pressure behavior and definition of shallow seams. *J. Chinese Journal of Rock Mechanics and Engineering* 21(8): 1174–1177.

Hou ZJ (1996) Preliminary study on control of roof strata under thick aeolian sand mining. *J. Coal Mining* (02): 50–53.

Li WP, Ye GJ, Zhang L, Duan ZH, Zhai LJ (2000) Study on the engineering geological conditions of protected water resources during coal mining action in Yu-Shen-Fu Mine Area in the North Shanxi Province. *J. J ournal of China Coal Society* 25(05): 449–454.

Liu YQ, Zhou HW, Li YH, Yi HY, Xue DJ (2015) Water-sand inrush simulation under shallow coal mining based on the particle flow code. *J. J ournal of Xi'an University of Science and Technology* 35(05): 534–540.

Shi PW, Hou ZJ (1996) Law of roof breaking movement of shallow seams in Shenfu mining. *J. Journal of Xi'an Mining Institute* 16(03): 204–207.

Wu Q, An YH, Liu WG, Zhang CF, Li JJ (2005) Water-soil environment issues and its controlling technology in Shendong Mining Field. *J. Coal Geology and Exploration* 33(03): 54–58.

Xue DJ, Zhou HW, Ren WG, Zhang BF, Liu YQ, Zhao YF (2015) Stepped shearing-induced failure mechanism and cracks propagation of overlying thin bedrocks in shallow deep coal seams mining. *J. Journal of China Coal Society* 40(8): 1746–1752.

Zhang J, Hou ZJ (2004) Physical simulation and analysis of the development law of shallow buried coal seam hydraulic fracture. *J. Mine Pressure and Roof Management* (04): 32–34.

Zhang JY, Hou ZJ, Tian RY (1998) The rule of rock pressure and overburden breaking in shallow buried mining field. *J. Mine Pressure and Roof Management* (03): 9–11.

Zhang XM, Hou ZJ (2007) Numerical simulation of overburden three zones for mining operation with thick overburden and less depth. *J. Coal Science and Technology* 02: 93–96.

Land Reclamation in Ecological Fragile Areas – Hu (Ed.)
© 2017 Taylor & Francis Group, London, ISBN 978-1-138-05103-4

Urban heat island effect of mining cities under mining disturbance: A case study of Wu'an

C.C. Mao & M.C. Fu
School of Land Science and Technology, China University of Geosciences, Beijing, China

ABSTRACT: Improper excavation, cover occupation and occupation of land caused by mining development change the thermal properties of city's underlying surface, which contributes to the Urban Heat Islands (UHI). This paper, firstly, estimates the spatial dynamics of Land Surface Temperature (LST) by single-channel algorithm based on the Landsat TM Thermal Infrared (TIR) bands in 1998, 2004, 2010, 2016 of Wu'an, and makes visual interpretation of mining boundary and partitions surface disturbance types of mining; secondly, analyzes the impact of mining on the urban heat island effect, the characteristics of temperature spatial distribution of different surface disturbance types and its contribution to the UHI effect from time and space perspective with spatial analysis method. The conclusions are that the average LST of Wu'an is on the rise with the mining scale expanding and the UHI effect is obvious and the LST of different surface disturbance types are obviously different.

1 INTRODUCTION

With mining development, the processes such as peeling of rocks and soil, accumulation, pressure, occupation and ground subsidence caused the surface water evaporation, transpiration cycle and surface heat capacity (Bian Zhengfu, 2009), which affected the land surface temperature and caused the change of microclimate in the mining area and surrounding areas (Peng Jian, 2005). The study of the UHI effect in the mining city will contribute to monitor the temporal and spatial distribution of the effects of mining on temperature, which serves the management of mining cities, land reclamation and restoration of ecological environment. In this paper, Landsat-5 TM and Landsat-8 OLI images are used to analyze the spatial distribution and dynamic changes of LST in the mining city of Wu'an in 2000, 2004, 2010 and 2016. Five main mining areas in Wu'an are choosen as the research object. Analyze the distribution of the surface temperature of the mining area, the dynamic change and the contribution degree of different disturbance type to the UHI effect, so that provides the basis for the study of mining development and land reclamation.

2 MATERIALS AND METHODS

2.1 *Overview of Wu'an*

This paper chooses the mining city of Wu'an as the research area, 1818 km^2. Wu'an belongs to monsoon climate of medium latitudes, cold and dry winter, hot summer rainy. Average annual temperature is 14.4°C, the maximum temperature is 37.6°C. Wu'an is rich in mineral resources, such as coal, iron, cobalt, aluminum, etc, which is an important energy base of Hebei province. The distribution of mineral resources is concentrated, with urban areas as the center, north-south distribution of iron ore resources; east-west distribution of coal resources. The mining development of Wu'an is dominated by open-pit mining and underground mining. There are any kinds of disturbances in the same period. It is typical for the study of temperature differentiation effect of mining surface disturbance.

2.2 *Data source*

In this study, three Landsat-5 images (2000-3-28, 2004-11-2, 2010-6-12), one Landsat-8 OLI images (2016-4-9) and Wu'an land use status map are used as data source. Use ENVI 5.1 and ArcGIS 10 to process remote sensing data and do spatial analysis.

2.3 *Land Surface Temperature (LST) retrieval*

The single-channel universality method proposed by Jime'nez-Mun'oz and Sobrino is used to retrieve the land surface temperature, and the empirical formula in the algorithm is parameterized by combining the channel response function of Landsat-8 thermal infrared channel, thus getting coefficient matrix applying to Landsat-8 (Yang Xuesen, 2015). LST retrieval includes three steps: land surface brightness temperature calibration, parameter estimation and land surface real temperature calculation (Sobrino JA, 2004; Jimenez-munoz JC, 2006). The spatial distribution data of LST are calculated, of which the mean temperature of the land surface is 0.8, 0.6, 0.5, –0.7°C differ from the measured temperature of the local weather station respectively. Retrieval results are within acceptable limits. According to the data of the meteorological website, four-stage temperature retrieval result are interpolated to the surface temperature of the same month to describe the time difference of temperature.

2.4 *Measurement of LST response in mine fields*

Use contribution rate to measure the contribution of the specific disturbance type to the temperature rise in the study area (Xie Miaomiao, 2011), ie, the ratio of the increase of the LST in specific disturbance type area to the increase in the land surface LST of all disturbance types. $C_{Rk} = \frac{(T_k - T_0)S_k}{(T_d - T_0)S_d}$, The larger the C_{Rk} value indicates that the effect of this disturbance type on the LST rise is more intense and contributes more to the regional warming, which is the key area of the local climate change.

3 INFLUENCE OF MINING DEVELOPMENT ON UHI EFFECT

The LST of Wu'an shows obvious temporal and spatial differentiation (Figure 1). On the time scale, with the expansion of mining development, the average LST of Wu'an is increasing. From 2000 to 2016, the average LST of Wu'an increased from 23.69°C to 41.09°C in April, which means range expansion and the increase in intensity of the UHI effect. The spatial distribution shows that high LST is mainly distributed in the central city and the main mining area.

Figure 2 shows the average LST of five main mining areas in Wu'an, and the city's average LST.

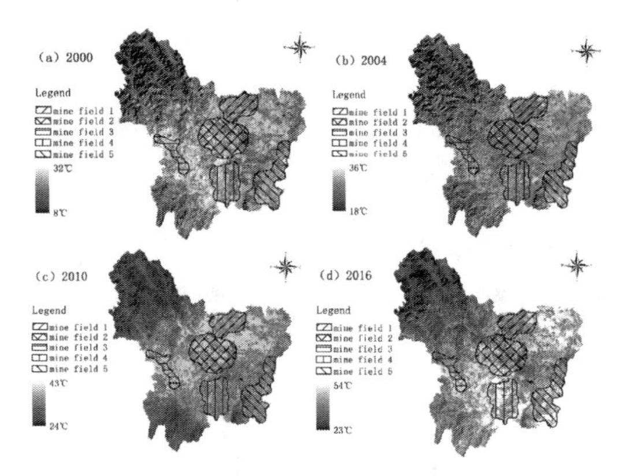

Figure 1. The distribution of LST and mine field.

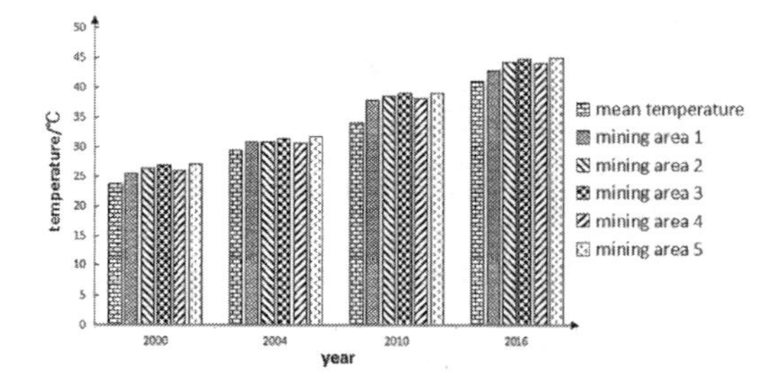

Figure 2. Average land surface temperature of different mine fields.

The average LST of the mining area was 5 ~ 7°C higher than the average LST of Wu'an, which means that the average LST of the mining area is increasing year by year. Due to the law of temperature diffusion, i.e., air diffuse from the high temperature region to the low temperature area, so that will promote the formation of UHI.

4 THE CONTRIBUTION RATE OF DIFFERENT DISTURBANCE TYPES TO HEATING

According to Wu'an land use status map, the mine field is divided into mining pit, slag-heap, tailings pile, ore dressing area and sinking area. After mining development disturbance, the LST of mining area shows a heterogeneous distribution in the space, and the mean LST of different disturbance types are quite different. Using the spatial analysis method, analyze the dynamic changes of the LST and the contribution rate on heating of different disturbance types in five major mining areas from 2000 to 2016 (Table 1).

The high LST area of mine field are slagheap and tailings pile, whose contribution rate to UHI effect is the biggest. The mean LST of slagheap in mine fields are about 9°C higher than that in the study area, while tailings piles' are about 8°C higher. The contribution of ore dressing area to the UHI effect is in the medium level. In 2000 and 2004, the mean LST of ore dressing area were 6.5°C higher than the mean LST of mine fields. After 2004, due to the remediation of mining in Wu'an, closing a small part of the mine, consolidating site of the ore dressing industry, which make the industry land more intensive, the contribution of ore dressing area to UHI effect also decreased. The low value area of the LST in mine fields are mining pit and sinking area, whose contribution rate to UHI effect is the least and almost didn't change in time. As mining technology each mineral is different, heating contribution value is also different. The mean LST of mining pits in iron ore

Table 1. The contribution rate to heating of different land disturbance types (unit: %).

	2000					2004				
	NO. 1	NO. 2	NO. 3	NO. 4	NO. 5	NO. 1	NO. 2	NO. 3	NO. 4	NO. 5
mining pit	12.60	12.78	11.47	11.87	7.62	12.47	13.10	12.63	12.56	8.07
slagheap	31.55	31.79	31.50	31.28	31.65	31.56	31.29	31.75	31.03	31.55
tailings pile	30.80	30.35	30.28	30.55	30.71	30.88	30.46	30.00	30.54	30.58
ore dressing area	25.05	25.08	26.75	26.30	20.25	25.09	25.15	25.62	25.87	20.06
sinking area	–	–	–	–	9.77	–	–	–	–	9.74
	2010					2016				
	NO. 1	NO. 2	NO. 3	NO. 4	NO. 5	NO. 1	NO. 2	NO. 3	NO. 4	NO. 5
mining pit	16.63	16.62	16.4	15.87	9.69	16.35	15.88	17.02	16.97	9.06
slagheap	31.54	31.79	31.25	31.43	33.68	31.42	31.65	31.26	31.13	33.45
tailings pile	30.86	30.53	30.61	30.85	32.05	30.88	30.93	30.77	30.58	32.1
ore dressing area	20.97	21.06	21.74	21.85	15.60	21.35	21.54	20.95	21.32	15.58
sinking area	–	–	–	–	8.98	–	–	–	–	9.81

area are 5.8°C higher than that in the study area, and the contribution rate is relatively high. In coal mine area, the mean LST of mining pits are 3.9°C higher than the mean LST of mine fields, while sinking areas' are 2.7°C higher.

5 CONCLUSION AND DISCUSSION

Wu'an Mining Development has drastically changed the land surface coverage, affecting the local climate, causing the LST change. The mean LST of mine fields are significantly higher than that of Wu'an urban. Mine fields are the main distribution area of the UHI. With the expansion of the mine development, the city's LST is increasing, the UHI effect is more and more obvious, the intensity of UHI increases.

The temporal and spatial heterogeneity of LST is closely related to type of land surface disturbance of mining development. Slagheap and tailings pile have obvious heating effect, which contribution rate to the UHI are the biggest. Contribution rate of ore dressing area to the UHI effect is at a medium level. Due to the difference in minescale, the LST of it has spatial and temporal heterogeneity. Ore dressing area in extensive mine fields have higher heating effect. After the adjustment of small mines, mining more intensive, contribution rate to heating has been reduced. The LST of mining pit and sinking area are relatively low in the whole mining area, the contribution rate to heating is low, and the promotion of the UHI effect is the weakest. Among them, contribution to heating is affected by minerals, mining technology and mining scale.

ACKNOWLEDGEMENT

This paper was supported by project Natural Science Foundation of China (41641008).

REFERENCES

Bian Zhengfu, Lei Shaogang, Chang Luqun, et al. (2009) Affecting factors analysis of soil moisture for arid mining area based on TM images. *J. Journal of China Coal Society*, 34(4):520–525.

Jimenez-Munoz JC, Sobrino JA. (2006) Error sources on the land surface temperature retrieved from thermal infrared single channel remote sensing data. *J. International Journal of Remote Sensing*, 27(5):999–1014.

Peng Jian, Jiang Yijun, Wu Jiansheng, et al. (2005) Eco-environmental effects of mining and related land reclamation technologies in China. *J. Progress in Geography*, 24(2):38–48.

Sobrino JA, Jimenez-Munoz JC, Paolini L. (2004) Land surface temperature retrieval from Lands-at-5 TM. *J. Remote Sensing of Environment*, 90(4):434–440.

Xie Miaomiao, Bai Zhongke, Fu Meichen, et al. (2011) Effects of land disturbance on surface temperature in large opencast coal mine. *J. Journal of China Coal Society*, 36(4):643–647.

Yang Xuesen. (2015) Research on the single-channel algorithm for land surface temperature retrieval from Landsat-8 data. Beijing: China University of Geosciences (Beijing).

Land Reclamation in Ecological Fragile Areas – Hu (Ed.)
 ISBN 978-1-138-05103-4

Application of fuzzy mathematics and GIS technology in the evaluation of mine environmental restoration

Q.T. Ma, G.F. Shang & S.H. Yu
Faculty of Graduate Studies, Hebei GEO University, Shijiazhuang City, Hebei Province, China

ABSTRACT: The mining industry has destroyed the original ecological environment of the mining area, which made the restoration of environment became increasingly important. Considering the gradual and fuzzy nature of environmental factors, this paper uses fuzzy comprehensive evaluation method to build the evaluation model, combined with GIS technology to analyze the evaluation results, and puts forward some reasonable suggestions. Take Zhangjiakou, Kangbao County, Beiguan brick environment clay ore mining area as an example. The evaluation results show that mine pit area is the third level land, the waste rock, the mound, bleachery area is the second level land, other areas are the first level land.

1 INTRODUCTION

With the rapid development of economy, mining projects are becoming frequent, which bring serious ecological damage to mine area. In order to solve problems and guide the coordinated development, we need to carry out environmental restoration evaluation.

Foreign environmental evaluation began in the 20th century, 60 years, comprehensive development in the 1970s, has now become an important branch of environmental science. A lot of environmental quality assessment was carried out in Europe, North America, Japan, Eastern Europe and the Soviet union that most of them is single factor environmental quality assessment method. At the beginning of 90s, Cendrero et al put forward a new method based on the natural unit classification system. In the 60s of last century, China began to study the geological environment of mining area, in the 90s, it turned on the adverse effects study of society and economy. In 2000, environmental assessment was carried out in succession, and various evaluation methods, such as Analytic Hierarchy Process (AHP), fuzzy comprehensive evaluation method and grey correlation degree method, were proposed. Fuzzy mathematics is a theory and method of dealing with fuzzy phenomena. This idea is very suitable to deal with the problem of environmental assessment with the characteristics of fuzziness and gradual change.

The mining environment of clay mine has the characteristics of high relative terrain height, the damage degree is uneven, and the characteristics of rock and soil are obvious. Therefore, this paper selects relative terrain height, slope, soil, rock and soil characteristics, vegetation coverage, the occupied and destoried land as evaluation index to build a fuzzy mathematical model. On the basis of the evaluation results, uses GIS to establish the evaluation chart, analyzes the factors that affect the environment, and puts forward some suggestions.

2 FUZZY COMPREHENSIVE EVALUATION METHOD

2.1 *Factor standardization*

According to the characteristics of geological environment, set up collection of evaluation factors: U = (relative terrain height, slope, soil, rock and soil characteristics, vegetation coverage, the occupied and destoried land). According to "General principles of Chinese regional environmental geological survey", the geological environment quality is divided into three grades. Creating remark collective V = (I, II, III). The maximum relative height of the study area is 10 m, so the relative height of the terrain is divided into three grades: less than 5 m, 5–10 m and more than 10 m; the characteristics of rock and soil is mainly divided into hard lump rocks, medium hard rocks, loose rocks of level three, quantization value is 1, 2, 3; the soil is divided into humus soil, sandy clay, debris soil in three categories, quantization value is 1, 2, 3; the slope of the study area is between 45 degrees–5 degrees, so the slope is divided into less than 20 degrees, 20 degrees–40 degrees, more than 40 degrees of three levels; according to the results of the survey, the vegetation coverage is divided into 70 degrees–100 degrees, 30 degrees–70 degrees, less than 30 degrees of three grades; the occupied and destoried land reflects the proportion of the destroyed land area in the evaluation unit, and divides it into three grades: less than 30%, 30%–80%, and more than 80% (Table 1).

2.2 *Determine factor weight*

The Analytic Hierarchy Process (AHP) can divide all the factors in the complex problem into an ordered hierarchy and construct the relative importance judgment matrix then use the maximum to determine the weight of each factor; Delphi method can make a quantitative evaluation of a large number of non technical and quantitative analysis factors. This paper uses Delphi–AHP method to determine the weight of each index. Mine environmental assessment is divided into three layers, including the target layer A (environment quality), criterion layer B (internal factors and external factors), index layer U (relative terrain height, slope, soil, rock and soil characteristics, vegetation coverage, the occupied and destoried land). Use the judgment matrix R to calculate the eigenvalues λ and eigenvectors A, then test the judgment matrix (Table 2).

2.3 *Membership functions*

The membership functions of the 3 levels corresponding to the incremental index factors and the decremental index factors were (1)–(6).

Table 1. Factor standardization table.

Factor	RH m	RS	Soil	Slop °	VC %	ODL %
I	0–5	Hard	Humus	<20	70–100	<30
II	5–10	Medium	Sandy	20–40	30–70	30–80
III	>10	Weak	Debris	>40	<30	>80

Note: RH represents the relative terrain height, RS represents the rock and soil characteristics, VC represents the vegetation coverage, ODL represents the occupied and destoried land.

$$R_1 = \begin{cases} 1 & X \le A_1 \\ \dfrac{(A_2 - X)}{(A_2 - A_1)} & A_1 \le X \le A_2 \\ 0 & X \ge A_2 \end{cases} \tag{1}$$

$$R_2 = \begin{cases} 0 & X \le A_1 \text{ or } X \ge A_3 \\ \dfrac{(X - A_1)}{(A_2 - A_1)} & A_1 \le X \le A_2 \\ \dfrac{(X - A_1)}{(A_2 - A_1)} & A_2 \le X \le A_3 \end{cases} \tag{2}$$

$$R_3 = \begin{cases} 0 & X \le A_3 \\ 1 & X > A_3 \end{cases} \tag{3}$$

$$R_1 = \begin{cases} 0 & 0 \le X \le A_2 \\ \dfrac{(A_2 - X)}{(A_2 - A_1)} & A_2 \le X \le A_1 \\ 1 & X \ge A_1 \end{cases} \tag{4}$$

$$R_2 = \begin{cases} 0 & X \ge A_1 \text{ or } X \le A_3 \\ \dfrac{(A_1 - X)}{(A_1 - A_2)} & A_2 \le X \le A_1 \\ \dfrac{(X - A_3)}{(A_2 - A_3)} & A_3 \le X \le A_2 \end{cases} \tag{5}$$

$$R_3 = \begin{cases} 0 & X \ge A_3 \\ 1 & X < A_3 \end{cases} \tag{6}$$

R_i represents the evaluation indicator's membership grade that corresponding to A_1, A_2, A_3. X represents the measured value of the evaluation index. A_1, A_2, A_3 is the standard value.

Table 2. The weight of each evaluation index to the target layer.

Target	Criterion	Weight	Index	Weight	Consistency check		Weight
Environment quality	Internal factors	0.3333	Relative terrain height	0.3337	$CI = 0.0269$	$\lambda_{max} = 3.0538$	0.1113
			Rock and soil characteristics	0.5247	$RI = 0.58$		0.1749
			Soil	0.1416	$CR = 0.0464 < 0.1$ pass		0.0472
	External factors	0.6667	Slope	0.1976	$CI = 0.0269$	$\lambda_{max} = 3.0537$	0.1317
			Vegetation coverage	0.3119	$RI = 0.58$		0.2079
			The occupied and destoried land	0.4905	$CR = 0.0464 < 0.1$ pass		0.3270

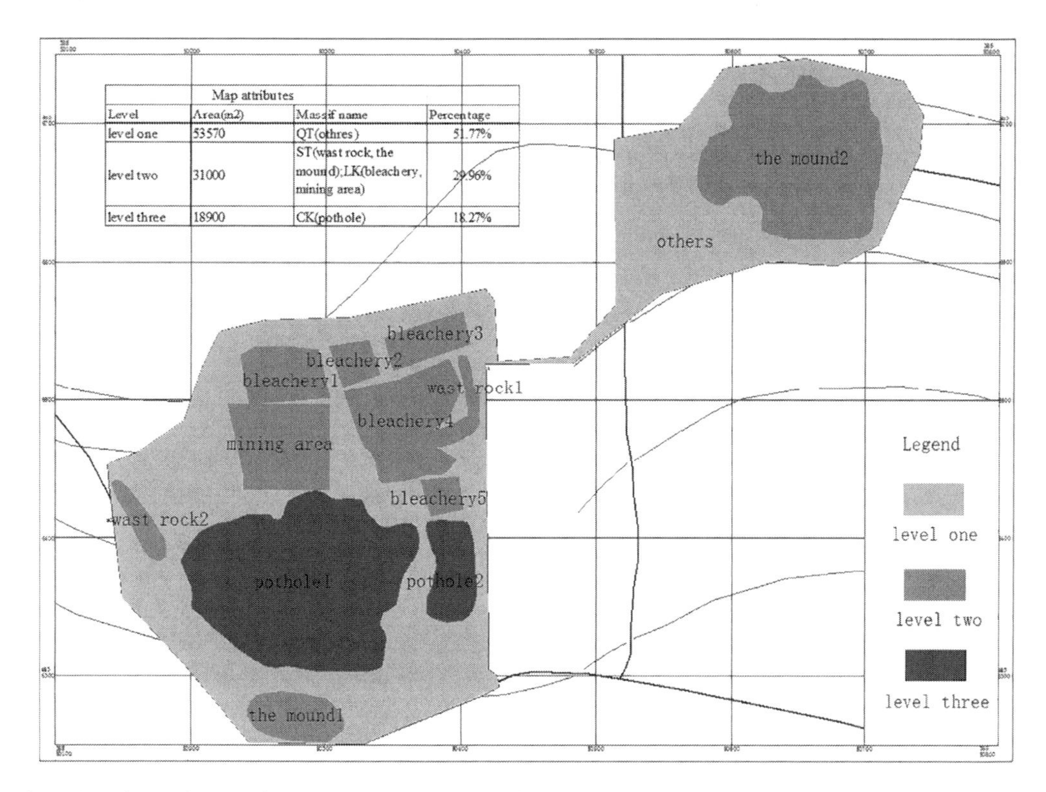

Figure 1. The zoning map is used to describe the level of destruction of different plots.

3 EVALUATION RESULTS

3.1 *The result of zoning evaluation*

Take the upper and lower limits of the interval of first and third, the middle of the interval of second as the standard value of the quantitative index. The quantization value was taken as the standard value of the qualitative index. The measured value of each index factor is obtained by field measurement. Put measured value and standard value into the fuzzy mathematical model and caculated by MATLAB, we can get the evaluation results *B*.

$$B_1 = [0.1, 0.22, 0.46] \quad (7)$$

$$B_2 = [0.08, 0.23, 0.13] \quad (8)$$

$$B_3 = [0.32, 0.38, 0.21] \quad (9)$$

$$B_4 = [0.65, 0.30, 0.00] \quad (10)$$

3.2 *GIS spatial analysis results*

Combining the results of fuzzy comprehensive evaluation with ArcGIS spatial analysis, we can get the evaluation zoning map Figure 1.

Light grey, grey and dark grey represent three quality levels of land environment. The evaluation results of CK (pothole), ST (wast rock, the mound), LK (bleachery, mining area), QT (others) are B1, B2, B3, B4 which show that pothole area is the third level land, the waste rock, the mound, bleachery, mining areas are the second level land, others are the first level land. With the application of ArcGIS to the analysis of geological environment zoning we find that the first level land has account for 51.77% of the total area, the second account for 29.96%, and the third account for 18.27%. The evaluation results are consistent with the actual situation.

4 CONCLUSION AND SUGGESTION

1. In this paper, the evaluation results are consistent with the actual situation of the study area, with strong objectivity and authenticity. Using the method of fuzzy mathematics model and ArcGIS spatial analysis to evaluate the environmental quality of mining area, and the evaluation result has a strong guiding role for the restoration and treatment of mine environment.

2. What affects the geological environment most is the potholes, which adviced to filled by wast rock and the mound. We can smooth the land by reducing relative terrain height and plot and increase vegetation coverage by planting trees and grass. Aiming at the influence factors to govern, so as to achieve the purpose of improving the overall quality of the environment.

REFERENCES

Bao Nisha, Ye Baoying, Bai Zhongke, Liu Xiaocui, Ma Xiao. (2009) Application of ArcGIS in land reclamation in mining area [J]. *Journal of Shanxi Agricultural University (Natural Science Edition)*, (06):501–504.

Chen Jianping. (2012) Study on Evaluation and Division of Geological Environment Quality in Xixiang County [D]. *Xian:Xi'an University of Science and Technology.*

Chen Jianping, Fan Limin, Ning Jianmin. (2014) Mine geological environment impact assessment based on fuzzy comprehensive evaluation and GIS Technology [J]. *Chinese Journal of Coal Geology*, (02):43–48.

Gupta RP, Joshi BC. (1990) Landslided hazard zoning using the GIS approach—a case study from the Ramaganga Catchment Himalalyas [J]. Engineering geology, 20(1–2):119–131, 125–135.

Raju, Prasada, Saibaba J. (1999) Landslide hazard zonation mapping using remote sensing and geolographic information system techniques—case study of Pithoragarh Area U.P. [J]. *International Geo-scienceand Remote Sensing Symposium (IGARSS),* 1:577–579.

Zhao Xinzhuo, Zhang Xueming. (2014) Application of Fuzzy Comprehensive Evaluation in Comprehensive Evaluation of Mine Environment [J]. *Science Innovation and Application, (01):293.*

Land Reclamation in Ecological Fragile Areas – Hu (Ed.)
© 2017 Taylor & Francis Group, London, ISBN 978-1-138-05103-4

Ecological footprint and carrying capacity of mining cities: A case study of Wu'an

Z.R. Wang & M.C. Fu
School of Land Science and Technology, China University of Geosciences, Beijing, China

ABSTRACT: The ecological footprint is an important method to measure the regional sustainable development level. This paper taking the mining city Wu'an as an example, amends the ecological footprint model and adds the pollution discharge account including waste water, waste gas and solid waste. The dynamic changing process and feature of the ecological footprint and ecological carrying capacity of the city from 2010 to 2014 are measured and discussed, and the sustainable development degree of Wu'an is evaluated. The study result shows that from 2010 to 2014, the per capita ecological footprint and per capita ecological carrying capacity of Wu'an showed a trend of increasing first and then decreasing, the per capita ecological footprint of Wu'an was much more than that of per capita ecological carrying capacity during the period of 2010–2014, and Wu'an is in a serious state of ecological deficit. This paper evaluates the existing problems and sustainable development of Wu'an in the process of social and economic development, with a view to provide scientific basis for mining city to correctly handle the relationship between population, resources, environment, economic and social development and formulate relevant policies.

Keywords: Mining cities; Ecological footprint; Land ecological carrying capacity; Wu'an

1 INTRODUCTION

The core of sustainable development is to strive to seek the harmonious development of man and nature. Meanwhile, we must link human development with resource consumption, environmental degradation, and ecological stress (Niu Wenyuan. 2012). However, with the development of China's social economy, the ecological and environmental problems are becoming more and more prominent. At present, the study of ecological footprint (Rees W E. 1992) has been widely used in many fields and aspects. In the aspect of model calculation, there are an ecological footprint comprehensive account model calculation framework (Rees W & Wackernagel M. 1996) and so on. In the field of application, land (Gerbensleenes PW. *et al.*, 2002), psychology (Verhofstadt E *et al.*, 2016) and other fields have been applied, but the research on the sustainable development level of energy-based cities is relatively few. This mining city faces more severe ecological and environmental problems than other cities (Zhang Jianjun *et al.*, 2007). Based on the ecological footprint theory, the paper revises the ecological footprint model. The ecological footprint and ecological carrying capacity of Wu'an from 2010 to 2014 were measured, and the sustainable level was evaluated objectively, so as to coordinate the contradiction between ecology and economic.

2 STUDY AREA AND DATA SOURCES

Wu'an lies in the south of Hebei Province, covering a surface area of 1819 km^2, located at 113° 45′–114° 22′E, 36° 28′-37 01′N. It is high in the west and low in the east, with a small basin in the middle. The whole city can be divided into three types: mountainous areas, low hilly areas, and basins. It belongs to the temperate continental monsoon climate. 2014 the total population of the city is 826,700, and the GDP of city reached RMB 60.65 billion. Wu'an is an important energy supplier, rich in coal, iron, cobalt, aluminum and others.

The data used in this paper are from *Wu'an Statistical Yearbook* (2010–2014) and *China Rural Statistical Yearbook*. Land use data are from the land use pattern of Wu'an City (2010–2014).

3 RESEARCH METHODS

In ecological footprint model, mainly includes the account of biological resource and energy resource. In this paper, increases the pollution discharge account to add the ecological footprint of waste. That of waste water and waste gas refers to the area of bio-productive land used to treat waste water and air pollutants. That of solid waste includes the area of land used for dispose.

$$EF = N \times ef = \sum r_j \times A_i = \sum (C_i / P_i) \quad (1)$$

$$EC = N \times ec = N \times r_j \times Y_j \times \sum a_j \quad (2)$$

where EF is total ecological footprint (hm^2); N is population (capita); ef is Per capita ecological footprint (hm^2/cap); A_i is per capita bio productive area of i- type goods (hm^2/cap); C_i is per capita consumption of i goods (t); P_i is average production capacity of i (t/hm^2); i is types; r_j is equilibrium factor. EC is total ecological carrying capacity (hm^2); ec is per capita ecological carrying capacity (hm^2/cap); a_j is per capita productive land area (hm^2/cap); r_j is equilibrium factor; Y_j is yield factor.

The ecological carrying capacity refers to the area of the biologically productive land that can be provided to humans. This paper uses the new equilibrium factors provided in *Living Planet Report 2012*. The yield factors used in this paper are 1.59 for cropland and build-up land, 3.86 for grassland, 2.96 for forest and 3.19 for water (Zhang Jianjun, 2007).

3.1 *Ecological footprint*

The paper adopted the method (Liu Lemian, 2009). According to the economic development of Wu'an City and the production and consumption of the main resources, selected items are as follows (Table 1). The calculation of biological resource consumption is based on the global average yield proposed by FAO. The various types of fossil energy conversion factors required take the global average energy footprint parameters proposed by Wackeragel. Based on the above data, the per capita ecological footprint of Wu'an (Table 2) is calculated.

3.2 *Ecological carrying capacity*

The calculation of ecological carrying capacity is the area of ecological supply area, means to figure out all kinds of eco-productive land area converted after the total value. According to the report of the World Commission on Environment and Development, at least 12% of the ecological capacity needs to be deducted for the conservation of biological diversity. Using the formula (2) to calculate out the per capita ecological carrying capacity of Wu'an from 2010 to 2014 (Table 3).

3.3 *Ecological deficit or surplus*

Through the difference calculation between ecological footprint and ecological carrying capacity, the per capita deficit of Wu'an (2010–2014) are as follows (Table 4).

Table 1. Biological resources, energy resources and pollution account of Wu'an.

Account	Land types	Consumption items
Biological resource	Cropland	Cereals, Beans, cotton, Oil crops, vegetables, Melons
	Forest	Wood, Walnut, Chestnut
	Pasture	Pork, beef, Lamb, poultry, Eggs, Milk, Garden fruit
	Water area	Aquatic products
Energy resource	Build-up land	Electricity
	Fossil energy land	Coal, Coke, Gasoline, Diesel, Gas
Pollution	Pollution absorption	The emissions of SO_2, NO_X, Dust and Sewage

Table 2. The per capita ecological footprint of Wu'an from 2010 to 2014 (hm^2/cap).

Year	Cropland	Pasture	Forest	Fossil energy land	Build-up land	Water area	Pollution absorption	Per capita ecological footprint
2010	0.3672	0.6378	0.0130	3.4369	0.2170	0.0220	0.4999	5.1938
2011	0.3793	0.6556	0.0132	5.3640	0.2511	0.0218	0.5580	7.2431
2012	0.3971	0.6390	0.0162	5.7707	0.2437	0.0231	0.6874	7.7772
2013	0.3868	0.6373	0.0084	2.6448	0.2691	0.0240	0.7025	4.6728
2014	0.3607	0.6717	0.0201	2.3074	0.2623	0.0247	0.9949	4.6418

Table 3. The per capita ecological carrying capacity of Wu'an from 2010 to 2014 (hm^2/cap).

Year	Cropland	Pasture	Forest	Build-up land	Water area	Biodiversity area (deduct)	Total supply area	Per capita ecological carrying capacity
2010	0.2865	0.0230	0.1054	0.1133	0.0081	0.0644	0.5363	0.4720
2011	0.2803	0.0226	0.0887	0.1075	0.0072	0.0607	0.5062	0.4455
2012	0.2771	0.0223	0.1052	0.1027	0.0077	0.0618	0.5151	0.4533
2013	0.2728	0.0222	0.1049	0.1116	0.0076	0.0623	0.5191	0.4568
2014	0.2674	0.0220	0.1017	0.1135	0.0074	0.0614	0.5120	0.4506

Table 4. 2010–2014 various types of land in Wu'an ecological deficit or surplus (hm^2/cap).

Year	Cropland	Pasture	Forest	Fossil energy land	Built-up	Water area	Pollution absorption	Per capita ecological deficit
2010	−0.0807	−0.6148	0.0925	−3.3725	−0.1037	−0.0139	−0.4999	−4.7218
2011	−0.0990	−0.6330	0.0755	−5.3033	−0.1436	−0.0146	−0.5580	−6.7976
2012	−0.1199	−0.6167	0.0890	−5.7089	−0.1409	−0.0154	−0.6874	−7.3240
2013	−0.1139	−0.6151	0.0965	−2.5825	−0.1575	−0.0164	−0.7025	−4.2160
2014	−0.0933	−0.6497	0.0815	−2.2459	−0.1488	−0.0173	−0.9949	−4.1912

4 EVALUATION OF ECOLOGICAL SUSTAINABILITY OF WU'AN FROM 2010 TO 2014

1) Analysis of Ecological Structure Demand. From 2010 to 2014, the average ecological footprint (ef) of Wu'an was 5.9057 hm^2, much higher than the global level (2.1 hm^2). The average ef of fossil energy land was 3.9048 hm^2, accounting for 66.12% of the total ef, It shows that the ecological consumption of Wu'an mainly comes from fossil energy land, which is closely related to the local iron, steel and the mining industry. Followed by pollution absorption 0.6886 hm^2, accounting for 11.66% of the total ef, it shows that Wu'an pollution is serious. Forest ef in 2012 dropped to 0.0887 hm^2, mainly caused by serious reduction of local walnut due to frost. 2010–2012, the per capita ef of Wu'an increased from 5.1938 hm^2 to 7.7778 hm^2, the growth rate was 49.75%. Among them, the ef of fossil energy land increased from 3.4369 hm^2 to 5.7707 hm^2 in 2010, accounting for 90.31% of the total growth. It was the main factor to stimulate the ef of Wu'an. From 2012 to 2014, the per capita ef of Wu'an decreased from 7.7778 hm^2 to 4.6418 hm^2, the ef of fossil energy land decreased from 5.7707 hm^2 to 2.3074 hm^2. The reason is mainly due to the impact of the global economy, coal prices plummeted, coal production decreased significantly, coupled with the control of iron and steel enterprises. It can be seen that the growth of ef in Wu'an mainly depends on fossil energy land. 2) Analysis of ecological supply. From 2010 to 2014, the average ecological carrying capacity (ec) of Wu'an was 0.4556 hm^2, which was much smaller than the global (1.8 hm^2), only 7.71% of the average ef, Among them, the average per capita ec of cropland is 0.2768 hm^2, accounting for 60.76% of the total ecological supply, which is the main part of the ecological supply of Wu'an. Cropland is the main factor affecting the ecological supply of Wu'an. It can be seen that the main reason for the change of ec is the change of ecological supply of cropland caused by non-agricultural build-up land and ecological returning farmland. 3) Ecological deficit (surplus) and sustainable development analysis. Wu'an was in a state of ecological deficit from 2010 to 2014, the average is 5.4501 hm^2, higher than the global level (3.3301 hm^2), grew first and then decreased, the trend is roughly the same as the per capita ef (Fig. 1). Build-up land was ecological surplus and shows an increasing trend gradually, besides, others are showing ecological deficit, showing that the urbanization of Wu'an occupied the biological production land to a certain extent. Forest is in a state of ecological surplus, mainly because forest consumption in Wu'an is smaller. The serious ecological deficit shows that nature of Wu'an is far beyond its own ecological carrying capacity, which is not in favor of the sustainable development. The main cause is large dependence of Wu'an on energy, this

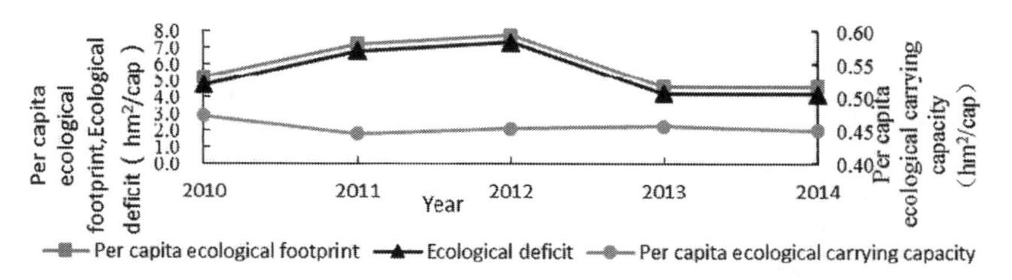

Figure 1. Dynamic change of ecological footprint (ef) in Wu'an from 2010 to 2014.

unreasonable use of resources structure is certainly lead to the region's resource depletion in the future.

5 CONCLUSION

In this paper, the ecological footprint and ecological carrying capacity of Wu'an (2010–2014) are measured. From 2010 to 2014, the per capita ecological footprint and ecological deficit of Wu'an showed synchronous trend, which first increased and then decreased. It mainly because the per capita ecological carrying capacity remained basically unchanged. The average of the per capita ef and ecological deficit were 5.9057 hm^2 and 5.4501 hm^2. The per capita ef reached a maximum (7.7778 hm^2) in 2012. The average ec is 0.4556 hm^2, which is far less than the global (1.8 hm^2), only 7.71% of the average ef of Wu'an, It shows that the sustainable development capacity of Wu'an has a rising trend, but the ecological environment is still facing great pressure, the contradiction of the rapid economic growth, ecological resources is still very prominent.

ACKNOWLEDGEMENT

This paper was supported by the National Natural Science Foundation of China (41641008).

REFERENCES

Gerbensleenes PW. Nonhebel S, Wpmf I. (2002) A method to determine land requirements relating to food consumption patterns. *J. Agriculture Ecosystems & Environment*, 90(1):47–58.

Liu Lemian (2009). Study on dynamic change and forecast of ecological footprint in Yanling County. Doctoral Dissertation, Changsha. Hunan Agriculture University, 2009.

Niu Wenyuan. (2012) Theory and Practice of China's Sustainable Development. *J. Bulletin of Chinese Academy of Sciences,* 27(3):280–290.

Rees W.E. (1992) Ecological footprints and appropriated carrying capacity: what urban economics leaves out. *J. Environment and Urbanization*, 4(2): 121–130.

Rees W, Wackernagel M. (1996) Urban ecological footprints: Why cities cannot be sustainable—and why they are a key to sustainability. *J. Environmental Impact Assessment Review*, 16(4–6):223–248.

Verhofstadt E, Ootegem LV, Defloor B, et al. (2016) Linking individuals' ecological footprint to their subjective well-being ☆. *J. Ecological Economics*, 127:80–89.

Zhang Jianjun, Fu Meichen, Liu Fuchang. (2007) Study on Ecological Footprint Assessment and Sustainable Development in Wuan. *J. China mining*, 16(4):14–18.

Land Reclamation in Ecological Fragile Areas – Hu (Ed.)

Landscape pattern change monitoring and predicting using remote sensing method—taking Jiaozuo as an example

P. Liu
School of Surveying and Mapping Land Information Engineering, Henan Polytechnic University, Jiaozuo, Henan

K. Li
School of Surveying and Mapping Land Information Engineering, Henan Polytechnic University, Jiaozuo, Henan
Yunnan Highway Development and Investment Co. Ltd., Kunming, Yunnan

R.M. Han, Y.M. Yang, X.Q. Cheng & C. Ma
School of Surveying and Mapping Land Information Engineering, Henan Polytechnic University, Jiaozuo, Henan

ABSTRACT: Monitoring LULC and landscape status using multi-temporal RSD is an important way for environmental impacts and ecological degradation monitoring caused by coal resource exploitation. In this research, multi–temporal RSD captured on 1994, 2004 and 2015, together with a vector map, are selected as raw data. On the basis of SVM classification and CA_Markov simulation results, 31 years' dynamic changing trend and the relationship between LULC change and landscape ecology are analyzed. The results demonstrate that the ability of CA_Markov model used for LULC changing trend simulation, and shown the changing trend of different land cover types, including water body, built-up area, farmland, bare land, etc. The signification of the research is that the effectiveness of RSD and CA_Markov model for landscape ecology changing monitoring and prediction are experimented and testified in mining area. Outcomes will certainly helpful to popularize the applicability of the strategies to environmental protection and utilization.

Keywords: Remote Sensing Data (RSD), Land Use and Land Cover (LULC) change, landscape ecology, CA_Markov, coal mining area

1 INTRODUCTION

As the main energy resource of China, coal resources provide fundamental supports for socio-economic development, but coal resource exploitation also results in serious damage to regional land cover, ecological system and human settlement environment, which makes mining arears become one of the most degraded and contaminated regions with fragile ecological environment. It is an important and efficient way to monitor and analyze the environmental impacts and ecological degradation caused by coal resources exploitation using multi-temporal RSD (Remotely Sensed Data) (Liu et al, 2015; Hu & Xie, 2005). The spatial-temporal of Land Use and Land Cover (LULC) restricted by remotely sensed data is useful for the quantitative research of ecological processes. Dynamic change of LULC over coal mining area was studied based on 22 scenes remotely sensed data and corresponding product, such as NDVI (Li, et al, 2016). Changes in land use related to mining in the north-western mountains of Spain during the previous 50 years are also analyzed based on aerial photographs and other materials (Redondo-Vega, et al, 2017). However, it will be more significant to predict landscape pattern changing trend on the basis of previous LULC status analysis.

This research was conducted in Jiaozuo coal mining area, located in northwestern part of Henan province, China. The rest of this paper is organized as follow. A brief introduces of data source and methodology selected is given in Section 2. In section 3, LULC change status and trends were analyzed. Finally, section 4 conclude this work with some remarks.

2 RESEARCH AREA AND METHODOLOGY

The research area is located at Jiaozuo city, which coverage is from 35°10'–35°21' in north latitude, and 113°4'–113°26' in east longitude. It borders the provincial capital of Zhengzhou to the south, Xinxiang to the east, Jiyuan to the west, Luoyang to the southwest. In this study, multi-temporal remotely sensed data captured on October 22,

1999, June 1, 2004 and December 3, 2015, together with the vector maps, are employed as source data. SVM (Support Vector Machine) classifier is used to character land cover types of research area. Furthermore, the land cover status in 2025 is simulated based on SVM classification results of 2004 and 2015, which make a probability for landscape pattern predicting in future. On the basis of SVM classification and CA_Markov simulation results, the dynamic changes in land cover types at different times and the land use condition of coal mining areas is obtained by analyzing the landscape pattern in landscape scale and class scale respectively. The relationship between LULC (Land Use and Land Cover) change and landscape ecology in this coal mining area is analyzed.

2.1 *LULC status obtain and changing analysis*

LULC status is obtained by SVM machine learning classifier. A Support Vector Machine (SVM) supervised classifier is adopted to extract LULC status, for its superior classification performances with respect to other classifiers in term of efficiency, accuracy and generalization of the results (Giocco et al, 2010). The mechanism of SVM is on the basis of statistical learning theory. In this study, the Radial Basis Function (RBF) of SVM is adopt, because it works well in most cases according to the result of previous research (Wu et al, 2004). The changing trend of different LULC types are analyzed using land use transfer matrix.

2.2 *Prediction of landscape pattern changing*

CA_Markov model, using cellular automata in combination with Markov chain analysis, is selected for landscape pater changing trend simulation and prediction. The process of landscape pattern prediction model selected in this research, CA_Markov model, can be described as follows (Sang et al., 2010; Liu et al., 2012): (1) obtaining landscape pattern with SVM classification; (2) determining transition rules on the basis of results from step (1). And using GIS spatial overlay analysis function to achieve transition probability matrix and transfer area; (3) using CA_Markov transition rule to predict future status of landscape pattern simulated. In this step, the most critical steps is determining start status, cell loop times and CA filters model.

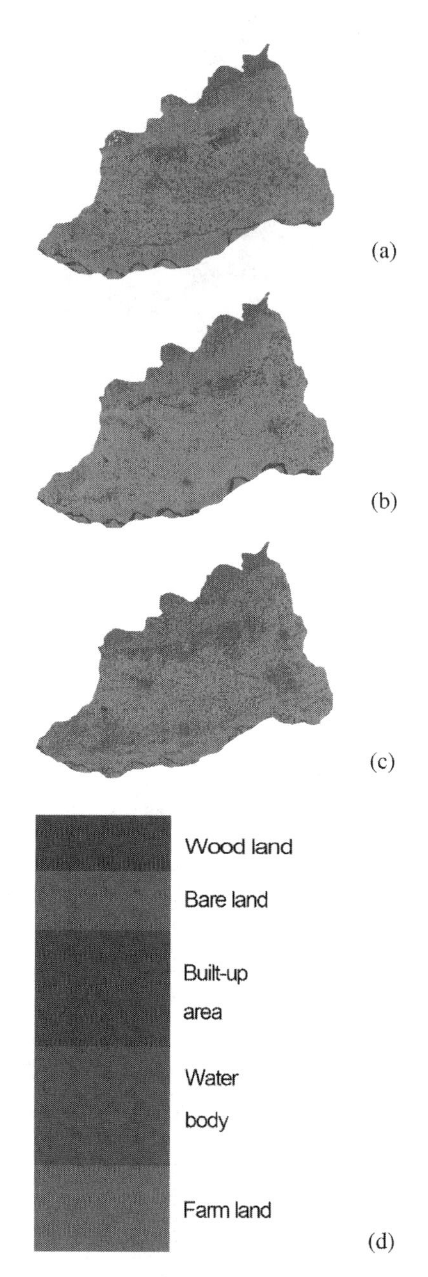

Figure 1. LULC classification results obtained in year of (a) 1994, (b) 2004, (c) 2015 respectively, and the classification accuracy analysis.

3 RESULTS AND ANALYSIS

3.1 *Landscape pattern status*

In this research, the parameter of SVM classifier are assigned as: 0.14 for gamma genes, 100-penalty parameter and 0 pyramid level. In this mining area, five LULC types is gathered, by considering real situation of the mining area and the requirement of monitoring. The classification land cover types are woodland, bare land, built-up area, water body and farmland, shown as Figure 1.

The classification accuracies are assessed from 4 aspects, they are Overall Accuracy (OA), User Accuracy (UA), Product Accuracy (PA), and kappa coefficient. UA and PA of different years

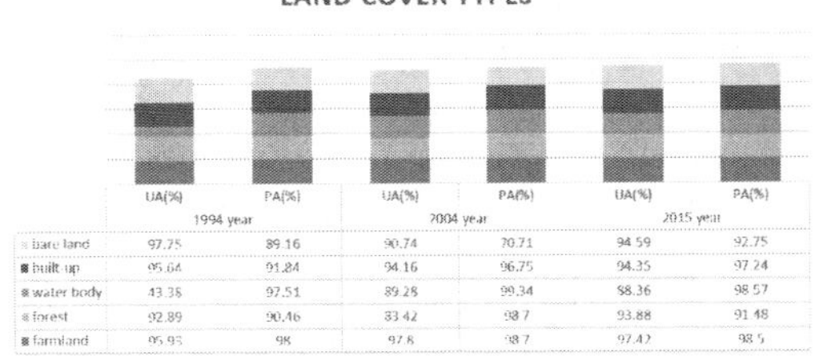

Figure 2. User and product accuracy of different land cover types in different year.

can be seen on Figure 2. The statistic results of OA in year of 1994, 2004, and 2015 are 96.61%, 96.23%, 97.63% respectively, and the corresponding kappa coefficient are 0.95, 0.94 and 0.96. From Figure 1 we can see that classification accuracy is satisfied the requirements of trend analysis of landscape pattern.

3.2 *Future status prediction*

In this stage, the CA_Markov model was selected to simulate landscape pattern status in future. The model is using initial state probabilities and determine trends probabilities at different times of stochastic processing to archive the purpose of predicting future trends (zhang, 2012). On the basis of landscape pattern status obtained in previous stage, conditional probability transition matrix is calculated. The simulation processing including three steps in gross: (1) landscape pattern status archived by SVM classifier or any other classifier you selected with a satisfied classification accuracy; (2) calculating transition probability and condition probability matrix using GIS analysis approach; (3) setting iteration times to simulate landscape pattern status in future. In this research, landscape pattern status both in 2015 and 2025 are simulated on the basis of results obtained from 1994 and 2004 RSD. The status map simulated in 2025 is analysis landscape changing trend in future, and the status map simulated in 2015 is used to comparing with status map obtained by SVM classifier. The error accuracy of simulation result is –0.013, 0.385, –0.022, 0.290 and 0.347 for farmland, forest, water body, built-up area and bare land respectively. The comparison analysis shown that there is an acceptable accuracy to simulate landscape status over times using CA_Markov model. So, the predicted landscape pattern status in 2025 is reliable for changing trend analysis in future.

3.3 *Future status prediction*

In this stage, 30 years landscape pattern changing trend from 1994 to 2024 was analyzed on the basis of SVM classification results and CA_Markov predicted outcomes. In order to monitoring, analyzing spatial-temporal changes of landscape pattern over mining area, landscape transition matrixes of different period are calculated (shown as Table 1, taking transition matrix from 1994 to 2004 as an example).

Table 1. Landscape transition matrix from 1994 to 2004 (km^2).

1994 \ 2004	BLT	WT	FM	FST	BAR	Total
BLT	85.95	1.40	63.58	2.74	8.49	317.8
B%	27.03	0.004	20.00	0.009	2.67	7.55
C%	18.22	1.95	2.43	1.32	1.00	
WT	0.93	13.53	11.22	0.37	11.25	93.15
B%	10.55	14.52	12.40	0.40	12.07	2.213
C%	0.19	18.80	0.42	0.18	1.33	
FM	79.44	3.58	763.3	1.57	75.66	1776.12
B%	4.47	0.20	42.97	0.08	4.26	42.205
C%	16.84	4.98	29.23	0.75	8.98	
FST	18.79	1.40	125.3	67.35	72.10	569.9
B%	3.30	0.24	22.10	11.83	12.69	13.54
C%	3.98	1.95	4.79	32.46	8.56	
BAR	42.80	16.95	394.2	33.29	266.5	1451.
B%	2.94	1.16	27.16	2.29	18.36	34.48
C%	9.07	23.56	15.09	16.04	31.68	
Total	471.6	71.93	2611.	207.4	842.0	4204.
OA%	11.21	1.71	62.10	4.935	20.20	100
CH%	48.35	–22.7	47.02	–63.5	–41.9	

Note: Where (*A*) is transition matrix between ***K*** and ***K* + 1** period, and value A_{ij} is area transformed; (***B***) is the proportion transformed between research periods; (***C***) is the percentage of contribution, means changing degree from ***K*** and ***K* + 1** period. BLT means building, WT means water, FM means farmland, FST means forest, BAR means bare-land, OA means overall accuracy, CH mean changing rate.

From transition matrix from 1994 to 2004, we can see that during the first 10 years, the changing trend of landscape patterns is that all land cover types were decreasing while built-up area was increasing. Between period from 2004 to 2025, built-up area is the most drastic change type the among different landscape types, the total changing rate is up to almost 70%. The outcomes of this research shown that there is a sustained downward trend in water body, while continues growth tendency in build-up area. And a wavelike change in farmland, forest and bare land, first increased and then decreased for farmland and forest, while in reverse for bare land.

4 CONCLUSIONS

Our research demonstrate that classification accuracy obtained by SVM classifier is satisfied the requirements of trend analysis of landscape pattern. CA_Markov model is able to predict LULC trends with good performance, and is an effective way to integrate RSD with spatial-temporal model for analysis of LULC change and corresponding environmental impact. The effectiveness of RSD and CA_Markov model for landscape pattern change monitoring and predicting are also experimented and testified in this paper. The outcomes of this research will certainly help to popularize the applicability of the strategies to environmental protection and resource utilization.

ACKNOWLEDGEMENT

The authors thank the support to this research from the National Natural Science Foundation of China under Grant No. 41601450, Henan Postdoctoral Foundation, The Key Research Project Plan of Colleges & Universities in Henan Province, the Ph.D. fund of Henan Polytechnic University (No. B2015-20), and the Fundamental Research Funds for the Universities of Henan Province (NSFRF140113).

REFERENCES

Giacco, F., et al. (2010) Uncertainty Analysis for the Classification of Multispectral Satellite Images Using SVMs and SOMs. *Ieee Transactions on Geoscience and Remote Sensing*, 48(10): p. 3769–3779.

Hu Zhenqi & Xie Hongquan. (2005) Study on land use /cover change of coal mining area based on remote sensing images. *Journal of China Coal Society*, 2005, 30(1): 44–48.

Liu P, Du PJ & Pang YF. (2012) Analysis and simulation of land cover and thermal environment in mining area based on remote sensing data and CA_Markov model. *Journal of China Coal Society*, 37(11): 1847–1854.

Redondo-Vega, JM, Gomez-Villa, A; Santos-Gonzalez & J. (2017) Changes in land use due to mining in the north-western mountains of Spain during the previous 50 years, *Catena*,149, 844–856.

Sang LL, Zhang C, Yang JY, et al. (2010) Simulation of land use spatial pattern of towns and villages based on CA_Markov model. *Mathematical and Computer Modelling*, 55(3–4): 938–943.

Wu TF, Lin CJ & Weng RC. (2004). Probability estimates for multi-classification by pairwise coupling. *Journal of Machine Learning Research*, 5: 975–1005.

Zhang, X. (2012) Prediction of land use based on CA Markov model in Gannan prefecture, Lanzhou University, Master thesis.

Land Reclamation in Ecological Fragile Areas – Hu (Ed.)

Monitoring of vegetation coverage by multi-temporal remote sensing images: A case study of Yulin coal mine

X.N. Chen, X. Yuan & Z.Y. Yang
College of Geomatics, Xian University of Science and Technology, Xi'an, Shaanxi, China
College of Geology Engineering and Geomatics, Chang'an University, Xi'an, Shaanxi, China

ABSTRACT: An increasing exploitation of mineral resources has made ecological destroyed and environment polluted, which has a serious impact on the ecosystem. In this paper, the change of vegetation cover is taken as an indicator of the terrestrial ecosystems. The Normalized Difference Vegetation Index (NDVI) data from 2000 to 2014 are employed to investigate the spatio-temporal variation of vegetation cover, and the effects of human activities and climate change on vegetation cover were also analyzed by using Sen-trend combined with the Mann-Kendall test and correlation analysis method. The results indicate that the vegetation coverage presented an increasing trend over Yulin, but there was no obvious change in some areas whereas the cover decreased largely in 2004. According to the analysis, climate change and human activity were the major factors for vegetation cover.

Keywords: Vegetation Coverage; NDVI; Correlation Analysis Method; Sen-trend; Mann-Kendall Test 1

1 INTRODUCTION

The vegetation, connecting soil with water and atmosphere, is the natural bond which can reflect overall condition of the ecological environment. Normalized Difference Vegetation Index is a ratio related to red and near-infrared bands about reflectivity in remote sensing images[1]. It goes hand in hand with vegetation coverage, biomass and leaf area index and is the most widely used index in variation of vegetation cover. This variation is a result of the interactions between the factors of human activities and climate change[2]. Previous studies indicate that the constantly warming climate significantly enhance vegetation activity. The vegetation cover has shown an increasing trend in the Yellow River, which is closely related to precipitation. Moreover, human activity should be taken seriously for their frequent actions[3]. Northern Shaan'xi is located in the Loess Plateau with poor water resources, weak ecology environment and serious water and soil loss. It is vegetation that provides for an efficient level of conservation effort[4]. With the implementation in ecological projects from 1999, there have been greater changes in regional vegetation. However, due to the increasing demand for higher mining, negative effects become prominent. Based on what discussed above, this article analyzed the space time structure of annual NDVI variation and it's influences from 2000 to 2014 by NDVI and meteorological data concerned, combined with the methods of correlation analysis and change detection. The comprehensive evaluation of the effect is made for not only a correct understanding, but also a further prediction.

2 MATERIAL AND METHODS

2.1 *Overview of the study area*

Northern Shaan'xi includes Yu'lin and Yan'an cities mainly, which is near northern Inner Mongolia, southern Tong'chuan city in Shaan'xi, western Gan'su and Ning'xia, and eastern Shan'xi (from 107.10°E to 111.4°E and from 34.49°N to 39.35°N). It covers 77191.087 km^2, and accounts for 40.11% of the total area in Shaan'xi. It represents the typical characters of temperate continental monsoon climate with mean annual air temperature of 9.2, mean annual precipitation of 500 mm, while mean evaporation of 1000 mm. The kind of physiognomy is complicated and various, including the transitional zone of aeolian landform, the hilly-gullied fragile region and the pimple mound region.

2.2 *Data sources*

The data is SPOT-VEGETATION (SPOT-VGT) S10, which was issued by Flemish Institute for Technological Research, VITO. It provide global coverage with the spatial resolution of 1 m and cover the period from 1998 to 2014 and with the radiation correction, atmospheric correction and geometric correction finished. The meteorological data ranging from 2000 to 2014 comes from China meteorological data sharing service system (http://cdc.cma.gov.cn).

2.3 *Methods of data processing*

2.3.1 *Time series analysis*

Common ways include principal components analysis, algebraic operation, regression analysis, wavelet transform and so on, while the integration of Sen-trend and Mann-Kendall is widely employed in the trend studies of vegetation cover. The calculating formula of Sen-trend is as follows:

$$\rho = median\frac{x_j - x_i}{j - i}, 1 < i < j < n \tag{1}$$

where x_j and x_i are time series records; $\rho > 0$ means the trend is up, vice versa. Mann-Kendall is used for mutation test, The testing model is expressed in the following equations:

$$Q = \sum_{i=1}^{n-1}\sum_{j=i+1}^{n} sign(x_j - x_i) \tag{2}$$

$$sign(s) = \begin{cases} 1(s > 0) \\ 0(s = 0) \\ -1(s < 0) \end{cases} \tag{3}$$

$$z = \begin{cases} \dfrac{Q-1}{\sqrt{Var(Q)}}(Q > 0) \\ 0(Q = 0) \\ \dfrac{Q+1}{\sqrt{Var(Q)}}(Q < 0) \end{cases} \tag{4}$$

where Q is test statistic; z is normalized test statistic; sample is indicated by n.

2.3.2 *Correlational analysis method*

Plant growth is the result of various factors, mutuality intensity among the factors is required. The degree that climate affects vegetation is deserve to study, and the calculating model is shown below:

$$r_{xy} = \frac{\sum_{i=1}^{n}\left[(x_i - \bar{x})(y_i - \bar{y})\right]}{\sqrt{\sum_{i=1}^{n}(x_i - \bar{x})^2 \sum_{i=1}^{n}(y_i - \bar{y})^2}} \tag{5}$$

2.3.3 *Effect of human activity*

With a regression analysis of the NDVI and multi-climatic factors, we can calculate the NDVI difference of the prediction values and factual values to obtain residual as human's influence. The calculating model of residual is as follows:

$$\varepsilon = NDVI_t - NDVI_p \begin{cases} > 0, positive \\ \approx 1, \\ < 0, negative \end{cases} \tag{6}$$

3 RESULTS AND ANALYSIS

3.1 *Annual variation of NDVI*

As is shown in Figure 1, NDVI in the researched region took on the feature of rising with fluctuations, but it was still an overall growth which showed an obvious effect in vegetation restoration of the northern Shaan'xi while there was a sharp cut in 2005. According to Figure 2, the areas of improvement were much larger than that of the degradation, which offered a further evidence for the increasing vegetation. And the areas of improvement were distributed mainly in the eastern regions, more than 84.65% of the whole range. The rate of improving decreased gradually from southeast to northwest.

3.2 *Effect of climate*

Based on Figure 3, with the arrival of growing season, NDVI showed the highest growth rate

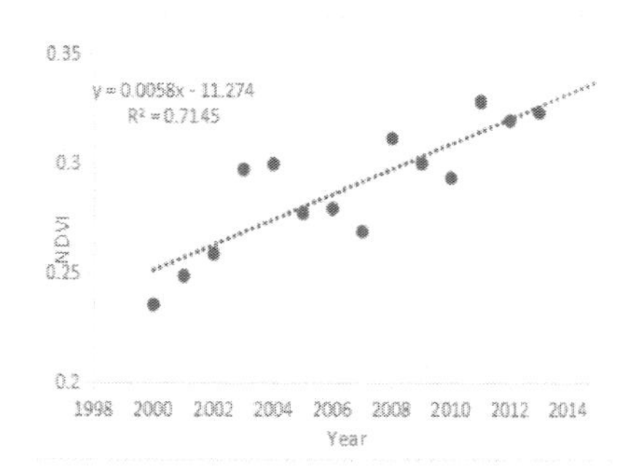

Figure 1. The variation tendency of annual average NDVI.

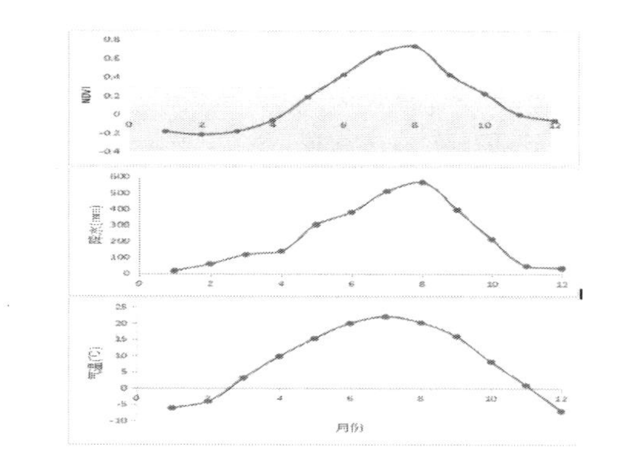

Figure 2. Spatial distribution of annual average NDVI of the northern Shaan'xi.

and peaked on August. The vegetation developed in the most vigorous stage at the moment, after which it weakened. The variation of temperature and precipitation is quite similar to that of NDVI. From the Figure 4, it was pointed out that there were significant spatial differences in the impacts that temperature and precipitation brought respectively.

3.3 *Human activity*

Figure 5 indicated that human activity made a considerable impact on vegetation coverage. The positive areas accounting for about 72.58% were in the east and the south. What boosted the growth of vegetation were whether afforestation and agricultural activity. Rather, the reasons for the

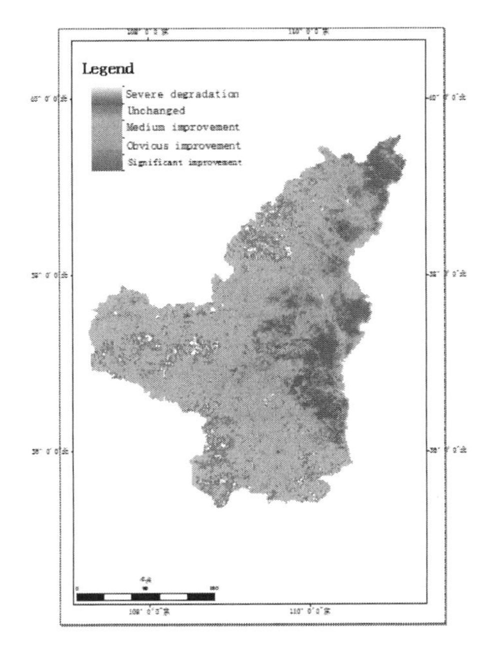

Figure 3. Intra-annual variation of monthly average of DNVI, precipitation and temperature.

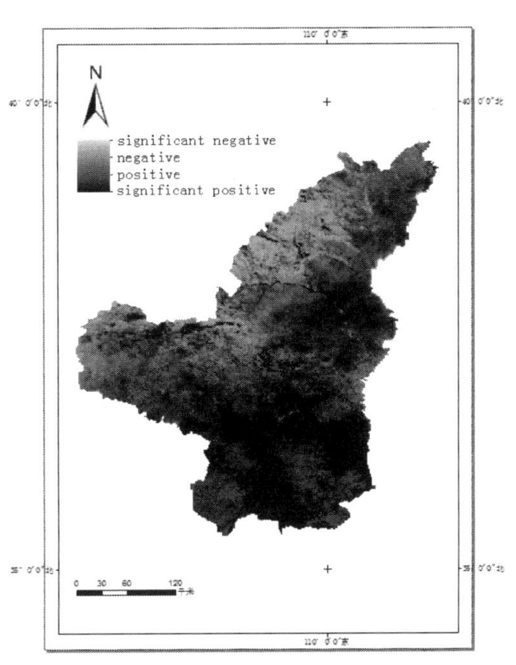

Figure 5. The characteristics of impact human activities on vegetation in northern Shaan'xi.

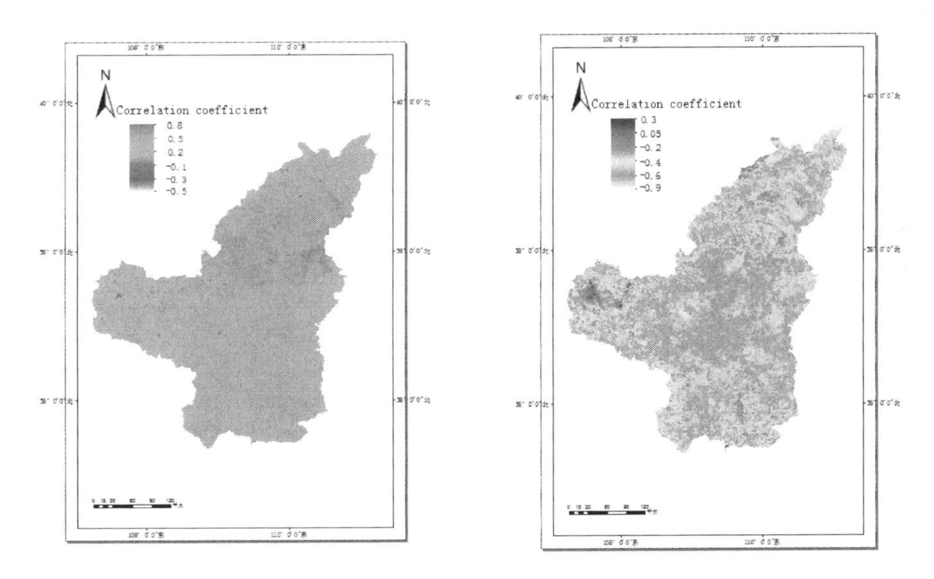

Figure 4. The spatial characteristics of correlation between NDVI and climate change.

negative were excessive urbanization and mineral exploration and the balance of nature and ecological environment had been broken due to what was discussed above.

4 CONCLUSION

1. The vegetation cover in the researched region exhibited an uneven distribution but still an overall growth in recent 15 years. As there was decline in some area, the extent of improvement were much larger than that of the degradation.
2. NDVI in northern Shaan'xi was significantly correlated with precipitation ($P < 0.01$). It was a pseudo effect that temperature effected on NDVI and there was not remarkable correlation between them.
3. Human activity had made a considerable impact on vegetation coverage, among which positive effect was much greater and the positive areas were mostly in the east and the south. In conclusion, vegetation coverage had an annual variation in northern Shaan'xi from the combined impact of the change of climate and human activity. The increasing of NDVI was due in large part to factors of precipitation and the areas with less vegetation relied more on it. Whether afforestation and agricultural activity were the main human factors that elevated vegetation cover, yet urbanization and mineral exploration affected the restoration and reconstruction of ecological environment as well. Consequently, the departments concerned are supposed to take appropriate measures to prevent ecological environment from negative influences of human activity.

REFERENCES

Lambin EF, Strahler AH. (1994). Indicators of land-cover change for change-vector analysis in multitemporal space at coarse spatial scales [J]. International Journal of Remote Sensing 5(10): 2099–2119.

Rouse Jr J, Haas R, Schell J. et al. (1974). Monitoring vegetation systems in the Great Plains with ERTS [J]. NASA special publication, 351: 309.

Townshend JR (1994). Global data sets for land application from the Advanced Very High Resolution Radiometer: an introduction [J]. International Journal of Remote Sensing 15(17): 3319–3332.

Tucker CJ, Townshend JR, Goff TE (1985). African land-cover classification using satellite data [J]. Science, 227(4685): 369–375.

Land Reclamation in Ecological Fragile Areas – Hu (Ed.)
 ISBN 978-1-138-05103-4

Remote sensing inversion of soil organic carbon in Jining city based on decision tree classification

W.J. Sun, X.J. Li & Q.C. Li
College of Resources and Environment, Shandong Agricultural University, Taian, China

ABSTRACT: This study used Landsat8 remote sensing data and the measured data of soil organic carbon, taking the carbon content of the surface 0–20 cm of soil in the study area as the research object, through multiple linear regression analysis of SPSS, building the prediction model: $soc = 24.476 + 51.503 \times b_1 - 49.824 \times b_4 - 11.297 \times b_6 - 21.778 \times b_7$, $R^2 = 0.747$, $P < 0.01$. Then, based on the Band Math and Decision Tree of ENVI, the soil organic carbon content and distribution in the whole study area were obtained. The results showed as follows: (1) The average soil organic carbon content in the study area was about 12.45 g/kg, which was in the middle even lower level. And the organic carbon content in some areas was close to 0 g/kg, the highest content was $22.5 \pm 4.5 g/kg$. (2) The organic carbon content was mostly concentrated in 12–18 g/kg, which area of about 6165.13 km^2, accounting for the total land area of 59.99%. It was the most widely distributed.

Keywords: Soil organic carbon; Remote sensing inversion; Decision tree classification; Distribution

1 INTRODUCTION

Carbon as an element of nature, moving in endless cycles, especially the form of CO^2 in the atmosphere and SOC in the lithosphere, closely related to human survival (Jin, et al., 2000). Soil organic carbon as part of the soil carbon pool, affecting the global climate and land productivity directly or indirectly. Therefore, it is of great significance to study the regional soil organic carbon status.

The research of soil organic carbon has been 200 years history. Since the 1970s, the popularity of statistical methods has facilitated the study of soil organic carbon. A large number of statistical methods were used to study the soil organic carbon content and vertical spatial distribution (Post et al., 1982. Foley, 1995. Xue, et al., 2011. Yuan, et al., 2008). Post & Emanuel (1982), using the soil profile and the Holdridge life-zone classification system to estimate the global soil organic carbon pool. Foley (1995) examed the global terrestrial carbon budget using a process-based model, DEMETER-1. Xue, Z.J. et al. (2011) studied the distribution of soil organic carbon in different depths by using soil sample data and geostatistics.

At the same time, remote sensing inversion based on RS and GIS software has increasingly become an important method to study the spatial distribution characteristics of soil organic carbon, spatial interpolation method and reflectance, DN value and other remote sensing information modeling method (Li, et al., 2005. Chen, et al., 2015) are commonly used inversion method. Dai, J.R. et al. (2015) achieved the visualization about the spatial distribution of surface soil organic carbon density in Shandong province by the analysis of the soil profile data and spatial analysis function of MapGIS; You, H.C. (2012) achieved remote sensing inversion mapping through the use of ArcGIS spatial interpolation, and then researched the spatial distribution characteristics of soil organic carbon of Shun Lin Country; Wang Qiong (2016) realized remote sensing inversion of soil organic carbon in cotton field in oasis area of northern Xinjiang by band operation.

Decision Tree as a scientific image classification method in remote sensing application, Chen Wen (2005) thought this method was not only suitable for large training data set, but the structure constructed easily, high efficiency and good classification accuracy. However, few studies have used the decision tree classification to carry out spatial inversion of soil organic carbon. This study took this as the starting point, study on soil organic carbon content in 0–20 cm tillage layer, remote sensing image reflectance and soil organic carbon content as the analysis factor for multiple linear regression analysis, and then established the organic carbon prediction model. Finally, the spatial remote sensing of soil organic carbon in the whole study area was studied by using the decision tree classification method.

2 MATERIALS AND METHODS

2.1 *General situation of study area*

Jining is located in the southwest hinterland of Shandong province. Scope of latitude and longitude is 34°27′ 11.83″ N-35° 58′ 42.27″ N, 115° 51′ 28.51″ E-117° 34′ 57.24″ E, 167 kilometers long in north and south direction, 158 kilometers long in east and west direction; The terrain is complex, mainly in plain depressions, mountains, hills, basins are distributed, the east higher than the west. It belongs to warm temperate monsoon climate, the average temperature is 13.3°C–14.1°C, annual precipitation in the 597–820 mm or so.

Jining City is both a agricultural city and a resource-based city, the total land area is about 11,187 square kilometers, the arable land area is about 6113.2 square kilometers, contains brown soil, cinnamon soil, shajiang black soil, fluvo-aquic soil, paddy soil and other soil types. Jining City is also one of the eight largest coal mining base, the main coal-bearing strata are more than 10 layers, recoverable thickness of about 10 meters.

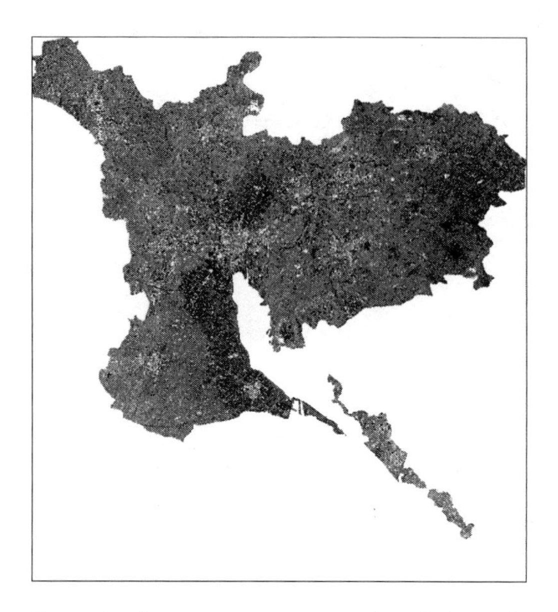

Figure 1. Image preprocessing results of the study area.

2.2 *Remote sensing image acquisition and processing*

Remote sensing images derived from Geospatial Data Cloud, Selected two landsat8 data images of October 2, 2015, corresponding to the sampling time. "Band number - Line number" are "122–35", "122–36".

Remote sensing images in practical applications, in order to satisfy the information required for the research project, it is necessary to carry out the preprocessing. And for different research content, the content of its pre-processing is not exactly the same. This study needed to obtain ground reflectance and geographic coordinates information as accurate as possible. Thus, the remote sensing image was mainly composed of five aspects: radiation calibration, atmospheric correction, geometric correction, mosaic and cutting. The result is shown in Figure 1.

2.3 *Soil sampling and analysis*

According to the farming mechanism of the study area, to avoid other objects to cover the ground, in early October 2015, totally 134 samples were collected on the basis of the design route, taking into account the uniformity and reliability of the samples, a total of 129 valid samples were screened for data statistics, each sample was positioned by hand-held GPS, the sampling depth is 0–20 cm above the surface, each soil sample was composed of five soil samples in the range of 30 m * 30 m in accordance with diagonal sampling method. The sample distribution is shown in Figure 2.

Figure 2. Distribution of soil sample.

Soil organic carbon determination methods are diverse, there are gravimetric methods, colorimetric methods and volumetric methods. In this study, potassium dichromate volumetric method was used—External heating method. After bringing the soil samples back to the laboratory, picked out the plant roots, stone bricks, small animals and other intrusions. After drying, air-drying, beating, grinding and sieving, then, began to do soil organic carbon determination.

2.4 *Modeling of soil organic carbon*

According to the organic carbon determination results, the sample points were divided into four groups which the soil organic carbon content were 5–10 g/kg, 10–15 g/kg, 15–20 g/kg, 20–25 g/kg. And then seven sample points were evenly extracted from each group, a total of 28 points as a verification point. The remaining 101 sample points were used as modeling points (as shown in Table 1).

This study used the modeling function of SPSS to analyze the relationship between the reflectance and the soil organic carbon content, and then took the soil organic carbon as the dependent variable and the reflectivity of each band as the independent variable to perform the multivariate linear stepwise regression analysis.

2.5 *Prediction of soil organic carbon content*

In order to obtain the range of soil organic carbon content of the study area, this study adopted two methods, one was Desity Slice which based on ENVI and Band Math, the other was Kriging Method which based on ArcGIS and the theory of statistics.

2.6 *Spatial inversion of soil organic carbon*

2.6.1 *Basic knowledge of decision tree classification*

Decision tree classification is based on remote sensing image data and other spatial data, through expert experience summary, simple mathematical statistics and inductive methods, to acquire classification regulation for remote sensing classification (Deng, 2010). Decision tree classification is a hierarchical classification method (Zhang, 2005) its classification process can be expressed very well with the decision tree structure, the decision tree comprise of a parent node (such as the top node in Figure 4) and a number of child nodes (the nodes except the top node in Figure 4), each node is a separate classification process, the end of each node represents a classification result, that is, a data category. Figure 4, Sequentially from the top node down progressively classified operation, a total of four nodes, that is, four levels, resulting in four categories.

Table 1. Soil organic carbon sample content parameters (unit: g · kg–).

Option	Sample points	The minimum value	The average value	The maximum value
Modeling sample set	101	5.13	13.61	24.98
Validation sample set	28	5.32	14.73	24.10

2.6.2 *The process of decision tree classification*

Firstly, based on the Desity Slice, the soil organic carbon content was divided into four content gradients: 0–5 g/kg, 5–12 g/kg, 12–18 g/kg, 18–27 g/kg. Then, constructing decision tree based on content gradient and prediction model (as shown in Figure 3). Next, executing decision tree and generated preliminary classification results which looked rough and was necessary to carry out Majority/ Minority Analysis as well as Clump and Sieve. Finally, converting it to vector file to import ArcMap for a map.

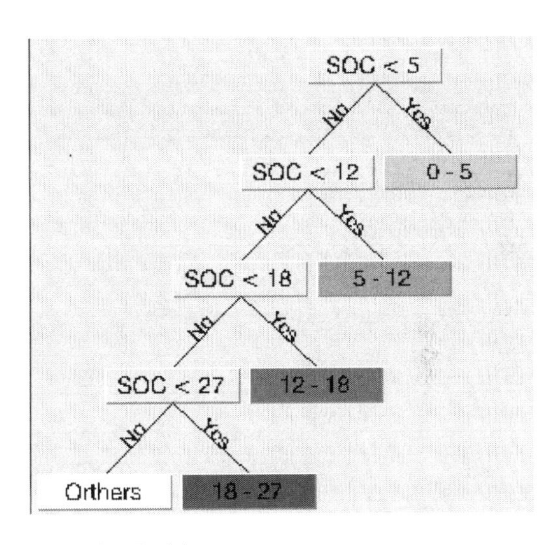

Figure 3. Decision tree structure.

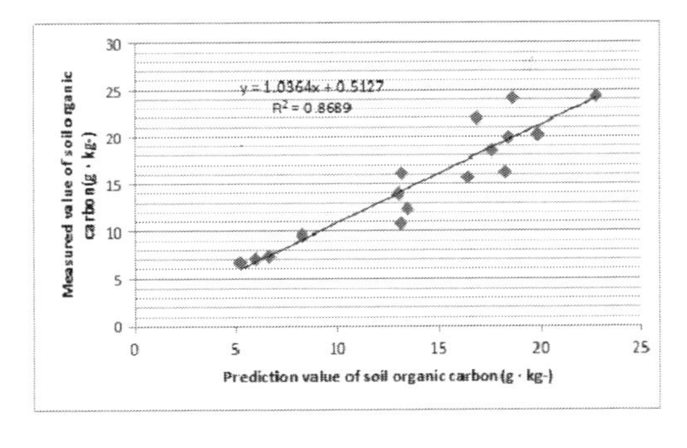

Figure 4. Reliability test of organic carbon inversion model.

3 RESULT ANALYSIS

3.1 *Model and verification of soil organic carbon retrieval*

The higher the correlation coefficient (R^2) of the model, the smaller the standard deviation (RMSE), the better the fit. Observed Table 2, found that the Mode 4 has maximum R^2 and minimum RMSE, but the significance (P) is high enough. Therefore, consider the model parameters such as R^2, P and RMSE, choose Mode 4.

That is:

$$soc = 24.476 + 51.503 \times b_1 - 49.824 \times b_4 - 11.297 \times b_6 - 21.778 \times b_7.$$

In order to verify the reliability of the model, this study used ENVI4.7 to extract the reflectance of each band corresponding to 28 verification points, substituted it into the inversion model and obtained the corresponding organic carbon prediction value. The predicted value was compared with the measured value by Statistical analysis, $R^2 = 0.8689$, there was a good linear relationship between the two (as shown in Figure 4), that was, organic carbon inversion model had good reliability, and it could be used to predict the soil organic carbon content throughout the study area.

3.2 *Prediction of soil organic carbon*

3.2.1 *Soil organic carbon content*

Table 3 shows that, since the Kriging interpolation method is based on the value of the adjacent variable (Niu, et al., 2001), that is, the sample value in this study, and then used the local factors to estimate the unknown value, with obvious geometric characteristics relative to the Density Slice. As a result, the range of organic carbon content is also closer to the organic carbon content of the collected soil samples. Density Slice method is based on the spectral characteristics of the image, through the band calculation (Band Math), each pixel is in accordance with the organic carbon prediction model to obtain a fixed value, in principle, higher accuracy, the predicted value is more reliable, however, due to the prediction model and the impact of ground objects, resulting in some data anomalies, need to be removed. Taking into account the actual situation, the organic carbon content can not be less than 0 g/kg, so the organic carbon content in the range of −1.62–0 g/kg calculation results deleted, at the same time as the maximum value exceeds the sample value too much, in the step-by-step test calculation, the first 34–39 g/kg "Aplay", found that the image was difficult to find this category by naked eye, so the corresponding data deleted, and then followed by 30–34 g/kg, 34–28 g/kg, 26–28 g/kg, 27–28 g/kg, 26–27 g/kg, eventually found the maximum close to 27 g/kg instead of 39.14 g/kg. That is, 27–39.14 g/kg range of data removed, and then calculate the average was about 12.45 g/kg. Therefore, the study found that soil organic carbon maximum content was about 27 g/kg, the lowest about 0 g/kg, the average value was 12.45 g/kg.

Table 3. Parameters of organic carbon under different statistical methods.

Option	The maximum value	The minimum value	The average value
Density Slice	27(39.14)	0(−1.62)	12.45
Kriging interpolation	23.57	3.44	14.83
Sample measured value	24.10	5.13	13.92

Note: Unit of content is g/kg; 39.14, −1.62 is the actual forecast value, and the value of 27 and 0 is adjusted according to the actual situation.

Table 2. Soil organic carbon prediction model and parameters.

Model	Expression	R^2	RMSE	P
1	$soc = 25.233 - 40.913 \times b_6$	0.690	4.404	0.000
2	$soc = 25.280 - 27.932 \times b_6 - 16.675 \times b_7$	0.709	4.105	0.02
3	$soc = 226.114 + 13.678 \times b_1 - 25.765 \times b_6 - 27.996 \times b_7$	0.726	3.882	0.05
4	$soc = 24.476 + 51.503 \times b_1 - 49.824 \times b_4 - 11.297 \times b_6 - 18.644 \times b_7$	0.747	3.553	0.05
5	$soc = 23.448 + 65.958 \times b_1 - 67.703 \times b_4 - 21.778 \times b_7$	0.744	3.561	0.01

Note: soc is the soil organic carbon content; bi is the i-band reflectivity; R^2 is the correlation coefficient; RMSE is the standard deviation; P is significant.

3.2.2 *Spatial distribution of soil organic carbon*

1. Analysis on the advantages of Decision Tree classification

The results of the three methods are shown in Figures 5 and Figures 6 and Figures 7.

As shown in Figure 5, the Kriging interpolation map had the obvious geometry features as it took samples as the divergence center. In other words, it was closely related to the number of samples. Thus, only in the case of dense distribution of sample points, could it maintain a certain degree of accuracy. Compared with the Kriging interpolation method, the density mapping method was based on the prediction model and Band Math, the content was dotted, the inversion effect was better, but the image looked rough (as shown in Figure 6).

Compared with the density segmentation, although the two are similar in classification basis and accuracy, but the classification process and drawing quality were different. The decision tree classification could be used to merge or remove the small patches on the inversion image with the help of the ENVI classification function, however, the density slice image was difficult to deal with, and the image was rough so that the classification effect was not as good as the former. Therefore, the decision tree classification was a good method of mapping in the spatial inversion of soil organic carbon, that was, to maintain the classification accuracy, but also to ensure a better visual effect.

2. Statistical analysis of spatial distribution of soil organic carbon

Using ENVI, the results of classification of decision tree were analyzed statistically. The results are shown in Table 4.

Table 4 shows that the organic carbon content in the study area is mainly concentrated at 12–18 g/kg, with an area of about 4850.56 km^2, accounting for

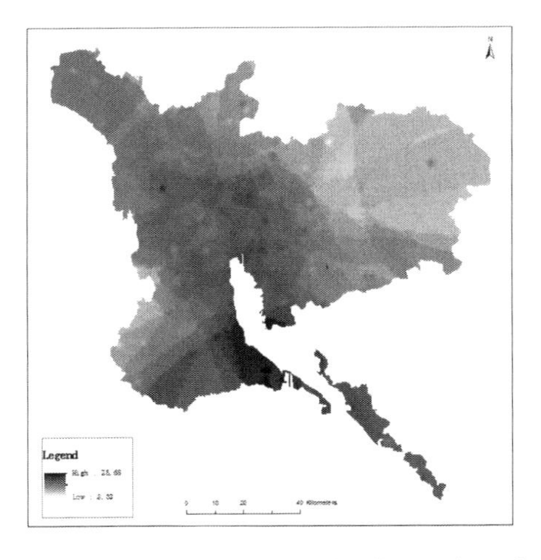

Figure 5. Spatial distribution of soil organic carbon based on Kriging interpolation.

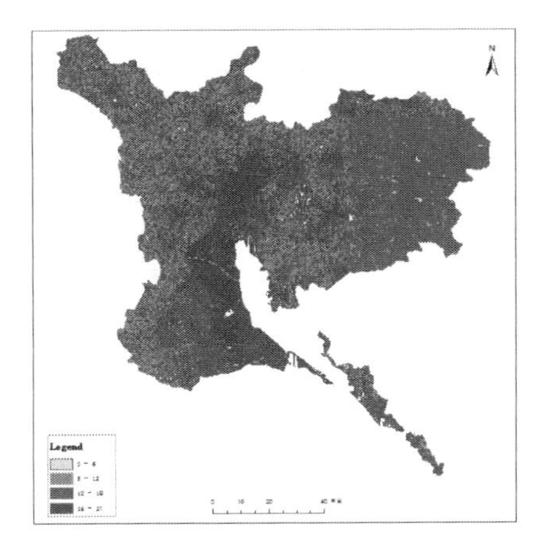

Figure 6. Spatial distribution of soil organic carbon based on density division.

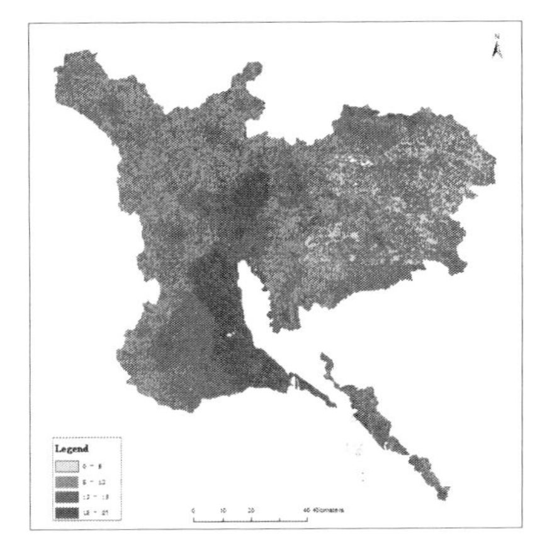

Figure 7. Spatial distribution of soil organic carbon based on density division.

Table 4. Results of soil organic carbon statistics.

Option (g/kg)	Area (km^2)	Proportion (%)	Distribution
0–5	501.43	5.17	distribution mainly in the eastern part of Jining City, and a small amount of distribution in the northern region
5–12	2396.20	24.68	distributing wildly outside Weishan County, Yutai County
12–18	4850.56	49.95	most widely distributed, even in each county
18–27	1972.24	20.20	concentrated in Rencheng District, Weishan County

49.95% of the total land area of the study area. The distribution is the most extensive and distributed even in all counties, indicating the overall level of soil organic carbon in Jining is not high; The distribution of the smallest 0–5 g/kg, an area of 501.43 km^2, less than 6% of the region, mainly scattered in the northeastern region; Soil organic carbon content of 5–12 g/kg, 18–27 g/kg of the two gradient, the area is quite, but the former distribution is more extensive, the latter is concentrated in the Rencheng district and Weishan County which almost all areas is in the city's highest level of soil organic carbon content, and its agriculture, forestry, animal husbandry and fishery has also been ahead of other regions. It can be seen that soil organic carbon plays an important role in land production.

4 CONCLUSION

1. Through the Band Math and Density Slice based on ENVI, the average soil organic carbon content of the tillage layer was about 12.45 g/kg which is at the middle level. Part of the organic carbon content close to 0 g/kg while the highest content in Weishan and Rencheng was about $22.5 \pm 4.5 g/kg$.
2. According to the results of remote sensing inversion and statistical analysis, the organic carbon content in the study area is mainly concentrated at 12–18 g/kg, the area is about 4850.56 km^2, accounting for 49.95% of the total study area. And its distribution is the most extensive, even in all the country.

REFERENCES

Chen, H.Y. et al. (2015) Remote sensing inversion of saline soil in the Yellow River Estuary based on improved vegetation index. *J. Transactions of the Chinese Society of Agricultural Engineering.* 36 (05), 107–112+114+113.

Chen, W. et al. (2005) Research on decision tree classification. *J. FujianComputer.* (8), 5–6.

Dai, J.R. et al. (2015) Spatial distribution characteristics of soil organic carbon density and its influencing factors in Shandong. *J. Research of Environmental Sciences.* 28 (9), 1449–1458.

Deng, S.B. (2010) Remote sensing image processing method. Beijing, Science Press.

Foley, J.A. (1995) An equilibrium model of the terrestrial carbon budget. *J. Tellus Series B-chemical & Physical Meteorology.* 47 (3), 310–319.

Jin, F. et al. (2000) Research progress of soil organic carbon storage and its influencing factors. *J. Soils,* 32 (1), 11–17.

Li, Y.H. et al. (2005) Inversion of ground relative reflectance based on TM/ETM + remote sensing data. *J. Shandong Agricultural University.* 36 (04), 545–551.

Niu, W.J. et al. (2001) Study of universal Kriging interpolation method. *J. Computer Engineering and Applications.* 37 (13), 73–75.

Post, W.M. & Emanuel, W.R. (1982) Soil carbon pools and world life zones. *Nature.* [Online] 298 (5870), 156–159. Available from: http://www.nature.com/nature/journal/v298/n5870/abs/298156a0.html [Accessed 8th Jaly 1982].

Wang, Q. et al. (2016) Spatial distribution pattern inversion of soil organic matter in cotton field based on HJ satellite. *J. Transactions of the Chinese Society of Agricultural Engineering.* 32 (1), 174–180.

Xue, Z.J. et al. (2011) Spatial heterogeneity of soil organic carbon on the scale of small watershed in hilly area of Loess Plateau. *J. Soil and Water Conservation.* 25 (3), 160–163.

You, H.C. (2012) Remote sensing retrieval of soil organic carbon and its spatial differentiation. [Online]. Available from: http://kns.cnki.net/kns/brief/default_result.aspx [Accessed 1st April 2012].

Yuan, F. et al. (2008) Estimation and spatial distribution of soil organic carbon storage in Jiangxi Province. *J. Ecology and Environment.* 17 (1), 268–272.

Zhang, Q. (2005) Study on Classification of Aerial Hyperspectral Remote Sensing Land Cover Based on Decision Tree Method—A Case Study of Yixing Experimental Area in Jiangsu Province. [Online]. Available from: http://kns.cnki.net/kns/brief/default_result.aspx [Accessed 1st May 2012].

Land Reclamation in Ecological Fragile Areas – Hu (Ed.)
© 2017 Taylor & Francis Group, London, ISBN 978-1-138-05103-4

Research of the spatial model of land use based on the CA-Markov model

C.X. Qiu, D. Han & Q.Q. Mao
College of Geomatics, Xi'an University of Science and Technology, Xi'an, China

ABSTRACT: Land use and its change is a reflection of the process of social and economic development, with a clear comprehensive and regional, regional and economic environment changes play a very important role. In order to study the law and the driving mechanism of land use change in loess hilly and gully region, taking the ArcGIS software as an example, ArcGIS software was used to integrate and overlay the data of land use in different periods, and the land use The time course and spatial pattern of land use change in Zhifanggou watershed before and after returning farmland were analyzed. Based on the land use pattern of 2013, the CA-Markov model of IDRISI software was used to simulate the land use change in 2023 and draw the conclusion.

Keywords: Returning cropland to forest; Land use/cover change; Zhifanggou watershed; ArcGIS; CA-Markov model

1 INTRODUCTION

China is facing the problem of sustainable development, such as population growth, shortage of resources and deterioration of ecological environment, which is closely related to land use. The complexity of land use change and the complexity of its problem structure determine the necessity of comprehensive land use research with value. In this paper, based on the data of the Zhifanggou watershed in Ansai County of Shaanxi Province, this paper, based on the years of location monitoring data, combined with topographic maps and satellite remote sensing images from 1990, 2000 and 2013, Platform and Remote Sensing (RS) technology to analyze the temporal and spatial dynamics of land use before and after returning farmland in Zhifanggou watershed, reveal the quantitative characteristics of land use, clarify the spatial and temporal pattern of land use change and the law of land use change, and the relevant information and IDRISI. This trend is simulated by CA-Markov model.

2 SURVEY AREA OVERVIEW AND DATA PROCESSING

2.1 *Research area profile*

Zhifanggou watershed (E 109°19′30″, N 36°51′30″) is located in the second sub-region of the hilly and gully region of the Loess Plateau. It is a tributary of the lower reaches of the Xingzi River,[4] which belongs to the typical loess hilly gully Area, belonging to Ansai County, Shaanxi Province.[5] The basin is warm temperate semi-arid climate zone, the basin within the hilly ups and downs, terrain broken gully density of 8.06 km/km^2, elevation 1100–1400 m.[6] The schematic diagram of the Zhifanggou watershed is shown in Figure 1 below.

2.2 *The specific workflow*

The specific workflow is shown in Figure 2.

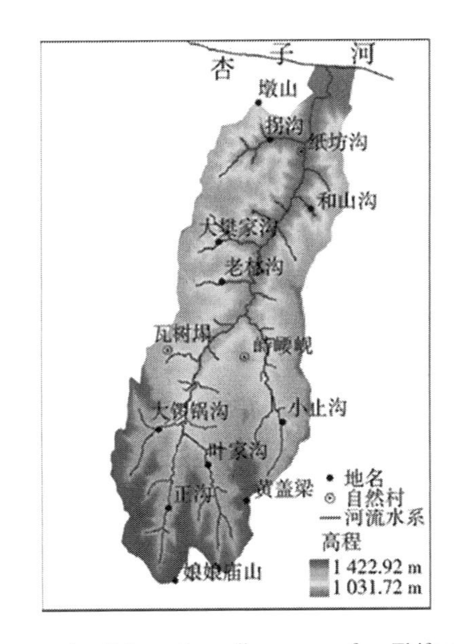

Figure 1. Schematic diagram of Zhifanggou watershed.

Figure 2. Technical flow chart.

2.2 *Data processing*

1. ArcGIS data preparation;
2. Data fusion;
3. Overlay analysis;[7]
4. Calculate the area and export the attribute table;
5. Export data processing;[8]

3 SPATIAL AND TEMPORAL EVOLUTION OF LAND USE

3.1 *Analysis of LUCC time process*

3.1.1 *Analysis of LUCC time process before returning Farmland (1990–2000)*

According to the calculation, the relative change rates of the main land use in the Zhifanggou watershed in Ansai County from 1990 to 2000 are shown in Table 1.

3.1.2 *Analysis of LUCC time process after returning Farmland (2000–2013)*

According to the calculation, the relative change rates of the main land use in the Zhifanggou watershed in Ansai County from 2000 to 2013 are shown in Table 2.

3.2 *LUCC spatial pattern analysis*

3.2.1 *Analysis of spatial pattern of LUCC before returning Farmland (1990–2000)*

The results show that the increase of rural residential area and orchard area is mainly due to the transfer of forest land area, the transfer amount is 0.2190 km^2 and 0.7854 km^2 respectively, and the probability of turning the forest land into the orchard is 24.31%. Most of the area was transferred to grassland and shrub, the probability of transfer was 19.93% and 14.86% respectively. The increase in cultivated land area is mainly due to the transfer of dryland and sparse forest land area into 0.1153 km^2 and 0.0989 km^2, respectively. Overall, the transfer rate of rural settlements is 90%, the conversion rate of orchards is 88%, and the transfer rate of reservoirs and forest land is the lowest. The area of shrub forest and orchard was the most obvious. In the period from 1990 to 2000, 0.9896 km^2, 0.8631 km^2 and 0.7854 km^2 were

Table 1. Relative change rate of major land use in Zhifanggou watershed from 1990 to 2000.

Land use type	1990 the total land area of various types	The total area of land in 2000	Added area	Relative change rate of land use
	m^2	m^2	m^2	m^2
Rural settlements	43315.79904	64697.02214	26381.2231	1.23
River surface	88370.35988	66989.13678	19225.5163	0.90
Reservoir water surface	10645.02185	25026.24495	32301.1152	2.25
dry land	107744.0591	66362.83602	445620.9716	10.77
orchard	120184.9693	241566.1924	15320.6541	0.13
woodland	97319.63124	72938.40814	16589.5028	0.68
Shrubbery	449902.8673	261284.0904	17256.1265	0.09
Natural grassland	1752439.4911	1762439.489	42056.2546	4.21
Saline land	399141.8641	187760.641	20134.9861	0.10
arable land	1226174.781	1076827.288	12071.3695	0.08

Table 2. Relative change rate of major land use in Zhifanggou watershed in 2000–2003.

Land use type	1990 the total land area of various types	The total area of land in 2000	Added area	Relative change rate of land use
	m^2	m^2	m^2	m^2
Rural settlements	64697.02214	103472.5069	55365.92072	1.43
River surface	66989.13678	71729.70324	59327.64578	12.51
Reservoir water surface	25026.24495	14953.5889	11107.01465	1.10
dry land	66362.83602	339715.6097	24114.76353	0.09
orchard	241566.1924	423872.5792	57114.165	0.31
woodland	72938.40814	1762613.315	7218.090882	0.00
Natural grassland	1598384.547	1500489.002	496133.4712	5.07
Artificial grass	164054.9426	71937.97897	61623.32893	0.67
Saline land	187760.641	166871.0906	118547.2083	5.67

Note: "0" indicates that the number of land use changes in a certain area during a study period is 0, that is, the relative change rate is zero.

transferred to dryland, shrub and orchard. The transfer probabilities were 30.63%, 29.81% and 24.31%, respectively.

3.2.2 *Analysis of spatial pattern of LUCC after returning Farmland (2000–2003)*

The results show that the increase of forest land area is mainly from the transfer of other types of land and open forest land, the transfer amount is 0.8605 km^2 and 0.1415 km^2 respectively, and the probability of grassland turning into shrub is 45.63% The probability of shrubs transferred to forest land was 35.78%, and the increase in cultivated land area was mainly due to the transfer of grassland and other types of land area into 0.410 km^2 and 0.3102 km^2 respectively. Overall, shrubs, arable land and dry land transfer rate of the highest, are 94%, orchards and forest land transfer rate followed by 86% and 83%, river water transfer rate of the lowest, 17% The Grassland, shrubs and cultivated land conversion between the most obvious, during the 1.0082 km^2 and 0.4104 km^2 of grassland were converted to shrubs and arable land.

3.3 *The time process and spatial pattern of land use change*

From the time point of view, the annual change rate of land use in Zhifanggou watershed is relatively small in the first 10 years of returning farmland to forest, and the change rate of reservoir water is the fastest, the annual rate of change is 10.39%, the rate of natural grassland change. The annual rate of change is 0.04%. The annual rate of change of other land use is 7.77% of the orchard, 3.80% of the rural settlements, –4.07% of the saline land, –3.22% of the shrubbery and 2.95% of the dry land, River surface –1.86%, cultivated land –0.94%. Saline soil, shrub forest, dryland forest land, river surface and farmland have a negative growth, that is, the area is reduced. Compared with the previous period, the relative change rate of land use types after returning farmland showed obvious difference. The relative change rate of river surface is the largest, the relative change rate is 12.51, the forest land is the smallest, the relative change rate is almost zero. The relative change rate of rural settlements, river surface, reservoir surface, natural grassland and saline land is R > 1, that is, the local land use change is larger than the whole change range; the relative land of dryland, orchard, forest land, shrub forest and artificial grassland. The rate of change is R < 1, which is less than the magnitude of the overall land use change.

From the spatial pattern, the relationship between the land use and the land use types after returning farmland is different. The transfer of land before and after returning from land to farmland, grassland and sparse forest land was the most significant. The land transfer after returning farmland occurred mainly between cultivated land, forest land and pasture land, which mainly occupied the area of cultivated land and the area of forest land. The area of land use in Zhifanggou watershed before returning farmland to forest is more and less. The area of forest land and grassland is large, and the land area of forest land, water area and other types is less. And the land use type of Zhifanggou watershed is reduced after the implementation of the project of returning farmland to forest. Many other types of land are converted into irrigated land, forest land and grassland, and the area of forest land and grassland gradually increases to the watershed. The vast majority of the area.

4 SIMULATION AND PREDICTION OF DYNAMIC EVOLUTION OF LAND USE CHANGE

4.1 *Prediction of land use distribution based on CA-Markov model in 2023*

1. The determination of the transfer probability
In the year, the change of land use type is divided into a series of discrete processes, which are averaged according to the average annual conversion rate of each type (the conversion area of each land type after 2000 to 2013 is divided by the age interval 13. The annual conversion area accounts for the percentage of the original area of the type) to determine the transfer probability of the land use type unit,[9] the rural residential area into other types of transition probability as the first line, the river surface into other landscape types of transfer probability as The second row, and so on, to establish the transition probability matrix, to determine the initial transfer probability matrix from 2000 to 2013.[10]
2. Dynamic simulation prediction
The transfer probability of the initial state is adjusted, the main adjustment is the transfer probability of cultivated land, grassland, forest land, residential area and water area, and the adjusted probability matrix of transition is shown in Table 3 below.

Table 3. Adjusted land use type transfer probability matrix.

	Settlements	woodland	Grassland	orchard	arable land	dry land	Saline land	Waters
Settlements	0.0657	0	0	0	0	0	0	0
woodland	0	0.0723	0.0297	0.0003	0.0252	0.0152	0	0.0273
Grassland	0	0.0041	0.0151	0.0015	0.0111	0.0011	0.0013	0
orchard	0	0.0370	0.0051	0.0182	0	0.0159	0	0
arable land	0	0.0051	0.0008	0.0407	0.0089	0.0085	0	0
dry land	0	0.0377	0.0114	0	0	0.0279	0	0
Saline land	0	0.0249	0.0016	0.0010	0.0009	0.0485	0.0486	0
Waters	0	0	0	0	0	0	0	0.0093

Table 4. Main land use type transfer matrix in Zhifanggou watershed.

	Settlements	Woodland	Grassland	Orchard	Arable land	Dry land	Saline land	Waters
	km^2	km^2	km^2	km^2	km^2	km^2	km^2	km^2
Settlements	553	0	0	1	79	13	0	0
woodland	190	16177	2190	14	1096	662	0	142
Grassland	162	10812	5578	549	4104	397	483	11
orchard	23	1162	159	571	0	499	0	0
arable land	19	846	350	2629	577	546	0	0
dry land	0	325	98	0	0	241	0	0
Saline land	0	609	38	24	21	0	1185	0
Waters	0	106	22	0	0	88	0	704
Total	947	30037	8435	3788	5877	2466	1668	857

3. Run the CA-Markov model

Based on the land use map of 2013, we input the probability matrix of land use area transfer in Zhaoyuan Gully from 2000 to 2013, then input the adaptive image set, and set the number of cycles of geo-cellular automata to 10, and use the land-The cellular automaton filter creates a weight factor with significant spatial meaning, and uses the weighting factor to act on adjacent grid cells to change the state of adjacent grids. In the solution, a 5×5 filter was used to obtain the land use prediction map of the Zhifanggou River Basin in 2023.

4. The number of area changes in the forecast

After the merger, the main land use type transfer matrix in Zhifanggou watershed from 2000 to 2013 is shown in Table 4 below.

5 DYNAMIC PREDICTION OF LAND USE CHANGE

According to the forecast, the area of the settlements, woodlands, grasslands, orchards, arable land, dry land, saline land and waters of Zhifanggou watershed will reach 131954.7 m^2, 4571225.3 m^2, 690310.7 m^2, 497044.2 m^2, 508629.6 m^2, 275068.8 m^2, 108581.163 m^2 and 75275.2 m^2. The results show that the land use change will continue to change from 2000 to 2013, and the area of forest land, orchard, rural settlements and dry land will increase as a whole, and the land use change will increase from 2013 to 2013. Year, the net increase in forest land was 1567525.3 m^2, an increase of 52.2%, mainly from the transfer of cultivated land and grassland area, while the area of cultivated land, saline land, grassland and waters continued to decrease, compared with the area of cultivated land. A net reduction of 79070.4 m^2, a decrease of 13.5%, mainly due to the transfer of woodland and orchard area. However, with the deepening of the project of returning farmland to forest, this increase and decrease will be reduced, because the area of returning farmland to forest is limited, the forest area changes over time will become saturated, and cultivated land can not be reduced, So the final forest land, grassland, arable land and other areas will remain basically stable, in a dynamic balance.

6 CONCLUSIONS

Based on the CA-Markov model, combined with the single land use dynamic degree, comprehensive land use dynamic degree, land use type relative rate of change, land use transfer matrix data and related indicators, this paper analyzes the policy of returning farmland to forest before (1990–2000) and after (2000–2013) the implementation of Ansai County. The time course and spatial pattern of land use change in Zhifanggou watershed are analyzed, and the CA-Markov model is used to simulate the change of land use structure in Zhifanggou watershed in 2023. So as to have some reference and reference value for returning farmland to forest project in Ansai County of northern Shaanxi.

REFERENCES

Chen Youqi. Based on GIS in China's land use change and its impact model [J]. Ecological Science, 2000,19 (3): 1–7.

Dong Qian-kun. Study on Spatial and Temporal Pattern of Land Use in Zhifanggou Watershed Based on GIS [J]. *J. Mining Survey*, 2015 (4): 59–64.

Effects of Vegetation Restoration Process on Mineral Nitrogen Accumulation in Soil Profile in Hilly and Gully Area of Loess Plateau—A Case Study of Zhifanggou Watershed Experimental Area [J]. *J. Journal of Soil and Water Conservation*, 2004, 18 (6): 58–62.

Journal of Geographical Sciences, 2008, 27 (6): 49–56. Li Xiao-hua, Li Xiu-hua, Li Tian-hong. Study on spatial scale effect of soil erosion in water erosion area of the Yellow River Basin [J].

Journal of Wuhan University (Information Science Edition), 2004, 29 (7): 575–579 (in Chinese with English abstract) [J]. *J. Journal of Wuhan University (Information Science Edition)*, 2004, 29 (7): 575–579.

Lei Min, Cao Ming-ming, Xi Jing. Comprehensive benefit evaluation and policy orientation of returning farmland to forest in Mizhi County [J]. *J. Journal of Soil and Water Conservation*, 2007, 27 (3): 151–156.

Li Xiubin. The core area of global environmental change research—International research trends of land use/land cover change [J]. *J. Acta Geographica Sinica*, 1996 (6): 553–558.

Lin Qing, Luo Ge-ping, Chen Xi. Study on the Research of LUCC Driving Force Model [J]. *J. Advances in Geography*, 2005, 24 (5): 79–87.

Liu Jiyuan, Bu and Ao Sier. Study on the Temporal and Spatial Characteristics of Modern Processes of Land Use Change in China—Based on Satellite Remote Sensing Data [J]. *J. Quaternary Sciences*, 2000, 20 (3): 229–239.

Yang Jing, Tong Jie, Zhang Shuai. Application of ArcGIS vector data spatial analysis in urban housing selection [J]. *J. Geospatial Information*, 2012, 10 (1): 119–120.

Land Reclamation in Ecological Fragile Areas – Hu (Ed.)
© 2017 Taylor & Francis Group, London, ISBN 978-1-138-05103-4

Analysis on dynamic change of vegetation coverage based on NDVI: A case study of Beijing-Tianjin-Hebei

Y. Mu, C.S. Zhang, X.Q. Wang, J.W. Zhang & X.Y. Zong
College of Geomatics, Xi'an University of Science and Technology, Xi'an, China

ABSTRACT: In order to study the condition of vegetation cover and its change trend influenced by human activities, by calculating NDVI of five-day synthetic data of geospatial data cloud and MODIS data of Beijing-Tianjin-Hebei in 2005, 2010 and 2015, the vegetation coverage is calculated by the Mixed-pixel dichotomy model, the changes of vegetation coverage at different periods and different data sources are compared. According to the change of vegetation coverage: from the time point of view, From 2005 to 2015, the area of vegetation coverage of middle coverage, low coverage and extremely-low coverage types increased, while the high and middle-high coverage types decreased in Beijing-Tianjin-Hebei; from space, the total area of vegetation coverage increased, while the area of vegetation coverage decreased in central urban area in Beijing-Tianjin-Hebei, which is closely related to the construction of the national "Three-North" Shelterbelt and attention to vegetation protection and rapid urbanization in recent years.

1 INTRODUCTION

Vegetation coverage is defined as the vertical projection area of vegetation within a unit area, vegetation cover is an important index to study the change of urban ecological environment quality. On the basis of vegetation coverage, it is the prerequisite and basis of ecosystem health assessment to study on temporal and spatial characteristics of vegetation. The change of land use and land cover change by vegetation index has become an important means to monitor the dynamic change of environment.

This experiment uses ENVI to calculate the normalized difference vegetation index of the three remote sensing images of Beijing-Tianjin-Hebei in September 2005, 2010 and 2015, the vegetation coverage is estimated by using the dimidiate pixel model and compared with that of the five-day synthetic data of the same geographical data.

2 GENERAL SITUATION OF STUDY AREA

Beijing-Tianjin-Hebei is located in the coast of North China, and its south and southeastern is close to Shandong and Henan Province. The study area is located in the semi humid and semi-arid transition zone, the main climate type is a warm temperate semi humid monsoon climate, in the northwest edge of temperate semi-arid climate.

3 DATA SOURCES AND PREPROCESSING

3.1 *Data sources*

The image data used in this experiment are MODIS image data and the NDVI data of five-day synthetic data of geospatial data cloud of Beijing-Tianjin-Hebei in 2005, 2010 and 2015, Three MODIS image data imaging at roughly the same time, September 18, 2005, September 11, 2010 and September 1, 2015. The NDVI of synthesis data of geospatial data cloud are 2005, 2010 and 2015, in September 11th.

3.2 *Data preprocessing*

MODIS raw data for PDS format, which is processed by EOS/MODIS satellite receiving and processing system, and the data of HDF format used in this experiment is generated.

1. Radiation correction: In the process of remote sensor observation target, because of the measured values include atmospheric reflection, scattering, radiation and other effects, resulting pixel value and the target spectral reflectance or spectral radiance and other physical inconsistencies, resulting in image distortion, through the radiation correction operation to eliminate such effects, get the correct reflection of the target and radiation characteristics.
2. Projection transformation: The geometric projection transformation is carried out after two images are cut by ENVI, and geometrically

registered and unified into the UTM/WGS84 coordinate system. The geometric error of the image after registration is controlled within one pixel.

3. Image cutting: In the use of shp format data in ENVI for cutting, shp format is the standard format of ArcGIS software. The SHP format automatically converted into AOI format data by ENVI, according to this data for image cutting.

Geospatial data cloud images for the national five-day of synthetic NDVI data, and the data range of [–1, 1], which is cut and threshold processed by ArcGIS.

1. Data Mask: The administrative boundary line shp format line data topology for the surface data by ArcGIS, and with the help of the mask function extract the NDVI image data.
2. Threshold processing: NDVI range of [–1, 1], negative values for the ground cover for the clouds, water, snow, etc., high reflection of visible light; zero indicates a rock or bare soil. In order to study the vegetation coverage in the area, the Con function of the raster calculator in ArcGIS is used to delete the NDVI information below zero and preserve NDVI value of the vegetation.

4 RESEARCH METHODS AND STEPS

4.1 *Vegetation index selection and calculation*

Vegetation index is a quantitative value of the vegetation of the earth's surface, which is extracted from the multi spectral remote sensing data. It is a comprehensive reflection of the vegetation type, coverage and growth in the unit pixel.

NDVI is defined as the ratio of the difference between the near infrared band and the visible red band and the sum of the two bands. Selected the first band (0.62~0.67 um), second band (0.841~0.876 um) of MODIS, namely the red and near infrared band, the formula is:

$$NDVI = (X_{nir} - X_{red}) / (X_{nir} + X_{red}) \quad (1)$$

In the formula: X_{nir} is the reflectance of the near infrared light band, and X_{red} is the reflectance of the red band. The higher of the plant height, the larger of the population and the leaf area coefficient, the greater of the NDVI.

4.2 *Vegetation coverage calculation*

Vegetation coverage refers to the ratio of the vertical projection area of vegetation canopy to the total land area, which is an important index to measure the surface vegetation condition. The method of remote sensing vegetation cover based on vegetation index includes empirical model method, vegetation index method, pixel decomposition model method and FCD model mapping method, the vegetation index method does not need to establish the regression model, more universal significance. According to the actual situation of the study area, the maximum and minimum values of vegetation coverage can be approximated by 100% and 0%, so the vegetation coverage formula based on the dimidiate pixel model:

$$f = (NDVI - NDVI_{\min}) / (NDVI_{\max} - NDVI) \quad (2)$$

In the formula: f for vegetation coverage, $NDVI_{\min}$ and $NDVI_{\max}$ represent the minimum and maximum values of NDVI. Based on the vegetation coverage gray-scale map of different phases by the above model in the study area and then reclassified.

4.3 *Vegetation coverage hierarchical graph generation*

Referring to the threshold value of vegetation identified in previous literature, combined with the characteristics of Beijing-Tianjin-Hebei, according to the land use map, the vegetation coverage in the study area is divided into 5 levels, as shown in Table 1.

The vegetation coverage map of MODIS data and geospatial data cloud synthesis data for each period of the study area is generated by ArcGIS according to the change of vegetation coverage, as shown in Figures 1–3. According to the classification results,

Table 1. Standard for classification of vegetation coverage in Beijing-Tianjin-Hebei.

Grade	Name	The state of vegetation coverage
1	High coverage	Vegetation coverage >75%
2	Middle-high coverage	Vegetation coverage 55%–75%
3	Middle coverage	Vegetation coverage 35%–55%
4	Low coverage	Vegetation coverage 10%–35%
5	Extremely low coverage	Vegetation coverage <10%

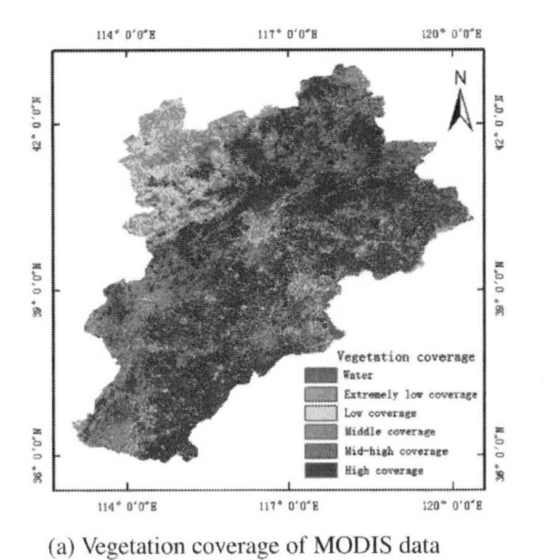

(a) Vegetation coverage of MODIS data

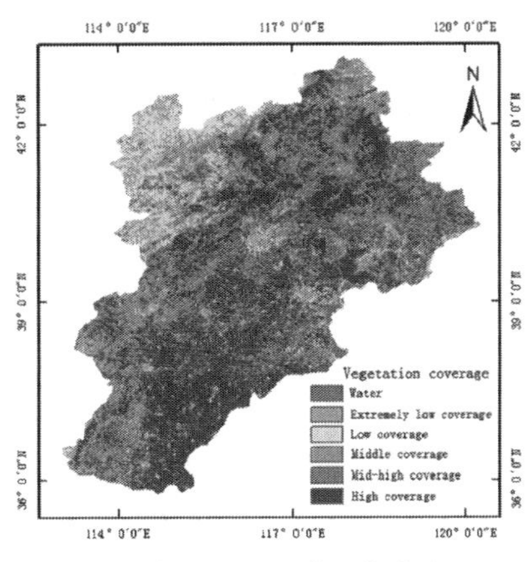

(b) Vegetation coverage of synthetic data

Figure 1. Vegetation coverage map in September 2005.

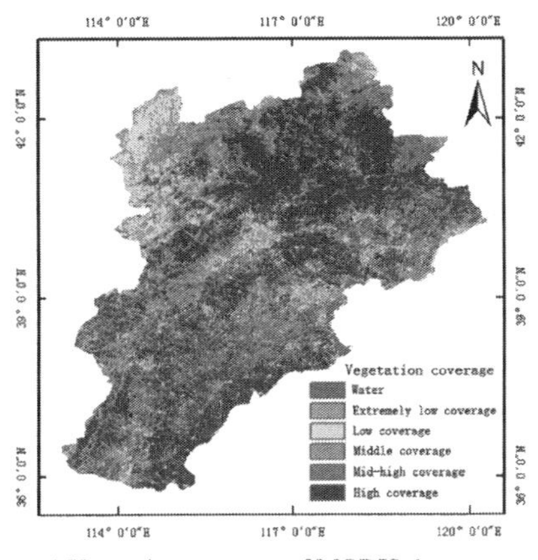

(a) Vegetation coverage of MODIS data

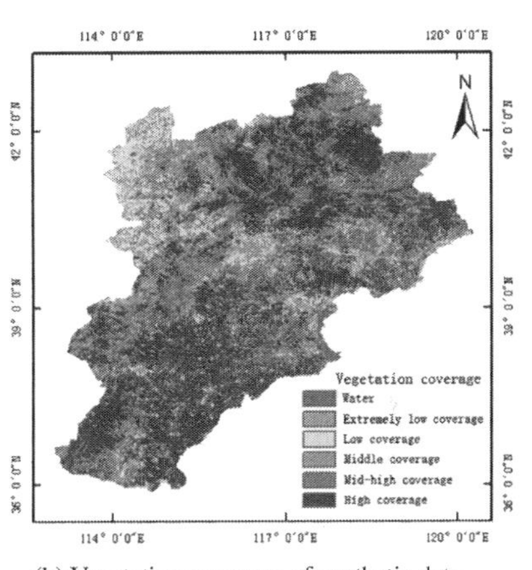

(b) Vegetation coverage of synthetic data

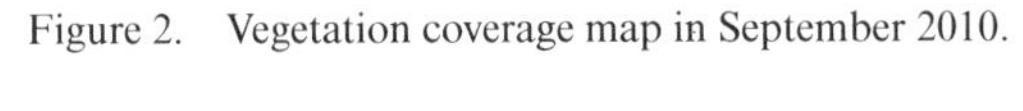

Figure 2. Vegetation coverage map in September 2010.

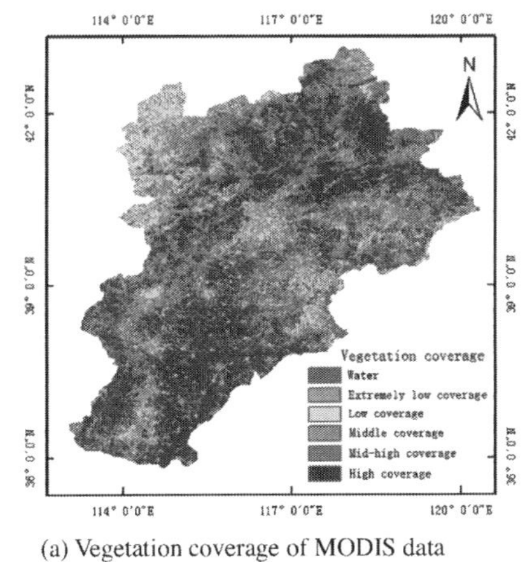

(a) Vegetation coverage of MODIS data

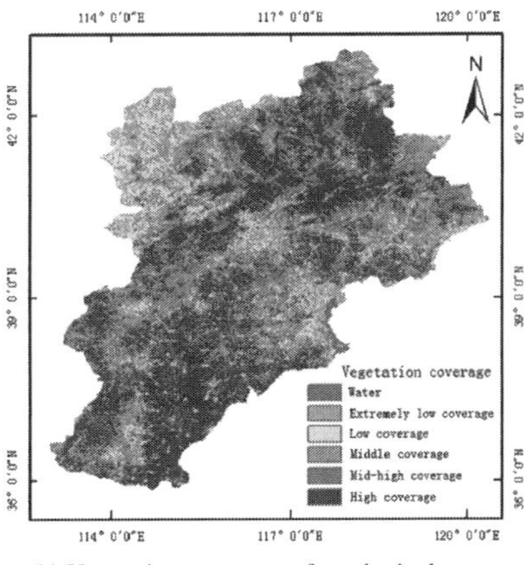

(b) Vegetation coverage of synthetic data

Figure 3. Vegetation coverage map in September 2015.

the grid numbers given by each of the classification results, the resolution of the original image and the number of grids corresponding to the classification results are calculated by multiplication, and the results of vegetation coverage in three periods are obtained.

5 RESULTS AND ANALYSIS

5.1 *The time variation characteristics of vegetation cover in the study area*

According to the vegetation coverage level of Beijing-Tianjin-Hebei from 2005 to 2015, MODIS and geospatial data cloud data synthesis of vegetation cover change area and change rate are calculated (Tables 2 and 3), the area of mid-high vegetation coverage is the largest, which accounts for 35% of the total; followed by the type of high coverage type; the proportion of middle coverage type is lower than that of high coverage type; and the proportion of low and extremely low coverage type area are lower.

Compared Table 2 and Table 3, the total vegetation coverage area of Beijing-Tianjin-Hebei has increased, and the coverage of vegetation has improved to some extent. The percentage of vegetation coverage area of the MODIS data showed a fluctuating change, decreased from 3.05% in 2005 to 2.75% in 2010, by the year of 2015 and increased to 3.63%, the overall trend of increase, the proportion of vegetation coverage area with extremely low vegetation coverage increased with the synthetic data; the area of low coverage and middle coverage increased to a certain extent; while the area of high coverage of vegetation decreased year by year; the proportion of mid-high coverage area in the vegetation is fluctuating and declining as a whole, the proportion of mid-high coverage area decreased in MODIS data; the results showed that the high and middle coverage types of vegetation in Beijing-Tianjin-Hebei are gradually degraded to middle, low and extremely low coverage types.

5.2 *Spatial distribution characteristics of vegetation cover in the study area*

From Figures 1–3 and Table 2 and Table 3, the total vegetation coverage of Beijing-Tianjin-Hebei showed an increasing trend; but the vegetation coverage of high and mid-high gradually decreased to middle, low and extremely low vegetation coverage; the vegetation coverage in the northwest of Zhangjiakou increased from extremely low and low coverage to middle coverage; the vegetation coverage decreased in the central city, which is degraded from the middle coverage type to the low

Table 2. Dynamic change of vegetation coverage of MODIS in Beijing-Tianjin-Hebei in 2005–2015.

Vegetation coverage	2005 area/km^2	Proportion/%	2010 area/km^2	Proportion/%	2015 area/km^2	Proportion/%	Change area/km^2	Change rate/%
Extremely low coverage	5068	3.05	4579.5	2.75	6049.75	3.63	981.75	19.37
Low coverage	14346	8.62	14717.75	8.84	16527.25	9.92	2184.25	15.23
Middle coverage	30927.25	18.59	33698.75	20.25	34683.5	20.82	3756.25	12.15
Middle-high coverage	61842.5	37.17	60933	36.61	57323.5	34.42	−4519	−7.31
High coverage	54211.5	32.58	52505.75	31.55	51977.25	31.21	−2234.25	−4.12
Total	166395.25	100	166434.75	100	166561.25	100	169	35.32

Table 3. Dynamic change of vegetation coverage of synthesis in Beijing-Tianjin-Hebei in 2005–2015.

Vegetation coverage	2005 area/km^2	Proportion/%	2010 area/km^2	Proportion/%	2015 area/km^2	Proportion/%	Change area/km^2	Change rate/%
Extremely low coverage	5121.5	3.08	5833.5	3.5	6008.25	3.61	886.75	17.31
Low coverage	14502.75	8.71	14999.75	9.01	16662.25	10.0	2159.5	14.89
Middle coverage	31012.5	18.64	31535.5	18.95	34456.75	20.68	3444.25	11.11
Middle-high coverage	61195.25	36.77	61541.25	36.97	57832	34.71	−3363.25	−5.5
High coverage	54580	32.8	52530.5	31.56	51674.5	31.01	−2905.5	−5.32
Total	166412	100	166440.5	100	166633.75	100	221.75	32.49

coverage type, especially in Beijing and Tianjin; the vegetation coverage of the northern mountainous area changed greatly, and the coverage area of vegetation is low-high-low fluctuation, the vegetation coverage in northern Chengde is degraded; the area of vegetation coverage in the eastern coastal area increased obviously.

5.3 *Driving force analysis*

The changes of vegetation coverage in Beijing-Tian jin-Hebei are mainly caused by human activities, which can be divided into the following aspects:

1. Rapid economic development, population explosion, from 2005 to 2015, the total population increased from 90000000 to about 110 millions, the area of urban construction land has been expanded, and the vast majority of farmland areas have become urban construction land in 2015, resulting in the expansion of low vegetation coverage area.
2. A large number of original wastelands and rural waste lands in Beijing-Tianjin-Hebei have been transformed into cultivated lands and woodlands, and the whole coverage of vegetation has been improved to a certain extent.
3. The construction of "Three North Shelterbelt", Beijing-Tianjin-Hebei surrounding green construction projects involving Beijing, Tianjin and Zhangjiakou, Tangshan, Qinhuangdao, Baoding five cities in 52 states (cities, districts), At present, the fifth phase of the project is under construction, which has improved the coverage of vegetation in some areas of Beijing-Tianjin-Hebei region to some extent.
4. With the gradual reduction of the coastal aquaculture industry, the original ponds and depressions in the coastal areas are gradually transformed into low vegetation covered beaches and grasslands.

The above-mentioned human activities are the main factors to promote the overall increase of vegetation coverage and the decrease of vegetation coverage in some areas, and the changes of vegetation coverage can be caused by temperature, soil, moisture and crop growth, but it will not cause significant changes in vegetation coverage in the short term, so it is not the main reason for the change of vegetation coverage.

6 CONCLUSIONS AND RECOMMENDATIONS

By calculating the NDVI of the MODIS data, and combining with the two pixel model of mixed pixels, calculating the vegetation coverage and comparing with the results of vegetation coverage of the NDVI of the five-day synthetic data of the spatial data cloud, the difference is not significant. This technology has certain feasibility, which can reflect the vegetation cover condition of the region, and has certain guiding significance for vegetation restoration and ecological environment construction in Beijing-Tianjin-Hebei.

With the increase of population density and the continuous development of social economy in Beijing-Tianjin-Hebei, the vegetation coverage is more and more vulnerable. In the future economic development, Beijing-Tianjin-Hebei should pay attention to the wasteland reclamation and the combination of returning farmland to forest and grass; formulate relevant laws and regulations, strengthen the supervision of vegetation protection; Under the premise of not destroying the ecological environment, the land use structure should be adjusted reasonably, occupy as little cultivated land and grassland as possible and pay attention to vegetation restoration in the mining area; formulate protection policy of regional vegetation, establish protected areas in the lake and garden; in the protection of arable land and economic development at the same time, increase the proportion of vegetation coverage appropriately, to create a beautiful, comfortable green city to provide ecological support.

REFERENCES

Chen, Y.J. (2009) Comparative Study on Dynamics of Vegetation of Stone Forest Scenic Spot based on NDVI. Forest Inventory and Planning, 34(4), 30–33. 327–333.

Cheng, H.F. Zhang, W.B. & Chen, F. (2008) Advances in Researches on Application of Remote Sensing Method to Estimating Vegetation Coverage. Remote Sensing for Land & Resources, (1), 13–18.

Luo, Y. Xu, J.H. & Yue, W.Z. (2005) Research on Vegetable Indices Based on the Remote Sensing Images. Ecologic Science, 24(1), 75–79.

Ma, Z.Y. Shen, T. & Zhang, J.H. (2007) Vegetation Changes Analysis Based on Vegetation Coverage. Bulletin of Surveying and Mapping, (3), 45–48.

Tang, S.H. Zhu, Q.J. & Zhou, Y.Y. (2003) A Simple Method to Estimate Crown Cover Fraction and Rebuild the Background Information. Journal of Image and Graphics, 8(11), 1304–1308.

Tian, Q.J. & Min, X.J. (1998) Advances in Study on Vegetation Indices. Advance in Earth Sciences, 13(4).

Wei, B. Zhang, A.D. Cui, Q.C. Zhang, L. & Li, S. (2014) Study on Dynamic Change of Vegetation Coverage in Dongying City Based on RS. Ludong University Journal (Natural Science Edition), 30(1), 69–72.

Wang, D.P. Li, C.A. & Li, H. (2009) Study on Dynamic Change of Vegetation Coverage in Wuhan Region Based on RS. Journal of Agricultural Mechanization Research, (7), 74–78.

Land Reclamation in Ecological Fragile Areas – Hu (Ed.)
© 2017 Taylor & Francis Group, London, ISBN 978-1-138-05103-4

Research of the comprehensive improvement effectiveness assessment of coal mine subsided land under the ecological civilization

Y.F. Li & Y.X. Cui
School of Environment Science and Spatial Informatics, China University of Mining and Technology, Xuzhou, Jiangsu, China

M.H. Zhang
School of Management, China University of Mining and Technology, Xuzhou, Jiangsu, China

ABSTRACT: Because of coal resource exploitation, different degrees of land disturbance and destruction are caused. But the land is a spatial carrier of the human society development, the protection and reasonable exploitation and utilization of the land are demands of ecological civilization based on sustainable green development. Therefore, it is a great significance that strengthen the comprehensive improvement of coalmine subsided land and ensure its effect. Based on the impacts of comprehensive improvement on the ecological civilization construction, this paper analyzes its improvement effect on the regional ecological environment and optimizes and build the effect of the comprehensive evaluation index system of coalmine subsided area through the factor analysis method. The fuzzy evaluation method is used to evaluate improvement effect to clear improvement significances in different mine life cycle stages, and to promote reasonable improvement of coalmine subsided land and regional ecological civilization construction.

Keywords: Coal mining area, Subsided land, Comprehensive improvement, Evaluation

1 INTRODUCTION

The interaction between human and nature is the most important content in human society development. Through the constant understanding of nature, the use of nature, the transformation of nature, and conscious or unconscious acceptance of natural punishment, human beings gradually realize that respect for nature, protect nature, and achieve harmony with nature is necessary for sustainable development when they transform nature and bring benefit themselves. The development and utilization of mineral resources is one of the important contents of human use of nature, which provides reliable basic material information for the development of human society and promotes the development of social civilization. The human beings should pay attention to give back to the nature when they take from nature. Make disturbed, even damaged nature repair reasonably and timely, which really embodies the harmonious coexistence between man and nature, a virtuous cycle and comprehensive development (Zhang Mi, 2013). Therefore, mining area is a place where the human makes bigger disturbances, should strengthen the construction of ecological civilization.

2 THE PRACTICAL SIGNIFICANCE OF COAL MINE SUBSIDENCE COMPREHENSIVE IMPROVEMENT

Coal resources are a stable component of strata endowment, and its exploitation will artificially break the stability of rock around the coal seam, which cause deformation, fault rupture and collapse of strata; the formation of subsidence basin, a reaction on the surface, affects the normal use of land. With the increasing strength and scale of coal resources exploitation, the subsidence area increases. According to incomplete statistics, as of 2013, China has about 1,500 coal mines, the mine goaf is about 700,000 hectares because of mining, of which more than 50% caused land collapse and damage (Sun Guishang and Li Jianzhong, 2014). A large number of coal mining subsidence concentrated in the country in Shanxi, Henan, Hebei and Heilongjiang and other regions has contributed to the construction of coal products. These places not only caused the destruction of cultivated land, resulted in imbalance of regional water cycle and damaged surface buildings, but also do they triggered a series of social problems, and target deviation to ecological civilization construction.

2.1 *Subsidence improvement in coal mining area and mine life cycle*

Rise of coal mining area due to the coal resources exploitation. But the coal resources are limited, the mine life cycle characteristics are significant. With the increasing intensity of coal resources exploitation, the land subsidence area gradually appeared and expanded, and the tasks of subsidence improvement appeared and intensified. At the same time, influenced by the spatial transfer of coal mining, the improvement of subsidence possesses typical characteristic of time and space, and it lags behind coal resources exploitation.

The improvement tasks of coal mining subsidence run through each stage of the mining life cycle and lags behind the mine life cycle. In the production stage [0, T1), the mine production gradually increased; but due to the beginning of insufficient mining, land subsidence area and the degree of subsidence are not significant, there are less tasks of subsidence improvement. In the reaching production stage [T1, T2), the output of the mining area increase rapidly until the design production capacity is reached. With the increasing intensity of mining, the degree and the range of the deformation land are intensified, and the comprehensive improvement tasks of the subsidence appear and increase rapidly. At this time, if timely improvement can be implemented according to the trend of subsidence, the subsidence improvement tasks and mining production can maintain a balance; but if the subsidence improvement lags, the expansion of the subsidence area and the cumulative increase in improvement will be caused inevitably. In the stable production stage, the intensity of the mine exploitation and the velocity of the surface subsidence tend to be stable. If the comprehensive improvement of the subsidence can be implemented with improved benefit of mining, which will promote the coordinated development of mining area; If the improvement lags, it will cause the rapid expansion of the subsidence area and the degree of subsidence will tend to maximum. After entering the aging period [T3, T4), the reserves of the mine area are reduced to less than 25% of the design reserves in the mining area, and the production of the mining area gradually decreases. However, the subsidence area will not be reduced immediately, which usually lags behind reduction of output 1–2 years. Even if the end of mineral resources mining, subsidence land will not immediately stagnate, the improvement task will not be completed immediately. (Figure 1)

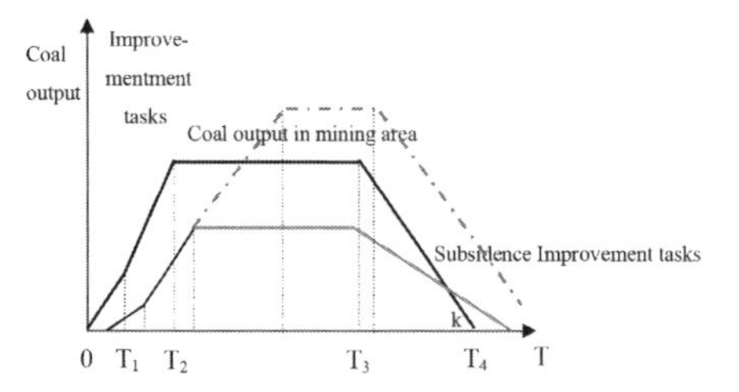

Figure 1. The relationship between production and subsidence improvement in mine life cycle.

2.2 *The necessity and reality of the subsidence governance in mining area*

The exploitation of coal resources provides the basic energy guarantee for the social and economic development, and promotes the construction of ecological civilization. However, if subsided land caused by the exploitation of coal resources is not timely governed, it may have a series of hazards. (Chen Xinsheng, Wang Qiaoni, etc., 2009; Liang Hailin, 2015)

The land subsidence caused by the exploitation of coal resources firstly shook the basis of ground production and directly affected the safety production of agriculture, transportation and construction within the mining area. Secondly, land subsidence will cause the soil and water loss and result in regional ecological environment degradation, which hinder the construction of ecological civilization; In addition, the increasing area and extent of the subsidence will endanger the infrastructure and buildings in the mining area, and cause a huge amount of dangerous source, which result in a large number of losses; finally, the increasing subsided land will lead to contradictions between enterprises and the region, and new social problems, which hinder the process of ecological civilization construction. Therefore, to strengthen the comprehensive improvement of mining subsidence have an outstanding practical significance.

Of course, subsidence improvement in coal mining area is a systematic project. Not only need a lot of money, but also need to improve the technical system for the improvement effect. At present, with the enhancement of comprehensive national strength, the degree of attention to the subsidence improvement is increasing, and the externalities of environmental costs are stressed. It is necessary to take effective measures to strengthen the comprehensive improvement of the subsidence in the process of coal resources exploitation. At the same time, strengthen the collaborative innovation of producing, teaching and research, improve the key technologies of subsidence improvement, to provide technical support for comprehensive improvement of subsided land.

3 EFFECT EVALUATION OF SUBSIDENCE COMPREHENSIVE IMPROVEMENT IN COAL MINING AREA

3.1 *Effect evaluation index system of subsidence comprehensive improvement*

Subsidence improvement in coal mining area is a systematic project with significant time and space characteristics; effect evaluation of subsidence comprehensive improvement is an evaluation afterwards, in order to objectively reflect the effect of subsidence comprehensive improvement, it is necessary to establish the system index system.

The design and selection of indexes in the evaluation index system should conform to the characteristics of the subsided land and meet the purpose of the effect evaluation (Qiao Li and Bai Zhongke, 2009; Cheng Jinhua, Chen Jun, etc., 2013). From the perspective of the construction of ecological civilization in coal mining area, subsidence comprehensive improvement should consider the economic effect, environmental effect and social effect (Lin Yanli, Yan Hongwen., 2011).

Based on this, follow the concept model (Formula 1) of effect evaluation of subsidence improvement in coal mining area, according to the different characteristics of the subsided land in the mining area, in order to ensure the objectivity and comparability of the evaluation and overcome the influence of the price fluctuation, taking availability of the indicators into account, the paper takes relatives indicators to reflect the specific improvement effect; to avoid the duplication of indicators and the impact of regional industrial development on the socio-economic development in the mining area, the factor analysis of the index system was carried out. Following the principle of not less than 85% of the overall sample representation, 20 indicators were selected to establish effect evaluation index system of subsidence comprehensive improvement in mining area (Table 1).

$$E = f(e_1, e_2, e_3) \quad (1)$$

In the formula: E is the effect of subsidence improvement in coal mining area; e_1 is the economic effect of subsidence improvement; e_2 is the environmental effect of subsidence improvement; e_3 is the social effect of subsidence improvement.

In order to objectively reflect the effect of subsidence improvement, these specific indicators take the fixed index, and look the beginning of the subsidence improvement as the base period, determine the indicator value through the actual information of evaluation period.

Table 1. Effect evaluation index system of subsidence comprehensive improvements in mining area.

Object level	Middle level	Indicator level	Notes
The effect of subsidence improvements in coal mining area	Economic Effect (e_1)	Reclamation rate of subsidence improvement (e_{11})	Positive
		Investment in unit improvement area (e_{12})	Negative
		Improvement rate land use (e_{13})	Positive
		Growth rate of output value in improvement area (e_{14})	Positive
		Growth rate of annual net income of operators in the management area (e_{15})	Positive
		Growth degree of agricultural productivity (e_{16})	Positive
		Growth rate of ecological service value per unit area (e_{17})	Positive
	Environmental effect (e_2)	Improvement rate of soil erosion control in improvement area (e_{21})	Positive
		Growth rate of vegetation coverage in the improvement area (e_{22})	Positive
		Growth rate of irrigation guarantee in the improvement area (e_{23})	Positive
		Improvement rate of natural disaster resilience in mining area (e_{24})	Positive
		Improvement degree of regional ecological landscape (e_{25})	Positive
		Land degradation rate (e_{26})	Negative
	Social effect (e_3)	Change rate of population fed by the land in mining area (e_{31})	Positive
		Completion rate of timely subsidence improvement (e_{32})	Positive
		The reuse ratio of subsidence improvement (e_{33})	Positive
		Improvement rate of road network density in subsidence area (e_{34})	Positive
		Improvement degree of living security in subsidence area (e_{35})	Positive
		Engel coefficient (e_{36})	Negative
		Growth rate of enterprise disputes (e_{37})	Negative

3.2 *The effect evaluation of subsidence comprehensive improvement in coal mining area*

The subsided area will transfer and expand with the operating space of coal resources exploitation. Therefore, the opportunity of subsidence improvement can be determined according to production layout, subsidence stability. Evaluation of subsidence comprehensive improvement can be implemented by stages and regions according to completion of subsidence improvement project.

In order to objectively reflect the subsidence improvement effect in coal mining area, combined with the establishment of the index system and its characteristics, this paper carries out the evaluation through fuzzy comprehensive evaluation method based on the AHP method. The AHP method is used in civilization construction of coal mining area to determine the subordinate degree of indicators of subsidence improvement by experts, the fuzzy comprehensive evaluation method is used to measure the effect of subsidence comprehensive improvement, which will reflect effect of subsidence comprehensive improvement quantitatively.

The measured effect value of subsidence comprehensive improvement is still need to be further evaluated, which needs a reasonable evaluation criteria. Traditionally, the objective of subsidence comprehensive improvement is to be available. But because of the ambiguity of this objective, the effect evaluation of subsidence comprehensive improvement lacks of operability. According to the design of comprehensive evaluation system, and the objective and connotation that "It is necessary to restore the land value and ecological environment, and ensure the land social effect.", of comprehensive improvement of subsided land. Based on the comprehensive evaluation value of the land around at the same period, determine the evaluation standard of effect of comprehensive improvement combined the earlier comprehensive evaluation value of background. Based on these, the effect of comprehensive improvement of coal subsidence is evaluated.

According to this evaluation system, a mining area constructed in 1996, meanwhile the design capacity of it is 5 million tons/year; and put into operation in 2000, of which output is 1.2 million tons; then achieve designed production capacity in 2004. The improvement situations of subsided land in two adjacent mining areas are showed in Table 2, the effect evaluation of improvement as shown in Figure 2.

It can be seen from the quantitative evaluation that the effect of subsidence improvement is not exactly the same in different evaluation areas, and the effect is different at different stages, which also affects the comprehensive improvement effect of the subsidence area. The mining area has a similar periodic rule of the subsidence improvement tasks in its life cycle, and the effect of comprehensive improvement also has certain regularity. Therefore, this evaluation model can ensure time and space range of evaluation according to needs, and it also can further find the weak link in the process of subsidence improvement according to evaluation results. Based on these, the corresponding measures are taken to ensure that subsided mine land improvement not only focuses on quantity, but also concerned about the quality of improvement, and effectively improve key weaknesses in process of the ecological civilization construction.

Table 2. The evaluation objects: the basic situation of mining subsidence area.

Evaluation object	211 mining area	213 mining area	The total in mining area
Area (hm^2)	308.2	426.5	8000
Improvement period	2012.3–2013.5	2013.2–2014.8	2000.5–2086.5
Investment (10^8 yuan)	2250	3168	
Area of cultivated land reclamation (hm^2)	49.3	64.2	
agricultural population in subsidence areas (person)	1368	1689	24566

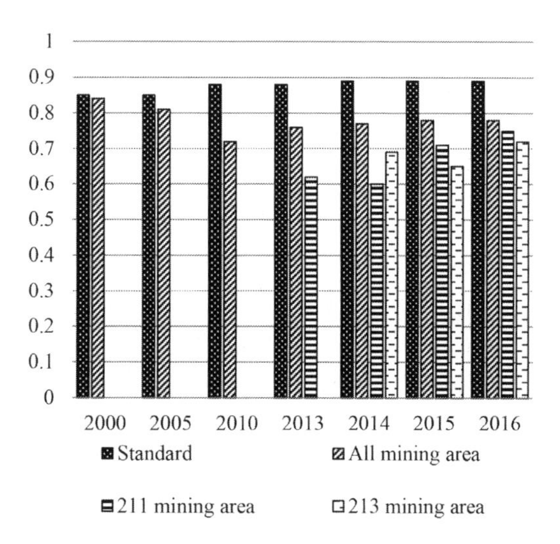

Figure 2. Effect evaluation of subsidence improvement in mining area.

4 SUMMARY

In the context of the ecological civilization construction in the whole society, the mining area provides important mineral raw materials and energy for economic and social development, meanwhile the mining area affected by human activities, is the key area of ecological civilization construction (Zhang Wei, Wang Jianfei, etc., 2014). Among them, the land in mining area is the carrier of regional ecological civilization construction, which must ensure its safe and sustainable use. Strengthening the subsidence comprehensive improvement is an important content of ecological civilization construction.

Coalmine subsidence improvement should pay attention to the scale of governance and the effect of improvement. Therefore, according to the characteristics of the mine life cycle, take corresponding improvement measures in different stages of life cycle.

Coalmine subsidence improvement should strengthen end-of-pipe treatment, and improve land use benefits. At the same time, it is necessary to strengthen the whole process of improvement, in aspect of time, in previous stage, reduce losses through reasonable planning; in the progress, control the destruction of strength through modern technology; in stable stages, implement comprehensive improvement timely. In aspect of space, we should strengthen the coordination of improvement planning and technology to ensure the improvement effect.

Strengthen the improvement responsibility and make the strict effect evaluation of improvement. In order to ensure the objectivity of the evaluation, the construction of index system and the evaluation standard can take the dynamic management way according to the actual; establish delayed evaluation system of subsidence to reflect effect of subsidence improvement; at the same time, clearing responsibility to avoid flashy subsidence improvement. Promote the construction of regional ecological civilization.

ACKNOWLEDGMENT

Supported by The National Key Research and Development Program of China (2016YFC0501 109).

REFERENCES

Cheng Jinhua, Chen Jun & Yi Xinghua. (2013) Research on Evaluation Indicators System of Ecological Civilization in Mining Area. *China Population. Resource and Environment*, 23(2), 1–10.

Chen Xinsheng, Wang Qiaoni & Zhang Zhiguang. (2009) Evaluation on Integrated Benefits of Reclamation Modes for Coalmine Collapse. *Journal of Geology*, 33(2), 174–178.

Liang Hailin. (2015) Effective Approaches in Comprehensive Treatment of Coal Mining Subsidengce. *Coal Engineering*, 47(12),71–74.

Lin Yanli & Yan Hongwen. (2011) Benefit Evaluation of Rural Land Comprehensive Consolidation in Theory. *Ludong University Journal (Natural Science Edition)*, 27(2), 164–167.

Qiao Li & Bai Zhongke. (2009) Research on the Evaluation Index System of Ecology Civilization in Minng Area. *Metal Mine*, (11), 113–118

Sun Guishang & Li Jianzhong. (2014) Research on Comprehensive Treatment for Mining Subsidence Land in China. *Contemporary Economics*. (21), 52–53.

Zhang Mi. (2013) Connotation Characteristics and realization path of socialist ecological civilization. *Studies on the Socialism with Chinese Characteristics*. (2), 84–87.

Zhang Wei, Wang Jianfei, Xiao Xin & Zhang Zhenguo. (2014) Polices and Regulations of Coal Mining Subsidence Area Consolidation: A comparison between China and USA. *Scientific and Technological Management of Land and Resource*, 31(4), 119–124.

Land Reclamation in Ecological Fragile Areas – Hu (Ed.)

Evaluation index system and ecological restoration of abandoned quarries

Q.Q. Zhang & T.Z. Zhang
School of the Environment, Tsinghua University, Beijing, China

ABSTRACT: The restoration and evaluation of degraded ecosystems is an important component in the sustainable development of "human-nature-economy" compound ecosystems. Based on an analysis of ecosystem evolution and structure-function-service integration, this research proposes that ecosystems can return to a dynamic balance of self-maintenance by enriching system elements, strengthening the relationships between different elements and diversified ecology processes, and actuating the restoration of internal functions, which includes the self-regulation of degraded ecosystems. In this paper, the degraded ecosystem structure and degradation characteristics of a quarry are combined with the major factors that influence recovery, and a reference frame for the restoration of the quarry ecological system is then selected to establish the key targets for ecological restoration. Three elements and 17 indexes are constructed from ecological processes, vegetation structures and ecological functions to characterize the ecological restoration, and these parameters can be used to evaluate and manage the ecological restoration of the abandoned quarry.

Keywords: quarry, ecological processes, vegetation structure, ecological function, ecological restoration, restoration evaluation, index system

1 INTRODUCTION

The excessive use of natural resources by human societies has increased the stress on natural ecosystems and resulted in climate change, environmental pollution, vegetation damage, biodiversity loss, etc. The constant acceleration of urbanization and industrialization worldwide, with the ensuing demand for mineral resources, means that newly-created quarries have largely destroyed many natural ecosystems (Wang et al., 2011). Quarrying results in significant visual and ecological impacts (Simón-Torres et al., 2014), not all of which have been identified yet (El-Taher et al., 2016). Quarrying causes drastic alterations. It destroys flora and fauna, thereby* reducing biodiversity and disrupting fundamental ecological relationships. Moreover, it extensively damages soil by modifying the original site topography and depleting and altering soil microbial communities (Corbett et al., 1996; Pinto et al., 2001; Milgrom, 2008; Mouflis et al., 2008; Simón-Torres et al., 2014). Because ecosystem degradation has increased in severity, investigations on ecosystem restoration are urgently required.

Ecological restoration is designed to restore the natural ecological system according to the laws of nature and recreate, guide or accelerate the process of natural evolution. Although human technology cannot restore natural systems, it can be applied to improve natural restorations by introducing important plants and animals into ecosystems to generate the basic habitat conditions and promote natural evolution and ecosystem recovery. Harper (1987) defined ecological restoration as the working process of assembling and testing communities and ecosystems. Jordan (1994) indicated that transforming ecological systems to prior or historical (natural or non-natural) states represents an ecological restoration. Cairns (1995) stated that an ecological restoration is the process of restoring the structure and function of a damaged ecosystem to the state prior to being disturbed.

Ecological restoration assessment takes specific target and system as reference, evaluate on the change of structure, function, quality, health and safety in the process of restoration (liu et al, 2014). Studies have indicated that ecological restoration evaluations have an important role in promoting restoration ecology as a scientific research area (Bradshaw 1993; Norton and Hobbs 1996; Lake 2001). Thus, a set of generally accepted criteria should be defined for use by ecologists and project engineers so that they can evaluate the success of ecological restorations in restoration projects (Alexander and Allan 2007) because a set of guiding criteria will greatly promote the evaluation of restoration projects and the reporting of recovery results (Yang and others 2013). The key to evaluating ecological restoration lies in the selection of the evaluation index and the construction of the index system. Methods of comprehensively and objectively selecting the evaluation index and scientifically designing the index system represent a hot topic in the field of ecological restoration and evaluation (Ding and Zhao 2014).

2 LITERATURE REVIEW

Diamond (1987) believes that ecological restorations can reconstruct a natural community, rebuild a self-maintaining community and maintain the continuity of the community, and his focus is on the restoration of vegetation. Egan (1996) (quoted in Hobbs etc, 1996) indicated that an ecological restoration is the process of reconstructing historical regional plant and animal communities and maintaining the sustainability of the ecological system and the traditional cultural functions of human beings. An ecosystem is affected by non-biological factors, such as solar energy, light, temperature, rainfall, wind, rock, soil, water, air, CO_2, O_2, N_2, inorganic salts, humus, proteins, and carbohydrates, as well as biological factors, such as producers, consumers and decomposers (Cai 2000). These factors are interrelated, and their roles constitute the entire functioning of the ecosystem and provide services for the environment and the functional maintenance of the ecosystem. The foundation for ecosystem restoration and construction is an increase in biodiversity, and plant diversity is particularly important. The development of plant diversity can increase the species diversity of ecosystems because high plant diversity can promote high productivity and provide a material basis for the ecological diversity of the ecosystem. Furthermore, different plant species can create a variety of heterogeneous habitats within an ecological system and accommodate more species assemblages. Moreover, different plants have multiple layers of roots that lead to the construction of various soil micro-habitats and accommodate diversified soil animals and microorganisms (Peng 2007).

According to the vertical structure of the plant community, the ecological system can be divided into several layers, such as trees, shrubs, herbs and the surface layer (moss, lichen) (Li 2011). The tree layer has tall stems and foliage, which perform photosynthesis and regulate gas exchange. Leaf transpiration can inhibit high temperatures and increase air humidity to adjust the microclimate. Trees, shrubs and herbs combine together to form a landscape, which provides scenic and recreation services for human beings. In the community ecosystem, pollination and seed dispersal for reproduction can be conducted via wind energy. Green plants primarily produce energy via photosynthesis and chemical energy bacteria to provide a variety of crops, fruit, prey and other resources for consumers (human beings and animals). The plant community is the primary producer, and it is also the habitat of animals. The ground layer (lichen or moss and other plants) can be used for water penetration and as an adsorbent for water conservation to achieve efficient water regulation. The soil is held in place by the root systems of plants, thereby preventing soil collapse and soil erosion. Microorganisms and fungi in the soil decompose biological debris to generate, store and accelerate the internal cycle of nutrients. The relationship between the structure and function of ecosystems and the services they provide can be described as Fig. 1 Structure of the plant community ecosystem and its function & service relationship:

Therefore, Natural ecological recovery process is essentially a process of soil and plant system synergy evolution. The degree of ecological recovery can be represented by different phase characteristics of soil and vegetation. Soil and vegetation conditions can most directly reflect the degree of ecological restoration (Zhang hua, 2013). As for evaluation indicators, first, it must be possible to measure and assess indicators repeatedly over time. Second, indicators should be variables that are sensitive to changes in community recovery status over time or within key ecosystem, which allows for the

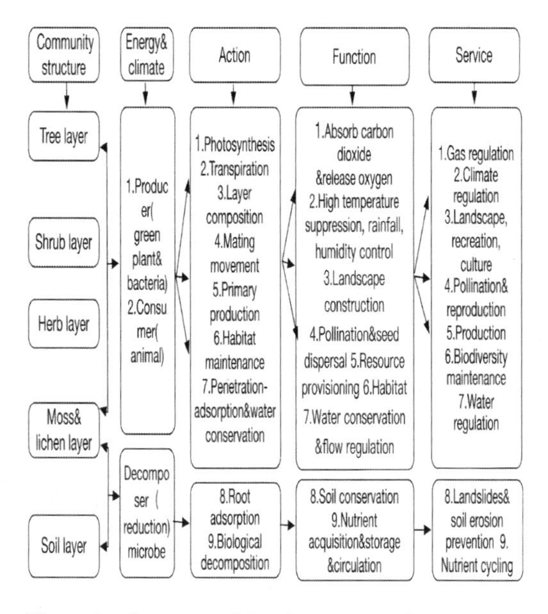

Figure 1. Structure of the plant community ecosystem and its function and service relationship.

exploration of interaction. Third, the effects of community- and individual-level experiences also should be considered concurrently (Jennifer Horney, 2017)

3 MATERIALS AND METHODS

3.1 *Materials*

Quarry is the site for production of sand and stone used for building materials. Sandstone ore used for construction is generally exposed to the surface, therefore, the vast majority of sand mining are open pit mining. So, the vegetation soil should be stripped off at first for sandstone excavation. In the process of vegetation stripping off, the whole ecosystem degenerates and disappears. Quarry ecosystems are primarily degraded because of man-made interference and damage to the ground vegetation. Anthropogenic deforestation and mining has resulted in the vegetation community structure destruction, biodiversity reductions, soil erosion and ecosystem degradation. Mining induces damage to vegetation, exposes the soil layers and leads to soil erosion and soil loss, and forest felling is another primary factor that causes vegetation degradation. Vegetation degradation is a dominant factor for soil erosion. Vegetation degradation and soil erosion overlap and drive the simplification of ecosystem elements and ecological processes. It is high risk of soil erosion. Therefore, it is urgent and important to carry out the ecological restoration of abandoned quarry (abandoned quarry means the depleted quarry) in time. Nevertheless, the evaluation of ecological restoration is the standard to define and evaluate the ecological success of restoration project.

3.2 *Methods*

3.2.1 *Developing the indicators*

Potential indicators were identified initially through a systematic review of the literature and categorized by Society for Ecological Restoration International (SER) or Core Capability. After aggregating the identified indicators, several methods were used to validate the final aggregated list, including a review of literatures, Quarry or mining reclamation plan, recovery plan of case studies from CNKI (the largest Chinese literature data base) and Web of science. Technology flow chart is as follows (see Figure 2).

A systematic review of ecosystem recovery literature was performed to identify research studies related to measuring the progress of recovery in order to extract potential recovery indicators and metrics. First, citations were obtained from mine and quarry reclamation and recovery management

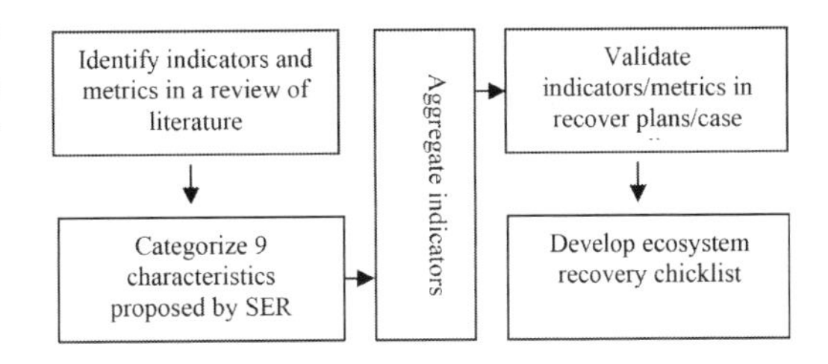

Figure 2. Methods used in the development and validation of recovery indicators for the Quarry Recovery Checklist.

scheme report; Second, a supplemental University of Tsinghua Libraries Articles and researches was conducted using the following keywords: 'quarry'; 'recovery'; and 'ecosystem'. In CNKI data base, in total, 26 peer-reviewed, scholarly articles published between 2010 and 2017 were retrieved. In addition, 15 conference papers published between 2006 and 2013 were retrieved. Involved in environmental science and resource utilization, mining engineering, building science and engineering, and agriculture disciplines, etc. Using the following keywords "ecological restoration quarry" search in Web of Science data base, in total, 60 meetings, 5 reviews and 128 articles published between 2010 and 2017 were retrieved. A total of 196 peer-reviewed publications and conference presentations were reviewed to determine if they incorporated recovery metrics or measures employed to gauge ecosystem recovery progress.

Ruiz Jaen and Aide (2005a) summarized and analyze the articles which published on "restoration Ecology" since it founded 11 years (1993–2003), reveal that species diversity, vegetation structure and ecological process are mainly as ecological recovery measures. Among species diversity, researchers usually take plant as ecological recovery index, which accounted for 79%; researchers who take arthropods as ecological recovery index accounted for 35%. Vegetation coverage, density, biomass, height is a common measure of vegetation structure index, which accounted for 62%, 58%, 39% and 39% respectively. In the research of ecological process, researchers who take biological interactions as ecological recovery index accounted (60%), followed by a depots (47%) and soil organic matter (39%) (Song, 2013).

3.2.2 *Validating the indicators*

As a first step in validating the indicators identified in the literature, quarry recovery plans, content-analysed for another project were reviewed to determine whether or not these indicators have been used in practice. 69 Recovery plans had been previously collected from all of the china: including quarries in Beijing city (Song, 2008; Zhang, 2013;

Table 1. Indicators identified in literature review, listing SER core capability and recovery focus area.

Core capability	Recover focus area	TILR
Soil physical & chemical properties	topsoil sickness, soil texture, organic matter	
	water content, PH value, compactness	48
	soil quality, available NPK, soil organic carbon	
Terrain	gradient, slope direction	
	slope damage/forming time	22
Climate	Rainfall, temperature	16
	irrigation condition	
	native plants, evergreen tree species ratio	
	vegetation coverage, coverage of woody plants	
Vegetation	evenness degree of trees and shrubs	51
	species diversity, species richness	
	litter structure	
	existing vegetation coverage	
	root system status, community structure	
	animal and plant species richness	
Biodiversity	the biology abundance	43
	existing species	

Zhang, 2013; Liu, 2011. etc) Chongqing city (Ning, 2003), Guangdong province, Fujian province, Zhejiang province (Yin, 2012), Guizhou province, Jiangsu province (Yu, 2012) and so on. Recovery plans from the original sample of 69 were included in this study if they scored above the overall mean score in the original plan quality assessment study (Berke et al., 2014). Each of these plan-based indicators was able to be categorized within aggregate indicators, illustrating that all plan-based indicators validated the existing literature-based indicators. According to core capability which proposed by Society for Ecological Restoration International (SER) and the indicators which have been validated, we consolidate indicators for ecosystem restoration evaluation as follows (see Table 2).

The plan based indicators were then reviewed to spot potential metrics to be added to the list of possible measurements. The case studies highlight a community's potential use of the recovery indicators as a means to evaluate the success of recovery, using actual recovery experiences. The validation of the indicators through case studies may elucidate potential sources for gathering data on a quarry's ongoing ecosystem recovery as well as its pre-disaster baseline status. In addition, the case studies may help to demonstrate further how local planning can be used to fulfil national recovery priorities, as outlined in Core Capabilities.

Table 2. Evaluation index system for quarry ecological restorations.

Elements	Indexes	Indication
Ecological process index	Top soil sickness	Bearing capacity
	Soil bulk density	Water/fertility retaining ability
	Soil available N,P,K	
	Soil organic matter	Soil productivity/ carrying capacity
	Soil enzyme activity	
Vegetation structure index	Species of arbor layer	System productivity; energy storage & conversion capacity; community structure & function; ecological niche diversity
	Species of shrub layer	
	Species of herb layer	
	Shannon diversity index	
	Ecological dominance index	
	Pioneer species	
	Constructive species	
Ecological function index	Soil microbial diversity	Niche diversity
	Soil microflora diversity	Diversification of decomposition process & material circulation
	Soil fauna functional group diversity	Material cycle diversification
	Biomass	

The above index system, except Pioneer species and Constructive species, other indicators are positive indicators, i.e., the value is bigger, the better. Pioneer species and Constructive species is moderate indicator, which need to be put into positive indicator, and then be nondimensionalized with standardization method. The calculation formula for moderate index transferring into positive index is:

$$X_i = \frac{1}{1+\left|-X_i\right|}, \quad (i=1,2,\ldots,n) \tag{1}$$

Establish observation matrix, using the method of Z – Score to calculate the original observation data and get its standardized matrix. Using the following formula to Standardize index:

$$Z_{ij} = \frac{\left(X_{ij} - \overline{X_j}\right)}{S_j} \tag{2}$$

$$\bar{X}_j = \frac{1}{n}\sum\nolimits_{i=1}^{n} X_{ij} \tag{3}$$

$$S_j^2 = \frac{1}{n-1}\sum\nolimits_{i=1}^{b}\left(X_{ij} - \bar{X}_j\right)^2 \tag{4}$$

Computing the correlation coefficient, then calculate the correlation coefficient matrix, the correlation coefficient was calculated by the following formula:

$$R_{ij}=\frac{1}{n-1}\sum_{i=1}^{n}Z_{ij}Z_{ji}\left(\mathrm{i},\ \mathrm{j}=1,2,\ldots,\mathrm{P}\right) \tag{5}$$

To calculate the characteristic roots and characteristic vectors, choose the main ingredients: select m principal components, usually takes the principal components which their cumulative contribution rate over 80% or more in practice and their corresponding characteristics of the root is greater than 1. That is:

$$\sum_{j=1}^{m}\left(\sum_{j=1}^{p}\lambda_j\right)^{-1} \tag{6}$$

Takes i principal component and its corresponding characteristic roots λ_j, which accounted for the proportion of the sum of extract characteristic root

$$W_i=\frac{\lambda_i}{\sum_{i=1}^{m}\lambda_k} \tag{7}$$

as the weight of i principal component.

After m main components were determined, then get the matrix of m principal component's contribution to the overall variance

$$\mathrm{A}=(\lambda_1,\lambda_{2,\ldots,}\lambda_m),$$

At the same time, get the contribution matrix that the original matrix in the previous m principal component

$$L=(L_1,L_{2,\ldots,}L_m).$$

The contribution matrix of each index to overall variance can be determined by following:

$$\mathrm{F=A.L}(f_1,f_{2,\ldots,}f_m).$$

The value of each element is the weight of corresponding indexes.

Finally, according to the comprehensive evaluation model to express the status of ecosystem recovery:

$$\mathrm{HI}=\sum_{i=1}^{m}\sum_{j=1}^{p}\left(w_jI_j\right)W_i \tag{8}$$

Among then, $\sum_{i}^{m}W_i=1,\ \ \sum_{j=1}^{n}W_j=1;\ W_i$ is the weight of i comprehensive index; W_j is the weight of j element of indicator among i comprehensive index. I_j is the status quo level of j indicator.

At last, weight sum of each index and get the comprehensive evaluation value.

4 DISCUSSION

The effects of an ecological restoration are usually measured by the attributes of ecological systems, such as the biodiversity, vegetation structure and ecological processes (Liu and Lu 2009). The quantity of biological species, the rate of biomass increase, the soil physical and chemical properties, etc, are among the recognized indicators of ecological restoration (Hao 2016, Zhong 2006). Because ecological processes are interrelated, an ecological system can achieve "structure-function-service" integration. Therefore, the ecological restoration intends to construct the evaluation index system from three aspects: vegetation structure, ecological process and ecological function.

Among the ecological process indicators, soil thickness, quality, parent material (lithology), and physical and chemical properties characterize the soil bearing capacity and soil fertility after restoring quarry waste dumps and the platform (or pithead) via soil dressing measures. The nutrient and enzyme activities characterize the diversification of soil material cycling process and the carrying capacity of land (Zhang, 2015). Changes in the soil properties, nutrient cycling and biological interactions are ecological processes that can reflect the success of an ecosystem restoration. The recovery of biological interactions is essential for long-term ecosystem functioning. Then, they just represent the 9 characteristics of ecosystem recovery which were put forward by SER. That is: ability of eliminate potential threats to ecosystems; ability of the ecosystem to recover from natural disturbances; self-supporting ability of the ecosystem.

In the vegetation structure index, the quantity of species refers to species quantities in a community or habitat. The Shannon diversity index is used to estimate the diversity of a community:

$$\mathrm{H'}=-\sum_{i=1}^{s}(\mathrm{p_i}\times log_2\mathrm{p_i}) \tag{9}$$

where s represents the total quantity of species and Pi represents the proportion of species i to the total quantity of species (Pielou 1974). When only one population is observed in the community, the Shannon index is equal to 0; when there are more than two populations in a community and only one member in each population, the Shannon index reaches the maximum value. The Shannon diversity index is a good method of characterizing the statistics of community diversity in an ecosystem restoration; therefore, it is suitable for evaluating ecological restorations. Dominant species have the highest number of individuals in each layer of the community and are the most important species in the layer because these species have the largest ecological role and determine the basic characteristics of the layer. Edificator is the dominant species in the upper layer of a plant community, and it is usually the community constructor. The constructive species determines the appearance of the community and restricts other components of the community (including plants, animals and microorganisms). Therefore, the species included in the restoration of ecological systems must be considered, and constructor species are important in the evaluation of the restoration scheme. The purpose of the restoration project should be to restore the first dominant species in the primary functional groups rather than a specific number of species, and then ecological redundancy should be restored via the functional traits of different additional species (Montoya 2012). This method is practical and can achieve the most important goal, which is achievable in most recovery projects. Pioneer species are species in the early stages or in the mid-stage of the succession of ecological communities. Pioneer species appear earlier and survive relatively easily, and they play a constructive role in the ecological restoration of communities in the ecological restoration framework. However, the maximum biological diversity method can be used to conduct artificial restorations, which eliminates the need to introduce pioneer species. However, in areas where recovery is relatively difficult or where the habitat conditions are not sufficient, the emergence of pioneer species plays an important role in the construction of a community. The vegetation structure index represents the community ecosystem productivity, the energy storage and conversion capacity, and the ecological niche diversity and its complicated functions in the community ecosystem. The restoration of vegetation communities is a prerequisite for the restoration of animal communities and ecological processes. Therefore, vegetation communities can be considered evaluation indexes, and they are easy to measure with low processing time.

Then, the index of vegetation structure can represent 9 characteristics of ecosystem recovery which were put forward by SER. That is: Whether the new community have similarity in the species diversity and community diversity to the reference site; is there any local species in new community? Is there any emergence of functional groups for the long-term stability of the ecosystem? Ability of the ecosystem to provide habitat for population reproduction.

Among the ecological function indexes, soil organisms represent a regulator of the ecological processes, such as decomposition, nutrient mineralization, etc. Microorganisms play an important role in the ecosystem by participating in nutrient cycling, organic matter degradation and energy flow. The generation cycle of arthropods is short; therefore, these organisms can reflect inter-annual variations in the recovered plots. Small arthropod species can effectively monitor subtle but important diversification factors that may affect the habitat quality. The flora community impacts the arthropod's food structure, habitat, natural enemy species, population dynamics, and fecundity, which then affects the diversity and richness of the plant communities. However, the diversity of the arthropod community also has an influence on the structure, function, stability and ecological processes of the ecosystem (Li and others 2008). Arthropods directly or indirectly use vegetation as a food item and habitat; therefore, they are sensitive to disturbances in the composition of the plant community. Changes in the diversity and complexity of the arthropod community could reflect habitat degradation; therefore, these organisms could play a role as a large-scale ecosystem biodiversity indicator in the evaluation of ecological restoration projects.

Then, the index of ecological function can represent 9 characteristics of ecosystem recovery which were put forward by SER. That is: the ability to maintain ecosystem functions; integrity of the ecosystem landscape.

5 CONCLUSIONS

The restoration of the ecological system of the abandoned quarry represents the restoration of self-maintenance and self-regulating functions within a human-nature-economic complex ecosystem, which is influenced by human disturbances and natural factors. The ecosystem management of quarries focuses on the recovery of the ecosystem "structure-process-function" integration, which can stimulate the self-repair function of the ecosystem and eventually generate an ecosystem that exhibits a dynamic balance and the relative stability of self-maintenance. The preliminary evaluation indexes described above were selected for the evaluation of degraded ecosystem restorations. However, to improve and test the indexes to determine their usefulness in restoration evaluations and monitoring system construction, the indexes must move beyond a theoretical discussion and be implemented in practical investigations to determine whether they can promote research on the evaluation, monitoring, early warning and restoration of degraded ecosystems caused by human interference.

REFERENCES

Alexander GG, Allan JD. 2007. Ecological success in stream restoration: Case studies from the Midwestern United States. Environmental Management, 40: 245–255.

Aronson J, Floret C, Le Floc'h E, Ovalle C, Pontanier R. 1993. Restoration and rehabilitation of degraded ecosystems in arid and semi-arid lands. II. Case studies in southern Tunisia, central Chile and northern Cameroon. Restoration Ecology 1: 168–187.

Bradshaw AD. 1993. Restoration ecology as a science. Restoration Ecology, 1: 71–73.

Cairns J Jr. 1995. Restoration ecology. Encyclopedia of Environmental Biology. 3:223–235.

Cai XM. 2000. Ecosystem ecology. Beijing: Science Press.

Corbett, E.A., Anderson, R.C., Rodgers, C.S., 1996. Prairie revegetation of a stripmine in illinois: fifteen years after establishment. Restor. Ecol. 4, 346–354, http://dx.doi.org/10.1111/j.1526-100X.1996.tb00187.x.

Dimond J. 1987. Reflections on goals and on the relationship between theory and practice. // Jordon W R III, Gilpin N, Aber J eds. Restorations Ecology: A Synthetic Approach to Ecological Research, Cambridge University Press, Cambridge, 329–336.

Ding JY, Zhao WW. 2014. Progress and prospects on evaluation of ecological restoration: a review of the 5th World conference on ecological restoration. Chinese Journal of Applied Ecology 25: 2716–2722.

El-Taher, A., García-Tenorio, R., Khater, A.E.M., 2016. Ecological impacts of Al-Jalamid phosphate mining, Saudi Arabia: soil elemental characterization and spatial distribution with INAA. Appl. Radiat. Isot. 107, 382–390, http://dx.doi.org/10.1016/j.apradiso.2015.11.019.

Griffiths CJ, Hansen DM, Jones CG, Zuël N, Harris S. 2011. Resurrecting extinct interactions with extant substitutes. Current Biology 21: 762–765.

Hao J, Guo D, Shangguan T. 2016. Ecological performance assessment on early plant reclamation in coal gangue yard. Acta Ecologica Sinica 36: 1946–1958.

Harper J L.1987. Self-effacing art: Restoration as imitation of nature. // Jordon W R III, Gilpin N, Aber J eds. Restorations Ecology: A Synthetic Approach to Ecological Research, Cambridge University Press, Cambridge, 35–45.

Hobbs R J, Norton DA. 1996. Towards a conceptual framework for restoration ecology. Restoration Ecology, 4: 93–110.

Hobbs RJ. 2003. Ecological management and restoration: assessment, setting goals and measuring success. Ecological Management & Restoration 4: S2–S3.

Jackson LL, Lopoukhine N, Hillyard D. 1995. Ecological restoration: a definition and comments. Restoration Ecology 3: 71–75.

Jennifer Horney, Caroline Dwyer, Meghan Aminto. Developing indicators to measure post disaster community recovery in the United States. [J]. Disasters, 2017, 41(1): 124–149.

Jordon W R III, 1994. "sunflower Forest": Ecological restoration as the basis for a new environmental paradigm. /Baldwin A D J ed. Beyond Preservation: Restoring and Inventing Landscape, University of Minnesota Press, Minneapolis, 17–34.

Lake PS. 2001. On the maturing of restoration: Linking ecological research and restoration. Ecological Management and Restoration, 2: 110–115.

Li Q, Chen YL, Zhou XY, Bei RT, Yin LH, Xiong ZP. 2008. Assessment of ecological restoration of degraded ecosystem and its biodiversity. Journal of Northwest Forestry University 23: 69–73.

Liu F, Lu L. 2009. Progress in the study of ecological restoration of coal mining subsidence areas. Journal of Natural Resources 4: 612–620.

Liu X F, Wang W J, Li J. Research advances and prospects of post-disaster ecological restoration assessment. Acta Ecologica Sinica, 2014, 34(3):527–536

Liu YG. Evaluation of artificial restoration effect of the abandoned Mines in Beijing. [D]. Beijing Forest University. Doctoral dissertation. Beijing, 2011.

Li zhenji CS. 2011. Synecology. Beijing: Meteorological Press.

Ma J, Liu S, Shi Z, Liu X, Miao N. 2010. A review on restoration evaluation studies of degraded forest ecosystem. Acta Ecologica Sinica 30: 3297–3303.

Milgrom, T., 2008. Environmental aspects of rehabilitating abandoned quarries: israel as a case study. Landscape Urban Plann. 87,172–179, http://dx.doi.org/10.1016/j.landurbplan.2008.06.007.

Montoya D, Rogers L, Memmott J. 2012. Emerging perspectives in the restoration of biodiversity-based ecosystem services. Trends in Ecology & Evolution 27: 666–672.

Mouflis, G.D., Gitas, I.Z., Iliadou, S., Mitri, G.H., 2008. Assessment of the visual impact of marble quarry expansion (1984–2000) on the landscape of Thasos island, NE Greece. Landscape Urban Plann. 86, 92–102, http://dx.doi.org/10.1016/j.landurbplan.2007.12.009.

Ning FS, You X, Yang HL. Countermeasures for Ecological and Landscape Restoration of Abandoned Quarries in Urban District of Chongqing City [J]. Bulletin of Soil and Water Conservation. Vol. 25, No.3, Jun., 2005

Peng S. Restoration ecology. Beijing: China Meteorological Press. 2007

Pielou E C. 1974. Population and community ecology: Principles and methods, Gordon and Breach, New York.

Ruiz-Jaen MC, Aide T M. Vegetation structure, species diversity, and ecosystem process as measures of restoration success. Forest Ecology and management, 2005a, 218:159–173.

SER (Society for Ecological Restoration International Science and Policy Working Group). 2004. The SER International Primer on Ecological Restoration. http//www.ser.org.

Simón-Torres, M., del Moral-Torres, F., de Haro-Lozano, S., Gómez-Mercado, F., 2014. Restoration of dump deposits from quarries in a Mediterranean climate using marble industry waste. Ecol. Eng. 71, 94–100, http://dx.doi.org/10.1016/j.ecoleng.2014.07.039.

Song BM. The studies on the ecological restoration in abandoned quarry in Xishan Mountain in Beijing: Process, characters and mechanism of natural restoration. [D] Shandong University Doctoral Disssertation. Jinan, 2008.
Wang, J., Li, Z., Hu, X., Wang, J., Wang, D., Qin, P., 2011. The ecological potential of a restored abandoned quarry ecosystem in Mt. Mufu, Nanjing, China. Ecol. Eng. 37. 833–841, http://dx.doi.org/10.1016/j.ecoleng.2010.12.026.
Yang ZP, Gao JX, Zhou KX, Zheng H, Wang Y, Li HM. 2013. Evaluation of ecological restoration: research progress. Chinese Journal of Ecology 32: 2494–2501.
Zhang H, Study on the Technology of Ecological Restoration in Huangyuan Village Quarry Loose Deposits of Beijing Fangshan Region. [D]. Beijing forest university. Beijing: 2013
Zhang HY. 2015. Study on the evaluation index and restoration of Karst degraded ecosystems based on the structure-process-function integration. Ecological Science 34: 205–210.
Zhang Y, Zhao TN, Shi CQ. Dynamic research on vegetation recovering in abandoned mine slopes in Beijing mountainous areas. Jounal of Arid Land Resources and Environment. Vol. 27, Jun. 2013.
Zhong S. 2006. Study on the theoretical system and evaluation method of ecological restoration of abandoned mine. Fuxin: Liaoning Engineering and Technical University.

Land Reclamation in Ecological Fragile Areas – Hu (Ed.)
© 2017 Taylor & Francis Group, London, ISBN 978-1-138-05103-4

Experimental investigation of synthetic (top) soils for mine rehabilitation: Column leaching study

Shashibhushan Biliangadi
IIT Bombay—Monash Research Academy, IIT Bombay, Mumbai, India

Mohan Yellishetty & Vanessa N.L. Wong
Monash University, Clayton, Victoria, Australia

Anil Kumar Dikshit
IIT Bombay, Mumbai, India

Suman Majumdar
JSW Group, Mumbai, India

ABSTRACT: Linz-Donawitz (LD) slag is one of the major by-products generated in the iron and steel industry. Alternative uses for waste by-products in the iron and steel industry are required due to increasing production, less land for disposal and stringent environmental regulations. Three different synthetic soils (M1 = 30% slag + 70% compost, M2 = 50% slag + 50% compost, and M3 = 70% slag + 30% compost) prepared from LD slag and sewage sludge compost were subjected to continuous column leaching studies for two months. The aim of the leaching study was to understand the variation in pH trends during leaching, mobility of metals and nutrients. Results showed that the pH of M2 and M3 synthetic soil and steel slag leachates remained in alkaline pH ranges. The pH for M1 was circumneutral until the liquid to solid ratio (L/S) = 0.5 and thereafter reached above pH 10. pH generally decreased with increasing percentage of compost. LD slag is a heterogeneous material comprising of 35–40% CaO and 6–8% MgO. Previous studies also suggest that with ageing and in presence of water, steel slag may form new mineral phases. Readily soluble minerals may dissolve in water more quickly however, minerals which are entrapped within complex heterogeneous phase may dissolve slowly. The highly alkaline pH trend during this leaching study may be due to presence of Ca-CO_3-OH type ion matrix in leachates. Therefore, when steel slag or synthetic soils are used as an artificial (top) soil, high pH leachates may have some negative ecological impacts.

1 INTRODUCTION

Iron ore is primarily extracted by open cut mining operations from hematite (Fe_2O_3), magnetite (Fe_3O_4), goethite ($HFeO_2$), siderite ($FeCO_3$), ilmenite ($FeTiO_3$), pyrite (FeS_2) are the minerals from which iron is extracted. Each of these minerals constitute varying amounts of iron (Yellishetty et al. 2012). In the iron and steel industry, every ton of steel produces around 2 to 4 Tonne (T) of solid, liquid and gaseous wastes. Slag is a major by-product generated during the iron and steel making process and is produced in Blast Furnace (BF) during pig iron making and in Steel Melting Shop (SMS)—either in Basic Oxygen Furnace (BOF) and/or Electric Arc Furnace (EAF). In 2016 steel production has reached to 1628 Million Tonne (MT) worldwide and 96 MT in India (Ministry of steel, 2017). In general, production of 1T of pig iron and steel leads to the generation of 300–540 kg of iron making/Blast Furnace (BF) slag and ~200 kg of steel slag respectively (Indian bureau of mines, 2016) which is high in oxides and silicates of calcium, magnesium, iron and aluminium.

Today, due to better understanding of material characteristics, most of the slag is being reused in cement making, pavement construction, as construction materials or as soil conditioners. The reuse of slag by other industries depends upon market demand but can be restricted by long transport distances (Drissen et al. 2009). The industries are concentrating towards zero-discharge strategy to meet environmental regulations as well to avail economic benefits from by-products. Steel slag utilization rates in USA, Japan, Germany and other developed countries is >95% mark (Yi et al. 2012). However, developing countries like India, China and others are still dispose of a large proportion of waste materials.

Table 1. Comparison of chemical composition of BF slag and BOF slag (Kumar et al., 2010).

Source	SiO_2 (%)	FeO (%)	Al_2O_3 (%)	CaO (%)	MgO (%)	MnO (%)	P_2O_5 (%)	S (%)
BF slag	33.00	0.40	19.00	32.00	10.00	0.20	–	0.80
BOF slag	14.22	23.17	0.93	47.80	9.73	0.57	2.34	0.03

Table 2. Chemical properties of steel slag and sewage sludge.

Material	Ca (mg/Kg)	Mg (mg/Kg)	Fe (mg/Kg)	Al (mg/Kg)	P (%)	C (%)	N (%)	S (%)
LD slag	291,081.31	50,453.02	202,123.75	11,967.05	8,938.44	0.56	0.03	0.05
SSC	25,327.50	5,574.00	14,250.00	7,450.00	5,439.00	20.47	1.52	0.42

Steel slag is generally alkaline with pH >10. The cement industry utilizes BF slag as it constitutes only 0.5–1.0% FeO, CaO, MgO, SiO_2 and Al_2O_3. Conversely, steel slag comprises a high iron (16–23%), magnesium and free lime concentrations which restricts its usage in the cement industry (Kumar et al. 2010). Table 1 shows the chemical composition of BF slag and BOF.

LD slag is used as a soil conditioner and fertilizer in acidic soils because of its high mineral and free lime content. LD slag with lower phosphorous concentrations is reused in blast furnace for recovery of metals like V, Mo and Ti, as coarse aggregates in roads and railway ballast. However, the majority of steel slag is generally dumped or disposed of in landfill due to low market demand and high transportation costs (Kumar et al. 2010).

Several elements which are present in slag such as Ca, Mg, Si, Mn, Co and Fe are necessary for plant growth and these minerals. These elements are useful when they are available in the required concentration and become toxic to plants as well as animal and humans if they exceed threshold concentration limits (Habashi 2011). Several studies have showed the benefits of the use of steel slag in agriculture. Application of steel slag to soils can treat Fe chlorosis in corn (Wang and Cai 2006).. Lopez et al. (1995) and Yi et al. (2012) stated several benefits of using steel slag as liming agents in acidic soils, to cover acidic landfills as free Ca and Mg can increase pH the presence of silicates assists in long term pH buffering, and steel slag can restrict fungal growth in soils.

Using steel slag to as a component of forming synthetic soils can assist in providing a source of material for minesite rehabilitation. Soil pH plays an important role in nutrient uptake by plants because at higher pH levels nutrients becomes unavailable for plants. The current long-term leaching experiment study was carried out to understand the variation in leachate pH during leaching of synthetic soils in contact with water. This study helps with predicting the approximate amount of time and water required to bring pH of synthetic soils to near neutral pH conditions.

2 MATERIALS

LD slag was supplied by JSW Vijaynagar plant, Toranagallu, Karnataka, India. Sewage Sludge Compost (SSC) was collected from Gippsland wastewater treatment plant, Victoria, Australia. Some important chemical properties of steel slag and SSC are given in Table 2.

3 METHODS

3.1 *Long-term leaching studies*

The column setup is shown in Figure 1. Columns used for the leaching experiment are made of transparent acrylic pipes of 10 cm Internal Diameter (ID) and are of total length 50 cm. A manually operable valve was fixed at the bottom of the column to collect leachates at regular intervals. A thin layer of glass wool was placed at the bottom of column to restrict the loss of material. About 6–8 cm of gravel bed was placed on top of the glass wool to improve water infiltration and movement of leachate. Each column was filled with 2.6 kg of synthetic soil over the gravel bed. The volume of de-ionised water equivalent to a liquid to solid ratio (L/S) = 0.25 (650 mL) was added into each column. Synthetic soil samples were watered once every ten days and leachate was collected after every three days. Leachate samples were collected at the bottom of column and immediately analysed for pH.

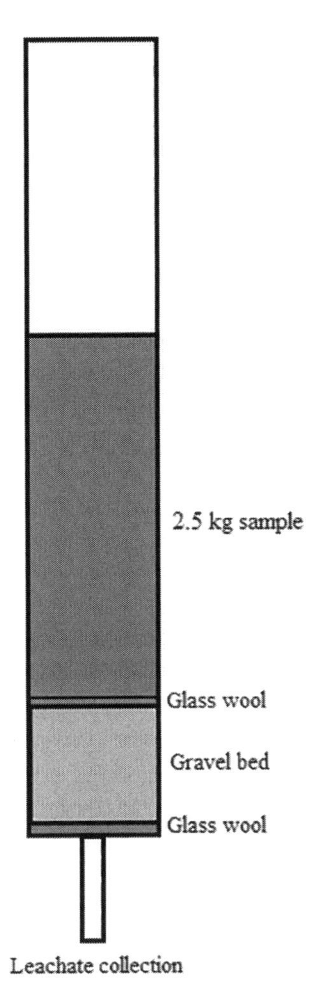

Figure 1. Column leaching set-up.

4 RESULTS AND DISCUSSION

pH variation with L/S during leaching study is shown in Figure 2. Results showed that the pH of M2 and M3 synthetic soil and steel slag leachates remained in the alkaline pH range. The pH for M1 was circum-neutral until the Liquid to solid ratio (L/S) = 0.5 and thereafter reached above pH 10. In the beginning, unstable organic acids present in compost might have reacted with water more rapidly than steel slag-leading to reduced pH conditions at the beginning of the experimental period.

LD slag is a heterogeneous material comprising of 35–40% CaO and 6–8% MgO. Previous studies also suggest that with ageing and in presence of water, steel slag may form new mineral phases. Readily soluble minerals may dissolve in water more quickly however; minerals which are entrapped within complex heterogeneous phase or micro-pores of slag may dissolve slowly compared to minerals which are present in macro-pores or onto the surface. The highly alkaline pH trend during this leaching study may be due to presence of Ca-CO_3-OH type ion matrix in leachates (Riley and Mayes 2015). Therefore, when steel slag or synthetic soils are used as an artificial soil, high pH leachate may have some negative ecological impacts such as death of aquatic species in nearby water streams, unavailability of nutrients to plants and so on.

pH variation with time during leaching study is given in Figure 3. During first few weeks of leaching study no biological colonies were observed in any of the synthetic soils samples. However, after 8 weeks of leaching (day 57 onwards). Growth of biological colonies/tissues were observed in case of M1 and compost samples. Initially biological colonies were observed only in one of the M1 columns where the pH was around 9.23 and thereafter bio-

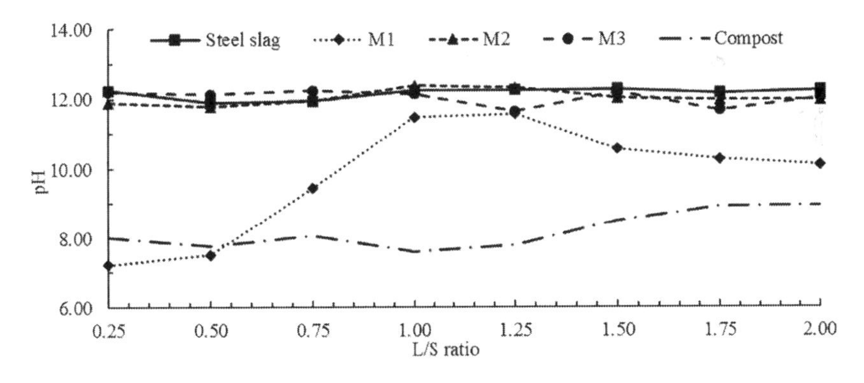

Figure 2. Variation in pH with L/S during long term leaching study.

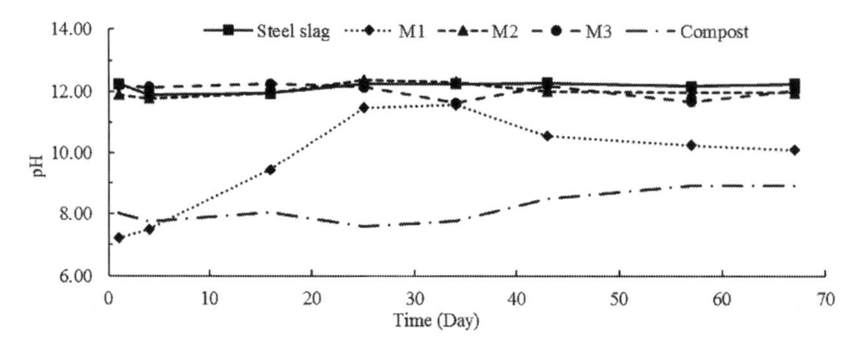

Figure 3. Variation in pH with time during long term leaching study.

logical colonies disappeared and remained in the samples. Similar kind of observations were made by (Hull et al. 2014) in streams affected by dumping of steel slag whose pH was buffered around 9.

5 SUMMARY

This column leaching study revealed that the pH of steel slag and synthetic soil samples remained in alkaline range. This may be due to the slow release of alkaline metals like Ca and Mg into the aqueous phase. Leaching of Ca and Mg in high quantities into streams or groundwater may increase the pH which may adversely impact aquatic environments. The alkaline pH trends during these leaching studies indicate that it will be difficult to decrease the pH down with leaching with water alone. Therefore, it is essential to neutralise the alkaline waste materials for artificial soil which is suitable for vegetation.

ACKNOWLEDGEMENT

We are thankful to JSW Steel for providing the LD slag from Vijaynagar works, Karnataka, India and supporting our research. We also extend our thanks to Gippsland wastewater treatment plant, Victoria, Australia for providing compost to carry out experiments.

REFERENCES

Drissen, P., Ehrenberg, A., Kühn, M. and Mudersbach, D. (2009) Recent development in slag treatment and dust recycling. Process Metallurgy, 737–745.

Habashi, F. (2011) Pollution problems in the metallurgical industry: A review. Journal of Mining & Environment, 2(1), 17–26.

Hull, S.L., Oty, U.V. and Mayes, W.M. (2014) Rapid recovery of benthic invertebrates downstream of hyperalkaline steel slag discharges. Hydrobiologia, 736(1), 83–97.

Indian bureau of mines. (2016). Slag-iron and steel. Nagpur: Ministry of mines, Government of India.

Kumar, D.S., Umadevi, T., Paliwal, H.K., Prasad, G., Mahapatra, P.C. and Ranjan, M. (2010) Recycling steelmaking slags in cement. World Cement, 11, 1–6.

Lopez, F., Balcazar, N., Formoso, A., Pinto, M. and Rodriguez, M. (1995) The recycling of Linz-Donawitz (LD) converter slag by use as a liming agent on pasture land. Waste Management & Research, 13(5), 555–568.

Ministry of steel. (2017). An overview of steel sector. Available from Ministry of steel, Government of India: http://steel.gov.in/overview.htm [Accessed 31st March 2017].

Riley, A.L. and Mayes, W.M. (2015) Long-term evolution of highly alkaline steel slag drainage waters. Environmental Monitoring and Assessment, 187(7), 463.

Wang, X. and Cai, Q.S. (2006) Steel slag as an iron fertilizer for corn growth and soil improvement in a pot experiment. Pedosphere, 16(4):519–524.

Yellishetty, M., Mudd, G., Mason, L., Mohr, S., Prior, T. and Giurco, D. (2012) Iron resources and production: technology, sustainability and future prospects.

Yi, H. Xu, G. Cheng, H. Wang, J. Wan, Y. and Chen, H. (2012). An overview of utilization of steel slag. Procedia Environmental Sciences, 16, 791–801.

Land Reclamation in Ecological Fragile Areas – Hu (Ed.)
 ISBN 978-1-138-05103-4

Dynamic monitoring and trend analysis of soil moisture in Shendong mining area based on MODIS

Y. Liu, H.R. Wang, W. Zhang & Z.Z. Meng
College of Geomatics, Xi'an University of Science and Technology, Xi'an, Shaanxi, China

ABSTRACT: Normalized Difference Vegetation Index (NDVI) and land surface temperature (T_s) data from the Multi-temporal Moderate Resolution Imaging Spectroradiometer (MODIS) with 250 m resolution from 2000 to 2015 were used to analyze the bi-parabolic NDVI-Ts space. The spatio-temporal features of soil moisture situation for 16a in Shendong mining area was evaluated based on the Temperature Vegetation Dryness Index (TVDI) obtained from bi-parabolic NDVI-T_s space. The future development trend of soil moisture was predicted using the unary linear regression method and rescaled range analysis (Hurst). The results show that soil moisture conditions in Shendong mining area increase gradually from northwest to southeast over the past 16a, and Hurst is higher in the middle of mining area, while it is low on the western. Combined with the slope and Hurst variation trend analysis in 16a of soil moisture, the drought conditions are slightly aggravated in future.

1 INTRODUCTION

Soil moisture is an important water resources in nature, it connects the atmospheric water, surface water and groundwater. Vegetation absorbs water mainly from precipitation, soil water, surface runoff and groundwater. However, atmospheric precipitation, surface runoff and groundwater must transfer into soil moisture which can be absorbed by the plant. The ecological environment of Shendong mining area is fragile and vegetation is given priority to with shrubs and herbs with the root system distributing in 0–5 m depth (Liu *et al.*, 2015). Thus, the growth of vegetation cannot absorb water from groundwater and then soil moisture has become the dominant factor in the mining area for vegetation growth and recovery. In this study, MODIS-NDVI and land surface temperature (T_s) were combined used to construct the bi-parabolic NDVI-T_s space, and a simplified Temperature Vegetation Dryness Index (TVDI) from bi-parabolic NDVI-T_s space was presented to evaluate the spatio-temporal features of soil moisture from 2000 to 2015 in Shendong mining area. The future development trend of soil moisture was predicted using the unary linear regression method and rescaled range analysis (Hurst).

2 MATERIALS

2.1 *Study area*

Shendong mining area (Figure 1) is located in southeast of the Ordos Plateau, northern edge of the Loess Plateau in northern Shaanxi and the South-East edge of Mu Us Desert, and is centered at (110° 18′30″E, 39° 11′30″N).

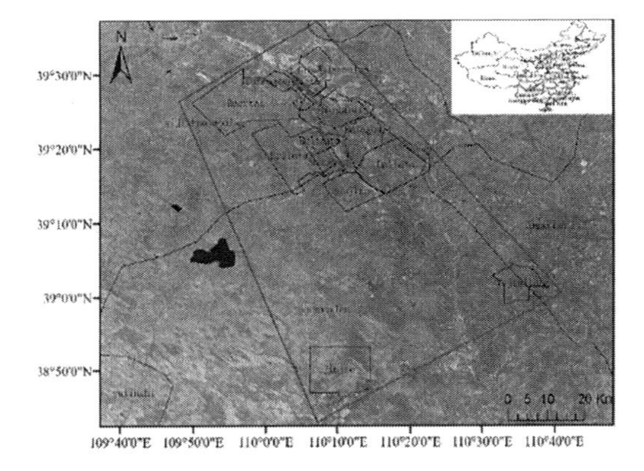

Figure 1. Map of the study area.

2.2 *Data*

The 8-day composite land surface temperature (MOD11A2) and the 16-day composite ground vegetation index (MOD13Q1) from 2000 to 2011 were used to analyze the scatter plot of the NDVI-T_s space. Ts data with 1 km resolution were obtained from MOD11A2 and resampled to 250 m resolution, whereas NDVI with 250 m resolution was obtained from MOD13Q1. Using the Maximum Value Composites method, the yearly maximum T_s and NDVI data from 2000 to 2015 were obtained to establish the NDVI-T_s space.

2.3 *Methods*

Sandholt et al. (2002) developed a simplified land surface dryness index based on the triangle NDVI-T_s space, i.e., TVDI (Figure 4), which can be expressed as follows:

$$TVDI = \frac{T_s - T_{s\min}}{T_{s\max} - T_{s\min}} \tag{1}$$

where

$$T_{s\max} = a_1 + b_1 \times NDVI$$
$$T_{s\min} = a_2 + b_2 \times NDVI \tag{2}$$

where T_s is the observed surface temperature at the given pixel; Ts_{min} is the minimum surface temperature in the triangle, defining the wet edge; and Ts_{max} is the maximum surface temperature observation for a given NDVI, defining the dry edge; a_2 and b_2 define the dry edge as a linear fit to the data; and a_1 and b_2 define the wet edge, which is the minimum surface temperature in the triangle.

In this study, the NDVI-Ts space is bi-parabolic. The fitting equation of the dry and the wet edges are as follows:

$$T_{s\max} = a_1 \times NDVI^2 + b_1 \times NDVI + c_1$$
$$T_{s\min} = a_2 \times NDVI^2 + b_2 \times NDVI + c_2 \tag{3}$$

where a_1, b_1, c_1, a_2, b_2 and c_2 are the fitting coefficients of the dry and the wet edges that can be obtained from the scatter plot of the NDVI-Ts space.

In this paper, we used Hurst Index to forecast change trend of soil moisture in future (Hurst, 1951), it based on three condition: 1) If H = 0.5, it shows that time series is a random sequence, there is no continuity; 2) If 0.5 < H < 1, the time series with long-range correlation and sustainable; 3) If 0 < H < 0.5, the time series with long-range correlation and anti-sustainable. Soil moisture change trend in 16a used a unary linear regression method, the formula are as follows:

$$Slope = \frac{n \times \sum_{i=1}^{n}(i \times M_{NDVI,i}) - \sum_{i=1}^{n} i \sum_{i=1}^{n} M_{NDVI,i}}{n \times \sum_{i=1}^{n} i^2 - (\sum_{i-1}^{n} i)^2} \tag{4}$$

3 RESULTS

3.1 *Scatter plot of NDVI-Ts space*

Figure 2 shows the scatter plot of the dry and the wet edges in the NDVI-T_s space of the study area.

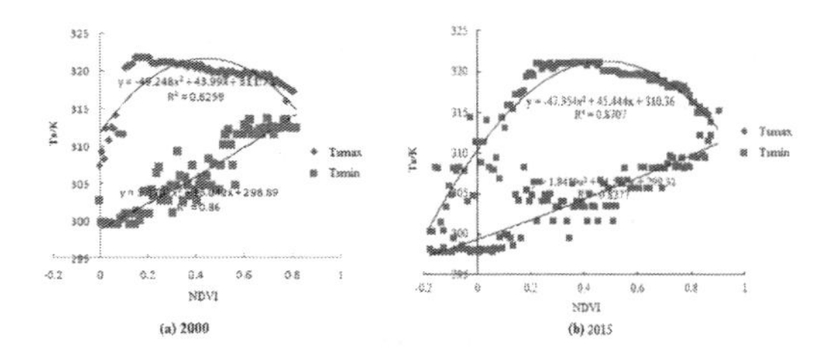

Figure 2. Scatter plots of the dry and the wet edges in the NDVI-Ts space in 2000 and 2015 of study area.

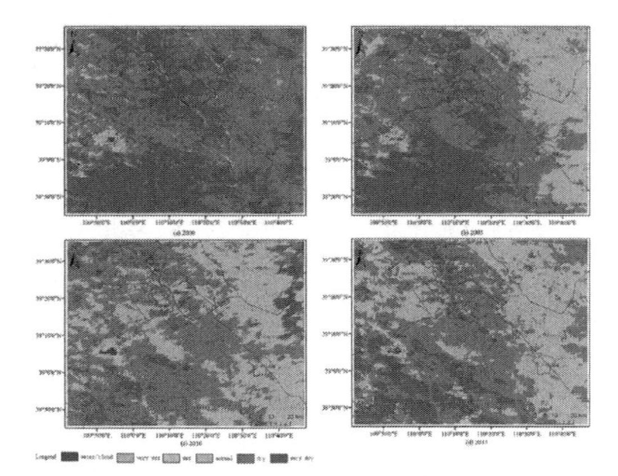

Figure 3. Soil moisture grade spatio-temporal variation in Shendong mining area from 2000 to 2015.

The figure shows that the NDVI-T_s space is a bi-parabola, the dry-edge position relationship has a high correlation coefficient of $R^2 > 0.62$, and the wet-edge position relationship has a correlation coefficient of $R^2 > 0.8$.

3.2 *Spatio-temporal variation analysis of soil moisture of Shendong mining area*

In space scale, soil moisture increases gradually from northwest to southeast from 2000 to 2015 (Figure 3). In 2000, most of the mining area is in a state of drought, the proportion is 96.03%, and in 2015, the drought area are mainly distributed in the southwest and northwest of study area, the percentage decline to 59.59%. Wet area are mainly distributed in the east and southeast of the mining area and the ratio is 4.41% in 2015, while wet area proportion was only 0.36% in 2000. In 2015, normal areas are mainly distributed in the eastern area, scattered in other areas, accounts for the largest share, at 35.87%, and the normal area was only accounted for 3.12% in 2000.

3.3 *Soil moisture forecast in future*

Hurst value in most area of Shendong mining area is between 0.35 and 0.45 (Figure 4), with weaker

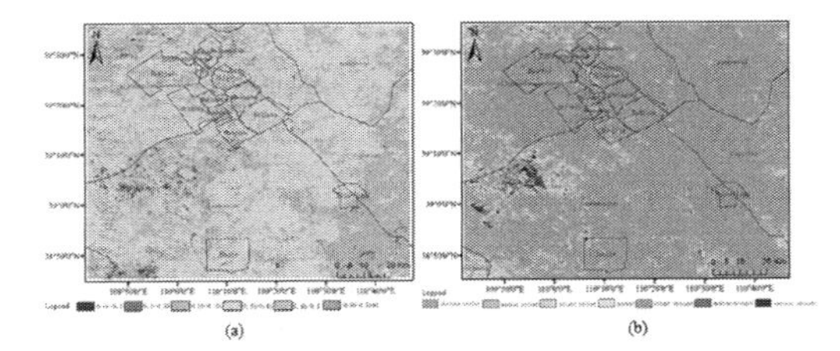

Figure 4. Soil moisture forecast based on Hurst (a) and Slope (b).

anti-sustainability, however, the drought was alleviated in most area of study area in 16 years, and thus soil moisture may be decrease in most area of Shendong mining area and the drought will aggravate in future. Hurst value in the core area and southwest of Shendong mining area is over 0.50, while soil moisture condition of these area is obvious relieved, and thus soil moisture in these areas may be improved in future.

4 SUMMARY

Dynamic monitoring the spatial pattern of soil moisture shows that the drought was eased in most area of Shendong mining area in past 16 years. Soil moisture in most area will be worse in future based on the analysis of Hurst index and unary linear regression method.

ACKNOWLEDGEMENTS

This work was supported by the National Natural Science Foundation of China (Grant No. 41401496).

REFERENCES

Hurst H E. (1951). Long-term storage of reservoirs: an experimental study. J. *Transactions of the American society of civil engineers, 116*:770–799.

Sandholta, I., Rasmussena, K., and Andersen, J. (2002). A simple interpretation of the surface temperature/vegetation index space for assessment of surface moisture status. J. *Remote Sensing of Environment, 79(2–3)*: 213–224.

Ying Liu, Lixin Wu & Hui Yue. (2015). Bi-parabolic NDVI-Ts Space and Soil Moisture Remote Sensing in an Arid and Semi-arid Area. J. *Remote Sensing, 41(3)*:159–169.

Land Reclamation in Ecological Fragile Areas – Hu (Ed.)
© 2017 Taylor & Francis Group, London, ISBN 978-1-138-05103-4

Regional difference analysis of ecological carrying capacity based on an emergy in Shaanxi province

B.H. Li, D.W. Jin, G.R. Ge & Y.D. Guo
Xi'an Research Institute of China Coal Technology and Engineering Group, Xi'an, Shaanxi, China

ABSTRACT: In order to reveal the differences and causes of regional sustainable development, the ecological footprint and ecological carrying capacity of thirteen cities in Shaanxi Province were calculated and analyzed. The results showed that the ecological footprint of Xi'an is 4.04×10^8 hm^2 and the biggest, Yangling is 0.30×10^7 hm^2 and the smallest; the ecological carrying capacity of Weinan is the largest reached to 6.00×10^7 hm^2 and Tongchuan, 0.77×10^7 hm^2 is the smallest; six cities appear ecological deficit, of which Xi'an is the most serious reached to 3.89×10^8 hm^2; five cities have ecological surplus, of which Hanzhong is the best having 2.65×10^7 hm^2. The economy development of Shaanxi Province mainly relies on the contribution of mineral resources and land resources. The rich ecological areas have relatively large land and mineral resources. The ecological surplus areas with the low development intensity are superior in natural conditions. The intensity of regional land and mineral resources development is the subjective cause, while the uneven distribution of natural conditions and resource reserves is an objective reason for regional differences.

1 INTRODUCTION

The theory of ecological footprint has set off a research boom since the Canadian scholars ReesWE and Wackernage first proposed and perfected. In the quantitative research process, many scholars found that the basic parameters in the inter-provincial of the national level are too large, especially when the deviation is greater. Therefore, the ecological footprint theory and calculation based on emergy are developed. Emergy theory is to link the various substances and energy with solar energy, unified for the solar value, and the unit is the sej. Subsequently, a series of in-depth studies were carried out, such as Liu Miao et al. used regional energy footprint to compensate for the lack of traditional ecological footprint theory. Zhao Zhiqiang et al. to break the traditional ecological footprint system closed limitations, and to introduce non-material system to the ecological footprint. Jindan et al. applied the ecological footprint model to the resource-based city for the first time by using the energy value theory to improve the regional emergy density and reduce the calculation bias. The studies were for a certain region or a certain period. It is rare about the regional composition and differential analysis. Therefore, in this study, the difference of ecological footprints in the urban areas of Shaanxi Province was studied by comparing and analyzing the internal, including to find out the ecological footprint consume regions and contribution factors, and to provide some ideas for more intuitively and specific gripe the over-development and maintenance of regional sustainable development.

2 MODEL APPLICATION

Shaanxi Province located in the southern plains and the southern Shaanxi Qinba mountain, which belonging to the continental monsoon climate. The precipitation in north and south differences obvious in the province. Twenty-eight kinds ecological footprint indicators of Six categories were selected, which in the calculation of ecological footprint of energy value, including twelve species agricultural products, eleven kinds animal products, one aquatic products, three energy consumption projects and construction land. three energy consumption projects are coal, oil and natural gas. The above indicators are conversion to the solar energy value by solar energy rate.

2.1 *Regional emergy density*

The global energy value benchmark set by Odum et al. is 15.83×10^{24} sej/a in 2000, and combined with the land area and location of Shaanxi Province, the total energy value is calculated as 1.59×10^{21} sej/a. the regional energy density is calculated as 2.73×10^{14} sej/hm^2, by the definition of energy value and emergy density in Shaanxi Province.

2.2 *Energy ecology footprint and ecological carrying capacity calculation*

The emergy ecological footprint Calculated by Equation 1. The results showed that ecological footprint and per capita energy footprint were the highest in Xi'an city, reached 4.04×10^8 hm² and 46.8 hm²/person, followed by Xianyang City, which reached 1.89×10^8 hm² and 38.2 hm²/person, and the smallest of ecological footprint is the Yangling city, that is 0.30×10^7 hm².

The ecological carrying capacity is closely linked to the renewable resources. Therefore, we are calculate the amount of renewable resources in Shaanxi Province. Yulin City, Ankang City and Baoji City ranked in the top three, respectively 63.5×10^{20} sej, 59.3×10^{20} sej and 53.1×10^{20} sej, See Table 1. On the other hand, soil fertility with the surface energy value e_2 is also included in the renewable resources, which calculated by the formula 3. The results shown in Table 2. The top three of the e_2 are Yulin City, Ankang City and Shangluo City, respectively 2.27×10^{23} sej, 2.14×10^{23} sej and 1.67×10^{23} sej. The ecological carrying capacity of the value is calculated by the formula 2 in Shaanxi Province, the largest is Weinan City, reaching 6.00×10^7 hm², followed by Hanzhong City, which is 5.67×10^7 hm², the smallest is Tongchuan City, which is 0.77×10^7 hm². See Table 3. For more intuitive differences display of urban ecological carrying capacity and ecological footprint, we are statistics the per capita ecological footprint and per capita ecological carrying capacity, see Figure 1.

According to the calculation results of regional ecological footprint and ecological carrying capacity, we can calculate the ecological ecology of the region in order to analyze the current situation of ecological environment in Shaanxi Province. The results showed that the most serious ecological deficit was xi'an city in Shaanxi Province, reached 3.89×10^{-8} hm², followed by Xianyang City, which was 1.76×10^{-8} hm². However, the most of ecological surplus was 2.65×10^{-8} hm² in Hanzhong City in Shaanxi Province.

3 FIGURES AND TABLES

3.1 *Figures*

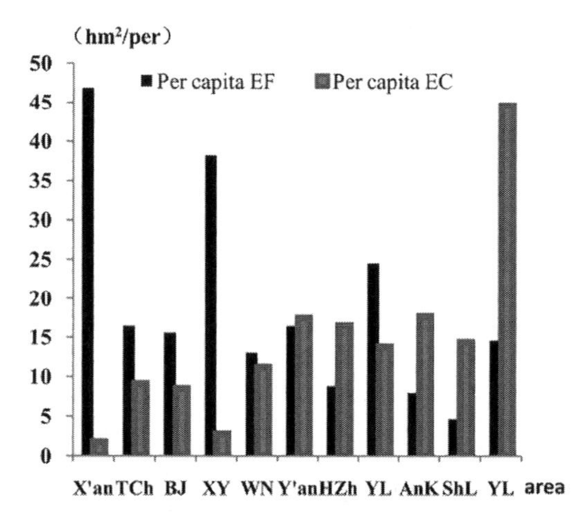

Figure 1. Emergy-based ecological footprint and ecological capacity in urban areas.

3.2 *Tables*

Table 1. Statistics of calculation for annual renewable resources emergy in urban areas (1019 sej).

Conversion rate (sej/j)	1	2450	17600	30500	46600	81000	57600	
Name	Solar	Wind	Rain potential	Rainwater chemistry	River potential	River chemistry	Earth rotation	Total
Xi'an	5.05	314	3.16	86.1	2.66	72.5	84.3	257
Tongchuan	1.94	121	3.28	89.3	0.17	4.5	32.4	133
Baoji	9.07	565	3.75	102	9.17	250	151	531
Xianyang	5.10	317	2.46	67	0.54	14.8	85	178
Weinan	6.52	406	2.45	66.7	43.8	1190	109	1430
Yan'an	18.5	115	6.09	166	0.29	7.97	309	519
Hanzhong	13.6	843	6.69	182	21.7	592	226	1050
Yulin	21.5	1340	3.57	97.4	3.51	95.5	358	593
Ankang	11.8	732	4.16	113	10.7	291	196	635
Shangluo	9.79	610	3.66	99.7	5.01	136	163	9.8
Yangling	0.07	4.14	4.22	115	4.43	121	1.1	246

Table 2. Statistics of e2 calculation in urban areas.

Name	Soil thickness (cm)	Crop area (km^2)	e_2 (sej)
Xi'an	30	7901.60	5.36×10^{22}
Tongchuan	30	3804.91	2.58×10^{22}
Baoji	30	16611.24	1.13×10^{23}
Xianyang	30	9665.53	6.56×10^{22}
Weinan	30	10281.96	6.98×10^{22}
Yan'an	30	26457.77	1.80×10^{23}
Hanzhong	30	24407.60	1.66×10^{23}
Yulin	30	33492.47	2.27×10^{23}
Ankang	30	31578.40	2.14×10^{23}
Shangluo	30	24551.51	1.67×10^{23}
Yangling	30	71.23	4.83×10^{20}

Table 3. Emergy-based ecological footprint and ecological capacity in urban areas.

Name	Emergy ecological footprint (10^7 hm^2)	Energy ecological capacity (10^7 hm^2)	Ecological surplus (10^7 hm^2)
Xi'an	40.4	1.53	−38.9
Tongchuan	1.39	0.77	−0.62
Baoji	5.84	3.19	−2.66
Xianyang	18.9	1.37	−17.6
Weinan	6.97	6.00	−0.98
Yan'an	3.64	3.88	0.23
Hanzhong	3.02	5.67	2.65
Yulin	8.29	4.67	−3.62
Ankang	2.10	4.68	2.58
Shangluo	1.10	3.39	2.29
Yangling	0.30	0.91	0.61

4 EQUATIONS

The regional emergy density is the ratio of regional energy consumption and regional energy density.

$$Eef = \frac{\sum_{i=1}^{n}\left[\left(Pd_i + \mathrm{Im}_i\right)\right] * Tr_i}{D} \tag{1}$$

Eef is the ecological footprint of energy, Pd_i is the number of items, Im_i is the input of consumption, Tr_i is the solar energy conversion rate corresponding to the consumption item. D is the regional emergy density.

$$Eec = \frac{e_1 + 0.03e_2}{D} \tag{2}$$

Eec is the ecological carrying capacity of eq, e_1 is the renewable energy value of the region, e_2 is the surface energy value, soil fertility can be used as renewable resource.

e_2 = surface soil thickness (generally 30 cm) × bio-productive land area × 10^{10} × soil bulk density (1.38 g/cm^3) × soil surface organic matter average content × typical surface Gibbs free energy (5.4 kcal/g) × 4186 (J/kcal) × Surface energy conversion rate (7.40×10^4 sej/J).

5 SUMMARY

According to the analysis results of regional ecological ridge, there are some differences in different regions of Shaanxi Province. Six cities such as Xi'an City, Xianyang City and Baoji City are ecologically deficit, and Yan'an City, Hanzhong City and Shangluo City are the ecological surplus. Through the deep analysis, the main factors causing the ecological red are also different. For Xi'an City, Xianyang City and Baoji City the main reason of ecological cash is the consumption of foreign fossil fuels. It can be seen that the consumption of coal, oil and natural gas in Xi'an City is far more than in other areas. In Xi'an City the reserves of resources or vacancies are coupled with the economy, and the population is relatively concentrated. This can also be illustrated by the ecological footprint of the absolute value and per capita value of ecological footprint. The actual demand has been far beyond of the carrying capacity. Xianyang City as another member of the integration of the West and the neighbor of Xi'an City carries the regional economic development of heavy industry. Behaving few energy reserves is the main reason of leading to the ecological deficit. For Yulin City, Tongchuan City and Weinan City the main reason of causing ecological cash is the excessive development of local mineral resources and the degradation of the ecological environment. The development of Hanzhong City, Ankang City, Shangluo and Yangling generally depends agricultural and where the natural conditions are relatively good. Especially Hanzhong City, Ankang City, Shangluo City being in the southern of Shaanxi Qinba mountainous have wet and humid climatic conditions, The Qinba mountainous areas mostly non-developed are rich in biological resources, which lead to relatively large ecological surplus and sustainable development conditions. It is obvious that economic development can be effected by the regional natural resource conditions and the main reason of changing regional ecological environment is also different.

The eco-footprint model is based on the fact that most of the resources are derived from the basic solar energy. In the calculation of ecological carrying capacity, only the solar energy, wind energy, rain energy, earth rotation energy, biological resources and soil are considered renewable resources under the current technical conditions. The coal, oil, natural gas and other fossil fuels and mineral resources can be more realistic to reflect the regional sustainable carrying level. It can be draw from the analysis of the ecological footprint of urban energy in Shaanxi Province that the ecological footprint model can accurately reflect the regional characteristics and find out the reasons for the high ecological footprint. For example, the ecological footprint of Xi'an (due to the consumption of fossil fuels) is 26 times of the ecological carrying capacity per capita, which is far from satisfying the regional development demand. This result also reflects the characteristics of the resource concentration caused by the economic development of the provincial capital city. In the agricultural development area, the ecological carrying capacity is generally greater than the ecological footprint, which has a direct relationship with the local natural conditions and the intensity of agricultural development. These areas tend to develop tourism and other industries.

REFERENCES

Bian Yousheng, Liu Yingkun, Bian Jing, The analysis on Output-input Ration of Artificial Supplementary Energy in Agro-eco-engineering[J]. *J. Engineering Science*, 2006,8(8):28–32.

Bian Zhengfu. Study on Spatial Variation of Agricultural Land Quality in Mining Area of Mining Area. *J. China University of Mining and Technology,* 2004, 33 (2): 213–218.

Brown M T, Arding J. Transform ities Working Paper. Gainesville: Center for Wetlands, University of Florida, 1991.

*Brown M T, Bar*di E. Handbook of Emergy Evaluation. Folio#3:Emergy of Ecosystems. Gainesville: Center for Environmental Policy, Environmental Engineering Sciences, University of Florida, 2001:61–62.

Jin Dan, Bian Zheng-fu. Effects of Ecological Footprint Model and Its Application in Resource-based Cities [J]. Acta Ecologica Sinica, 2010, 30 (7): 1725–1733.

Li Lianlu, Huang Yuzhu, Han Chunru, Study on the transforming coefficient for industrial energy input of agro-ecosystem[J]. *J. Rural Ecological Environment,* 1989, (4):32–36.

Liu Miao, HU Yuan-Man, Chang Yu. Improvement of Ecological Footprint Method Based on Emergy Theory [J]. *Journal of Natural Resources*, 2008, 23 (03): 447–457.

Liu QP, Lin ZS, Feng NH, Liu YM. A modified model of ecological footprint accounting and its application to croplandin Jiang su, *J. China. Pedo sphere*, 2008, 18(2):154–162.

Liu Xunhao, The preliminary study on the efficiency of energy conversion of farmland in different regions of China[J]. *J. Journal of Beijing Agricultural University*, 1982, (1):1–20.

Odum HT. Environmental Accounting: Emergy and Environmental Decision Making. New York: John Wiley, 1996:51, 73–206, 298.

Tang Yuxian, Tan Shiwen, Zhang Boquan and other translations. Soil science principles [M]. Beijing: *J. Agricultural Press*, 1984.

Wen Dazhong, Research methods of agro-ecosystem (III) [J]. *J. Rural Ecological Environment*, 1986(2):48–51.

Zhao Zhi-qiang, LI Shuang-cheng, Gao Yang. Environmental footprint model of open system based on emergy improvement and its application [J]. Acta Ecologica Sinica, 2008, 28 (5): 2220–2231.

Land Reclamation in Ecological Fragile Areas – Hu (Ed.)
 ISBN 978-1-138-05103-4

Estimation and analysis of the Vegetation Coverage of "Three-North" based on MODIS data

H. Zhang, C.S. Zhang & X.Q. Wang
College of Geomatics, Xi'an University of Science and Technology, Xi'an, China

ABSTRACT: Based on the MODIS the Vegetation Coverage (VC) of the Three-North Shelterbelt Project in 2005, 2010 and 2015 was deduced by the binary model based on the normalized vegetation index. The results show that: 1) The VC of the three-North project area in 2005 is 34.93%, in 2010 is 39.88%, in 2015 is 42.24%. There was a significant increase in the VC of the Three-north Shelterbelt in 2005–2015, but the growth rate was decreased. 2) The VC in the northwest of the Three North Shelterbelt is the lowest, and the VC in most areas is below 40%, which is low coverage type. The VC in the northeastern region is over 60%, which is the high coverage type. 3) The VC in the Mengxin area was the lowest, the northern part of North China is the highest. The VC's growning of the Loess Plateau is the fastest in the hilly and gully region.

Keywords: MODIS, NDVI, dimidiate pixel model Vegetation Coverage

1 INTRODUCTION

Being the main body of terrestrial ecosystem Vegetation is an important indicator of environmental change, as the link between soil, atmosphere and water. Vegetation Coverage (VC) refers to the percentage of the vertical projected area of vegetation (including leaves, stems, branches) on the ground as a percentage of the total area of the statistical area (Gitelson et al., 2002). Vegetation coverage is an important index reflecting the coverage of surface vegetation cover. It is an important basic data to study the hydrological, meteorological and ecological aspects of the study area. Therefore, it is of great practical significance to establish a quick and accurate method for calculating vegetation coverage.

Vegetation coverage extraction methods mainly includes the traditional ground measurement and emerging remote sensing measurement. The accuracy of traditional ground measurement is high but not suitable for large and medium scale research area. Remote sensing measurement method can quickly and accurately monitor large-scale range and can provide reliable data in real time, so the remote sensing measurement method to extract vegetation coverage is widely used. The current method of estimating vegetation coverage is based on the relationship between vegetation spectral information and vegetation coverage, which can be divided into statistical model and physical model. According to the different models the statistical model method is divided into vegetation index method, regression model method and pixel decomposition method. Regression model method depends on the measured data of the specific area. In a small range this method has a certain degree of accuracy, but there is still a lot of restrictions about its promotion. Vegetation index method does not need to establish the regression model, the vegetation index used is generally verified, and has a good relationship with the coverage; and hardly rely on the surface measured data, so using of vegetation index to estimate vegetation coverage is better, Which is more general than the regression model method and can be extended to a wide range of areas to be the common vegetation coverage calculation method.

In this paper, In this paper, using the advantages of vegetation index to estimate vegetation coverage fully, and the coverage of the three-north shelter forest project area is calculated by combining the simple and practical pixel dichotomy according to the vegetation index to analyze the spatial and temporal characteristics of vegetation coverage in the area, which provided scientific basis for the ecological benefits of the Three-north Shelter forest and the further guidance of the construction of the Three-north Shelter forest.

2 RESEARCH AREAS AND RESEARCH METHODS

2.1 *Research area overview*

The Three-north Shelter forest project area is a specific regional concept, which is a national project of our country in order to fundamentally change the ecological and environmental problems such as sandstorm hazard and soil erosion in northwest China, north China and northeast China. "Three North" shelter forest system east of Heilongjiang Binxian, west of Xinjiang Uzbe Bishan, north of the northern border, the southern coastal river, Yongding River, Fen River, Weihe River, Tao River downstream, Lakun Mountains, including Xinjiang, Qinghai Gansu Province, Inner Mongolia, Shaanxi, Shanxi, Hebei, Liaoning, Jilin, Heilongjiang, Beijing, Tianjin and other 13 provinces, municipalities and autonomous regions of the 559 counties, with a total area of 406.9 million square kilometers which is 42.4% of the China's land. According to the natural characteristics of Three-north area, the fragility of ecological environment and the type of disaster, it is divided into four protected forest system areas in Mengxin area, Loess Plateau region, northern North China and northeast region. The three-North Shelterbelt project is a large area, and is long-term afforestation activities which will not only affect the ecology of afforestation areas, but also have a certain influence on the surrounding areas. Therefore, the object of this study is not limited to afforestation area but also the whole vegetation state of the project area.

2.2 *Remote sensing data*

MODIS as the world's new generation of "map of one" optical remote sensing equipment, with 36 spectral bands to observe the Earth's once surface every one to two days, the spatial resolution is 250 m (1 ~ 2 band), 500 m (3 ~ 7 band) and 1 km (8 ~ 36 band) 3, with a high time resolution and coverage of a wide (scanning width of 2330 km) and so on. In addition, MODIS data can be obtained from the data sharing platform website in real time, which is conducive to a wide range of vegetation, land and other information dynamic monitoring and evaluation. The data used in this study is a 16-day synthetic MODIS-NDVI product with a spatial resolution of 1 km and a time span of January 2005 - December 2005, January 2010 - December 2010, January 2015 - December 2015. The characteristics of MODIS data can meet the requirements of time and space in the vegetation coverage of the Three North Shelterbelt project area. In order to reduce the cloud and the impact of the sun-target-sensor geometric angle, we use the 16-day synthetic MODIS-NDVI product as the area of the cloud. Vegetation coverage monitoring of the basic data. In this paper, when the remote sensing images were selected, the vegetation coverage was synthesized from January 2005 to December 16, 2005, and the MODIS-NDVI data of the 225–240 days of vegetation growth was selected. August vegetation growth status), and the average vegetation coverage of the entire Three-north Shelter Project in August was calculated.

2.3 *Research methods*

2.3.1 *Pixel binary model*

According to the principle of the pixel dichotomy model, the information S observed by the remote sensing sensor can be divided into the information Sv provided by the green vegetation part and the information Ss provided by the bare soil part,

$$S = Sveg + Ssoil \tag{1}$$

And the vegetation is composed of two parts of the mixed pixels, vegetation coverage f is calculated as the proportion of green vegetation part of the soil cover area ratio of 1-f, assuming that all the vegetation covered by pure pixels, the remote sensing information For Sveg. The information Sv contributed by the vegetation part of the mixed pixel can be expressed as:

$$Sv = f * Sveg \tag{2}$$

Similarly, assuming that all the pure pixels covered by the soil, the resulting remote sensing information for the Ssoil mixed pixel soil contribution information Ss can be expressed as:

$$Ss = (1\text{-}f) * Ssoil \tag{3}$$

so we can get the information of the mixed pixel:

$$S = f * Sveg + (1\text{-}f) * Ss \tag{4}$$

The formula for transforming the available vegetation coverage:

$$f = (S (Sveg\text{-}Ssoil) \tag{5}$$

where Ssoil and Sveg are fixed parameters, so the vegetation coverage can be estimated by using the remote sensing information according to the formula.

2.3.2 *Estimation of vegetation coverage using NDVI*

Vegetation index (vegetation index), refers to the remote sensing sensor to obtain multi-spectral data, linear and non-linear combination has a certain significance of the various values for the vegetation. It is an indirect indicator of the vegetation growth, coverage, biomass and vegetation planting characteristics calculated from the characteristics of the vegetation reflectance band. It has been proved that vegetation index has a good correlation with vegetation coverage, and it is very suitable to calculate vegetation coverage with vegetation index. Among them, NDVI is the most widely used vegetation index of Normalized Vegetation Index (NDVI). It has many advantages such as plant growth status and vegetation spatial distribution. For example, the sensitivity of vegetation detection is high and the detection range of vegetation coverage is wide.

NDVI is calculated as: NDVI = (NIR-R) / (NIR + R) (6) where NIR is near infrared and R is red. The NDVI value of each pixel can be expressed as the information NDVIveg contributed by the green vegetation part, and the information NDVIsoil, which is contributed by the bare soil part, satisfies the condition of the formula (5), thus substituting NDVI f = (NDVI-NDVIsoil) / (NDVIveg-NDVIsoil) (7), in the equation NDVIveg represents the NDVI value of the pure vegetation pixel, while NDVIsoil is the NDVI value that is completely in the bare soil or no vegetation coverage area.

2.3.3 *NDVIsoil and NDVIveg values*

For most types of bare ground surfaces, NDVIs should theoretically be close to zero and do not change over time and space. However, due to soil moisture, soil type, roughness, etc., the NDVIsoil value of bare ground surface changes with time and space And NDVIs and NDVIv can not take a fixed value, usually by virtue of the experience to judge for each of the changes in the vegetation, the NDVIs and NDVIv can not take a fixed value, The NDVI value of the NDVI data is calculated. The NDVIveg and the NDVIsoil are calculated according to the frequency statistics table. The 5% of the soil element is the value of NDVIsoil, and the cumulative frequency is 95% in the land as the value of NDVIveg.

2.3.4 *Estimation of Vegetation Coverage in Three-north Shelterbelt*

The generalization of the vegetation coverage estimation process in the Three-north Shelterbelt is as follows:

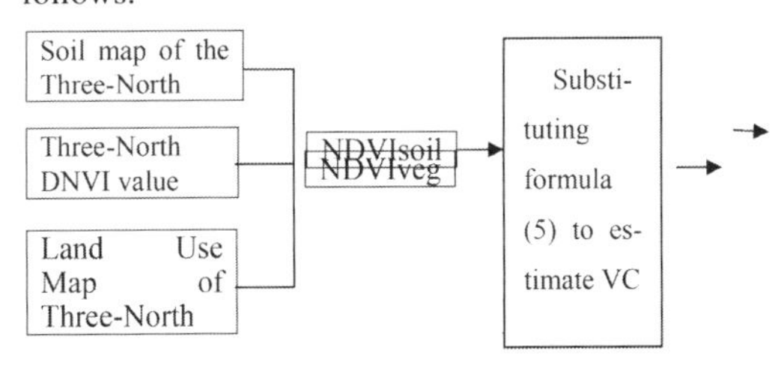

3 RESULTS AND ANALYSIS

This can be calculated from August 2005, August 2010 and August 2015 vegetation coverage, using equal spacing re-classified as 6, that is, 0%, 0% <f < = 20%, 20% F < = 40%, 40% <f < = 60%, 60% <f < = 80%, 80% <f < = 100% (Figure 1, Figure 2, Figure 3).

From the spatial scale analysis, it can be seen that the vegetation coverage in the northwest of the Three-North Shelterbelt project area is the lowest, most of the vegetation coverage is below 40%, the coverage is low; the vegetation coverage in the northeastern region is more than 60% And the vegetation coverage in the central region was between the two, and the coverage rate was the highest. The average vegetation coverage in the Mengxin area was the lowest, and the Loess Plateau was higher than the Mengxin area, but lower than that in the North China and North China and northern North China. The average vegetation coverage in the Loess Plateau is the fastest, especially in the hilly and gully areas with severe soil erosion in the

Figure 1. Vegetation coverage in 2005.

Figure 2. Vegetation coverage in 2010.

Figure 3. Vegetation coverage in 2015.

Loess Plateau. From the time scale, the average vegetation coverage of August North Shelterbelt in August 2005 was 34.93%. The average vegetation coverage in August 2010 was 39.88%. The average vegetation coverage in August 2015 was 42.24%. The results showed that there was a significant increase in the VC of the Three-North Shelterbelt in 2005–2000, but the growth rate was decreased.

4 CONCLUSION

Based on the MODIS-NDVI data from August 2005, 2010 and 2015, the vegetation coverage of the Three-north Shelterbelt area was estimated, and the temporal and spatial distribution characteristics of the Three-north Shelterbelt were analyzed. In 2005–2015, the overall level of vegetation coverage in the Three North Shelterbelt Area appeared upward trend, the coverage vegetation of the majority provinces have increased. Therefore, it can be considered that the vegetation in the region has been well recovered under the combined effects of policies and natural conditions. Vegetation changes are affected not only by climate but also by human activities. To a certain extent, climate plays a leading role in the change of vegetation cover, and economic, technical and social factors such as temporal and spatial changes of vegetation cover are often decisive. influences. Such as Beijing, Tianjin and other cities due to the expansion of large areas of the city, which led to reduced vegetation coverage.

Using NDVI vegetation index to estimate the vegetation coverage, the method is simple and easy, and it is helpful to make quantitative assessment of the overall situation of vegetation in the area. Therefore, this study has laid a foundation for the long-term dynamic monitoring of vegetation coverage in the whole Three-north Shelterbelt Area, which has great significance to the scientific evaluation of the ecological environment in the Three North Shelterbelt.

REFERENCES

Carlson TN, Ripley DA. (1997). *On the relation between-NDVI, Fractional vegetation cover, and leaf area index*. Remote Sensing of Environment, 62: 241–252.

Cheng Hongfang, Zhang Wenbo & Chen Feng (2008) *Research on Remote Sensing Estimation of Vegetation Coverage*—Remote Sensing of Land Resources, 75 (1): 13–18.

Dun can J, Stow D & Franklin J (1993) *Assessing the relationship between spectral vegetation indices and shrub cover in the Jornada Basin, New Mexico*. International Journal of Remote Sensing, 14: 3395–3416.

Liu Lvliu, Xiao Fengjin (2006) *Spatial and temporal variation of vegetation NDVI and temperature and precipitation in the Yellow River Basin*, Chinese Journal of Ecology, 25 (5): 477–481.

Three North Construction Bureau of the Ministry of Forestry 1993. *China's Three North Shelterbelt System Overall Planning Program*. Yinchuan: Ningxia People's Publishing House.

Qin Wei, Zhu Qingke & Zhang Xuexia (2006) *Vegetation coverage and its measurement method of progress*. Northwest A & F University (Natural Science Edition), 34 (9): 163–169.

Land Reclamation in Ecological Fragile Areas – Hu (Ed.)
© 2017 Taylor & Francis Group, London, ISBN 978-1-138-05103-4

Research on the relation between soil micromorphology and landuse in loess tableland around Xi'an

Z.X. Zhou
College of Geomatics, Xi'an University of Science and Technology, Xi'an, Shaanxi, China

J.L. Pang
College of Tourism and Environment, Shaanxi Normal University, Xi'an, Shaanxi, China

ABSTRACT: For studying the soil micromorphological characteristics under different land use modes, showing the micromorphology indicating significance of land use change, this paper choose loess tableland around Xi'an as a study region. The paper studied the soil micromorphology characteristics on different land uses, found the relation of soil micromorphology and land use, and the accurate interpretation of micromorphology to land use change instructions. So, this study is of importance and relevance significance. The innovation of the thesis are as following: make out the soil micromorphology composite index and access methods, and explore human activities and natural factors effect on the soil micromorphology. And describe the specific characteristics of soil micromorphology in response to land use change.

1 INTRODUCTION

Since the beginning of the 20th century, some scholars tried to apply petrography method to soil science research. It is the microscopic observation of the soil under the polarizing microscope, related with distribution pattern of soil material, shape and size, material composition, and various shape and size of space, and the ratio between its components. Soil micromorphology record the information of forming processes (such as mineral composition, material movement, land use, soil development direction, etc. (Villagran, Gianotti, 2013; Müller, Thiemeyer, 2010; Aldeias, et al., 2014; Rachman, et al., 2005). At present, the soil micro-morphological study is becoming more and more important, gradually formed an independent branch of soil science: the soil micro-morphology, which is the study of the micro morphological characteristics and particles of the soil with the applicates of microscopy. The study of sediments and soils has been shown to be an essential component of environmental reconstruction which may concern either landscapes and human impact at a regional scale, or site formation processes ruled by both natural factors and human activities (Stein, Farrand, 1985). Different land use patterns and management patterns often lead to changes in the physical and chemical properties of soil and soil microstructure, so that some of the characteristics of the soil change (Adejuwon, Ekanade, 1988; Chatre, Bresson, 1994; Dorronsoro, 1996; Attou, et al, 1998; Heidari, Asadi, 2015). Although soil micromorphological study has made great progress, but its research content is still few the application to the study on land use changes and few application to study the assessment of the ecological effect of landuse changes.

2 STUDY AREA

The city of Xi'an is located in the central Guanzhong Basin in the southern Loess Plateau with a semi-humid and continental monsoon climate. In the study region, loess landforms in Eastern District of Xi'an city, the loess topography is diverse, including four geomorphic units: Daiwang-Mae tableland, Bailuyuan tableland, Bahe river terrace and the loess hill, and the agricultural land use patterns in space and time effects on soil. The land cover is given priority to with artificial cultivation of plant community, mainly in field crops such as wheat, orchards and other types.

3 STUDY METHODS

3.1 *Sample collection*

Based on field investigation in key areas, typical sites were selected as samples, There are a few main landuse types to be choose as samples, including the forest land, grassland, garden land and cultivated land, and the cultivated land or garden of different cultivated period. At a sample land,

Table 1. The environment conditions of sample sites.

Sample cite	Location	Serial number	Elevation	Slope	Aspect	NDVI	Landuse type	Planting time
I	N34°24′45.5″ E109°19′30.8″	1	462 m	9°	65.33	52.44	cornfield	5 years
II	N34°25′27.4″	2	488 m	6.5°	334.14	53.26	cornfield	10 years
	E109°21′10.4″	3	486 m	7°	334.14	53.67	cornfield	20 years
		4	482 m	6°	334.14	55.52	cornfield	30 years
III	N34°22′11.9″ E109°26′8.7″	5	678 m	20.1°	277.41	47.92	cornfield	20 years
IV	N34°18′25.4″ E109°23′45.8″	6	1018 m	10.5°	135	52.64	cornfield	8 years
V	N34°16′47″ E109°22′38.8″	7	1099 m	7°	26.57	21.22	trees	60 years
VI	N34°12′57.4″	8	849 m	10.7°	203.83	47.61	grass	3 years
	E109°23′21.6″	9	860 m	18°	203.83	53.98	cornfield	60 years
VII	N34°8′24.4″	10	662 m	15.3°	55.02	54.39	orchard	10 years
	E109°17′36.5″	11	668 m	18.9°	55.02	56.03	cornfield	50 years

with similar slope and the basically same slope aspect, the profile of topsoil is divided into 5 layers (0–10 cm, 10–20 cm, 20–30 cm, 30–40 cm, and 40–50 cm). In each soil layer, 3–5 clods of soil were taken, then air-dried and made into thin slices (the thickness of Soil slice is 30 μm) for micromorphological analysis.

3.2 *Image acquisition*

The micromorphology was observed under a petrographic microscope and its image was quantificationally measured by Leica Qwin Standard V2.6 image processing system (Pang, et al., 2007). Based on analyzing soil micromorphology quantitative data, and different enlarged multiple digital image effect on data analysis, the images in 10 × 10 times are most suitable for quantitative statistical analysis in Leica Qwin Standard V2.6 image processing systems.

3.3 *Soil micromorphology integrated factors*

With factor analysis, it was found that 10 integrated factors represent the vast majority of soil micromorphology information, including: coarse particle shape, coarse particle largest size, pore degrees, pore shape, the number of coarse particle, minimum width of coarse particle, round degrees, the total number of coarse particle, minimum area of coarse particle. According to the contribution rate of 10 factors variables, measured micromorphology composite index, which reflects the total differnce of micromorphology.

3.4 *Soil micromorphology index*

The results obtained by factor analysis are often used for comprehensive evaluation. According to the 10 factors of soil micromorphology, with the contribution rate (two decimal places) as the weight, the soil micromorphology index are calculated by the following comprehensive judgment formula:

$$A = \sum a_i \times F_i \quad (1)$$

where A is the soil micromorphology index; ai is the contribution rate of the ith factor, Fi is the score of the ith factor.

4 RESULT

According to observering and quantificationally describing soil micromorphology in different soil profile under the different types, patterns and history of land use, the rules are summed up about the soil micromorphology response to LUCC as described below: (1) The different land uses have different characteristics of soil micromorphology: soil micromorphology in woodland and grassland have similarities, and soil micromorphology in the cropfields and orchards have significant similarities. (2) The micromorphology in different soil layers is different: with depth changes, the soil profile changes are obvious and its soil micromorphology characteristics have regularity. (3) Along with the increase of planting time, the overall trend is soil sponge micromorphology development, soil cling degree growth, iron dyed crack phenomena and reunion phenomenon becoming increasingly evident. This is because the development process is artificial soil mature process. (4) Therefore, there is a corresponding relationship

Table 2. The correlation between the composite index of soil micro morphology and the non-zonality natural factors.

Soil horizon	0–10 cm	10–20 cm	20–30 cm	30–40 cm	40–50 cm	Average absolute value
Elevation	−0.31	0.46	−0.1	−0.55	−0.33	0.35
Slope	−0.26	−0.63	0.59	−0.62	−0.09	0.44
Aspect	−0.39	0.02	0.004	0.36	0.27	0.21
NDVI	−0.1	−0.45	0.2	0.39	0.19	0.27
Landuse type	0.36	0.38	−0.07	−0.25	−0.9	0.39
Planting time	−0.56	−0.45	0.54	−0.5	0.8	0.57

between the certain soil characteristic and land-use individual situation: the micromorphology of sponge-shaped is the most prominent farming soil micromorphology; and at 10 cm below the surface the coarse particles such as feldspar, source from loess manures in artificial fertilization process, except wind-blown sand dust; there are clay-binding or porous coated with $CaCO_3$ cementation, and bacteria filamentous or nodular secondary calcite which reveal that soil development in semi-dry climate, or in field with dry farming. Because the study area precipitation in the 25–35 cm depth, if iron dyed clay coating in 25–35 cm, which is the significant symptoms of illuvial horizon of forest soil in loess plateau region. (5) Overall human landuse changes impact on soil more obviously than non-zonality natural factors: land-use time > slope > land-use type > elevation > land-cover change > aspect. As described in Table 2.

5 SUMMARY

Based on the Leica digital image processing system and the analysis of the landuse change effects on soil micromorphology, this paper researched human activities and non-zonality natural factors effects on soil development by a smaller scale of the space environment. This paper provided new ideas to combines both micro and macro level analyses that link theory with practice in land use studies field, and provided vital scientific theory on combating ecological problems and improving soil ecological functions. Through field trips to determine the wild sampling locations, the selected soil profiles are typical. It is an important experience, using Leica Qwin Standard v2.6 image processing system, to quantitatively analyzes the relationship between the soil micromorphology and the land-use change. The soil micromorphology index data are available to explanation of the specific characteristics of soil development in response to land use change.

REFERENCES

Adejuwon J.O. & Kanade O (1988). Comparison of soil properties under different land use types in a part of the Nigerian cocoa belt. *Catena*, 15: 319–333.

Aldeias, V., P. Goldberg & HL Dibble (2014). Deciphering site formation processes through soil micromorphology at Contrebandiers Cave, Morocco. *Journal of Human Evolution*, 69(1):8–30.

Attou, F., A. Bruand & Y.L. Bissonnais (1998). Effect of clay content and silt-clay fabric on stability of artificial aggregates. *European Journal of Soil Science*, 49: 569–577.

Chatre L.M. & Bresson (1994). Micromorphological indicators of anthropogenically induced soil structural degradation. *Transactions of 15th World Congress of Soil Science*. Mexico.

Dorronsoro (1996). Interactive computer programme for demostration of micromorphological aspects of calcification processes in soil. In: Soil Micromorphology. *10th International Working Meeting on Soil Micromorphology*, Moscow, pp165.

Heidari A. &. Asadi (2015). Micromorphological characteristics of polluted soils in Tehran petroleum refinery. *Journal of Agricultural Science & Technology*, 17(4):1041–1055.

Müller S. & H. Thiemeyer (2010). Potentials of reconstructing the formation and transformation of slope deposits by the use of soil micromorphology, *Journal of Economic & Social Research*, 12:3–4.

Pang J.L., Qiu H.Y & Huang C.C (2007). Comparison of Micromorphological Features of two Agricultural Cultivated Soils in Guanzhong Areas, Shaanxi Province. China. *Agricultural Sciences in China*, 6(9); 1089–1098.

Rachman, A., S.H. Anderson &. J. Gantzer (2005). Computed-tomographic measurement of soil macroporosity parameters as affected by stiff-stemmed grass hedges. *Soil Science Society of America Journal*, 69:1609.

Stein J.K. & Farrand W.R (1985). Archaeological sediments in context, *Center for the Study of Early Man, Institute for Quaternary Studies, University of Maine at Orono*, 18(6):677–679.

Villagran X.S. & C. Gianotti (2013). Earthen mound formation in the Uruguayan lowlands (South America): micromorphological analyses of the Pago Lindo archaeological complex. *Journal of Archaeological Science*, 40(2): 1093–1107.

Land Reclamation in Ecological Fragile Areas – Hu (Ed.)
 ISBN 978-1-138-05103-4

Dynamic monitoring of land desertification in coal mining districts in the north of Shaanxi Province

H. Yue
College of Geomatics, Xi'an University of Science and Technology, Shaanxi, China

ABSTRACT: Shendong mining area is one of the major coal production centers in China, belonging to the arid and semi-arid desert mine area. In order to seek the harmonious development of Northern Shaanxi Province and make up the impacts of modern coal mining on land desertification and geology environment at the junction of desert and loess in the north of Shaanxi Province, this paper use MODIS NDVI data, dynamic analysis the changes of vegetation cover in 2000–2014, the results indicated that vegetation coverage in North of Shaanxi Province area showed an overall increasing trend in 15a, most of the surface vegetation coverage has improved, the degree of desertification declined, very severe desertification area and severe desertification area had dropped from 59.81% to 28.28%. The fragile ecological environment and arid climate are main natural driving fact for land desertization, while underground mining activities are also a noticeable factor.

1 INTRODUCTION

The coal resources are rich in northern Shaanxi province of China, coalfield proven reserves is as high as more than 1339 t, as one of the world's seven big coal field (Wu *et al.*, 2009). However, mining area belongs to the temperate continental climate, geographical environment for the Mu Us Desert and the loess plateau border areas, lack of water resources, vegetation sparse, fragile ecological environment, and natural environment quality is on the decline (Liu *et al.*, 2011). Since the 1980s, the country started large-scale investment in the development of the north of Shaanxi province. Mostly coal seam covering very thin strata, coal bed mining subsidence will until the earth's surface, not only cause huge loss of water resources, but also cause wind-blown sand flow on the earth's surface (Zipper *et al.*, 2011). The fragile ecological environment and the influence of human activity in the north of Shaanxi leads to become the one of the important source of desertification and sandstorm.

2 STUDY AREA AND DATA

2.1 *Study area*

Study area is located at the central part of Loess Plateau, which include Yulin city and Yan'an city of Shaanxi province, with a total area of 92521.4 Km^2.

2.2 *Data*

This research adopts the Earth Resources Observation Science centers in the United States in 2000–2014 (Earth Resources Observation and Science Center, EROS) provide the spatial resolution of 1 KM which contains the Shaanxi province of MODIS Vegetation index (MOD13A3) products (MODIS/Terra Vegetation Indices or L3 Global 1000 m SIN Grid V005).

2.3 *METHOD*

This paper use the maximum synthesis (MVC) to extract the maximum NDVI of the year, also Fractional Vegetation Coverage (FVC) is used to reflect the desertification (Gu *et al.*, 2009). FVC is defined as on the ground of vertical projection area of percentage of the total area of statistical area, is a comprehensive quantitative indicators of plant community covering the surface condition. FVC is widespread use in hydrology, climate, environment and the assessment of land degradation, desertification, etc. FVC formula described as below:

$$FVC = (NDVI - NDVIsoil) / (NDVIveg - NDVIsoil) \quad (1)$$

$$NDVI_{soil} = \frac{VFCmax * NDVImin - VFCmin * NDVImin}{VFCmax - VFCmin} \quad (2)$$

$$\text{NDVIveg} = (1 - \text{VFCmin}) * \text{NDVImax} - \frac{(1 - \text{VFCmax}) * NDV\text{Imin}}{\text{VFCmax} - \text{VFCmin}} \quad (3)$$

When the area approximately take $VFC_{max} = 100\%$, $VFC_{min} = 0\%$, FVC can be defined as formula (4)

$$\text{FVC} = (\text{NDVI} - \text{NDVImin})/(\text{NDVImax} - \text{NDVImin}) \quad (4)$$

This paper take a confidence level of 95% and 5% in the study area as $NDVI_{max}$ and $NDVI_{min}$ separately.

3 RESULTS

3.1 *Vegetation dynamic change analysis*

For average NDVI change trend year by year in northern Shaanxi province showed in Figure 1.

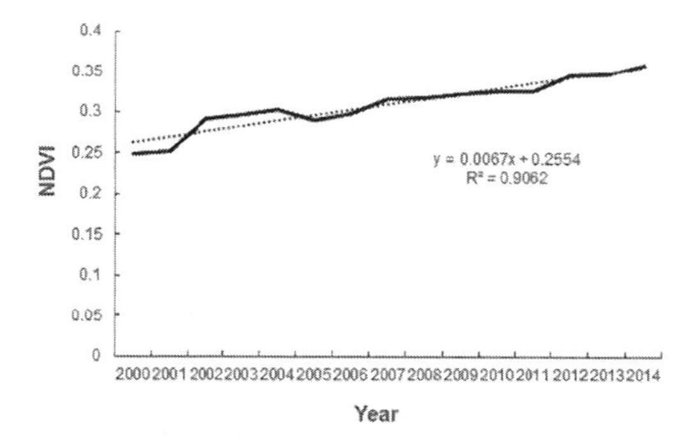

Figure 1. Annual maximum NDVI trend.

In the period of 2000–2014, the mean NDVI although there are fluctuations, but overall trend is on the rise ($R^2 = 0.906$). In each year of 12 months NDVI value raised after a fall trends, and rising faster than falling. A maximum NDVI from 0.389 to 0.573, which in July through September; Minimum value from 0.147 to 0.194, which in January to March. It showed as an obvious seasonal change.

3.2 *Desertification degree of temporal change analysis*

According to the different vegetation coverage the desertification degree could be divided into micro, mild, moderate, severe and extreme (Table 1). Statistical analysis the area and proportion of each desertification degree which showed in Table 2. Extremely severe desertification area of the overall trend tended to decrease year by year, extremely severe desertification area is 22267 km^2 in 2000, which declined to 11885 km^2 in 2014, and area accounts for the proportion was decreased from 18.84% in 2000 to 10.06% in 2014. Extremely severe desertification and severe desertification transferred to moderate, mild, micro degree of desertification grades.

3.3 *Analysis of characteristics of spatial distribution*

According to the division of vegetation coverage level of land desertification in the study area, and get space distribution of land desertification in northern Shaanxi in 2000–2014 by using ArcGis (Figure 2).

The most serious desertification regions distributed in transition of the Inner Mongolia plateau and Mu Us Desert. Serious desertification area (red) are mostly concentrated in the north of

Table 1. Classification of desertification degree in northern Shaanxi.

Land desertification grade	Extremely	Severe	Moderate	Mild	Micro
FVC	<0.1	0.1–0.3	0.3–0.5	0.5–0.7	0.7–1
NDVI	<0.19	0.19–0.34	0.34–0.51	0.51–0.78	0.78–0.83

Table 2. Land desertification area and percentage of northern Shaanxi.

	Extreme		Severe		Moderate		Mild		Micro	
Year	Area/km²	pct./%	Area/km²	pct. /%	Area/km²	pct. /%	Area/km²	pct. /%	Area/km²	pct. /%
2000	22267	18.84	48418	40.97	21971	18.59	8102	6.86	17416	14.74
2005	14699	12.44	40399	34.19	29198	24.71	11133	9.42	22745	19.25
2010	13190	11.16	27686	23.43	35493	30.03	15238	12.89	26567	22.48
2014	11885	10.06	21529	18.22	30296	25.64	26863	22.73	27601	23.36

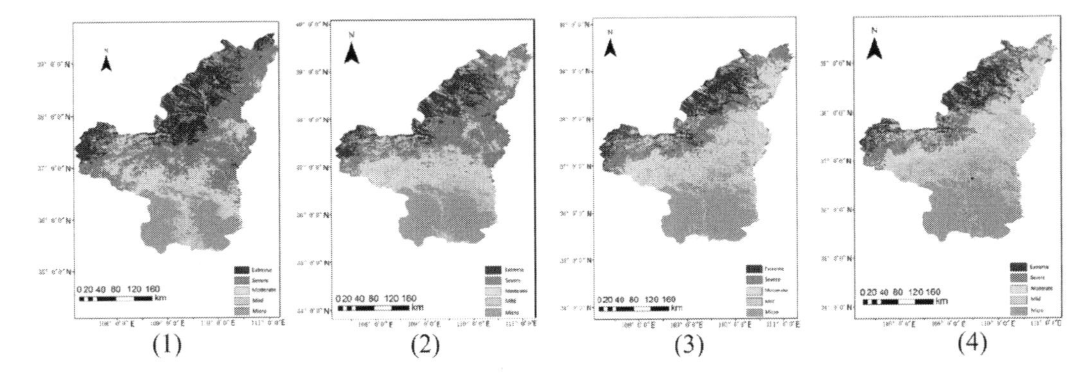

Figure 2. Vegetation coverage map in 2000–2014.

relatively high latitude, severe desertification areas are mainly in Yulin city of Fugu county, Shenmu county, Yuyang district and Hengshan county and Dingbian county.

4 CONCLUSION

This paper using the MODIS Terra satellite Normalized Difference Vegetation Index (NDVI) data during 2000–2014, the desertification land classification combined with linear regression analysis method, vegetation coverage and dynamic change of land desertification in northern Shaanxi province were analyzed. The results showed that over the past 15 years, most of the surface vegetation cover was improved. The primary reason was the input and the attention of the environmental protection in Shendong mining area.

ACKNOWLEDGEMENT

This work was supported by the National Natural Science Foundation of China (No. 41401496).

REFERENCES

Gu Z J, Zeng Z Y, Shi X Z, et al. (2009). Assessing factors influencing vegetation coverage calculation with remote sensing imagery. J. *Int. Remote Sens, 30 (10)*: 2479–2489.

Liu Ying, Wu Li-xin, Ma Bao-dong, et al. (2009). Soil Moisture Monitoring in Shendong Mining Area and Bi-parabola NDVI-Ts Space. J. *Science & Technology Review, 29(35)*:39–44.

Wu Li-xin, Ma Bao-dong, Liu Shan-jun. (2009). Analysis to vegetation coverage change in Shendong mining area with SPOT NDVI data. J. *China Coal Society, 34(9)*:1216–1222.

Zipper CE, Burger JA. Skousen JG, et al. (2011). Restoring forests and associated ecosystem services on appalachian coal surface mines. J. *Environmental Management, 47(5)* 751–765.

Land Reclamation in Ecological Fragile Areas – Hu (Ed.)
 ISBN 978-1-138-05103-4

Hyperspectral estimation model of plumbum concentration in soil of mining areas based on wavelet transform and random forests

J. Lv, X.M. Li & J. Kang
College of Geomatics, Xi'an University of Science and Technology, Xi'an, Shaanxi, China

ABSTRACT: Heavy metal pollution of tailings is one of the serious problems in environmental pollution, accurate estimation of soil heavy metal content is very important for the mine soil pollution monitoring. Taking Jinduicheng mine tailings in Shaanxi as the study area, soil spectral were measure with ASD spectrometer, Plumbum element content of soil samples were obtained by laboratory analysis. The wavelet transform was applied to the soil hyperspectral data for noise reduction, and the noise reduction of soil spectrum is studied by using the first derivative spectral transform and the continuum removal method. Plumbum content in the mine tailing soil were estimated by random forests, inversion results were compared with the original high spectral data and the noise reduction spectral data. The results showed that: the estimation model on the spectral data set after noise reduced by wavelet transform achieved a correlation coefficient R^2 of 0.774, and the root mean square error of *RMSE* is 249.125, the prediction accuracy is better than the original hyperspectral data. The results provide a theoretical basis for exploring the characteristics of soil hyperspectral data extraction, and has important significance for the heavy metal pollution monitoring of tailing soil in the mining areas.

Keywords: Wavelet transform; Soil; Plumbum; Random forest; Hyperspectral

1 INTRODUCTION

Heavy metal pollution is one of the most serious problems of environmental pollution in the mining area. Therefore, how to estimate the content of heavy metals in the tailings soil is one of the important subjects in the environmental monitoring.

The method of monitoring and identifying the heavy metal pollution in soil is usually used to carry out chemical analysis from the field sampling to the laboratory. Chemical analysis of the general requirements of the destruction of soil samples, from a large number of soil samples collected, dried, weighed, grinding to test, in the large-scale monitoring of heavy metals in the soil is time-consuming, laborious.

Hyperspectral remote sensing is the electromagnetic spectrum of visible light and near infrared, mid infrared and thermal infrared range, get a lot of very narrow spectrum continuous image data technology, the imaging spectrometer can collect hundreds of very narrow spectral information [1]. Hyperspectral remote sensing can provide a detailed feature rich spectral feature description, so as to improve remote sensing image recognition ability for object [2], hyperspectral remote sensing the fine recognition ability for mine tailing soil heavy metal pollution monitoring guarantee.

At present, scholars have used multiple stepwise regression partial least squares regression method, artificial neural network, support vector machine method of soil reflectance spectra and soil heavy metal content inversion model was established and the quantitative estimation of soil heavy metal content. Because the spectrometer can produce the system noise in the measurement of the soil spectrum, and the spectrum of the soil sample is easy to be affected by the background noise. Therefore, it is necessary to enhance the reflectance spectrum of the soil, reduce the noise and improve the quality of the soil spectrum before the establishment of the model for estimating the content of heavy metals in soil.

Wavelet analysis is the original signal is decomposed into different scale efficient tool for signal processing, wavelet is a mathematical function data into different frequency domains, and each domain has its frequency and scale feature resolution ratio of [3]. The wavelet shrinkage and thresholding method can remove the noise due to the sparse representation of the wavelet function.

Effect of spectral information of soil heavy metal pollution is weak, so it is necessary to deal with the soil spectral denoising of soil reflectance spectra in this paper, and the original soil reflectance, derivative spectral transform, envelope removal methods were analyzed using random forest to construct quantitative lead content of mine tailing Hyperspectral Estimation model.

2 MATERIALS AND METHODS

2.1 *Study area*

The study area is located at Jinduicheng mining area, Huashan county, Shaanxi province. Jinduicheng molybdenum mine located in the South East of Qinling Mountains, Huashan County of Shaanxi province in Jinduicheng, the area of the mining is 4.5 km^2, with an altitude of 1211 m. Jinduicheng molybdenum is a large molybdenum deposit in China, and has proven molybdenum reserves of 1011461.22 tons.

2.2 *Field campaign data*

60 sampling sites were selected during July, 2012 in the research areas. The surface soil (0–20 cm) were collected, then they were dried at 20°C for 3 days. The soil were crushed with 2 mm polyethylene sieve in order to remove gravel, pebbles and plant debris, they were grinded after the polyethylene 0.15 mm sieve. The soils were divided into two parts, one for analysis and testing of Pb concentration in soil, one for spectral measurement of soil.

2.3 *Random forests*

Random forests is a statistical machine learning method, which is created by Breiman [4]. Random forests can achieve comparable results with boosting algorithms and support vector machines. Random forests has been applied in a large number of remote sensing researches for image classification of hyperspectral data [5–6], SAR data [7], LiDAR [8–9].

A random forest is a classifier consisting of a collection of tree-structured classifiers {h(x, Θk), k = 1,...} where the Θk are independent indentically distributed random vectors and each tree casts a unit vote for the most popular class at input x.

3 RESULTS AND ANALYSIS

3.1 *Selection of optimal wavelet bases*

Using SNR and RMSE to evaluate the effect of wavelet denoising. The signal to noise ratio is calculated as follows:

$$SNR = 10\log\sum_{1}^{N}\frac{x_i^2}{(x_i^2-\overline{x}_i)^2} \quad (1)$$

It can be seen from the table, when the soil spectral reflectance decomposition is 6, which selected the Haar6 wavelet, The SNR value is the smallest, and the RMSE value is the largest, the results show that the Haar6 wavelet has better noise reduction effect on the soil spectral reflectance, therefore, the wavelet spectrum of soil lead content was selected by Haar6 wavelet denoising.

3.2 *Analysis of hyperspectral estimation of lead content in soil*

Using random forest respectively on the original soil spectral data of soil spectral data set, Haar6 set, wavelet denoising after soil spectral data the first derivative transformation set, soil spectral data envelope after removal of the set of quantitative estimation model of Pb content in soil hyperspectral, inversion of soil lead content. The results showed that the estimation of Pb content in soil in soil Haar6 spectral data after wavelet denoising is set, the correlation coefficient between the estimated and measured values is R^2 0.774, the Root Mean Square Error (*RMSE*) is 249.125, and the result is better than that of the original hyperspectral data, the first derivative data set and the data set after the envelope removal (Table 2). The results show that the Hyperspectral Estimation model constructed by the random forest on the soil spectral data set based on the Haar6 wavelet denoising has a higher precision.coordinate expression the accuracy of the first order derivative data set in the data set of random forest construction is higher than that of the original spectral data set, the results show that the first derivative transformation and the envelope removal method have obvious effect on the soil spectral pretreatment.

The content of Pb in soil Hyperspectral Estimation Model Inversion in different spectral data on lead content accuracy comparison as shown in Figure 1, Contrast discovery: The simulated and measured values of random forest on the Haar noise reduction spectral data sets are better, the accuracy was significantly higher than that of lead content in original hyperspectral data, First

Table 1. The de-noising effect of Harr1~Harr6 wavelet.

Wavelet basis	Haar1	Haar2	Haar3	Haar4	Haar5	Haar6
SNR	79.632	78.984	79.892	80.678	80.321	80.854
RMSE	276.912	250.126	264.759	280.145	273.764	249.125

Table 2. Accuracy comparison of random forests inversion of Pb concentration based on different datasets.

Data set	R^2	*RMSE*
Raw data set	0.683	273.552
Haar6 wavelet denoising dataset	0.774	249.125
First derivative dataset	0.697	268.590
Data set after dislodging enveloping line	0.771	249.972

derivative data set and the envelope after removal of the data set on the estimation accuracy. But in the larger lead content and lead content measured deviation of 4 sampling points and the sampling point 14, sampling points16 and the sampling point 18, the lead content predicted by the estimated model is larger than the measured lead content, there may be doping that several of the soil and some other heavy metals, lead to these sampling points of soil spectral Pb spectral response is affected by them.

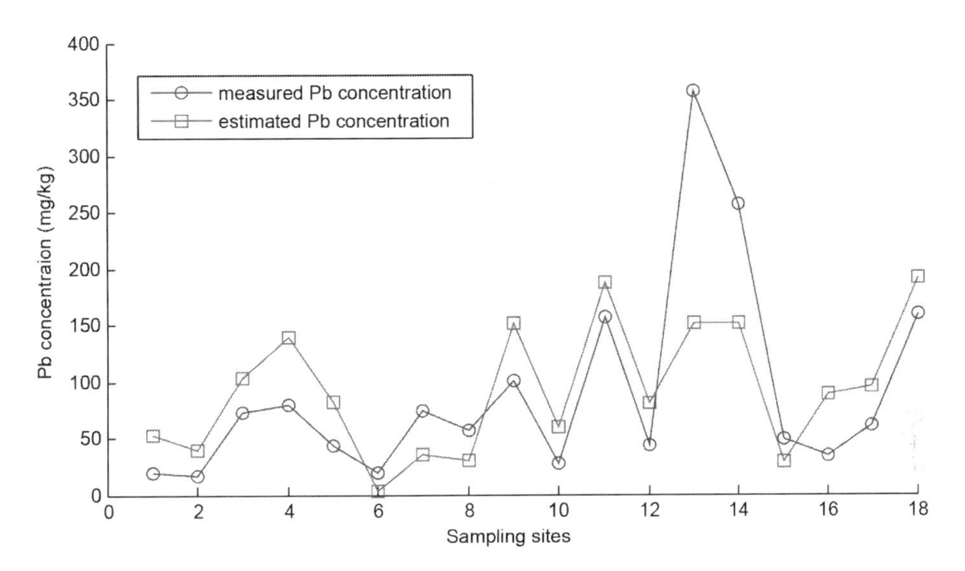

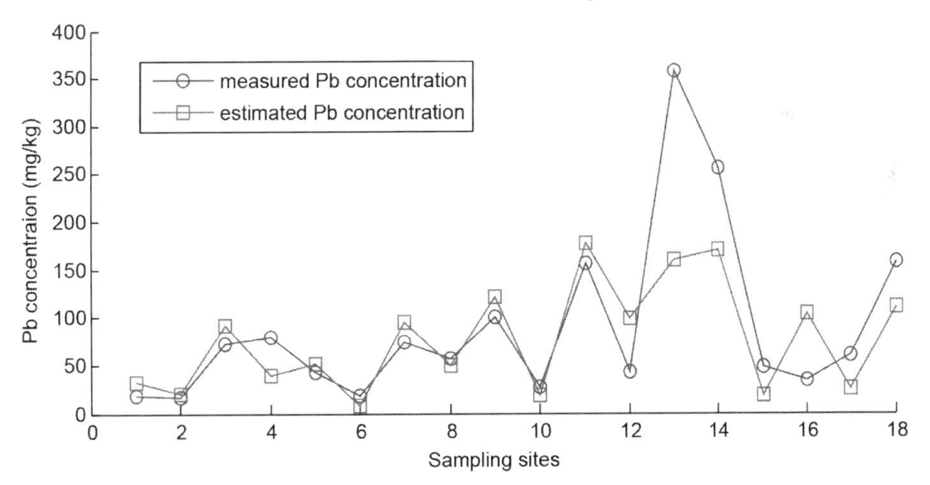

Figure 1. Inversion of Pb on test sample with different spectral datasets.

4 CONCLUSIONS

Isomap manifold learning method can effectively reduce the dimensionality of hyperspectral data, in the premise of maintaining the original characteristics of the spectrum, as far as possible to reduce the data redundancy, extracting useful information. In this paper, Shaanxi JinDuiCheng mine tailings soil as the research object. The results show that the random forest on soil spectral data Haar6 wavelet denoised sets constructed Hyperspectral Estimation Model for estimating soil lead content mining to obtain the high accuracy inversion model to predict the outcome of R^2 is 0.774, RMSE is 249.125. This method makes full use of the advantage of the heavy metal content extraction of hyperspectral remote sensing technology, it has important practical significance for the future of heavy metals pollution monitoring, but also provide technical support for environmental monitoring in mining area.

ACKNOWLEDGEMENTS

This work was funded by Natural Science Basic Research Plan in Shaanxi Province of China (Program No. 2015JQ4105), Open Research Fund of State Key Laboratory of Information Engineering in Surveying, Mapping and Remote Sensing (16E02), and Scientific Research Program Funded by Shaanxi Provincial Education Department (Program No. 16JK1496).

REFERENCES

Breiman L. Random Forest [J]. Machine Learning, 2001, 45:5–32.

Cao L, Coops N C, Innes J L, et al. Tree species classification in subtropical forests using small-footprint full-waveform LiDAR data [J]. International Journal of Applied Earth Observation & Geoinformation, 2016, 49:39–51.

Goetz A F H. Three decades of hyperspectral remote sensing of the Earth: A personal view. Remote sensing of environment, 2009, 113 (S1):5–16.

Goetz A F H, Vane G, Solomon J E, Rock B N. Image spectrometry for earth remote sensing. Science, 1985, 228 (4707):1147–1153.

Luo S, Wang C, Xi X, et al. Fusion of airborne LiDAR data and hyperspectral imagery for aboveground and belowground forest biomass estimation [J]. Ecological Indicators, 2017, 73:378–387.

Meyer H, Lehnert L W, Wang Y, et al. From local spectral measurements to maps of vegetation cover and biomass on the Qinghai-Tibet-Plateau: Do we need hyperspectral information? [J]. International Journal of Applied Earth Observation & Geoinformation, 2017, 55:21–31.

Strang G. Wavelets and Dilation Equations: A Brief Introduction [J]. Siam Review, 1989, 31(4):614–627.

Sun W, Shi L, Yang J, et al. Building Collapse Assessment in Urban Areas Using Texture Information From Postevent SAR Data [J]. *IEEE Journal of Selected Topics in Applied Earth Observations & Remote Sensing*, 2016:1–17.

Land Reclamation in Ecological Fragile Areas – Hu (Ed.)

Quality monitoring and regional differences of ecological restoration in a mining city: A case study of Wu'an City, China

Y.Y. Ru & M.C. Fu
School of Land Science and Technology, China University of Geoscience Beijing, Beijing, China

ABSTRACT: Ecological restoration quality monitoring is important for implementing ecological restoration on mining cities. Based on RS and GIS techniques, the present study uses LANDSET satellite data to calculate NDVI (Normalized Vegetation Index), EVI (Enhanced Vegetation Index) and NDMI (Normalized Water index) for years 2007, 2010, 2013, 2014 and 2016. Through comparative analysis, this study select NDVI to invert FVC (Fractional Vegetation Coverage). The results reveal that: (1) NDMI correlates with NDVI better than EVI; (2) Unban area's FVC has been staying a low level in Wu'an and there ecological condition needs to be improved; (3) During years 2007–2010, the FVC of Wu'an changed a little generally; (4) In general, ecological status of the city enhanced from 2013 to 2016. This paper shows that after the government's closure of some small mines and with the help of reclamation measures and the Taihang Mountain greening projects, the ecological restoration goes well.

1 INTRODUCTION

In a long run, the ecological environment of mining city influences its development. And ecological restoration is an important means to improve its environment. Ecological restoration quality monitoring can support the conduction of ecological restoration project.

FVC is a useful indicator reflecting the quality of environment (Cheng Hongfang et al., 2008), which can be quickly obtained by RS technology (Chen Qiao et al., 2015; Wang Jingrui et al. 2015). For example: Guo Hui et al. used LANDSET data to invert FVC aiming at quantifying the effect of Tarim river transportation on the vegetation coverage of its surrounding place (Hui Guo et al., 2017). Wang Xiaoxia et al. used MODIS data to draw the FVC chart and used it to describe the development of the region along Hei River (Wang Xiaoxia et al., 2016).

This paper aims to use LANDSET data to analyze the ecological restoration condition of Wu'an City from 2007 to 2016. In the first step, because of NDMI's excellent functions to invert he soil and vegetation moisture, the study analyzes the correlation between NDVI and NDMI as well as that between EVI and NDMI. And the vegetation index having the best correlation with NDMI will be considered as the basis of inversion for FVC. Then, getting the FVC data of years 2007, 2010, 2013, 2014 and 2016, the paper studies the quality and regional differences of ecological restoration in Wu'an City, which aims to help its future ecological restoration work.

2 SITE DESCRIPTION

Wu'an City, Hebei Province, China is a typical mining city, which locates between east longitude 113°45′ to 114°22′, and north latitude 36°28′ to 37°01′. Having abundant resources and a large number of mineral enterprises, the mining city holds 1818 km^2

3 MATERIALS AND METHODS

The remote sensing image data used in this study are from LANDSET satellites. Its source is www.gscloud.com.cn. Because of the long research period, data of years 2013, 2014 and 2016 is from the Landset8 satellite, data of years 2010, 2010 from the Landset5 satellite. The spatial and temporal resolution of the two satellites mentioned above are both 30m and 16d.

Data preprocessing includes geometric correction, radiometric calibration, image mosaic, atmospheric correction, image cropping and so on. Considering the error of cloud cover, the annual NDVI, EVI and NDMI were all synthesized with MVC (maximal synthesis method).

There are a variety of inversion methods for FVC. In this paper, we choose the Pixel

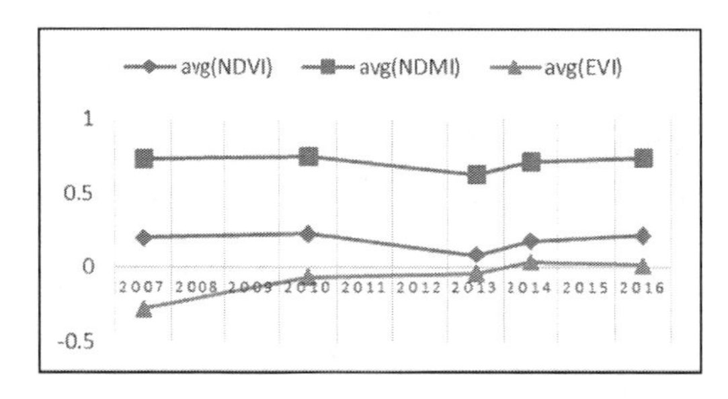

Figure 1. 2007–2016 NDVI, NDMI, EVI mean distribution of the line chart.

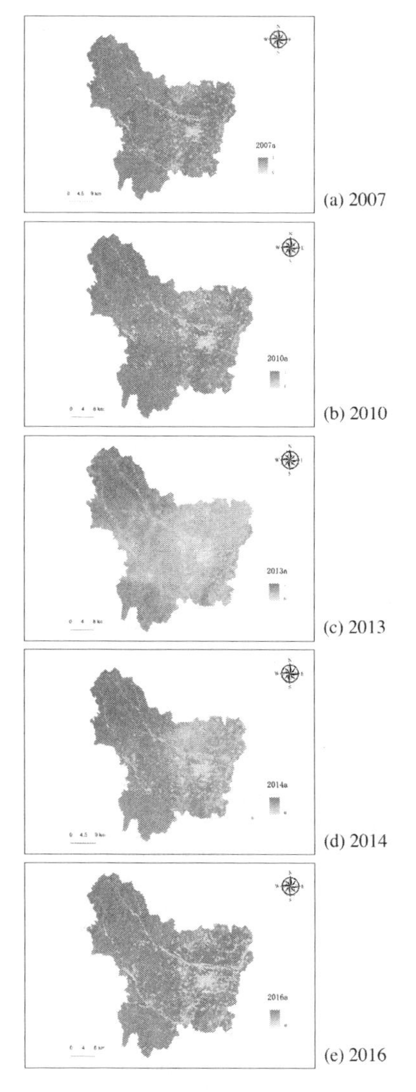

(a) 2007

(b) 2010

(c) 2013

(d) 2014

(e) 2016

Figure 2. Frequency of FVC for years 2007, 2010, 2013, 2014, 2016.

binary model based on mixed pixel method (Liu Guangfeng et al., 2007; Zhong Lina et al., 2013).

4 RESULT

4.1 *Correlation analysis of vegetation indexes*

In this study, the average of NDVI, NDMI and EVI in 5 years is plotted as a line chart, as shown in Fig. 1. It can be seen that NDMI's correlation with NDVI is much better than that of EVI. Therefore, NDVI is chosen for inverting FVC.

4.2 *FVC*

For the inversion of FVC, we define the confidence degree as 1 (Li lin, 2008). Figure 2 - Figure 6 shows the FVC images of year 2007–2016. FVC values are equally divided as 5 sections (0–0.2, 0.2–0.4, 0.4–0.6, 0.6–0.8 and 0.8–1.0). Table 1 shows the distribution of FVC values in the above five sections. In Fig. 2, we grant different sections different kinds of green color. The bigger, the darker.

5 DISCUSSION

5.1 *Correlation between NDVI and NDMI*

Fig. 1 shows that NDVI strongly correlates with NDMI. The two polylines are very similar. But the correlation between EVI and NDMI is weaker. That is why we choose NDVI to invert FVC. In addition, the data shows that during the years 2010–2013, the mean of NDVI displays an obvious decreasing. But it slowly increases both during the years 2007–2010 and 2013–2016.

5.2 *Temporal and spatial difference of FVC*

We define the pixel whose FVC value between 0–0.2 as the bare ground. Table 1 implys that the bare land area increased slightly during the years 2007–2010, but the overall FVC changed a little. From 2010 to 2013, however, with the bare land

Table 1. Wu'an city's frequency of FVC values for the years 2007, 2010, 2013, 2014, 2016.

Section	2007a/%	2010a/%	2013a/%	2014a/%	2016a/%
0–0.2	46.4343	46.8424	47.6622	46.7671	46.4245
0.2–0.4	3.0945	2.8978	17.1881	5.3998	1.4095
0.4–0.6	8.5443	7.6492	21.7909	13.822	7.6641
0.6–0.8	39.2995	40.7451	10.9139	24.2444	17.8788
0.8–1.0	2.6274	1.8655	2.4449	9.7667	26.6231
Total	100	100	100	100	100

area increasing slightly, nearly 75% of land whose FVC values are between 0.6 and 0.8 turned into lower vegetation-covered land. In combination with Figure 1, two possible reasons should be considered. The first reason is that landset5 and landset8 have a large difference in sensing pixels whose FVC values in [0.6, 0.8]. And the second possible reason is that too much land degradation occurred during the period. The exact answer needs further study. In addition, Table 1 also shows that during years 2013–2016, bare land area gradually decreased, and the overall vegetation density increased.

Fig. 2 is the images of geographical distribution of FVC values during the research period. It can be seen that the distribution of FVC changed a little from 2007 to 2010. During the years 2013 to 2016, it is obviously that the vegetation coverage changed hardly in the northwest and southwest of Wu'an city. And FVC values of the east part of the city increased. In addition, the overall vegetation coverage of the city has increased, which corresponds to the Table 1. The following are two possible reasons. One is that the government of Wu'an City closed a lot of small mines in recent years, which not only promotes the intensive use of land, but also reduce the land-destructing behaviors. The eastern part of the city is coal area. Closing the small mines would strength ecological restoration undoubtedly. Another possible reason is the promotion of Taihang Mountain greening project. Wu'an city is an important part of Taihang Mountain greening project, the national project would undoubtedly enhance the overall ecological quality of the city. Besides, urban area FVC values always stay at a low level within the research period. This mean that urban vegetation cover changed a little. And planting in urban area was always at a low level.

6 CONCLUSION

This study find that NDVI is strongly correlated with NDMI. And using it to invert FVC in Wu'an city. Conclusions are as following:

6.1 *Significant regional differences of quality of green space*

During the study period, FVC values of the urban area were generally stay at a low level. However, that of the northwest and southwest places display high FVC values. And FVC in the eastern region of Wu'an has increased significantly.

6.2 *Gradually improved quality of green space*

During the years 2007–2010, Wu'an city's FVC changed little. But from 2013 to 2016, FVC maintained a high level or gradually increased in most areas of the city, and the overall ecological quality of Wu'an City has improved.

ACKNOWLEGEMENT

The authors acknowledge the support provided by Natural Science Foundation of China (41641008).

REFERENCES

Cheng Hongfang, Zhang Wenbo, Chen Feng. (2008) Advances in remote sensing estimation of vegetation coverage. *Remote Sensing for Land & Resources,* (01):13–18.

Chen Qiao, Chen Yongfu, Chen Xingliang et al. (2015) Monitoring of change of forest land resources in forest changing area based on TM remote sensing image—A Case Study of Wu'ning County, Jiangxi Province. *Journal of Northwest Forestry University*, (02), 166–171.

Hui Guo, Guli Jiapaer, Anming Bao et al. (February 2017) Effects of the Tarim river's middle stream water transport dike on the fractional cover of desert riparian vegetation. Ecological Engineering, 99, 333–342.

Li Lin, Tan Bingxiang, Feng Xiaolan. (2008) Study on vegetation cover age and its dynamic change of Beijing suburbs by remote sensing. Agriculture Network Information, (06), 38–41.

Liu Guangfeng, Wu Bo, Fan Wenyi et al. (2007) Extraction of vegetation coverage in desertification regions based on the dimidiate pixel model. Research of Soil and Water Conservation, (02), 268–271.

Wang Jingrui, Shen Wenjuan, Li Weizheng et al. (2015) Performance comparison of remote sensing retrieval model of plantation biomass based on rapid eye. *Journal of Northwest Forestry University*, (06), 196–202.

Wang, Xiaoxia, Jia, Kun, Zhang, Yuzhen et al. (2016) Fractional vegetation cover estimation method through dynamic Bayesian network combining radiative transfer model and crop growth model. *IEEE* Transactions on Geoscience & Remote Sensing, 54(12), 7442–7450.

Zhong Lina, Zhao Wenwu. (2013) Detecting the dynamic changes of vegetation coverage in the loess plateau of China using NDVI data. Science of Soil and Water Conservation, (05), 57–62.

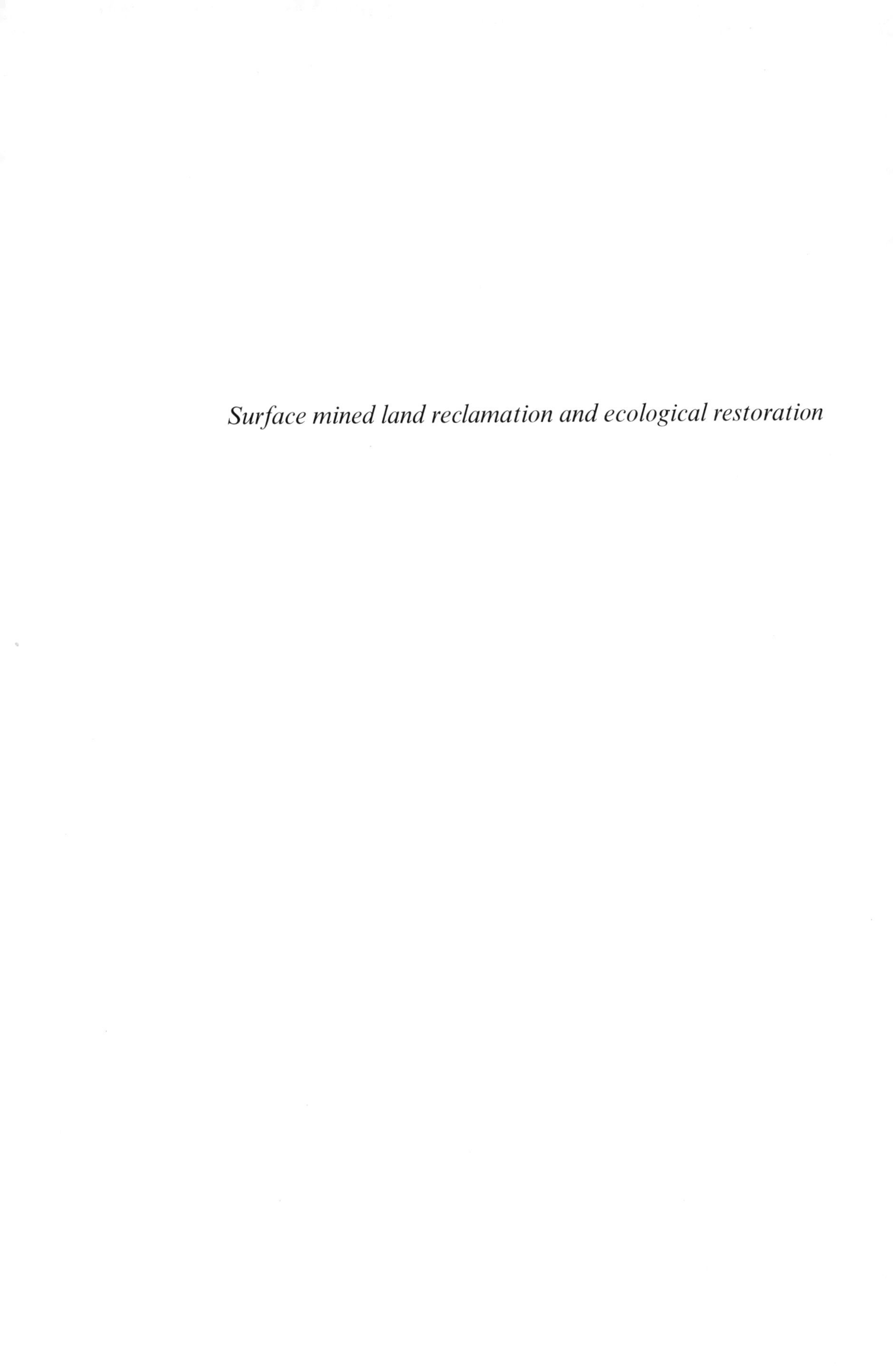

Surface mined land reclamation and ecological restoration

Land Reclamation in Ecological Fragile Areas – Hu (Ed.)
 ISBN 978-1-138-05103-4

Application of a mixture of fly ash and topsoil for rehabilitation in open-pit coal mines in South East Asian countries

A. Hamanaka, S. Matsumoto, K. Murakami, H. Shimada & T. Sasaoka
Department of Earth Resources Engineering, Kyushu University, Fukuoka, Japan

ABSTRACT: Rehabilitation of post-mine land in open-pit coal mines is necessary for sustainable development of coal. Although topsoil containing much nutrition should be secured for plant growth, there was the shortage and the erosion of topsoil by heavy rain in the tropical climate, especially in South East Asian countries. Therefore, in this study, the application of Fly Ash (FA) to topsoil was evaluated in terms of the growth of plants with *Acacia mangium*, aiming at securing topsoil in open-pit coal mines. As the result of the plant experiment, the inhibition of plant growth attributing to the dissolution of Aluminum (Al) from FA was observed. The growth rate of the plants was decreased at the mixing ratio of more than 40% of FA. Thus, topsoil can be secured by mixing FA in topsoil in consideration of the mixing ratio, contributing to the successful rehabilitation of post-mine land in open-pit coal mine.

1 INTRODUCTION

1.1 *Rehabilitation in coal mining*

Rehabilitation of post-mine land is necessary for sustainable development of coal. Coal is mined with the destruction of surface structure, such as forest and grasses, in open-pit coal mining. Thus, rehabilitation of post-mine land is required at the end of the development of coal for environmental conservation. In open-pit mining, topsoil, which is formed from the surface down to approximately 1.0 m depth, is stored during the excavation of coal, followed by the placement in surface layer in post-mine land during rehabilitation since they contain much nutrition that is useful for plant growth (Sheoran et al. 2010).

However, the shortage of topsoil attributing to the loss while hauling soil results in the difficulty of rehabilitation in some cases. Furthermore, soil erosion caused by heavy rain in the tropical climate in South-East Asian countries causes the loss of topsoil as shown in Figure 1, leading to the failure of rehabilitation (Sheoran et al. 2010). Therefore, it is important to secure enough amount of topsoil for rehabilitation in coal mining in South-East Asian countries.

Figure 1. Soil erosion due to heavy rain in a rehabilitation area in a coal mine in a South East Asian country.

1.2 *Utilization of Fly Ash (FA) for securing the topsoil*

In order to secure the amount of topsoil, Fly Ash (FA), which is considered industrial waste after thermal electric power generation with coal was mixed in topsoil in the past research (Hamanaka et al. 2014). The amount of topsoil was successfully increased by mixing FA in topsoil without soil erosion by rainfall (Gorman et al. 2000, Matsumoto et al. 2016). Although FA has been utilized with the aim of increasing the amount of topsoil in several cases, the effects of metals in FA and the high alkalinity are concerned in terms of the inhibition of plant growth (Mahlaba et al. 2012, Cheng 2003). The effects have to be elucidated for securing the amount of topsoil by mixing FA.

In this study, the effects of FA in topsoil on the growth of plants were investigated, aiming at securing topsoil by mixing FA for effective rehabilitation in coal mining in South-East Asian countries.

2 METHODS

2.1 *Samples*

Topsoil was prepared by mixing the decomposed granite produced by the Gulin Kita-Kyu Co., Ltd., and the Kyushu Bentonite produced by the Shinagawa Yogyo Co., Ltd. based on the soil composition which was reported in post-mine land in open-pit coal mine in Indonesia (Matsumoto et al. 2016). The soils were classified as sand, silt, and clay after screening, followed by the preparation of the topsoil. Besides, Fly Ash (FA) taken in the coal fired power plant in Japan was mixed in the topsoil at the mixing ratio of 0%, 20%, 40%, 60%, and 80%, aiming at increasing the amount of topsoil. The topsoil with FA were homogeneously mixed by the cone and quartering method. They were labeled as FA0%, FA20%, FA40%, FA60%, and FA80%, and utilized for plant experiment after physical and chemical analysis of the topsoil in addition to FA as described as below.

2.2 *Plant experiment*

In this study, *Acacia mangium* which inhabits tropical forest in South East Asian countries was planted on the topsoil mixed with FA as indicated in section 2.1 in order to elucidate the effects of FA in topsoil on the growth of plants: the seeds were obtained in Japan. *Acacia mangium* has ever been successfully applied in post-mine lands for bauxite, copper, coal, and iron in the world, and widely utilized for a primary reclamation of post-mine land in many cases as shown in Figure 2 (Leon et al. 2013). It also can grow under the various conditions, such as in compacted soils, dry area, and humid area.

Acacia mangium was planted on the prepared topsoil for 133 days until a clear distinction is observed in the phytotron glass room G-9 in Biotron Application Center, Kyushu University under the conditions as follows: at 30°C and 70% of relative humidity assuming the local climate in the coal mine in Indonesia. In this test, 5 plants were planted in pots by the mixing ratio of FA, and the height was measured every week. The growth rate was calculated with the change of height and the initial height (8.0 cm) of *Acacia mangium* every week. 500 mL of water was supplied to the pots every 3-4 days. The liquid fertilizer HYPONeX-R (N-P-K = 6-10-5) diluted to 1,000 mg/L with deionized water was, moreover, added to them weekly to promote the growth. The leachate from the bottom of the pots was sampled to report the change of pH. At the end of the experiment, the plants were taken to supply to the dissolution analysis with acids as described in section 2.4 for the purpose of understanding the effects of metal concentration in plant body on plant growth. In this study, the leaves, stem, and root were mixed and supplied to the dissolution analysis since it was unable to separate the samples in each part.

Figure 2. Acacia mangium in a rehabilitation area in a coal mine in Indonesia.

2.3 *Physical analysis*

The topsoil was supplied to the particle size distribution test (ASTM D422-63 2007), and the constant head permeability test (ASTM D 2434-68 2006) after the application of FA in the topsoil by reference to the standard of ASTM in order to understand the effects of physical conditions of the topsoil on the growth of *Acacia mangium*.

2.4 *Chemical analysis*

The FA was supplied to X-ray Fluorescence (XRF) analysis to measure the metal content in FA. The result was compared to that of FA produced in the coal fired power plant in the Indonesian mine.

The plant samples were taken at the end of the plant experiment, and washed with deionized water with the washing process using a sonication (UT-106H, SHARP) at room temperature so as to remove soil particles. They were dried at 60°C for 72 hours and pulverized using mortar and pestle. 5 mL of 61% nitric acid (HNO_3) and 35% hydrochloric acid (HCl) mixed at volumetric ratio of 3:1 was added into 0.25 g of the plant samples by reference to the past study (Quadir et al. 2011). After the dissolution process, the solution was supplied to DigiPREP Jr. (SCP Science, Quebec, Canada) at 110°C until the samples were completely dissolved. In the case of obtaining undissolved samples in acids after the above process, 1 mL of the acids was added and the dissolution process was repeated. The volume of the solution was adjusted to 20 mL by adding deionized water, followed by the measurement of metal content using ICP-AES (VISTA-MPX ICP-OES (Seiko Inst., Japan)) after the filtration with 0.45 μm of membrane filter. The concentration of Al, As, B, Fe, Mn, S, and Zn was measured, and it was calculated with mg per dry unit weight (mg/g) in this study.

3 RESULTS AND DISCUSSION

3.1 *Physical and chemical properties of the simulated topsoil mixed with fly ash*

The FA utilized in this study were composed of 67% of Si, 19% of Al, 4% of Fe, and other elements such as Ca and K on the basis of the results of XRF analysis. Comparing the results with the metal content in FA which was taken in the coal fired power plant in the Indonesian mine, the content of Al in FA showed a similar value: the content of Al was 13–15% in FA in the Indonesian mine. It can be seen that the content of Al in the topsoil with FA rises with the increase of the mixing ratio of FA for the results. Considering that the inhibition of the growth of plants by aluminum was reported in the past research, the content of Al in FA may affect the growth of *Acacia mangium* in the plant experiment (Kochian et al. 2005).

Soil texture of the simulated topsoil is drawn at different mixing ratio of FA in Figure 3, and the permeability is shown in Figure 4. There is a close connection between a permeability of soil and the elongation of root of plants, and a permeability of soil is one of the important indicators to evaluate the effects of soil conditions on plant growth (Tangahu et al. 2011). Whereas the ratio of silt gradually rose with the increase of the mixing ratio of FA in the topsoil, the proportion of sand decreased in Figure 3. Additionally, the permeability of the topsoil decreased with the increase of the mixing ratio of FA. This was attributed to the increase of the composition of silt which showed low permeability compared to that of sand with the mixture of FA. Although the significant change of permeability of soils can affect the growth of plants, the growth of plants was not affected by the physical conditions, such as soil texture and permeability, in this study due to the minor change of the permeability: the change was reported within 3.0×10^{-4} cm/s.

In short, there was not significant effect of physical conditions of the topsoil on the growth of *Acacia mangium* during the plant experiment. It can be, therefore, expected that the difference of the growth rate of *Acacia mangium* was attributable to the mixing ratio of FA.

3.2 *Plant growth of Acacia mangium on the simulated topsoil mixed with fly ash*

Figure 5 shows the change of growth rate of *Acacia mangium* for 133 days at different mixing ratio of FA in the plant experiment. The growth rate decreased with the increase of the mixing ratio of FA in the results. There was stagnation in the change of the growth rate at the mixing ratio of

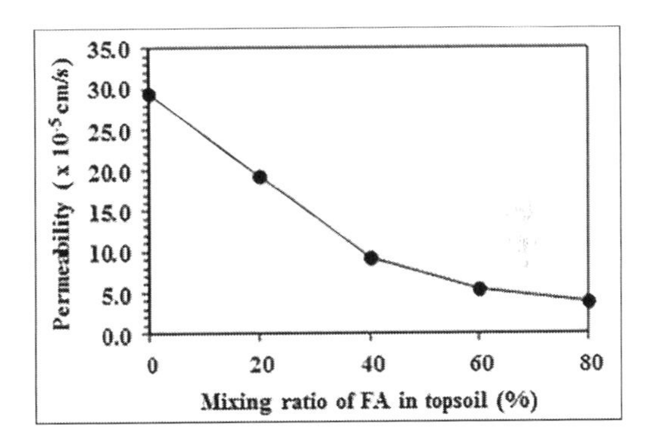

Figure 4. Permeability at different mixing ratio of FA in the topsoil.

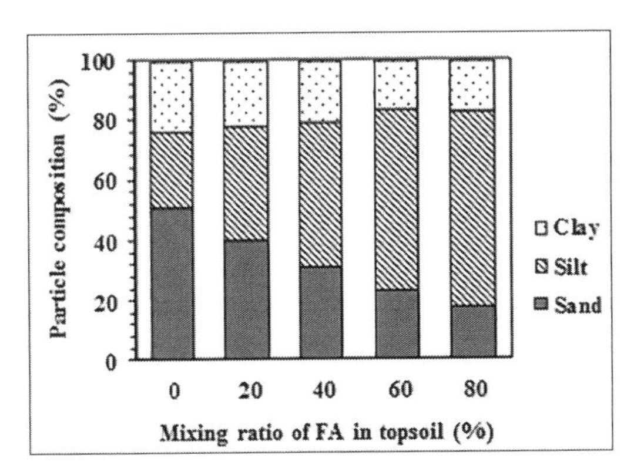

Figure 3. Soil texture at different mixing ratio of FA in the topsoil.

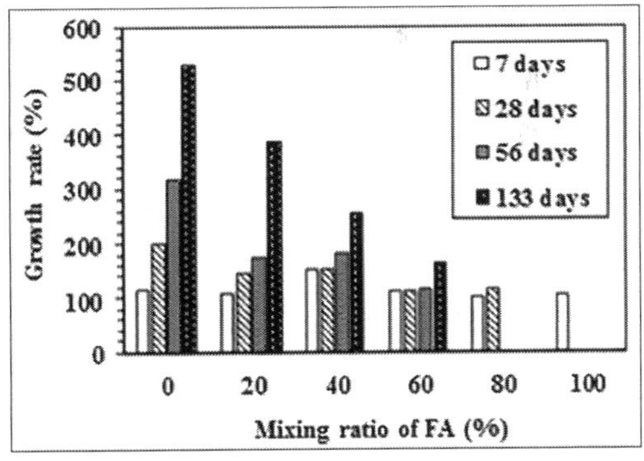

Figure 5. Growth rate of Acacia mangium at different mixing ratio of FA.

over 40% of FA. Besides, all of *Acacia mangium* were died after 28 days at the mixing ratio of over 80% of FA, resulting from the mixture of FA in topsoil.

In addition, the pH in leachate was approximately 9.0 in the topsoil with the mixture of FA, although it was 7.5 without FA. The pH in leachate continued to show ca. 8.0 in topsoil with FA after 50 days, indicating that the mixture of FA in topsoil resulted in the alkali conditions in the leachate in consideration of the pH = 7.5 in the topsoil without FA. FA has a high alkalinity in the form of CaO (Pradhan & Deshmukh 2008), and the content of Ca was observed in XRF analysis, resulting in the alkali conditions. Moreover, Figure 6 describes the concentration of Al in the body of *Acacia mangium*, which was measured by the dissolution analysis with acids. The results showed that the concentration of Al was more than 1.8 mg/g at the mixing ratio of more than 40% of FA, whereas it was less than 0.6 mg/g in the topsoil at the mixing ratio of less than 20%. Aluminum exists as $Al(OH)_3$ which is an insoluble mineral at neutral condition (ca. pH = 7.0); meanwhile, it is released as $Al(OH)_4^-$ at alkali condition: aluminum dissolves as $Al(OH)_4^-$ at ca. pH = 9.0 (Matsumoto 2000). Aluminum inhibits the elongation of roots of plants (Kochian et al. 2005, Kikui et al. 2005), suggesting that Al in the topsoil derived from FA dissolved under the alkali condition and were concentrated in the body of *Acacia mangium* along with the inhibition of the growth at the mixing ratio of over 40% of FA. In FA20% sample, the concentration of Al in the plant body was, however, lower than that of FA40%–80% since the amount of dissolved Al from FA was not higher than that of FA40%–80% due to the low mixing ratio of FA. For these results, the mixing ratio of FA with the aim of increasing the amount of topsoil without the inhibition of plant growth should

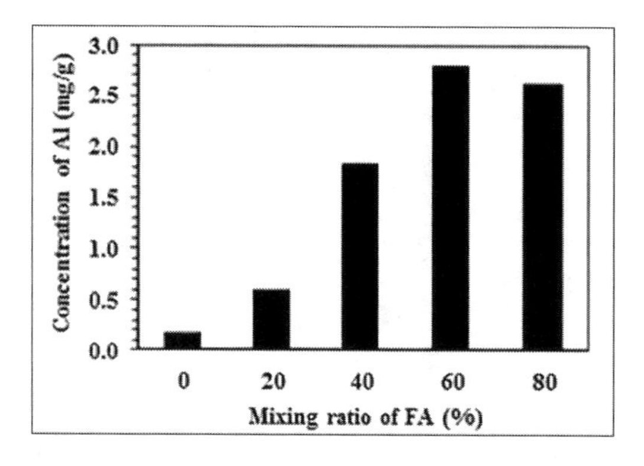

Figure 6. Concentration of Al in the body of Acacia mangium after the plant experiment at different mixing ratio of FA.

be adjusted at the mixing ratio of less than 20% of FA in this study in terms of the concentration of Al in the body of plants. Furthermore, the content of Al in FA in this study showed a similar value with that of FA which is produced in the Indonesian mine, indicating that the utilization of FA for the purpose of increasing the amount of topsoil may result in the inhibition of plant growth due to the effects of Al in open-pit coal mines in South East Asian countries.

However, since the tolerance to Al in plants varies depending on the plant species and the chemical composition of FA depends on the quality of coal, additional experiment with other species of plants and coal is required in order to establish the guideline of the utilization of FA in topsoil to secure the enough amount of topsoil.

4 CONCLUSIONS

In this study, the application of Fly Ash (FA) to topsoil was evaluated in terms of the effects of FA on the growth of *Acacia mangium*, aiming at securing topsoil in open-pit coal mines. The results obtained by the experiments are summarized as follows:

1. The FA in this study was composed of a large amount of silt. Therefore, the permeability of the topsoil decreased with the increase of the mixing ratio of FA in topsoil. However, there was not significant effect of physical conditions of the topsoil on the growth of plants due to the minor change of the permeability: the change was reported within 3.0×10^{-4} cm/s.
2. The FA in this study consisted of 67% of Si, 19% of Al, 4% of Fe, and other elements such as Ca and K.
3. There was stagnation in the change of the growth rate at the mixing ratio of over 40% of FA. Besides, all of *Acacia mangium* were died after 28 days at the mixing ratio of over 80% of FA, resulting from the addition of FA in topsoil.
4. The concentration of Al was more than 1.8 mg/g at the mixing ratio of more than 40% of FA, whereas it was less than 0.6 mg/g in the topsoil at the mixing ratio of less than 20%. Considering that the pH in leachate was approximately 9.0 in the topsoil with the mixture of FA, Al in the topsoil derived from FA dissolved under the alkali condition and were concentrated in the body of *Acacia mangium* along with the significant inhibition of the growth at the mixing ratio of over 40% of FA. Therefore, FA has to be utilized in topsoil at the mixing ratio of less than 20% of FA in order to increase the amount of topsoil in this

study in terms of the concentration of Al in the body of plants.
5. The content of Al in FA in this study showed a similar value with that of FA which is produced in the Indonesian mine, indicating that the utilization of FA for the purpose of increasing the amount of topsoil may result in the inhibition of plant growth due to the effects of Al in open-pit coal mines in South East Asian countries.

ACKNOWLEDGEMENT

The FA samples were obtained in the coal fired plant in Japan, and the authors are grateful to the plant for providing the samples. We also would like to express our appreciation to the Biotron Application Center, Kyushu University for plant experiment.

REFERENCES

American Society for Testing and Materials (ASTM). (2006) *ASTM Standard D 2434-68*, Standard Test Method for Permeability of Granular Soils, ASTM.

American Society for Testing and Materials (ASTM). (2007) *ASTM Standard D422-63(2007)e2*, Standard Test Method for Particle-size Analysis of Soils (withdrawn 2016), ASTM.

Cheng, S. (2003) Effects of heavy metals on plants and resistance mechanisms. A state-of-the-art report with special reference to literature published in Chinese journals. *Environ. Sci. Pollut. Res. Int.* 10(4), 256–264.

Gorman, J.M. & Sencindiver, J.C. & Horvath, D.J. & Singh, R.N. & Keefer, R.F. (2000) Erodibility of fly ash used as a topsoil substitute in mineland reclamation. *J. Environ. Qual.*, 29(3), 805–811.

Hamanaka, A. & Inoue, N. & Matsumoto, S. & Shimada, H. & Sasaoka, T. & Matsui, K. & Miyajima, I. (2014) Rehabilitation of Open Cut Coal Mine with Paper Mulberry (*Broussonetia papyrifera*) in Indonesia. *J. of the Polish Mineral Engineering Society* 15(2), 159–163.

Kikui, S. & Sasaki, T. & Maekawa, M. & Miyao, A. & Hirochika, H. & Matsumoto, H. & Yamamoto, Y. (2005) Physiological and genetic analyses of aluminium tolerance in rice, focusing on root growth during germination. *J. of Inorganic Biochemistry* 99(9), 1837–1844.

Kochian, L.V. & Pineros, M.A. & Hoekenga, O.A. (2005) The physiology, genetics and molecular biology of plant aluminum resistance and toxicity. *Plant Soil* 274, 175–195.

Leon, J.D. & Castellanos, J. & Casamitjana, M. & Osorio, N.W. & Loaiza, J.C. (2013) Hai, R. (ed.), *Alluvial gold-mining degraded soils reclamation using Acacia mangium plantations: an evaluation from biogeochemistry*, 155–176. New York: Nova Science.

Mahlaba, J.S. & Kearsley, E.P. & Kruger, R.A. (2012) Microstructural and mineralogical transformation of hydraulically disposed fly ash—implications to the environment. *Coal Combustion and Gasification* 4, 21–27.

Matsumoto, H. (2000) Plant responses to aluminum stress in acid soil molecular mechanism of aluminum injury and tolerance. *Kagaku to Seibutsu* 38(7), 425–458.

Matsumoto, S. & Ogata, S. & Shimada, H. & Sasaoka, T., Kusuma, J.G. & Gautama, S.R. (2016) Application of coal ash to postmine land for prevention of soil erosion in coal mine in Indonesia: utilization of fly ash and bottom ash. *Advances in Materials Science and Engineering* 2016, 1–8.

Pradhan, A. & Deshmukh, J.P. (2008) Utilization of fly ash for treatment of acid mine water. *J. of Environmental Research and Development* 3(1), 137–142.

Quadir, Q.F. & Watanabe, T. & Chen, Z. & Osaki, M. & Shinano, T. (2011) Ionomic response of *Lotus Japonicus* to different root-zone temperatures. *Soil Science and Plant Nutrition* 57, 221–232.

Sheoran, V. & Sheoran, A.S. & Poonia, P. (2010) Soil reclamation of abandoned mine land by revegetation: a review. *Int. J. Soil, Sediment and Water* 3(2), 1–21.

Tangahu, B.V. & Sheikh Abdullah, S.R. & Basri, H. & Idris, M. & Anuar, N. & Mukhlisin, M. (2011) A review on heavy metals (As, Pb, and Hg) uptake by plants through phytoremediation. *Int. J. Chem. Eng.* 2011, 1–31.

Land Reclamation in Ecological Fragile Areas – Hu (Ed.)

 ISBN 978-1-138-05103-4

Effect of acid soil properties on soil erosion at a dumping site in an open-pit coal mine, Indonesia

H. Shimada, S. Matsumoto, A. Hamanaka, T. Sasaoka & G.J. Kusuma
Geological Survey of Japan, AIST, Ibaraki, Japan
Faculty of Mining and Petroleum Engineering, Institute Technology of Bandung, Bandung, Indonesia

ABSTRACT: Soil erosion which easily progresses due to heavy rain in the tropical climate is a serious issue for mined-land reclamation in Indonesia since it causes plant death. Moreover, acidifi-cation of soils occurs because of sulfide minerals in mining area. The physical properties of soils are changed by the acidification of soils according to previous studies. Thus, since the erosion rate of soils greatly depends on physical characteristics of soils, the relation between the change of physical properties of soils caused by soil acidification and soil erosion has to be considered for erosion control in mining area. In this study, physical properties of the soil samples in which the soil-pH was adjusted with H_2SO_4 were measured, followed by the artificial rainfall test with the samples in order to understand the effect of acidification of soils on erosion rate. The results showed that the decrease of cohesive strength among soil particles caused by the decline of consistency limit with soil acidification resulted in the increase of erosion rate. Hence, soil acidi-fication should also be prevented with the measures against soil erosion for successful mined-land reclamation. Based on DPSIR model and considered previous studies and realities, this essay chooses a typical case, Wu'an, in order to build an ecological security evaluation system. With reference to relevant standards, eco-logical security level was determined. There is a want of developmental driver of urban system, pressure of habitat overload and obviously insufficient social response. In order to get out of "resource curse", Wu'an should make the innovation-driven improvement and increase the investment in environmental protection.

1 INTRODUCTION

Indonesia is one of major coal producers and exporters in the world. Both of the coal production and the export have gradually expanded with increase of world demand for coal. The coal is mined by open-pit mining method in Indonesia, and a broad post-mining area is built after the mining operation as shown in Figure 1. The broad area has to be revegetated in terms of environmental conservation.

It is difficult to conduct revegetation in such a post-mining area due to the poor condition for plant growth Therefore, topsoil which is stored during excavation process is backfilled in the post-mining area in many open cast coal mines since soil organic matter and soil microbe in topsoil can promote the growth of vegetation.

In many coal mines, acid sulfate soils or rocks are mined as waste materials with excavation of coal. These soils or rocks consist of several kinds of minerals such as silicate and sulfide minerals (e.g. pyrite: FeS_2). This sulfide mineral leads to Acid Mine Drainage (AMD) by the reactions when they are exposed to oxygen and water during excavation process as follows:

Figure 1. Post-mining area in an open cast coal mine in Indonesia.

$$FeS_2 + 7/2O_2 + H_2O \rightarrow Fe^{2+} + 2SO_4^{2-} + 2H^+ \quad (1)$$

$$Fe^{2+} + 1/4O_2 + H+ \rightarrow Fe^{3+} + 1/2H_2O \quad (2)$$

$$Fe^{3+} + 3H_2O \rightarrow Fe(OH)_3 + 3H^+ \quad (3)$$

$$FeS_2 + 14Fe^{3+} + 8H_2O \rightarrow 15Fe^{2+} + 2SO_4^{2-} + 16H^+ (4)$$

These waste materials are, generally, backfilled in deep in the ground to prevent the contact of sulfide minerals to oxygen and water. However, they are mixed in topsoil and backfilled in surface layer in post-mining area in some cases.

Additionally, withered plants due to acidic condition were observed during revegetation process after the construction in construction site in Japan. This case indicates that the mixing of such waste materials consisted of sulfide minerals in topsoil affects revegetation process in mining operation. There is, though, not a clear guideline to evaluate topsoil for effective revegetation in consideration with soil acidification in open cast coal mines in Indonesia.

For the reasons, field conditions in post-mining area in the open cast coal mine in Indonesia were investigated through field investigation and laboratory vegetation test with Acacia Manguim on acid soil, aiming at the establishment of proper assessment guideline of acid-topsoil for effective revegetation.

Soil erosion is one of serious environmental issues in terms of land degradation as shown in Figure 2 (Holz et al., 2015; George et al., 2009). Soil erosion which is caused by heavy rain is also a serious issue for mined-land reclamation in Indonesia in the tropical climate since it causes plant death. As physical characteristics of soils have a close connection with soil erosion, soil management and measures against rainfall play an important role to prevent soil erosion (Chow et al., 2003).

In the waste dump in mines, soil acidification and soil erosion progress simultaneously. Soil acidification is caused by the exposure of sulfide in waste rocks to atmosphere with excavation, and it has been reported in the world (Lin et al., 2005; Munksgaard et al., 2012). In open-cast mine, this situation easily progresses due to the large scale mining. Moreover, physical properties of soils, such as shear strength and Atterberg limits, change under acidic conditions according to past studies.

Figure 2. Soil erosion in the western United States (Holz et al., 2015).

Gratchev and Towhata indicated that acidic conditions changed the strength of soils (Gratchev and Towhata, 2013). Furthermore, Hasan et al. concluded that soil acidification affected the change of Atterberg limits of soils (Hasan et al., 2013). For the reasons, the changes of physical properties of soils through soil acidification possibly affect soil erosion. There are, however, few studies which investigate the relationship between soil acidification and soil erosion. Thus, it is important to elucidate the effects of soil acidification on soil erosion for taking measures against soil erosion in mines.

In this research, artificial rainfall test was conducted with simulated soil samples, which were adjusted to soil properties in the waste dump in open-cast coal mine in Indonesia and to different pH with H_2SO_4 in order to understand the effects of acidic conditions on soil erosion. The measures against soil erosion were examined by comparing the results with the field conditions in open-cast mine in Indonesia.

2 METHODS

Simulated soil samples were prepared based on soil properties in the waste dump in A coal mine in Indonesia. The physical properties of the simulated soil samples were investigated after adjusting soil pH to acidic conditions, followed by the artificial rainfall test with the simulated soil samples. The measures against soil erosion in consideration of soil acidification were examined by comparing the results with field conditions in A coal mine in Indonesia.

2.1 *Sample preparation*

Simulated soil samples were prepared by mixing sand, silt, and clay on the basis of the soil properties in the waste dump in A mine as shown in Table 1. After the sample preparation, soil pH of the simulated soil samples was adjusted to different pH values ranging from 2.0 to 6.0 with H_2SO_4 and deionized water. The water content of the adjusted samples was adjusted to 1.5–2 times of liquid limit of the simulated soil samples. Five samples were prepared in this preparation: soil pH was 2.0, 3.0, 4.0, 5.0, and 6.0, respectively.

2.2 *Physical properties of soils*

Major minerals in the simulated soil samples were investigated with X-Ray Diffraction (XRD) analysis using wide angle goniometer RINT 2100 XRD under the following conditions: radiation CuKα; operating voltage 40 kV; current 26 mA; divergence slid 1 deg; anti-scatter 1 deg; receiving slit

Table 1. Soil properties in the waste dump in A mine in Indonesia and that of simulated soil samples.

Sample	Sand (%)	Silt (%)	Clay (%)	Texture
Soils in A mine	2.7~48.9	26.6~57.8	24.4~39.6	Clay Loam Light Clay Silt Clay
Simulated soil	36.7	26.7	36.6	Light Clay

0.3 mm; step scanning 0.050°; scan speed 2.000°/min; and scan range 2.000°–65.000°.

Atterberg limits test was, additionally, conducted with simulated soil samples which were adjusted to different soil pH based on the standard of ASTM D4318-05 in order to understand the physical characteristics of soils under acidic conditions (ASTM, 2005). The plastic index (I_P) was calculated by subtracting with plastic limit (W_P) from liquid limit (W_L) in the test. Furthermore, particle size distribution of the adjusted soil samples was analyzed on the basis of the standard of ASTM D422-63 (ASTM, 2007). The effect of acidic conditions on soil characteristics was investigated by performing the particle size distribution analysis under different pH conditions. Zeta potential of the simulated soil samples was, moreover, measured in 10^{-3} M KCl solution using Zetasizer Nano ZS (Malvern) under different pH conditions which was adjusted with HCl. 5 mg of fine-grained sample was added to 20 ml of the solution which was adjusted to different pH values ranging from 2.0 to 6.0 with HCl. The effect of acidic conditions on physical properties of soils was investigated by comparing the results with the change of particle size distribution of soils under different pH conditions.

2.3 *Artiicial rainfall test*

Soil erosion rate of simulated soil samples which was adjusted to different pH was measured in artificial rainfall test. Figure 3 shows the schematic view of the artificial rainfall equipment. In this equipment, rainfall was supplied through raindrop needles to soil samples for an hour. Annual soil erosion H was calculated with the equation as shown below:

$$H = \frac{(R \times E)}{(A \times G \times I)} \tag{5}$$

H is annual soil erosion (cm/year), R is annual rainfall (mm/year), E is soil erosion per an hour (g/hour), A the section area (cm^2), G is the dry density of sample (g/cm^3), I is rain intensity (mm/hour).

The rainfall intensity was set at 80 (mm/hour) based on the observed values in Indonesian mines (National Environment Agency, 2011; Hamanaka et al., 2013). Water content of the soil sample was set at 15%, and slope angle was set at 35° by reference to field conditions in the waste dump in A mine. The annual rainfall R was set at 3000 (mm/year) which is the average value in Indonesia (Dina, 2009). Simulated soil samples were set at 240 mm by 155 mm of the case in the equipment at the same filling rate. Surface stream and percolation water were separately collected and they were dried at 105°C in an oven. The weight of the residue was measured to calculate the amount of eroded soil per an hour E. The calculated amount of annual soil erosion was compared to the Indonesian standard of soil erosion (Hamanaka et al., 2013). In addition to the comparison, the measures against soil erosion were studied by considering field conditions which were observed in open-cast coal mines in Indonesia.

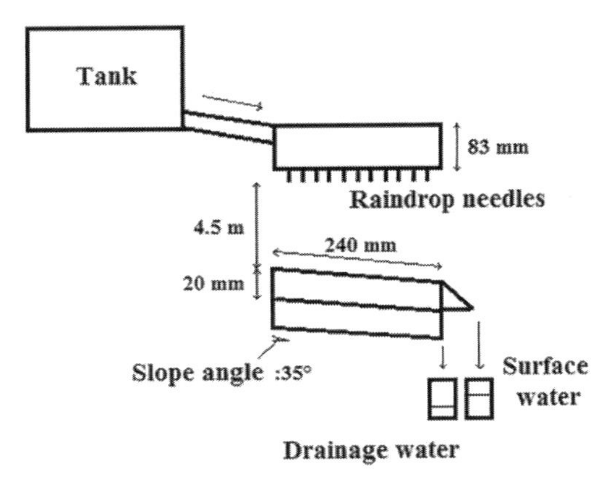

Figure 3. Schematic view of artificial rainfall equipment.

3 RESULTS AND DISCUSSION

3.1 *Changes of physical properties of soils under acidic conditions*

Table 2 shows Atterberg limits of soils in the waste dump in A mine and simulated soil samples. There was not significant difference in WP, WL, Ip, and soil pH between the samples. Moreover, simulated soil samples mainly consisted of quartz, kaolinite, montmorillonite, and illite according to the result

Table 2. Soil properties in the waste dump in A mine in Indonesia and that of simulated soil samples.

Sample	Liquid limit W_L (%)	Plastic limit W_p (%)	Plasticity index I_p (–)	Soil pH
Soils in A mine	26.9~53.9	19.1~24.3	7.9~26.9	3.4~5.7
Simulated soils	43.2	24.1	19.1	5.91

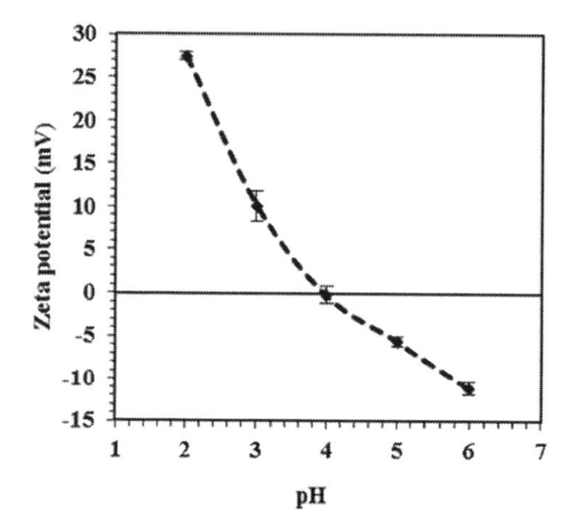

Figure 4. Zeta potential of simulated soil samples at different pH value.

of XRD analysis. Due to the content of these clay minerals, simulated soil samples were categorized into light clay as shown in table 1. These results indicated that simulated soil samples showed similar characteristics with soils in the waste dump in A mine.

As a next step, simulated soil samples which were adjusted to different soil pH with H_2SO_4 was used for Atterberg limits test. The result showed that W_P, W_L, and I_p declined with the decrease of soil pH from 6.0 to 2.0. Besides, particle size of simulated soil samples increased with the decrease of soil pH based on the result of particle size distribution analysis. Thus, the decrease of soil pH led to the increase of particle size of simulated soil samples along with the decline of Atterberg limits. The decrease of Atterberg limits with soil pH reduction is consistent with the result in the past study (The Geotechnical Society of Japan, 2004).

Figure 4 shows the change of zeta potential of simulated soil samples under different pH conditions. Zeta potential of simulated soil samples increased with the decrease of pH value from 6.0 to 2.0. This suggested that pH-dependent charge of soil particle edges became positive, resulting in the attraction between positive and negative permanent charge (The Clay Science of Japan, 1987; Iwata and Kida, 1997). The aggregation of soil particles due to the attraction caused the increase of particle size of simulated soil samples. Moreover, it would appear that dissolved Al^{3+} derived from clay minerals in simulated soil samples caused aggregation of soil particles as an aggregating agent, leading to the increase of particle size of simulated soil samples. In short, the decrease of soil pH caused the change of physical properties of soils with the decrease of Atterberg limits and the increase of particle size.

3.2 *Soil erosion with the change of pH*

Figure 5 shows annual soil erosion rate of simulated soil samples under different soil pH conditions in artificial rainfall test. Red lines indicate the standard of soil erosion in Indonesia (Hamanaka et al., 2013). The risk is very low when soil loss (cm/year) is < 0.15, low when it is 0.15–0.9, moderate when it is 0.9–1.8, and high when it is 1.8–4.8.

Annual soil erosion in the artificial rainfall test gradually increased with the decrease of soil pH in the result. Since the cohesive strength between soil particles was low due to low consistency limit caused by aggregation of soil particles as described above, simulated soil samples were easily eroded by simulated rainfall. The risk of soil erosion was classified into moderate when soil pH was 6.0; however, the risk was classified into high with the decrease of soil pH. This suggested that soil

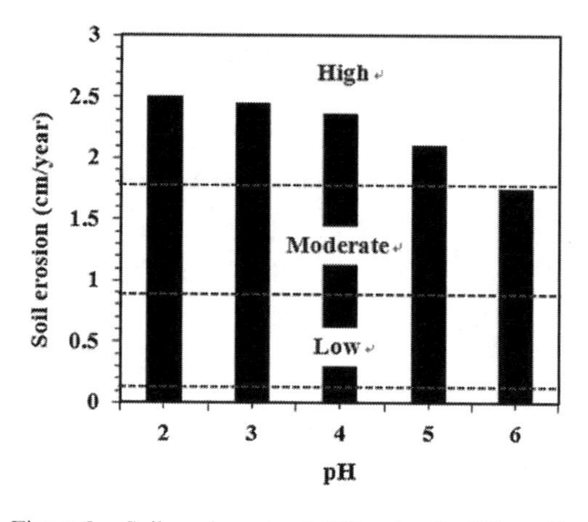

Figure 5. Soil erosion rate at different soil pH in artificial rainfall test: red lines indicate the Indonesian standard of soil erosion: classification is very low when it is < 0.15; low when it is 0.15–0.9; moderate when it is 0.9–1.8; high when it is 1.8–4.8 (Hamanaka et al., 2013).

erosion may occur over time with soil acidification even if the risk of soil erosion is categorized into low. In regards to acid generation from sulfide minerals in waste rocks in mines for a long term, soil acidification over time may result in soil erosion during the operation (Shea, 1995). Therefore, regular monitoring for soil conditions is required to prevent soil erosion in the area where soil acidification possibly occurs such as in the waste dump and in pit in mines.

In open-cast coal mines in Indonesia, soil erosion was observed in the waste dump where acidic water with pH = 2.1 occurred as shown in Figure 6. This suggested that acidification of the soils and soil erosion progressed simultaneously in the waste dump in mines. Thereby, waste materials causing soil acidification through a weathering process of rocks have to be preferentially placed within the waste dump in order to prevent soil erosion in mine area. Since both of positive charge of soil particles and aggregation of soil particles caused by Al^{3+} possibly progress with soil acidification in mines, it can be said that soil erosion is accelerated in the waste dump in the mines. Besides, Figure 7 describes the soil erosion on the slope and acidic water in pit. When the slope in pit is exposed to rainfall after excavation of coal, acidic water occurs and the slope surface is eroded. Moreover, non-reacted face is exposed to atmosphere with soil erosion, resulting in the progress of the occurrence of acidic water. Consequently, waste rocks which cause soil acidification should not be left on the slope in pit for prevention of soil erosion as well as the occurrence of acidic conditions. For these situations in open-cast mines in Indonesia, the measures against soil acidification contribute to prevention of soil erosion. This can, additionally, result in successful reclamation of the waste dump in mines.

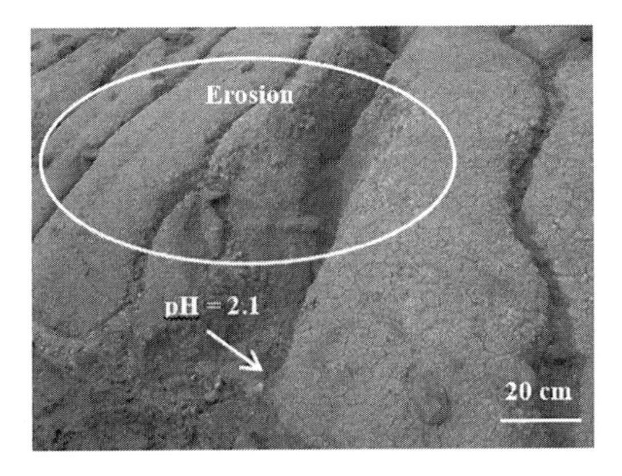

Figure 6. Soil erosion under acidic conditions in the waste dump in A mine: pH of the drainage water was 2.1.

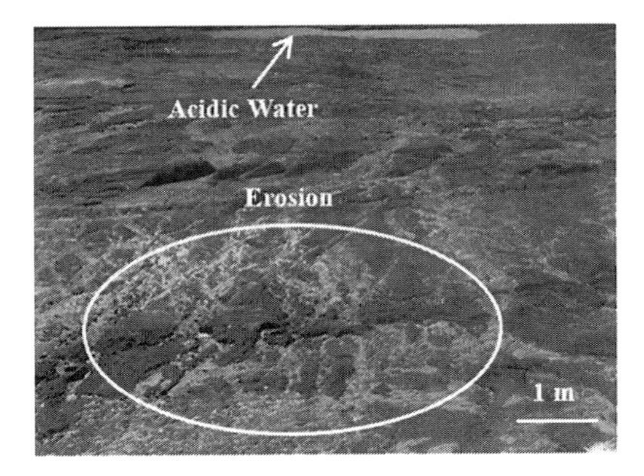

Figure 7. Soil erosion and acidic water on the slope in pit in A mine.

4 CONCLUSIONS

Since the erosion rate of soils greatly depends on physical characteristics of soils, the relation between the change of physical properties of soils caused by soil acidification and soil erosion has to be considered for erosion control in mining area. In this study, physical properties of the soil samples in which the soil-pH was adjusted to acidic conditions with H_2SO_4 were measured, followed by the artificial rainfall test with the samples in order to understand the effect of acidification of soils on erosion rate. The main results are summarized as follows:

1. With the decrease of soil pH, Atterberg limits of simulated soil samples sharply declined and the particle size increased. Positive charge of soil particle edge with soil acidification caused the attraction between positive and negative permanent charge, leading to the results. Moreover, dissolved $Al3^{+}$ derived from clay minerals in simulated soil samples under acidic conditions possibly caused aggregation of soil particles as an aggregating agent, resulting in the increase in particle size of simulated soil samples.
2. Since the cohesive strength between soil particles decreased due to low Atterberg limits by aggregation of soil particles with the decrease of soil pH, simulated soil samples were easily eroded by simulated rainfall in artificial rainfall test.
3. Soil acidification and soil erosion progress simultaneously in mines; therefore, waste materials causing acidic conditions should be preferentially placed within the waste dump and not be left on the slope in pit for prevention of soil erosion as well as soil acidification.

ACKNOWLEDGEMENT

The authors are grateful to the Green Asia Program for financial support. The authors also would like to express their appreciation to the mines for kind assistance with field investigation.

REFERENCES

American Society for Testing and Materials (ASTM). (2005) *ASTM Standard D 4318-05*, Standard test methods for liquid limit, plastic limit, and plasticity index of soils, ASTM.

American Society for Testing and Materials (ASTM). (2007) *ASTM Standard D422-63(2007)e2*, Standard Test Method for Particle-size Analysis of Soils (withdrawn 2016), ASTM.

Chow, T.L. & Rees, H.W. & Fahmy, S.H. & Monteith, J.O. (2003) Effects of pulp fiber on soil physical properties and soil erosion under simulated rainfall. *Can. J. Soil Sci* 83(1): 109–119.

Dina, L. (2009) Prospect of keprok orange in kutai timur regency. *EPP* 6(1): 36–43.

George, A. & Pierret, A. & Boonsaner, A. & Christian, V. & Planchon, O. (2009) Potential and limitations of Payments for Environmental Services (PES) as a means to manage watershed services in mainland Southeast Asia. *Int. J. of the Commons* 3(1): 16–40.

Gratchev, I. & Towhata, I. (2013) Stress–strain characteristics of two natural soils subjected to long-term acidic contamination. *Soils and Foundations* 53(3): 469–476.

Hamanaka, A. & Shimada, H. & Sasaoka, T. & Matsui, K. & Miyajima, I. & Ichinose, M. (2013) Assessment of soil erosion at the rehabilitation area in Indonesian coal mine. *Proceedings of the Spring Annual Symposium on the Mining and Materials Processing, MMIJ Spring meeting*, 28–30 March 2013, Tokyo, Japan, pp.370–371.

Hasan, M.R. & Hasan, M.H. & Islam, M.R. & Razi, K.A.A. & Alam, S. & Abdullah, T. (2013) Changes in geotechnical properties of soil with pH in household and industrial waste dump site. *Int. J. of Applied Sciences and Engineering Research* 2(2): 119–127.

Holz, D.J. & Williard, K.W.J. & Edwards, P.J. & Schoonover, J.E. (2015) Soil erosion in humid regions: a review. *J. of Contemporary Water Research & Education* 154: 48–59.

Iwata, S. & Kida, T. (1997) Tsutinokannkyoukenn, Fujitekunosisutemu Co., Ltd.

Lin, C. & Tong, X. & Lu, W. & Yan, L. & Wu, Y. & Nie, C. & Chu, C. & Long, J. (2005) Environmental impacts of surface mining on mined lands, affected streams and agricultural lands in the dabaoshan mine region, southern China. *Land Degradation & Development* 16: 463–474.

Munksgaard, N.C. & Lottermoser, B.G. & Blake, K. (2012) Prolonged testing of metal mobility in mining-impacted soils amended with phosphate fertilisers. *Water Air Soil Pollut.* 223: 2237–2255.

National Environment Agency. (2011) Annual weather review 2011. http://www.nea.gov.sg/training-knowledge/publications/annual-weather-review-2011.

Shea, C.W. (1995) Regulating for the Long Term: SMCRA and Acid Mine Drainage. *J. of Natural Resources amp Environmental Law* 10(2): 193–216.

The Clay Science of Japan. (1987) Handbook of clay and clay minerals, Gihodoshuppan.

The Geotechnical Society of Japan. (2004) Geotechnical hand-book in Japanese, Japan, The Japanese Geotechnical Society.

Land Reclamation in Ecological Fragile Areas – Hu (Ed.)
 ISBN 978-1-138-05103-4

Trend change of soil nutrients in the green slope of an open pit coal mine dump

Q. Zhu, Z.Q. Hu, X.R. Liu, Y. Zhao & Y. Cai
China University of Mining and Technology, Beijing, China

ABSTRACT: This research chosen Baorixile Opencast Mine, Inner Mongolia in the grassland area as the study area, take samples in the reconstructed soil (1~5 years) of the dump slope to reveal the trend of soil nutrients variation in the reconstructed soil. Based on the determination the available potassium, available nitrogen, available phosphorus, TN and organic matter, the Principal Components Analysis (PCA) has been applied to calculate the Soil Nutrient Integrate Score (IFI) to reflect the change of soil nutrients with time. The results showed that the soil nutrients contents decreased in the first 3 years and then increased, and the variation range was 57.91 ~ 81.04. The IFI in the reconstructed soil of the dump slope was affected by the soil erosion and the surface accumulation of plants root. The accumulation rate of phosphorus and potassium in soil is accelerated by vegetation restoration, and then enhanced the soil IFI, which exceeded the original level after 5 years soil reconstruction.

1 INTRODUCTION

Opencast coal mining is an important form of coal production. China's large-scale opencast coal mines are mainly located in the northwest arid and semi-arid grassland area. And the process of opencast mining will inevitably and strongly influence the soil environment and the surrounding grassland eco-system, which caused a series of secondary environmental problems and have attracted the attention of global researchers (Liu Zhiyong 2011, Song Ziling 2007). Practically, the restoration measure of the dump slope is usually soil reconstruction followed with revegetation to increase the green space, enhance the dump slope stability and reduce dust (Liu Chunlei 2011, Tai Peidong 2002). The main characteristics of the opencast coal mine dump slope including: the precipitation in grassland area is low and concentrated; drought and strong winds in spring and autumn; short frost-free period (plants can only grow in June to September); slope angle of dump is usually close to natural repose angle. The study on soil nutrients changes in sloping field is mainly concentrated on the slope farmland in the southeast China recently. *Xin Yan* compared the effects of different tillage methods on soil nutrient loss in slope farmland in Liaoning province, and suggested that the soil erosion can be effectively decreased by reducing the surface disturbance, increasing the surface roughness and reducing the runoff intensity (Xin yan 2013). *Xu Zhiyou* monitored the soil nutrients of conversion of cropland to grassland in the loess hilly area of southern Ningxia province found that the content of soil nutrients increased with the decrease of position on the same slope (Xu Zhiyou 2010). *Li Jianqiang* analyzed the content and composition of slope farmland soil nutrients in the upper reaches of the *Yangtze river,* and concluded that planting shrub or shrub-grass hedgerow on slope along contour line significantly improved soil nutrient content and mitigated soil erosion (Li Jianqiang 2010). The existing research rarely studied on changes of soil nutrients in the soil of the dump slope in the opencast coal mine.

Principal components analysis is usually used for soil quality evaluation which can eliminate the errors caused by auto-correlation and get the comprehensive evaluation score to evaluate the quality of soil nutrient accurately (Wu Yuhong 2010). Therefore, this research determined the available potassium, available nitrogen, available phosphorus, TN and organic matter in the topsoil of dump slope in *Inner Mongolia Baorixile* Surface Mine, which has been reconstructed 1 to 5 years and used Principal Components Analysis (PCA) to calculated the Soil Nutrient Integrate Score to reflect soil nutrients changes with time, to reveal the trend of nutrients variation in the reconstructed soil in the opencast coal mine dump slope in grassland area and provide theoretical basis for opencast coal mine dump revegetation.

2 RESEARCH AREA PROFILE

Hulun Buir is located in the northern temperate which has continental climate. *Baorixile* mining area is located in *Hailar* District of *Hulun Buir*

City whose area is about 220 km^2, which belongs to the high plain area of northern *Inner Mongolia*, and the terrain is relatively flat and open, slightly undulating, the relative elevation is 60 ~ 106 m, its soil mainly composed are Quaternary sand and silty sand. The average annual temperature is –1.5 to 6.5°C, the coldest month (January) the average temperature is –18 to 30°C, the hottest month (July) the average temperature is 16 to 21°C. Average annual rainfall is 190 ~ 406.3 mm, the precipitation of Winter and spring is about 40 to 80 mm, while that of summer is about 200 to 300 mm, accounting for 65% to 70% of annual precipitation. The covering soil is the original soil and then planted Elymus and Agropyron on it. The water required for plant growth entirely from natural rainfall after 5 years except extremely dry weather.

3 RESEARCH METHODS

3.1 *Sample collection*

The topsoil samples of dump slope of 2011, 2012, 2013, 2014 and 2015 were collected in May. The sampling lines were laid at the bottom, middle and top of the slope, samples were collected every 50 m along the sampling line, and the coordinates of sampling points were recorded by portable GPS. Set three sampling points in the original soil stacking area and collected 3 samples at each sampling point and mixed them. Soil samples with depth of 0 to 10 cm were collected in sealed bags.

3.2 *Test methods*

Remove the stones and grass roots of the samples, and then air-dried and grinded the samples, screened the soil with a 100 mesh, and tested available phosphorus, available potassium, total nitrogen, available nitrogen and organic matter. The available potassium was extracted by neutral ammonium acetate-atomic spectrophotometry. The determination of available phosphorus was carried out by using HCl-H_2SO_4 extraction-spectrophotometry. The total nitrogen was measured by sulfuric acid-accelerator, and the determination of organic matter was determined by $FeSO_4$ solution after oxidation with potassium dichromate-H_2SO_4.

3.3 *Data analysis*

The data were processed and plotted with Excel 2010, The correlation between the indexes of soil nutrient was analyzed by SPSS 20.0, and then used factor analysis to screen out the principal components, calculated the actual contribution rate of each soil nutrient index (Huang An 2014).

Principal components analysis is widely used in the comprehensive evaluation of multiple indicators. This method is based on the idea of dimensionality reduction, which can eliminate the correlation between the evaluation indicators and the work of selecting synthesis method, descript the relative status of the sample objectively (Zhang Zilong 2013). The method of comprehensive evaluation of soil nutrient content can be summarized as follows: (1) Standardize the original index data; (2) Determine the number of principal components; (3) Explain the meaning of principal component; (4) Calculate the Soil Nutrient Comprehensive Score.

$$IFI = \lambda_1 \cdot F_1 + \lambda_2 \cdot F_2 + \cdots + \lambda_n \quad (1)$$

F represents a single principal component score; λ represents the contribution rate of the principal component.

The soil quality can be measured according to the scores of soil samples on a principal component and comprehensive index.

4 RESULT ANALYSIS

4.1 *Descriptive analysis*

Sample test results in Table 1 showed that the nutrient contents of different soil samples were different. The variation coefficients of five soil nutrient indexes: available phosphorus> total nitrogen> available potassium> available nitrogen> organic matter. The variation coefficients of available phosphorus is 0.74, which is far more than others. The degree of

Table 1. Soil nutrient content.

Value	Available K/(mg/kg)	Available P/(mg/kg)	TN/(g/kg)	Available N/(mg/kg)	Organic matter/(g/kg)
Maximum	377.85	30.17	3.25	135.94	3.69
Minimum	119.13	2.13	0.53	28.90	1.04
Mean	194.31	7.96	1.28	89.10	2.54
Standard deviation	63.83	5.85	0.48	25.02	0.62
Variation coefficient	0.33	0.74	0.37	0.28	0.24

variation of the five indicators was graded according to the size of the variation coefficients: $Cv \leq 0.1$ is weak variability, $0.1 < Cv < 0.3$ is medium variability, $Cv > 0.3$ is strong variability. The table showed that available phosphorus, available potassium, total nitrogen are strong variability; alkali nitrogen, organic matter Cv was a medium variability which is lower than 0.3. The results show that the covering soil nutrient content has changed greatly with time.

4.2 *Principal component analysis*

The key factor of comprehensive analysis of principal component analysis is the actual contribution rate of each participating factor to the total score, which is the main factor to determine the weight of each factor. The actual contribution rate of each factor depends on the principal component load obtained by principal component analysis, and the comprehensive score of each point is calculated by the formula.

4.2.1 *Correlation analysis*

Principal component analysis requires correlation between the indicators. Therefore, it is necessary to test the correlation of soil nutrient index before the principal component analysis.

It can be seen from Fig. 2 that there was a significant correlation between organic matter, available nitrogen and total nitrogen ($P < 0.01$), and available phosphorus were positively and very significantly correlated with available potassium ($P < 0.01$), which was consistent with the prerequisites for factor analysis.

This study tested the correlation between variables. The result showed that the KMO value was 0.691, which was less than 7 and Bartlett sphericity test showed P is less than 0.001, which indicated that variables was correlated so factor analysis can be carried out.

4.2.2 *Principal component analysis*

The total variance of 5 soil nutrient index correlation matrix showed that in which whose characteristic root is greater than 1 is main component. As shown in Table 3, there was 2 principal components whose cumulative contribution rate was 90%, which means the 2 principal components contained 90% of the total amount of information provided by the original data. Therefore, the first 2 principal components were used as the comprehensive variables to evaluate the soil nutrient content.

This paper used 0.5 as the threshold value of the load coefficient which in Principal Component Analysis (PCA) usually is 0.3 ~ 0.5. The variance contribution of the first principal component, which represented the integral effect of soil nitrogen and carbon, was 63.9% and was the highest and the most important factors. That was result from the high load of total nitrogen, available nitrogen and organic matter. The variance contribution of the second principal component was 25.6%, which was the secondary factor, and the principal component of the total potassium and available phosphorus was more than 0.5, which could be regarded as the main component of the comprehensive factor of P and K in soil.

The 2 principal component scores coefficient matrix (Table 5) was obtained by calculating the rotating load, which can be used to obtain the linear equation of 2 principal components.

$$F_1 - 0.399 \times Z_1 + 0.357 \times Z_2 + 0.370 \times Z_3 + 0.133 \times Z_4 + 0.009 \times Z_5 \tag{2}$$

$$F_2 - 0.159 \times Z_1 + 0.012 \times Z_2 + 0.045 \times Z_3 + 0.567 \times Z_4 + 0.542 \times Z_5 \tag{3}$$

Table 3. The characteristic root and the contribution of variance of each indicator.

	Initial eigenvalue		
Indicator	Characteristic root	Contribution of variance /(%)	Cumulative contribution /(%)
1	3.197	63.949	63.949
2	1.279	25.589	89.537
3	0.273	5.458	94.995
4	0.217	4.340	99.335
5	0.033	0.665	100.000

Table 2. Correlation coefficient of soil nutrients.

	TN	Available N	Organic matter	Available K	Available P
TN	1				
Available N	0.788**	1			
Organic matter	0.766**	0.963**	1		
Available K	–0.218	–0.400*	–0.337*	1	
Available P	–0.248	–0.451**	–0.421*	0.770**	1

**Significant correlation at .01 level (bilateral).
*Significant correlation at .05 level (bilateral).

Table 4. Principal components load.

	The first principal component	The second principal component
TN	0.794	0.449
Available N	0.938	0.262
Organic matter	0.914	0.305
Available K	–0.627	0.701
Available P	–0.677	0.652

Table 5. The component score coefficient matrix.

	Component	
	1	2
TN	0.399	0.159
Available N	0.357	0.012
Organic matter	0.370	0.045
Available K	0.133	0.567
Available P	0.009	0.542

Taken the variance contribution rate of each principal component as the weight, then the scores of the soil samples in the 2 principal components can be calculated according to Formula 1. The specific evaluation model is:

$$IFI = 0.639 \times F_1 + 0.256 \times F_2 \tag{4}$$

4.3 *Change characteristics of soil nutrient IFI over time*

The general statistics of comprehensive score of 45 soil samples showed that:

Table 6 shows that the comprehensive score of soil nutrient is: 5a >3a > original soil >1a >2a >4a; the covering years of the soil nutrient comprehensive score of the overlying soil from 1 to 5 years decreased firstly and then increased except for that of 3 years. The soil comprehensive score of 3 years was significantly higher than that of other years. It was probably because the slope angle of covering soil for 3 years was smaller than that of other slopes, and its Soil nutrient loss rate was slow.

Table 6 and Figure 1 shows that the comprehensive score of slope topsoil decreased slowly after covering soil for 1 to 4 years, no significant difference with the topsoil. And the comprehensive score of slope surface soil after covering soil for 5 years significantly higher than the original soil, which means that planting plant on slope soil can gradually restore and even improve the original soil fertility after 5 years. The soil comprehensive score of the second main components was also increased, which indicated that soil nutrients improvement was mainly caused by the increase of phosphorus and potassium content in soil.

It can be seen from Figure 2 that the change of the first principal component score is small in fifth years, and there is no significant difference between

Table 6. The average soil integrate score (IFI) in Different reclamation years.

	The first principal component	The second principal component	IFI
Original soil	70.85a	77.94b	70.80ab
1a	66.06a	100.49b	67.94ab
2a	55.60ab	92.63b	59.24b
3a	65.59a	127.71ab	74.61ab
4a	50.91b	99.14b	57.91b
5a	59.76ab	167.39a	81.04a

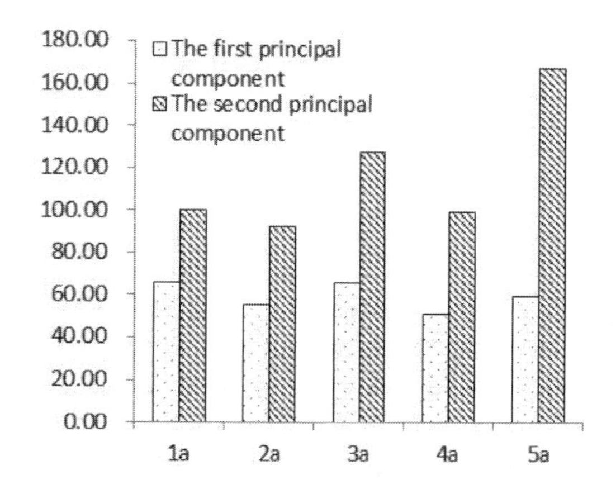

Figure 1. Soil IFI in different years.

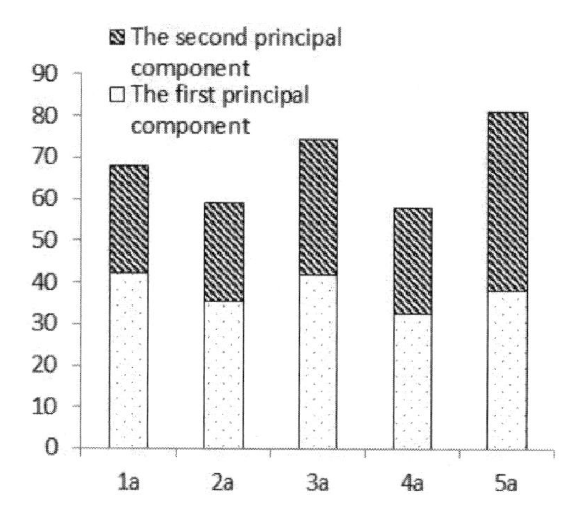

Figure 2. The score of each principal component.

the two years except for the external covering soil in the last fourth years. The results showed that the content of carbon and nitrogen in soil remained stable, and the change of the first principal component score was stable in five years after covering soil. The second principal component score significantly decreased in the first 2 years after covering soil and then increased, which means that phosphorus and potassium elements in soil lost and accumulated in the same time, and the accumulation rate is greater than the loss rate after covering soil 2 years.

Zhang Aiqun believes that plant roots secrete organic acids to activate Al-P, Fe-P and Ca-P in soil, which improved soil phosphorus availability and promoted the accumulation of soil O-P, and lead to soil available phosphorus content increasing, which was called the soil surface accumulation of available phosphorus. The Al-P, Fe-P and O-P in the soil would convert to water-soluble phosphorus and easily loss in the rain leaching condition. The soil nutrient content decreased in the first 1 to 3 years after soil covering, which was because of not completely recovered Vegetation, weak Secretion of organic acids and low fertility of vegetation and low soil microbial activity. Vegetable fertilizer function gradually restored with the extension of the covering ages. Meanwhile, the ability for soil to retain and maintain exogenous phosphorus enhanced, and the organic phosphorus in the flora and fauna caused by vegetation restoration can promote the soil phosphorus content by the microbial decomposition and mineralization, which lead the soil available phosphorus content increased rapidly (Zhang Aiqun 2009). Available potassium has a similar effect to available phosphorus, but the potassium salt can dissolve in water and it would lose with the erosion of rain, so the increase rate of that behind the available phosphorus.

5 CONCLUSION

1. The nutrient contents in soil had obvious differences in different soil cover ages, the variation coefficients showed: available phosphorus > total nitrogen > available potassium > available nitrogen > organic matter, the available phosphorus variation coefficient was 0.74 which was the maximum, the other indexes were closed to the strong variability threshold (Cv = 0.3).
2. The soil nutrient comprehensive score which was calculated by PCA showed that the main component score and the comprehensive score of slope topsoil changed significantly in different soil cover years, no significant change between adjacent years, which means that soil nutrient changes slowly.
3. The slope topsoil nutrient comprehensive score (IFI) decrease firstly and then increased, the soil score after covering 4 years was 57.91 which was minimum among the five, and it significantly increased to 81.04 after 5 years which exceed that of original soil. The comprehensive factors of carbon and nitrogen changed small, and that of phosphorus and potassium decreased slowly and then increased rapidly. The accumulation of phosphorus and potassium in covering soil increased with time and vegetation restoration, the accumulation rate gradually exceeds the loss rate, then the soil nutrient increased.

REFERENCES

Huang An, Yang Lian'an, Du Ting, et al. 2014. Comprehensive Assessment of Soil Nutrients Based on PCA[J]. *Arid Zone Research*, 31(5):819–825.

Li Jianqiang, Zhang Hongjia, Cheng Jinhua, et al. 2010. Effects of hedgerow intercropping with different species on soil nutrients and its distribution on steep land in upper reaches of Yangtze River[J]. *Ecology and Environmental Sciences*, 19(11):2574–2580.

Liu Chunlei, Wang Jinman, Bai Zhongke, et al. 2011. Analysis of land reclamation technologies for surface coal mine in arid grassland[J]. *Metalmine*, 5: 154–157.

Liu Zhiyong & Pei Guodong, 2011. The present situation and development of outdoor coalmine enterprise in our countrys[J]. *Science and Technology Information*, 4(2): 102.

Song Ziling. Study on the Design Theory and Method for Modern Surface Mining[D]. Fuxin: Liaoning Technical University, 2007.

Tai Peidong, Sun Tieheng, Jia Hongyu, et al. 2002. Restoration for refuse dump of open-cast mine in steppe region[J]. *Journal of Soil and Water Conservation*, 16(3): 90–93.

Wu Yuhong, Tian Xiaohong, Tong Yan'an, et al. 2010. Assessment of integrated soil fertility index based on principal components analysis[J]. *Chinese Journal of Ecology*, 29(1): 173–180.

Xin yan, Wang Xuan, Qiu ye, et al. 2013. Effects of Different Tillage Modes on Soil, Water and N,P Nutrient Loss on Sloping Croplands in Liaoning Province[J]. *Journal of Soil and Water Conservation*, 27(1):27–30.

Xu Zhiyou, Yu Feng, Gao Hongjun, et al. 2010. Variation Laws of Soil Nutrients of Returning Cultivated Land to Woodland or Grassland at Different Slope Positions of Semi-Arid Gullied Rolling Loess Area[J]. *Soil and Water Conservation in China*, 08:39–41+68.

Zhang Aiqun, He Liyuan, Zhao Hui'e, et al. 2009. Effect of organic acids on inorganic phosphorus transformation in soils and its readily available phosphate[J]. *Acta Ecologica Sinica*, 29(8):4061–4069.

Zhang Zilong, Wang Wenquan, Miu Zuoqing, et al. 2013. Application of Principal Component Analysis in Comprehensive Assessment of Soil quality under Panax Notoginseng Continuous Planting[J]. *Chinese Journal of Ecology*, 32(6):1636–1644.

Land Reclamation in Ecological Fragile Areas – Hu (Ed.)
 ISBN 978-1-138-05103-4

Collaborative efforts towards ecological habitat restoration of a threatened species, Greater Sage-Grouse, in Wyoming, USA

Peter D. Stahl & Michael F. Curran
Wyoming Reclamation and Restoration Center, University of Wyoming, Laramie, Wyoming, USA

ABSTRACT: Wyoming is the least populated state in the U.S. with ~ 250,000 km^2 of relatively undisturbed ecosystems and wildlife habitat as well as an abundance of natural resources. Increasing demand for natural resources has resulted in more resource extraction activities, including surface mining and oil & gas drilling, and associated disturbance in these ecosystems, which, along with wildfire, invasive plants, and infrastructure development have lead to decreases in both the amount of habitat available and the quality of remaining habitat. These are thought to be the primary factors responsible for 50% population loss of Greater Sage-Grouse (*Centrocercus urophasianus*) in its historical range. The U.S. Fish and Wildlife Service (USFWS), in 2010, determined the Greater Sage-Grouse was warranted for Endangered Species listing but precluded because other species were under more severe threat. Inclusion of the Greater Sage-Grouse as an Endangered Species would likely result in considerable negative economic impacts in the state of Wyoming which depends on extractive natural resource production for economic health. To conserve the Greater Sage-Grouse and it's habitat in Wyoming, the state developed a natural resource policy titled "Sage-Grouse Core Area Strategy and Governor's Executive Order for Sage-Grouse" which involves protecting existing wildlife habitat and ecological restoration of habitat affecting by resource extraction and other disturbance factors mentioned above. The policy being carried out was developed by a collaborative team of state and federal agencies, industry, non-governmental organizations, and private landowners known as the Sage-Grouse Implementation Team. Specifics of restoration efforts are discussed.

1 WYOMING ENVIRONMENT, ECOSYSTEMS, ECONOMY AND GOVERNANCE

Wyoming is the least populated state in the U.S. with ~ 250,000 km^2 of relatively undisturbed ecosystems and wildlife habitat as well as an abundance of natural resources. The state is located in the Rocky Mountain region of the northern United States and includes several large mountain ranges separated by high elevation intermountain basins. The entire state is located over 940 meters in elevation with an average elevation of 2040 m. Wyoming has a semi-arid climate with precipitation ranging from 15 to 40 cm (6–16 in) annually but averages 37 cm (14.5 in) per year. Large amounts of snow fall in the mountains making the Wyoming a cold, dry state with long winters and short growing seasons from 42–145 days. Precipitation is the most common factor limiting ecosystem restoration efforts.

Mountainous areas of Wyoming support large coniferous forests dominated by Ponderosa Pine, Lodgepole Pine, Douglas Fir and Subalpine Fir but vegetation in the basins is dominated by large expanses of sagebrush-grasslands, also know as sagebrush steppe. The most important plant species in these areas are the various subspecies of the aromatic shrub Big Sagebrush (Basin Big Sagebrush, Mountain Big Sagebrush, and Wyoming Big Sagebrush) and grasses such as Western Wheatgrass, Green Needlegrass, Thickspike Wheatgrass, Buffalograss and Indian Ricegrass. Many wildlife species such as Pygmy Rabbits, Sagebrush Lizards, Golden Eagles, Badgers, Pronghorn Antelope, and Elk thrive in the sagebrush grasslands. Population numbers of several wildlife species occurring in sagebrush grasslands such as Greater Sage-Grouse and Mule Deer, however, for various reasons are declining. The two most common land uses in Wyoming are wildlife habitat and domestic livestock grazing. For this reason, the majority of ecosystem restoration work is restoring native sagebrush-grassland plant communities back to their original condition.

Wyoming's economy is dominated by natural resource production, particularly mined products like coal, uranium, and sodium carbonate. Wyoming leads the nation in production of these materials and also produces large amounts of oil and natural gas. Extraction of all of these materials from the earth causes disturbance of the ecosystems on the earth's surface lying above them legally requiring restoration. Another very important industry in Wyoming is tourism. People from across the United States and the world travel to Wyoming to visit famous National Parks like Yellowstone, hike in the mountains and watch the abundant wildlife.

Lands and natural resources in Wyoming are governed by both the state of Wyoming and the United States Federal Government. The U.S. Federal Government owns and controls 48% of the land within Wyoming borders and primarily regulates natural resource development and extraction as well as land reclamation and habitat restoration on those lands. The state of Wyoming Department of Environmental Quality regulates natural resource extraction and reclamation on state and private lands. Surface coal mining is regulated by the state of Wyoming within guidelines set by U.S. federal law (the Surface Mine Control and Reclamation Act of 1977). A number of other agencies are involved in natural resource production, environmental quality, and wildlife management in Wyoming; including, at the federal level, the U.S. Bureau of Land Management, U.S. Forest Service, U.S. Fish and Wildlife Service, and at the state level the Wyoming Oil and Gas Conservation Commission, the Wyoming Department of Game and Fish, and the Wyoming State Lands Department.

2 LOSS OF SAGEBRUSH-GRASSLANDS WILDLIFE HABITAT

The greater sagebrush ecosystem covers much of the western United States and remains one of the largest relatively intact biomes in the U.S. but once covered over 60 million ha. Huge tracts of this native vegetation type remain but large portions of the sagebrush grassland have been lost to agricultural conversion, sagebrush eradication, urbanization, wildfire, invasive species and natural resource extraction. Sagebrush-grasslands provide habitat to a large number of species including the Greater Sage-Grouse, pronghorn antelope, the Sage Thrasher, mule deer and pygmy rabbits. Some of these species are sagebrush dependent and cannot live without this shrub. A number of these species are in decline, which is thought to be related to loss of quality sagebrush-grassland habitat

The population of Greater Sage-Grouse (Centrocercus urophasianus) has decreased in the western United States from it's historically high numbers. The range of this species has declined by half primarily due to human activities reducing the amount of quality sagebrush habitat (Shroeder et al., 2004; Harju et al., 2010). Because of facts like these, the U.S. Fish & Wildlife Service is considering placing the Greater Sage-Grouse on the Endangered Species List to protect the species. Natural resource development such as mining and oil & gas drilling have had significant negative impacts on sage-grouse habitat and sage-grouse populations in a number of western states, including Wyoming (Holloran, 2005; Aldridge and Boyce, 2007). Listing of the Greater Sage-Grouse as an endangered species would most likely result in additional federally mandated restrictions on disturbances in sage-grouse habitat which would curtail the most important economic activities in Wyoming, natural resource production, and have very serious economic impacts.

Current state and federal regulations requiring reclamation of sites disturbed by natural resource extraction has reduced the amount of habitat negatively impacted by this activity. Wyoming has been a leader among western states in developing strategies to better manage natural resource development and sagebrush-grassland wildlife habitat that result in benefits for both.

3 SAGE-GROUSE IMPLEMENTATION TEAM

The State of Wyoming created a Sage-Grouse Implementation Team (SGIT) in 2014 to review data and make recommendations to the Governor regarding actions, funding, and regulatory actions necessary to maintain and enhance sage-grouse populations and sage-grouse habitats in Wyoming. The SGIT was formed to broadly represent the major interests found in the state of Wyoming and includes representatives of the Agricultural Industry, Oil & Gas Industry, Mining Industry, Wind Generation and Transmission Industry and Conservation or Sportsmen's Groups. Also represented are the following Wyoming State agencies: Wyoming Game and Fish Commission, Wyoming Department of Agriculture, Wyoming Department of Environmental Quality, Wyoming Wildlife and Natural Resource Trust Fund Board, Wyoming Oil and Gas Conservation Commission, Wyoming Office of State Lands and Investments. Wyoming House of Representatives, Wyoming Senate. The SGIT has a mandate to work with the following Federal Agencies which also have sage-grouse conservation as a high priority issue and regularly attend the monthly SGIT meetings: United States Bureau of Land Management, United States Fish and Wildlife Service, United States Forest Service, and the United States Natural Resource Conservation Service. A number of special interest groups are also participating in the statewide sage-grouse conservation effort and include and include the Wyoming Stockgrowers Association, the Petroleum Association of Wyoming, the Wyoming Mining Association, The Nature Conservancy, the Thunder Basin Grassland Prairie Ecosystem

Association and the Audubon Society. A number of local Sage-Grouse Working Groups were also formed around the state to address local issues regarding sage-grouse conservation.

4 SAGE-GROUSE EXECUTIVE ORDER

The state of Wyoming has developed a strategy to conserve sage grouse and minimize human disturbance to their habitat in key areas of the state that also provides for continued production of natural resources. Implemented in 2008, the Governor's Sage-Grouse Executive Order (SGEO) involves a comprehensive system for limitation and management of land surface disturbing activities within designated sage-grouse core population areas (State of Wyoming, 2008). Land disturbance is tightly regulated and disturbed lands can account for no more than 5% of the total land area in a sage-grouse conservation core area, no disturbances are permitted or allowed within 1.0 km of a sage-grouse lek and disturbance density should not exceed one per square mile (one per 640 acres; State of Wyoming, 2011). Sage-grouse populations not located in core areas are not protected by the same conservation policies (Gamo, 2013). In non-core areas, sage-grouse and their habitat are safeguarded by a 0.4 km (0.25 mile) preserved area around leks and a 3.3 km (2 mile) seasonal buffer zone during the spring mating season when leks are active.

The SGEO also gives general stipulations designed to maintain existing sage-grouse habitat by permitting development activities in core areas in a way not causing declines in sage-grouse populations (State of Wyoming, 2011). These stipulations include requirements for native plant community restoration of disturbance associated with natural resource extraction activities and wildfire on public lands. Specific stipulations are given to Oil & Gas producers as well as Miners and include limitations on density of well pads and mine operations (one per square mile) as well as amounts of suitable habitat disturbed (State of Wyoming, 2011).

5 SPECIFIC EXAMPLES OF COLLABORATIVE HABITAT RESTORATION EFFORTS

The Douglas Core Area Restoration Team (DCA RT) was formed by the Wyoming Governor's Office in 2013 to reduce the high level of disturbed land in the Douglas Core Sage-Grouse Conservation Area (DCA). Due to large wildfires in the area in the past 25 years a substantial amount of land in this area does not have sagebrush and is considered disturbed. There has also been heavy pressure to drill oil wells in the DCA as there are known reserves underground. Because the SGEO requires Sage-Grouse Conservation Core Areas to maintain disturbance levels under 5% of the total land area, the DCA RT was charged with restoring disturbed sage-grouse habitat within the DCA to "suitable" sage-grouse habitat as defined by the SGEO. The overall goal of the DCA RT is to Identify and implement projects that will maintain, enhance or restore sage-grouse seasonal habitats within the DCA to maintain sage-grouse populations in the DCA.

The DCA RT was assembled by the Wyoming Governor's Office and currently consists of 27 members representing State and Federal government agencies, an oil company operating in the area, special interest groups such as the Stock Growers Association, the Audubon Society, and the Nature Conservancy. All of these groups participate in the choice, planning and conduct of the projects implemented by the DCA RT. One of the most important activities so far executed are sagebrush plantings in burned portions of the DCA. Over 30,000 sagebrush seedling have been planted in the DCA to restore suitable sage-grouse habitat and serve as seed sources for further sagebrush reestablishment. The DCA RT is also working to control invasive species, especially cheatgrass, where they are problematic and testing new sagebrush seeding technologies.

Another example of a collaborative habitat restoration project is the Wyoming Oil and Gas Reclamation Database (WOGRD). The database was developed by the Wyoming Reclamation and Restoration Center at the University of Wyoming with the cooperation and involvement of 20 oil and gas companies, private environmental consulting firms, the Petroleum Association of Wyoming, as well as state and federal government agencies. The database collects and stores data on land disturbance associated with oil and gas drilling in Wyoming as well as progress on reclamation efforts and the methods used to mitigate the disturbance. This database has been crucial to evaluate the environmental impacts of and assess reclamation and restoration efforts associated with the recent oil and gas boom in Wyoming. The US Fish and Wildlife Service used the WOGRD in their process of coming to a non-warranted Endangered Species Act listing decision for the Greater Sage-Grouse.

REFERENCES

Aldridge, C.L. & Boyce. M.S. (2007) Linking occurrence and fitness to persistence: habit-based approach for endangered sage-grouse. *Ecological Applications*, 117:508–526.

Gamo, S.R. (2016) *Effectiveness of Wyoming's Sage-Grouse Core Areas in Conserving Greater Sage-Grouse and Mule Deer*. Dissertation, University of Wyoming, Laramie.

Harju, S.M., Dzialak, M.R., Taylor, R.C., Hayden-wing, L.D. & Winstead, J.B. (2010) Thresholds and time lags in effects of energy development on Greater Sage-Grouse populations. *Journal of Wildlife Management*, 74:437–448.

Holloran, M.J. (2005) *Greater Sage-Grouse (Centrocercus urophasianus) population response to natural gas field development in western Wyoming*. Dissertation, University of Wyoming, Laramie.

Schroeder, M.A., Aldridge, C.L., Apa, A.D., Bohne, J.R., Braun, C.E., Bunnell, S.D., Connelly, J.W., Deibert, P.A., Garnder, S.C., Hilliand, M.A., Kobriger, G.D., McAdam, S.M., McCarthey, C.W., McCarthy, J.J., Mitchell, D.L., Rickerson, E.V., & Stiver, S. J. (2004) Distribution of sage-grouse in North America. *Condor*, 106:363–376.

State of Wyoming. (2008) Office of Governor Freudenthal. *State of Wyoming Executive Department Executive Order. Greater Sage-Grouse Area Protection*. 2008–02.

State of Wyoming. (2011) Office of Governor Mead. *State of Wyoming Executive Department Executive Order. Greater Sage-Grouse Area Protection*. 2011–05.

Land Reclamation in Ecological Fragile Areas – Hu (Ed.)

Technology and application for ecological rehabilitation on self-maintaining vegetation restoration

Y. Zhao
State Environmental Protection Engineering Center for Ecological Restoration on Destroyed Surface, Beijing, China
Louis Ecological Engineering Co. Ltd., Beijing, China

L.J. Chai
State Environmental Protection Engineering Center for Ecological Restoration on Destroyed Surface, Beijing, China

J. Chen
Inner Mongolia Geological Environment Monitoring, Hohhot, China

X.M. Wang
State Environmental Protection Engineering Center for Ecological Restoration on Destroyed Surface, Beijing, China
Louis Ecological Engineering Co. Ltd., Beijing, China

ABSTRACT: The technology for self-maintaining vegetation restoration is based on the restoration or reconstruction of soil habitat system, vegetation community system and substance cycle system for destroyed surfaces. The technology is used widely to rehabilitate damaged vegetation of highway and railway slopes, mines in the northern freezing-thawing, semi-arid and southern humid regions. After the applications of the technology in various regions, the vegetation community becomes higher compatibility with the surrounding native vegetation, and vegetation coverage of demonstration projects could reach more than that of native vegetation. A self-maintaining vegetation system can be derived from the damaged vegetation without artificial aftercare.

1 INTRODUCTION

Vegetation plays an important role in soil and water conservation, carbon sequestration process in ecosystem (Li, 2004). Nowadays, Artificial restoration of vegetation is considered as one of the most effective approaches to improve ecosystem health. The composition of plant community is different on different restoration stages and the diversity would increase as restoration time increasing (Shen & Zhang, 2015). However, unreasonable artificial vegetation restoration under specific environments would induce vegetation degradation and plant diversity decreasing. Additionally, the effects of artificial vegetation restoration reported could not be self-sustaining.

The paper will introduce a new technology for vegetation restoration, which refers to the reconstruction of soil habitat system, vegetation community system and substance cycle system. As restoration time increasing, soil physicochemical, microbial properties and plant diversity show an initial increasing at early stage and reach a stable status at late stage, and a self-maintaining vegetation system can be derived from the damaged vegetation system without artificial aftercare.

2 SELF-MAINTAINING VEGETATION RESTORATION

2.1 *Reconstruction of soil habitat system*

The soil is a loose layer of porous material on the surface of the land, and is direct supplier of medium and nutrients for plant growth (Paniagua & Kammerbauer, 1999). The reconstruction of soil habitat, is mainly based on the use of functional materials, which can enhance soil nutrients, water-stable and the growth of plants. The reconstructed soil habitat system including humus layer, topsoil layer and subsoil layer, not only can be stably attached to the bedrock layer, which has many advantages such as a certain mechanical strength, no crack and anti-erosion performance, but also supply sufficient nutrient and acid-alkali buffering capacity. Due to complex and various soil conditions in China, it is necessary for soil characteristics to improve soil quality by using targeted materials. The modified planting layer, which is constructed on the destroyed surfaces to form functional habitat of plant growth by spraying technology, could be consistent with the surrounding natural soil, and also has some engineering mechanical properties.

2.2 *Reconstruction of vegetation community system*

Vegetation can greatly influence above-ground plant community structure and diversity during restoration process. Meanwhile, it can impact the properties of the below-ground ecosystem via enhancing litter and root input; increasing soil nutrients accumulation, improve soil physical structure and enhancing soil microbial biomass and activity during vegetation restoration process. The vegetation community is constructed by basing on ecosystem function and natural vegetation community structure (Li, 2004), which contains the diversity of plant species, characteristics of specific seasonal features, zonal and dynamic succession, horizontal and vertical vegetation community structure (such as the diversity and combination of arbor-shrub-grass) and seasonal succession of vegetation community (such as pioneer species, native species, and niches of each species).

Different vegetation types have different impacts on soil physicochemical and microbial properties because of their different growth patterns (Yin & Yin, 2005). The distribution and growth of plants are closely related to the natural conditions, such as climate and soil types. The types of mine landforms are complex and diverse, and the differences of vegetation communities in different mines are obvious. Therefore, the self-maintaining vegetation restoration uses the selected native plants to emphasize the adaptation of the regional climate and soil. Taking full account of intraspecific and interspecific competition of plant species, native arbor-shrub-grass combinations are to recreate a climax community, and restore naturally self-growing and self-maintaining.

2.3 *Reconstruction of substance cycle system*

Soil micro organisms are decomposers and transformers of soil organic materials (White *et al.*, 2005). Organic materials provided by various micro organisms, which decompose residues of animals and plants (such as litter), is main energy source for plant growth (Zak *et al.*, 2003). Microorganisms can absorb and decompose organic materials to enhance material circulation and energy flow between vegetation community system and soil habitat system. Additionally, the physiological activities of micro organisms also can affect soil formation and development to improve soil physical and chemical properties (Gros *et al.*, 2004). For example, some kinds of soil micro organisms can transform nitrogen in the air into the fixed state of nitrogen which plant can directly utilize.

The construction of substance cycle system is based on the study of distribution of native soil microbial community. The applications of functional microbial products can fix nitrogen, release phosphorus and potassium, and then consistently supply nutrient for the growth of plants. While the growth and prosperity of plants can continually provide carbon source to facilitate soil micro organisms, plants and micro organisms establish mutually beneficial substance circulation system, and continuously improve soil fertility.

3 APPLICATIONS OF SELF-MAINTAINING VEGETATION RESTORATION

An ecological restoration demonstration project of Jingxin Expressway is at the beginning of Hanjiaying (Pile No. K0+105.258) and the end of Chahar of Wulanchabu (Pile No. K72+050) with a total length of 70.67 km and an engineering area of 187,250 m^2. There is a typical Mongolian plateau continental climate with an annual average rainfall of 384 mm and evaporation of 2,400 mm. The soil of sunny slopes has poor permeability and water stability, and there was very sparse vegetation. Bare and shallow soil is further easy to cause soil erosion. The technology of self-sustaining vegetation restoration was applied to the sunny slopes under fully considering slope stability, landscape, safe and comfortable driving.

The best combination of plant species is *Robinia pseudoacacia*, *Ulmus pumila*, *Salix psammophila*, *Lespedeza bicolor*, *Agropyron cristatum*, *Melilotus suaveolens* and other six kinds of plant species. According to the reconstruction of soil habitat, vegetation community and substance circulation system, vegetation coverage of the sunny slope increases from 70% to 90%, and the diversity of plant species show an upward trend from 2013 to 2016 (Figure 1). Four kinds of sunny soil enzyme activity (sucrose, catalase, urease and celluase) are better than those of shady slope (Figure 2). For example, sunny soil sucrase activity increases from 6.91 mg/g·24 h to 10.12 mg/g·24 h, which is higher than 9.68 mg/g·24 h of shady soil in 2015. Road landscape of the demonstration project is remarkable.

Figure 1. The succession of a self—maintaining vegetation system of the demonstration project.

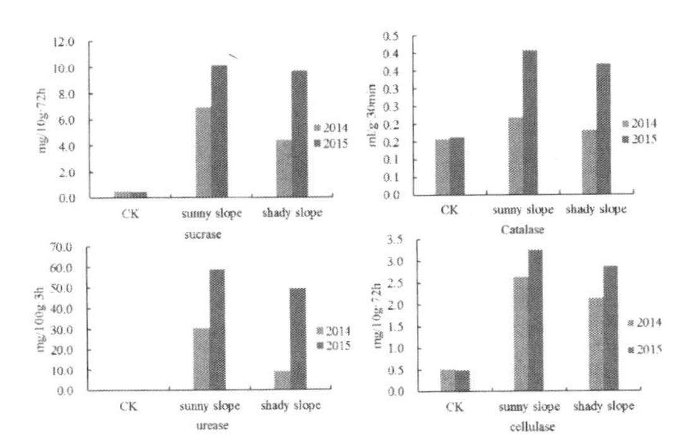

Figure 2. Four kinds of enzyme activity of the sunny soil and shady soil.

4 SUMMARY

Nowadays, artificial restoration of vegetation is considered as the most effective approaches to improve ecosystem health. Taking full consideration of ecological principles, landscape, slope stability and cost-effect, the self-maintaining vegetation restoration technology was applied to many projects. The effects of many vegetation restoration projects were obviously remarkable. The technology can enhance stability of destroyed surfaces, and increase soil nutrients accumulation and vegetation coverage. The ecosystem damaged finally develops to be free after-care vegetation ecosystem the same as grassland or forest ecosystem.

This study was sponsored by the National Key Research and Development Program of China (2016YFC0501109) and the program of Inner Mongolia Bureau of land and resources (Technical requirements on project of mine geo-environmental ecological restoration in Inner Mongolia).

REFERENCES

Gros, R., Monrozier, L.J., Bartoli, F., Chotte, J.L. & Faivre, P. (2004). Relationships between soil physicochemical properties and microbial activity along a restoration chronosequence of alpine grasslands following ski run construction. J. *Applied Soil Ecology, 27*: 7–22.

Li, W.H. (2004). Degradation and restoration of forest ecosystems in China. J. *Forest Ecology and Management, 201*: 33–41.

Paniagua, A., Kammerbauer, J., Avedillo, M. & Andrews, A.M. (1999). Relationship of soil characteristics to vegetation successions on a sequence of degraded and rehabilitated soils in Honduras. *J. Agriculture, Ecosystems and Environment, 72*: 215–255.

Shen, Y.C., Zhang, X.Z. & Wang J.S. (2015). Ecological Restoration in the Typical Areas. *J. Contemporary Ecology Research in China*, 265–374.

White, C., Tardif, J.C., Adkins, A. & Staniforth, R. (2005). Functional diversity of microbial communities in the mixed boreal plain forest of central Canada. J. *Soil Biology and Biochemistry, 37*: 1359–1372.

Yin, R.S. & Yin, G.P. (2005). China's Ecological Restoration Programs: Initiation, Implementation, and Challenges. J. *An Integrated Assessment of China's Ecological Restoration Programs*, 1–19.

Zak, D.R., Holmes, W.E., White, D.C., Peacock, A.D. & Tilmanm, D. (2003). Plant diversity, soil microbial communities, and ecosystem function: are there any links?. J. *Ecology, 84(8)*: 2042–2050.

Land Reclamation in Ecological Fragile Areas – Hu (Ed.)
© 2017 Taylor & Francis Group, London, ISBN 978-1-138-05103-4

Preliminary analysis and study on land reclamation and vegetation restoration—a case study of HuaXin coal mine in TongChuan city

X.H. Li & W.H. Ma
Country College of Geology and Environment, Xi'an University of Science and Technology, Xi'an, Shanxi Province, China

ABSTRACT: The coal resources are still one of the most important energy sources in China, but the exploitation of coal mine makes damage to the land resources and the ecological environment. In the context of China's shortage of land resources and degradation of environment, the study of land reclamation and vegetation restoration has important practical significance. This paper takes Shaanxi province Tongchuan Huaxin coal mine as an example and uses the basic theory of land reclamation, through the analysis of the main ecological problems on the land reclamation and vegetation restoration, terraced reclamation, scheme of groundwater balance and biological engineering design for reconstruction of mining land reclamation and vegetation restoration are rebuild in detail, which restore useful land and improve the ecological environment in mining and finally can realize the mine ecological environment protection, land resource utilization and sustainable development of local economy.

1 INTRODUCTION

The land is a valuable natural resource, but with the exploitation of mineral resources, land resources will be severely damaged, and will have a serious impact on people's lives, production and ecological environment. In particular, a large number of Chinese population, lack of land resources and dwindling arable land and the rapid development of economic construction is becoming more and more unprepared. Land reclamation has become an important part of land development and utilization activities, and is the basic policy of implementing the policy of "cherishing and using land rationally, practically protecting cultivated land". Based on the requirement of sustainable utilization of land resources and ecological environment construction, land reclamation must be an important way to supplement cultivated land.

2 SURVEY OF MINING AREA

Huaxin mining area is located in Tongchuan city of Shaanxi Province Huang Bao Zhen Du Jia Yuan Zi Jia Po Wangyi District in, the terrain of the coal mine is strongly cut by the water system, and the dendritic gully is extremely developed, ravines aspect, topography, geomorphology old. According to the scene investigation, the phenomenon of ground subsidence and ground fissure is not currently appear in mining area, but the surface vegetation is sparse, land is mainly used for farming, therefore, soil erosion is easily appear, and slope is mostly the collapsibility of loess collapse in the area, its stability is poor. At present, the main environmental problem is soil erosion, coal mining caused underground aquifer draining, groundwater level decline, the phenomenon of unreasonable stacking about a large number of abandoned stone (recrement) as well as industrial and domestic sewage emissions.

3 MAIN ECOLOGICAL PROBLEMS IN MINING AREA

3.1 *Soil erosion status*

There are two main types of soil erosion: water erosion and gravity erosion, Huaxin mining area is given priority to with water erosion, the main types are surface erosion, gully erosion, raindrop erosion, flash floods and so on. The surface erosion is widespread in the whole area, and the distribution of gully erosion is also common. During the mining process, the waste rock and the waste residue discharged have destroyed some of the land and the vegetation on the bottom of the ditch. In addition to the other temporary occupation of land, will inevitably destroy the natural vegetation and disturbance of the original surface is relatively stable, make the soil loose, bare ground area, and cause additional soil erosion.

3.2 *Changes of soil physical and chemical properties*

The ground surface forest, grassland has the function of soil and water conservation vegetation encroachment, after the destruction of the exposed surface, even without being washed, the surface humidity range increased, soil physical and chemical properties are bad. One of the most obvious change is to strengthen the role of the decomposition of organic matter, reduce soil organic matter content, is not conducive to the cultivation of other re vegetation, in addition, due to the construction damage and mechanical excavation, soil enrichment process is blocked, effects of biological and material exchange between the soil.

3.3 *Destruction of surface vegetation*

Influence of mine construction on vegetation mainly occurs in the construction period of the road, station and auxiliary systems engineering construction etc. These construction activities are to clear vegetation, excavation of the surface and ground construction, resulting in the direct construction of the local surface vegetation completely destroyed, the construction area within a certain range of vegetation will be subject to varying degrees of damage. Mining area will cause surface subsidence area due to coal mining, on the edge, as a result of surface cracks, subsidence area terrace, the surface soil is loose, water level is reduced and local vegetation is damaged, and it can also cause shrubs tilt, lodging etc, affect the growth of vegetation.

3.4 *The effect on land-use*

The impact of the mining area on the local land-use is mainly the site excavation, road construction and auxiliary systems and other construction sites, The land-use of these facilities is basically permanent about these facilities, the land lost its original and ecological function, so it has certain influence for the local land-use. The temporary area will produce a temporary effect, arable land can restore the original land-use function within 1 year in general, and shrubs need 2–3 years when the construction is over. Therefore, the adverse effects that temporary land occupation on land-use and economy in the whole region is limited in construction period.

4 PREVENTIVE CONTROL AND LAND RECLAMATION TECHNOLOGY

4.1 *The principles of land reclamation and ecological reconstruction in mining area*

According to exploitation plan of HuaXin coal mine in TongChuan city, combined with the degree of land destruction, in line with the principle of "prevention first, adjust measures to local conditions, prevent from complex combination", we should take appropriate control measures to prevent the harmful reclamation in the mining process, and prevent it in advance.

4.2 *Restoration of cultivated land*

The main type of cultivated land is dry land in the whole mine by subsidence area, the fracture often occurs in the mining area boundary pillar gob zone and transition zone, as well as the transition zone of the different subsidence depth, due to the increase of slope, cultivated land will lose the ability to maintain soil and water, soil nutrient loss, land production capacity reduction or loss of function. Most of the land in the Ida collapse area is mainly cultivated land, so the main form of reclamation is to construct or repair the terraced fields. Terraced reclamation construction mainly includes the topsoil, subsoil and flat ridge building links, according to the characteristics of cultivated land area is relatively small and the land is mainly dry land in subsidence area, we recommend the use of raw mix pile soil reclamation, and deep applicate manure and fertilizer, we can choose suitable for local planting crops and varieties, and use advanced runoff agriculture technology to ensure the stability of agricultural production.

4.3 *Restoration of vegetation*

Because there is most gentle plateau surface in HuaXin coal mine, surrounding farmland, the design of the wellhead and the industrial square is located in south side of the new Sichuan alum water, the main task that engineering construction is facing ecological reconstruction and restoration include mine industrial site, air space, explosive magazine, the gangue field and around road greening vegetation, in order to compensate for the loss caused by project construction.

The greening of industrial sites should take into account the requirements of beautifying the environment and protecting the environment, it is considered to beautify the environment in front of the factory and choose suitable ornamental tree species, grass irrigation, etc. Around the workshop should be considered to protect the environment green and choose suitable tree species, grass irrigation that have function of noise reduction and dust removal. The green space should be mainly choose greening tree species of noise reduction function. According to the local climate and soil conditions, the street trees of mine road on both sides are planted with evergreen and deciduous trees, we should choose the suitable tree species small evergreen tree and deciduous shrub juniper trees that the tree is beautiful, drought, cold, and lateral roots were developed, wind resistance is also strong; tree species should be selected locust, the French phoenix tree, drooping willow, willow etc. Green area of 3080 m^2, designed for 1*1 m, deep hole is 0.7 m, and the plant spacing was 4 m for planting 140 trees.

5 SUMMARY

The status of China's coal as the main energy will not change for a long period of time in the future, and the amount of coal mining on land destruction has been ranked first in all industrial sectors, so the coal mine land reclamation will continue to focus on land reclamation in our country. According to the terrain and topography of HuaXin coal mine in TongChuan city, put forward the destruction of land reclamation goals for cultivated land and woodland. Combined with the degree of land destruction, we must put forward specific reclamation measures and preventive control. Effective controlling of soil erosion, improving the reclamation of soil, improving the productivity of the soil and the ecological environment of the mine, achieve the protection of the ecological environment of the mine, land use and sustainable development of local economy.

REFERENCES

Gong, C. & Ding, Y. & Hou, S.Z. (2008) Recovery and harnessing of the mine ecological environment. *Coal Technology*. 27(10), 3–4.

Gong, F.Q. (1999) (2014) Study on land reclamation and vegetation restoration of silica ore. *Environmental Science and Management*. 39(7), 158–160.

Guan, X. (2016) Analysis on the recovery and management of mine ecological environment in inner Mongolia area—a case study of Shendong coal group. *The Chinese Journal of Geological Hazard and Control*. 27(3), 1003–8035.

Jiang, F. (2010) Geological environment in the mining area and its development trend in Fujian province. *The Chinese Journal of Geological Hazard and Control*. 21(2), 44–49.

Li, J.C. & Bai, Z.K. & Zhang, L.C (2000) *Land reclamation and ecological reconstruction in open pit coal mine*. Beijing: Science Press.

Xu, H.C. (2006) The main problems and Countermeasures of coal mine land reclamation. *Shanxi Coking Coal Science and Technology*. 35(1), 7–8.

Zhang, D.F. & Bai, Z.K. & Ye, B.Y. (2007) Evaluation method of reclaimed land in mining area. *Resource Development and Market*. 27 (7), 685–687.

Zhang, J. & Li, S.C. (2002) Study on land reclamation and ecological restoration in mining area. *Opencast Coal Mining Technology*. 1(14), 36–39.

Zhang, Y. & Liu, B. (2002) Effect of different vegetation types on soil erosion by water. *Acta Botanica Sinica*, 45(10), 1204–1209.

Land Reclamation in Ecological Fragile Areas – Hu (Ed.)

Study on root distribution and mechanical properties of *pinus tabulaeformis* in dumps of an opencast coal mine, Shanxi, China

C.J. Lv
College of Resources and Environmental Science, Shanxi Agriculture University, Taigu, China

W.G. Chen
College of Life Sciences, Shanxi Agriculture University, Taigu, Shanxi, China

D. Chen & X.X. Guo
College of Resources and Environmental Science, Shanxi Agriculture University, Taigu, China

ABSTRACT: The study area was the dumps of maximum opencast coal mine in Shanxi, the three reclamation substrates of thin loess, thick loess and soil-rock mixture were chosen as sample plots, the root distributions of *pinus tabulaeformis* growing in which were surveyed by profile excavation method and local specimens method. The mechanical properties of roots and water-holding capacity of litter of *pinus tabulaeformis* were tested. The results showed that:(1) In the thin loess, the root diameter of *pinus tabulaeformis* was 0–6.5 mm, the root weight density and root length density distributed evenly and were wavy with the increasing diameter in each soil layer. In the thick loess, the root diameter was between 0–3.5 mm, the root weight density and root length density were the biggest in deep soil layers. In the soil-rock mixture, the range of root diameter and root weight density and root length density were all the smallest, and roots tended to develop in surface soil. (2) The root diameters tested successfully ranged from 1.49 to 3.31 mm, the tensile force increased with increasing diameter according to a power function, the tensile strength decreased with increasing diameter at the level of a power function. The root stress-strain curve showed the single-peak curve with the characteristics of elastic-plastic. (3) The water-holding rate of litter increased with socking time at a power function, and there was a significant difference in three reclamation substrates ($P < 0.05$). The reclamation substrate has a great influence on the root distribution and soil reinforcement.

Keywords: *Pinus tabulaeformis*; reclamation substrate; root distribution; tensile properties; water-holding rate of litter

1 INTRODUCTION

The mining waste land is a special kind of degraded ecosystem, which is formed by the artificial accumulation. In the mining area soil erosion is serious and ecological environment is fragile. The large-scale mining activities have accelerated the degradation of the ecosystem, and formed unique damage characteristics (Hu Zhenqi, 2010). The slope of waste dump forms repose angles Angle (about 36°), with steep, long and loose. The reclamation conditions in China is also very different from abroad (BaiZhongke et al., 2012). Most studies have shown that in the early stage of reclamation, vegetation restoration rapidly is an economical and effective measure that can effectively reduce the soil erosion of mining waste land (Akihiro et al., 2015).

Over the past 30 years, the concept of reinforcing slope with plant has been more and more welcomed both at home and abroad (Coppin, 1990; Gray D H, 1996; Norris J E, 2005; Xiong Yanmei et al., 2007; Chiatante, 2003). It has been recognized that the vegetation roots can increase the shear strength of soil and reinforce slope by the anchoring of coarse roots (C.Bergeron et al., 2009), or the tensile properties of fine roots (Schmidt, 2001; Chen Lihua et al., 2008). The root function of soil reinforcement is affected by many factors such as root density, distribution depth and tensile strength soil thickness (Gray et al., 1982; Stokes et al., 2009). The root distribution is also affected by the soil thickness and topography environment (Farrish, 1991; Guillermo et al., 2016; Li Shaocai et al., 2005). Generally in a flat landform, root system distribution is symmetrical; but on the steep slope, root system develops a special structure of bilateral fan-shape with lateral roots growing preferentially upslope (Madhusudan B et al., 2000; Chen Lihua et al., 2008) to ensure the plant stability and adapt to the environment.

Mine soil is a heterogeneous mixture of rock and soil, the reclamation substrates both in the horizontal direction and the vertical profile are very complex. Compared with the soil developed from the original landform, mine soils have a wider range of particle size. In these special artificial landscape, it has not been clear that root is how to distribute and how to reinforce the soil. Therefore the study will explore the root of *Pinus tabulaeformis* distribution characteristics and root mechanical properties restoring for 9 years in three reclamation substrates of thin loess covered, thick loess covered and soil–rock mixture, to provide the basis for the vegetation configuration and soil profile construction in mining area.

2 MATERIALS AND METHODS

2.1 *Study sites*

Pingshuo opencast coal mine is one of the largest and most modern coal mine in China, which is located in Shuozhou city in the north of Shanxi province, the eastern part of the Loess Plateau, China. The geographic coordinates is 112°10′E~113°30′E longitude, 39°23N′~39°37′N latitude, In this ecological fragile zone area, soil erosion is strong. It is a typical temperate semi-arid continental monsoon climate, with strong wind in spring and winter, with cool temperate of 4.8–7.8°C averagely in summer. The annual average precipitation is about 450 mm. Anjialing opencast mine is the study area, the three largest area substrates of thin loess, thick loess and soil-rock mixture were selected as sample plot, with restoration for 9 years *pinus tabulaeformis*. The altitude is from 1345–1349 m, the slope degree is from 34–38° (Table 1).

Table 1. Basic situation of sample plot.

Sample plot	Reclamation years (a)	Canopy density (%)	Plant height (m)	Thickness of covering soil (cm)	Slope degree (°)
Thin loess	9	90	1.5	30	34
Thick loess	9	93	1.8	60	38
soil-rock mixture	9	90	1.9	No soil	38

Note: The sample plot of thin loess, its surface layer was covered by thin loess of 30 cm, its lower was the mixture of soil and weathering stones in diameter 3–5 cm. The sample plot of thick loess, its surface layer was covered by thick loess of more 60 cm with a few stones. The sample of soil-rock mixture means that the surface layer of 0–15 cm was weathering gravel, the lower was grit of 4 cm in diameter.

2.2 *Root morphology investigation*

The root distribution survey was carried out by profile excavating method and local specimen method. In May 2015, in the selected standard plot, the soil profile was excavated at 20 cm of the standard tree downslope and sampled every 15 cm until without root. Each soil sample size were 30 cm × 10 cm × 15 cm (length × width × height). The roots were separated from the soil by dry sieve method. The root diameters were measured with the vernier caliper (0.01 mm), then were classified 6 levels according to the diameter: <0.5 mm, 0.5 ~ 1.5 mm, 2.5 ~ 3.5 mm, 3.5 ~ 4.5 mm, 4.5 ~ 5.5 mm, 5.5 ~ 6.5 mm. The root total length were measured and the root biomass were weighed at 85°C drying.

2.3 *Mechanical properties of Pinus tabulaeformis roots*

Root tensile was tested using the portable tensile tester in situ. In the soil profile, along the root extending direction, dug gently the root to around 15 cm long, measured the diameter with vernier caliper, then clamped the root and pulled at a constant speed of 10 mm/min along the root extension direction, until the root fractured. Finally recorded the root tensile force F, and calculated the tensile strength P.

$$P = 4F / \pi D^2 \quad (1)$$

where P is the root tensile strength, MPa; F is the maximum tensile force, N; D is the root mean diameter, mm.

2.4 *Litter water-holding properties*

Water-holding capacity and water absorption rate were determined by indoor immersion method. Put the air dry litter 100 g into the yarn bag of 0.2 mm aperture, and then immersed into tap water, and weighted the litter after 1, 2, 5, 10, 15, 20, 30, 40, 50, 60, 390, 600, 630, 1380, 1410, 1440 min respectively. According to the water-holding rate (g/g)

$$R = (W_{abs} - W_{air}) / W_{air} \quad (2)$$

It is generally believed that the water-holding capacity is maximum after immersing 24 hours (Liu Shirong et al., 1996). Where R is the water-holding rate, g/g; W_{abs} is the litter weight after absorption, g; W_{air} is the litter weight of air dry, g. Maximum water—holding capacity (t/hm^2) = maximum water-holding rate (g/g) × litter storage (t/hm^2).

3 RESULTS

3.1 *Distribution characteristics of root weight density*

Root weight density refers to the root dry biomass per unit volume, g/dm^3.

1. In the thin loess (F1-a), the root diameter was between 0 and 6.5 mm, root weight density ranged from 0.27 to 1.07 g/dm^3. Different root diameter distributed basically in each soil layer, but the root weight density has no obvious regularity with increasing root diameters and was irregular wavy.
2. In the thick loess (F1-b), the root diameter was between 0 and 3.5 mm, root weight density varied from 0.27 to 1.07 g/dm^3. The root weight density of the surface layer was minimum. The distribution of root diameter was different in different soil layers. The range of root diameter in 15–30 cm was 0–3.5 mm, while that in 0–15 cm soil layer was 0–1.5 mm. The root weight density decreased with the increasing root diameter in 0–15 cm and 30–45 cm, but the root weight density showed wave with the increasing root diameter in the 15–30 cm.
3. In the rock-soil mixture (F1-c), the root diameter range was only 0–2.5 mm in the profile, the root weight density range was 0.03–0.30 g/dm^3. The root weight density in 30–45 cm soil layer was the least, the root developed better in the shallow soil layer than in the deep layer.
4. In the reclamation substrates of thin loess, thick loess and soil-rock mixture, the average root weight density of *Pinus tabulaeformis* was 2.90, 0.74, 0.36 g/dm^3 respectively. The root density increased first, then decreased with soil depth in thick loess and soil-rock mixture; and the root density increased with soil depth in thin loess. The root density difference of three reclamation substrates was significant ($F = 15.158$, $P = 0.005$) and it was no significant in different soil layers ($F = 1.857$, $P = 0.236$).

3.2 *Distribution characteristics of root length*

Root length density refers to the root total length per unit volume, g/dm^3.

As can be seen from Figure 2(a-c), in the three reclamation substrates of thin loess, thick loess and soil-rock mixture, root length density showed a decreasing trend with the root diameter increased each soil layer. The root length density had no obvious regularity in different soil layers in thin loess plot, but In the thick loess plot, the root length density of 0–15 cm soil layer was the smallest, and the root length density tended to grow in the deep soil layer, in the soil-rock mixture plot, the root length density of 30–45 cm soil layer was the smallest, and the root tended to grow in the shallow soil layer.

Figure 2(d) showed that, the root length density increased and then decreased with soil depth. The root length density in three soil layers and three reclamation substrates both had no significant difference ($F = 2.008$, $p = 0.215$; $F = 2.106$, $p = 0.203$).

3.3 *Root tensile force and tensile strength*

Pinus tabulaeformis tested was 9 years, located in the wind erosion and water erosion area, and the growing substrates was not good, there was not too much root biomass, the root diameter tested successfully was mainly in 1.49–3.31 mm.

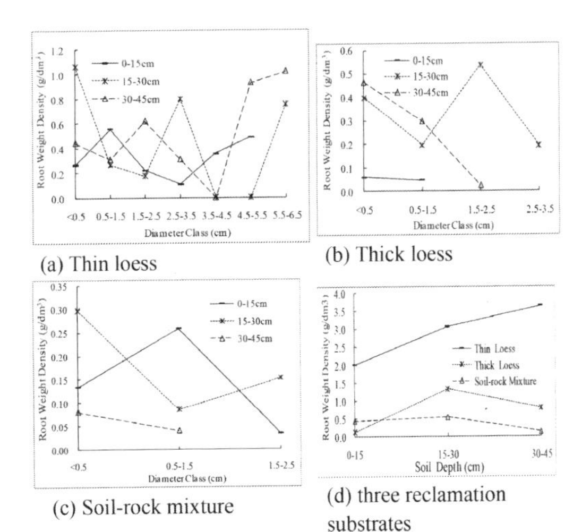

(a) Thin loess (b) Thick loess (c) Soil-rock mixture (d) three reclamation substrates

Figure 1. Distribution characteristics of root weight density in three reclamation substrates.

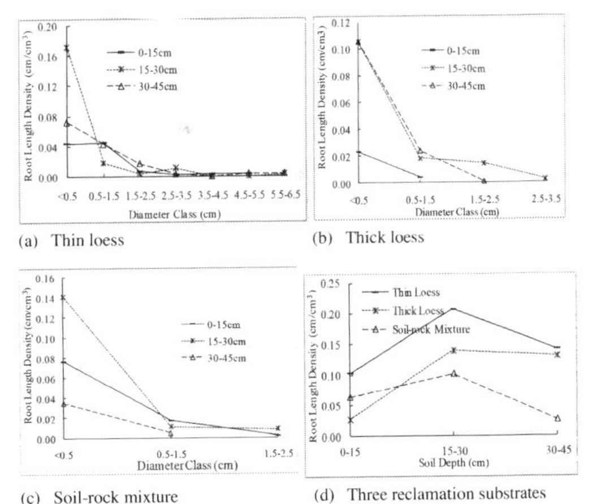

(a) Thin loess (b) Thick loess (c) Soil-rock mixture (d) Three reclamation substrates

Figure 2. Distribution characteristics of different root length.

The tensile force was 24–100.3N, the average was 55.4N. The tensile strength was 10–18 MPa, the average tensile strength was 13.8 MPa. Tensile force increased with the increasing diameter followed by the power function, $F = 13.586D^{1.7119}$ $R^2 = 8282$. The tensile strength decreased with the increasing diameter in the level of the power function, $P = 16.295D^{0.238}$ $R^2 = 0.11$.

3.3.1 *Stress-stain curve of roots*

As is known to all that vegetation roots can reinforce the slope stability because of the strong tensile properties and deformation. So the root stress-strain curve is the foundation of the root tensile process (Fig. 4).

The root stress-strain curve was single peak curve without obvious yielding and necking, the whole curve showed elastic-plastic characteristics (Fig. 4). According to a large number of test data(Lv Chunjuan et al., 2011), the stress-strain curve of thinner roots in diameter showed a non smooth curve (Figure 4), that of larger diameter root was relatively smooth. The results indicated that the thin roots had stronger buffering capacity than the thick roots. The third-order parabolic function can reflect well the basic characteristics of stress-strain curve.

3.4 *Water absorption characteristics of litter*

In the three kinds of reclamation substrates of thin loess, thick loess and soil-sock mixture, the thickness of litter was 3 cm, 2.17 cm, 2 cm, and the litter volume was 5.13 t/hm², 6.97 t/hm², 10.89 t/hm², the natural water content was 18.40%, 11.79% and 6.63% respectively.

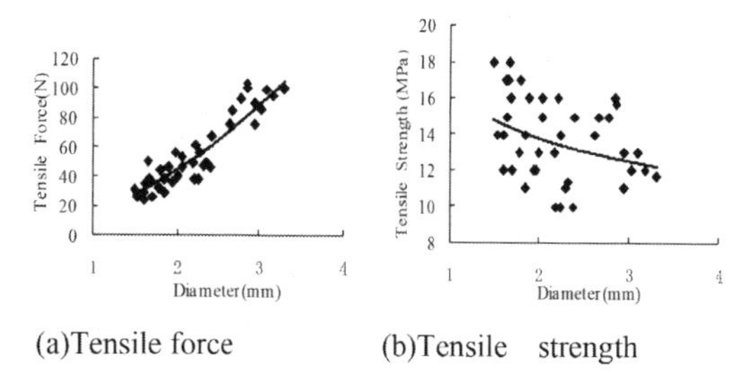

(a)Tensile force (b)Tensile strength

Figure 3. Variation of Root tension resistance with diameter.

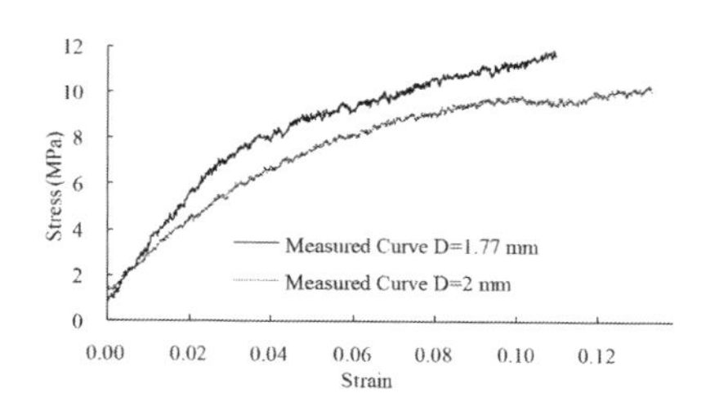

Figure 4. Root stress-strain curve (D = 1.77 mm, $y = 2 \times 10^{-8} x^3 - 4 \times 10^{-5} x^2 + 0.0347x$, $R^2 = 0.993$; D = 2 mm, $y = 8 \times 10^{-9} x^3 - 2 \times 10^{-5} x^2 + 0.0254x$, $R^2 = 0.9843$ where y is stress, MPa; x is strain.)

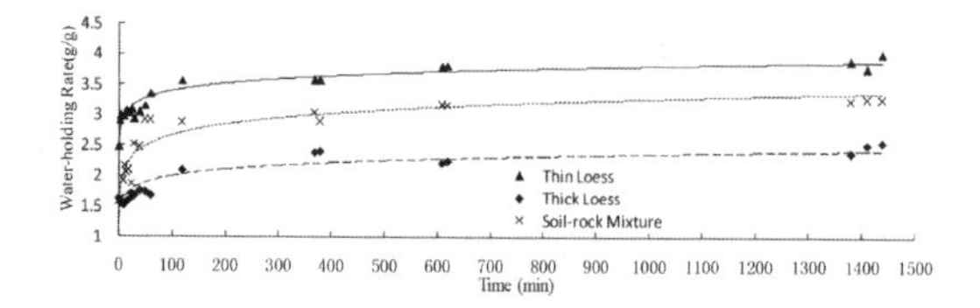

Figure 5. The water holding rate of Pinus tabulaeformis litter in three substrates.

According to Figure 5, The saturated water holding capacity of litter was 4.0195, 2.5656 and 3.2684 g/g in thin loess, thick loess and soil-sock mixture, respectively. The water holding capacity of litter in the three kinds of reclaimed substrates increased with soaking time, and the best fit function was log function ($y_{thin} = 0.1807\ln(x) + 2.563 R^2 = 0.9212$; $y_{thick} = 0.1572\ln(x) + 1.2737$ $R^2 = 0.8386$; $y_{mixture} = 0.2622\ln(x) + 1.4585$ $R^2 = 0.8944$. where y is water holding rate of litter, g/g; x is soaking time of litter, min.)There was a significant difference in water holding capacity of litter in three reclamation substrates ($P < 0.05$).

4 CONCLUSION AND DISCUSSION

In three reclamation substrates of thin loess, thick loess and soil-rock mixture, we found that the tensile force increased with the increasing diameter, and the tensile strength decreased with the increase of root diameter. The results were consistent with the other tree species in literature. Root stress-strain curve was peak without obvious yielding and necking, showing elastic-plastic characteristics. But for thinner roots, the curve was no smooth and had buffering to the tensile force.

Reclamation substrates and the three-dimensional structure in profile had important effects on the development and distribution of roots in the soil layers. In thin loess substrate, the thickness of covering soil was only 30 cm, the lower was soil and weathered rock which restricted the development of fine roots. So more roots grew in the covering soil layer and evenly distributed. In thin loess substrate, root diameter range was the widest of 0–6.5 mm, root length density and root weight density were wavy with increasing diameter in three soil layers. In the thick loess substrate, the entire substrate was relatively homogenous and loose, the roots followed the law of natural growth, tended to deep soil layer. In the soil-rock mixture substrate, the surface of 10 cm was weathering tiny sand, the lower was big size of stones with a little of weathered detritus, so the roots more distributed on the surface. Even so, the fine root was also hard to survive; root hairs had to evolve into clumped distribution. In the soil-rock mixture substrate, the root diameter was the narrowest of 0–2.5 mm, and the root tended to top soil layer. In short, the root length density had no significant difference, but the root weight density and water-holding rate of litter had significant difference in three substrates.

REFERENCES

Akihiro Hamanakakaa, Naoya Inoueaa, Hideki Shimadaaa, et al. (2015). Design of self-sustainable land surface against soil erosion at rehabilitation areas in open-cut mines in tropical regions [J]. International Journal of Mining, Reclamation and Environment. 29(4), 305–315.

Bai Zhongke, Lv Chunjuan, Wang Jinman, et al. (2012). Control of rock erosion and utilization of water resources in mining area [M]. Beijing: China land press.

Bergeron, C., J.-C. Ruel, J.-G. Elie et al. (2009). Root anchorage and stem strength of black spruce (Picea mariana) trees in regular and irregular stands[J]. Forestry. 82 (1), 29–41.

Chen Lihua, Yu Xinxiao, Song Weifeng, et al. (2008). Root soil mechanics [M]. Beijing: Science Press, 10–11, 44.

Chiatante, D., S.G. Scippa, A. Di Lorio, et al. (2003). The influence of steep slopes on root system development[J]. J. Plant Growth Regul. 21, 247–260.

Coppin, N.J., Richards, I.G. (1990). Use of vegetation in civil engineering[M]. London: C.I.R.I.A. Butterworths. 5–10.

Farrish, K.W. (1991). Spatial and temporal fine-root distribution in three Louisiana forest soil [J]. Soil Sci. Soc. Am. J. 55, 1752–1757.

Gray, D.H., Leiser, A.T. (1982). Biotechacal Slope Protection and Erosion Control[M]. New York: Van Nostrand Reinhold Co., 271.

Gray, D.H., Sotir, R.B. (1996). Biotechnical and soil bioengineering Slope stabilization: a practical guide for erosion control[M]. Chichester: John Wiley and Sons, 64–70.

Guillermo Tardío, Alejandro González-Ollauri, Slobodan B. Mickovski (2016). A non-invasive preferential root distribution analysis methodology from a slope stability approach[J]. Ecological Engineering. 97, 46–57.

Hu Zhenqi (2010). The opportunity and challenge of land reclamation and ecological reconstruction in Shanxi coal mine area [J]. Shanxi agricultural science. 8(1), 42–45, 64.

Li Shaocai, Sun Hailong, et al. (2006). Mechanical properties of the interaction between the root system and rock mass of slope protection [J]. Chinese Journal of rock mechanics and engineering. 25(10), 2051–2057.

Liu Shirong, Wen Yuanguang, et al. (1996). Ecological and hydrological functions of forest ecosystem in China [M]. Beijing: China Forestry Press, 3–7.

Lv Chunjuan, Chen Lihua, Zhou Shuo, et al. (2011). Agricultural engineering tensile mechanical property [J]. Journal of roots of different tree species. 27 (S1), 329–335.

Madhusudan, B., Shrestha, M., Horiuchi Y., et al. (2000). A study on the adaptability mechanism of tree roots on steep slopes. A. Stokes (Ed.), In the Supporting Roots of Trees and Woody Plants: Form, Function and Physiology. Developments in Plant and Soil Sciences, Dordrecht: Kluwer Academic Publishers, 51–57.

Norris, J.E. (2005). Root reinforcement by hawthorn and oak roots on a highway cut-slope in Southern England[J]. Plant Soil. 278, 43–54.

Schmidt, K.M., Roering, J.J., Stock, J.D., et al. (2001). Root cohesion variability and shallow landslide susceptibility in the Oregon CoastRange[J]. Can. Geotech. J. 38, 995–1024.

Stokes, A., Atger, C., Bengough, A.G., et al. (2009). Desirable plant root traits for protecting natural and engineered slopes against landslides[J]. Plant and Soil, 324(1/2), 1–30.

Xiong Yanmei, Xia Hanping, et al. (2007). Study on the effect and mechanism of plant root system on slope erosion resistance [J]. Chinese Journal of Applied Ecology. 18 (4), 895–904.

Land Reclamation in Ecological Fragile Areas – Hu (Ed.)
© 2017 Taylor & Francis Group, London, ISBN 978-1-138-05103-4

Study on the ecological security of Wu'an based on DPSIR

Z. Xu, J.J. Zhang & Z.Y. Li
School of Land Science and Technology, China University of Geosciences (Beijing), Beijing, China

ABSTRACT: Based on DPSIR model and considered previous studies and realities, this essay chooses a typical case, Wu'an, in order to build an ecological security evaluation system. With reference to relevant standards, ecological security level was determined. There is a want of developmental driver of urban system, pressure of habitat overload and obviously insufficient social response. In order to get out of "resource curse", Wu'an should make the innovation-driven improvement and increase the investment in environmental protection.

1 INTRODUCTION

Mining cities have created economic growth while environmental risk is increasing. Therefore, it turns into the focus of government and academia. The sustainable development of a region won't be threatened regardless of whether environment and living space are damaged or not. That's ecological security (Jiang Yong *et al.*, 2011). The ecological security evaluation has been developed such as PSR model, ecological footprint method etc. DPSIR, developed by the European Environmental Committee from PSR and DSR models (ZHANG Yanet al., 2014), reflect the interrelationship among economy, environment and resources (HE Chun-lan et al., 2010); and it's an orderly way to explore causality (YU Bo-hua & LU Chang-he, 2004). Thus, this paper constructed an ecological security evaluation system based on DPSIR, aiming to quantify the ecological security degree, finally to support local transformation evidently.

2 OVERVIEW AND DATA SOURCES

Wu'an is located in the south of Hebei Province (113°45′E~114°22′E and 36°28′N~37°01′N). Despite its solid industrial foundation, it outputs negative growth of major products with a still huge total quantity because of resource constraints and industrial reformation. All data originates from the Wu'an Statistical Yearbook (2011–2015).

3 METHODOLOGY

3.1 *Ecological security evaluation system*

This system was constructed by typical, comprehensive and available principles. Previous researches and realities were considered and index is weighted by coefficient of variation.

3.2 *Ecological security evaluation model*

Ecological security is driven by five factors including driving force, pressure, state, influence and response (Zhang Ji-quan et al., 2011). Formula of ecological security value is as follows:

$$E_i = \sum_{j=1}^{m} W_{ij} \times P_j \quad (1)$$

E_i is the ecological security value; P_j is the jth index weight; W_{ij} is the normalized value of the *jth* index of the *ith* year. According to previous studies (QIU Wei et al., 2008) (LI Yu-ping. & CAI Yun-long, 2007) and reality of Wu'an, safety level was divided as follows:

4 RESULTS AND ANALYSIS

4.1 *Studies on the factors of ecological security*

Factor scores were obtained (Figure 1). Driving forces can reduce ecological security. Human activities constantly change and ultimately shape the structure and function of the city system. The figure shows that the driving force index has the most dramatic changes in the five years. The highest value is the lowest value of nearly 4 times. Driving force soared may promote prosperity, but also increase risk. Pressure is the direct cause of habitat changes. When minimum of pressure appears in 2013, driving force reaches its peak, which indicates that environment was overloaded by urgent economic demands. State is the result of the interaction between driving force and pressure. During study period, value decreased nearly 6 times from 0.188

Table 1. Wu'an ecological security evaluation system.

Index I	Index II	Weight	Attribute
Driving force	Per Capita GDP (yuan)	0.047	+
	Natural population growth rate (‰)	0.128	+
	Mining investment accounted for whole society	0.054	+
Pressure	Industrial SO_2 Emissions (t)	0.049	–
	Industrial enterprises water intake (m^3)	0.070	–
	Population density ($person/km^2$)	0.042	–
State	Whole society electricity consumption (100 million kW·h)	0.083	–
	Per capita arable land area (mu/person)	0.048	+
	Per capita water consumption (m^3)	0.105	–
Impact	Number of secondary and tertiary industry practitioners (person)	0.066	+
	Per capita park green area (m^2)	0.068	+
	Urbanization rate (%)	0.075	+
Response	Urban greening coverage (%)	0.043	+
	Environmental pollution controlaccounts for GDP	0.061	+
	Urban sewage treatment rate (%)	0.061	+

Table 2. Ecological safety level and division criteria.

Interval	Level	Safety degree	Characteristics
(0.9–1.0]	I	Ideal safety	Perfect structure, healthy habitat
(0.75–0.9]	II	Good safety	Structure is basically perfect, good habitat
(0.6–0.75]	III	General safety	Structure is still complete, general habitat
(0.45–0.6]	IV	Critical safety	Structure has deteriorated trend, sensitive habitat
(0.3,0.45]	V	Risk alert	Structure has been damaged, habitat warning
[0,0.3]	VI	Extremely unsafe	Structural damage, ecological disaster formation

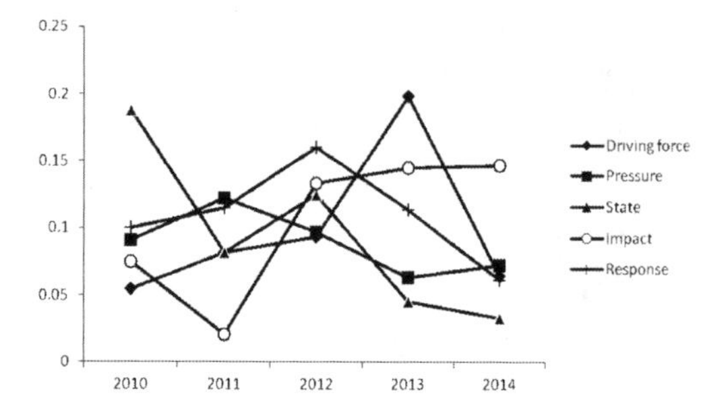

Figure 1. The scores of each factor in the DPSIR model.

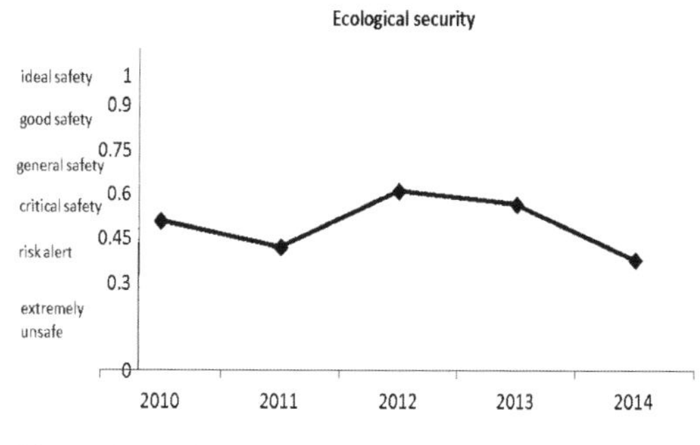

Figure 2. Wu'an ecological environment safety evaluation results.

to 0.033. That means system is unstable and self—regulating ability is poor. The objects of impact are human and society. Wu'an has strong economic foundation, thus it can weaken adverse effects of state. Response is measure carried out by government and others. Value increased slowly in the first three years because government was affected by the downturn market. Greenery coverage and pollution control investment dropped to the lowest, further exposing response is not sufficient in 2014.

4.2 *Ecological security grade*

Security grades were obtained (Figure 2). Except 2012, 2010–2013 are in critical security. Wu'an has been at risk alert in 2014. The level of ecological security is low and the trend is declining.

5 SUMMARY

Ecological security was low in Wu'an, which shows that the local economy has increased the environmental pressure, resulting in declining tolerance of habitat. The traditional industrial structure cannot balance the relationship between the economic construction and ecological protection. And due to lack of response, Wu'an city system showed impairment of structure and function, regulation and anti pressure ability degradation, serious ecological problems and other features. Although economy is still steady, Iron-oriented industry cannot stimulate sustainable growth. It should emphasis on current development model which had "insecure" factors, increase environmental protection investment, and finally ensure ecological health. DPSIR reveals relationship of urban system components and analysis underlying logic of environmental damage. However, there are three aspects lack necessary norms and convincing instructions so far, such as construction of index system, setting of weight, and classification of ecological security level standard, which worth studying further.

ACKNOWLEDGMENTS

This paper is Contribution No. 41571507 of the National Natural Science Foundation of China.

REFERENCES

He Chun-lan, Rao Hui & Zhao Xiao-qing (2010) Advances in ecological security assessment research in china. *YunNan geographic environment research*, 22(03), 104–110.

Jiang Yong, Fu Mei-chen, Du Chun-yan & Wu Xian-bing (2011) A case study on Yong Qing country: dynamic assessment of ecological safety based on DPSIR model. *Resources and industries*, 13(01), 61–65.

Li Yu-ping. & Cai Yun-long (2007). Security Evaluation of Land Ecology in Hebei Province. *Acta Scientiamm Natumlium Universitatis Pekinensis*, 43(06), 784–789.

Qiu Wei, Zhao Qing-liang, Li Song & Zhang Cheinchi (2008). Ecological Security Evaluation of Heilongjiang Province with Pressure-State-Response Model. *Environmental Science*, 29(04), 1148–1152.

Yu Bo-hua & Lu Chang-he (2004). Application of DPSIR framework for analyses of sustainable agricultural development. *China population, resources and environment*, 14(05), 70–74.

Zhang Ji-quan, Yi Kun-peng, Hiroshi Tani, Wang Xiufeng, Tong Zhi-jun & Liu Xing-peng (2011). Ecological security assessment of Baishan City in Jilin Province based on DPSIR. *Chinese Journal of Applied Ecology*, 22(01), 189–195.

Zhang Yan, Shen Lei & Yu Wen-jia (2014). Evaluation of regional energy security based on the DPSIR model: a case study of Guangdong province. *China Mining Magazine*, 23(07), 32–37.

Land Reclamation in Ecological Fragile Areas – Hu (Ed.)

Study on the landscape design in southern Shaanxi mining area based on similar natural terrain reconstruction

H.H. Chen
College of art, Xi'an University of Science and Technology, Xi'an, Shaanxi, China

ABSTRACT: A large number of high strength of the development of all types of mining to the South Shaanxi caused serious damage in the mountainous terrain structure, natural geology, hydrology and soil and other ecological environment. In order to protect the environment and resources in Qinba mountainous area, to restore the natural ecological landscape, green mountains and rivers, the digital correlation analysis of the characteristics of the terrain, soil and vegetation growth of the adjacent undisturbed watershed and abandoned mining area is analyzed in this paper. On this base, to minimize the damage to the natural environment in the mountain area of southern shaanxi, we try to explore the landscape design method of similar natural terrain for remodeling the mountainous area and try to give a guidelines in trench remodeling, soil reconstruction, vegetation and landscape remodeling.

1 INTRODUCTION

The southern part of Shaanxi is located in the hinterland of Qinling Mountains, includes most areas of Hanzhong, Ankang and Shangluo. Here is rich in gold, molybdenum, iron, manganese, copper, mercury, sulfur, phosphorus, fluorite, quartz, asbestos, marble, limestone, slate and other metal minerals. But most minerals are deeply buried in the vein, except a few minerals can be mined open-pit, the vast majority mining ways are admitted, inclined and vertical[1]. The surrounding rock of mining area is mostly hard bedrock, which with strong mechanical strength. Although the ground subsidence and fissure caused by mining is rare, but because of the extensive development, mining rich abandoned the poor, so the damage to the environment in the mining area is very serious[2].

In recent years, with the enhancement of China's rapid economic development and environmental protection, and reference experience from the United States, Germany and other developed capitalist countries in the experience of landscape ecological restoration of mining areas, our country has obtained considerable development and progress in research on ecological reconstruction, such as the mining area terrain restoration, land reclamation and vegetation restoration, etc. Mining land reclamation and landscape restoration also made a variety of attempts and exploration, part of them is made of the earth art for sightseeing, part of them is made of the art park and playground for the public use, there are also part of the exhibition hall, the museum, so that the abandoned mining land have the value of follow-up. But the southern Shaanxi area relatively remote, the first choice of ecological restoration is to revert to the original appearance as far as possible, that is using the near natural reconstruction landscape design idea to restore the local ecological environment[3].

"Similar nature" literally means "close to nature". To restore the natural condition should be placed in the sustainable development of mining wasteland is preferred. The use of natural processes, take the natural succession method is an important means, including the use of native species, vegetation restoration and succession community, improve soil quality, natural river and water natural recovery process, to improve the productivity and stability of the natural ecological system[4].

2 SIMILAR NATURAL REMODELING LANDSCAPE DESIGN PRINCIPLES

2.1 *Similar natural remodeling landscape design connotation*

2.1.1 *Cultural connotation*

Natural is the highest realm of spiritual pursuit of the Chinese people. In "Dao de jing" Laozi said: "people learn from earth, earth learn from universe, universe learn from law, law learn from natural". This philosophy is rooted in the Chinese people's thinking, impact on China's landscape architecture, building construction, and even in national governance, and they think That all things are born in nature. Chinese people look at nature, not only in the living environment, but deep in the life of the return to the origin of the spirit of desire and demand.

Chinese traditional culture always lays stress on theory that man is an integral part of nature, they believe that human life is limited, transient, but the life of the natural universe is infinite. If one's life is to be perfect and lasting, it must be connected with the life of the universe, and draw wisdom and strength from nature.

2.1.2 *Sustainable development connotation*

The idea of sustainable development is to make full use of the self regulating ability of the nature, so that mankind can seek the best balance among the mine development, economic growth and environmental protection. The idea of similar natural landscape design is to give full play to the power of nature, so that the original land function can be restored, and the purpose of sustainable utilization can be achieved[5].

2.1.3 *Landscape aesthetics connotation*

Restoration of mining wasteland should be coordinated with the adjacent undisturbed landscape, and the surrounding landscape should form a complete basin with the upper and lower reaches. The damaged landscape should restore to ecological productivity and visual appeal landscape structure. Using the relatively stable landscape structure as the standard, the restoration planning and design of the mine waste land fully embodies the beauty of nature.

2.2 *The concept of similar natural remodeling landscape design*

Similar natural governance is the first proposed for rivers governance, it is a kind of management scheme which is close to nature, economic and practical, and maintains the beauty of landscape[6]. Yang Cuixia, Beijing Forestry University, in her doctoral thesis put forward the concept of "near natural terrain reconstruction" of Mining Reclamation,"Similar natural remodeling landscape design" is based on this concept of landscape design for the restoration of abandoned mining areas in hills. This theory of landscape design is the comprehensive utilization of topography, landscape ecology, landscape aesthetics principle, when we are planning the restoration of landscape design similar to unperturbed region in mining area.

The landscape design of abandoned mine land is different from the commercial landscape design. For the mining area has its own ecological cycle system, planning and design should be carried out from inside to outside. In order to meet the basic conditions of the ecological cycle, the visual appearance is consistent with the surrounding natural environment. Therefore, in the process of landscape remodeling, taking into account the local unique natural conditions and environment, and thinking about the concept, methods, steps and implementation process when we taking similar natural remodeling landscape design.

2.3 *The content of the similar natural remodeling landscape in an abandoned mining area*

2.3.1 *Similar natural remodeling of topography and geomorphology*

The first step of mining landscape restoration is the restoration of terrain. The similar natural remodeling of topography has a positive effect on the recovery of the local hydrological and ecological space, and is also the premise of the later landscape planning. This reconstruction in mining area is a kind of terrain regulation model which can simulate the natural form of the river basin, or similar to the natural form, It is on the base of the survey which using basin geomorphology theory, environment theory and technology foundation of regional landscape ecological and geographical environment of the mining area, taking river basin as research unit, using computer simulation technology, taking the hydrological morphological characteristics of adjacent undisturbed as the design reference target, design of erosion in the process of the formation of a terrain near the natural environment and in accordance with the landform, and the implementation method of governance[7].

2.3.2 *Similar natural remodeling of vegetation*

In the remodeling process of natural vegetation, ecological sustainable development is the key, the purpose is to reproduce as much as possible similar to the undisturbed region vegetation ecosystem. We should consider the following two issues in the design and implementation process:

1. Give priority to local plant species.

Native tree species have adapted to local soil and climatic conditions and have higher survival rates when grown. In addition, we should according to the Seasonal characteristics of different plant species growth when planting, the mountain elevation is different, different plant species should be different.

2. Phasic planting

Landscape restoration of abandoned land in mining area is a long process. In the process of transformation, the ecological environment restoration needs to consider many aspects of the ecological elements, should be guaranteed in the environmental carrying capacity to promote the update measures with a controlled and phased manner. Improvement can be achieved by a number of small projects to gradually promote the whole ecosystem restoration and landscape function[8].

3 THE LANDSCAPE RESTORATION DESIGN IN A SOUTHERN SHAANXI MINING AREA

3.1 *The ecological status of the southern Shaanxi mining area*

Mineral resources in southern Shaanxi in the main metal mineral resources, especially the well-being of the region of Xunyang County, is one of the four famous mercury mine base in China.

It is said that the Lintong Qin tombs underground palace constructions river landscape with mercury, its Ore material come from Southern Shaanxi mining area. Here is the rolling about 1500 meters above sea level of the "M" shaped mountain, many metal mineral resources is distributed in bedrock rocks. In recent decades, due to the quick success, extensive development, ecological environment has been severely damaged in mining area of southern Shaanxi. At present, the most prominent problem has three points: 1) Destruction of mountain. Due to the inconvenience of traffic, it can not be ordered mining, most mining areas are mined by simple blasting, and this way of mining damage to the structure of the mountain, leaving a large number of steep loose bare landslide. 2) Unreasonable stacking of tailings and gangue occupy and destroy mountain vegetation. In a multi metal mine survey in mining area, waste residue, gangue is piled up naturally with the hillside. The hillside with natural rolling without the protection of the natural rolling ore is very unstable and easy to cause secondary disasters such as landslides[9]. 3) The pollution of surface water. The toxic and harmful substances in the fall in the river mine waste residue such as heavy metal ions in high content, easy to form pollution under the erosion of rain and stream, and it is also damage to river surface water. Liu Yue (Shanxi Datong University) evaluated the environmental quality of surface water in a lead-zinc mining area in southern Shaanxi, the result is V, heavy metal pollution is seriously. The results of this test are related to the sustained release of heavy metal elements in the tailings. The channel water has not reached the water quality requirements of farmland irrigation, and can not be used for farmland irrigation, this phenomenon is in urgent need of governance[10].

3.2 *Topographic restoration and ecological reconstruction*

Aiming at the problem of environment in mining area, the landscape design content is as follows:

3.2.1 *Topography and geomorphology reconstruction*

The landscape is reconstructed by the method of subsection management, which is divided into upper, middle and lower parts. Each section uses different landscape restoration techniques.

The upper part of mountain mostly is a steep with hard rock, and it is the slope hard to keep the soil. We can use the method of filling the crater to shaped terrain, and planting hardy native plants for greening. The middle of the mountain mostly is accumulated with natural rolling gangue, masonry wall can be used to strengthen the gangue, on this basis, we can also plan the terraces and use mounding planting green vegetation. The lower part of the mountain mostly are rolling huge waste rocks, and they maybe block the river. If necessary, the tunnel can be diverted by the way of excavating the tunnel in the mountain, so as to prevent the water from washing slag and cause surface water pollution.

3.2.2 *Vegetation remodeling*

The mountain area restore green in southern Shaanxi mining area, can not plant only one kind of grass or plant in all of the whole mountain, it should be classified according to the altitude of the mountain to construct timing, and hierarchical landscape. Hanzhong city, Yang County highest elevation of 3071 meters in the mountainous area in southern Shaanxi, Pingli, Xunyang, Baihe area is less than 1500 meters of the Mountain hilly area. The lowest point is located in the right bank of Han River which Baihe county AnKang city junction with Hubei province, 170 meters above sea level[11].

According to Dr. Zhang Qiaoming of the Northwest Agriculture and Forestry University in Qinling Mountains and the middle section of South Main flora species diversity study, that the southern slope middle of Qinling Mountains Vegetation changes about 1000 m to the top of 2868 m, the plant diversity index increase with the altitude increases, reaches the maximum in middle altitude, and then decreased gradually[12]. The vegetation in the mountainous area is mainly forest, and the coastal gradient shows a vertical distribution pattern, from the bottom to the top are deciduous oak forest, deciduous birch forest and subalpine coniferous forest[13]. Therefore, the landscape reconstruction should be fully demonstrated, respect for natural changes, the selection of suitable tree species.

Sequential landscape planting, that is the planting of the mountain was carried out before and after the plant growing time. In the process of ecological restoration and landscape reconstruction, not for the pursuit of rapid greening, we should ensure that the environment under the control of the carrying capacity of a limited, phased planting, so that the plants grow up orderly.

4 CONCLUSION

Landscape ecological restoration in southern Shaanxi mountain mining area is urgent, this is the key to the ecological security and sustainable development of mining industry. The design concept of landscape restoration of similar natural landscape reconstruction has a very important practical significance for the rapid restoration of vegetation growth, the reduction of soil and water pollution in the abandoned mining area mountainous region of Southern Shaanxi. Similar natural remodeling can minimize the difference between the reclaimed land and the natural environment. It is the best way to reconstruct the landscape and the special environment of the mountain area.

The core of Similar natural remodeling is the terrain and vegetation restoration. Terrain restoration, firstly, the topography of Undisturbed mining area should be analyzed, these data are used to guide the landform and physiognomy restoration of the abandoned mining land. The restoration of vegetation should be based on the elevation of the mountain, and the hierarchical classification, the first choice is local vegetation and using the timing of cultivation techniques to remold.

REFERENCES

Chen Hehu, Chen Xiongjun. *Close to the traditional dwellings in southern Shanxi* [M]. Shaanxi: Xi'an Jiao Tong University Press, 2016.05.

Gao Jarong, Near Natural Control:Torrent Control Engineering Based on the Landscape Ecology[J]. *Journal of Beijing Forestry University*. 1999, 21(1):80–85.

Hu Zhenqi, Yang Xiuhong, BAO Yan, et al. On the Restoration of Eco-Environment [J]. *Science & Review Technology*. 2005, 1:38–41.

Li Bin, Chen Yuehua, Tong Fangpin, et al. Study on the Vegetation Restoration and Create a Sustainable Landscape: As an Example Hunan Lengshuijiang Antimony Mine [J]. *Chinese Agricultural Science Bulletin*. 2010, 26(9):273–276.

Li Naiying, Guo Cailing, XI Zhende, et al. Study on the Exploitation and Utilization of Mineral Resources and Regional Sustainable Development in Qinling-Daba MTS. Of Southern Shaanxi.[J]. *Resources Science*. 1999, 21(4):31–34.

Liu Xiaoqing. *The Research on Landscape Planning of Abandoned Mine Lands Based on Ecological Restoration*[D], Jangsu: Nanjing Tech University, 2014, 06.

Liu Yue, Guo Wenqian. Evaluation of Surface Water Quality Based on Fuzzy Comprehensive Evaluation Method: A Case Study in Southern Shaanxi Lead-Zinc Area[J]. *Journal of Eastern Liaoning University*. 2015, 22(2)116–118.

Tang Zhiyao. Altitudinal Patterns of Plant Species Diversity in Niubeiliang Qinling Mountains[J]. *Biodiversity Science*. 2004, 12(1):108–114.

Xiang Maoxi, HE Weizhong, LI Yonghong. An Analysis of Geological Environment Problems of Mines in Shaanxi[J]. *Ground Water*. 2014, 36(4):89–91.

XI Zhende. The Enviornmental Problems in the Exploitation and Utiliization of Mineral Resources in Southern Shaanxi and some Countermeasures for Their Control[J]. *Journal of Northwest University (Natural Science Edition)*, 1998, 25(1):68–71.

Yang Cuixia, Yang jianying, Zhang Chengliang, et al. Design principle and method of Near-natural restoration of mine wasteland [M]. Beijing: SciencePress, 2015.06:20–25.

Zhang Qiaoming. *Patterns of Plant Community and Biodiversity on South Slope in the Middle of Qinling Mountains*[D]. Shaanxi: Northwest A&F University, 2012.05.

Zhong Yuanchun. *Research on landscape Ecology Maintenance and Restoration on Mining Waste Lands*. [D]. B.E.: Hunan Agricultural University, 2005.

Land Reclamation in Ecological Fragile Areas – Hu (Ed.)

 ISBN 978-1-138-05103-4

Varied soil liable organic carbon contents among reclamation patterns in Pingshuo opencast coal mine, China

Y. Yuan
School of Land Science and Technology, China University of Geosciences, Beijing, P.R. China

Z.Q. Zhao
School of Land Science and Technology, China University of Geosciences, Beijing, P.R. China
Key Laboratory of Land Consolidation and Rehabilitation, Ministry of Land and Resources, Beijing, P.R. China

Y.Z. Wang, P.T. Zhu, Y. Chen & S.Y. Niu
School of Land Science and Technology, China University of Geosciences, Beijing, P.R. China

ABSTRACT: Six types of reclaimed forest stands (three types of monoculture forests and three types of mixed forests) were identified in Pingshuo opencast coal mine, China, to determine Total Organic Carbon (TOC) and Soil Liable Organic Carbon (SLOC) in reclaimed RMSs. Adjacent, unreclaimed plot (UR) was selected as references as well. The results showed that after 21-year reclamation, concentrations of TOC and SLOC varied among forests, among which mixed forests had the highest fractions of TOC and SLOC concentrations. Contents of TOC and SLOC were observed decreased as soil depth increased. At the same depth, the Efficient Ratio (ER) of three SLOC varied in the order low SLOC (16.35-24.75%) > middle SLOC (4.14-12.15%) > high SLOC (0.12–1.37%). Carbon Management Index (CMI) varied significantly among forests as well. Overall, TOC and SLOC could be significantly improved after reclamation, and mixed forests especially R. pseudoacacia-P. tabulaeformis could be the most effective reclamation pattern.

Keywords: Soil liable organic carbon; Opencast coal mine; Reclamation; Pingshuo

1 INTRODUCTION

Surface mining, the most common technique used for mining of coal, often causes great degradation to our terrestrial ecosystems (Ussiri and Lal, 2005). Notably, soil profile structure was thoroughly deteriorated with 80% of Total Organic Carbon (TOC) lost in the process of opencast coal mining (Ghose, 2001). The new soils after reclamation are called Reclaimed Mine Soils (RMSs) with poor soil nutrients and immense potential to sequestrate TOC, which provides a platform to observe the accumulation of total and liable SOC from scratch.

Accumulation of TOC is vital for the formation of RMSs, because TOC is an important factor that determines many other soil biological properties in RMSs (Frouz, et al., 2009). In addition, increasing the storage of C in soils has substantial potential to mitigate increases in atmospheric carbon dioxide (CO_2) concentrations. The Soil Liable Organic Carbon (SLOC), accounting for small proportion of TOC, is highly important as it fuels the soil food web, thereby strongly influencing nutrient cycling with accompanying effects on soil quality and productivity. Moreover, its oxidation drives the flux of CO_2 from soils to the atmosphere (Chan et al., 2001). SLOC has a much shorter turnover time (generally less than 10 years) and thus is affected much more rapidly by management-induced changes in organic matter inputs or losses. (Janzen et al., 1997). It has, therefore, been suggested as an early indicator of the effects of soil management and cropping systems on TOC quality (Yang, 2012). Nevertheless, the storage of TOC, especially the SLOC were rarely elucidated in reclaimed mine land.

The specific objectives of this study were to: (1) explore the TOC sequestration and SLOC under reclaimed forest and cropland ecosystems, (2) detect the SLOC Efficient Ratio (ER) Carbon Management Index (CMI) under mixed and monoculture reclamation ecosystems in Pingshuo opencast coal mine, China.

2 MATERIALS AND METHODS

2.1 *Study area*

Pingshuo opencast coal mine (112°10′–113°30′E, 39°23′–39°37′N), the largest opencast coal mine in China, is located in the north of Shanxi Province, east of the Loess Plateau. The ecological environment is fragile and it has an arid to semi-arid continental monsoon climate. The total annual precipitation averages between 428.2 to 449.0 mm and the average annual effective evaporation ranges from 1786.6 to 2598.0 mm. The annual mean temperature is 6.2°C and the dominant vegetation is hay.

2.2 *Sampling plots design*

RMSs of forest plots were identified from 5 predominant types of reclaimed forest with different reclamation patterns in Pinshuo mine, including 3 types of monoculture forest and 2 types of mixed forest: Plot UM: *U. pumila* monoculture; Plot PM: *P. tabulaeformis* monoculture; Plot RM: *R.pseudoacacia* monoculture; Plot RP: *R.pseudoacacia-P. tabulaeformis*; Plot RUA: *R.pseudoacacia-U. pumila-A. altissima*.

Adjacent, undisturbed forest (*Populus simonii* monoculture) was selected for reference, and hereafter called "undisturbed forest (UF)". In addition, a new dump (N39°30′13″, E112°20′14″) without reclamation was selected as "unreclaimed plot (UR)" to provide a baseline for comparison of changes in the TOC and SLOC in RMSs. The unreclaimed plot, with elevation of 1482.7 m, was established in the year 2013.

2.3 *Sampling and processing*

A 100-cm soil profile was dug in each plot after the leaf litter and humus layer was removed. Composite soil samples from 0–20, 20–40 and 40–60 cm depth were collected during August 2014. In each plot, soil samples were collected randomly at three sampling locations after stripping the litter and fermentation on the horizon of topsoil. In each sampling location, five sub-samples were obtained within a range of 10 × 10 m to make the composite sample. The composite soil samples were sealed in plastic bags and transported to the laboratory.

2.4 *Experimental analyses*

After roots and other plant debris were removed, all samples were air-dried and then passed through 0.2 mm meshes for TOC and SLOC analysis. The TOC concentrations were determined following the dry combustion method (Kalembasa and Jenkinson, 1973) using a CHN analyzer. Low, middle and high SLOC were determined using 333, 167 and 33 mM $KMnO_4$ according to Blair et al. (1995).

2.5 *Calculation of ER and CMI*

ER was calculated using the following equations:

$$ER = LOC / TOC \tag{1}$$

CMI was calculated following Blair et al. (1995) using the following equations with and soils in unreclaimed plot as reference.:

$$L = \frac{C\ in\ fraction\ oxidized\ by\ KMnO_4}{C\ remaining\ unoxidized\ by\ KMnO_4} = \frac{C_L}{C_{NL}} \tag{2}$$

$$LI = \frac{Liability\ of\ C\ in\ sample\ soil}{Liability\ of\ C\ in\ references\ soil} \tag{3}$$

$$CPI = \frac{Sample\ total\ C}{Reference\ total\ C} = \frac{C_r\ Sample}{C_r\ Refererence} \tag{4}$$

$$CMI = CPI \times LI \times 100 \tag{5}$$

where L was liability of C, LI was liability index, CPI was carbon pool index and CMI was carbon management index.

2.6 *Statistical analysis*

The significant difference between the mean values of TOC and SLOC in all plots was analyzed using analysis of variance (ANOVA) with LSD post-hoc test at $P < 0.05$ using SPSS 16.0 (SPSS Inc. Chicago, USA).

3 RESULTS AND DISCUSSION

3.1 *TOC concentration*

TOC contents varied significantly among forests ($P < 0.05$) at the same soil layer (Fig. 1). RP was observed the highest fraction among the 7 forest types at all soil layer (0–20, 20–40 and 40–60 cm), indicating that *R. pseudoacacia-P. tabulaeformis* mixed forest was the most effective reclamation pattern to restore TOC and in Pingshuo opencast coal mine (Yuan et al., 2016). UF, however, had a lower

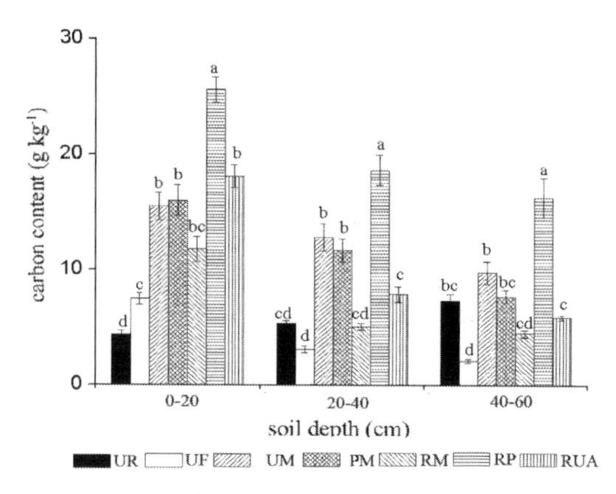

Figure 1. TOC concentrations of 7 plots in Pingshuo opencast coal mine.

TOC concentration compared to reclaimed forests at all soil layers. It demonstrated reclamation facilitated C sequestrating in mine soils. Generally, TOC contents decreased as the soil depth increased (Fig. 1), during which contents in 0–20 cm were significantly higher than that of 20–40 cm and 40–60 cm (P < 0.05), indicating that reclamation restored TOC dominantly in the surface of soil profile. Note that TOC concentrations were observed in an increasing order as soil depth increased in UR, which could be due to procedure in soil layer reconstruction in Pingshuo dumps—fertile soils were discarded under the coverage of poor soils.

3.2 *Low, middle and high SLOC concentration*

Similar to the distribution of TOC among 7 forest types, low, middle and high SLOC varied significantly among forests (P < 0.05) (Fig. 2). The highest fraction of low, middle and high SLOC were found in RP and RUA at 0–20, 20–40 and 40–60 cm (*P* < 0.05), indicating that mixed forests facilitated SLOC accumulating compared to monoculture forests and the undisturbed forest. SLOC contents in UF were observed lower than that of reclaimed forests and SLOC contents decreased as soil depth increased, except those of UR. The higher proportion of SLOC in mixed forest was in accordance with a higher litter or organic material input in RP and RUA.

3.3 *ER of SLOC*

A higher value of ER represents that SLOC is easier to be decomposed by microorganism and absorbed by plants (Janzen et al., 1997). The data presented herein (Fig. 3) showed low, middle and high SLOC ER among forests. Generally, ER of low and middle SLOC did not varied among forests (P > 0.05) except that UR had the least ER. ER of high SLOC, however, varied among forests (P < 0.05). A lower fraction of SLOC ER in UR could be attributed to few organic carbon input and indicating scarce of SLOC could be available for plant to absorbed. ER of low, middle and high SLOC were found decreased as soil depth increased (P > 0.05) in Fig. 3, which could be due to more organic material accumulated in the top soils rather than deep soils. Moreover, ER of low SLOC (16.35–24.75%) exceeded that of middle (4.14–12.15%) and high SLOC (0.12-1.37%), which was in favor of Xu et al. (2013).

3.4 *CMI of SLOC*

CMI indicates the influence that forest type exerts on soil organic carbon pools. The data presented herein (Fig. 4) showed the low, middle and high SLOC CMI among forests. Generally, CMI of low, middle and high SLOC varied significantly among

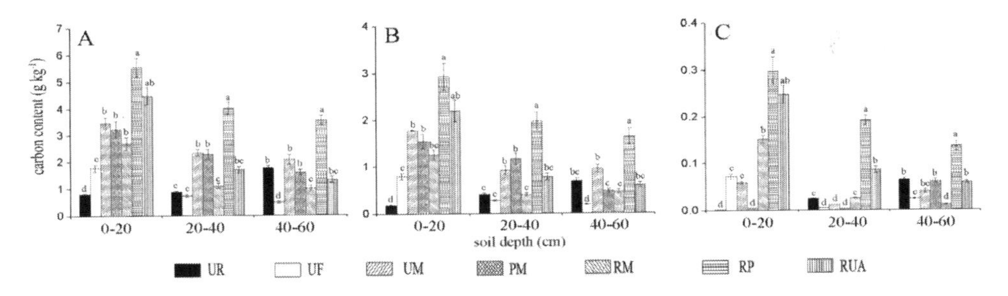

Figure 2. Low, middle and high SLOC concentrations of 7 plots in Pingshuo opencast coal mine. A: low SLOC; B: middle SLOC; C: high SLOC.

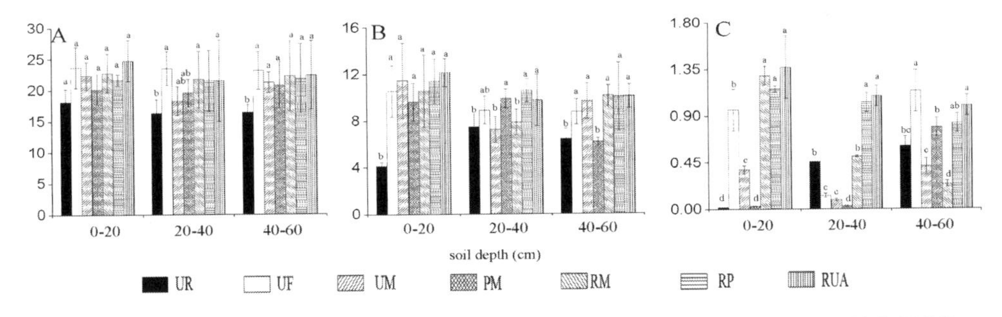

Figure 3. ER (%) of 7 plots in Pingshuo opencast coal mine. A: low SLOC; B: middle SLOC; C: high SLOC.

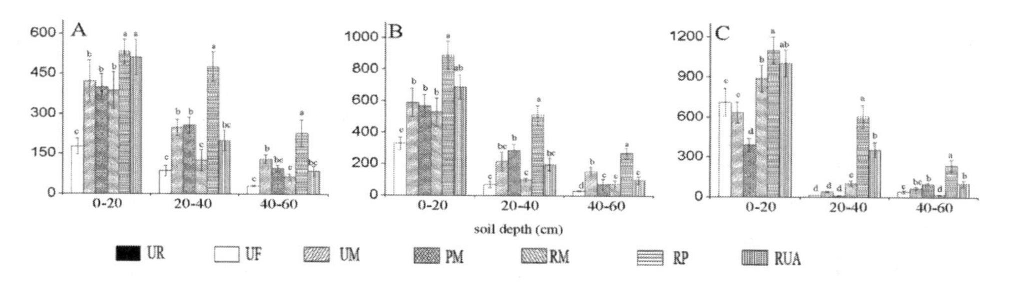

Figure 4. CMI of 7 plots in Pingshuo opencast coal mine. A: low SLOC; B: middle SLOC; C: high SLOC.

forests ($P < 0.05$). Moreover, CMI of low, middle and high SLOC were found decreased as soil depth increased ($P < 0.05$). The variance of CMI in 7 plots in our study area demonstrated reclamation patterns had significantly influence on TOC and SLOC especially on the topsoil.

4 CONCLUSIONS

Concentrations of TOC and SLOC varied among forests, among which mixed forests had the highest fractions of TOC and SLOC concentrations. TOC and SLOC contents were observed decreased as soil depth increased. The variance of ER and CMI among forests indicating reclamation patterns exerted significantly influence on TOC and SLOC in RMSs. Overall, TOC and SLOC could be significantly improved after reclamation and *R.pseudoacacia-P. tabulaeformis* and *R.pseudoacacia-U. pumila-A. altissima* were recommended as favorable reclamation patterns in Pingshuo opencast coal mine.

ACKNOWLEDGMENTS

This study was supported by the Chinese Universities Scientific Fund (grant no.2652016115).

REFERENCES

Chan, K.Y., Bowman, A. & Oates, A. (2001). Oxidizible organic carbon fractions and soil quality changes in an oxic paleustalf under different pasture leys. *J. Soil Science,* 166, 61–67.

Frouz, J., Pizl, V., Cienciala, E. & Kalcik, J. (2009). Carbon storage in post-mining forest soil, the role of tree biomass and soil bioturbation. *J. Biogeochemistry*, 94, 111–121.

Ghose, M. (2001). Management of topsoil for geo-environmental reclamation of coal mining areas. *J. Environment Geology,* 40, 1405–1410.

Janzen, H.H., Campbell, C.A., Brandt, S.A., Lafond, G.P. & Townley Smith, L. (1992) Light fraction organic matter in soils from long term crop rotations. *Soil Science Society of J. American Journal,* 56, 1799–1806.

Kalembasa, S.J. & Jenkinson, D.S. (1973). A comparative study of titrimetric and gravimetric methods for the determination of organic carbon in soil. *J. Science, Food and Agriculture,* 24, 1085–1090.

Ussiri D.A.N. & Lal, R. (2005) Carbon sequestration in reclaimed mine soils. *J. Critical Reviews in Plant Science,* 24, 151–165.

Xu, P., Jiang, C.S., Hao, Q.J. & Zhu, D. 2013 Effects of the different land use on soil labile organic matter and carbon management index in Junyun Mountain. *J. Environment Science*, 34(10), 4009–4015.

Yang, X.Y, Ren, W.D., Sun, B.H. & Zhang, S.L. (2012) Effects of contrasting soil management regimes on total and labile soil organic carbon fractions in a loess soil in China. *J. Geoderma,* 177–178, 49–56.

Yuan, Y., Zhao, Z.Q., & Zhang, P.F. (2017) Soil organic carbon and nitrogen pools in reclaimed mine soils under forest and cropland ecosystems in the Loess Plateau, China. *J. Ecological Engineering*, 102, 137–144.

Land Reclamation in Ecological Fragile Areas – Hu (Ed.)
 ISBN 978-1-138-05103-4

Soil available water in texturally different soils from overburden of an open pit mine in Xilinhot

X. Wang, Y.L. Zhao, H.F. Liu & S.S. Tian
Institute of Land Reclamation and Ecological Restoration, China University of Mining and Technology, Beijing, China

ABSTRACT: Soil available water content is a vital index of assessing soil drought resistance and soil water holding capacity. Basic physical parameters (texture, particle size, bulk density and total porosity) and soil water characteristic curves of different soil layers which taken from overburden of an open pit mine in Xilinhot were measured. The results showed that the soil water holding capacity from high to low is L5 > L1 > L2 > L3 > L4. The present researches indicated that the soil available water, field capacity, permanent wilting point showed defined relation with clay and silt content, and this relation turned out that an increase volumetric water retention occurred with increasing of clay and silt and/or bulk density. In general, L5 had a good water holding capacity than others, and its thick soil layer make it suitable as the topsoil alternatives of soil profile reconstruction in land reclamation.

1 INTRODUCTION

Coal mining often makes soil structure destroying and soil degrading, combining with the local ecological fragile and desertification. Hence, mine soils has the characteristics of poor structure, low nutrient availability, low water holding capacity and low biomass productivity. In land reclamation, chemical, physical and biological properties are always selected as the main soil quality indicators, however among all of the factors that affect local vegetation growth, soil water content is the most important. Soil water availability is very crucial for plants because their growth principally depends on the soil water storage. Thus, measurements of soil water content and soil Available Water (AW) of different soils is an important field for research and it can conduct as an important basis used to choose what kind of soils that are more suitable for land reclamation. However, there is little information about the AW condition for a range of soils in Xilinhot (Inner Mongolia) yet. Therefore, the objectives of this study were: (1) to determine the Soil Water Characteristic Curve (SWCC) of five soil layers; (2) to calculate and compare the soil Available Water (AW), Field Capacity (FC), Permanent Wilting Point (PWP) among these soils (3) to analysis soil water holding capacity of various soil types in this region.

2 MATERIALS AND METHODS

2.1 *The study site and soil properties*

Soil samples were collected from five layers of four profiles in Xilinhot (Inner Mongolia). Laboratory basic physical test and particle size distribution test were taken to these soil samples. The selected soils were mainly different in soil texture (i.e. clay content). The region has a semi-arid climate, with mean annual temperature and precipitation of 1.7°C and less than 300 mm, respectively.

2.2 *Soil water retention measurements and modeling*

The SWCC is defined as the relationship between water content and water potential for the soil (Williams 1982). The soil samples were brought to the state of full saturation with water (SW), 0 kPa. Next, soil water retention was determined at water potentials of 0, 1, 10, 30, 50, 100, 300, 500, 1000 and 1500 kPa using porous ceramic plate apparatuses. The SWCC model development by van Genuchten (1980) is widely used and selected for fitting:

$$\theta = \theta r + (\theta s - \theta r)\left[1/1 + (\alpha h)^{n}\right]^{m} \quad (1)$$

where θ ($cm^3\ cm^{-3}$) is the soil water content, θs ($cm^3\ cm^{-3}$) and, θr ($cm^3\ cm^{-3}$) are saturated and residual soil water content, respectively. The parameters η and α are fitting coefficients, and h is the water potentials.

2.3 *Calculations of FC, PWP and AW*

The soil available water can be defined as the difference between water content at Field Capacity (FC) and Permanent Wilting Point (PWP):

$$AW = FC - PWP \quad (2)$$

The soil water contents at FC and PWP were obtained from the calculated SWCC and the water content values under the water potentials of 33 kPa and 1500 kPa are common choices for FC and PWP, respectively. The soil water retention data were used for determination and calculation of FC, PWP and AW and corresponding water potentials: gravitational water GW = SW – FC, water available for plants AW = FC – PWP, water unavailable for plants UW = PWP.

3 RESULTS AND DISCUSSION

Some physical properties of the studied soils are shown in Table 1. The soil bulk density values ranged from 1.22 to 1.72 g cm^{-3} with an order L3 > L2 > L4 > L1 > L5 and the L3 is the most compacted soils (Table 1). Table1 shows that the soil samples taken from this region had a high proportion of sand and a low proportion of silt and clay, thus the soil textures were characterized as sandy loam, loamy sand, sand or silt loam. The trends of porosity did not exhibit any notable changes related to soil layers.

3.1 *Measurement and fitting of SWCC*

Fig. 1 shows the fitted SWCC. According to the fact that the calculated R^2 values were in the range of 0.9867–0.9995, the imitative effect of the applied model to the soil water retention data were fine.

Soil available water depends in a complex way on soil structure and composition, consequently the shapes of the SWCC illustrated the water holding capacity and the condition of water content of the studied soils. The SWCC plotted for five soils had visible difference revealing the variance in retention capabilities, porosity and grain composition. The SWCC showed that the maximum water holding capacity attained the highest value in L5, medium-in L1, L2, and the lowest in L3, L4. And retention of AW for plants was visible diverse between the studied soils.

Table 1. Soil particles distribution and basic physical properties of the studied soil.

Sample	Clay (%)	Silt (%)	Sand (%)	Bulk density (g/cm^3)	Porosity (%)
L1	4.55	37.73	57.72	1.23	0.51
L2	4.43	34.04	61.54	1.59	0.39
L3	3.32	21.80	74.87	1.72	0.33
L4	0.15	2.69	97.16	1.54	0.37
L5	7.44	71.72	20.83	1.22	0.51

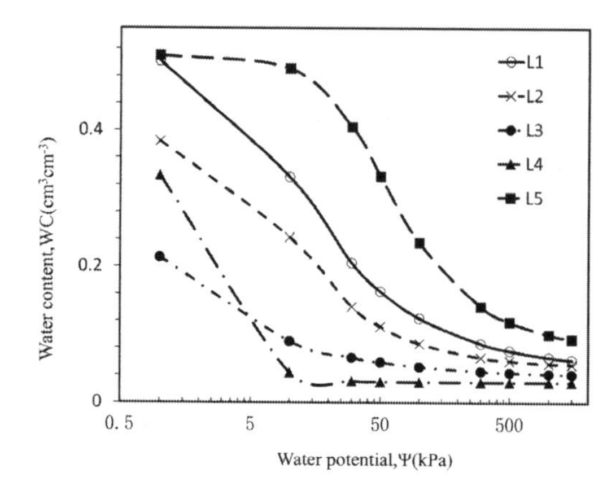

Figure 1. Fitted soil water characteristic curves.

Table 2. Water retention categories and water constant of the studied soils.

Parameter	GW, m^3 m^{-3}	AW, m^3 m^{-3}	UW, m^3 m^{-3}	FC, m^3 m^{-3}	PWP, m^3 m^{-3}
L1	31.5	13.3	6.3	19.6	6.3
L2	25.6	7.8	5.6	13.4	5.6
L3	26.5	2.4	4.1	6.5	4.1
L4	32.7	0.1	3.1	3.1	3
L5	11.8	29.9	9.3	39.2	9.3

3.2 *FC PWP and AW values for plants in the studied soils*

The volumetric values for FC, PWP, AW, GW, and UW are presented in Table 2. There were significant differences in these values among the studied soils. In general, the highest value of GW were found in L4 and its lowest value was obtained in L5. The highest value of UW were observed in L5, and its lowest values were seen in L4. On the other hand, the highest values of FC PWP and AW were all obtained in L5, while the values of other soils decrease along with depth of soil layers, the difference of which is extremely significant among different layers. And the changes were extremely similar to that of clay and silt, in contrast with sand. The highest values of FC PWP and AW were observed in L5 (Table 2) with high clay and silt content and low bulk density (Table 1). The lowest values of FC PWP and AW were determined for L4 (Table 2) with highest sand content and relatively high bulk density (Table 1). This relation indicates that an increase volumetric water retention occurred with increasing of clay and silt and/or bulk density.

4 CONCLUSION

1. Soil bulk density in the L3 are the highest among the five types of studied soils, while in the L5 is the lowest. The soil of texture, bulk density, porosity did not exhibit significant laws along with soil layer depth.
2. L5 had a better water holding capacity than other soil samples along with a highest clay content. Soil water holding capacity in other soils decrease with soil layer depth. The present researches indicated that the AW, FC and PWP showed defined relation with clay and silt content.
3. According to the survey, the reserves of soil L5 was more sufficient than other soils, in addition, the fine water holding capacity make it suitable as the topsoil alternatives of soil profile reconstruction in land reclamation.

ACKNOWLEDGEMENT

This study was supported by the National Key Research and Development Program of China (2016YFC0501103) and the ShenHua Group Corporation Limited. The experiment was carried out in China University of Mining and Technology (Beijing).

REFERENCES

Beata Kolodziej, Maja Bryk, Anna Slowinska-Jurkiewicz, Otremba Krzysztof & Miroslawa Gilewska (2016). Soil physical properties of agriculturally reclaimed area after lignite mine: a case study from central Poland. Soil & Tillage Research, 163, 54–63.

Horne, D.J., & Scotter, D.R. (2016). The available water holding capacity of soils under pasture. *Agricultural Water Management*, 177, 165–171.

Kołodziej, B., Bryk, M., Słowińska-Jurkiewicz, A., Otremba, K., & Gilewska, M. (2016). Soil physical properties of agriculturally reclaimed area after lignite mine: a case study from central Poland. *Soil & Tillage Research*, 163, 54–63.

Wang, Yates, S., R., Lowery, van Genuchten, & Th, M. (1998). Estimating, soil hydraulic properties using tension infiltrometers with varying disk diameters. Soil Science, 163(5), 356–361.

Webster, R., & Oliver, M.A. (1992). Sample adequately to estimate variogram of soil properties. *European Journal of Soil Science*, 43(1), 177–192.

Land Reclamation in Ecological Fragile Areas – Hu (Ed.)
© 2017 Taylor & Francis Group, London, ISBN 978-1-138-05103-4

Study on the treatment method of stripping soil in an opencast mining area based on trade background

Y.Y. Li, W. Zhou & J.H. Tong
Department of Land Science and Technology, China University of Geosciences, Beijing, China

ABSTRACT: Land is a scarce resources, so rational use and protect is very important. This paper based on the trade context studied a new method for treating peeling soil in opencast mine. Research purposes: Through established the trading system of stripped topsoil resources protect ecological environment in stripped zone and coverage zone, increase the available land area. Research content: This paper analyzes the differences about mines reclamation between domestic and foreign, puts forward the trading system of soil stripping, and demonstrates its economic, social and ecological feasibility. Expected research results: Establishing the basic framework of the soil trading system; Maximizing use of stripped soil, increasing the available land area, and easing land conflicts; Providing the basis for the government formulate relevant policies and other scholars carry out land-related research.

1 INTRODUCTION

Soil stripping in open mining area refers to strip material from the open-pit mine to the designated site during the opencast working, that the amount of stripping can be several times to the amount of mining and it need a large number of land for storage. Land is covered more than 450000 hm^2 by opencast mining excavation and occupation in China (Bian, 2005).

2 ANALYSIS OF RECLAMATION DIFFERENCES AT HOME AND ABROAD

The United States and Germany is the first to start reclamation in 1970s and 1980s (Hu & Zhao 2004). Next, other developed countries formulate relevant legal, started earlier than the domestic; The United States, Germany and other focus on reclamation ecological goals (Jin & Bian, 2009), while China's early reclamation aimed at increase the number of land (Hu & Zhao 2004); The United States, Canada major research soil reconstruction and improvement, vegetation reconstruction and erosion control. The United Kingdom is mainly to research pollution reclamation and waste reclamation (Hu, 1996), In the domestic, Bai & Zuo (2001) study mining planning and design, ecological restoration (Bai, 1999) and others researches. With the development of new technologies, Zhou (Zhou & Bai, 2009), Zhang (2012) used GIS, RS, GPS, etc. in planning and design, reclamation area monitoring and other aspects.

3 AN OVERVIEW OF THE STRIPPING SOIL TRADE SYSTEM

The stripping soil trade system is the coordination system of government, enterprises and individuals. The specific transaction process is as follows:

4 FEASIBILITY ANALYSIS

4.1 *Market demand analysis*

1. The number of cultivated land decreased year by year (Figure 1), the proportion of high quality cultivated land is small only 29.40%, the responsibility of reclamation is arduous.
2. The demand of urban construction land is increasing (Figure 2).
3. The government attaches importance to stripping land reclamation work.

4.2 *Engineering technology analysis*

1. Soil stripping technology caused widespread concern. Fu & Chen (2004), Yu & Yuan (2016), Dou et al. (2014) study the stripping process, stripping mode. Che (2008) study on the stripping of construction machinery. Xu et al. (2011), Lin et al. (2015), Zhang et al. (2015)

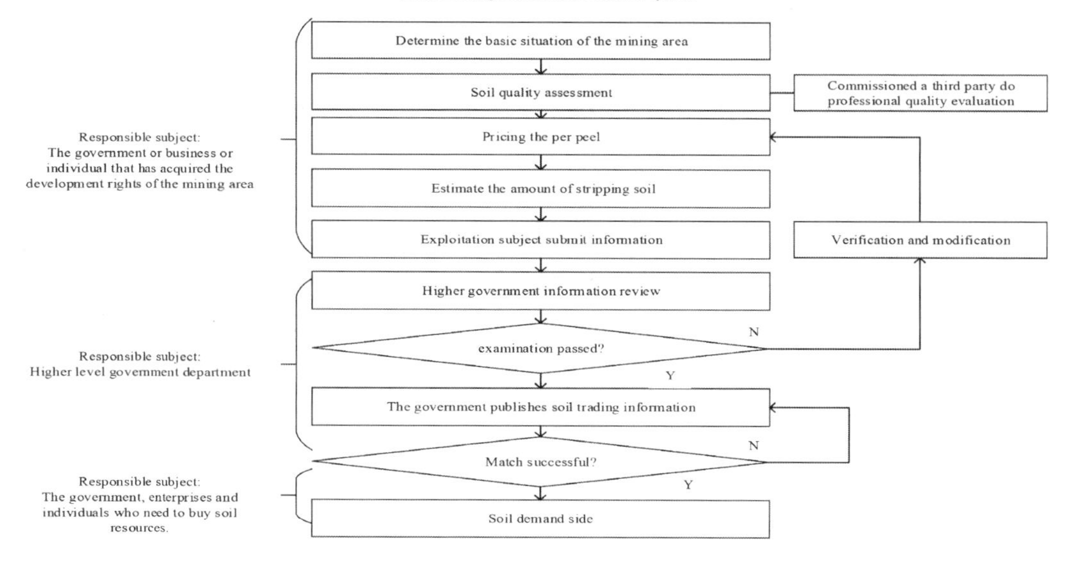

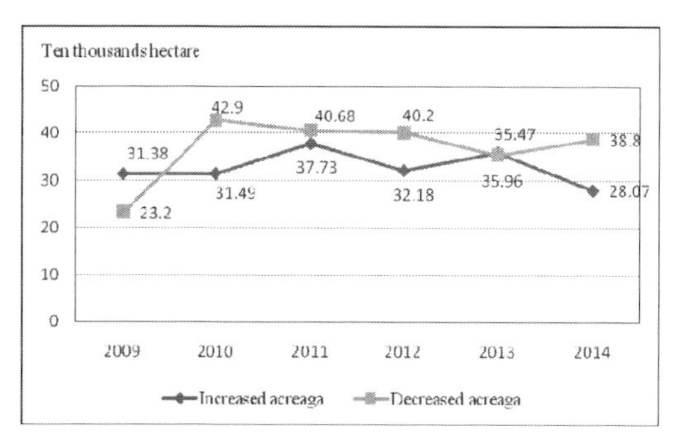

Figure 1. Change of cultivated land from 2009 to 2014.

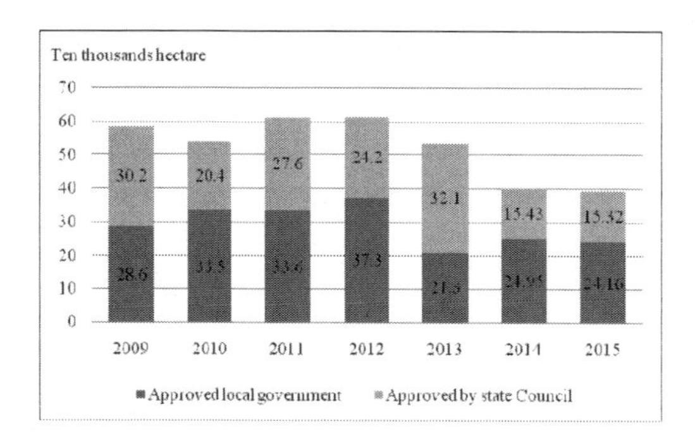

Figure 2. 2009 to 2015 the number of approved construction land changes.

study on the cover reconstruction technology of the different targets.

2. The amount of stripping soil obtained by the mine DEM images through using ARCGIS.

4.3 *Comprehensive benefit analysis*

4.3.1 *Economic benefits*

1. Farming layer soil sale with price, resulting in direct economic benefits.
2. Soil quality has improved, crop yield increased, so bringing direct economic benefits.

4.3.2 *Social benefits*

System can alleviate contradiction between cultivated land protection, construction land supplying, ameliorate low-yielding fields, effectively increase supply social agricultural products and solve the problem of construction land supplying (Zhang et al., 2015).

4.3.3 *Ecological benefits*

Topsoil stripping is better than the terms of mixed reclamation and it can improve soil quality and sustain for soil microbial growth. Fu et al. (2004) proved that topsoil stripping can improve the soil quality quickly (Tables 1 and 2), and Zhang et al. (2002) proved it can effectively increase the number of soil microbial.

5 SUMMARY AND PROSPECT

This paper establishes basic framework of soil trading system, demonstrates the feasibility of soil trade system from three aspects: economic benefit, social benefit and ecological benefit. But there are still shortcomings, reflected in the following: the study of targeted, lack of the overall concept; research lacks data support; national legal policy

Table 1. Effect of different technologies of reclamation on content of soil nutrients.

Reclamation method	Soil depth/cm	Organic matter/g/kg	Total nitrogen /kg	Total phosphorus /kg	Available potassium /kg
Topsoil stripping	0–20	18.58	0.70	0.56	157.5
Mixed reclamation	0–20	14.55	0.68	0.50	108.0

Table 2. Effect of different technologies of reclamation on soil properties.

Reclamation method	Bulk Density/cm	Organic matter/g/kg	Capillary moisture capacity/kg	Total porosity/kg
Topsoil stripping	0–20	1.53	27.30	43.77
Mixed reclamation	0–20	1.17	37.04	55.85

support is light. To achieve rapidly development, we should deal with the next problems: strengthen practical research and provide strong data support; according to different conditions, flexibly hand the stripping, covering and trading problems; government, business and personal should coordinate with each other; national and local governments need to strengthen policy support and legal norms.

ACKNOWLEDGMENT

Fund project: National Natural Science Foundation of China. "Spacial heterogeneity and indicative significance of reclamation soil in loess opencast coal mining (41570108311)."

REFERENCES

Bai, Z.K., Zuo, X., (2001) Case Study on Land Reclamation Planning of Large Opencast Coal Mine. *Journal of Soil and Water Conservation*, 15(4), 118–121.

Bian Z.F. (2005) Study on Land Reclamation and Ecological Reconstruction in China's Coal Mining Area. *Resources & Industries*, 7(2), 18–24.

Bai, Z.K., Wang, W.Y., Li, J.C. etal. (1999) Research on Basic Theory and Method of Ecological Reconstruction in Mining Area. *Energy Environmental Protection*, (1), 10–14.

Che, Z.X. (2008) Study on Key Technology Equipment of Semi-continuous Mining in Topsoil Separation of Open-pit Coal Mine. *China Coal*, 34(4), 47–48.

Dou, S. Dong, x. Dong, L.J. et al. (2014) Technical System of Topsoil Stripping in Song Liao Plain—A Case Study of Song Yuan City. *Journal of Jilin Agricultural University*, 36(2), 127–133.

Fu, M.C., Chen, Q.J. (2004) Top soil stripping and its process in ecological reclamation of mining area. *Metal Mine*, 2004(8), 63–65.

Hu, Z.Q. (1996) New Progress of Land Reclamation in Foreign Countries. *China Land*, (10), 41–42.

Hu, Z.Q., Zhao, Y.L., Cheng, L.L. (2004). Land Reclamation Goal and Connotation Extension in China. *China Land Sciences*, 18(3), 3–8.

Jin, D., Bian Z.F. (2009) Comparison and Reference of Land Reclamation Policies and Regulations at Home and Abroad. *China Land Sciences*, 23(10), 66–73.

Lin, S.Y. Zhang, L. Che, T.T. et al. (2015) Effects of Different Topsoil Thickness on Soil Nutrients and Corrosion Resistance. *Acta Agriculturae Universitatis Jiangxiensis.* (3), 556–563.

Xu, Y. Zhang, F.R. Zhao, H.F. et al. (2011) Discussion on the necessary condition of soil peeling for soil fertility in tillage layer. *China Land Sciences*, (11), 93–96.

Yu, D., Yuan, S.G. (2016) A Discussion on the Technology of Constructing the Partial Stripping of Cultivated Land. *Chinese Journal of Agricultural Resources and Regional Planning*, 37(8), 47–51.

Zhou, W., Bai, Z.K. (2009) Analysis on the Evolution of Ecological Environment in the Combined Mining Area of Pingshuo Coal Mine. *Journal of Shanxi Agricultural University (Natural Science Edition)*, 29(6), 494–500.

Zhang, J.M., Ren, Y.Q., Fu, K.J. et al. (2012) Application of Remote Sensing and Gis Technology in Large Dump Site Location. *Metal Mine*, 49(9), 111–113.

Zhang, F.R., (2015) Technical and Economic Analysis on the Utilization of Cultivated Land Topsoil in the Black Soil Area for Rural Reclamation. *Chinese Journal of Soil Science*. (5), 1034–1039.

Zhang, J.E. Liu, W.G. Hu, G. (2002) Relationship between soil microbial quantity and soil fertility under different land use patterns. *Ecology and Environmental Sciences*. 11(2), 140–143.

Land Reclamation in Ecological Fragile Areas – Hu (Ed.)

Peatland restoration after open pit mining

Aleksandr Mikhailov
Department of Mechanical Engineering, Saint Petersburg Mining University, Saint Petersburg, Russia

Arkadii Kim
Department of Water Management and Ecology, Saint Petersburg State University of Architecture and Civil Engineering, Saint Petersburg, Russia

Oleg Prodous
ENCO-Engineering LLC, Saint Petersburg, Russia

ABSTRACT: For design of peat excavation technology on the peat deposit "Orshinsky moss" of the Tver region, Russia technically feasible and economically expedient solutions were proposed. This peat deposit is located in 30 km from Tver city. This peat deposit has the area over 35 km^2. Peat resources are estimated at approximately 28.0 million tons. The geological structure of a peat deposit allows to conduct working off of a pit in the open way. The production capacity of the enterprise for peat with conditional moisture of 70% – 545 760 m^3/year. The project provided excavation and drying of peat raw materials to moisture content of 70%, its transportation to the plant for processing, technical land reclamation. The reclamation direction of the pit is a nature protection pond by natural filling and overgrowing for the purpose of repeated bogging of the territory.

Arrangement and formation of a landscape of future pond depend on a form of open-pit development and level of ground waters bedding. During creation of a pond it is given a rectangular shape with depth – 3–4 m. In the center of this pond there is an area (island) 40 × 30 m in size for nesting of waterfowl.

1 INTRODUCTION

Wetlands cover about 5 to 8% of the world's land surface, and over 50% of the world's wetlands are peatlands. These percentages extrapolate to between 386 to 409 million ha of peatlands. (Stefanie, 2004). Peatlands are natural systems performing local, regional and often global functions. Peat is a biogenic deposit which, consists of about 90–95% water and about 5–10% solid material. Various extraction techniques exist for the peat excavation from a peat deposit. There are some principles of the pit mining: 1) principles for the wise use of peatlands; 2) peat is produced in a sustainable way; 3) in peat production primarily earlier drained peatlands are used; 4) In the production there must be technically and economically best available environment protection technology in use.

The Code of Practice for Responsible Peatland Management defines the principles for responsible management for the first stages of the production chain including site selection and preparation, peat extraction, and after-use of peatlands (Code, 2011). The basic condition for peat extraction is the naturally accumulated peat deposit. Positive experiences have also been gained on creating wetlands on bog foundations. A lot of documented information is available on different alternatives (Selin, 1999) which can be used in planning the after-use.

After-use may include rehabilitation of the peatland ecosystem, alteration of land use to forestry, agriculture, recreation or urban development or a combination of different land use forms. Where peatland rehabilitation is the chosen after-use strategy, member companies shall aim for a high level of conservation of biodiversity and peatland ecosystem function.

The important part of any restoration project is definition of the target ecosystem. The restoration scheme assumes a further increase in the proportion of near-natural restoration in peat mined areas; mostly the creation of a gentle lake coastline (including island) with favorable conditions for shallow water vegetation (Rehounkova et al., 2011).

Wise after-use at terminated peat cutover areas should consider a number of important issues such as landscape and biodiversity conditions. One such option is wetland rewetting. Effects of this relate to vegetation, biology. Vegetation and limnic life show a great part of the success for biodiversity. Previous studies (Lundin et al., 2008) have indicated that, prior to the creation of cutaway lakes, considerable on-site management work should be carried out. Depending on the lake depth, size, aquatic chemistry, vegetation and other properties, the created lakes can be used for recreation purposes or can be important sites for wildlife, especially waterfowl conservation (Lamers et al., 2002).

Creation of artificial lakes can go together very well with wetland restoration but may require further considerations and additional actions. Lake creation is frequently accompanied by the development of marsh vegetation and open water habitats which provide food, nesting sites, and protection for fish, waterfowl and wetland animals (Kavanagh, B. 1998).

The creation of the lake should start by leaving a 30–60 cm layer of peat to support invading aquatic vegetation. It is desirable to create an uneven surface on the bottom of the lake in order to create islands and variations in water depth (Vikberg, 1996). In recent years, creation of shallow lakes (1–2 m depth) in territories of cutaway peatlands is also considered as a valuable management activity. The islands should be at least 10 m² in size and need to be protected against erosion (Hörnsten, 1992).

The project provided development of a peat deposit in the open-pit mining, drying of peat raw materials to moisture content of 70%, its transportation for processing on plant, technical land reclamation. This paper provides an overview of the current research specifically aimed at the regeneration of peatland after open pit mining through creation of nature conservation pond.

2 BACKGROUND

The study area is located in the Tver region. Tver city is 167 km away from Moscow. The peat deposit "Orshinsky moss" is located 30 km east of Tver city (56°55′05,13″ N, 36°24′19,54″ E). The site is a natural bog with Sphagnum peat. Peat thickness – 6 m. Type of peat (Sphagnum peat). Degree of decomposition (von Post scale) – H2-H4.

The climate of Tver region is moderately continental, i.e. displaying some features transitional from the continental climate to a damper one of the north-western areas. Average yearly precipitation is 653 mm, 132 mm of which falls as snow. The warmest time of year in Tver is in July when it is 17.5°C on average, but could get up to 22.7°C maximum. The coldest time of year in Tver is in January when it is −10.7°C on average, but could get down to −14.2 °C minimum. Average rainy days −131. Average snowy days −107. The vegetative period - 170 days.

The region is divided into two forestry areas: the south taiga and the mixed coniferous-broad-leaved forest. Thus, the considered territory is in a zone of excess moistening. Depth of frost penetration in soils - 135 cm. The area is characterized by high relative humidity of air during the whole year. Winds western and southwest prevail. Small speed of wind is noted in the fall and in the winter. The considered territory is a part of the Volga-Tveretsa lowland which is a part of the Upper Volga lowland. The southern and northern parts of the area a flat relief with absolute marks from 135 to 140 m. Biases of a surface change from 0,5–1% to 3–4%. Surface of peat bogs equal strongly overgrown with a sedge, grass and small shrubby vegetation.

3 PROJECT PURPOSE

For design of peat excavation technology on the peat deposit "Orshinsky moss" of the Tver region, Russia technically feasible and economically expedient solutions were proposed. This peat deposit has the area over 35 km². Peat resources are estimated at approximately 28.0 million tons. This raised bog peat lying in a hollow on a watershed of the Volga and Medveditsa Rivers.

For the choice of the site of extraction of peat three sites F1, F2, F3 on the nature of drying network, extent of drawdown of a peat deposit and a condition of its surface, availability and infrastructure are allocated on earlier drained area (Figure 1).

Now more and more significant role of peat will be shown in the field of its agro-industrial use. The following real prospects development of peat resources are outlined:

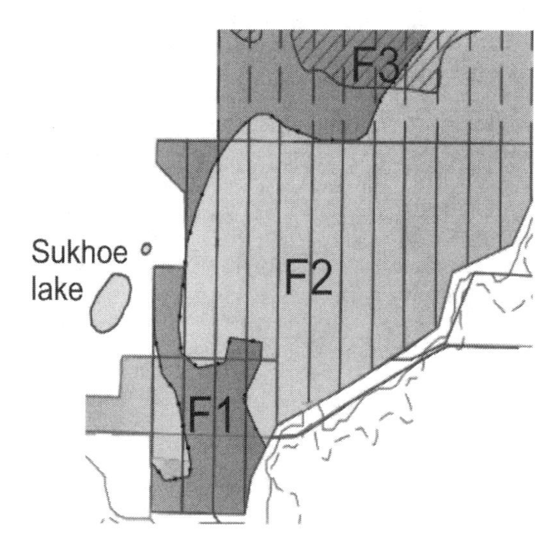

Figure 1. Plan of peat deposit.

- extraction of peat for preparation of organic and organ and mineral fertilizers;
- production of soil for hothouse and greenhouse farms, etc. products for vegetable growing of the protected soil;
- fishery, etc. the directions of use of the developed areas of peat fields.

Peat is perfectly granulated. The granulated peat is good fuel and product for agriculture and horticulture and will be highly profitable business.

The following constraints and goals were therefore set:

- Minimizing the environmental impacts (eliminating dust, noise and water impacts, reducing greenhouse gas emissions);
- Maximizing the utilization of solar energy and minimizing the weather sensitivity;
- Improving the peat product quality.

The advantages and disadvantages of the peat production method:

- Minimizing the environmental impacts;
- Extension of the production season (April - September);
- Optimal utilization of the weather conditions;
- Production efficiency 20 times higher than current peat production;
- Increased amount of exploitable peat resources;
- Rapid restoration of the production area back to a carbon sink.

Analyzing the technological level of peat production manufacturing processes, we selected two key directions to enhance the degree of energy potential use (Mikhailov, 2013):

- magnification of peat production season duration;
- intensifying the processes of peat excavation and drying.

4 PROJECT DESCRIPTION

Working off of a pit can happen all the year round. Overburden acts and are stored separately, and further is used at reclamation of lands. The project accepted transport, one-onboard longitudinally cross system of development of a peat deposit. System of development—continuous, when quarrying mountain preparatory work during operation is absent. The technological scheme is characterized by use of the mining-transport equipment of cyclic action: excavator (hydraulic shovel) and tractor and transport units.

Development of a peat deposit conduct on all depth of a peat deposit lower than the level of the parking of the excavator a side face with loading of peat in vehicles and transporting on a surface out of pit limits on the place of peat dehydration and storage.

Excavation depending on thickness of a peat deposit changes from 1,5 to 6,0 m. The peat deposit is excavating by one ledge. The summer technology is provided in the period May-October. The axis of the excavator driving is combined with a peat dredging axis. Transport trailers for loading give the forward course and establish at distance 1 m from a ledge brow with such condition that the average angle of excavator arrow rotation did not exceed 70°. The constructive sizes are presented on Figure 2.

This technology has several different components, which can be modified to adapt to site-specific conditions. The site needs to be accessible by the heavy equipment needed to conduct the excavation activities. There should be a readily available source of suitable fill material for backfilling the excavation. Excavation and transport can be conducted in any climate; however, actual fieldwork is generally seasonally dependent. The hydrogeology at the site should be considered, and excavation may not be appropriate if the groundwater is shallow.

Excavation and transport technologies have the following advantages:

Figure 2. Location of the study sites on the peat deposit.

- permanent;
- immediate result for the removal of exposure pathways;
- easily implemented;
- limited long-term monitoring and institutional controls;
- flexible.

The mining process consisted of excavating the contaminated ground and hauling it to an on-site lined unit via bulldozer and front-end loader. Stockpiles of contaminated material will be constructed around the open-pit storage for the future using for reclamation plan. Dust was not an issue as the material was reportedly wet and solid. The dried peat is loading on semitrailer with the front-end loader and transporting to the pellet plant. The hauling distance is 12 km.

In the conditions of this pit the accepted excavation system has to provide safe, economic and fullest excavation of peat raw materials at observance of measures for environmental protection.

The body of a peat deposit of this site on a hade belongs to horizontal. The peat deposit is tracked on depth to 6 m from the surface of the field (ordnance-datum 144 m), at the same time the level of ground waters is in winter time on ordnance-datum 140 m. On this classification sign the longitudinal system of open-pit mining is accepted (one-onboard) when moving the front of excavation works perpendicular to the gross channel.

In the project, it is accepted technological schemes of open-pin mining by the excavator and tractors complex applied to mining works in all working zone of a pit. According to the plan of pit reclamation, in the center of a pit the area the sizes of 40 × 30 m under the natural island in the center of the artificial pool after pit reclamation is left untouched.

Advantage of the scheme: a possibility of fast input of a pit in operation, high mobility.

The types of equipment used and the environmental conditions will affect the man- and machine-hours required to complete a given amount of work. Each piece of equipment is specifically designed to perform certain mechanical tasks. For this project, the excavator is the most important piece of equipment required for peat removal and handling. It is efficient, relatively low cost, can move high tonnages of materials quickly.

The production capacity of the enterprise for peat raw materials with conditional moisture content – 545 760 m^3/year. The project provided excavation and drying of peat raw materials to moisture content of 70%, its transportation to the plant for processing, technical land reclamation. The reclamation direction of the developed pit is a nature protection pond by natural filling and overgrowing for the purpose of repeated bogging of the territory.

Extraction of peat can be made before formation of snow cover on the peat field taking into account duration of standing of snow cover and existence of seasonal permafrost during the spring period.

5 RECLAMATION PLAN

There are multiple possibilities for wise after-use of terminated peat extraction areas (Joosten & Clarke, 2002). Wetland restoration is one and relates to restoration of the hydrology to achieve a functioning wetland ecosystem. Indicators of success are water quality, vegetation development and freshwater biology such as bottom fauna. There are several examples of creating lakes in depleted peatlands in Europe and especially in Russia after peat hydraulicking or excavating by multibucket excavator of fifty years ago. It is important to keep in mind that pools are aquatic habitats that do not need to be beautiful to meet fauna requirements. These pools were created following simple rules [Quinty & Rochefort, 2003]:

- *Size and shape*: It was suggested to create pools about 60 × 200 m. This represents a minimum range of size to be useful for wildlife such as birds during migration.
- *Depth*: Pools have a depth that allows the presence of permanent standing water all summer. It should be between 2 and 6 meters. The presence of some sort of structure in and out of the water is important for the establishment of many species. It is also important not to reach the mineral substrate (the need to have 0.5 m of peat to avoid reaching the mineral substrate).
- *Slopes*: The creation of a gentle slope on one side and an abrupt slope on the opposite side should help increase the diversity. It is also the easiest way to create pools with hydraulic excavator.

According to water and physical characteristics peat belongs to potentially fertile soil suitable for growth of plants and to types suitable for biological reclamation and to overgrowth. The required direction of reclamation is a reservoir of nature protection appointment by natural filling and overgrowing.

At reclamation of the developed peat pit the following requirements have to be fulfilled:

- carrying out reclamation right after the end of development of a peat pit;
- area planning and cleaning from technogenic wastes.

Pool after peat excavation belongs to superficial – 5–6 m, and on the area to medium-sized – 10–15 hectares. Planning of restoration areas (the site of topsoil, the temporary areas, a strip for the movement of the construction equipment) is carried out by method of topsoil moving by bulldozer T10MB and loader Amkodor 342P. The total area of annual land reclamation makes 11.0 hectares, including: a pool (a dredging bottom) – 9.1 hectares, slopes of a pit – 0.7 hectares, the planned surrounding territory –1.2 hectares. Making of slopes is provided with a dip (1:3) that excludes development of erosive processes on slopes. The minimum depth of a reservoir has to make not less than 2 m at the minimum seasonal level of fluctuation of water.

From conditions of peat production, the pool has rectangular forms in the plan with maximum width 220 m and maximum length 465 m. Arrangement and formation of a landscape of future pond depend on a form of open-pit development and level of ground waters bedding. During creation of a pond it is given a rectangular shape with depth – 3–4 m. In the center of this pond there is an area (island) 40 × 30 m in size for nesting of waterfowl.

For creation of fertile (potentially fertile) layer of earth around the pool soil of overburden soil which prior to mining works was removed and complex in dumps is used. The territory is planned on all area around the created pool. Ground water in the flooded ponds can be considered pure. Hydrology is a key factor in peatlands, and restoration of an appropriate hydrological regime is a major goal along with the re-establishment of peatland. After completion of dredging it passes into the category of surface or lake water. The first part of the restoration was blockage of the drainage system. The plan of reclamation and arrangement of the pool formed in dredging is shown in the Figure 3.

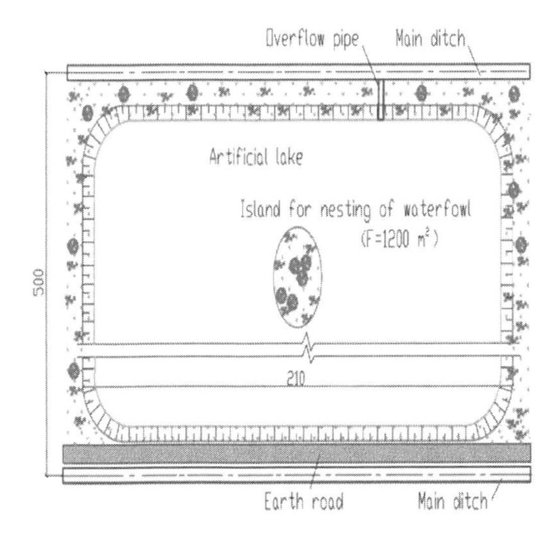

Figure 3. Reclamation plan of the mined-out quarry.

The restoration procedure should drive this mined peatland towards a functional and 'typical' peatland ecosystem. Most species colonizing the restored zones were typical peatland species (Mikhailov, 2012).

Long-term observations showed that, birch and pine can develop around of the pool. The restoration techniques that were used fifty years ago, seem very effective for restoring the mined peatland (Figure 4).

By good rewetting conditions due to inflowing ground water and surface waters the main species are *Glyceria fluitans L.*, *Juncus effuses*, *Potamogeton natans*, *Phragmites australis*, *Carex rostrata* and occurrence of *Equisetum limosum*, *Alisma Plantago-aquatica*. The water bodies host a large number of amphibians: *Lissotriton vulgaris*, *Triturus cristatus*, *Pelophylax lessonae* and are important for dragonflies (*Leucorrhinia pectoralis*). Water bodies are also nesting sites for several birds (*Fulica atra, Anas crecca, A. platyrhynchos, Podiceps nigricollis, Aythya fuligula, Bucephala clangula*).

Aquatic plants produce directly and indirectly food for algae, macroinvertebrates, fish, and avifauna. The average depth of fish lakes should be 1.5–2.0 m with some deeper holes in places, possibly under overhanging trees. Initial colonization of bottom fauna and vegetation showed high species number but turned more trivial after some years and the new wetland conditions resembles ordinary small shallow lakes in the landscape.

Pumping might be needed in sites where there is not enough water available either to create the lake and/or to maintain it during drier periods. The depth of the lake should be adjusted according to the bird species one would like to favor.

Figure 4. Artificial pool after peat hydraulicking.

Thus, summing up told, it is necessary to notice that on the majority of moss bogs of wood zone it is possible to carry out few-scale extraction of renewed vegetative resources without serious fears for causing of a notable damage to peatland ecosystems.

6 SUMMARY

The project provided development of a peat deposit in the open-pit mining, drying of peat to moisture content of 70%, before transportation for processing on plant, technical land reclamation. The after-use phase of the bog area can only start when a sufficiently large unbroken area is formed of the pit which are no longer used. There are several alternatives for using the bog area after the peat production is finished. One of them is the creation of artificial pool.

Thus, summing up told, it is necessary to notice that after open-pit mining of peat it is possible to carry out big-scale extraction of renewed peat resources without serious fears for causing of a notable damage to peatland ecosystems.

REFERENCES

Code of Practice for Responsible Peatland Management. Available from: http://www.epagma.eu/sites/default/files/documents/epagma_code_of_practice_revised_september_2011. pdf. [Accessed 20th January 2017].

Hörnsten, L., 1992. Efterbehandling av torvtäkter utbrutna med djupbrytningsteknik – en literaturstudie – Treatments of peat bogs harvested by deep digging technique. Rapport 1992:36, Vattenfall Research, Värmeteknik. Vällingby, Sweden. 64 pp.

Joosten, H. and Clarke, D. (2002) Wise use of mires and peatlands. Background and principles including a framework for decision-making. International Mire Conservation Group and International Peat Society. UK. 303 pp.

Kavanagh, B. (1998) Cutaway boglands: a new landscape for birdlife. *In: Anon. (ed) The Future Use of Cutaway Bogs. Proceedings from the first comprehensive conference on cutaway bog rehabilitation*. Ferbane. Brosna Press. pp. 34–40.

Lamers L.P.M., Falla S.J., Samborska E.M., Van Dulken I.A.R., Van Hengstum G, Roelofs J.G.M. (2002) Factors controlling the extent of eutrophication and toxicity in sulfate-polluted freshwater wetlands. Limnology and Oceanography, 47(2), 585–593.

Lundin, L.; Lode, E.; Strömgren, M.; Nilsson, T. (2008). Wetland: Wise after-use at terminated peat cuttings. *Proceedings of the 13th International Peat Congress, Volume I: After Wise Use - The Future of Peatlands, Tullamore, Ireland, 8–13 June 2008. Ed. Farell, C.; Feehan*, J. International Peat Society, pp. 430–433.

Mikhailov A. (2012) Restoration of peatlands after selective white peat excavation. *Book of Abstracts of the 14th International Peat Congress. Peatlands in Balance Stockholm, Sweden June 3–8, 2012*. p. 448.

Mikhailov, A. (2013) Peat surface mining methods and equipment selection. *Mine Planning and Equipment Selection. Proceedings of the 22nd MPES Conference, Dresden, Germany, 14th–19th October 2013. Drebenstedt, Carsten, Singhal, Raj (Eds.)* 2014, XXVII, Volume 2, pp.1243–1249.

Quinty, F. and Rochefort, L. (2003) Peatland restoration guide, 2nd ed. Canadian Sphagnum Peat Moss Association and New Brunswick Department of Natural Resources and Energy. Québec, Canada, 106 pp.

Rehounkova K., Rehounek J., Prach K. (2011) Near-natural restoration vs. technical reclamation of mining sites in the Czech Republic. Faculty of Science, University of South Bohemia in České Budějovice Available from: http://www.calla.cz/data/hl_stranka/ostatni/Sbornik_anglicky.pdf. [Accessed 10th January 2017].

Selin, P. (1999) Turvevarojen teollinen käyttö ja suopohjien hyödyntäminen Suomessa/Industrial use of peatlands and the re-use of cut-away areas in Finland. University of Jyväskylä, Jyväskylä, Finland, 239 p.

Stefanie Leupold (2004) After use of cutaway peatlands —an overview of options and management planning. Available from: http://www.gret-perg.ulaval.ca/fileadmin/fichiers/fichiersGRET/pdf/Doc_generale/Stencil108.pdf. [Accessed 10th January 2017].

Land Reclamation in Ecological Fragile Areas – Hu (Ed.)
© 2017 Taylor & Francis Group, London, ISBN 978-1-138-05103-4

Effect of different vegetation restoration years on soil fertility in the Antaibao open pit

X. Wang, J.C. Li, J.Y. Yue, C.Y. Guo, N. Lu, Y.H. Wang & S.Q. Yang
Shanxi Biology Institute, Taiyuan, Shanxi Province, China

ABSTRACT: The environment pollution and ecological destruction caused by mining wasteland have made serious crisis to the ecologically sustainable economic development. This paper summarized the changes of soil physical properties and chemical properties and biological properties during the progress of vegetation restoration. The results showed that the physical and chemical properties, enzyme activity and microbial diversity of soil significant varied between different forest age in mine reclamation and control plots. With increasing stand age, soil physical and chemical properties, enzyme activity and microbial diversity were different form control plots. After vegetation restoration, the bulk density and water content decreased in mine area which is different form control plots; however, other indexes increased in mine reclamation of surface mine reclamation area. In addition, the soil fertility of 20 years was better than that of 15, 10 years and control plots. The interaction between vegetation development and soil biological activity. With the process of vegetation restoration, soil structure has been improved and soil fertility has been improved.

1 INTRODUCTION

Soil fertility is the basic and essential characteristic of soil. It is the ability of soil for plants' growth supply and coordinating nutrient, water, air and heat. It is a comprehensive reflection of soil physical, chemical and biological properties (Zhao Q G, 1996). Many natural and man-made ecological processes such as vegetation succession, climate fluctuation, land use change and so on have affected the temporal and spatial evolution of soil fertility significantly (Hajabbasi M A, Jalalian A and Hamid R K, 1997). However, in recent years, it has paid more attention to vegetation restoration technology, plant community evolution trend and ecological benefit research in the field of vegetation restoration at home and abroad (Ren X X, Cai T J, Wang X F, 2010), while the overall research on the evolution of soil fertility is relatively small, and lack the study on characteristics of different vegetation restoration years on soil fertility at the abandoned land of open pit coal mine. Based on the reclamation area of Antaibao open pit in Shuozhou, through the investigation and statistical analysis of soil nutrient index and biological index of man-made forest in different planting years, this study has understood and grasped the soil fertility status after the abandoned land renewed in the mining area, analyzed the characteristics of soils in vegetation restoration from fertility perspective, and laid a foundation for establishing scientific soil fertility control system and improving soil management level.

2 OVERVIEW OF THE STUDY AREA

Pingshuo mining area belongs to Shuozhou City, northern of Shanxi Province. Its geographical coordinates is 112°11′ to 113°30′ of east longitude, 39°23′ to 39°37′ of north latitude. After more than 20 years of reclamation, the current mining ecosystem has been effectively restored, and has been formed the multi-level plant layout structure of Arbor-Bush-Grass which gives priority to *Robinia pseudoacacia*, *Chinese pine*, *Sea-buckthorn*, and *Elm* (Yuan Y, Zhao Z Q and Bai Z K, 2016).

2.1 *Research methods*

2.1.1 *Sample survey and soil sampling*

In the Early July of 2014, this study chooses four different samples as objects: the mixed plantation of *Robinia pseudoacacia-Caragana* planted 20 years (20 a), the mixed plantation of *Robinia pseudoacacia-Sea-buckthorn* planted 15 years (15 a), the mixed plantation of *Robinia pseudoacacia-Caragana-Elaeagnus angustifolia* planted 10 years (10 a) and the natural restoration grass enclosed 20 years (CK).

2.1.2 *The determination of soil properties*

In order to evaluate the quality of soil fertility after vegetation restoration, three types of soil indicators were selected as the factors to evaluate soil fertility quality, which were indicators of soil physical properties (soil bulk density, soil water holding capacity and soil water content), the indicators of soil nutrient traits (pH, soil organic matter content, soil total nitrogen content, soil available nitrogen content, soil total phosphorus content and soil available phosphorus content) and the indicators of soil biological traits (the activities of sucrase, urease, catalase, polyphenol oxidase and the quantities of bacteria, fungi, actinomycetes, nitrogen-fixing bacteria and the community diversities of bacteria and fungi). (Lao J C, 1988, Guan S Y, 1986, Department of Micrology, Institute of Soil Science, CAS, 1985, Zhou J, Bruns M A, Tiedje J M, 1996, Krsek M, Wellington E M H, 1999).

2.1.3 *Data analysis*

The data were processed and analyzed by SPSS 19.0 statistical analysis software. Analysis of DGGE electrophoretic patterns by Quantity One 4.31. In this study, Soil Quality Index (SQI) was selected as the comprehensive evaluation index when evaluating soil fertility quality. (Mato R E, Chhonkar P K, Singh D, 2008, Andrews S S, Karlen D L, Mitchell J P, 2002, Bastida F, Moreno J L, Hemandez T, 2006).

3 RESULTS AND ANALYSIS

The order of soil indexes under different vegetation types is not the same, that is, the improvement ability of different vegetation types to different soil indicators is different. The order of single index of different vegetation types can not reflect the difference of soil comprehensive fertility quality of different vegetation types, so the study introduced the Soil Quality Index (SQI) to make comprehensive evaluation of soil fertility quality of different types. According to the results of principal component analysis, synthetic analyzed the sensibility of soil properties' indicators and the relationship between each indicators, selected soil organic matter, total phosphorus, sucrase, urease, polyphenol oxidase, actinomycetes, nitrogen-fixing bacteria and denitrification as the main indicators to evaluate the soil quality of vegetation restoration in abandoned land of Antaibao open pit. According to the formulas 1 and 2, the results of soil quality index under different forest ages were 20 a(2.547)> 15 a(2.061)> 10 a(1.303)> CK(0.509). It can be seen that the soil quality index fluctuates between 0.509 and 2.547. In the process of vegetation restoration, the soil quality index showed corresponding changes with the vegetation restoration years. The results showed that 20 a had higher soil fertility in the four samples, followed by 15 a, but the soil fertility quality of the three artificial vegetation restoration samples was superior to that of natural restoration sample.

4 DISCUSSION

The results showed that after the vegetation restoration in Antaibao open pit, the soil physical properties, soil nutrient and soil microenvironment have been improved, the number of microbes have been increased, the community diversity has been enhanced, and the soil fertility quality has been advanced. The comprehensive evaluation results of soil fertility indicate that soil fertility will change with the extension of vegetation restoration. With the increase of recovery time, litter account, the amount of soil moisture, the soil surface temperature is effectively reduced, which greatly improved the vegetation growth environment, promoted the growth of various plants in the forest, affected and promoted the soil microbial species and quantity, the activity and quality of enzyme, so as to improve the decomposition rate of litter in forest. The soil nutrients are returned, and finally soil fertility is improved. The artificial vegetation planting samples with a long history have more soil hydrolytic enzymes, catalase and microbial content, and have higher enzyme activity than the samples with short history (Fang X, Tian D L, Qin G X, 2009). At the same time, the litter of samples is increasing with increased recovery time. It is more conducive to nutrient return with the increasing broadleaf species and soft decomposition litter and broadleaf forest has the ability of return litter earlier and increase soil nutrients (Xiang W H, Tian D L, Yan W D, 2003). The emergence of this phenomenon is the result of the combined effects of multiple ecology processes. In the three samples of different vegetation restoration years, the soil fertility could be better preserved in the sample of 20 years, while the effect of maintaining fertility of planted forest in 15 years and 10 years is reduced. It can be seen that the improvement of soil condition is different from that of different vegetation planting years.

5 SUMMARY

The plantation is more close to the top community in the region for more than 20 years, with more complicated community structure, more abundant species composition, more types and amount of soil microbes and enzymes. In this artificial forest, the litter is more easily decomposed, the better nutrient return and nutrient conservation capacity lead to higher quality of soil fertility.

REFERENCES

Andrews S S, Karlen D L, Mitchell J P. (2002) A comparison of soil quality indexing methods for vegetable systems in Northern California. Agric Ecosyst Environ, 90: 25–45.

Bastida F, Moreno J L, Hemandez T, et al. (2006) Microbiological degradation index of soils in a semiarid climate. Soil Biol Biochem, 38: 3463–3473.

Department of Micrology, Institute of Soil Science, CAS, M. (1985) Soil microbiology research. Beijing, Science Press.

Fang X, Tian D L, Qin G X, Xiang W H. (2009) Nutrient contents and enzyme activities in the soil of Cunninghamia lanceolata forests of successive rotation and natural restoration with follow after clear-cutting. Scientia Silvae Sinicae, 45(12): 65–71.

Guan S Y, M. (1986) Soil enzyme and study method. Beijing, Agricultura Press.

Hajabbasi M A, Jalalian A, Hamid R K. (1997) Deforestation effects on soil physical and chemical properties, Lordegan, Iran. Plant and soil, 190: 301–308.

Krsek M, Wellington E M H. (1999) Comparison of different methods for the isolation and purification of total community DNA from soil. Journal of Microbiologial Methods, 39: 1–16.

Lao J C, M. (1988) Manual of analysis of agricultural soil. Beijing, Agricultura Press.

Masto R E, Chhonkar P K, Singh D, et al. (2008) Alternative soil quality indices for evaluating the effect of intensive cropping, fertilization and manuring for 31 years in the semi-arid soils of India. Environ Monit and Assess, 136: 419–435.

Ren X X, Cai T J; Wang X F. (2010) Effects of vegetation restoration models on soil nutrients in an abandoned quarry. Journal of Beijing Forestry University, 32(4):151–154.

Xiang W H, Tian D L, Yan W D, Kang W X, Fang X. (2003) Biomass dynamic and nutrient accumulation of natural restoration at early stage after fallow in clear-cutting forestland of Chinese fir plantation. Acta Ecologica Sinica, 23(4): 695–702.

Yuan Y, Zhao Z Q, Bai Z K, et al. (2016) Niche characteristics of dominant herbaceous species under different land reclamation patterns in Antaibao opencast coal mine. Chinese Journal of Ecology | Chin J Ecol, 35(12):1–8.

Zhao Q G. (1996) Modern soil science and sustainable development of agriculture. Acta Pedologica Sinica, 33(1):1–12.

Zhou J, Bruns M A, Tiedje J M. (1996) DNA recovery from soils of diverse compostition. Applied and Environmental Microbiology, 62:316–322.

Land Reclamation in Ecological Fragile Areas – Hu (Ed.)

The open-pit mining dust characteristics in the eastern Junggar coal field and its impact on the surrounding plants

J.J. Yang & Y.E. Cao
College of Resource and Environmental Science, Xinjiang University, Urumqi, China
Key Laboratory of Oasis Ecology, Urumqi, China

J. Wang
216 Investigation Team of Nuclear Industry, Urumqi, China

G. Lu
The Xinjiang Uygur Autonomous Region Soil and Water Conservation Ecological Environment Monitoring Station, Urumqi, China

G. Wang
216 Investigation Team of Nuclear Industry, Urumqi, China

L.P. Zhao & X.Y. Zhang
College of Resource and Environmental Science, Xinjiang University, Urumqi, China
Key Laboratory of Oasis Ecology, Urumqi, China

ABSTRACT: In this study, the dust amount produced in waste dump, coalyard and by transport in Eastern Junggar coal field was analyzed, meanwhile, the contaminated plant physiological indices were measured by photosynthesis meter. The above researches were used to estimate environmental pollution by mining. The results show that the coalyard dust was significantly related with wind speed, and the simulation value was far beyond that of waste dump, but dust concentration in the dump was higher. The amount of dust and the concentration of the paving road were respectively 42.35 ($kg \cdot km^{-1} \cdot truck^{-1}$), 0.04021 ($PM_{10}$), while those of the non-paved road were respectively 42.35 ($kg \cdot km^{-1} \cdot truck^{-1}$) and 0.04021 ($PM_{10}$). The net photosynthetic rate and stomatal conductance of Ulmus pumila & Haloxylon ammodendron decreased significantly for the influence of coal dust. Reasonable and effective management measures can reduce the maximum extent of dust pollution and damage to the surrounding vegetation.

1 INTRODUCTION

The dust produced in the process of coal mining is not only large, but also complex in composition, especially in the open pit coal mine with large disturbance and few control measures, its influence is more extensive. The dust starting mechanism and dust removal of the open pit yard and its influencing factors are also concerned by scholars (Yang, 2007; Chang, 2009; Cong et al., 2010; Ruan et al., 2016). The dust has an inestimable damage to the metabolism, morphology and development of vegetation, and even can affect the plant community structure (Haung, 2007; Fan et al., 2013; Yang et al., 2016). The study on dust pollution and its effect on vegetation are prevalent in the dust amount of city greening tree species and photosynthetic response (Pang et al., 2009; Yao, 2009; Cai, 2010; Lv, 2011), however, there are few studies on the effects of dust and photosynthetic response of plants in arid areas (Li et al., 2012a; Gao, 2014), and there are few reports on the influence of dust and its influence on Vegetation in open pit mines in arid areas (Naidoo and Chirkoot, 2004).

The vegetation in arid area is rare and ecological environment is very fragile, and the exploitation of open pit mine is the biggest influence on the surface and the surrounding environment. In order to make sure the ecological environment carrying capacity in the study area and ensure its stability, in this study, the dust emission caused by opencast coal mining and its influencing factors in arid region are quantitatively described through the model and particle concentration monitoring, at the same time, the photosynthetic response of vegetation to dust was studied to reveal the influencing degree and the tolerance limit of vegetation in the study area.

The study area is located in the southern Gurbantunggut Desert, the northern desert of Jimsar County of China, It has the typical extreme dry continental climate. It is low in annual rainfall, abundant in sunshine, rich in heat. The average temperature is 7°C, the average annual precipitation is 183.5 mm while the average annual evaporation is 2042.3 mm. The dominant wind direction in the study area is northwest wind, the average wind speed is 2 m/s, the maximum wind speed is 16 m/s.

2 MATERIAL AND METHOD

2.1 *Dust calculation*

2.1.1 *Calculation of dust discharge in dump of open pit mine*

Zhundong Wucaiwan opencast coal mine of Shenhua (M1), Red Sand Springs coal mine (M2), Dacheng coal mine (M3) are selected as the main research object, and its Dump (D) surrounding is investigated. Dust amount of dump is calculated by the Japanese MITSUBISHI Heavy Industries Ltd. of Nagasaki Institute formula, the formula is as follows:

$$Q_P = \beta\left(\frac{W}{4}\right)^{-6} \times U \times A_P \qquad (1)$$

where Qp is dust emission of dump (mg/s), β is empirical coefficient, W is the surface moisture content of dump (%), U is the average wind speed of dump (m/s), Ap is the area of dump (m^2).

2.1.2 *Dust calculation of opencast coal yard*

The survey projects and survey area of opencast coal yard is the same as those of dumps'. The dust emission is calculated by the formulas of the Architectural Institute of Xi'an metallurgy, the formula is as follows:

$$Q_P = 4.23 \times 10^{-4} \times U^{4.9} \times A_P \qquad (2)$$

where Qp is dust emission of opencast coal yard (mg/s), U is the average wind speed of opencast coal yard (m/s), Ap is the area of opencast coal yard (m^2).

2.1.3 *Calculation of road dust*

The road is divided into paved road (R_1) and non paved road (R_2), and the empty passing is marked as K. The estimation of the amount of dust in the road on the road with scattered materials is carried out by the empirical formula of Shanghai environmental protection center and Wuhan Institute of water transport engineering, the formula is as follows:

$$Q = 0123 \times \frac{V}{5} \times \left(\frac{M}{6.8}\right)^{0.85} \times \frac{P}{0.5} \times 0.72L \qquad (3)$$

where Q is the amount of dust (kg/car) for the car driving, V is the vehicle speed (km/h), M is the vehicle load (T), P is the road surface material (kg/m^2), L is road length (km).

2.2 *Photosynthetic response of vegetation*

In October 2016, three samples (P1, P2, P3) are selected near coal chemical factory Zhundong opencast mining area, the samples were obviously polluted by fly ash and dust, Ulmus pumila is selected as the greening tree species and Haloxylon ammodendron is selected as the widely distributed species. On the day before the test, the sample plant is rinsed, and half of the plant is treated with fly ash, the treatment method is that the fly ash is evenly distributed on the leaves of the plant until the blades are evenly covered, and the treatment is marked as S, ensure that the other half of the plant from the ash pollution and marked as K. In case of rain, repeat the steps above. The net photosynthetic rate (Pn, $\mu molCO_2 \cdot m^{-2} \cdot s^{-1}$) and stomatal conductance (Gs, $molH_2O \cdot m^{-2} \cdot s^{-1}$) of the samples were measured with the LI6400 photosynthesis measurement.

3 RESULTS

3.1 *Status of dust in the study area*

The dust rate (shown in Table 1.) of dump and coal yard is calculated with the formula (1) and (2). As can be seen from the table, the coal yard dust is generally higher than the amount of dump. By field investigation can be known, except for special reasons, the air particle concentration of coal yard is higher than that of coal heap dump. Compared

Table 1. Amount of raised dust from Dump and Coal pile.

site	area (km^2)	water content %	average wind speed ($m \cdot s^{-1}$)	dust amount ($kg \cdot h^{-1}$)	concentration of $PM_{2.}$	particle number of $PM_{2.5}$	concentration of PM_{10}	particle number of PM_{10}
M2D1	0.52	4.52	2.4	0.13	0.002	28.857	0.007	9.143
M2D2	0.91	4.52	2.4	0.23				
M1D	5.07	5.23	3.7	0.82	0.017	299.143	0.101	138.571
M3D	0.19	4.57	2.8	0.05	0.002	34.571	0.007	7.714
M2H	0.05	1.99	2.5	7.83	0.001	25.143	0.005	5.714
M1H	0.03	2.61	3.5	20.68	0.009	148.500	0.030	53.500
M3H	0.02	1.63	2.8	3.92	0.005	86.400	0.017	20.800

Table 2. Dust amount of roads.

road	length (km)	inventory ($kg{\cdot}m^{-2}$)	vehicle speed ($km{\cdot}h^{-1}$)	automotive load (t)	dust amount ($kg{\cdot}km^{-1}{\cdot}$ $cars^{-1}$)	concentration of $PM_{2.}$	particle number of $PM_{2.5}$	concentration of PM_{10}	particle number of PM_{10}
R1	391.96	1.10	80	40	14.02	0.009	59.048	0.016	13.476
R1 K	–	–	80	15	6.09	–	–	–	–
R2	75.65	4.90	30	80	42.35	0.011	170.458	0.040	66.602
R2 K	–	–	40	15	13.61	–	–	–	–

to the dump, fly ash of coal yard is more easily spread into the air by the wind because of its small particles. Table 1 also shows that the main factors affecting coal yard dust is the average wind speed of the coal yard and therefore general coal yards have higher windscreen dust control measure. The main factor that affects the amount of dust is the area of the dump, wind speed and other factors due to the different performance of the model did not show a larger role, but particulate matter concentration in the air of the dump shows that when the average wind speed is high (3.7 $m{\cdot}s^{-1}$), $PM_{2.5}$ particle number can be as high as 299, in addition, coal yard also showed the same characteristics, it can be concluded that the main factor influencing the dust amount of dump and coal yard are average wind speed, while the other factors such as water content, area and so on, have different contributions to the dust amount due to different calculation methods.

The road arranged in a crisscross pattern in the mining area. Except that the main thoroughfare is paved road, the other access road of the mine is off road. Although the off road is only a small part of the length of the road, but the dust caused by it is much higher than the paved road, and the reason for this phenomenon is that there are no more than 5 times the amount of material on the off road, and the fine particulate matter contained in the material is an important component of dust. In addition, because that the unevenness of the off road and the local limit the speed and quantity of the vehicle, therefore to some extent, it limits the great pollution caused by the off road dust in the study area.

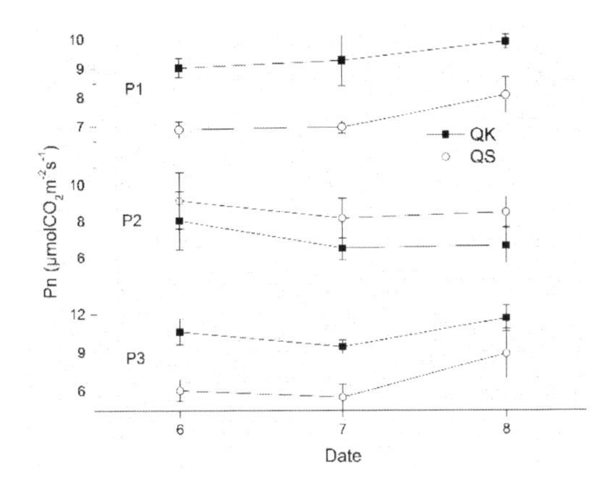

Figure 1. Net photosynthetic rate of Ulmus pumila.

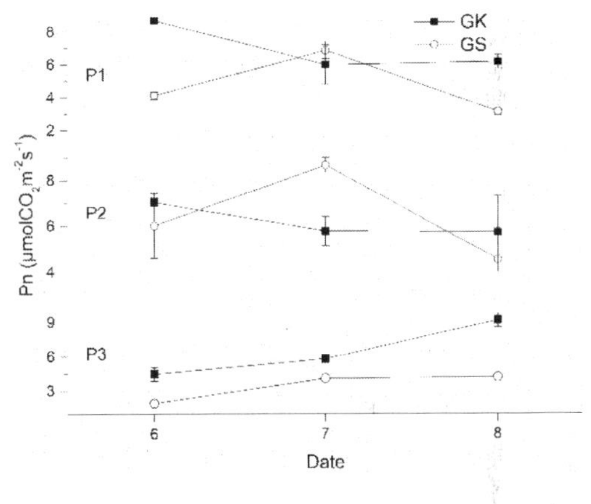

Figure 2. Net photosynthetic rate of Haloxylon ammodendron.

3.2 *The photosynthetic response of vegetation on coal dust*

3.2.1 *Effect of fly ash treatment on net photosynthetic rate*

The study set up the ash treatment test of Ulmus pumila and Haloxylon ammodendron, and the photosynthetic response was analyzed. The study found that, for Ulmus pumila, the changes of net photosynthetic rate in the experimental group and the control group were not obvious during the monitoring period. Except P2 point, the net photosynthetic rate of P1, P3 in control group was significantly higher than that of the experimental group, and it can be inferred that the ash treatment has a certain degree of influence on the net photosynthetic rate of Ulmus pumila, and the ash will weaken the photosynthetic capacity of Ulmus pumila.

As for Haloxylon ammodendron, net photosynthetic rate in the experimental group and the control group did not show obvious regularity. The comparison between Figure 1 and Figure 2 shows that Ulmus pumila has higher photosynthetic rate than that of Haloxylon ammodendron. But by the influence of ash, Haloxylon ammodendron did not show a consistent rule, this may be due to the special physiological structure and the adaptability to the stress environment. Therefore, to some extent, ash treatment on Ulmus pumila photosynthetic inhibition was greater than that of Haloxylon ammodendron.

3.2.2 *Effect of coal ash treatment on stomatal conductance*

The study compared the changes of stomatal conductance in Haloxylon ammodendron and Ulmus pumila, it can be found that the effect of coal ash treatment on stomatal conductance was not obvious, comparatively, the assimilating branches of

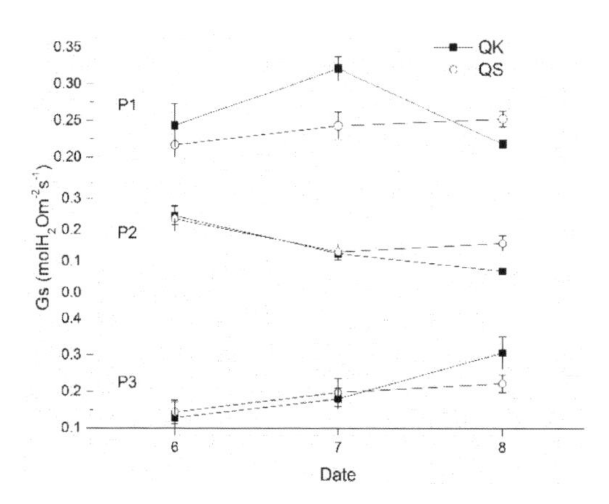

Figure 3. Stomatal conductance of Ulmus pumila.

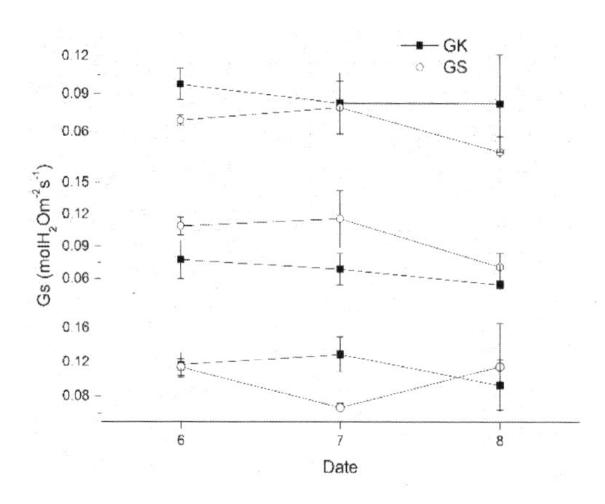

Figure 4. Stomatal conductance of Haloxylon ammodendron.

Haloxylon ammodendron was more sensitive to ash treatment. It can be inferred that, because of the lager leaves and both sides of the blade are provided with an air hole for gas exchange, therefore, through long-term adaptation, one time manual processing does not change the original adaptation mechanism to a great extent, the stomatal conductance of Ulmus pumila leaves in response to ash treatment is not obvious. For Haloxylon ammodendron with assimilating branches, ash can affect the larger leaves area for photosynthesis, combined with its relatively low intensity of photosynthesis, therefore, ash treatment has certain effect on the stomatal conductance of assimilating branches in Haloxylon ammodendron.

3.2.3 *Comparison of photosynthetic difference under the treatment of fly ash*

To further clarify whether the ash treatment has significant influence on net photosynthetic rate and stomatal conductance of Ulmus pumila and Haloxylon ammodendron, variance analysis was used to compare the difference of photosynthetic responses of two plant species. Both control and experimental groups, the net photosynthetic rate of Ulmus pumila was significantly higher than that of Haloxylon ammodendron ($P < 0.01$), which shows that in the process of photosynthesis, the plant with leaves are more efficient than the plant with assimilating branches. Fig. 1 shows that in stress conditions, Haloxylon ammodendron with assimilating branches has a stronger ability to adapt to stress, therefore, in the arid areas with poor environment, the two plants have different survival strategies to deal with different habitats.

No matter the Ulmus pumila or Haloxylon ammodendron, different treatments did not have a significant impact ($P > 0.05$) on the stomatal conductance of two plant species, this may be due in a short period of time, plant stomatal conductance did not response timely to the environment, therefore, we can infer that in a short period of time, the decrease caused by plant photosynthesis was not due to the change of stomatal conductance.

4 SUMMARY

This study calculated the dust amount of opencast coal mine dump, coal yard and road in the eastern Junggar coalfield, and explored the impact of ash on the vegetation photosynthesis, the conclusion is as follows: i. The coal yard has a higher dust emissions than that of coal mine dump, while the main influencing factor is the average wind speed in the region. ii. The off road dust is the main source of road dust pollution, and surface material and small particles are important causes of a large number of dust. iii. Ash treatment has certain inhibition effect on photosynthesis of Ulmus pumila and Haloxylon ammodendron, in contrast, Haloxylon ammodendron has higher resistance to stress circumstance than that of Ulmus pumila. iv. In a short period of time, the main factor of the plant photosynthesis inhibition caused by ash treatment was not the change of stomatal conductance.

REFERENCES

Cai, Y., H (2010). Study on dust-retention effect and photosynthetic characteristics of urban keynote tree. Fujian Agriculture and Forestry University.

Chang, T (2009). Research on coal dust starting and dust control in open-air coal storage yard. Shanxi University.

Cong, X., C., Chen, Z., L., Zhan, S., F (2010). Experimental study of static dust emission from a coal pile in the open air yard. *Journal of China University of Mining & Technology,* **39**(6):849–853.

Fan, Q., Li, S., Y., Guan, T., L., Wu, X., X., Wang, R. & Ren, L., J (2013). Then ecological effect on plant and soil around opencast coal mine from the mineral dust. *Northern Environment,* **25**(9):104–108.

Gao, D (2014). Effect of coal dust on the growth of plants in grassland near coal mining in arid region. Moogolia Agricultural University.

Huang, F (2007). Effect of highway dust pollution on photosynthesis in plant leaves. Wuhan University of Technology.

Li, S., Y., Gu, F., Qiu, Y., Z., Jin, Z., Z., Li, Y., G & Gao, X., Y (2012). Evaluation on Dust-holding Capabilities of Photosynthetic Organs of Ten Shrub Species in Hinterland of the Taklimakan Desert. *Arid Zone Research,* **29**(6):1022–1031.

Lv, D.P (2011). Study on dust catching ability and effectiveness of dust pollution on photosynthetic parameters of three vertical greening plants. Nanjing Forestry University.

Naidoo, G. & Chirkoot, D (2004). The effects of coal dust on photosynthetic performance of the mangrove, Avicennia marina in Richards Bay, South Africa. *Environmental Pollution* **127**: 359–366.

Pang, B., Zhang, Y.L & Wang, D (2009). The present and further research of dust-retention by urban plants. *Shandong Forestry Science and Technology,*(02):126–130.

Ruan, W.G., Mo, Z.Z., Zhang P. & Li, Y.C (2016). Analysis on dust production process and pollution control measures in opencast mining. *Environmental Science and Management,* **41**(6):96–99.

Yang, D (2007). Research on dusting and migration rule of the dust at open-air coal yard. Liaoning Technical University.

Yang, H.L., Wei, L.L., Ye, X.H., Liu, G.F., Yang, X.J & Huang, Z.Y (2016). Effects of caol dust deposition on seeding growth of *Hedysarum laeve* Maxim., a dominant plant species on Ordos Plateau. *Acta Ecologica Sinica,* **36**(10):1–8.

Yao, J (2009). The physiological and ecological impacts of Urban typical afforestation trees polluted by dust. Nanjing Forestry University.

Subsidence land reclamation and ecological restoration

Land Reclamation in Ecological Fragile Areas – Hu (Ed.)
© 2017 Taylor & Francis Group, London, ISBN 978-1-138-05103-4

Reclamation of prime agricultural farmlands: A retrospective 40 years after reclamation

Robert G. Darmody & Kevin McSweeney
Department of Natural Resources and Environmental Science, University of Illinois, Urbana, Illinois, USA

ABSTRACT: Illinois was an important location for innovation and evaluation of mine reclamation regulations, techniques, and approaches. Active surface and underground mining for coal is not what it was in the latter half of the previous century and before, but the reclaimed areas are a testimony to the efforts of the early pioneers in restoring mined and subsided land to their former agricultural productivity. Illinois has some of the best agricultural soils in the world, and underneath of much of that area is coal, thus promoting an enhanced sensitivity to good quality mined land reclamation. The University of Illinois was the lead agency in mine land reclamation research and established research plots to test various reclamation techniques and systems. Many of those plots remain in agriculture and the long term success of reclamation will indicate that the post mining improvements in the soils and landscapes are permanent and not a result of the intense management applied initially.

1 INTRODUCTION

The Surface Mining Control and Reclamation Act (SMCRA) was a revolutionary response to a perceived permanent threat to the soils and landscapes of the U.S. (Public Law 95–87) At the time it was passed, 1977, the environmental movement was in full swing in the U.S. and opposition to mining, in particular coal mining, was mounting. Previous to SMCRA, mining operators were not required to make any major reclamation efforts after extraction of the resource which they owned regardless of the ownership situation of the surface. This was, in part, in keeping with the common cultural concept in the U.S. at the time that property owners had the right to do anything they wanted to with their property, including leaving a "moon scape" after coal was surface-mined (Table 1).

This presented a classic case of individual rights conflicting with community rights. SMCRA changed all that and mining companies had, in Illinois, for example, to restore the agricultural productivity to the lands that they disturbed by surface mining. Likewise, coal companied that subsided the surface due to high extraction mining technologies had to restore the land use of the areas that they subsided as part of their mining plans. The problem with these new requirements and regulations was that mining companies did not have the expertise or the research base to make the intentions of SMCRA realized. Thus, a reclamation research program was established at the University of Illinois, and elsewhere in the U.S. to determine appropriate technologies to facilitate effective reclamation. In Illinois it was known as the Reclamation Research Program for surface mining and for planned subsidence mining it was the Illinois Mine Subsidence Research Program. These two research programs were largely supported by grants from the federal and state governments, along with support from the Illinois Farm Bureau, Illinois Coal Association, and the Illinois Clean Coal Institute. It involved field trials of various reclamation approaches and evaluation of crop yield responses to the reclamation of mined land and mitigation of subsided land. Experiments and research plots were developed and monitored for several years initially, but the question remains regarding the true long-term effectiveness and effects of the reclamation efforts. Consequently, the University of Illinois has proposed revisiting the research sites to assess the legacy effects.

Table 1. Attributes of surface mined soils, pre and post SMCRA.

Pre law	Post SMCRA
Rocky	Rock-free
Short steep slopes	Graded
Loose	Compacted
Chemically reactive, FeS_2, $CaCO_3$	Chemically Stable

2 METHODS AND MATERIALS

2.1 *Planned mine subsidence mitigation research*

Longwall mining: Coal in Illinois is found in a formation known as the Illinois Coal Basin, a spoon-shaped deposit that is surfaced mined on its outer edges and deep mined towards its interior. Underground mining in Illinois was first recorded in 1810, and began on an industrial scale in the1890's. In pre SMCRA times, the underground workings were small scale and subsidence was generally unplanned and limited to small areas. As mining technology advanced, longwall machines were brought into the industry in the 1970s and the resultant extensive planned subsidence generated strong opposition from citizen groups. In Illinois, longwall mine subsidence relates to about 70% of the extracted coal seam thickness, or about 1.2–1.5 m drop at the center of the mine panels (Darmody et al., 1989).

IMSRP: The Illinois Mine Subsidence Research Program (IMSRP) was initiated to assess the probability of mitigating the damage associated with longwall mining (Darmody et al., 1992). The research covered a period of about nine years. One of the aims was to assess the impacts of subsidence on crop yields and the effectiveness of subsidence mitigation in the return to pre-subsidence crop yields and land use. IMSRP research methodology involved the development of an agricultural impact assessment method utilizing infrared aerial photos taken in the spring over longwall mines to detect increased soil wetness associated with mine subsidence. The extent and the location of wetter areas were recorded and the locations were visited in the following fall to collect crop samples from the impacted areas and local unmined reference areas. Results were based on the recently subsided and mitigated areas and the overall impact on crop yields above longwall mines varied with precipitation and was estimated to depress crop yields by about 5% (Table 2) (Darmody et al., 1992; 1994).

Table 2. Overall longwall mining subsidence—induced reduction in corn yields, includes entire mine foot print.

Corn yield reduction (%)			
Year 1	Year 2	Year 3	Avg.
7.4	4.2	2.4	4.7

Findings indicated that, at least initially, mitigation efforts were successful in essentially restoring crop yields (Table 3).

However, the short term mitigation success involved the installation of the infrastructure necessary to remove excess water, the problem in Illinois with its low relief, high rainfall situation (Figs. 1, 2). It is unknown if the successful restoration of crop yields as the initial results showed, have been sustained over the following decades. It is our intent to develop a research program to revisit the research sites to evaluate the long-term effects of subsidence and the success of subsidence mitigation on agricultural systems (Figs. 3, 4).

2.2 *Mine reclamation research in Illinois*

Surface Mining: The coal industry found it economic to mine coal by removing the overburden as equipment improved and demand grew and this technique became significant in the 1920's. Because of the cost of reclaiming was high, and the legal requirements few, reclamation of surface mined land was insignificant until passage of SMCRA, as we have noted above. The Surface Mining Reclamation Research Program at the University of Illinois was established to investigate methodologies to comply with SMCRA requirements to restore crop yields on Illinois prime farmlands. The mining companies had no expertise in this area and their need to replace topsoil in compliance with SMCRA and their 24/7 operations involve hauling material during wet conditions lead to what was early identified as the culprit, soil compaction (Dunker and Jansen, 1987; Dunker et al., 1992). Additionally, the bond release time frame (10 yr.) precluded use of successional strategies for soil building. Conventional soil handling techniques involved movement and placement of topsoil via scrapers and end-dump trucks that drove on the soil thus compacting it (Figs. 5, 6) (Hooks et al., 1992). We realized that topsoil replacement strategies needed to be improved to prevent surface compaction and 'insulation' of subsoil from plant root exploitation. Prevention of soil compaction would

Table 3. Success rate of initial mitigation attempts on longwall undermined subsided farmland in Illinois.

	Year 1	Year 2	Year 3	Year 4	Mean
Crop	Yield difference on mitigated subsidence fields (% of ref.)				
Corn	101	93	71*	70*	81*
Soybean	96	124*	86*	81*	93

*Significant at 5% level, includes a "snap shot" of mitigation success, some sites would be re-mitigated. IMSRP research sites were those identified as severally impacted and in need of mitigation, thus the results are biased towards the worse cases.

Figure 1. Subsided mitigation: road grade raised and drainage culvert positioned to remove surface water from field.

Figure 2. Installation of drain tile to remove excess water from subsided field, southern Illinois.

Figure 3. Subsided Illinois farmland, before mitigation.

Figure 4. Subsided Illinois farmland, after mitigation.

Figure 5. Scraper placing topsoil; compaction will result from the applied load.

Figure 6. Placement of soil on graded overburden with an end-dump truck driving on soil and generating damaging compaction.

be the best alternative and methodologies to that end were investigated. In addition, deep tillage was believed to be a useful tool for compaction alleviation if it could not be avoided. Impediments to long term success of soil building centered on efficient and speedy imperatives of reclamation operations, as incorporated into the overall mine plans.

Reclamation Research Experiments: Conventional surface mining relied on immense power shovels that moved the overburden and soil materials, but one innovated company invested in a bucket wheel excavator to do that job (Figs. 7, 8). The main advantage with this methodology was that the soil materials were moved on a conveyor belt and placed in the reclaimed area without the need to drive high ground pressure vehicles on the rebuilt soil, thus avoiding compaction (Fig. 9). Examination of the resultant soil revealed a unique structural composition to the rebuilt soil we termed "Fritted" referring to the soil spheres developed as the materials rolled along on the conveyor belt before gentle placement.

Figure 7. Conventional power shovel removing overburden at surface coal mine in Illinois.

Figure 8. Bucket wheel excavator removing overburden and transporting it across the pit at a surface mine in Illinois.

Figure 9. Belt conveyor placing soil without compaction as would be the case with dump trucks.

Table 4. Captain Mine mixed soil materials research plot treatments, replaced topsoil over fritted soil over subsoil.

Treatment	Topsoil Thickness (cm)	Fritted soil thickness (cm)
A/3	40	8
A/10	40	25
A/15	40	38
A/20	40	51
M10	0	25
M20	0	51
Control—Cisne Soil	Native pre-mining local soil	

Captain Mine Experimental Research Plots: To assess the possible benefits of the bucket wheel handled soil materials a series of research plots were developed incorporating different amounts of topsoil replacement over soil built from various depths of bucket-wheel excavated soil materials (Table 4) (McSweeney et al., 1981; 1984; 1987).

Corn (*Zea mays*) and soybean (*Glycine max*) crops were grown on the plots and results indicated that crop yields could be as good or better than crops grown on local, native soils (Table 5).

This evaluation of the rebuilt mine soils occurred shortly after the soils were reconstructed, so the long-term success has yet to be evaluated—a proposed research project. We recently made a preliminary revaluation of the soil profiles on the Captain Mine research plots. After the mine closed, the area including the plots was repurposed as a state park and since then has not been

Table 5. Initial Captain Mine mixed soil materials research plot crop yields (kg/ha).

Treatment	Corn	Soybean
A/3	7,086 a[†]	1,819 a
A/10	6,835 a	1,693 ab
A/15	6,961 a	1,693 ab
A/20	6,146 b	1,693 ab
M10	6,271 b	1,505 b
M20	6,396 b	1,568 ab
Control—Cisne Soil	7,023 a	1,693 ab

† The same letter within a column indicates no significant difference at the 0.05 level.

Figure 10. Excavation of a soil test pit to examine soils at the Captain Mine ~35 years after reclamation.

managed for crop production but the vegetation has been allowed to grow unmanaged. We excavated soil pits to observe the changes in the soil structure and plant rooting behavior in the 30+ years post mining (Fig. 10). Images of the soil profile taken immediately after soil reconstruction show the fritted structure developed consequently by the bucket wheel/conveyor belt handling as well as the compaction/massive structure resulting from topsoil replacement (Figs. 11, 12). Upon recent reexamination, the fritted soil structure is still evident, prismatic structure has begun to develop in some formerly dense horizons, yet compaction persists in other horizons (Figs. 13, 14, 15). At the surface, granular structure has developed, (Fig. 16) but in compacted horizons, roots are confined to the replaced topsoil (Fig. 17). A more extensive evaluation of the long-term success of this reclamation methodology awaits additional funding.

Burning Star Mine Experimental Research Plots: The Captain Mine experiments involved a specialized, and unique, materials handling methodology.

Figure 11. Reclaimed soil pit face exhibiting fritted structure from conveyor moved soil materials, shortly after soil placement.

Figure 12. Reclaimed soil pit face exhibiting the combination of fritted structure and massive physical condition, shortly after placement.

Conventional mining was unlikely to adopt that expensive, specialized approach. Compaction, given the conventional technologies, was something less easily prevented and more likely mitigated after the fact. In keeping with this reality, several methods

Figure 13. Thirty-forty years of soil development in a reclaimed mine soil—fritted structure persists.

Figure 15. Thirty-forty years of soil development in a reclaimed mine soil—compaction persists.

Figure 14. Thirty-forty years of soil development in a reclaimed mine soil (0' mix—no topsoil)—prismatic structure in dense horizon.

Figure 16. Thirty-forty years of soil development in a reclaimed mine soil—reformation of granular structure in the surface horizon.

for lessening the deleterious effects of compaction by tillage were proposed. The Burning Star Mine research plots were developed to assess several tillage machines in their ability to promote better soils as evaluated by crop yields (Hooks et al., 1987; Dunker et al., 1995). Among the tillage implements evaluated was the DMI Deep Ripper plow that was rated to operate to a depth of 122 cm (Fig. 18).

Figure 17. Thirty-forty years of soil development in a reclaimed mine soil—roots largely confined to replaced topsoil.

Figure 18. DMI deep plow, compaction can be mitigated to a depth of 122 cm, one of the tillage treatments at the Burning Star Mine research plots.

Table 6. Burning Star reclaimed minesoils tillage experimental research plot treatments and corn yields.

Treatment	Tillage tool	Plow depth (cm)	Corn yield (kg/ha)
CHS	Chisel Plow	23	4,582 d†
TG2	DMI Tiger II	36	4,206 d
RM1	Vibratory Shank	81	5,461 bc
TLG	Vibrate Cut & Lift	81	5,147 c
DM3	Deep Plow	96	5,587 b
DM1	Deep Plow	122	7,030 a
DM2	Deep Plow	122	7,093 a
Control—Cisne Soil	Conventional	23	7,281 a

† The same letter within a column indicates no significant difference at the 0.05 level.

Two versions of this plow mitigated the deleterious effects of soil compaction as evidenced in the corn (*Zea mays*) yields on plots with this treatment not significantly different from local, native soils used as a reference (Table 6).

However, we have yet to evaluate the persistence of the benefits of deep tillage subsequent to the passage of three decades since deep tillage and there is some concern that the beneficial effects might not persist. This is another potential research project that awaits new grant support.

3 CONCLUSIONS

Coal mining in Illinois, as elsewhere, fluctuates with the demand for coal and the industry's response to economic, social, and environmental constraints. Currently, coal mining is at a low ebb in Illinois and elsewhere. However, there are legacy issues with previous, let alone active, coal mining that SMCRA in the U.S. and other regulators internationally need to address. Our evaluation of subsidence mitigation and surface coal mining reclamation indicates that in the short term, say less than a decade, agricultural productivity can be largely restored, and in some rare cases, improved, as a consequence of high quality reclamation. However the issue remains to be confirmed on the persistence of these initial results to assure high quality, productive agricultural soils in perpetuity.

REFERENCES

Darmody, R.G., I.J. Jansen, S.G. Carmer, and J.S. Steiner. 1989. Agricultural impacts of coal mine subsidence: Effects on corn yields. J. Environ. Qual. 18:265–261.

Darmody, R.G., R.T. Hetzler, and F.W. Simmons. 1992. Coal mine subsidence: effects of mitigation on crop yields. International J. of Surface Mining and Reclamation 6:187–190.

Darmody, R.G. 1994. Effects of coal mine subsidence on agricultural soils and hydrology. p. 130–141. *In* B.A. Trent, R-A. Bauer P.J. Demaris, and N. Kawamura (ads.) Guidelines resulting from the Illinois Mine Subsidence Research Program, 1985–1993. Final report to the Illinois Clean Coal Institute. Illinois. State Geological Survey, Champaign, IL.

Dunker, R.E. and I.J. Jansen. 1987. Corn and soybean response to topsoil replacement and irrigation on surface mined land in western Illinois. J. Soil and Water Cons. 42:277–281.

Dunker, R.E., C.L. Hooks, S.L. Vance, and R.G. Darmody. 1992. Rowcrop response to high traffic vs low traffic soil reconstruction systems. Proceedings, 1992 National Symposium on Prime Farmland Reclamation, University of Illinois, Urbana, IL, August 10–14, 1992. Pp 1 l–18.

Dunker, R.E. C.L. Hooks, S.L. Vance, and R.G. Darmody. 1995. Deep tillage effects on compacted surface-mined land. Soil Sci. Soc. Am. J. 59:192–199.

Hooks, C.L., I.J. Jansen, and R.W. Holloway. 1987. Deep Tillage Effects on Mine Soils and Row Crop Yields. National Symposium on Mining, Hydrology,

Sedimentology and Reclamation, Springfield, Illinois. December 6–11, 1987.
Hooks, C.L., R.E. Dunker, S.L. Vance, and R.G. Darmody. 1992. Rowcrop response to truck and scraper hauled root media systems in reconstruction of surface mine soils. Proceedings, 1992 National Symposium on Prime Farmland Reclamation, University of Illinois, Urbana, IL, August 10–14, 1992. Pp 19–24.
McSweeney, K., I.J. Jansen, and W.S. Dancer. 1981. Subsurface horizon blending: An alternative strategy to B horizon replacement for the construction of post-minesoils. Soil Sci. Soc. Am. J. 45:795–799.
McSweeney, K., and I.J. Jansen. 1984. Soil structure and associated rooting behavior in minesoils. Soil Sci. Soc. Am. J. 48:607–612.
McSweeney, K., I.J. Jansen, C.W. Boast, and R.E. Dunker. 1987. Row crop productivity of eight constructed minesoils. Reclamation and Revegetation Research. 6: 137–144.
Public Law 95–87. 1977. Surface Mining Control and Reclamation Act. U. S. Code Vol. 30, Sec 1265.

Land Reclamation in Ecological Fragile Areas – Hu (Ed.)
© 2017 Taylor & Francis Group, London, ISBN 978-1-138-05103-4

Concurrent mining and reclamation for underground coal mining subsidence impacts in China

Yoginder P. Chugh
National Academy of Inventors, USA
Southern Illinois University Carbondale, Carbondale, Illinois, USA

ABSTRACT: Large scale underground mining of coal resources in China using longwall mining has resulted in ecological and environment problems, including surface subsidence that is considered serious due to competing interests of prime agricultural lands, food security, and regional economic development. The subsided lands must be rehabilitated soon after mining to be agriculturally productive to minimize loss of farmland. Similarly, precious water resources must also be managed during and after mining to protect this natural resource. Toward these goals, the concept of "Concurrent mining and subsidence reclamation (CMR)" was proposed by Professor Hu of the China University of Mining and Technology, Beijing (CUMTB). Over the last two decades CMR concepts have evolved and successfully applied in the field in different parts of China. This innovative technology has increased available farmland during the mining process, and provided better land protection and food security in mining areas even with high groundwater table. The technology has been used in 5 of the 14 large coal bases in China. This paper describes the technology concepts, design and guiding principles for planning with two case studies from different regions to enhance its application both in China and in other countries.

Keywords: Mining subsidence, farmland and water resources protection, concurrent mining and subsidence reclamation planning

1 INTRODUCTION

China is currently the largest producer and consumer of coal in the world. According to the BP Statistics, 2015, China produced about 3.75 billion mt of coal that is about 47.7% of the global coal production. It consumed about 3.94 billion mt of coal, or about 50% of the total consumption in the world. Coal mining can result in land, water, and air impacts related to waste disposal, toxic gas emissions, and land subsidence with associated changes in surface topography, and damage to land and structures (Hu et al 2016a). Land subsidence has created major ecological problems in China because 90% of the coal is mined from underground mining using high extraction longwall mining systems, in contrast to several other large producers of coal around the globe (USA, India, and Australia) who produce a significant amount of their coal production using surface mining methods (BP Statistics). Land subsidence impacts in China are more prominent in the east and northeast China (Hu and Luo 2006; Hu et al 2006)., which are the main coal and agricultural production regions (called "Overlap Regions") These regions cover about 40% of the total farmland in China, produce about 45% of the national food production, and mine about 58% of the national coal production. Furthermore, ground water table is relatively high in many areas of the Overlap Regions, which result in large scale water-logged agricultural areas after mining. The farmland and water resources can be highly impacted by underground mining. Therefore, technologies must be developed to make subsided lands agriculturally productive soon after mining to minimize loss of farmland and their pre-mining productivity level. Similarly, water resources and regional economic development must also be managed during and after mining to sustain regional economies. Toward the above goals, China's land reclamation science and technology has made considerable progress since the mid 1980's to restore subsided and subsiding lands to minimize impacts to natural resources (Hu 1994a; Hu 1994b). Some of the outstanding technology developments and applications until early 1990s are summarized below (Hu and Xiao 2013).

- "Digging Deep to Fill Shallow": This technology divides the subsidence-prone areas into two parts: deep and shallow. The deep areas are dug deeper to develop fish ponds, or water recreational areas and the excavated soil is used to fill the shallow areas to reclaim lands for agriculture. Hydraulic dredging and pumping equipment are often used for this.

- "Direct Rehabilitation": If the shallow subsidence areas are not water-logged and the Ground-Water Level (GWL) is also not very high, direct rehabilitation of subsided land may be possible through regrading with or without terracing to support drainage and minimize soil erosion.
- "Backfilling low lying areas-not currently being used for agriculture- for rehabilitation as agricultural land": It involves landfilling low lying areas with coal wastes, construction materials, and fly ash to develop agriculturally productive lands.
- "Subsurface Water Drainage": This technology establishes a system of drains below the subsurface water level/s to drain the impounded water and lower the subsurface water level so that the subsided land can be cultivated for agricultural production.

Reclamation methods with and without backfilling have been extensively practiced in China. However, most rehabilitation is generally done after subsiding lands have achieved the final subsidence basin or final settlement condition after mining. However, with all these approaches over 50% of the agricultural land may be already submerged into water, including some with highly fertile topsoil. Furthermore, subsiding lands are not agriculturally productive. These approaches are post-mining land rehabilitation technologies that typically result in a low percentage of farmland reclamation and high reclamation cost. Furthermore, these approaches may not allow efficient protection of water resources.

2 STATEMENT OF THE PROBLEM

China mines coal resources in several regions (Shandong, Anhui, Henan, Hebei, and Jiangsu) where mining, agricultural, water resources, and regional economic development have competing economic interests. The developments above can now be extended to develop mining and reclamation technologies that would: 1) Minimize the topographic changes related impacts of subsidence on agricultural production and productivity, and water resources quality and quantity; and, 2) Synergize the different industries operations in a manner that loss of revenues from different industries are minimized during and after resources extraction. The goal is to maximize the revenue for all industries involved during and subsequent to mining of coal resources.

Currently, subsided areas after mining are generally reclaimed upon completion of mining over a large region to take advantage of the economies of scale and mining sequence for different longwall panels. This is termed here as traditional reclamation or TR or "end of pipe solution" where land grading, soil placement, and agricultural production are performed mostly after the land has stabilized (Hu et al., 2013; Hu and Xiao, 2013b) after mining. TR approaches may satisfy only the first reclamation goal identified above, but they would most likely not satisfy the second goal. TR approaches may not develop efficient development and utilization of the coal, farmland and water resources, and regional economic development. Over the last 10–15 years, researchers at CUMTB are developing alternate CMR approaches to reclaim subsided land as mining progresses. They are similar to concurrent mining and reclamation concepts used in surface mining but are more complex. Examples include research by Li et al. (1999) who designed a pre-reclamation plan for Liuqiao coal mine based on land damage analysis. They concluded that the best time for pre-reclamation is after skipping mining of the No. 6 coal seam and before mining the central isolated working face. Furthermore, the layout of the pre-constructed water pond must correspond with the underground mining layout. Zhao and Hu (2008) developed an "appropriate reclamation timing model" for subsiding land that incorporates planning for reclamation. Hu et al (2013) proposed CMR concepts for longwall mining to reclaim high value land resources (Hu and Chen 2016), which were successfully demonstrated in the field in Anhui Province, China. In laying out mining plans, subsidence development during various stages were analyzed and pertinent factors such as vertical subsidence, post-mining slope, water area, and land use condition were considered in reclamation planning. Hu et al (2016b) discussed a CMR design case study that considered the value of both land and water resources. Limited CMR research to date has focused primarily on reclamation goal 1 above and there is room for additional development even there to better protect farmland and water resources. Over the last five years, the Institute of Land Reclamation and Ecological Restoration at the CUMTB and Southern Illinois University Carbondale (SIUC in USA) researchers are working collaboratively not only to minimize the impacts to farmland and water resources but in some cases, enhance utilization of all resources while simultaneously minimizing negative economic impacts. The goals of these development efforts are to minimize loss of and/or to enhance: 1) Farmland use for agriculture during and after the entire mining process, 2) Surface and ground water resources for agriculture, and economic development; and 3) Revenue from utilization of all resources.

3 CHARACTERISTICS OF MINING REGIONS IN CHINA

Coal resources distribution and subsidence related surface damage characteristics: Proven coal reserves are estimated as 1.53 trillion mt as of 2015 and rank third in the world (Xiao et al. 2014). Their distribution is shown in Figure 1. The degree of subsidence-related damage in different mining regions varies due to the geological and hydrogeological differences. In eastern China, where ground water table is near surface, subsidence results in large lakes and wetlands impacting about 85% of the pre-mining farmland land use. In western China, an ecologically fragile area susceptible to soils erosion, subsidence leads to increased soil erosion, depletion of water resources, lowering of ground water table, reduction in vegetation area, and land desertification. In southern China, with warm and humid climate area and hilly terrain, subsidence may cause landslides and endanger public safety. Therefore, CMR land reclamation strategies should consider regional or even site specific differences.

4 CONCURRENT MINING AND RECLAMATION (CMR) CONCEPTS TO MINIMIZE SUBSIDENCE IMPACTS

Critical scientific planning elements of CMR: Figure 2 shows the proposed four scientific design

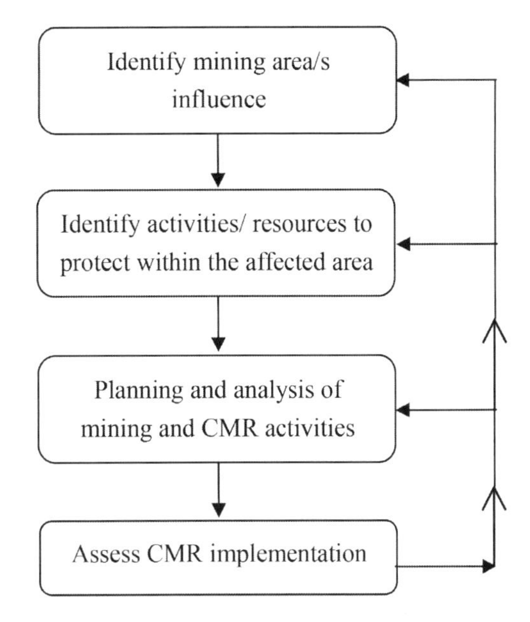

Figure 2. Critical scientific planning elements of CMR.

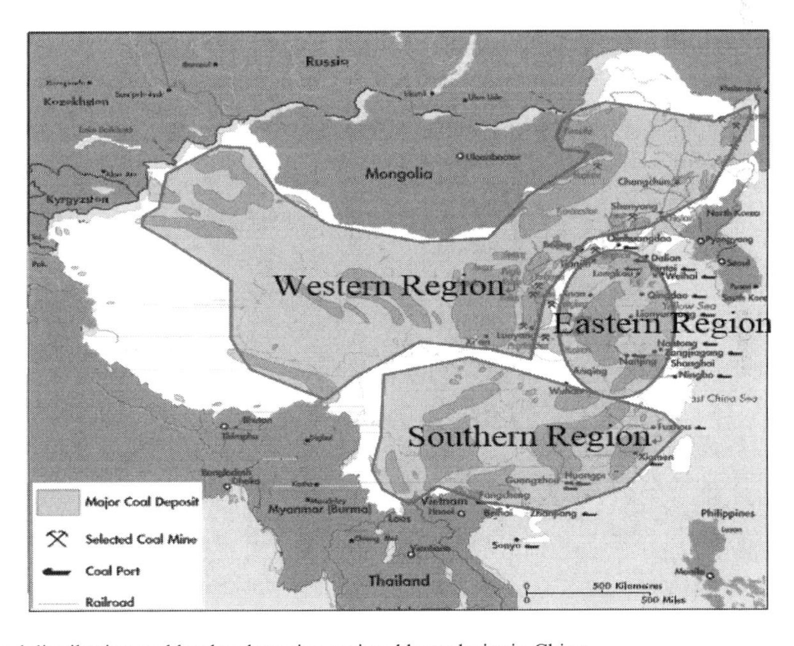

Figure 1. Coal distribution and land reclamation regional boundaries in China.
Source: Modified from US DOE. EIA. Country Energy Profile. China. February 1995.

elements for CMR planning. An excellent application of the above elements with a case study was provided in Hu et al., 2016. This paper presents an additional case study and summarizes the previous one to illustrate them.

1. Identify mining area influence on the surface and subsurface: For a selected mining plan, this element estimates the affected area on the surface and/or subsurface due to mining over a select period of time. It utilizes characteristics of the coal seam/s and associated underlying and overlying strata (geologic, hydro-geologic, and geometric such as seam pitch, anomalies), mining operations (panel width, panel length, mining rate, etc.), regional subsidence characteristics (angle of draw, angle of break, tensile and compressive strain, etc.), and surface topography to achieve this. The goal is to identify the nature and extent of the expected influence and/or damage to the variable/s of interest (farmland, aquifer, forest land, overlying or underlying coal seam, etc.) over the area so that CMR activities can be planned to minimize the impacts within the affected area and adjoining areas as mining progresses. This design element requires good data on all of the relevant characteristics listed above. A knowledge of subsidence engineering is essential for this exercise and good software programs are available to perform the design activities.
2. Identify activities and/or resources for protection within the affected areas: This element identifies resources (farmland, water resources, forests, structures) that need to be protected. For example, we may have to harvest the crops from the affected area before it goes under water due to surface subsidence, or we may have to remove structures, or move inhabitants away from the affected areas, or remove subsoil or topsoil from the affected areas. This element may also help identify areas where topsoil and/or subsoil may be stored temporarily to minimize CMR cost. This design element is a prerequisite to planning CMR activities.
3. Planning and analysis of mining and CMR activities within the identified areas: This element uses output of elements (1) and (2) above to plan CMR activities that would help protect desired resources, minimize CMR cost, and/or maximize the objective function which may be revenue from reclaimed areas through conservation and protection of identified resources, and regional and community development activities. This may require pursuing design elements (1) and (2) iteratively to achieve the desired objective function. It may consider an alternative mining plan, alternate reclamation technology or an alternate regional or community development plan. Since the near-term impact of CMR activities can be small, it is necessary to perform engineering economic analyses over an extended period of time (10 years or more) to provide a realistic view of the benefits.
4. Assess CMR implementation on identified objectives: This analysis element provides feedback at appropriate intervals on the performance of planned CMR activities and need for alternative planning, if needed. Just like design element 3 above, this element may also require iterative analysis of alternate mining plans, alternate reclamation approaches, and regional and community development activities.

5 CMR PLANNING CONSIDERATIONS

In achieving CMR planning goals (1) and (2) identified earlier in the statement of the problem section, one must consider current or pre-mining land and water utilization and future (post-mining) or proposed land and water resources utilization potential from regional economic development point of view. The CMR planning exercise should look at multiple alternatives that would enhance the regional economic development with sustainable future. For example, CMR may consider development of larger agricultural area, a new township or a new recreational area based on expected changes in surface topography and associated damage to existing structures. CMR planning must also consider alternate, creative regional economic development opportunities to support sustainable communities and to maximize explicit and implicit revenues. Mining companies should cooperatively work with local, regional, and national community planning groups to maximize return on land rehabilitation and water resources utilization opportunities.

CMR planning requires consideration of several important geographic, geologic, and current land use and hydrologic variables such as: 1) Surface topography and spatial distribution of land slope overlying proposed mining areas, 2) Geologic characteristics of soils and weak overburden overlying the bedrock; 3) Agricultural productivity characteristics of soils and their variation with depth below surface; 4) Surface and subsurface structures that may require subsidence damage consideration (highways, aquifers, existing reservoirs and major drainage patterns), and their future reconstruction costs; and 5) Pre-mining land and water use patterns.

The future potential for alternate uses is affected by important mining related variables that include: 1) Seam thickness, seam pitch, seam pitch orientation and their variation, 2) Areal extent and shape of the mining property (E-W or N-S), 3) Overall size of the reserve and expected subsidence-influence area, 4) Mining depth, 5) Panel size, 6) Mining advance rate, 7) Multiple coal seams, and 8) Mining width to depth ratio. Any surface or mining-related variables that directly or indirectly affect surface deformations must be considered. There is generally very little control on most of these variables. However, there may be some flexibility in panel orientation, panel width to depth ratio, and mining advance rate to control surface deformations and optimize CMR planning. Similarly, the presence of pre-existing water storage areas and water drainage patterns may allow some flexibility in CMR planning.

The pre-mining land use patterns vary locally and regionally, and future regional development needs (farmland, water resources, residential and commercial development needs, etc.) for economic development should be considered during the mine life of 20–30 years and could have significant impact on CMR planning. Another intuitive important factor would be the local costs for different types of reclamation (farmland, water storage and fisheries, recreation, pasture land, etc.) to achieve desired post-mining land use and water management structures and facilities. The goal of CMR planning should be to look at multiple alternatives that would enhance subsidence affected areas for the regional economic development with sustainable future.

6 FIELD DEMONSTRATION OF CMR PLANNING THROUGH CASE STUDIES IN CHINA

6.1 *Case study I (Anhui Province, Years 2007 to 2015)*

Description of the case study area: The study area is located in east China (Figure 3), and is the alluvial plain of the Huaihe River. The terrain is almost flat with ground slopes not exceeding 5°, with many ponds and rivers in the area. The surface elevations vary from +21.6 m to +25.4 m above the mean sea level or MSL (Figure 3), with the average of +23.1 m. It is located in the semi-humid warm temperate continental climate with four seasons. Most rain falls in summer (from June to August) with the average value of 926 mm. The GWL is about 1.5 m below the ground surface.

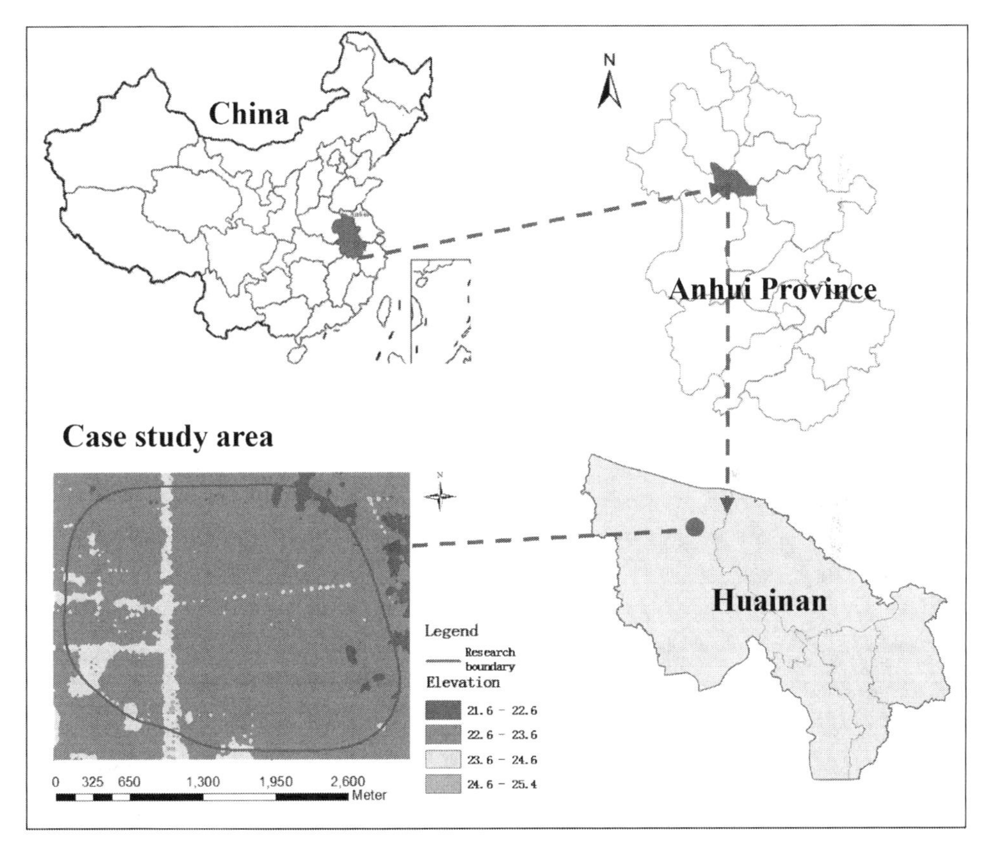

Figure 3. Location and pre-mining surface topography of — Case study I.

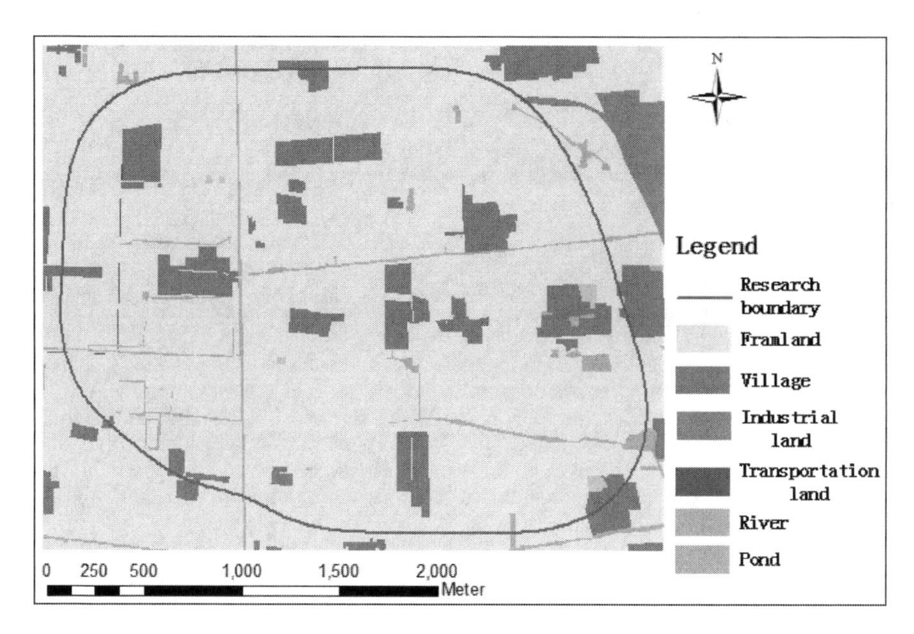

Figure 4. Pre-mining land use distribution in the case study area- Case study I.

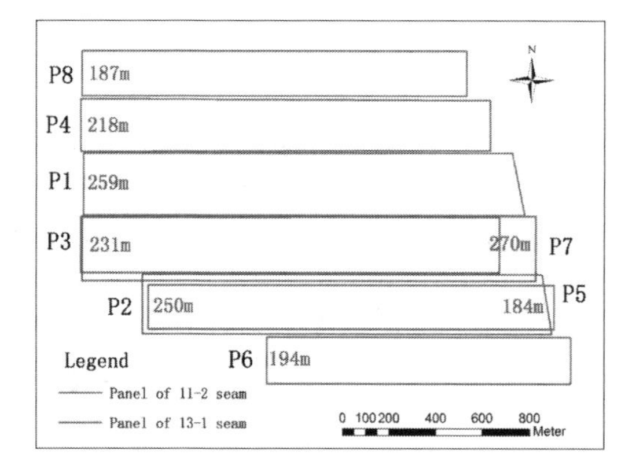

Figure 5. Layout of longwall panels in the case study area- Case study I.

Table 1. Relevant subsidence prediction parameters for Case study I and II.

Parameters	Case Study I	Case Study II
Subsidence factor for initial mining	1.11	0.8
Subsidence factor for repeated mining	0.91	/
Tangent of main influence angle	2.1	1.8
Horizontal displacement factor	0.31	0.35
The displacement distance	50 m	0.07H*
Influence propagation angle (deg.)	87.6	87

*H is the mining depth.

The mine began production at the end of 2007 and extracted mainly No. 11-2 and 13-1 seams.

Pre-mining land use is a very high-yield farmland area and covers about 86.79% of the entire mining area (Figure 4). It is a typical rice cultivation area in China with planting season from October to June. Corn is another important crop that usually grows from early June to late September. The farmland and water resources need to be protected in the area.

Geology and mining plan: There are eight (8) longwall panels (Figure 5) in the area with the mining sequence of P1-P2-P3-P4-P5-P6-P7-P8. Among them, six (6) panels mine 11-2 coal seam and the left two (2) panels mine 13-1 seam. The average thickness of 11-2 and 13-1 seam are 2.49 m and 3.56 m. The average face advance rate is about 5 m per day for both faces. So it requires about 12-months to mine each panel and the case study covers about eight (8) years of mining. The near surface alluvial materials vary from 346 m to 563 m in thickness. The relevant subsidence prediction parameters are given in Table 1.

Mining impacts simulation and analysis: The entire mining was divided into eight (8) stages based on the mining layouts: Stage 1- Completion of face P1; Stage 2- Completion of faces P1and P2; Stage 3- Completion of faces P1, P2 and P3; Stage 4- Completion of faces P1, P2, P3 and P4; Stage 5- Completion of faces P1, P2, P3, P4 and P5; Stage 6- Completion of faces P1, P2, P3, P4, P5 and P6; Stage 7- Completion of faces P1, P2, P3, P4, P5, P6

Table 2. Mining influence, water-logged and farmland areas in each stage- Case study I.

Stage	Max subsidence (mm)	Influenced area (ha)	Water area (ha)	Water area (%)
1	1105	290.58	0.00	0.00
2	1150	418.12	0.00	0.00
3	2239	426.67	90.27	21.16
4	2301	479.46	125.49	26.17
5	3302	486.23	163.62	33.65
6	3698	527.21	183.79	34.86
7	5779	531.24	209.36	39.41
8	5779	576.46	242.49	42.07

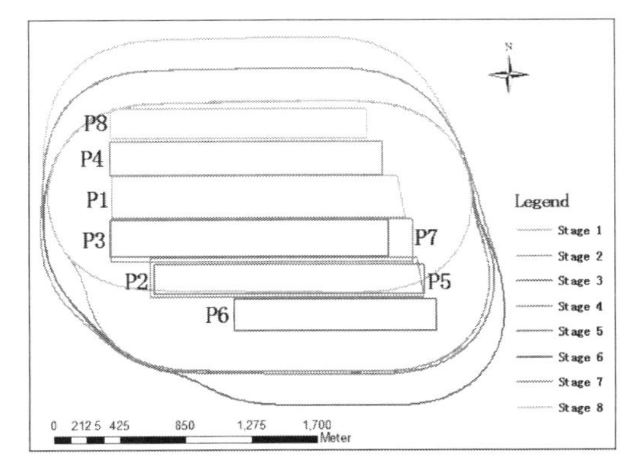

Figure 6. Subsidence-influenced boundaries after each mining stage- Case study I.

and P7; and Stage 8- Completion of faces P1, P2, P3, P4, P5, P6, P7 and P8. Surface displacements for each mining stage were predicted using the probability integration method and Knothe time function (Xiao et al. 2013). The subsidence-influenced region and maximum subsidence in different mining stages are shown in Table 2 and Figure 6. Incremental surface subsidence after each mining stage with a consideration of original surface topography is shown in Figure 7.

Farmland protection planning: Analyses indicated that if the resulting surface elevation after mining subsidence was less than +21.6 m, the land would submerge under water and it could not be used as farmland (Figure 8). The post-mining land use after each mining stage is discussed below.

Stage 1: During this stage the mining influence area is not very large, and there is no farmland that submerges into water; Stage 2: Since P2 is not adjacent to P1, the subsidence-influenced region extends over a wide area, but there are still no water-logged areas and farmland use is not impacted; Stage 3: Since P3 is located between P1 and P2, subsidence-influenced-region does not change much. However, the maximum subsidence increases to 2239 mm from 1150 mm in Stage 2 and some farmland submerges under water; Stage 4: P4 is next to P1 and subsidence-influenced region and water-logged areas extend towards north and more farmland submerges under water; Stage 5: The P5 is located above P2 and after its extraction, subsidence impact areas related to these two panels are superimposed and the maximum subsidence increases significantly to 3302 mm; Stage 6: P6 is adjacent to P2, and the subsidence-influenced region and water-logged areas increase gradually and extend to the south side; Stage 7: P7 is mined above P3 and therefore maximum subsidence increases significantly to a value of 5779 mm. Since P4 is the last panel on the south side, the subsidence impacts reach their final boundary on the south side; Stage 8: The subsidence impact region and water-logged areas further expand towards north side due to P8 and one big pond appears around the center of the TR study area.

Comparison of TR and CMR reclamation plans

TR Plan: All reclamation is planned during the ninth year after P8 mining. The regions which submerge under water would be dug deeper to get about 0.66 million m^3 of soil. The areas adjacent to the edges of the influence area will be graded as farmland and another 0.89 million m^3 of soil will be obtained and used as fill material. Since the average elevation of original topography (H_0) is +23.1 m. The designed elevation of reclaimed farmland (H_R) is +22.6 m. So some soil will be obtained from the edge of basin where the elevation is more than +22.6 m after subsidence (Figure 9). The dug-out soil will be used to fill low lying areas to the design elevation value of +22.6 m to grow crops. One large pond will be constructed around the farmland (Figure 10). Overall, the reclaimed study area will have 399.19 ha of farmland and 177.27 ha of water pond. Thus, it will result in cultivating about 69.25% of the case study area for farmland in the future (Table 3).

CMR Plan- Several guiding principles were used to develop the CMR plans. These included: 1) Local geographical environment, pre-mining land use patterns, and social and economic needs, 2) Minimize cost to excavate and transport the soils, 3) Dig deep, if necessary, to fill the shallow land areas to protect and enhance farmland, 4) Grade the disturbed land as soon as possible and develop contour ditches to protect farmland and enhance farm productivity; 5) Balance cutting and filling to minimize need for additional soil; 6) Consider future subsidence of reclaimed land; and 7) Maximize farmland and water resources protection and productivity of agriculture.

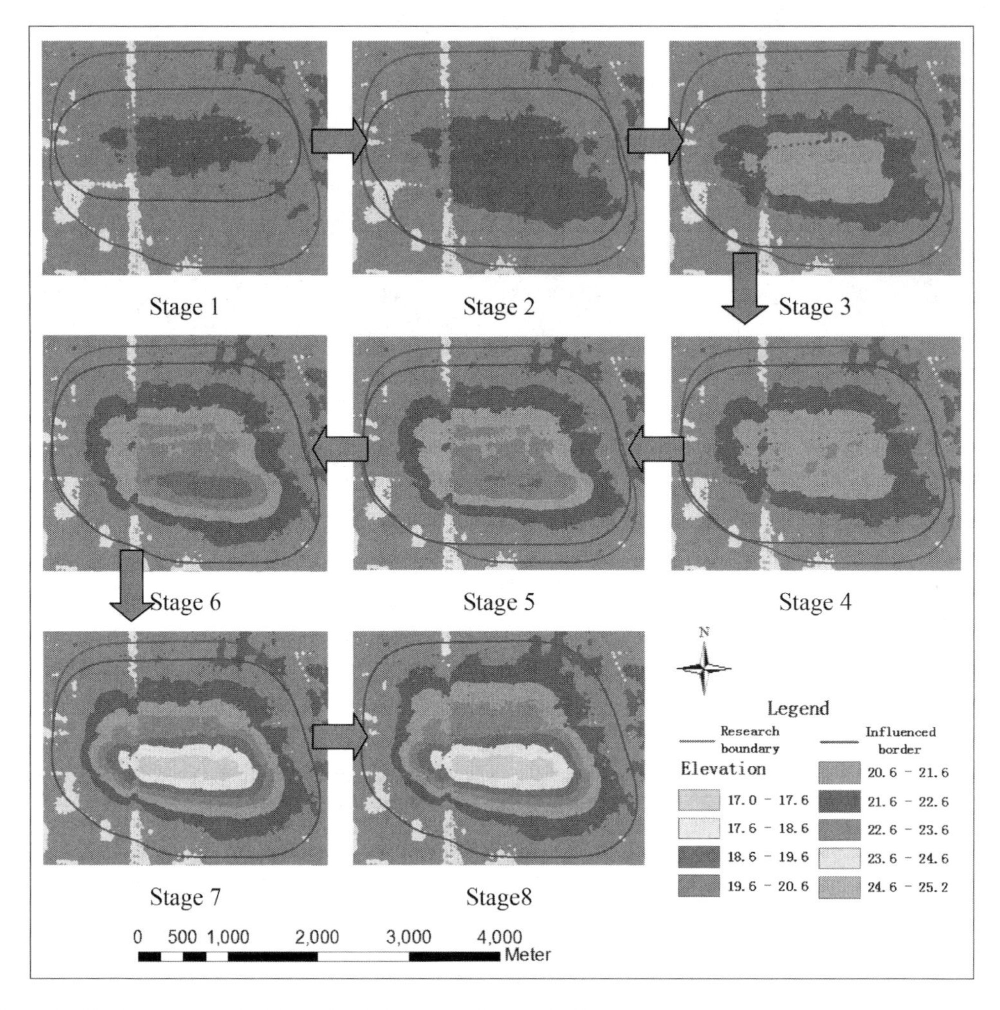

Figure 7. Surface topography after each mining stage- Case study I.

Since farmland and rivers and ponds protection were the primary goals here, it was decided that the surface elevation of the final reclaimed farmland elevation should not be less than +22.6 m The CMR activities for each phase are described below and are shown in Figure 11.

Stage 1: Only a small part of the land is disturbed by mining subsidence and there are no water-logged areas; Stage 2: After mining P3, large amount of land will submerge into water and therefore advance reclamation planning is necessary. It is proposed to dig about 0.96 million m^3 of soil dug from the deep region above P2 to form a small reservoir. With the consideration of future subsidence, the dug-out soil is used as filling material to raise the elevation of the surrounding regions in south and west sides so that land can be used as farmland; Stage 3: In order to save the precious soil, reservoir would be extended towards the P3 direction, and about 0.84 million m^3 of soil will be dug-out to fill the northeast area which would otherwise be water-logged due to mining subsidence.

Stage 4: With the extraction of P4, the region above P4 would submerge under water. So the reservoir will be further enlarged to the north side to obtain 0.38 million m^3 of filling soil and the shallow region in the northwest side is reclaimed as farmland; Stage 5: The reservoir is projected to reach its final boundary. About 0.29 million m^3 of soil will be dug-out from the reservoir to fill the east side region, and contour ditches are constructed to help transport water to the reservoir; Stage 6: Around the western edge of the final subsidence-influenced area, land will be regraded for farmland. About 0.30 million m^3 of soil will be

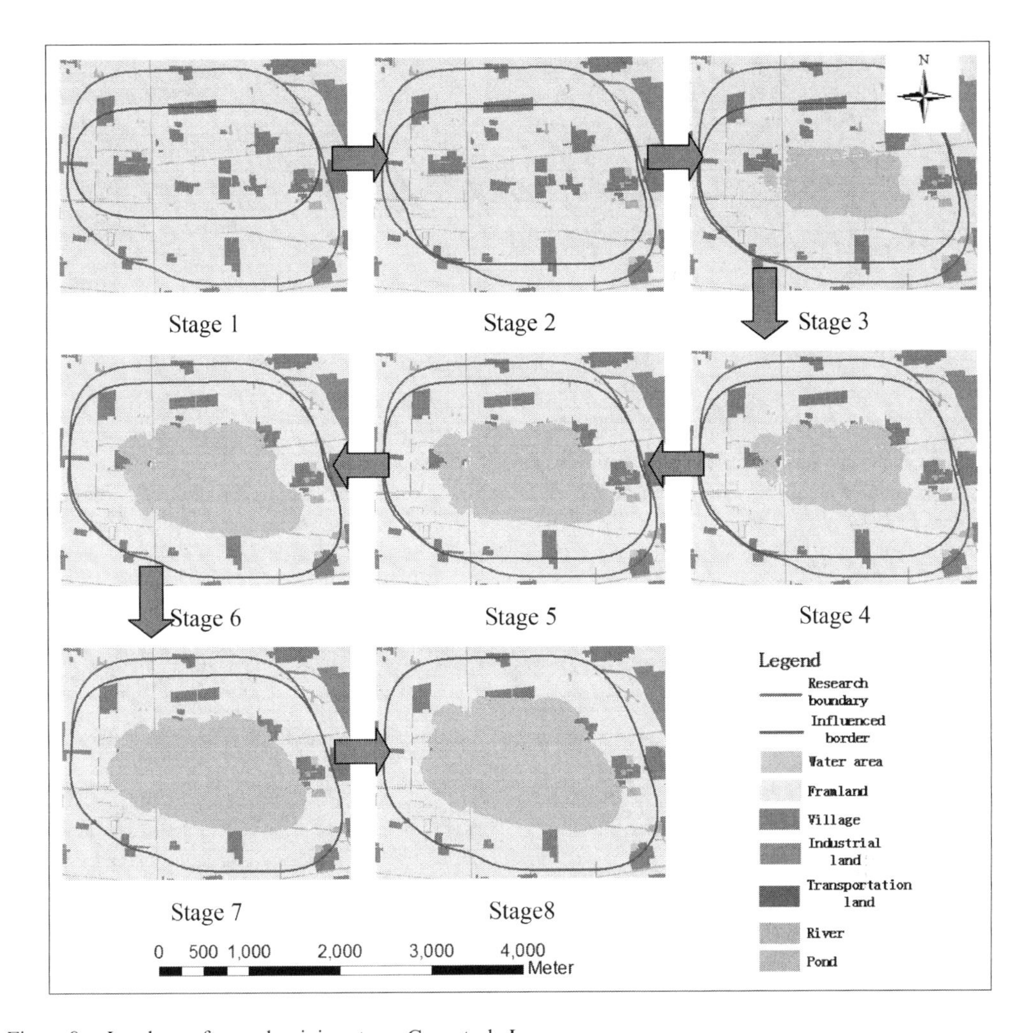

Figure 8. Land use after each mining stage- Case study I.

obtained from the region where surface elevation exceeds +22.6 m after subsidence, and the shallow areas around it will be filled to above +22.6 m level to cultivate crops; Stage 7: Due to the original surface topography and mining subsidence, only 0.29 million m^3 of soil can be obtained from the large region in the south and east of the subsidence basin, and is used to fill the area adjacent to it; Stage 8: The last panel on the north side is proposed to be mined, and 0.30 million m^3 of soil would be obtained from the relatively high areas to fill the shallow areas to be cultivated as farmland. The large reservoir is proposed to be developed as a Water Park to improve eco-environment and local economy. It will also provide water resources for farmland irrigation and local area industries at the same time. The resulting reservoir area and cultivated farmland cultivated in each mining stage based on the developed CMR plan is shown in Table 4 with the estimate that about there would be 477.48 ha will be reclaimed as farmland which accounts for 82.83% of the total study area.

Comparison of TR and CMR plans

Farmland utilization: The percentages of farmland in TR plan are 50.80% to 72.87% lower than that in the CMR plan during years 3–9. Even upon completion of all reclamation activities the percent reclaimed farmland in TR is about 69.25% as compared to 82.83% in the CMR Plan (Table 5).

Water resources development: The data for the two plans is shown in Table 6. The area in TR is larger than in the CMR plan except for Years 1 and 2. However, the water volumes in TR plan are smaller than in the CMR plan since steps were taken to protect water resources as mining progressed to use them efficiently in future.

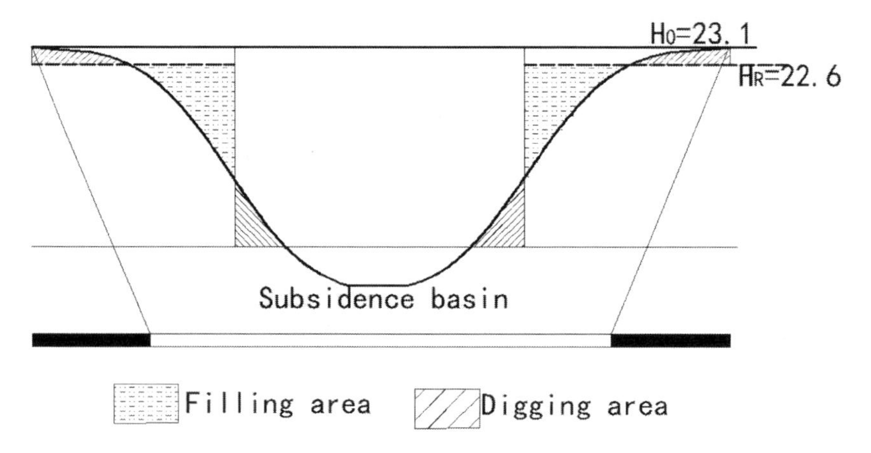

Figure 9. Schematic diagram of filling and digging areas.

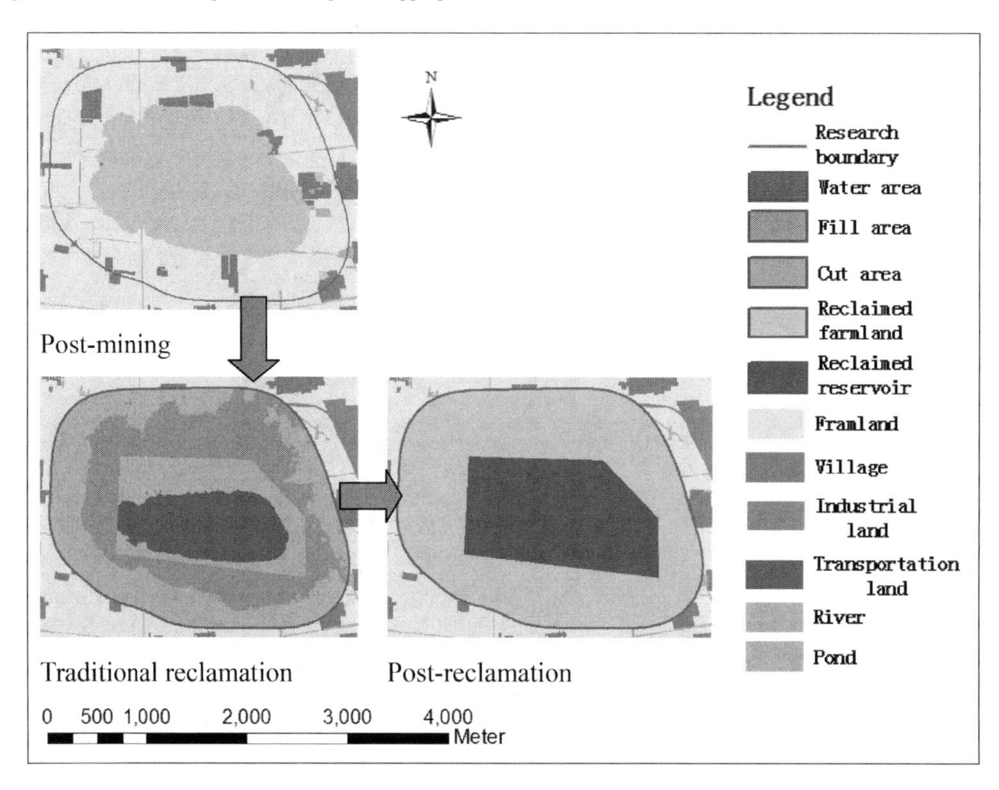

Figure 10. TR plan reclamation layout- Case study I.

Table 3. Relevant reclamation data for TR plan- Case study I.

Plan	Farmland (ha)	Water area (ha)	Water volume (million m^3)	Earthwork (million m^3)
TR	399.19	177.27	4.78	1.55
CMR	477.48	98.98	5.56	3.36

This case study has used all design and guiding principles discussed earlier to demonstrate that CMR plans have both short-term and long-term benefits for sustainable mining, with farmland use and protection and regional economic development. It represents analysis of only one CMR plan which may not be optimum. Alternate CMR plans could be developed that provide even more benefits than this one.

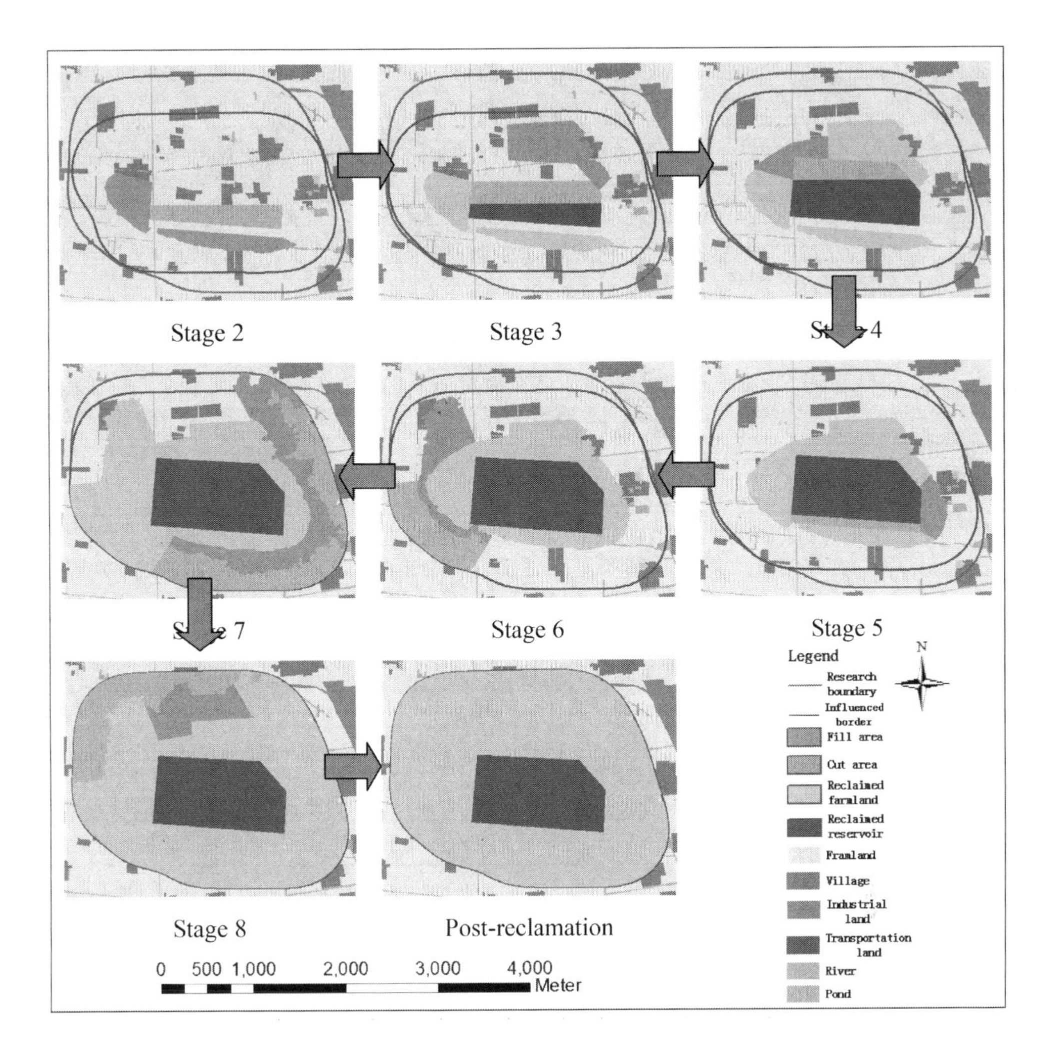

Figure 11. CMR plan reclamation layout- Case study I.

Table 4. Reclaimed farmland and water reservoir areas during different mining stages- Case study I.

	Farmland		Water Reservoir	
Stage	Area (ha)	Percentage (%)	Area (ha)	Percentage (%)
1	500.33	86.79	0.00	0.00
2	476.10	82.59	29.23	5.07
3	471.99	81.88	62.23	10.80
4	452.65	78.52	86.57	15.02
5	445.24	77.24	98.98	17.17
6	450.24	78.10	98.98	17.17
7	463.24	80.36	98.98	17.17
8	477.48	82.83	98.98	17.17

6.2 *Case study II (Shandong Province, Years- 2009 to 2014)*

This case study was recently published by Hu et al. (2016) and represents the state-of-the-art in CMR planning techniques. It is included here only in summary form to strengthen some of the principles discussed in earlier sections.

Description of the case study area: The study area (Figure 12) is located in southwest of Shandong Province with relatively high GWL of 2.5 m, The highly productive farmland area covers about 86.60% of the entire mining area (Figure 13). Corn and wheat are the most important crops and the multiple cropping index is 200%.

Geology and mining plan: No. 3 coal seam, 9 m thick is mined in the study area and the seam

Table 5. Available farmland for cultivation for TR and CMR plans- Case study I.

	TR Plan		CMR Plan		
Year	Area (ha)	% of total	Area (ha)	% of total	Difference (%) (TR-CMR)
1	500.33	86.79	500.33	86.79	0.00
2	500.33	86.79	476.10	82.59	4.20
3	420.06	72.87	471.99	81.88	–9.01
4	389.84	67.63	452.65	78.52	–10.90
5	356.71	61.88	445.24	77.24	–15.36
6	341.54	59.25	450.24	78.10	–18.86
7	320.97	55.68	463.24	80.36	–24.68
8	292.84	50.80	477.48	82.83	–32.03
9 (after reclamation)	399.19	69.25	477.48	82.83	–13.58

Table 6. Water resources areas and volumes in TR and CMR plans- Case study I.

	TR Plan		CMR Plan		Difference (TR-CMR)	
Year	Area (ha)	Volume (m. cu.m)	Area (ha)	Volume (m. cu.m)	Area (ha)	Volume (m. cu.m)
1	0.00	0.00	0.00	0.00	0.00	0.00
2	0.00	0.00	29.23	0.58	–29.23	–0.58
3	90.27	0.47	62.23	2.15	28.04	–1.68
4	125.49	0.74	86.57	2.74	38.92	–2.00
5	163.62	1.57	98.98	3.72	64.64	–2.15
6	183.79	1.98	98.98	3.97	84.81	–1.99
7	209.36	4.18	98.98	5.55	110.38	–1.36
8	242.49	4.51	98.98	5.56	143.52	–1.05
9 (after reclamation)	177.27	4.78	98.98	5.56	78.30	–0.78

pitches average of 5°. Relevant subsidence prediction parameters are given in Table 1. There are five (5) longwall mining panels (Figure 14) in the area with the mining sequence of P1-P2-P3-P4-P5, and face advance rate of 5 m per day.

Subsidence impacts analysis: Similar to case study I, mining of the area was divided into five (5) phases based on the mining layout: Phase 1- Completion of face P1; Phase 2- Completion of faces P1and P2; Phase 3- Completion of faces P1, P2 and P3; Phase 4- Completion of faces P1, P2, P3 and P4; and Phase 5- Completion of faces P1, P2, P3, P4 and P5. In the end about 36.57% of the land would submerges under water and cannot be used for farming (Table 7).

Comparison of CMR and Traditional Reclamation (TR) plans

Similar to in Case study I, farmland availability with TR (Figure 15) and CMR (Figure 16) plans are summarized in Table 8. The differences in each phase vary from 3.01% to 37.14%. Upon completion, the reclaimed farmland by CMR is higher (88.5%) with long-term revenue benefits as compared to TR (75.56%).

Water resource volume data during mining (Years 1–5) and after TR completion (Year 6) are shown in Table 9. The available water resources volumes with CMR are much higher than TR. Final water volume by CMR is 8.80 million m^3, which is 15.01% more than TR. CMR planning measures can protect water resources in the subsided areas through stripping topsoil before it sinks into water and is used as filling material in low-lying areas. Soil erosion is controlled and water resources are channeled into reservoirs where they can be used in future. Water storage areas can be effectively used as fish ponds with good water management. It can also be used to design as water or wet land park which should benefit the local people and eco-environment.

7 UNDERGROUND MINING AND RECLAMATION PRACTICES FOR PRIME AGRICULTURAL LANDS IN ILLINOIS-USA

Industry overview: Over 70% of the State of Illinois is underlain by coal with the worlds' larg-

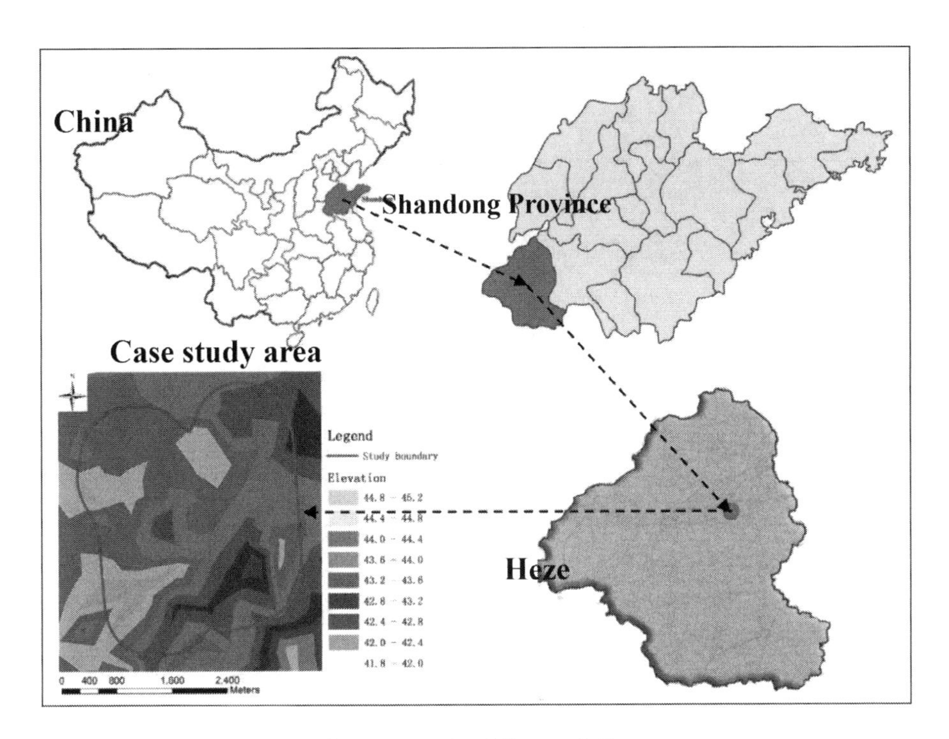

Figure 12. Location and pre-mining surface topography of Case study II.

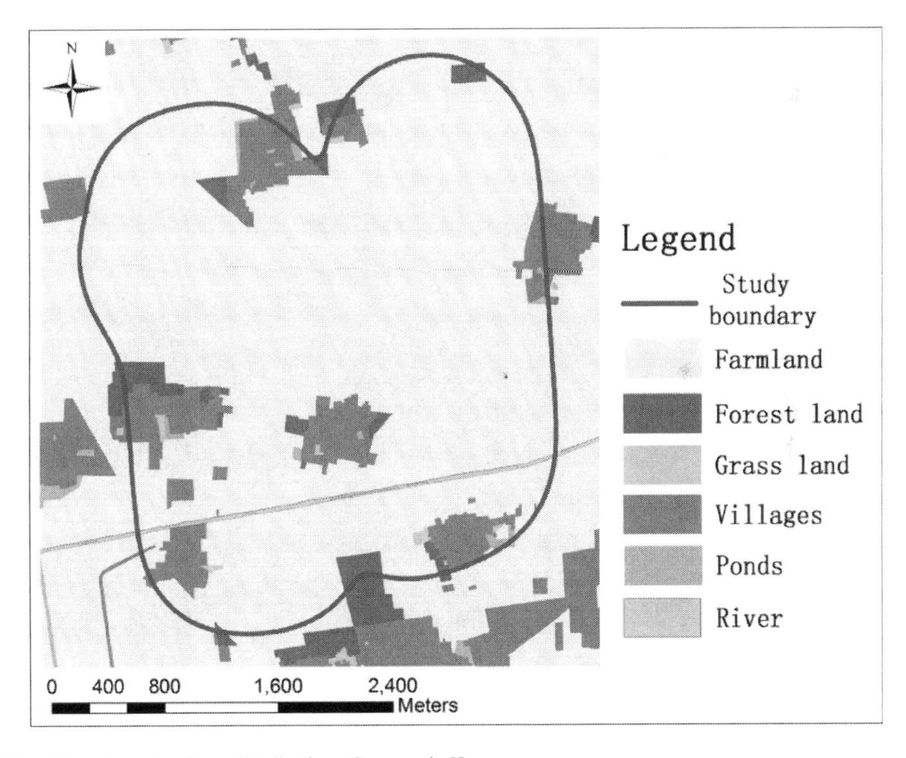

Figure 13. Pre-mining land use distribution- Case study II.

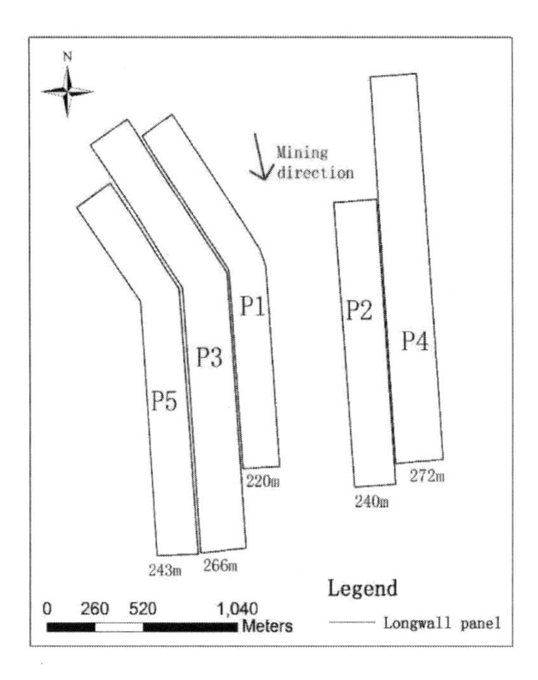

Figure 14. Layout of longwall panels- Case study II.

est bituminous coal reserve base (about 85 billion tons of demonstrated reserves). In 2014 Illinois produced about 58 million mt of coal employing 4100 persons. About 90% of coal was mined using underground mining methods, and about 80% of it using retreat longwall mining methods. The average productivity for all underground mines was 7.35 mt per hour while productivity from longwall mines ranged 10–12 mt per hour (Illinois Office of Mines and Minerals Annual Report, 2015).

Geology and mine characteristics: Almost all the coal was mined from two coal seams (No. 6 or Herrin, and No. 5 or Harrisburg). They vary in thickness from 1.3 m to 2.1 m and are mined at depths varying 36 m to 270 m. The seven longwall faces operate at depths varying from 120 m to 220 m. The coal seams are overlain by weak shales, sandstones, and limestones (20 MPa to 60 MPa uniaxial compressive strength or UCS), and underlain by weak claystone and shales (5 MPa to 15 MPa UCS). Caving of roof strata typically occurs immediately behind the face shield supports.

Regional surface and climate characteristics: Most of the coal in Illinois is mined in southern Illinois. The surface topography in the region is flat to gently rolling with surface slopes of less than 5% but with

Table 7. Subsidence influence area, water area and farmland in each mining phase- Case study II.

		Water area		Farmland	
	Influence area (ha)	Area (ha)	Proportion (%)	Area (ha)	Proportion (%)
Phase 1	355.04	44.40	4.88	749.39	82.32
Phase 2	549.42	102.28	11.24	692.49	76.07
Phase 3	721.53	197.16	21.66	598.81	65.78
Phase 4	846.07	278.74	30.62	517.65	56.87
Phase 5	910.31	332.93	36.57	467.58	51.36

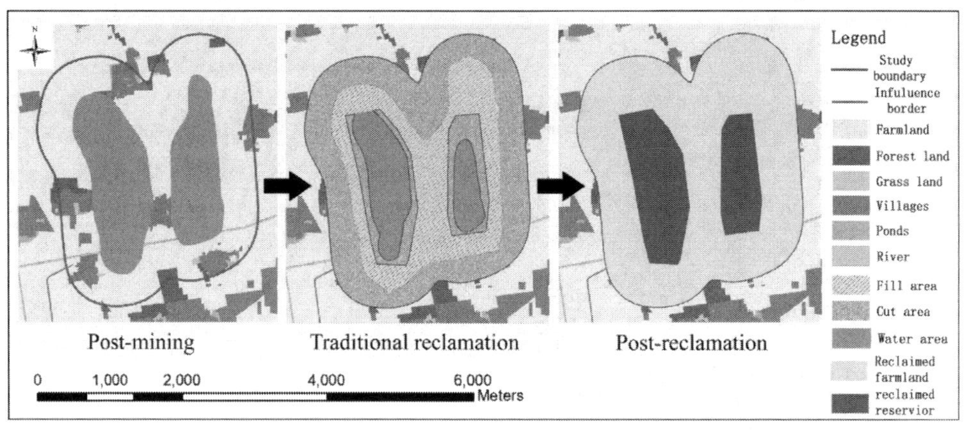

Figure 15. Traditional Reclamation (TR) plan- Case study 2.

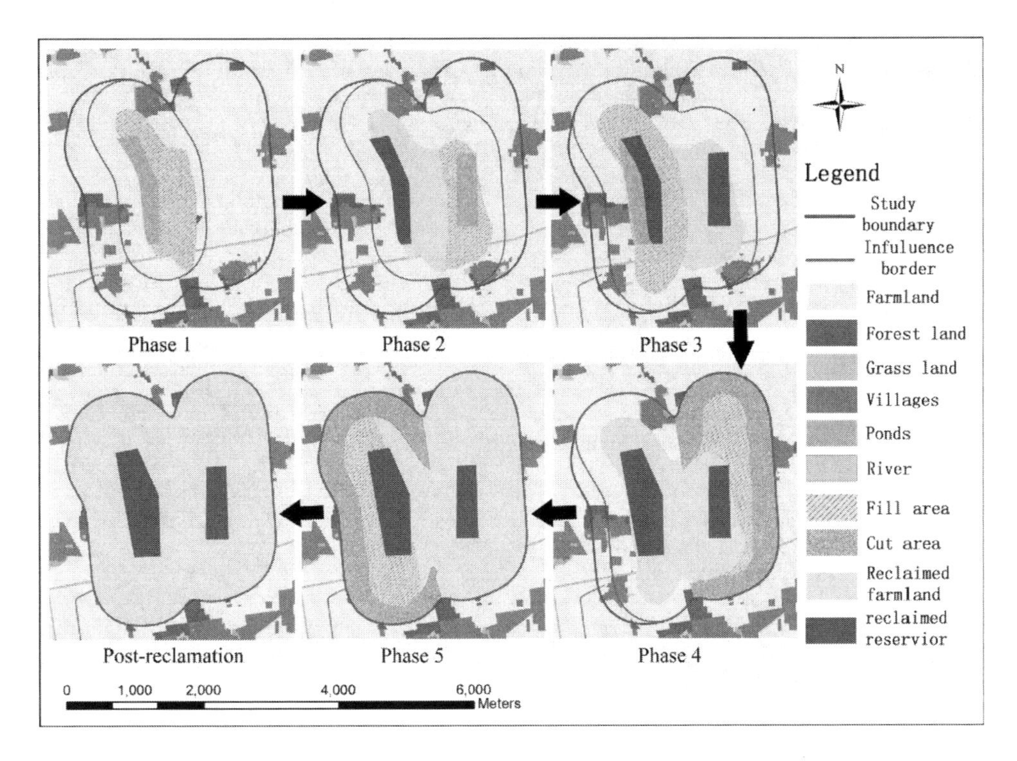

Figure 16. CMR plans during different mining phases- Case study 2.

Table 8. Farmland utilization in TR Plan and CMR Plans- Case study II.

	TR Plan		CMR Plan		
Year	Area (ha)	Proportion (%)	Area (ha)	Proportion (%)	Difference (%) (TR-CMR)
1	749.39	82.32	776.74	85.33	−3.01
2	692.49	76.07	759.12	83.39	−7.32
3	598.81	65.78	735.57	80.80	−15.02
4	517.65	56.87	771.37	84.74	−27.87
5	467.58	51.36	805.62	88.50	−37.13
6	687.83	75.56	805.62	88.50	−12.94

Table 9. Comparison of water resources in TR Plan and CMR Plan- Case study II.

	TR Plan		CMR Plan		
Year	Area (ha)	Volume (m. cu.m)	Area (ha)	Volume (m. cu.m)	Difference (m. cu.m) (TR-CMR)
1	44.40	0.22	39.85	1.81	−1.60
2	102.28	0.68	73.96	3.44	−2.77
3	197.16	3.03	104.69	6.61	−3.59
4	278.74	4.95	104.69	7.57	−2.63
5	332.93	7.04	104.69	8.80	−1.77
6	222.48	7.66	104.69	8.80	−1.15

local slopes reaching 10–15%. Illinois also is blessed with some of richest quality farmland in the world termed "Prime Farmland". Corn and soybeans are the primary crops that grow on poorly-drained soils (Darmody et al. 2014). Temperatures from April to August range 15–25 C^0 and the average rainfall of 80–120 mm occurs during the above period.

Mining operations and subsidence characteristics: All the seven longwall face operate super-critical faces with width varying from 300 m to 360 m; with some faces proposed to be 450 m wide. The face lengths vary 4500 m to 9000 m. Three-entry development entries are practiced with the width of the development entry pillar system varying from 70 m to 100 m depending upon seam underlying floor characteristics. The face advance rate ranges 27 m to 36 m per day. Thus typical mining life for a panel ranges 0.5 to 1.0 years. The angle of draw and subsidence factor values within the subsidence basin typically vary 15^0 to 25^0, and 0.55 to 0.7, respectively. Maximum subsidence values around the center of the panel range 1.2 m to 1.8 m. The chain pillars of development entries typically subside about 300 mm. About 80% of the maximum subsidence occurs behind the face after the longwall face has advanced about 100 m, with the remaining residual subsidence occurring within 3–4 months after mining.

Subsidence impacts analysis: With above discussion of surface and mining characteristics, subsidence impacts typically include: 1) Shallow water-logged areas, 2) Changes in drainage around the edges of subsidence basin, 3) Breaking of drain-tiles around the edges of the subsidence basin where tensile and compressive strains can be high, 4) Short-term flow changes in perennial and intermittent streams, 5) Damage to temporary and permanent structures, submergence of topsoil and subsoil in water if it is not removed in a timely fashion; and damage to infra-structures (roads, culverts, bridges, etc.). Since land is privately controlled, or owned by organizations (parks, forests, etc.) the coal company must interface with individual land owners and organizations to deal with subsidence impacts.

Subsidence reclamation regulations: Farmland mitigation requirements are given in Section 1817.121 (Subsidence Control) of the 62 Ill. Adm. Code 1817.121(c) (1), and are effective June 27, 2005. These are written to protect the land and water resources overlying the underground-mined areas. It requires that the mining company must have an approved reclamation plan prior to issuance of mining permit. The philosophy behind subsidence reclamation requirements was best summarized by Barkley (2017). Surface mining of prime agricultural lands requires classifying the pre-mining farmland into low capability, high capability, and prime based on productivity. Prior to issuance of mining permit, the coal company must post a reclamation bond which will be released only after the company has demonstrated the post mining yields to be equal to or greater than pre-mining yield capability. The State of Illinois is the only state in the USA that requires low and high capability designations.

Surface impacted areas due to underground mining however do not require to sub-classify the land similar to above and there are no yield-based performance requirements for reclaimed subsided lands. The primary goal of subsidence reclamation is to ensure appropriate drainage since agricultural soils drain poorly and shallow water-logged areas are likely to develop. The permitting agency hypothesizes that with drainage restoration the land should have similar yields as before subsidence. This hypothesis was developed based on extensive studies performed by Illinois researchers (Darmody et al., 2014) under the Illinois Mine Subsidence Research Program over a period of ten years (1985–1995) that showed that subsidence impacts on agricultural productivity are small as long as drainage is maintained. It also considers that farmland productivity depends upon how farming is done.

8 CONCURRENT MINING AND RECLAMATION OF SUBSIDENCE IMPACTS IN CHINA AND IN ILLINOIS, USA: SOME THOUGHTS

CMR is a complex planning tool that can enhance the value of land and water resources reclamation as demonstrated through multiple case studies included and referred to in this paper. However, it requires significant cooperation among land owners, water resources companies, and regional economic development agencies. Therefore, it can be commercially applied in countries which have more centralized control. It is an excellent technology that can make mining more sustainable with both short-term and long-term benefits. Some of the prerequisites for its success are not present in the USA. For example, mining alternate panels far away from each other, or mining alternate panels in multiple seams is not common. Similarly, the face advance rates are 10–20 times higher than in China. So, dynamic reclamation planning becomes an extremely important unit operation that can negatively impact operations productivity. Mining companies may not want to risk negative impact to mining operations. The maximum subsidence is much smaller than in China that can be easily mitigated through relatively easy regrading activities. Furthermore, since mining companies are operating super-critical panels, the areas around the panels that experience higher ground slopes are a small percentage of the mined-out panel area, the regrading efforts to ensure proper drainage for farmland cultivation are generally small and can be easily managed by high capacity equipment during the summer months when the soils are dry. Further development of CMR and its application should continue to make mining more sustainable in areas where there are competing interests between population, land use, and water resources use.

9 AREAS FOR FURTHER DEVELOPMENT OF CMR

The development of CMR was motivated by maximum utilization of farmland resources during the period when recovering valuable coal resources. However, as its development has gown and it is better understood, it could be used to minimize loss of or enhance any variable that would minimize subsidence impacts and improve regional economic and community development. Furthermore, it could consider multiple variables during the optimization process using principles of operations research and systems analysis. For example, protection of water, forest, land structures, or other critical resources could be interest variables. For this development, one would need good relative data on cost of associated reclamation and resulting revenue or economic enhancement for each variable considered. Such data are currently not available and should be developed. This area of development may also include order of extraction of multiple coal seams to maximize the objective function. Water resources, particularly potable water resources should be of immediate interest in China since the cost is very high. So, authors strongly recommend CMR development to protect and manage water resources in the near future. Synergizing CMR planning to protect multiple resources of national interest will require considerable cooperation among several agencies and may require central oversight and review. Further development of CMR should also include development of alternate, lower cost mining, backfilling, and reclamation technologies to take full advantage of CMR planning.

10 CONCLUDING REMARKS

This paper provides an overview of the "Concurrent Mining and Reclamation or CMR" technology. CMR technology studies have demonstrated that farmland utilization and water resources development can be minimized and/or enhanced through CMR planning as compared to traditional reclamation approaches. It is an advanced technology in subsidence land reclamation, and a leading technology in the world. The application of CMR in China is explained in this paper through two case studies from different parts of China with different geologic conditions, surface topographic and hydrologic characteristics. Most of the CMR applications to date have focused to minimize the loss of farmland use for agriculture during and after the entire mining process without a consideration of regional economic planning. A recent research study has also attempted to quantify impacts of CMR planning on surface hydrologic resources. That has identified another important research area of CMR for minimization and/or enhancement of water resources quantity and quality related to underground mining of coal. Since water resources represents one of the global mega challenges, this area of research has very high potential. Furthermore, CMR planning studies require at a minimum sub-optimization studies, application of principles of systems analysis and operations research need to be incorporated. CMR planning has the potential to develop and sustain emerging economic communities while simultaneously recovering coal as a valuable energy resource. The concepts and technologies are scientifically sound with potential for applications around the global.

Additional CMR studies are currently in planning stages to document and reinforce the developed concepts here. The CMR planning will be facilitated through development of engineering and economic analyses software and is being considered. These tools will allow a better understanding of the effect of different CMR variables on development of concurrent mining and subsidence reclamation plans and their impact on cost effectiveness.

REFERENCES

Barkley, D (2017) "Private communication from March 1 to March 17, 2017."

BP statistical reviews of world energy, 2015, www.bp.com/statisticalreview.

Darmody, R.G, R. Bauer, D. Barkley, S. Clarke, D. Hamilton (2014) Agricultural impacts of longwall mine subsidence: the experience in Illinois, USA and Queensland, Australia. International Journal of coal Science and Technology. 1(2): 207–212.

Hu ZQ (1994a) The technique of reclaiming subsidence areas by use of a hydraulic dredge pump in Chinese coal mines. Int. J. Min Reclam. Environ 8(4):137–140.

Hu ZQ (1994b) Land reclamation plan for mining subsidence. Mining survey 2:32–34.

Hu ZQ, Chen C, Xiao W, et al. (2016a) Surface movement and deformation characteristics due to high-intensive

coal mining in the windy and sandy region. International Journal of Coal Science and Technology, 3(3):339–348.
Hu ZQ, Chen C. (2016) Impact of underground coal mining on land ecology and its restoration in windy and sandy region. Journal of Mining Science and Technology, 1(2):120–130.
Hu ZQ, Li J, Zhao YL (2006) Problems, reasons and countermeasures for environmental quality and food safety in the overlapped areas of crop and mineral production. Sci. Technol. Rev. 24(3):21–24.
Hu ZQ, Luo YM (2006) Suggestions on environmental quality and food safety in overlapped areas of crop and mineral production. Sci. Technol. Rev. 24(3):93–94.
Hu ZQ, Xiao W, Wang PJ, Zhang YL. (2013) Concurrent mining and reclamation for underground coal mining. Journal of China Coal society. 38(2):301–307.
Hu ZQ, Xiao W. (2013a) Optimization of concurrent mining and reclamation plans for single coal seam: a case study in Northern Anhui, China. Environmental Earth Sciences, 68(5): 1247–1254.
Hu ZQ, Zhang RY, Y.P. Chugh and Jia J. (2016b) Mitigating Mine Subsidence Dynamically to Minimize Impacts on Farmland and Water Resources: A Case Study, Int J Environ Pollu 59(2):169–186.
Hu ZQ and Xiao W. (2013b) New Idea and New Technology of Mine Land Reclamation: Concurrent Mining and Reclamation, Coal Sci Technol, 41(9), 178–181.
Illinois Office of Mines and Minerals Annual Report (2015). https://www.dnr.illinois. gov/ mines/ Pages
Li TQ, Qi JZ, Zhou JH and Jiang, S. (1999) Method for pre-reclamation of dynamic subsidence area in Liuqiao No. 2 Coal Mine. Mining survey 2:57–58.
Xiao W, Hu, ZQ, Zhang, RY and Zhao, YL (2013) 'A simulation of mining subsidence and its impacts to land in high ground water area- An integrated approach based on subsidence prediction and GIS', Disaster Advance, 6(S4): 142–148.
Xiao W, Hu ZQ, Chugh YP, Zhao YL (2014). Dynamic Subsidence Simulation and Topsoil Removal Strategy in High-Groundwater Table and Underground Coal Mining Area- A Case Study in Shandong Province. International Journal of Mining, Reclamation and Environment, 28(4): 250–263.
Xiao W, Hu ZQ, Fu, YH (2014) Zoning of land reclamation in coal mining area and new progresses for the past 10 years. International Journal of Coal Science and Technology 1(2):177–183.
Zhao YL and Hu ZQ. (2008) Proper time model for prereclamation of unstable subsidence. Journal of China Coal Society. 33(2):157–161.

Land Reclamation in Ecological Fragile Areas – Hu (Ed.)
 ISBN 978-1-138-05103-4

Research on mining technology of extra-thick coal seam under rich water aquifers in a bitter cold steppe region

J. Fang
State Key Laboratory Coal Resources and Safe Mining, China University of Mining and Technology, Beijing, China
Science and Technology Development Department, Shenhua Group Corporation Limited, Beijing, China
State Key Laboratory of Groundwater Protection and Utilization in Coal Mining, Shenhua Group Corporation Limited, Beijing, China

D.Z. Gu, Q.S. Li & Z.G. Cao
Science and Technology Development Department, Shenhua Group Corporation Limited, Beijing, China
State Key Laboratory of Groundwater Protection and Utilization in Coal Mining, Shenhua Group Corporation Limited, Beijing, China

ABSTRACT: The Mindong Mine, which belongs to the Shenhua group, China, is located in the bitter cold steppe region of the eastern part of Inner Mongolia, where the ecological environment is seriously restricted by underground water. In order to reduce the destruction of extra-thick coal seam mining to underground aquifer and the fragile surface grassland ecosystem, field engineering detections, theoretical analysis and simulation experiments were adopted to study the reasonable mining method of extra-thick coal seam under rich water aquifer in Mindong Mine. Through the research on mining geological conditions and hydrogeological condition of mine, three extra-thick aquifers were detected over the roof of 16-3up coal seam, the first mining coal seam of mine. The upper water-enriched region was detected by the joint geophysical prospecting methods combined ground transient electromagnetic method and underground DC electrical method. The height of water-flowing fractured zone and its ratio to the mining height were measured through numerical simulation and grout loss of boreholes. By adjusting the mining height of coal seam to avoid the height of water flowing fractured zone reaching the upper aquifer, the aim of preventing groundwater system from destroying was achieved. Field practices indicate that the underground water is stable and the grassland ecosystem of mine is well protected.

Keywords: Bitter coldsteppe region; Ecological function impairment; Water-flowing fractured zone; Partition and limiting thickness; Water protection mining

1 INTRODUCTION

The MengDong grassland located in the hulun buir is an important ecological barrier in north China, with a cold and semi-arid, soil barren, ecologically fragile characteristics, such as annual rainfall less than 400 mm, the minimum temperature of 47.5°C, and the thickness of the overburden only 30 cm. The region is also the important coal production base for China, and made much important contribution to protect our country northeast energy supply. In recent years, with the strength of the coal excavation going up, the increasingly serious ecological damage caused by mining, resulting in a decline in groundwater level and soil erosion, vegetation destruction, etc (State Coal Industry Bureau, 2000; Wu et al., 2014). Therefore, it is urgently to be solved in the region which how to furthest protect the ecological environment with coal excavating.

The Mindong mine of the Shenhua, one of the large-scale production of mine in MengDong prairie, located in the middle and lower reaches of the YiMin River, is a hilly area located in the northeast corner of YiMin coal field belongs to the western slope of Great Khingan Mountains. The exploiting method is underground mining and the mining technology is fully mechanized top coal caving. It leads to water leakage and water inrush disasters (see chapter 1) which is the large thickness of coal seam occurrence, formation aquifer water in the mining. It is the main direction for ecological impairment and the practice of water-preserved mining How to balance the protection of water resources and the control of coalmine water damage.

Existing research shows that it is the key point to solve the problem of water protecting due to the coal mining that controlling the development of the overlying strata water flowing fractured zone reasonably, reducing or avoiding the damage of water flowing fractured zone to aquifer water (Ma et al., 2008; Wang et al., 2012). It is well known that mining height has directly effect on the development degree of overlying strata water flowing fractured zone, and the higher the mining height, the greater the fracture zone height is larger, the broader the range of influence (Chen et al., 2006; Gao et al., 2014; Ren et al., 2004; Zhang et al., 2001). For that, it is effective way that reasonable adjusting the mining height to reduce the height of fractured zone for Mindong mine which is mining the thick coal seam under water aquifer. So, it is studied that the development degree of the overlying strata water flowing fractured zone caused by mining the thick coal seam based on the physical condition of Mindong mine; According to the actual situation of aquifer occurrence, reasonable adjustment of coal seam mining scheme, to protect groundwater resources and the realization of the aim of reducing depletion of mining.

2 WORKING FACE MINING CONDITIONS

The Mindong mine affiliated with Dayang of shenhua group, constructed in 2008 with 5 million t/a of design production capacity, uses the vertical shaft way develops with the long arm of fully mechanized caving mining technology. The working face (01) is located in the panels numbered south one where the coal numbered 16-3 top coal seam. The initial working face that it is 195.6 m width, has a design length of excavating numbered 1300 m with the caving mining technology, the average designing mining thick is 10 m, with the largest mining thick 14 m and the minimum mining thickness of 3 m. The 01 surface mass production started in February 7, 2012, the initial water inflow is approximately 60 m^3/h, the overlying strata water flowing fractured zone brings communication with rich water aquifer as the greater of the mining effect; the mine was forced to shut down because of 1 the sudden increase of water inflow with 650 m^3/h when the working face excavating to 16 m. After it was shut down, the 95 drill hole is constructed, with around 950 m^3/h of drain water inflow in working face and goaf, moreover, the total yield is 1055 m^3/h. thus it can be seen that it has been the main problem that have much influence on the mine safety and green mining due to the damage of the aquifer and the water inrush disasters caused by mining; it is particularly important to research and form proper coal mining plan and the countermeasures of water-preserved mining. The hydrogeological exploration shows that geologic structure of the working face roof aquifer from top and bottom is quaternary system aquifer, 15 coal seam at the top of the aquifer (No. I aquifer), 16 coal seam roof aquifer (No. II aquifer), 16 coal seam and coal seam between aquifer (No. III aquifer), it exists a water-resisting layer between the four aquifers. It can be founded that in addition to the No. III aquifer is poor of water, the watery of the other three layers of aquifer are strong, which is the main water filling source in the underground working face, is also the main aquifer needed to be protected in the process of coal seam mining. It is the primary problem during the application of water-preserved mining must to be solved that which aquifer is communicated for the development of the overlying strata water flowing fractured zone.

3 ENGINEERING DETECTION OF THE DEVELOPMENT HEIGHT OF OVERLYING STRATA WATER FLOWING FRACTURE ZONE

In order to ensure the smooth extraction of the south one panels of Mindong mine, it is very critical to achieve the data of the development of the overlying strata water flowing fractured zone on condition of full-mechanized caving mining technology. Therefore, it is selected to carry out the engineering detection about the development height of the overlying strata water flowing fractured zone which the first 02 face located in the east wing of the south one panels.

3.1 *Engineering detection scheme*

According to the actual mining geological conditions, this paper measured way is constructing the geology drilling in 02 surface of the earth's surface before and after the mining respectively, comparing two borehole flushing fluid loss consumption to judge original formation fracture development degree and the development degree of water flowing fractured zone, and the caving zone. The observation of the borehole flushing fluid loss is according to the coal industry standard (MT/T 865-2000). The height of water flowing fractured zone of borehole flushing fluid loss observation method (State Coal Industry Bureau, 2000).

This measurement is in the central 02 face, which is 718.36 m distance to the open-off cut, 32 m to the transport gateway, the hole before mining is T1 hole, after mining is T2 hole, and the distance between two holes is 5 m (Figure 1). The depth of T1 hole is 350 m, the final hole horizon is in the 16-3 coal floor; the depth of T2 hole is 319 m, the final hole horizon is 16-3 coal seam roof. The hole T1started on May 6, 2014, it only applies simple hydrology observation when the depth is more than 150 m, no loss observation, and when the depth is larger than 150 m, using the drive pipe fix hole, change the radius to make a fresh start drilling

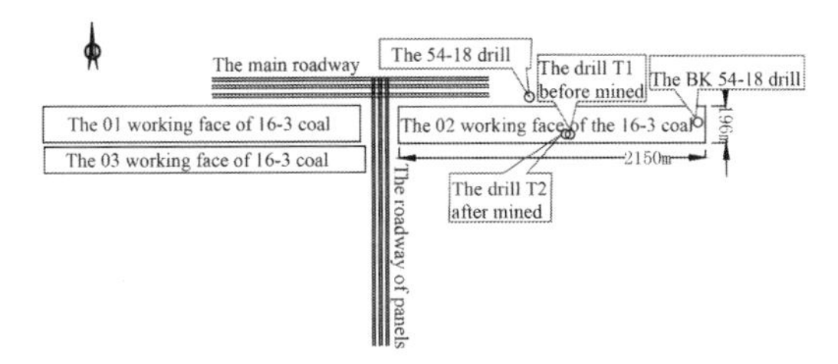

Figure 1. Overburden height damage measured borehole layout plan.

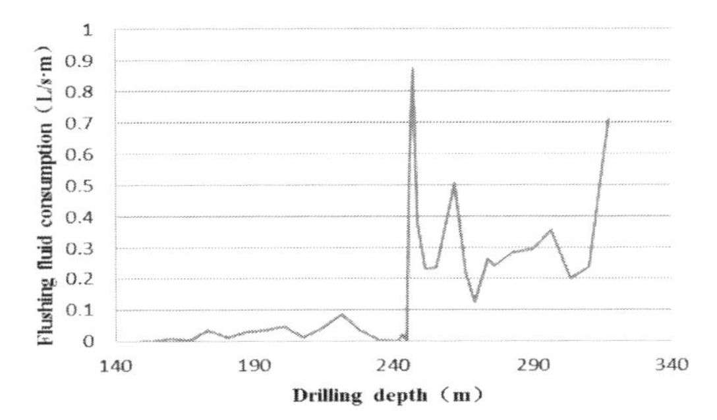

Figure 2. The curve of flushing fluid consumption of T2 hole.

hole. It starts to core when the depth is bigger than 240 m, and stops when it is final.

3.2 *Borehole flushing fluid consumption*

During the whole observation for T1 hole, the rinses loss quantity per unit time is reduced to 0–0.625 L/s, average 0.043 L/s, unit time footage rinses leakage quantity reduced to 0–1.042 L/s· m, average 0.073 L/s· m. During the whole observation for T2 hole, unit time loss is 0–8.40 L/s with 0.902 L/s on average. The loss with unit penetration and unit time is 0–14.00 L/s· m, the average is 1.538 L/s· m. Both unit time loss and unit penetration and unit time loss, which has a great amplification than before hole on the order of magnitude, and has the trend of going larger with the increase of drilling depth .it showed that Affected by mining activities, it occurs the overlying strata damage, deformation, stress redistribution, creating new cracks. The rinses consumption is greater than the original formation.

When the T2 hole continue drilling to the 246.22 m, it occurs that the interruption of flushing fluid circulation. After plugged, the grout did not come back when keep downhole, and the grout was leakage all (Figure 2). Verified by drilling coring, the core was broken with borehole water table dropped down quickly this place. It is speculated that this place is the water flowing fractured zone vertex. When T2 hole drilling to 293.35 m, the consumption which is 14 L/s· m is the maximum value of the whole process of observation, in view of the above, it can be concluded that this place is the vertex of caving zone.

The hole T2 rinses consumption observation results show that the vertex of the water flowing fractured zone is in the hole depth of 246.22 m, and the vertex of caving zone is in the hole depth of 293.35 m. The depth of coal roof in the location of drilling which is 330 m, so, it could be calculated that the height of measured water flowing fractured zone is 83.78 m, of caving zone is 36.65 m.

4 SCHEME OPTIMIZATION WORK FACE MINING

The research result that combined with the conclusion of field engineering practice showed that the ratio of the caving and mining is 5, and the ratio of the crack and mining is 11. It can be used to calculate the development height of the water flowing fractured zone of the work face under different ratio of mining and caving. When all the 14 m thick coal was mined, the height of the water fractured zone was up to 154 m; when the mining height was 3 m, minimum height of the water fractured zone is 33 m. Based on the analysis of the above, it could be a criteria to adjust reasonably the thickness of the caving and mining, control the development height of the water flowing fracture, achieve the goal of protection of aquifers that the subsequent working face mining aquifer's actual occurrence. Considering the strata No. III aquifer water is weaker, therefore, it was allowed that the No. III aquifer has been damaged by the fracture, and it is the main protection objection that the overlying No. II aquifer and its above aquifer.

Taking the subsequent 03 working face as an example, according to No. 2 aquifer occurrence situation in the mining range of working face, the three sections can be divided along the strike, as shown in Figure 3. In accordance with the relevant provisions of the national security procedures, if the mining height is less than 3 m, the height of waterproof protective layer is 2 times of that, when it is fully mechanized caving, the waterproof protective layer thickness in accordance with 1.3 ~ 2 times than the height of mining height. Thus, it can adjust and get the reasonable mining height, according to the actual distance of each section aquifer to coal seam.

The range of the section I is in 530 m from the work face cut, since the face cut that section No. II aquifer from coal seam 40.1 ~ 49.7. Therefore, the design mining height is 3 m with only mining and no caving; correspondingly, the height of the water flowing fractured zone is expected to 33 m, the thickness of the waterproof protective layer is 7.1–16.7 m, the requirement of water preservation can be satisfied. The range of the section II is from

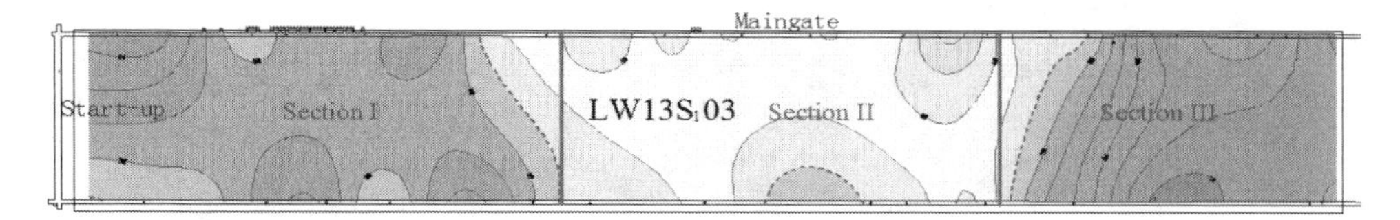

Figure 3. Partition graph of work face mining method.

the distance of 150 m to the cut to the distance of 1005 m to the cut, the range of the No. 2 aquifer is from 63.1–85.7 m. the thickness of the Waterproof coating is the 1.5 times of the mining height, the total thickness of caving and mining is 5 m. The range of the section III is from the distance of 1005 m to the cut to the location of stopping line, the range of the No. 2 aquifer is from 102.9–110.7 m. In a similar way, the thickness of the Waterproof coating is the 1.6 times of the mining height, and the total thickness of caving and mining is 8 m.

Field production practice results show that after using the above optimized mining plan in the 03 working, the overlying water aquifer is effectively protected, the water inflow of working face is greatly decreased, and mining with water protected is successfully achieved.

5 CONCLUSION

1. Respectively using the method of field measurement and numerical simulation, the research on the development height of the caving zone and water fractured zone of the 02 work face of Mindong mine under the condition of full-mechanized caving of weak strata was achieved, the comprehensive analysis showed that the range of caving belt height is 36.65–39.68, the development height range of the water fractured zone is 83.78–93.12 m, the ratio of the caving and mining is 5, and the ratio of the crack and mining is 11.The scientific basis was provided for the coal work face.
2. Using the FLAC^{3D} to make three-dimensional numerical simulation about the 02 work face of Mindong mine, about the work face damage with the advancing. The development height of the water fractured zone at different stages was analyzed by using the method of plastic zone distribution. It was got that as the advancing distance increasing from 100–200 m, the development height of the water fractured zone got "mutations"; when the working face advancing distance increase to 600 m, the water fractured zone height keep basically stable, and it has a guiding significance for prevention and control of water work of roof.
3. according to precise calculation about the height of water flowing fracture zone to adjust production plan, the work face was divided into 3 different sections which were 3 m, 5 m and 8 m, to achieve the aim that protecting the water resources in the coal exploration area and ensuring the coal recovery rate and recovery efficiency at the same time.

ACKNOWLEDGEMENTS

Financial supports from the State Key Research Development Program of China (2016YFC0501109) is greatly appreciated.

REFERENCES

Chen R.H., Bai, H.B. & Feng, M.M. (2006) Determination of height of water flowing fractured zone in overburden strata above fully-mechanized top coal caving face. *J. Mining & Safe ty Engineering*. 23(2), 220–223.

Gao, B.B., Liu, Y.P. Pan, J.Y., & Yuan, T. (2014) Detection and analysis of height of water flowing fractured zone in underwater mining. *J. Rock Mechanics and Engineering*. 33(1), 3384–3390.

Ma, Y.J., Wu, Q. Zhang, Z.Y., Li, T., & Meng, F.Z. (2008) Research on prediction of water conducted fissure height in roof of coal mining seam. *Coal Science and Technology*. 5, 59–62.

Ren F.H., Cai, M.F. & Lai X.P. (2004) Monitoring and analysis of the damage height of over burden rock mass on the mined out area. *J. Journal of University of Science and Technology Beijing*. 26(2), 115–117.

State Coal Industry Bureau (2000) *Regulations on buildings, water bodies and railways and the main shaft and this coal pillar and press coal mining*. Beijing, Coal industry publishing house.

Wang, L.G., Wang, Z.S. & Huang, J.H. et al (2012) Prediction on the Height of Water-Flowing Fractured Zone for Shallow Seam Covered with Thin Bed rock and Thick Windblown Sands. *J. Journal of Mining & Safety Engineering*. 29(5), 607–612.

Wu, Q., Zhao, S.Q. & Dong, S.N (2013) *The manual of Coal mine water prevention and control*. Beijing, Coal industry publishing house.

Zhang, J.Q., Liao, G.H. & Huang, Z.W. et al (2001) Dynamic principles of development of strata separation by fully mechanized top coal caving. *J. Journal of University of Science and Technology Beijing*. (6), 492–494.

Zhang J.Q., Liao, G.H., Huang, Z.W., Feng, E.J. & Liu, Q.Z. (2005) Practices and application of coal mining technology under water body in Zhaopo Mine. *J. Journal of University of Science and Technology Beijing*. 33(4), 29–31.

Land Reclamation in Ecological Fragile Areas – Hu (Ed.)
 ISBN 978-1-138-05103-4

Technique of long-distance reclaiming of subsidence land with Yellow River sediments

L.H. Duo, Z.Q. Hu, Y.X. Du, K. Yang & T. Xu
Institute of Land Reclamation and Ecological Restoration, China University of Mining and Technology, Beijing, China

ABSTRACT: In order to explore the new technology of long-distance filling reclamation with Yellow River sediments, 17 mining subsidence in the Northern of Jining city, Shandong province, are used as the research object to introduce the technology. The technology includes 3 parts: predicting mining subsidence, extracting and long-distance transporting Yellow River sediments, and reclaiming subsidence land to be the cultivated land with filing measures. Three sediment fields are laid in the diversion gate of Chengai. Suitable sediments-extracting equipment and pipelines are selected to extract and transport the sediments. Four pump stations are set on the competent route. Different filling methods are chosen according to the condition of the subsidence land. Finally, making the planning of farmland, farmland water conservancy and field road and making the design of shelter forest, to establish a complete farmland ecological system.

Keywords: long distance; coal mining subsidence; sediments; filling reclamation

1 INTRODUCTION

Coal has always been among the main sources in China, accounting for about 70% of primary energy (Hu et al. 2013, 2015). In 2015, national coal production reached 3.68 billion tons, accounting for 46.9% of the world's coal production (BP Statistical Review of World Energy, 2015). Well mining inevitably causes land subsidence in china (Hu et al., 2013; Wang et al., 2015). The plain mining area of high water level in the east is an important grain producing area and coal producing base, so the land subsidence is not only damaging the cultivated land, but also causing a large area of water accumulated (Hu et al., 2006; Yang et al., 2014), which increase the difficulty of land reclamation. The traditional method of "dig deep to pad shallow" in mining subsidence land often leads to poor quality of reclaimed soil and low productivity (Hu et al., 2008; Hu, 1997). At present, the most common filling methods are coal gangue filling, fly ash filling and lake mud filling. However, because of the insufficiency of filling materials and the risk of secondary pollution, the using range of coal gangue and fly ash filling is limited (Xu et al., 2014; Hu et al., 2003; Wang & Dong, 2009; Wang et al., 2006; Zhang et al., 2008). Similarly, because of the thick mud layer of the lake mud filling, the drainage and consolidation time are relatively long, causing cultivation begin in two or three years (Xue et al., 2006; Zou et al., 2009; Shi, 2010). In order to solve the problem that recovery rate of cultivated land is low in the area of high water level, Hu put forward the concept of filling reclamation using Yellow River sediments, and establish a test field in the Liangshan County, in the north of Jining City, Shandong province to verify the feasibility of using Yellow River sediments as filling materials (Hu et al., 2015; Wang et al., 2014). The test field is located 7 km away from the Yellow River sediments filed, which belongs to the short-distance filling. Nevertheless, because of the complex process, high initial investment and difficulty in production management, the application of long-distance filling technology (>30 km) (Hu et al., 2015) is rarely reported.

In this paper, technique of long-distance reclaiming subsidence land with Yellow River sediments is proposed, and it can be used in 17 coal mining subsidence areas in the north of Jining City, Shandong Province.

2 GENERAL SITUATION OF THE STUDY AREA

Jining is located in the southwest of Shandong Province. The geographical coordinates of Jining are 116°25′14″–116°59′57″ east longitude and 35°05′–35°38′17″ north latitude. The study area is coal mining subsidence land in the north of Jining, including 17 coal mines in the north of Rencheng district, north of Yanzhou district, Jiaxiang county, Wenshang county and Liangshan county. The total mining area is 734.12 km^2. The accumulated explored reserves are 4 billion tons and the annual yield is 22.95 million tons (Liu, 2013). Among 17 coal mines, 13 have been put into production including the mines in Liangbaosi, Xuchang, Tangkou, Daizhuang, Geting, Yangcheng, Yunhe, Xinhe, Luxi, Xinyi, Yiqiao, Tangyang and Hegang. Four mines are under construction namely the mines in Yangying, Yineng, Longxiang and Hongqi. The subsidence area of study area in 2009 was 3540.7 hm^2. The area of accumulated water for the whole year is 415.51 hm^2. The subsidence depth was 1.2~5.1 m.

The study area is mainly plain depression. The ground elevation is 36.00–40.00 m. The north and east are slightly higher while the south and west are slightly lower. The ground slope is normally 1/3000-1/6000. The study area belongs to the warm temperate monsoon continental climate with four distinctive seasons and abundant rain. It has much draught in spring, much rain in summer, draught in autumn. The winter is dry and cold with less snow. The average rainfall for many years is 666.3 mm. The annual maximum rainfall is 1285 mm and minimum rainfall is 437 mm. Average evaporation capacity is 1751.7 mm for many years. The frost season is quite long and lasts for around 140 days each year. The normal control farmland near the study area has high quality, which is the result of Yellow River sediment deposition more than 40 years ago. The planting structure is the form of harvest twice a year, and wheat is planted in winter and maize in summer.

3 TECHNIQUE OF LONG-DISTANCE FILLING WITH YELLOW RIVER SEDIMENTS

3.1 *Technical process*

Technique of long-distance filling with Yellow River sediments is proposed in this paper, which includes three parts: predicting mining subsidence, extracting and long-distance transporting Yellow River sediments, and reclaiming subsidence land to be the cultivated land with filing measures.

Step I: Predicting and analyzing mining subsidence land: According to the present situation of subsidence land and mining plan, the future developing trend of the subsidence land is predicted. For the mines under construction and to be constructed, with the help of subsidence prediction software, the total subsidence area and volume could be predicted, and then determine the amount of sediment required.

Step II: Extracting and transporting the water-sediment mixture for long distance: This part is the focus of this paper. The key point of sediments extraction is to determine the location of extracting point, the time and the equipment. The key point of sediments transport is to determine the pipeline material and the pipeline layout. According to the subsidence location and amount of sediments demanded, combined with distribution of beach area in the Yellow River, the location of extracting point and the suitable equipment are selected. The method of pipeline transportation is used to transport sediments from the Yellow River to the subsidence land, and the pipelines are divided into main pipes and branch pipes. According to power of the pump and the transmission distance of sediments, pressurizing pumps are set up on the main line. Because of the long-distance transport, the mixed relay transmission mode is adopted, which is used to transport the sediments to the subsidence. A pump station is set in each relay pump, and the size of the pump station is determined by the size of mud pump.

Step III: Reclaiming subsidence land to be the cultivated land with filing measures: For the land which will to be subsidence land, according to the topography of the subsidence land, the filling direction is determined and the filling strip is divided. Stripping topsoil and subsoil of each strip and piling up around to form a mound. Filling each strip and draining out the water to accelerate the sedimentation. After the consolidation of sediments, backfilling subsoil and topsoil. Finally, leveling the land to restore the subsided land to cultivated land. For the subsidence land in which the water depth is less than two meters, the best way is drainage in dry season. Other operations like dividing strips, filling, drainage, sediment consolidation, backfilling subsoil and topsoil, land leveling are conducted. For the mining subsidence

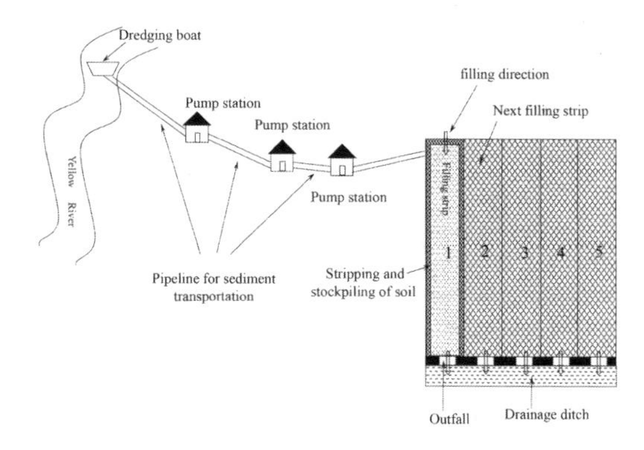

Figure 1. Diagram of long-distance reclaiming subsided land with Yellow River sediments.

in which the water depth is more than two meters, directly filling the subsidence with Yellow River sediment to removing the water, and backfilling soil from other areas to form the reclaimed land.

3.2 *Features and advantages of technical process*

The technical process has the following features and advantages:

1. High efficiency of sediments extraction: Sediments field and equipment can be selected to take Yellow River sediments with enough concentration. The efficiency of sediments extraction is high.
2. Sediments can be transported for long distance: Long distance transportation (>30 km) can be realized by reasonable combination of pipeline transportation and relay pressure pump station. It can transport for 90 km in the future.
3. Soil source can be ensured: Stripping the topsoil and subsoil in advance, can ensure sufficient amount of soil to backfill.

4 APPLICATION OF THE TECHNIQUE OF LONG-DISTANCE RECLAIMING OF SUBSIDENCE LAND WITH YELLOW RIVER SEDIMENTS

4.1 *Forecast and analysis of coal-mining subsidence land*

Based on the investigation of the current situation of coal mining subsidence in the north of Jining, the range of subsidence land in the next few years will be predicted. This paper focuses on large-area and future subsidence forecast. It researches on the change trend of spatial and temporal pattern in coal-mining subsidence area. The precision requirement of this research is different from ordinary engineering application. According to the mining plan, analyzing the mining data in a detailed way and selecting appropriate forecast parameters. The mining area composed of the same coal seam and the same mining method is the basic unit of division, to predict the future development trend of the subsidence. For the mines under construction and to be constructed, the mining plan can't be collected. So the MSPS software can be used to predict the volume of the subsidence land. Prediction results of subsidence volume of the filling area with Yellow River sediments in northern Jining are shown in Table 1.

4.2 *Extraction and transport of the water-sediment mixture for long distance*

4.2.1 *Optimization of the location, time and equipment for sediments extracting*

The spatial distribution of subsidence land and the temporal and spatial variation of sediment concentration should be considered. According to the optimal cost and the best ecological demand, the optimal selection of the location, time and equipment for sediments extracting are given, which providing support for long-distance filling reclamation.

Table 1. Subsidence prediction results of coal mine in northern Jining.

Mine	2020		Mine closure		The distance away from Yellow River/km
	Subsidence area/km^2	Subsidence volume/10^8 m^3	Subsidence area/km^2	Subsidence volume/10^8 m^3	
Xuchang	13.00	0.20	33.73	0.86	86
Daizhuang	21.53	0.34	62.27	1.59	80
Geting	7.60	0.13	14.67	0.37	71
Tangkou	17.53	0.24	70.53	1.8	77
Yunhe	7.07	0.11	8.47	0.22	74
Luxi	6.07	0.1	17.93	0.46	75
Hegang	2.80	0.04	7.80	0.2	82
Xinhe	1.40	0.02	2.20	0.06	79
Liangbaosi	10.80	0.17	70.60	1.8	52
Longxiang	2.00	0.03	16.87	0.43	54
Hongqi	2.00	0.03	12.27	0.31	52.5
Yangcheng	11.20	0.17	42.27	1.08	41
Tangyang	5.80	0.09	15.73	0.4	66
Yiqiao	4.53	0.07	12.13	0.31	66
Yineng	2.00	0.03	25.13	0.64	71
Yangying	5.40	0.07	39.53	1.01	8.5
Xinyi	8.73	0.14	39.93	1.02	78
Total	129.53	1.94	492.00	12.53	

1. Optimization of the location for sediments extracting

When choosing the optimal location of sediments field, some principles must be observed. The location for sediments extracting can't affect the node engineering arranged. The location of sediment field is arranged outside of the production dyke, and avoid the water intake of the Yellow River. The first canal of sediments extracting locates the upstream of diversion gate in Chengai of Liangshan county. This section is located in transition section from Gaocun to Taochengpu. The number of levee on the right bank is 312+600~333+000. The length of watercourse is 30.74 km. Spacing of levee is 3.7–8.5 km. The width of channel is 0.27~0.6 km. The width of beach land is 0.5–6.45 km. Average top rake is 0.148‰.

There are three sediment fields around the watergate for water diversion from the Yellow River in Chengai. The size of field No. 1 is 450 × 3000 m. The designed average digging depth is 3 m and the sediments reserve of the field is about 4.04 million m^3. The size of field No. 2 is 800 × 2800 m. The designed average digging depth is 3 m and the sediments reserve of the field is about 6.52 million m^3. The size of field No. 3 is 800 × 2300 m. The designed average digging depth is 3 m and the sediments reserve of the field is about 5.52 million m^3.

2. Optimal time of sediments extraction

Based on the data of water resources in Shandong section of the Yellow River in recent years, analyzing the flood season and the time of high water diversion rate to determine the sediments extraction time. Sediments extraction can't be carried out in the period when water diversion and sediments extraction are forbidden and sediments concentration is low. According to the calculation of the water demand of the filling reclamation, determining the time of sediments extraction to ensure the successful implementation.

Figure 2. Location of the sediment field in Chengai.

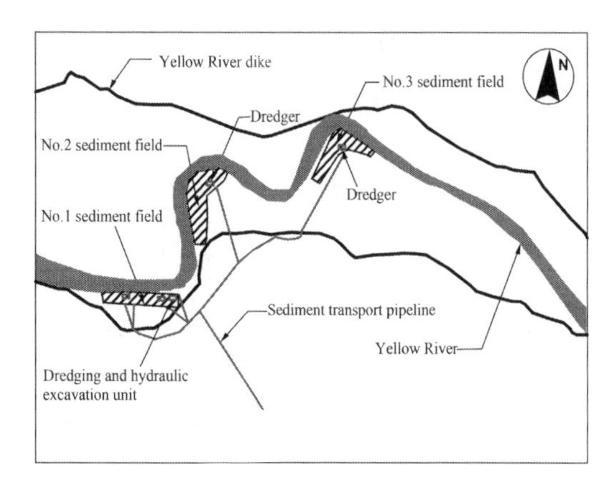

Figure 3. Sketch map of the location of the sediment field.

The water amount of the Yellow River in Shandong section is distributed unevenly. There is much water in summer and autumn while little in spring and winter. The water from upstream from July to October every year accounts for around 45% of water in the whole year. The coming water from March to June accounts for 32% of water in the whole year. The water is minimum from November to February next year due to ice flood, accounting for nearly 23% of the water in the whole year. According to data analysis, the flood season of the Yellow River in Shandong section is from July to October. The water-diversion amount is the largest in spring (March to June) and flood season (July to October), accounting for 83% of the whole year. According to the experience, freeze-up period in winter should not be water diversion period. Sediments are forbidden to be extracted from July to August. The average water-diversion days for many years is 253 d. When the guarantee rate is 50%, the number of days for water diversion is 268 d. When the guarantee rate is 75%, the number is 193 d. When the guarantee rate is 95%, the number is 163 d.

3. Optimal equipment of sediments extraction

The equipment of electrical dredger, DPB electric pump platform and hydraulic excavator pump combination are selected. The electrical dredger cuts sediment by rotation. The sediment particles can blend with water to form mud. DPB pump platform and hydraulic pump dig sediment and form mud through several hydraulic machines. Then the mud is absorbed by mud pump and delivered to sediment-transport pipeline through mud-discharge pipe. Finally, the sediment is transported to subsidence land.

Considering the actual efficiency of the equipment and large work area, which will influence the efficiency to some extent, the equipment is configured as 3920 m^3/h. Six electrical dredgers of YS-200 type are applied to work in the channel. The designed flow rate of each dredger is 350 m^3/h and the total flow rate is 2100 m^3/h. Three electric pump platforms of DBP500-35-3 type are applied to work in the edge area of channel. The designed flow rate for each is 500 m^3/h and for all is 1500 m^3/h. One mud pump of NL125-20 type is used with the flow rate of 200 m^3/h and it is matched with four sets of 80-50-2050 high-pressure water pumps with the flow rate of 50 m^3/h. One mud pump of NL100-28 type is used with the flow rate of 120 m^3/h, matched with two Z80-50-200B high-pressure water pumps with the flow rate of 60 m^3/h.

4.2.2 *Sediments transport by pipeline*

Pipeline laying should follow certain principles. Pipeline laying needs to meet the requirements of water conservancy construction management and to occupy cultivated land as less as possible. Pipeline laying also needs to coordinate with water conservancy, traffic and other public facilities. The laying route is better to be straight and try to minimize the use of pipe fittings so as to reduce resistance of pipe network. The laying should not obstruct nor damage the buildings and the pipeline should be laid along one side of the road and channel. Straight underground laying is applied.

Because of the friction between the sediment and the pipeline in the process of sediment transport, the pipeline must use wear-resistant materials. At present, the pipeline which can be used for materials delivery mainly include FRP pipe, ordinary steel pipe, composite steel pipe lined with ceramic, polyethylene pipe with ultra-high molecules, X65 pipeline steel pipe, composite steel pipe lined with cast stone and composite pipe with steel framework. According to the on-site condition and experience, the main pipeline uses composite steel pipe lined with cast stone which is wear-resistant and has thin wall. The diameter of the main pipe is 800 mm. The branch pipelines use ordinary steel pipes with the diameter of 350–600 mm. The pipeline begins near the right bank of the Liangji canal. Then the water-sediment mixture is pressurized through pump station and transported to subsidence land through the pipes with the diameter of 800 mm.

4.2.3 *Pressure pump layout*

The pressure pump station plans to use oil isolation pump with the type of 2DGN-300/4. Wear-resistant oil isolation pump has high lift and is widely used for tailings discharge by mine enterprises and for transporting coal in power industry. The oil isolation pump of 2DGN-300/4 type is used after comparison. Pump room is set for each relay pump. The size of pump room is determined by the size and quantity of mud pump.

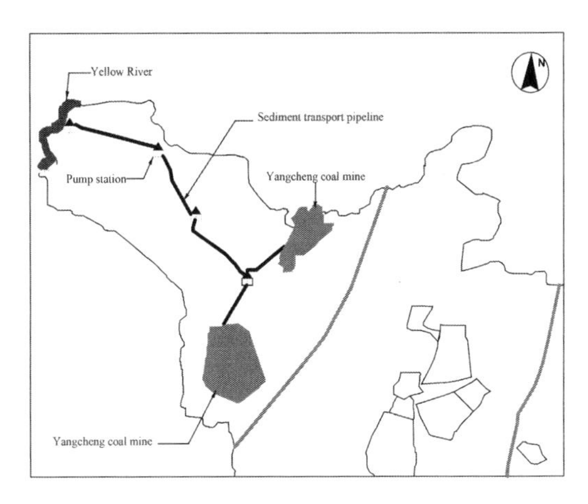

Figure 4. Layout of pipelines and pump stations.

Four pump stations are set and are distributed with the spacing of 14 to 19 km. They are labeled from No. 1 to No. 4 pump station from southwest to northeast. The pump stations of No. 1 and No. 2 are located in Liangshan county. The pump station of No. 3 is located in Wenshang county and No. 4 is in Rencheng district. Water-sediment mixture is extracted by equipment and then transported through pipeline to No. 1 pump station where water-sediment mixture is pressurized. Along the main pipeline, the mixture is transported to No. 2 pump station. Then the main pipeline is connected to No. 3 pump station which goes south to connect with No. 4 pump station. The main pipeline is 50.53 km long which forms the main pipeline system. Then the mixture is transported through the branch pipeline which is 13.88 km long to subsidence land in Yangcheng. Then the mixture is transported to subsidence land in Liangbaosi through the branch pipeline of 9.22 km. Because of the long distance of sediment transport, the hybrid relay transmission is applied. The relay of pump stations is used to transport the water-sediment mixture to subsidence land.

4.3 *Reclaiming subsidence land to be cultivated land with filing measures*

Different filling methods are chosen according to the condition of the subsidence land. There are 3 kinds of subsidence land. They are the land which will to be subsidence land, the subsidence land in which the water depth is less than two meters and the subsidence land in which the water depth is more than two meters. For the land which will to be subsidence land, according to terrain condition, the filling direction and strips are determined. Strips are always divided as regular rectangle shapes. Stripping topsoil and subsoil of each strip and then filling subsidence land with Yellow River sediments. Using perfect drainage system to drain away the water. Backfilling the subsoil and topsoil after the sediment consolidation. Finally, leveling the land to make the subsided land to be cultivated land. For the subsidence land in which the water depth is less than two meters, using the original or new drainage system to drain the water in the subsidence land, and then using the operation steps as described above to divide strips, strip the topsoil and subsoil, fill with sediments, drain out water, consolidate sediments, backfill the topsoil and subsoil, level the land and etc. For the subsidence land in which the water depth is more than two meters, filling directly the subsidence with Yellow River sediments and removing the water accumulated in subsidence land. Then using soil removed from other areas to cover the subsidence land to form the reclaimed land.

After the reclamation, taking engineering and biological measures to restore the land to an available state. The farmland irrigation and drainage, field road and shelter forest are respectively designed. Land leveling is mainly based on topography and terrain conditions. Based on the layout of roads and drainage ditch, dividing farmlands into parts. Meanwhile, the filed direction should ensure that the crop can absorb as much light as possible from the sun. Considering the characteristics of terrain, requirements of farming mechanization and water management in farmland, the field is designed of 800 meters in length and 200 meters in width. According to the actual situation and the traditional irrigation method, wells are located in the reclaimed land. Each well can control the average area of 8–12 hectares. The water yield of each well is 60 m^3/h. Field roads link the transport and land farming between village and village. The roads are 5–6 meters in width, constructed with concrete pavement. Based on the farmland layout, the production road is designed in the combination of the needs of farming. The production road is 3 meters in width and 15 cm in thickness. The subgrade is plain soil. Tree species, which are not easy to make the crop diseased, are selected as shelter forest species, such as willow and fast-growing poplar. Also, such tree species should be commonly selected by the local and widely accepted. So far the system of farmland ecosystem is constructed.

5 CONCLUSION

1. Technique of long-distance filling with Yellow River sediments is proposed in this paper, which includes three parts: predicting mining subsidence, extracting and long-distance transporting Yellow River sediments, and reclaiming subsidence land to be the cultivated land with filing measures.
2. The technology can be applied in 17 coal mining subsidence land in the north of Jining City, Shandong Province. Three sediment fields are arranged in the diversion gate of Chengai. The equipment of electrical dredger, DPB electric pump platform and hydraulic excavator pump combination are selected for sediments extraction. The main pipeline of sediments conveyance uses composite steel pipe lined with cast stone which is wear-resistant. The diameter of the main pipeline is 800 mm. The branch pipeline uses ordinary steel pipes with the diameter of 350–600 mm. Because of the long distance, four pump stations are established. There are 3 kinds of subsidence land. Depending on different conditions of the subsidence land, different filling methods are chosen.
3. It is an exploration of long-distance filling reclamation with Yellow River sediments. The technique can be widely applied to reclaiming subsidence area around the Yellow River basin. The technique is valuable to be popularized.

ACKNOWLEDGEMENTS

Funding: This work was supported by the National Key Technology Research and Development Program (2012BAC04B03) during the Twelfth Five-Year Plan Period.

REFERENCES

Hu, Z.Q., Li J. & Zhao Y.L. 2006. Problems, reasons and countermeasures for environmental quality and food safety in the overlapped areas of crop and mineral production. *Science & Technology Review*, 24(3): 21–24.

Hu, Z.Q., Qi, J.Z. & Si, J.T. 2003. Contamination and assessment of heavy metals in fly ash reclaimed soil. Transactions of the Chinese *Society of Agricultural Engineering*. 19(2), 214–218.

Hu, Z.Q., Wang, P.J. & Shao, F. 2015. Technique for filling reclamation of mining subsidence land with Yellow River sediment. *Transactions of the Chinese Society of Agricultural Engineering*. 31(3), 288–295.

Hu, Z.Q., Wang, P.J., Ji, R.Q., Zhao, Y.L. & Shao, F. 2015. Method of filling reclamation in mining subsidence with the Yellow River sediments for long distance. *Chinese patent. 103255762B.*

Hu, Z.Q., Xiao, W., Wang, P.J. & Zhao, Y.L. 2013. Concurrent mining and reclamation for underground coal mining. *Journal of China Coal Society*, 38(2): 301–307.

Hu, Z.Q. 1997. Principle and method of soil profile reconstruction for coal mine land reclamation. *Journal of China Coal Society*, (06):59–64.

Hu Z.Q., 2008. Land reclamation and ecological reconstruction. *China University of Mining and Technology Press,* Xuzhou.
Liu, L.F. 2013. Study on the subsidence government and ecological reconstruction planning of Jining mining area. *Fudan University*, Shanghai.
Shi, H.C. 2010. The application of reclamation technology in mining subsidence with lake mud in Datun company. *China High Technology Enterprises*. (3): 69–70.
United Kingdom. 2015. *BP Statistical Review of World Energy*.
Wang, P.J., Hu, Z.Q., Shao, F., Jiang, Z.D., Qiao, Z.Y., Liu, D.W. & Chen, Y.K. 2014. Feasibility analysis of Yellow River sediment used as the filling reclamation material of mining subsidence land. Journal of China Coal Society. 39(6), 1133–1139.
Wang, P.J., Shao, F., Liu J.T., Li, X.Y., Hu, Z.Q. & Russell S. Yost. 2015. Simulated experiment on drainage and fine sediment retention effects of geotextiles in land reclamation with Yellow River sediments. Transactions of the Chinese Society of Agricultural Engineering (Transactions of the CSAE), 31(17): 72–80.
Wang, Y. & Dong, J.H. 2009. Potential ecological risk assessment of filling reclaimed soils polluted by heavy metals in Xuzhou mining area. Journal of China Coal Society, 34(5): 650–655.
Wang X.Y., Yang J. & Guo H.X. 2006. Study on heavy metals in soil contaminated by coal waste rock pile. Journal of China Coal Society, 31(6): 808–812.
Xu, L.J., Huang, C., Zhang, R.Q., Liu, H.P., Yan, J.P., Helmut, M. & Lutz, M. 2014. Physical and chemical properties and distribution characteristics of heavy metals in reclaimed land filled with coal gangue. Transactions of the Chinese Society of Agricultural Engineering (Transactions of the CSAE), 30(5): 211–219.
Xue, S.X. 2006. Application in land reclamation of filling subsidence with lake mud in Zhang Village. Energy Technology and Management. (5): 58–59.
Yang, G.H., Hu, Z.Q., Zhao Y.L., Yang, Y.Q. & Yu, Y. 2014. Proposals on countermeasures of reclamation control in coal mining subsidence land with high underground water level. Coal Engineering, 2014, (06): 91–95.
Zhang, L., Han, G.C., Chen, H., Ma, M.G. & Guo, H.D. 2008. Study on heavy metal contaminants in soil come from coal mining spoil in the Loess Plateau. Journal of China Coal Society, 33(10): 1141–1146.
Zou, Z.Y., Shi, H.C. & Sun, G.Q. 2009. Application of lake mud stowing technology in the reclamation of subsided areas caused by coal mining. China Coal, 35(12):105–106, 122.

Land Reclamation in Ecological Fragile Areas – Hu (Ed.)
© 2017 Taylor & Francis Group, London, ISBN 978-1-138-05103-4

Design of an ecological damage restoration scheme in a mining subsidence area—taking the Xuzhou Dawu mining area as an example

X.M. Yao & X. Cui
China University of Geosciences, Beijing, China

ABSTRACT: In this thesis, the subsidence area of *Dawu* mining area in Xuzhou City, Jiangsu Province is chosen as the research object, and the ecological damage and restoration scheme were studied, using MAPGIS6.7 software. This thesis evaluates the suitability of ecological restoration in the subsidence area of *Dawu* mining area. Finally, with the aid of soil reconstruction, water system repair and other technologies and means, embankment construction, land leveling, irrigation and drainage, farmland protection and other aspects of ecological restoration program design.

1 INTRODUCTION

In recent years, with the development of China's economy, a large number of production and construction projects continue to damage the development of land resources, the production activities have had a greater impact on the ecological environment, social and economic stability. At present, China's damaged land restoration is generally based on reclamation, lack of planning from the perspective of land remediation, and can't guarantee the sustainable use of land resources. This thesis chooses the subsidence area of *Dawu* mining area in Xuzhou City, Jiangsu Province as the research object, based on the theory of sustainable development and location theory, This thesis analyzes the present situation of land use and coal mining subsidence in *Dawu* mining area, and puts forward the scheme of ecological damage repair.

2 OVERVIEW OF MINING AREA AND PRESENT SITUATION OF COAL MINING SUBSIDENCE

2.1 *Dawu mining area profile*

Dawu mining area is located in the *Huang-Huai-Hai* Plain. The thesis focuses on the planning and design of land reclamation in the coal mining subsidence area. *Dawu* District, 35 km from the main city of Xuzhou City, latitude 34°17′~34°32′, longitude 117°17′~117°42′. Its coal field is an important part of Xuzhou coal field, The coal area of the *Dawu* coalfield is 38.28 square kilometers, all over the *Dawu*, *Qingshanquan* and other towns.

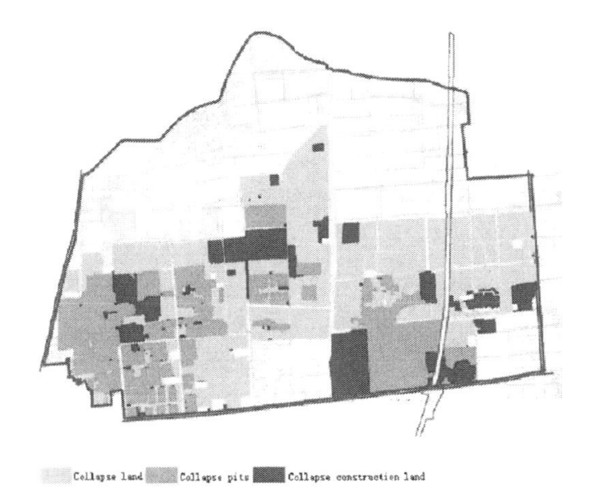

Figure 1. Dawu mining subsidence area space layout.

2.2 *The status quo of coal mining subsidence*

The total area of the coal mining subsidence area is 1050.31 hectares. Including 897.44 hectares of agricultural land, construction land 106.35 hectares, unused land 46.52 hectares. The spatial layout of the area is shown in Fig. 1.

3 SUITABILITY EVALUATION OF COAL MINING SUBSIDENCE AREA

3.1 *Evaluation unit and evaluation index*

In this study, the land use map of the measured topographic map is used as the evaluation unit, according to the evaluation unit. In terms of evaluation indicators, the influencing factors of cultivated land suitability evaluation include external conditions, soil condition, water level, irrigation and drainage conditions, terrain slope, land pollution condition and land use status.

3.2 *Determine the weight of the evaluation index*

In this thesis, the regional cultivated land suitability analysis is divided into four grades: Highly suitable (100 points), Moderate fit (80 points), Basic suitable (60 points), Not suitable (0 points). In this study, the Delphi method was used to quantify the grading and weight of each index. The expert group is mainly composed of experts and engineers of *Dawu* District Land and Resources Bureau, *Dawu* District Agriculture Bureau, *Dawu* District Water Conservancy Bureau, which has long been involved in the land remediation practice of coal mining subsidence area, and university professors build the analytic hierarchy process, after establishing the judgment matrix, and through the consistency test. The factor weights for each suitability direction are shown in Table 1:

Table 1. The single factor weight of the suitability direction.

Evaluation factor	Weights
External condition	0.050
Soil conditions	0.192
Water level	0.168
Irrigation and drainage conditions	0.171
Terrain slope	0.113
Land pollution	0.234
Original land use	0.071

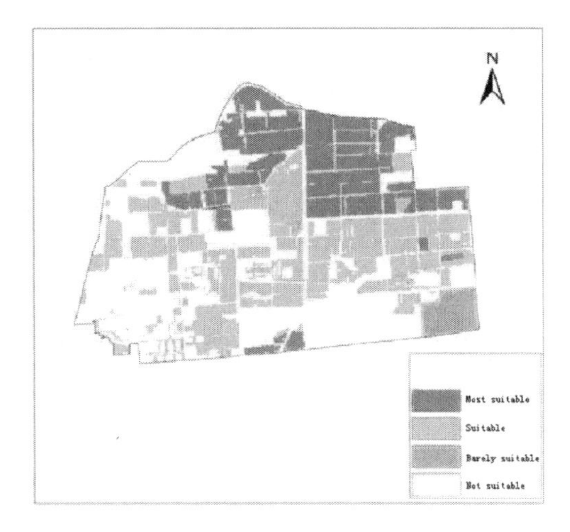

Figure 2. Adaptability evaluation results.

3.3 *Evaluation results*

According to the results of index quantification rating and weight determination, using MAPGIS6.7 software, and the weighted sum method is used to calculate the total score of the attributes of each evaluation unit, we can get the spatial distribution of each suitability direction.

In Figure 2, the strongest cultivated land is distributed in the northeastern part of the subsidence area; it is in the edge of the collapse of the basin. The areas of moderate appropriateness of cultivated land are mainly distributed in the subsidence area, the surface subsided significantly, but mostly as a whole sinking, seasonal water is more serious, and it is necessary to rehabilitate normally.

4 DESIGN OF ECOLOGICAL RESTORATION SCHEME FOR COAL MINING SUBSIDENCE AREA

4.1 *Regional ecological restoration zoning*

The ecological restoration program in this area is characterized by ecological remediation to improve the agricultural production and living conditions, enhance the ecological level as the theme. The functional positioning also meets the requirements of the area. As show in Fig. 3:

4.2 *Ecological restoration program design*

4.2.1 *Embankment design*

Agricultural areas of low-lying, the surface elevation is 27.0*m*-29.0*m*, ecological lake flood level is 30.0*m*, to ensure that ecological lake is not into the agricultural area, this design uses the polder method, constructed embankment on the border

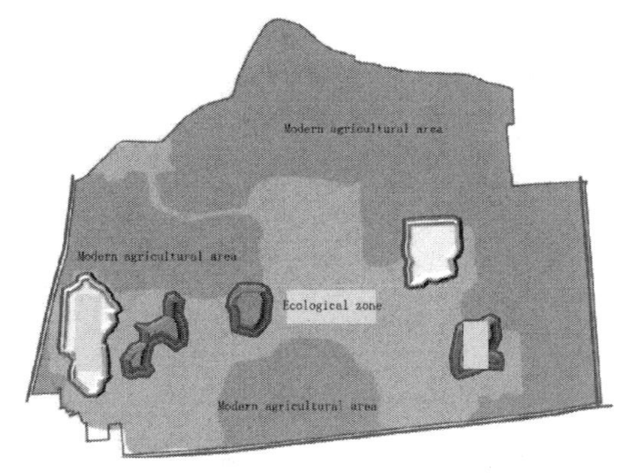

Figure 3. Regional ecological restoration zone.

of the plots, the width of the embankment is 6*m*, the top width of the clay core wall is 2 m, Slope coefficient is 1.5, The water diversion and the bottom is protected by M10 pulp.

4.2.2 *Land leveling design*

1. Topsoil stripping backfill: According to the area collapse, design topsoil stripping backfill, the tillage layer is 18 cm, the thickness of the stripped topsoil is 20 cm, after stripping soil, provide specialized site stacking.
2. Land formation earthwork calculation and soil reconstruction: Earthwork calculation using grid method, calculation application CASS software auxiliary completion, and the basic unit of earthwork calculation is flattening unit. The soil reconstruction design consists of three parts: Filling materials: use of fly ash, coal gangue and lake mud to fill, and take a different land configuration. Isolation layer: A 5 *cm* clay layer was placed on the upper part of the fly ash filling layer, finally covered with topsoil. Overburden thickness: because the main crops are wheat, corn, soybean in the region, combined with the high diving area of farmland drainage drop the actual needs, the thickness of soil reconstruction is 50 *cm*.

4.2.3 *Shelter forest design*

The area windstorm is not a major disaster, but due to surface damage caused by environmental damage, surface soil erosion is more serious, so dredging farmland forest network is designed, Choose sophora japonica, willow, privet and other trees along the main road on both sides, use Joe, irrigation two-story structure, row spacing 2 m * 2 m, A total of the use of sophora japonica and willow 24232 trees, privet 23554 trees.

4.2.4 *Irrigation and drainage design*

According to irrigation requirements, the area uses the new irrigation pumping station to raise water, in the agricultural area of new dry, two-stage irrigation pipeline to continue irrigation, pipeline water utilization factor is 0.92. Concrete pipe include 600 *mm*, 500 *mm*, 400 *mm*, 300 *mm*; PVC pipe include 250 *mm*, 200 *mm*, 160 *mm*, 110 *mm*.

5 SUMMARY

This thesis puts forward the key contents of the planning and designs of land ecological restoration in *Dawu* coal mining subsidence area based on the commonly used repair technology. Land repair is an extremely complex project. This thesis was described on some of the typical representative of the technology and it lacks the actual data support and construction plan. In practical application, the land repair needs to choose specific and reasonable method according to the situation.

REFERENCES

He S J, Guo H C, Wei C Y, et al. (2008). Land restoration in coal mining fields in China. Acta Geographica Sinica. 28(04), 172–175.

Jian-feng Zhang, & Qi-xiang Sun. (2005). Causes of wetland degradation and ecological restoration in the Yellow River Delta Region. Forestry Studies in China, 2005(02), 77–78.

Runkle. (1999). Effect of longwall mining on surface soil moisture and tree growth. Proceedings of 3rd subsidence workshop due to underground mining, 1999(11).

Szczepanska, J. & I. Twardowska. (2008). Distribution and environmental impact of coal-mining wastes in Upper Silesia, Poland. Environmental Geology, 17(5), 147–148.

Land Reclamation in Ecological Fragile Areas – Hu (Ed.)
 ISBN 978-1-138-05103-4

Physical and chemical properties of soil at different mining subsidence areas in windy and sandy regions

Y. Rong, Z.Q. Hu, Y.J. Wu & Y.M. Yuan
Institute of Land Reclamation and Ecological Restoration, China University of Mining and Technology, Beijing, China

ABSTRACT: In order to research the effect of mining subsidence on soil physical and chemical properties, the subsidence areas of Da Liuta mining district were divided into pelvic floor area and marginal area. The results showed that:1) The influence of mining subsidence on soil physical properties had obvious zoning characteristics, the influence on marginal zone was greater than pelvic floor area. Soil moisture content, bulk density, porosity in pelvic area showed a self-recovery phenomenon 1a after mining, while not in marginal area. The influence of mining subsidence on physical properties of topsoil were higher than those of deep soil. 2) The influence of mining subsidence on soil chemical properties was relative slight and had partition feature, the influence on marginal zone was greater than pelvic floor area. Mining subsidence reduced the content of soil total nitrogen and available potassium, but had no effect on soil available phosphorus or organic matter.

1 INTRODUCTION

As an important basic energy and raw material, coal resources play a strategic role in the long-term development of China. Coal mining inevitably causes a certain degree of damage to the ecological environment, while contribute to the economic development. Western windy and sandy region is located in the fragile eco-environment belt of the wind-water erosion crisscrossing region, and the high-intensity coal mining has worsened the fragile ecological environment. The development and utilization of coal resources have resulted in improper large scale excavation, subsidence, and farmland cover occupation, which leads to a chain reaction of various factors such as geology, soil, water and other factors. It is of great significance to explore the influence of mining subsidence on soil physical and chemical properties, which are one of the direct factors that reflect the change of ecological environment in mining area. The influence of mining subsidence on soil has been extensively studied by many scholars (Wang et al., 2014; Liu et al., 2014; Wang et al., 2006; Zhang et al., 2003; He et al., 2007; Chen et al., 1999; Huang et al., 2003). The influence of coal mining subsidence on soil physical and chemical properties could be short-term, long-term or irreversible. Most of the previous research adopted the method of substitution of space for time and made the static analysis at a certain time. However, there has been little research on the influence the types of subsidence land and regional differences on soil physical and chemical properties.

In view of the question, profits from the existing research results, the study carried out a 2-year dynamic monitoring of soil physical and chemical properties within 1.2 m of different region sites of subsidence basin in the western windy and sandy region. The study explored the characteristics of soil physical and chemical properties in different subsidence areas and their self-repair process. This study intends to provide feasible suggestions for land reclamation and ecological restoration in the western windy and sandy region.

2 MATERIALS AND METHOD

2.1 *Experimental site characteristics*

Daliuta coal mine is located on the north of Shaanxi province. This area is a transition zone of Maowusu sand land and loess hill, which belongs to Yanhe stratum of Erdos section of North China formation. The studied gangue field is surrounded by active, fixed and semi-fixed sand dunes and is located in a valley 4.8 km away from the Wulanmulun River in the east. The unconfined aquifer of this area consists of Quaternary residual sand, aeolian sand and lacustrine silty-fine sand. The mean annual rainfall is 440.8 mm, 71% of which falls from June to September every year. The total evaporation is 2163 mm per year.

2.2 *Plot setting*

The working face's length is 2881.3 m, width is 280.5 m, with a coal thickness of 7.07–7.7 m and an average depth of 234 m, the thickness of the overlying bedrock of 100–185 m. This working face adopts the long wall mining and management roof caving mining methods. The working face began to coal resources mining on September 3, 2013 with advance rate 10 m per day. According to mining subsidence theory, the study divided the coal mining subsidence basin into two areas: the pelvic floor area (PD) and the marginal area (BY). PD refers to the uniform settlement area in the middle of the working face, with almost no additional slope, and the dynamic cracks have the characteristics of rapid closure (Hu et al., 2014); BY is the non-uniform settlement area near the upper boundary of the working face, resulting in additional slope, existing long-term edge cracks (He et al., 1991; Lei et al., 2009). In order to avoid fortuity, six monitoring points were set repeatedly. And the unexplored area (CK) was taken as the control which is 120 m away from the open-off cut, three monitoring points were arranged. Hand-held GPS are used for sampling in 2014 years (1a after mining) and 2015 years (2a after mining).

2.3 *Sampling method*

Three parallel soil samples in three selected plots were collected to measure the moisture content, and the remaining soil samples were packed in zip-lock bags, which brought back into the room to analyze the soil physical and chemical indexes after being naturally dried and sieved. Meanwhile, three samples were collected from each layer of 0–20 cm, 20–40 cm, 40–60 cm, 60–80 cm, 80–100 cm, 100–120 cm with ring sampler respectively. the soil bulk density were weighed and calculated after the samples were dried in the temperature of 105°C. After the natural drying and sieving of soil samples from each layer correspondingly, physic-chemical property analysis were carried out in laboratory.

2.4 *Analytical methods*

The soil physical and chemical properties were measured according to soil agricultural chemical analysis methods compile d by the Chinese Soil Society (Bao et al., 2015). Soil physical properties, including Soil moisture content, bulk density, porosity and soil chemical properties, including organic matter, total nitrogen, available phosphorus, available potassium were determined for each sample. The methods were adopted as follows respectively: the oven drying technique, cutting ring method, pycnometer method, Kjeldahl method, $NaHCO_3$ extraction-Mo-Sb colorimetry, NH_4Ac extraction—flame photometric and potassium dichromate oxidation.

Excel 2013 was used to calculate the standard deviation. The software SAS 8.1 was used for analysis of variance. Treatment differences were tested using the Duncan's method of multiple comparisons. Sigma Plot 12.5 was used graph and present the data.

3 RESULTS AND DISCUSSION

3.1 *Effects on soil physical properties*

3.1.1 *Soil Moisture Content*

Windy desert region has characters of bare ground and fragile ecosystem with scarce rainfall and strong evaporation intensity. The lack of water typically affects and slows down the growth process and the rate of the plant roots, whose distribution is the most important indicating parameter of the region's development and utilization to some extent. Fig. 1 shows that the variation of Soil

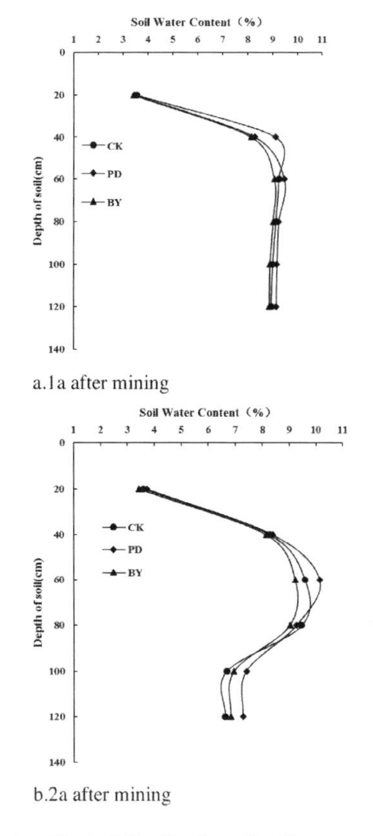

Figure 1. Vertical distribution of soil water content of different subsidence areas.

Moisture Content (SMC) in the 0–120 cm soil layers of different subsidence areas in the last 1a after mining and 2a after mining.

The SMC was divided in 60 cm depth of CK, showing two different kinds of characteristics in the last 1a after mining (Fig. 1a): the SMC increased with the constant deepening of the soil in the 0–60 cm soil layers, while showed an opposite character in the 60–120 cm soil layers within a narrow range. The SMC increased with the constant deepening of the soil in the 0–60 cm soil layers of PD, and the increase rate was 0.10%/cm, which was almost as the same as 0.09%/cm of the CK, but the SMC in the 60–120 cm soil layers was slightly decreased at a steady rate. The SMC in the 0–60 cm soil layers of PD was smaller than that of CK, but in the 60–120 cm soil layers the SMC was higher than that of CK, showing a certain recovery phenomenon. The SMC in the 0–60 cm soil layers of BY increased with the constant deepening of the soil and remained stable in the 60–120 cm soil layers. The SMC in all soil layers of BY was still lower than that of CK. It indicated that the SMC of BY did not show a recovery trend.

The vertical distribution of the SMC in the last 2a after mining was significantly different from 1a after mining (Fig. 1b). The SMC in all three regions increased in the 0–60 cm soil layers and then reduced significantly in the 60–100 cm soil layers, while remained gradually stabilizing trend in the 100–120 cm soil layers. There may be several times light rain and moderate rain within two months before the monitoring, which supplemented replenishment of the surface soil content. The SMC of PD was slightly higher than that of CK except in 80 cm, indicating that the recovery phenomenon was obvious. The SMC of PD in the 10–80 cm soil layers was still smaller than that of CK, and in the 80–120 cm soil layers was higher than that of CK, showing a certain recovery phenomenon.

On the basis of above study, the SMC of PD in the last 1a after mining has showed self-repair, while the SMC of BY in the last 2a after mining has restored in a certain soil layers. The major reason was that the dynamic cracks of PD have the characteristics of rapid closure. After the closure of the cracks, the SMC would restore with time. However, the marginal cracks of BY were still exist after the mining (Wang et al., 2015; Li et al., 2012; Zhang et al., 2013; Hu et al., 2014), which would rely on sandy landfill later.

3.1.2 *Soil Bulk Density*

Soil Bulk Density (SBD) refers to the weight of solid particle in per unit volume of soil, which value depends on the factors such as soil texture, structure and density. The SBD has an important effect on the soil permeability, water holding capacity, nutrient migration and anti-erosion ability, so it can be seen as a comprehensive characterization of soil quality condition. Fig. 2 shows the variation of the SBD in the 0–120 cm soil layers of different subsidence areas in the last 1a after mining and 2a after mining.

Fig. 2 shows that the SBD of PD and BY were lower than that of CK in the last 1a after mining and 2a after mining, but the difference was not significant between these treatments ($P > 0.05$) and the difference was not significant between the corresponding depth of these treatments ($P > 0.05$).

Overall, the SBD showed a tendency of CK > PD > BY in the last 1a after mining (Fig. 2a), expect in the 120 cm depth, which further demonstrated that uneven settlement and cracks of the surface had led to soil structure change in the early stage of mining disturbance. The SBD of PD was almost the same as that of CK in the last 2a after mining (Fig. 2b), but the SBD of BY was still lower than that of CK. The above analysis showed that mining disturbance had a significant impact

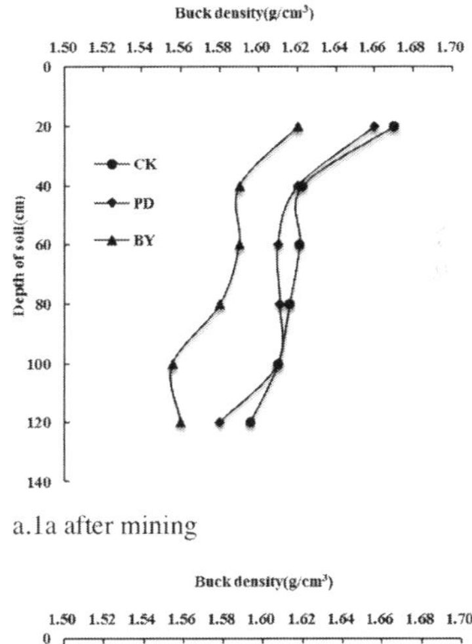

a.1a after mining

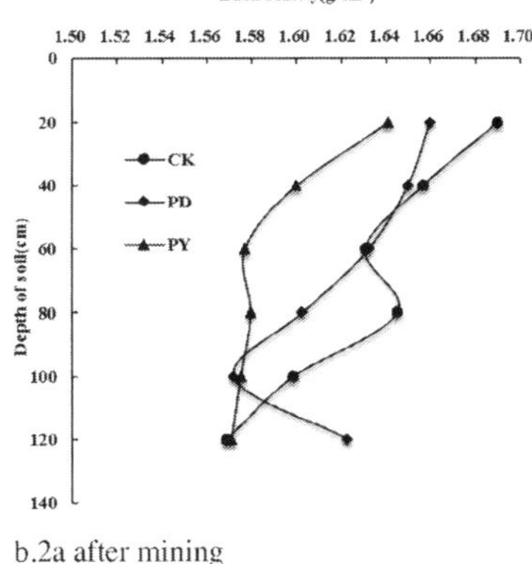

b.2a after mining

Figure 2. Vertical distribution of soil buck density of different subsidence areas.

on the SBD in the short term, but there was little impact in the long term. Moreover, the SBD of PD would restore to the pre-mining state 2a after mining, while the SBD of BY hadn't occurrence this phenomenon.

3.1.3 *Soil Porosity*

Soil Porosity (SP) is generally refers to the proportion of the soil pore volume in per unit volume under the natural state and it's an important indicator of soil fertility. Fig. 3 shows the variation of the SP in the 0–120 cm soil layers of different subsidence areas in the last 1a after mining and 2a after mining.

It can be seen from Fig. 3 that the SP has increased because of mining disturbance, but a less than 6% increase. The SP of PD and BY were higher than that of CK in the last 1a after mining and 2a after mining. But the difference was not significant between these treatments (P > 0.05) and the difference was not significant between the corresponding depth of these treatments (P > 0.05).

Overall, the PD showed a tendency of CK > PD > BY in the last 1a after mining (Fig. 2a), expect in the 120 cm depth, which further demonstrated that uneven settlement and cracks of the surface had led to soil structure change in the early stage of mining disturbance. The SP of PD was almost the same as that of CK in the last 2a after mining (Fig. 2b), but the SP of BY was still higher than that of CK. The above analysis showed that mining disturbance had a significant impact on the SP in the short term, but there was little impact in the long term.

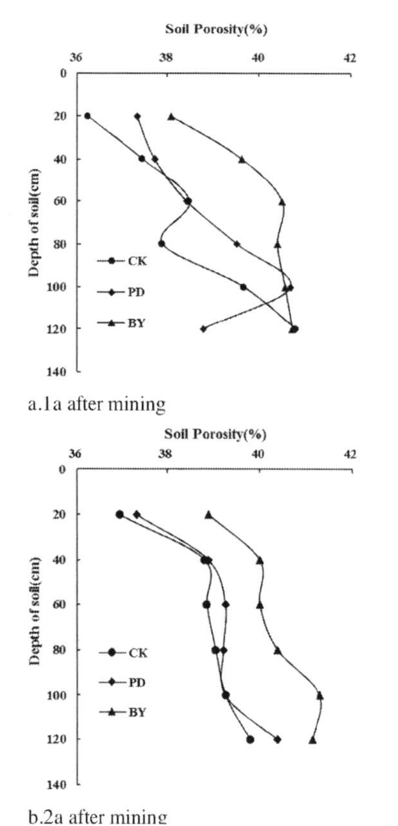

Figure 3. Vertical distribution of soil porosity of different subsidence areas.

3.2 *Effects on soil chemical property*

Soil nutrients characterize the quality and production capacity of soil. In the aeolian sandy area, coal mining subsidence changed the soil physical property, and this change may in turn affect the soil chemical property. The quantity of soil nutrients is an important indicator of soil fertility, which have a certain impact on the growth of surface vegetation.

3.2.1 *Soil Organic Matter*

Soil Organic Matter (SOM) is the most important indicator of soil quality indicators, which plays the most crucial role in the process of soil formation and fertility evolution. Fig. 4 shows the variation of the SOM in the 0–120 cm soil layers of different subsidence areas in the last 1a after mining and 2a after mining.

It can be seen from Fig. 4 that the SOM of PD, BY and CK were mainly concentrated in the 0–40 cm soil layers. The SOM of three treatments decreased in the 0–60 cm soil layers with depth in the last 1a after mining and went its minimum level in the 60–80 cm soil layers, but increased in the 80–120 cm soil layers, which might be related to the roots of vegetation. The SOM in the 0–40 cm soil layers in the last 2a after mining was higher than that in the last 1a after mining, because of more rainfall and reduced soil evaporation under tree canopies.

In general, the SOM of CK was higher than that of the subsidence area, but the difference was not significant (P > 0.05), which indicated that there was no significant effect of coal mining on the SOM of windy and sandy region.

3.2.2 *Soil Total Nitrogen*

Soil Total Nitrogen (STN) is an important nutrient for crop growth, which contains organic nitrogen and inorganic nitrogen. Many literatures indicated that soils in aeolian sandy land were low in the STN and mainly comes from plant foliage and animal debris. Fig. 5 shows the variation of the STN in the 0–120 cm soil layers of different subsidence areas in the last 1a after mining and 2a after mining.

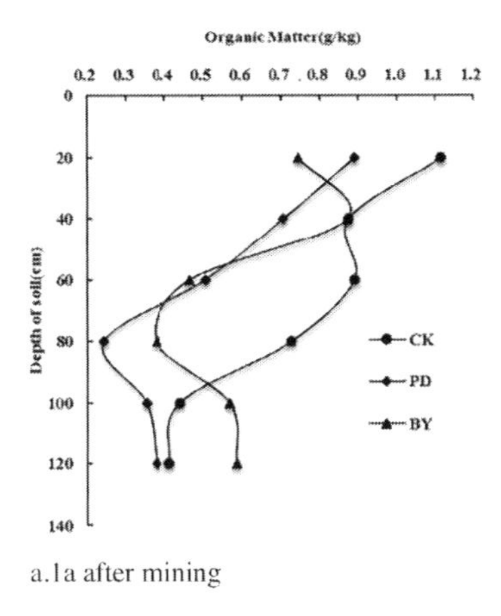

a.1a after mining

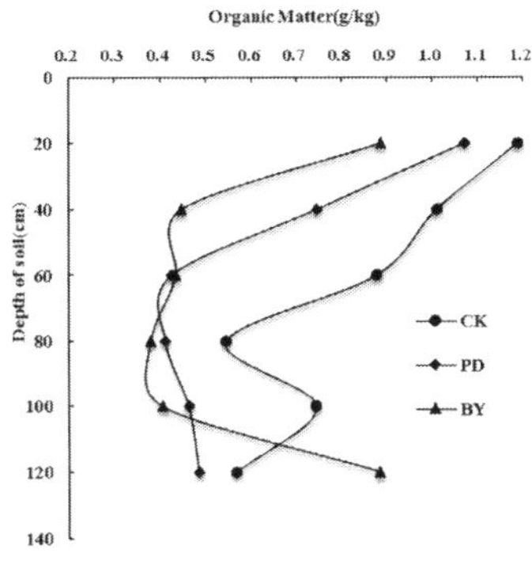

b.2a after mining

Figure 4. Vertical distribution of soil organic matter of different subsidence areas.

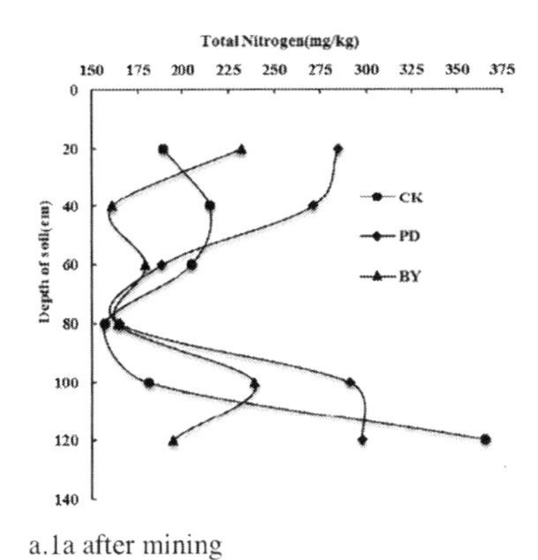

a.1a after mining

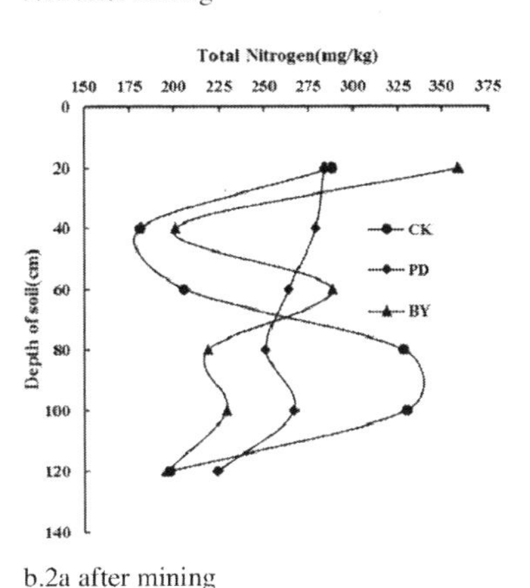

b.2a after mining

Figure 5. Vertical distribution of soil total nitrogen of different subsidence areas.

As is shown in Fig. 5, the variation trend between subsidence area and control area was basically the same in the last 1a after mining, namely, the STN decreased and then increased with depth. The STN of PD were higher in the 0–40 cm and 80–120 cm soil layers than that of CK, but the STN of BY were lower in the 0–40 cm and 80–120 cm soil layers than that of CK. The STN of the subsidence area in the 40–80 cm soil layers was lower than that of CK. There was a significant difference ($P < 0.05$) between PD in the 40–80 cm soil layers and CK in the 0–40 cm and 80–120 cm soil layers. The STN of CK increased and then decreased with depth in the last 2a after mining, while the STN of the subsidence area showed a gradually decreasing trend. The STN of the subsidence area in all soil layers were lower that of CK, except in the 0–40 cm soil layers, and the STN of BY were higher than that of CK. In general, the influence of mining subsidence on the STN had partition feature. The STN of PD had a certain self-repairing ability; but CK didn't show this phenomenon because of fissures increased the nitrogen loss, and the conclusion is consistent with the study of Zang Yin-tong (Zang et al., 2010).

3.2.3 *Soil Available Phosphorus*

Soil Available Phosphorus (SAP) determines the soil phosphorus supplying capacity directly, and is also an important indicator of soil fertility. Fig. 6 shows the variation of the SAP in the 0–120 cm soil layers of different subsidence areas in the last 1a after mining and 2a after mining.

It can be seen from Fig. 6 that the variation of the SAP was 2.1–4.12 mg/kg, which indicated that soils in aeolian sandy land were low in the SAP. The SAP of the subsidence areas and CK decreased and then increased with depth in the last 1a after mining, and the SAP of CK were higher than that of PD and BY. The statistical analysis of the SAP of three treatments was done determine if there are statistically significant differences between them. The results showed that there was no significant difference ($P > 0.05$) between CK, PD, BY and at

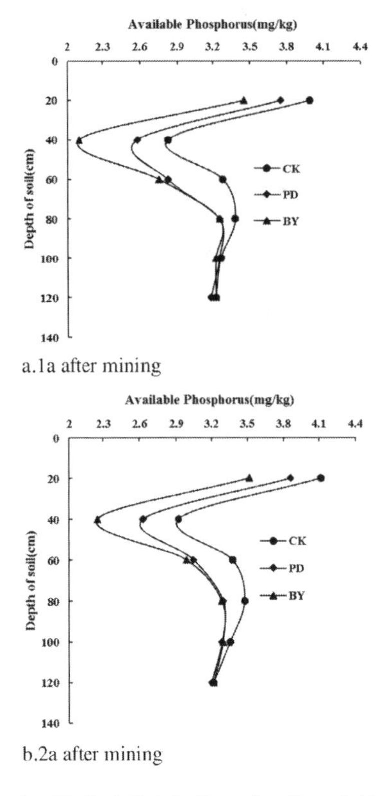

Figure 6. Vertical distribution of soil available phosphorus of different subsidence areas.

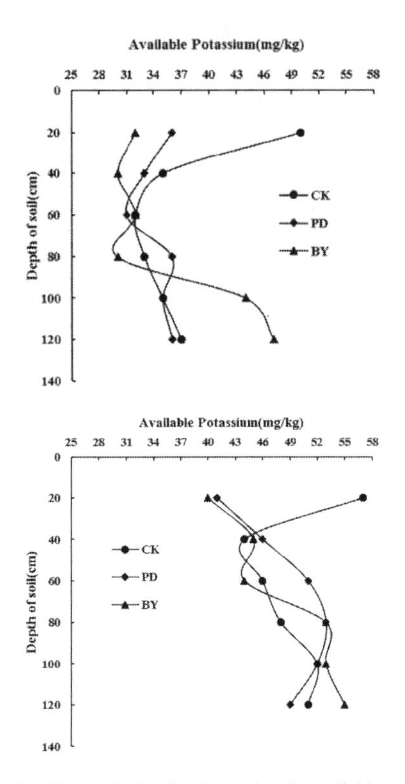

Figure 7. Vertical distribution of soil available potassium of different subsidence areas.

corresponding position of these areas. The variation of the SAP in the 0–120 cm soil layers of different subsidence areas in the last 2a after mining has the same features as that of the last 1a after mining and the SAP of BY was lower than that of PD and CK. Mining subsidence had an effect on the SAP, but only to a small extent. Compared with PD, mining subsidence has a relatively high impact on the SAP of BY, because of densely distributed of the edge cracks. The SAP of the topsoil would leach into the deep soils accompanied with leaching of rains. Besides, the sandy soil had the poor ability to fix and buffer phosphorus in the arid area and this negative impact was difficult to eliminate naturally in the short term.

3.2.4 *Soil Available Potassium*

Soil Available Potassium (AP) is an important component of soil available nutrient, which is the characterization parameter of soil current potassium capacity. The AP is affected by factors such as soil parent material, leaching of rains and surface vegetation, and is generally associated with the ability of plants to absorb potassium. Fig. 7 shows the variation of the AP in the 0–120 cm soil layers of different subsidence areas in the last 1a after mining and 2a after mining.

It can be seen from Fig. 7 that the AP increased of different subsidence areas in the last 1a after mining and then decreased with depth and lower than that of CK. The AP of different subsidence areas in the 0–40 cm soil layers decreased by 16 percent in comparison with CK and reached the lowest levels in the 40–80 cm soil layers. The AP of BY in the 100–120 cm soil layers was significantly higher than that of CK and PD ($P < 0.05$) and there was a significant difference ($P < 0.05$) between topsoil and substrates of BY, which indicated that mining subsidence have an impact on the AP of BY. This variation continued in the last 2a after mining, but the contents of the AP in soil had changed little. Compared with 1a after mining in the 40–80 cm soil layers, the AP of subsidence were higher than that of CK. The AP of PD and CK in the 0–20 cm soil layers were significantly different from those of CK ($P < 0.05$). From the above analysis, coal mining subsidence had an effect on the AP of BY and although the AP has improved 2a after mining, the negative effect was difficult to be eliminated in a short term.

4 CONCLUSIONS

On the basis of analysis on soil physical and chemical properties of the different subsidence areas, the following conclusions can be drawn:

1. The influence of mining subsidence on soil physical properties had obvious leveling and zoning characteristics, the mining subsidence's influence on the marginal zone was greater than the pelvic floor area. The soil moisture content, bulk density, porosity in the pelvic area showed a self-recovery phenomenon 1a after mining, while not in the marginal area. The influence of mining subsidence on physical properties of topsoil were higher than those of deep soil, which couldn't be restored in a short time.
2. The influence of mining subsidence on soil chemical properties was relative slight and had partition feature, the mining subsidence's influence on the marginal zone was greater than the pelvic floor area. Coal mining subsidence reduces the content of soil total nitrogen and available potassium, but has no effect on content of soil available phosphorus or organic matter.
3. While performing land reclamation and ecological rehabilitation in the west windy and sandy coal mining subsidence land, we should implement sub area management. In this way, not only can it reduce the possibility of secondary disturbance brought by the artificial restoration measures, but it can reduce the cost of ecological restoration.

ACKNOWLEDGEMENT

This research was supported was provided by Shenhua Group Corporation Limited key support project of the coal joint fund (U1361203). Thanks are also due to some participants for rendering assistant cooperation during studies.

REFERENCES

Bao Shi-dan. (2015). Soil agriculturalization analysis. China Agriculture Press, Beijing.

Chen Long-qian, Deng Ka-zhong, Xu Shan-kuan. 1999. Space variation law of physical characteristics about farmland soil due to mining subsidence. *J. Journal of China Coal Society*, 24(6):586–590.

He Guo-qing, Yang Lun, Ling Geng-di. (1991). Coal mine subsidence. China University of Mining and Technology Press, Xuzhou.

He Jin-jun, We I Jiang-sheng, He Xiao. 2007. Ground subsidence caused by mining affected to soil physics features in loess hills. *J. Coal Science and Technology*, 35(12):92–96.

Huang Luo-ming, Luo Yong-ming. 2003. Land remediation and ecological restoration of mined land. *J. Acta Pedologica Sinica*, 40(2): 161–169.

Hu Zhen-qi, Long Jing-hua, Wang Xin-jing. 2014. Self-healing, natural restoration and artificial restoration of ecological environment for coal mining. *J. Journal of China Coal Society*, 39(8):1751–1758.

Hu Zhen-qi, Wang Xin-jing, He An-min. 2014. Distribution characteristic and development rules of ground fissures due to coal mining in windy and sandy region. *J. Journal of China Coal Society*, 39(1): 11–18.

Lei Shao-gang. (2009). Monitoring and Analyzing the Mining Impacts on Key Environmental elements in Desert Area. China Mining University, Xuzhou.

Li Quan-sheng, He An-min, Cao Zhi-guo. 2012. Study on surface self-healing under modern coal mining technology in Shendong. *J. Coal Engineering*, (12): 120–122.

Liu Zhe-rong, Yan Ling, He Xiao. 2014. Effects of mining subsidence on physical and chemical properties of soil in the subsided land of the Daliuta Mining Area. *J. Journal of Arid Land Resources and Environment*, 28(11):133–138.

Wang Jian, Gao Yong, Wei Jiang-sheng. 2006. Influence of Mining Subsidence on Physical and Chemical Properties of Soil in Windy Desert Area. *J. Journal of Soil and Water Conservation*, 20(5):52–55.

Wang Qi, Quan Zhan-jun, Han Yi. 2014. Effects of mining subsidence on soil properties in windy desert area. *J. Journal of Soil and Water Conservation*, 11(6):118–122.

Wang Xin-jing, Hu Zhen-qi, Hu Qing-feng. 2015. Evolution and self—healing characteristic of land ecological environment due to super-large coalface mining in windy and sandy region. *J. Journal of China Coal Society*, (9): 2166–2172.

Zang Yin-tong, Wang Ji, Ding Guo-dong. 2010. Study on environmental variability of soil moisture in windy desert area after mining subsiding. *J. Journal of Arid Land Resources and Environment*, 47(2):262–269.

Zhang Fa-wang, Hou Xin-wei, Han Zhan-tao. 2003. Impact of Coal Mining Subsidence on Soil Quality and Some Protecting Technique for the Soil Quality. *J. Geography and Geo-Information Science*, 19(3):67–70.

Zhang Jian-min, Li Quan-sheng, Hu Zhen-qi. 2013. Study on ecological restoration mode of ultra wide fully-mechanized coal mining in west chain Aeolian sand area. *J. Coal Science and Technology*, 41(9): 173–177.

Zhang Li-juan, Wang Hai-lin, Hu Bin. 2007. Analysis of the soil nutrition and enzyme activity and their correlations in the coal mining subsidence area of Jiao Zuo City. *J. Environmental Science and Management*. 32(1):126–129.

Land Reclamation in Ecological Fragile Areas – Hu (Ed.)
© 2017 Taylor & Francis Group, London, ISBN 978-1-138-05103-4

Using loess and wind-blown sand in cementitious material for backfill mining to control land subsidence

X.D. Wang, G.G. Xu, S.B. Zhu, H. Wang, K.K. Yuan, B.Q. Wu & F. Wang
Xi'an Research Institute, CCTEG, Xi'an, Shaanxi, China

ABSTRACT: Wind-blown sand is widely distributed on the ground surface and can be used as backfill material. Mining with backfill is a proven mining practice in the coal industry and using cementitious backfill material of high concentration is an important development. The usual backfill material is composed mainly of wind-blown sand, fly ash, and cement. My team experimented with adding loess to the backfill mix to improve its characteristics. We determined the optimum proportion of loess to be added to backfill material to produce a high-quality, sand-rich cementitious material. In assessing the material, important parameters are compressive strength, bleeding rate, degree of stratification, setting time, and slump. When these were tested, results showed that the proper proportion of loess could improve performance. As the proportion of loess increased, the strength initially decreased but then increased; the bleeding rate, degree of stratification, and setting time decreased; and the slump rose. Loess is an inexpensive and widely available material that can improve the quality of cementitious material rich in wind-blown sand. The optimal mix proportion must be determined by test.

1 INTRODUCTION

According to different cementitious backfill materials, backfilling mining technology is sorted into several types, and an important type is backfilling mining technology by cementitious materials of high concentrations (Cui *et al.*, 2010 and Yang *et al.*, 2014) or paste backfilling material (Chang et al., 2011 and Sun et al., 2013). As a roof management method, cementitious materials should satisfy the strength requirement. Wind-blown sand, as natural and low-cost substance, is not usual raw material to solve the problem of ecology and geological environment after mining in northwest region of China, and just few scholars explored the engineering properties of wind-blown sand (Liu et al., 2014). A large amount of laboratory and field tests on the properties of wind-blown sand in cementitious backfill materials were done by our team. We try to take advantage of native industrial materials to improve combination property of cementitious materials of wind-blown sand. Working performance of cementitious backfill materials is controlled by multi-factors (Wei et al., 2009 and Sun et al., 2012). In consideration of the fine particle of wind-blown sand and the mineral composition and particle size distribution of loess (Fan et al., 2011, Shi et al., 2005 and Dang et al., 2009), engineering properties of cementitious backfill materials with the addition of loess and wind-blown sand will be explored.

2 MATERIALS AND EXPERIMENTAL METHODS

In this paper, the bulk density of wind-blown sand was 1.53 t/m^3. Moisture content was 0.4%. Mud content was 1.76%. The density of P.O. 42.5 cement was 3.1 t/m^3. The secondary fly-ash was chosen with bulk density of 0.92 t/m^3, The basic chemical components of raw materials of cementitious materials of wind-blown sand are shown in Table 1. Figure 1 shows particle size distribution of raw materials by laser particle size analyzer. From Figure 1(a), it could be known that natural wind-blown sand belongs to ultra-fine sand and badly graded sand. Loess is primarily air-dried and sieved through a 0.5 mm sieve. And chemical composition and granularity distribution of loess are respectively shown in Table 1 and Figure 1.

The initial concentration of samples are all 83.3% in the test. The raw materials and cementitious materials were taken to conduct physical and mechanical property tests in laboratory, including unconfined compressive strength test, Scanning Electron Microscope (SEM), slump, coursing degree of mortar, Setting time, bleeding rate, and so on.

Table 1. Chemical composition of raw materials of cementitious material of wind-blown sand.

Sample	Test results/%												
	SiO_2	Al_2O_3	TFe_2O_3	MgO	CaO	Na_2O	K_2O	P_2O_5	MnO	TiO_2	Loss of ignition	FeO	SO_3
Wind-blown sand	78.14	12.35	0.96	0.23	1.22	3.26	2.99	0.02	0.01	0.16	0.80	0.32	0.088
Cement	22.08	5.59	3.37	1.67	58.43	0.39	0.71	0.25	0.25	0.36	4.16	0.02	1.77
Fly ash	53.75	27.82	5.20	0.83	6.32	0.56	1.37	0.20	0.05	1.15	1.68	1.33	0.67
Loess	66.10	11.51	3.47	1.58	5.39	2.04	2.22	0.11	0.05	0.59	6.31	0.92	0.055

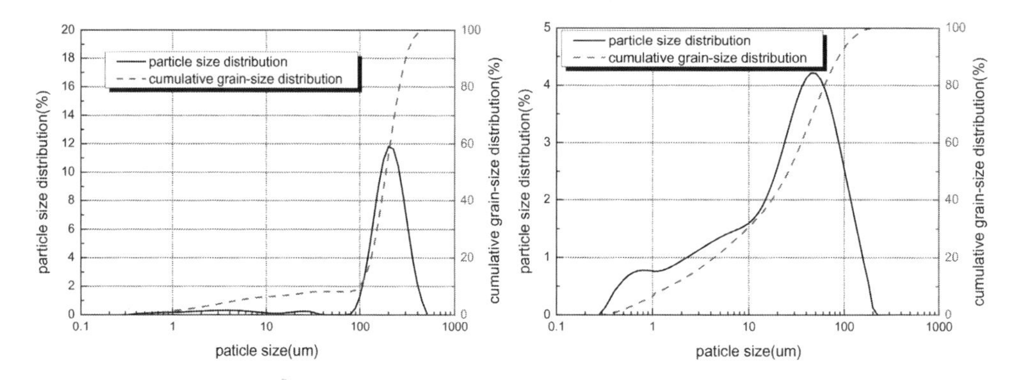

Figure 1. Granularity distribution of raw materials of cementitious material of wind-blown sand.

3 RESULTS

3.1 *Response of strength*

Compressive strength is one of important evaluation indexes of engineering properties of cementitious materials. It concerns the support effect of backfilling materials. The ratio of loess quality and solid total quality increases from 5.9% to 29.4%. Table 2 is the relationship between compressive strength and loess proportion. It shows that the strength of cementitious materials is firstly decreased and then increased slightly with the increase of loess addition. The macroscopic liquidity of cementitious materials decreased with the increase of loess addition. Compared with the samples of no mixing loess, the strength of cementitious materials is greatly affected by small admixture in a short time after mixing the loess. And the strength of cementitious materials gradually increases with the increase of curing period and the mixing amount. This law is more obvious for the large mixing amount of loess. From Table 2, it is positive for strength of the cementitious materials when the loess was mixed in cementitious materials. The largest ratio of loess quality and Solid total quality is about 30%, so the strength of the cementitious materials needs further verification when increasing the quantity of mixing loess.

3.2 *Response of bleeding rate*

Bleeding rate determines water-retaining property of cementitious materials, the particle gradation and workability of mixture. If bleeding rate is large, cementitious materials have poor water-retaining property, and water is easy to separate out. Figure 2(a) is the relationship between bleeding rate and loess proportion. It shows that bleeding rate of cementitious materials increases with the increasing of time and finally trends to be a constant. Bleeding rate of samples with addition of loess goes through three stages: high speed and well-distributed bleeding stage (ab stage), low speed and progressive decreasing bleeding stage (bc stage) and ending bleeding stage (cd stage). Bleeding rate of cementitious materials decreases with the increasing of the quantity of mixing loess. Bleeding rate suddenly decreases when the quantity of mixing loess is greater than 11.8%. At the same time, bleeding rate and the speed of bleeding of samples without loess are larger than others when the time is 90 min. As time goes on, the disciplines tend to be similar. It is important to choose the proper time to control the bleeding

Table 2. Mixing proportion of samples of cementitious material and compressive strength.

No.	Mass ratio				Loess proportion/%	Compressive strength/MPa		
	Wind-blown sand	Loess	Cement	Fly ash		3D	7D	14D
S0	13	0	1	3	0	0.87	1.67	2.85
S1	12.5	0.5	1	3	/	/	/	/
S2	12	1	1	3	5.9	0.65	1.42	2.71
S3	11	2	1	3	11.8	0.86	1.60	2.77
S4	10	3	1	3	17.6	0.69	1.69	2.94
S5	9	4	1	3	23.5	0.82	1.70	2.95
S6	8	5	1	3	29.4	0.89	1.75	3.14

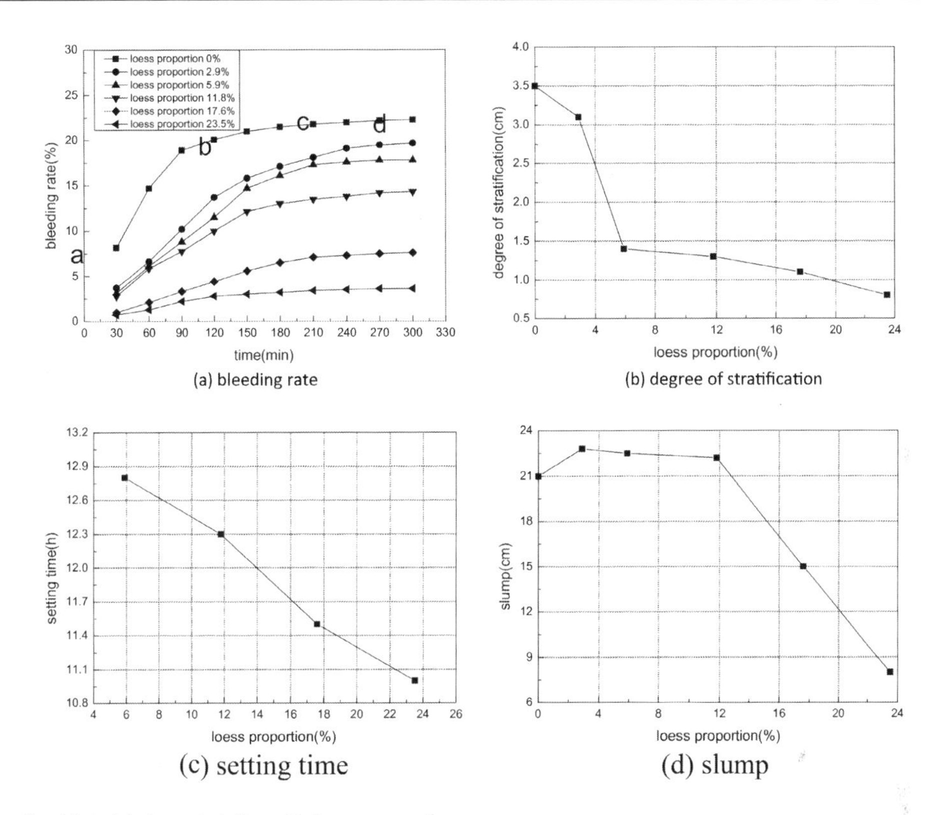

Figure 2. Material characteristics with loess proportion.

rate within limits, ensure material workability and complete the filling after preparing cementitious materials.

3.3 *Response of degree of stratification*

Water retention refers to the ability of mortar to preserve moisture. If the mortar water retention is not good, in the static release, the transmission process will produce segregation, bleeding and other issues. Stratification leads to gradient of particle gradation in the vertical direction of cementitious materials. The vertical direction of the water holding capacity of cementitious materials are different, resulting in work performance deterioration and reduction of comprehensive strength. Figure 2(b) shows the stratification of cementitious materials is decreasing with the increase of loess content. But when loess proportion is at the range from 2.9% to 5.9%, the degree of stratification appears pelter from 31 mm to14 mm. However, with the further increase of loess content, the decrease of stratification will tend to be gentle slowly. And when the content of loess increases to 23.5%, the stratification is reduced to 6 mm.

3.4 *Response of setting time*

Setting time is an index that measures the time of the cementitious material flowing from flow to plastic. The cementitious material need maintain sufficient fluidity or workability before it is pumped in the working face completely, and it cannot concrete too fast. But the cementitious material must be concrete as soon as possible after backfill completely, so the setting time cannot be too long. As shown in Figure 2(c), the influence of loess content on the setting time of cementitious material is also obvious. Under the same conditions, the setting time of cementitious material is decreasing linearly with the increase of the loess content. The linear law can be described as follows:

$$T = -0.1\omega + 13.06 \tag{1}$$

where, T represents the setting time, h; ω represents the content of the loess, %. As shown in Eq. (1), the sensitivity of loess content to condense of cementitious material is 10%, and there is no obvious mutation point. The influence on the characteristic is relatively moderate and stable.

3.5 *Response of slump*

The slump is an important index to reflect the fluidity of the cementitious material with large aggregate. The slump constant which reflects the flowability and frictional resistance in the pipeline is an important parameter for cementitious material. The content of cement, fly ash and initial mass concentration of cementitious material are 5.9%, 17.6% and 83.3%, respectively, in the experiments. As shown in Figure 2(d), the slump of the slurry material increased with the increase of the amount of loess when the amount of cement and fly ash maintain constant and the amount of the loess is between 2.9% and 11.8%. The slump in that condition is larger than that when the amount is 0. The slump declines sharply when the content is larger than 11.8%.

4 CONCLUSIONS

1. The appropriate amount of loess incorporation can effectively improve the performance of the material, but the sensitivity of each performance to loess content is different. Bleeding rate, slump and stratification are more sensitive, but the setting time is relatively weak.
2. With the increase of loess, the compressive strength of cementitious material decreases first and then increases, but the variation range of overall strength is small. The samples doped with loess have a greater residual strength because of the presence of strong frictional forces in addition to the mechanical bite force between the particles after compression failure.
3. With the increase of loess, the bleeding rate decreased significantly, and the bleeding termination of three stages which are high-speed uniform bleeding, low-speed diminishing bleeding and bleeding termination is obvious. The stratification is decreasing continuously, and after the loess content is more than 5.9%, the decrease of the stratification will tend to decrease slowly. The slump showed a first increase and then decreased, and when the content was more than 11.8%, the slump was linearly decreasing.

REFERENCES

Chang Qing-ling, Zhou Hua-qiang, Bai Jian-biao, et al. (2011) Stability Study and Practice of Overlying Strata with Paste Backfilling. Journal of Mining & Safety Engineering, 28 (2): 279–282.

Cui Zeng dil & Sun Heng-hu. (2010) The preparation and properties of coal gangue based sialite paste-like backfill material. Journal of China Coal Society, 35 (6): 896–899.

Dang Ya-ai, Li Shi-qing, Wang Guo-dong, et al. (2009) Fractal characteristics of soil particle composition for typical types of soil profile on Loess Plateau. Transactions of the CSAE, 25(9): 74–78.

Fan Henghui, Gao Jian-en, Wu Pu-te, et al. (2011) Analysis of influence factors for solidified soil strength based on change of physicochemical properties of loess. Rock and Soil Mechanics, 32 (7): 1996~2000.

Liu Peng-liang, Sun Kai-hua & Shang Pei. (2014) Cemented Filling Mining Process Applied in Continuous Coal Mining Face. Coal Engineering, 46(7): 52–54.

Shi Yu-xin, Dai Xue-rong, Song Zhi-guang, et al. (2005) Characteristics of Clay Mineral Assemblages and Their Spatial Distribution of Chinese Loess in Different Climatic Zones. Acta Sedimentologica Sinica, 23 (4): 690–695.

Sun Qi, Zhang Xiangdong & Yang Yu. (2013) Creep constitutive model of cemented body used in backfilling mining. Journal of China Coal Society, 38 (6): 994–1000.

Sun Qing-weil, Zhu Han & Cui Zheng-long. (2012) Preparation and Performance Study of Cemented Filling Material Based on Fly Ash and Coal Gangue. China Safety Science Journal, 22 (11): 74–80.

Wei Xiu-quan, Sun Heng-hu & Wang Ying-ying. (2009) Research on Microstructure and Physical, Mechanical Characteristics of Paste-Like Binging Material. Bulletin of the Chinese Ceramic Society, 28 (S1):37–40.

Yang Bao-gui, Han Yu-ming, Yang Peng-fei, et al. (2014) Research on Ratio of High Concentration Cementation Stowing Materials in Coal Mine. Coal Science and Technology, 2014, 42 (1): 30–33.

Land Reclamation in Ecological Fragile Areas – Hu (Ed.)

Effect of the exploitation of mineral resources on the ecological environment in Jiaozuo

L.J. Li & S.Y. Li
Henan Polytechnic University, Jiaozuo, Henan, China

ABSTRACT: Mining is the base of economic development and important property in Jiaozuo. It is very urgent and important to protect and govern ecological environment in the process of the resource exploiture. This paper discusses the formation of the geological disasters, ecological destruction and environmental pollution problems after exploitation of mineral resources in Jiaozuo, using the construction of Mountain stitches park and Forest park as an example, and puts forward some countermeasures and suggestions.

Keywords: mineral resources; exploit; environment influence; measure

1 INTRODUCTION

Resources and environment are the two big problem restricting the sustainable development of society, We need both environment and resources (Wang Heng Chen. 2015). Mining induced geological hazard, ecological destruction and environmental pollution has become hot spots.

2 JIAOZUO OVERVIEW

Jiaozuo is located in the northwest of henan province (Figure 1). Located in the transitional belt of the taihang mountains and plains of the north, the general topography is lower from north to south. Jiaozuo through census has more than 40 kinds of mineral resources, proven reserves of coal, limestone, bauxite, etc. More than 20.

Because of the mineral resources development, processing and utilization of unreasonable, has damaged to the mine and its surrounding environment and induced a variety of geological disasters.

3 THE ENVIRONMENTAL IMPACT OF RESOURCES DEVELOPMENT AND UTILIZATION

3.1 *The environmental impact of mineral resources development*

To the 1980s-1990s, Phoenix mountain area have 12quarries, south of the mountain was stripped

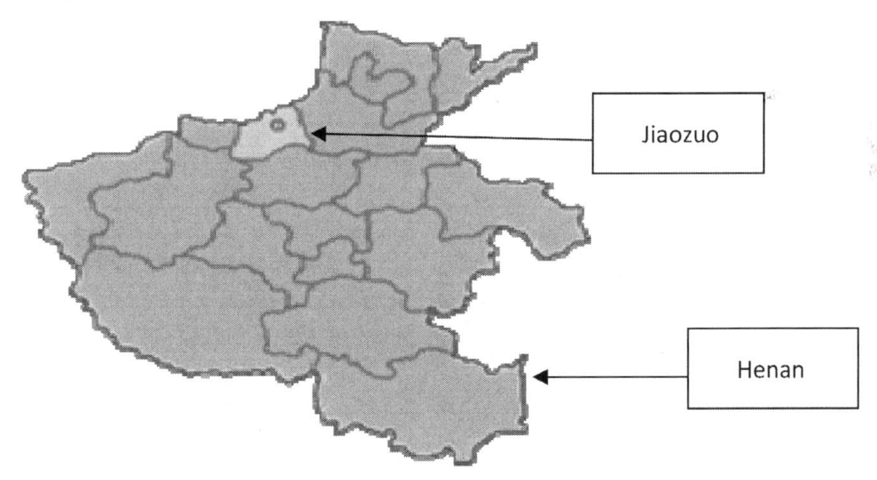

Figure 1. Jiaozuo location map.

the half. Make the original lush green mountain to the holes in the barren hills. Slag, mining pit takes up a lot of forest land. High and steep slope and debris bred the collapse, landslides and other geological hazard, Serious damage to the Jiaozuo ecological environment. Before 1999, Jiaozuo urban air quality level 2 less than 20 days.

In 1999, Jiaozuo cross the rubicon phoenix mountain. The municipal land and resources department compiled the mineral resource planning in the city, Could be divided into forbidden mining area, the limit of mine mining and exploration area. Shut down all the quarries near phoenix mountain, now Jiaozuo seam mountain national mine park is set up.

In zhongzhan district zhucun of Jiaozuo, sag and aperture are produced in the mining area will cause a lot of soil element loss as the surface runoff, further reducing the strength of the soil and cause widespread collapse. soil around zhucun the coal gangue has suffered different degree of heavy metal pollution (Hu Bin, 2004).

3.2 *The environmental impact of tourism resources development*

3.2.1 *Jiaozuo seam mountain national mine park*

Jiaozuo seam mountain national mine park is located in the north of jiaozuo city, with a total area of 0.9 square kilometers, is a show in coal mining sites, limestone mining sites, surface collapse sites governance landscape as the main body, ecological restoration measures show as the core, and the fusion of ancient kiln sites, modern studio and other human landscape in a body's comprehensive national mine park, is a perfect complement to yuntai mountain tourism resources.

Jiaozuo investment about 100 million yuan, to the ecological environment comprehensive improvement of quarry, The bare slope above the steep cliff flat into terraced fields, planting cypress, slope zone of the lower slower slope planting trees and grass, and watering supply by the pipe implements. Bare rock slope in the past, have grow a verdant grass irrigation combined with plant community, mining legacy destruction of ecological environment in the past, has become a green shrub garden now. Nowadays, seam mountain national mine park has built the subject sculpture, cliff waterfall (by residual rock transformation), xizi lake (by irrigation reservoir reconstruction), the star plaza, the bottom garden (converted quarries), parent-child paradise garden (converted mine spoil area)...(Figure 2).

"Mountain stitches" landmark sculpture implies the mountain as a living objects, human friend, like surgery, to close to the mountain mining "wound", make its restore "health", benefit to humanity.

Figure 2. Jiaozuo seam mountain national mine park.

Figure 3. Jiaozuo forest park.

Jiaozuo forest park is located in zhongzhan district zhucun mining coal mining subsidence, the terrain ups and downs. On the basis of birds garden scenic spot, as a whole, the people's park zoo move in jiaozuo forest park. Built north biggest, complete functions, distinctive forest ecological zoo (Figure 3).

Jiaozuo forest park, known by the general public as the "forest in the city, city people harbor", Jiaozuo propaganda protect to Wild animals and plants, display important platform of man and nature in harmony.

4 COUNTERMEASURES AND SUGGESTIONS

The resource-based cities to complete the resource-based type to the integrated type, the mining area to the ecological type, self service to the regional center of the transformation (Zheng Zhong, 2006). To reduce geological disasters, restore and protect the ecological environment, promote the sustainable development of mineral resource-exhausted city, are as follows:

1. Highlight key areas, with a strict legal system to protect the ecological environment (Lu Hao. 2016). Jiaozuo by strengthening the construction of legal system of environmental protection, norm all kinds of behavior of the environmental resources development, utilization, protection and management, effective control of ecological destruction and environmental pollution, sustainable production, consumption and management mode.
2. Establish responsibility system of environmental quality administrative leadership, strengthen the administrative supervision and law enforcement, environmental protection division management, strictly implement the system of construction project environmental assessment.
3. Used in a variety of comprehensive measures, formulate and implement the development and utilization of natural resources compensation fees, resource processing tax policy, resolutely eliminate high energy consumption and pollution heavy and hopeless backward production techniques.
4. Strengthen the propaganda and education of environmental protection, to enhance the consciousness of the people's environmental protection and participate, more effective and timely to provide services for environmental management and pollution control, urban development is not more than the bearing capacity of the natural and environmental self-purification ability.

5 SUMMARY

Mining has adverse impact on the ecological environment, through the integration of various resources, the reasonable development and utilization to protect the environment, synchronized planning, synchronous implementation, synchronous development.

ACKNOWLEDGMENT

Fund project: Henan Polytechnic University Dr fund (B2013-017).

REFERENCES

Hu Bin, Ren Yu-fen, Fang Yuan-yuan & Li Dong-yan (2004). Analysis of heavy metals in soil around zhucun

gangue dump in Jiaozuo. *Energy Environmental Protection*, 18, 53–56.
Lu Hao (2016). Set sail for green development life ecological civilization, environmental and resources protection committee of the National People's Congress in 2015 to work. *The People's Congress of China*, 02, 12–13.
Wang Heng Chen (2015). Construction of contemporary natural environment aesthetics—on the relationship between environment and resources. *Journal of Zhengzhou University*, 48, 107–110.
Zheng Zhong (2006). Transformation of resource-based city regeneration and regulation and management research—the domestic and foreign resources city transformation practice project and approach. *Resource Industrial Economy*, 01, 8–10.

Land Reclamation in Ecological Fragile Areas – Hu (Ed.)
 ISBN 978-1-138-05103-4

Maximizing ecological services value of abandoned mine land using integrated simulation models

L.P. Zhang, Y.F. Huang & M. Cao
College of Resources and Environmental Sciences, China Agricultural University, Beijing, China
Key Laboratory of Arable Land Conservation (North China), Ministry of Agriculture, Beijing, China
Key Laboratory of Agricultural Land Quality, Ministry of Land and Resources, Beijing, China

H.Y. Zhang
College of Science, China Agricultural University, Beijing, China

ABSTRACT: Large scale mining has resulted in great numbers of Abandoned Mine Land (AML). Transforming AML into other land use categories (i.e., forest land or cultivated land) and achieving higher Ecological Services Value (ESV) have become a hot issue for the ecologically sustainable development in destroyed mining regions. This paper combined the Linear Programming model and the Conversion of Land Use and Its Effects at Small Region Extent (CLUE-S) model to simulate AML shift in the Mentougou District in Beijing, China in 2020, which aims at realizing maximal ESV in post-mining landscape. The Linear Programming model was concerned with optimizing the demands for land-use categories, while the CLUE-S model translated these demands into spatial location of land use conversions. The results showed AML was mainly converted into forest, thus maintaining greater ecological service function. This study's findings offer an approach to simulate AML dynamics and provide rehabilitation suggestions for decision makers.

1 INTRODUCTION

Intensive mining activities have resulted in large areas of Abandoned Mine Land (AML), which means disturbed or contaminated land after mining activities and couldn't be used without ecological restoration (Hu *et al.*, 2015; Kostarelos *et al.*, 2015). Converting AML into farmland or forest land with high Ecological Service Value (ESV) can protect regional environment safety and achieve maximum ESV (Soltanmohammadi *et al.*, 2010; Sullivan and Amacher, 2013; Bonta *et al.*, 1997; Alday *et al.*, 2011.) Some specific land-use types are of higher ESV (Costanza *et al.*, 1998; Xie *et al.*, 2010), such as forest and grassland, for those land-use categories can improve water holding capacity, supplying more oxygen, breaking winds and fixing sands. So the core is to improve the areas of ecological land, which means converting AML to forest as many as possible. The purpose of AML restoration is to shift them into some other land use categories. However, limited research is conducted about how to optimize the spatial allocation of AML to realize higher ESV.

In this paper, Linear Programming model is used to obtain future land-use dynamics via the objective function under a series of land-use conversion constraints. After getting future land-use demands, we adopt the CLUE-S model (Verburg *et al.* 1999, Verburg *et al.* 2009, Verburg *et al.* 2014) to analyze the land use transformations with the target of achieving maximum ESV. We took the Mentougou District in Beijing, China as a case study. The objectives of the present study were to (1) design land-use change characteristics with Linear Programming model for achieving higher ESV; (2) simulate the spatial distribution of AML in 2020 using the CLUE-S model.

2 DATA

2.1 *Case region*

The Mentougou District (39°48′–40°10″N, 115°25′–116°10″E) is in the west of Beijing, China. In recent years, a great numbers of mines were closed due to newly released environment protection policy, leading to 4130 ha AML. The situation of the research area are shown in Figure 1. Eight land use types were included: cultivated land, garden land, forest land, grassland, construction land, AML, water, and unutilized land, with areas of 831 ha, 1678 ha, 30165 ha, 3686 ha, 2733 ha, 3573 ha, 905 ha, and 3063 ha, respectively (Figure 2).

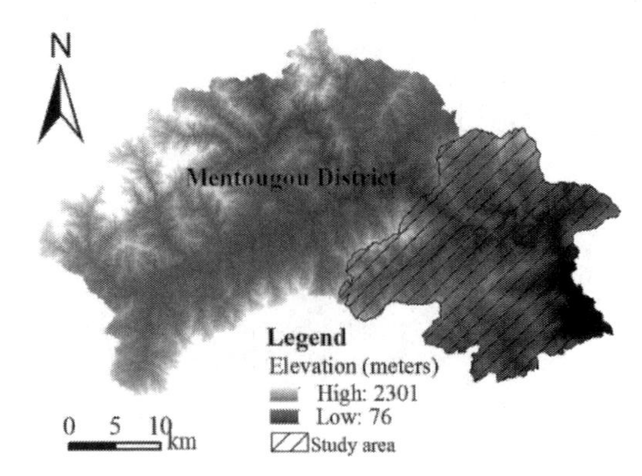

Figure 1. The situation of the research area.

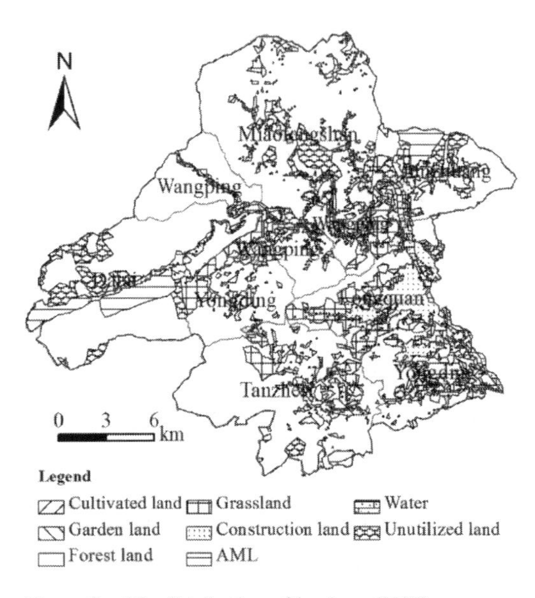

Figure 2. The distribution of land use (2007).

2.2 *Data processing*

The land-use map was from the Bureau of Land Recourses, Beijing. Elevation and slope of research area were extracted from DEM. The contents of Soil Organic Matter (SOM) was collected from the local soil map. The distance to the nearest river and road were measured in ArcGIS 10.0 software to depict transportation situation. The DEM; local soil map; distribution of river and road were obtained from the Beijing Digital Soil System. Population density, per capita income, and annual afforestation areas of 2007 were from the Statistical Yearbooks (2008). All data were transformed to the same projection system with a unified grid size of 100 × 100 m.

3 METHODS

3.1 *Abandoned mined land suitability analysis*

Abandoned mined land suitability is one of key factors influencing the transformation of AML. In this study, the Simple Limitation Method is used to assess reclamation suitability using eight factors: slope, surface material composition, soil texture, SOM, soil layer thickness, irrigation and drainage situation, transport situation, and land damage conditions (Zhang *et al.*, 2016). The results of suitability analysis can be one constraint in Liner Programming model.

3.2 *Liner programming model*

Two modules, i.e., non-spatial and spatial modules, are included in the CLUE-S model. Liner programming model was used to calculate the demands for all land use categories in the non-spatial section with the objective of achieving higher ESV. The ESV for all land use categories involved in this study are shown in Table 1.

From Table 1, we can see that more areas of reclaimed forest land mean higher ESV. Equation (1) is the objective function and the according constraints can be seen Equation (2).

$$Z' = \max\begin{pmatrix} ESV_1 \times (x_1 + x_9) + ESV_2 \times (x_2 + x_{10}) \\ + ESV_3 \times (x_3 + x_{11}) + ESV_4 \times x_4 \\ + ESV_5 \times x_5 + ESV_6 \times x_6 + ESV_7 \times x_7 \\ + ESV_8 \times x_8 \end{pmatrix} \tag{1}$$

$$\begin{cases} x_1 + x_2 + x_3 + \ldots + x_{11} = 46634 \\ x_9 + x_{10} + x_{11} = 3072 \\ x_6 = 501 \\ x_9 \le 85.3 \,\&\, x_{10} \le 176.1 \,\&\, x_{11} \le 1729.0 \\ x_1 + x_9 \ge 1656 \\ x_2 + x_{10} \ge 1678 \\ x_3 + x_{11} \ge 29738 \\ x_5 \ge 6688 \end{cases} \tag{2}$$

In Equation (1), Z' is the objective function. ESV_1 to ESV_8 describe the ESV of eight land-use categories. x_1-x_8 are the areas for eight land-use categories,. x_9, x_{10}, and x_{11} means the areas of AML which are suitable for transformation into cultivated land, garden land, and forest land, respectively. The *xs* can be calculated through the constraints of the Linear Programming model in Equation (2). In Equation (2), the first three equations represent the areas of the study area, reclamation areas of AML, and areas of AML without any restoration measures in 2020, respectively. The fourth equation is the results for mined land suitability analysis. The last four in equations

Table 1. ESVs for all land use categories in the study area (CNY ha^{-1}) (Xie et al., 2010).

	Cultivated land	Garden land	Forest land	Grassland	Construction land	AML	Water	Unutilized land
ESV	3296.98	7516.28	11735.57	4870.35	0	0	18926.32	580.10

Note: The ESV of garden land equals the average ESV of cultivated land and forest land.

represents the minimum areas of cultivated land, garden land, forest land, and construction land in the case region of 2020 issued by the *General Plan for Land-Use in the Mentougou District* (2006–2020) for ecological reservation and farmland protection.

3.3 *CLUE-S model*

CLUE-S model translates these demands into land use conversions due to the probabilities and rules of different land-use types (Verburg *et al.* 2014). The spatial module comprises a spatially explicit allocation procedure. The spatial distributions of the land-use types are quantified using a binomial logit model with the percentages of the types as dependent variables and geomorphologic, transportation accessible, soil-related, and socioeconomic driving factors as independent variables.

Considering data relevance, suitability, and acquisition, eight driving factors were selected: elevation (X_1), slope (X_2), distance to the nearest river (X_3), distance to the nearest road (X_4), SOM (X_5), population density (X_6), per capita income (X_7), and annual affore station areas (X_8). Specifically, X_1 and X_2 are used to describe the terrain conditions, and X_3, and X_4 are used to describe transportation accessibility. X_5 is one of the most important indicators of soil properties, which is essential for the reclamation of AML for cultivation. X_6 and X_7 are the key socioeconomic factors influencing the social development of the towns. X_8 is the indicators reflecting climatic conditions.

Conversion between the different types determines the changes that will eventually take place. The specific conversion settings of the eight types affect the temporal dynamics of the prediction, which are composed of two parameters: conversion elasticity (ELAS) and transition matrix. The first parameter, which ranges from 0 (easy conversion) to 1 (irreversible change), is determined on the basis of expert knowledge and observed behavior in recent years. The value for the second parameter, transition matrix, is 0 (irreversible transition) or 1 (easy conversion) and indicates the possible conversions for each type. Any land-use type was not allowed to transfer into AML. Water and construction land are restricted to shift into other categories due to their stable attributes (Zhang *et al.*, 2016).

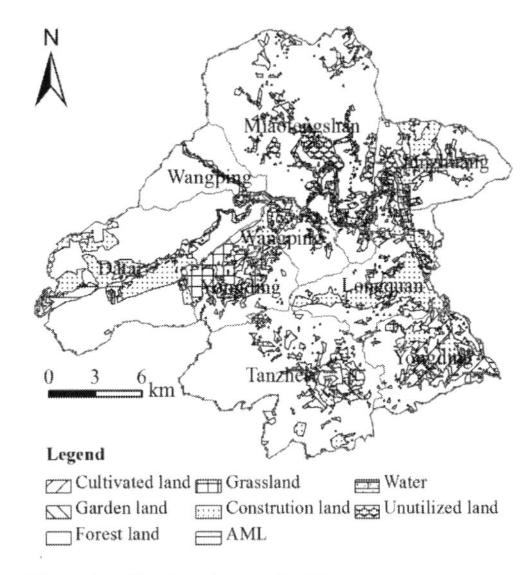

Figure 3. Simulated map of 2020.

4 RESULTS AND DISCUSSION

4.1 *Simulation accuracy of combined models*

The results of logit regression were tested by the Receiver Operating Characteristic (*ROC*), which should be higher than 0.7. *ROC*s for the eight land-use types were 0.854, 0.839, 0.825, 0.947, 0.900, 0.895, 0.861, and 0.837, respectively. The results of regression coefficients of the driving variables were then applied to regression analysis. The simulation accuracy can be assessed by *Kappa* indices (Gobin *et al.*, 2002). We compared the simulated map of 2013 with the true land use map in ENVI 5.0 software and found the *Kappa* index was 0.89, meeting the demands.

4.2 *Simulation results in 2020*

The simulated land use categories in 2020 were presented in Figure 3. The areas for the eight different land use categories were 1670 ha, 1681 ha, 29805 ha, 2659 ha, 6689 ha, 498 ha, 950 ha and 2682 ha, respectively. An increasing trend can be observe in the number of cultivated land, forests, water, and construction land (Figure 3). However, the decline of grassland and unutilized land can be

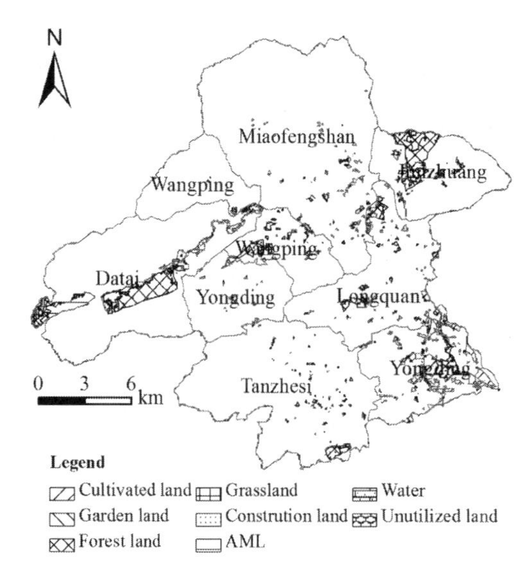

Figure 4. Special distribution of AML patches in 2020.

obvious. The number of garden land was more or less the same. In order to extract the shift of AML for 2020 in space, we overlapped the simulated map with the land use map of 2007. In Figure 4, a great shift of cultivated land and forest from AML can be found. The new cultivated land was observed in Yongding Town because of the flat terrain and fertile soil. Almost no spatial distribution of cultivated land was found in the other towns, which were hampered by the low levels of driving factors. Assuming forest development followed the trend shown in the non-spatial module, some AML can be converted to forest land. The newly formed forest basically emerged in Yongding, Miaofengshan, Longquan and Junzhuang.

5 CONCLUSION

In order to enlarge the ESV, the simulated land use dynamics are obtained through the combined Linear Programming and CLUE-S models for the case region. The combined model passed the examination of *ROC* and *Kappa* indices. Transforming more AML to forest land can improve the ESV, meeting the demand of higher ESV in the post-mining landscape.

ACKNOWLEDGMENT

The authors would like to acknowledge the financial support provided by the Foundation for Public Welfare of the Ministry of Land and Resources of China (no. 201411017).

REFERENCES

Alday, J. G., Marrs, R. H. & Martínez-Ruiz, C., (2011) Vegetation succession on reclaimed coal wastes in Spain: the influence of soil and environmental factors, *J. Applied Vegetation Science*, 14(1), 84–94.

Bonta, J. V., Amerman, C. R., Harlukowicz, T. J. & Dick, W. A., (1997) Impact of coal surface mining on three Ohio watersheds-surface-water hydrology, *J. Journal of the American Water Resources Association*, 33(4), 907–918.

Costanza, R., d'Arge, R., de Groot, R., Farber, S., Grasso, M., Hannon, B., Limburg, K., Naeem, S., O'Neill, R V., Paruelo, J., Raskin, R G., Sutton, P. & van den Belt, M. (1998) The value of the world's ecosystem services and natural capital, *J. Ecological Economics*, 25(1), 3–15.

Gobin, A., Campling, P. & Feyen, J. (2002) Logistic modelling to derive agricultural land use determinants: A case study from southeastern Nigeria, *Agriculture, J. Ecosystems & Environment*, 89(3), 213–228.

Hu, Z. Q., Fu, Y. H., Xiao, W., Zhao, Y. L. & Wei, T. T. (2015) Ecological restoration plan for abandoned underground coal mine site in Eastern China, *J. International Journal of Mining, Reclamation and Environment*, 29(4), 316–330.

Kostarelos, K., Gavriel, I., Stylianou, M., Zissimos, A. M., Morisseau, E. & Dermatas, D. (2015) Legacy soil contamination at abandoned mine sites: Making a case for guidance on soil protection, *J. Bulletin of Environmental Contamination and Toxicology*, 94(3), 269–274.

Soltanmohammadi, H., Osanloo, M., & Aghajani Bazzazi, A., (2010) An analytical approach with a reliable logic and a ranking policy for post-mining land-use determination, *J. Land Use Policy*, 27(2), 364–372.

Sullivan, J. and Amacher, G. S., (2013) Optimal hardwood tree planting and forest reclamation policy on reclaimed surface mine lands in the Appalachian coal region, *J. Resources Policy*, 38(1), 1–7.

Verburg, P. H., Soepboer, W., Veldkamp, A., Limpiada, R., Espaldon, V. & Mastura, S. S. A. (2014) Modeling the spatial dynamics of regional land use: The CLUE-S model, *J. Environmental Management*, 30(3), 391–405.

Verburg, P H., van de Steeg, J., Veldkamp, A. & Willemen, L. (2009) From land cover change to land function dynamics: A major challenge to improve land characterization, *J. Journal of Environmental Management*, 90(3), 1327–1335.

Verburg, P H., Veldkamp, A. & Fresco, L O. (1999) Simulation of changes in the spatial pattern of land use in China, *J. Applied Geography*, 19(3), 211–233.

Xie, G. D., Zhen, L., Lu, C. X., Xiao, Y. & Li, W. H. (2010) Applying value transfer method for eco-service valuation in China, *J. Journal of Resources and Ecology*, 1(1), 51–59.

Zhang, L. P., Zhang, S. W., Huang, Y. J., Cao, M., Huang, Y. F. & Zhang, H. Y. Exploring an ecologically sustainable scheme for landscape restoration of abandoned mine land: scenario-based simulation integrated linear programming and CLUE-S model, *J. International Journal of Environmental Research and Public Health*, 2016, 13, 354.

Land Reclamation in Ecological Fragile Areas – Hu (Ed.)

Characteristics of runoff coefficient of abandoned mine land using the storm water management model

X.D. Li, H. Yang, J. Wang, L.J. Guo & Z. Wang
School of Mechanics and Civil Engineering, China University of Mining and Technology, Beijing, China

ABSTRACT: Utilizing Storm Water Management Model (SWMM) as the auxiliary means, the paper takes a-century-old coal mine named Fengfeng as the typical example to analyze the runoff process of Abandoned Mine Land (AML). Based on the meteorology, soil and vegetation, topography and other environmental conditions, the article chooses rainfall, slope, underlying surface types of AML as the research variables. The results show that (1) the rainfall intensity is the direct factor influencing runoff coefficient, which plays a critical role in runoff speed; (2) coal waste's slope has a lower influence than its stacking life to the rainfall runoff; (3) AML's runoff coefficient of abandoned factories tends to 1; (4) The underlying surface type most closely affected by rainfall intensity in AML is the soil pavement.

1 INTRODUCTION

Due to digging, subsidence, and occupation, Abandoned Mine Land (AML) has been generated in the process of mining, excavating and smelting and is damaged, idle, abandoned or unutilized land, which needs remediation before reusing. China is a big coal mining production and consumption country. China has a large amount of AML, and it is distributed widely. AML's regeneration has significant influence to the sustainable development of resource-based cities.

For a long time, the research subjects of AML are mainly concentrated in the field of ecological restoration. In recent years, it is developing from technical means to comprehensive evaluation, diversified land use pattern, landscape reconstruction, and so on. However, there are few studies focusing on the characteristics of AML's Runoff Coefficient (RC). According to *the guideline for Environmental Performance Evaluation of City Ecological Construction*, the RC is a vital evaluation index of AML's regeneration. RC is utilized to evaluate the impact on the regional hydrological environment by changing the structure and properties of underlying surface. It is necessary to study characteristics of RC that is beneficial to solve the problem of soil erosion, site pollution, landscape reconstruction, and reducing the influence of AML on surroundings.

This paper takes a-century-old coal mine named Fengfeng as the typical example to study the characteristics of AML's RC. Based on the meteorology, soil and vegetation, topography, and other environmental conditions, the article chooses rainfall, slope, underlying surface types of AML to simulate the runoff process.

2 STUDY AREA DESCRIPTION

The study area (36°27′58″N; 114°13′07″E) is located in Fengfeng mining area, Hebei province. With a 187-meter average altitude of the site, there is a coal waste over 30 meters high. The annual average precipitation is 549.4 millimeters. The original topography of the site has been almost destroyed, and the vegetation coverage is low. The soil is mainly based on cinnamon soil and coal waste matrix soil.

3 MATERIALS AND METHODS

Due to the complexity of storm runoff process, the factors influenced are various. Among the factors, rainfall conditions and surface conditions are the two basic ones influencing RC. The article chooses rainfall, underlying surface type, and slope to analyze the characteristics of RC.

3.1 *Rainfall design*

Based on the storm intensity equation and rainfall data of Handan city (Equation 1), the simulation scene employs Chicago Hydrograph Model to synthesize different rainfall scenes with different return periods. The suggested criterion for rainfall peak is 0.4, and the rainfall interval is 5 min. Refer to *the Outdoor Drainage Design Code* (GB50014-20062014 version), the return Period

(P) values are 2, 5, 10, 20, 50, and 100. And the value of rainfall duration is 120 min.

$$q = \frac{3953.190(1+0.997\lg P)}{(t+16.393)^{0.852}} \quad (1)$$

where q, P *and* t refer to average rainfall intensity (L/(s.hm^2)), duration of rainfall (min), storm return periods (a), respectively.

3.2 *Selection of underlying surface type*

The study selects abandoned factories area, coal waste area, soil pavement area as experimental objects. The research area are 119 thousand square meters, 29 thousand square meters, and 132 thousand and 600 square meters, respectively. Besides, the abandoned factories area accounts for 33.2% of the total area, the coal waste area accounts for 8.1%, and the soil pavement area accounts for 58.7%, respectively.

The soil condition and slope of coal waste are the two main factors influencing runoff yield. Besides, the soil condition of the weathering layer is closely related to the pile up time. Therefore, according to the pile up time the coal waste is divided into the new coal waste (8–15 years) and the old coal waste (more than 30 years). Because of the frequent disturbance, the slope of new coal waste is sharper, which is in the range of 40 to 60 degrees. In this study, we select 50 degrees and 60 degrees to represent the new coal waste's slope, and 30 degrees and 40 degrees to represent the old coal waste's slope.

3.3 *Runoff simulation*

SWMM (Storm Water Management Model) is utilized in this study. It inputs rainfall time series, soil preconditions and underlying surface type data to simulate the runoff process of AML.

The study selects Horton infiltration model to simulate regional precipitation infiltration process. According to the existing research results, the infiltration parameters of the Abandoned Factories (AF), Coal Waste (CW) and Soil Pavement (SP) are as follows (Table 1).

Table 1. Infiltration parameters of different underlying surface types.

Parameter	AF	SP	CW (8–15 years)	CW (≥30 years)
f_c	3	6.6	8	13
f_0	17.13	76.2	13.8	17.5
K	5.81	4	4.2	4.5

F_c, f_0, and k refer to stable infiltration rate, initial infiltration rate, empirical value, respectively.

Table 2. Parameters of SWMM model.

Parameter	Slope	DS (mm)	MC
AF	–	1.27	0.011
SP	–	2.54	0.15
CW	30	3.96	0.1007
CW	40	3.36	0.0934
CW	50	2.88	0.0875
CW	60	2.52	0.083

The depression storage, Manning coefficient, maximum infiltration rate and minimum infiltration rate are parameters that SWMM required. And there is a linear regression relationship between the slope and the depression storage of coal waste, and the depression storage and the Manning coefficient of coal waste, respectively (Equation 2 and 3).

$$y = -0.51\ln(x) + 2.72, \quad R^2 = 0.99 \quad (2)$$

$$y = 20.337x - 1.0589, \quad R^2 = 0.823 \quad (3)$$

According to characteristics of the site, the Depression Storage (DS) and Manning Coefficient (MC) of this area are as follows (Table 2).

4 RESULTS AND ANALYSIS

4.1 *Runoff coefficient of coal waste*

Through 30 rainfall experiments, the study analyzes the relationship among RC, rainfall, slope and the pile up time (Table 3). The coal waste's RC is high. It is distributed between 0.504–0.838, and more than 50.4% of the rainwater was lost during the rain. That is because the topsoil is sandy soil. The soil is loose and the unstable resulting in weak resistance. At the same time, lacking of nitrogen and phosphorus and others, the coal waste has low plant coverage and vegetation succession. The bare gangue matrix accelerates rainfall runoff.

4.1.1 *Rainfall intensity influence on the runoff process of coal waste*

By changing the rainfall intensity, the RC of coal waste is simulated. The results are shown in Fig. 1. In the same slope condition, the RC gradually increases with the rain intensity. Taking the slope of 40 degrees as an example, the RC gradually increases from 0.516 to 0.758, which changes 46.9%. It indicates that the magnitude of rainfall intensity is a direct factor affecting the RC, which plays a key role in the runoff of coal waste. Besides, from Fig. 1 we can figure out that the variation of coal waste's RC is affected by coal waste's stacking age.

Table 3. Runoff coefficient of coal waste.

P	Rainfall	Slope 30°	40°	50°	60°
P = 2	56.27	0.504	0.516	0.646	0.653
P = 5	73.44	0.594	0.603	0.719	0.725
P = 10	86.43	0.645	0.653	0.757	0.762
P = 20	99.41	0.685	0.692	0.787	0.791
P = 50	116.58	0.727	0.733	0.816	0.82
P = 100	129.57	0.752	0.758	0.834	0.838

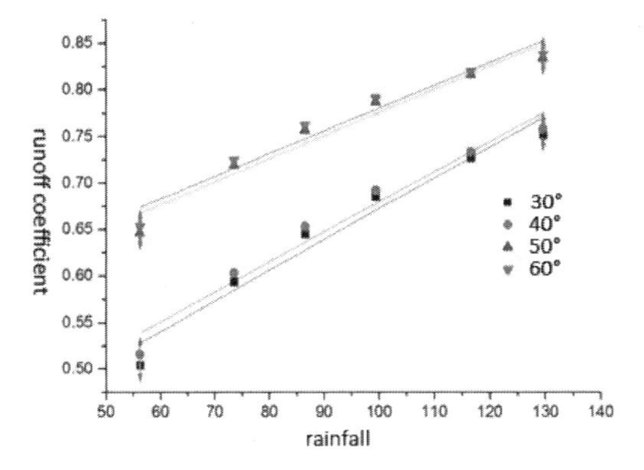

Figure 1. Rainfall intensity influence on RC of coal waste.

4.1.2 *Slope influence on the runoff process of coal waste*

By changing the slope, the RC of coal waste is shown as Fig. 2. The RC of 40 degrees and 50 degrees slope are obviously changed due to the different pile up time. In different stacking years, the maximum infiltration rate and the minimum infiltration rate of coal waste matrix are diverse, which is an important factor influencing the confluence of coal waste. In the same stacking age, the RC has a linear relationship with the slope, but the increase of the RC is not obvious as the slope increases. With the increase of coal waste's slope, the RC increased from 0.504 to 0.516, and the rate of change was only 2.38%. When the rainfall return period increases to 100 years, the RC decreases with the slope rate to 0.80%, which indicates that slope has low impact on rainfall runoff.

4.2 *Characteristics of abandoned factories and soil pavement*

Using SWMM to simulate the flow confluence process in abandoned factories area, it can be seen that the RC almost approached 1 (Fig. 3). With the increase of rainfall intensity, the RC

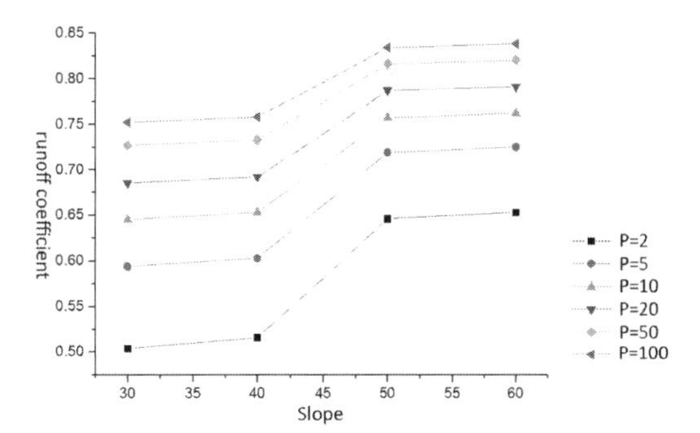

Figure 2. Slope influence on RC of coal waste.

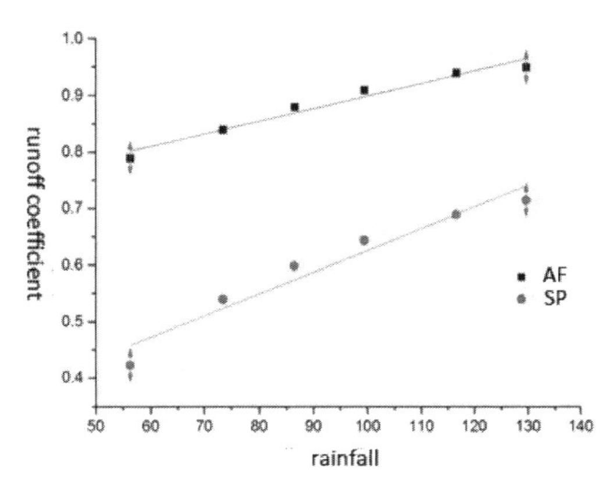

Figure 3. Relationship between rainfall and AF and SP.

Table 4. Characteristics of RC in AML.

Underlying surface types	RC
CW	0.504–0.653
AF	0.79–0.838
SP	0.423–0.715

approaches 1 quickly. This is due to impermeable ground blocking the rain directly into the ground. Soil pavement is another important component of AML, accounting for 58.7% of the total area. With the increase of rainfall intensity, the RC of soil pavement increased rapidly from 0.423 to 0.715 (Fig. 3), and the rate of change is 69.03%. It is the largest underlying surface types influenced by rainfall intensity of AML.

To sum up, the RC of each component in the abandoned mine land are as follows (Table 4).

5 CONCLUSION

Mining significantly impacts water environments with increased runoff and degradation of water quality. The outcomes of this research study presented in the paper, which simulated relationships between runoff coefficient and rainfall, slope, underlying surface types parctical guidance in the planning the future AML regeneration. It shows that rainfall intensity is a direct factor affecting the AML's RC, which plays a key role in the runoff of AML. The effect of slope on runoff of coal waste is lower than that of the pile up age. Under the same stacking age, the RC has a linear relationship with the slope, but the increase of the RC is not obvious as the slope increases. Under different stacking years, the RC of coal waste changed markedly. The study also found that the RC of the abandoned factories area almost approaches 1. Soil pavement is another important component of AML. With the increase of rainfall intensity, it rapidly increases from 0.423 to 0.715. it is the largest underlying surface types affected by rainfall intensity in AML.

REFERENCES

Cai Jianbo, Lin Ning, Du Xiaosong, Feng Yungang, (2011) Effect analysis of low-lying greenbelt to reduce urban runoff depth and runoff coefficient. *Journal of Urban Roads Bridges & Flood Control*, (06), 119–122+318.

Liang Yuting, (2014) Study on the factors of rainfall-runoff coefficient. *Journal of Soil & Water Conservation*. A thesis submitted in partial satisfaction of the Requirements for the degree of Architecture and Civil Engineering Municipal Engineering in the Graduate School of Hunan University.

Shi Mingxin & Wu Faqi & Tian Guocheng, (2015) Experimental study on effect of surface roughness on overland flow velocity. *Journal of Hydroelectric Engineering*, (06), 117–124.

Tao Wanghai & Wu Junhu, (2016) Study on numerical simulation of slope runoff and sediment yield rule. *Journal of Soil and Water Conservation*, 30 (1), 54–57.

Zang Yajun, (2008) Characteristics of rain infiltration and its stability research of coal waste dump in mountains area. A thesis submitted to Chongqing University in partial satisfaction of the Requirements for the degree of doctor of Engineering.

Land Reclamation in Ecological Fragile Areas – Hu (Ed.)
 ISBN 978-1-138-05103-4

Distribution and formation characteristics of karst collapse in major cities of China

L. Yang, Q. Wu, C.L. Liu, H.B. Hou & S.Q. Liu
College of Geoscience and Surveying Engineering, China University of Mining and Technology, Beijng, China
Urban Geology Department, Institute of Hydrogeology and Environmental Geology, CAGS, Shijiazhuang, Hebei, China

ABSTRACT: Based on the assessment results of the environmental and geological investigation of major cities in China and through comprehensive study on karst collapse, this paper summarizes the distribution and growth laws in topography, geological structure and distributions in different geographical areas. It points out that karst collapse is mainly distributed in low terrain, obviously subject to stratum structure. In view of quantity, karst collapse is more distributed in the medium-sized cities of Western region. The formation of karst collapse is influenced by dual effects of both natural and human activities, among which groundwater movement is the most active factor.

1 INTRODUCTION

The karst collapse refers to the collapse of the soil in the karst area, the collapse of the karst bedrock and the collapse of the overlying soil along with the underlying bedrock (Qian J P, 2007). karst collapse is a typical sudden geological hazard in karst area. Due to the concealment and suddenness of karst collapse, human life and social and economic construction have suffered serious losses. China's soluble rock distribution area covers 3.65 million Km^2, accounting for 1/3 of whole territory making China one of the countries with the most developed karst area (Lu Y R, 2010). To reduce the harm caused by karst collapse, we must first understand its distribution, formation conditions and influencing factors.

2 DISTRIBUTION CHARACTERISTICS OF KARST COLLAPSE

2.1 *Distribution characteristics of karst collapse in topography and tectonics*

The karst collapse is mainly distributed in the karst area with frequent human activity. In view of topography, the karst collapse mainly occurs in the karst negative landform peaks, the isolated peak plain and the first terrace; in view of terrain, the karst collapse is mostly distributed in the low-lying belt and high and low terrain transition zone (Liu B C et al., 2000).

The distribution of karst collapse is also more prone to occur in area of the tectonic belt and anticline, diagonal axis and wings. For example, Hechi City, Laibin City, karst collapse occurred in the tectonic zone (Zhu SH Z et al., 2000). The distribution of karst collapse is also closely related to groundwater activity.

2.2 *Distribution of karst collapse in different geographic regions in China*

2.2.1 *Distribution of karst collapse in major cities in various provinces and municipalities*

From the national perspective, karst collapses exist in 69 major cities with a total number of 1332. Collapses in Guangxi, Hunan, Guangdong, Shandong, Jiangxi and other provinces in China are the most developed (Table 1). The number of collapses in the above five provinces has reached 1101, accounting for 82.6% of the total.

2.2.2 *Disribution of karst collapse in cities of different regions*

China's urban distribution area is divided into four regions, namely "Eastern, Middle, Western, Northeastern". The east includes 10 provinces and cities namely Beijing, Tianjin, Hebei, Shanghai, Jiangsu, Zhejiang, Fujian, Shandong, Guangdong and Hainan. The middle includes 6 provinces namely Shanxi, Anhui, Jiangxi, Henan, Hubei and Hunan. The west includes 12 provinces namely Inner Mongolia, Guangxi, Chongqing, Sichuan, Guizhou, Yunnan, Tibet, Shaanxi, Gansu, Qinghai, Ningxia

Table 1. Distribution of Karst Collapse in different provinces of China.

Province	Cities with karst collapse	Total number
Guangxi	Guili, Hezhou, Yulin, Nanning, Liuzhou, Guigang, Hechi, Laibin, Chongzuo, Qinzhou	775
Guangdong	Guangzhou, Foshan, Yunfu	78
Shandong	Taian, Linyi, Jinan, Zaozhuang	58
Hunan	Yongzhou, Chenzhou, Xiangtan, Loudi, Changsha, Zhuzhou, Zhangjiajie, Yueyang, Shaoyang, Huaihua	135
Jiangxi	Jiujiang, Yichun, Ji'an, Shangrao, Jingdezhen, Pingxiang, Xinyu	55
Guizhou	Zunyi, Guiyang, Liupanshui, Anxun, Tongren, Bijie, Xingyi, Kaili	29
Anhui	Tongling, Huainan, Anqing, Huaibei, Wuhu	37
Hubei	Huangshi, Enshi, Xianning, Ezhou	35
Yunnan	Qujing, Kunming, Zhaotong, Lijiang, Yuxi	34
Hebei	Tangshan	21
Zhejiang	Hangzhou, Shaoxing, Jinhua, Huzhou	30
Fujian	Longyan, Sanming	26
Jiangsu	Xuzhou, Nanjing, Wuxi, Suzhou, Zhenjiang	16
Xizang	Lhasa	3

Note: The table "Yuxi (1)" said Yuxi's karst collapse is one.

and Xinjiang. The Northeast includes Liaoning, Jilin and Heilongjiang provinces.

The karst collapse mainly occurs in the western cities of our country, more in frequency than the central and eastern parts, the northeastern region has no karst collapse disaster, see Table 2.

2.2.3 *Distribution of karst collapse in cities of different sizes*

Major cities in China are divided into three levels as large, medium and small. Municipalities, municipalities, provincial cities belong to large cities. Cities with a population of more than 500,000 belong to medium-sized cities. Cities with a population less than 500,000 belong to small cities.

The karst collapse is mainly distributed in medium-sized cities. The distribution of karst collapse in large and medium-sized cities is relatively small, as shown in Table 3.

Table 2. Distribution of Karst Collapse in cities of the four regions.

Region	Cities with karst collapse	Total number
Eastern	Guangzhou, Longyan, Foshan, Tai'an, Yunfu, Tangshan, Tangshan, Linyi, Xuzhou, Jinan, Zaozhuang, Shaoxing, Nanjing, Huzhou, Wuxi, Suzhou, Jinhua, Sanming, Zhenjiang	229
Middle	Yongzhou, Chenzhou, Xiangtan, Tongling, Jiujiang, Yichun, Huangshi, Huainan, Shien, Jian, Shangrao, Loudi, Changsha, Anqing, Xiangning, Zhuzhou, Ezhou, Jingdezhen, Zhangjiajie, Pingxiang, Huaibei, Yueyang, Xinyu, Huaihua, Wuhu, Shaoyang	262
Western	Guilin, Hezhou, Yulin, Nanning, Liuzhou, Guigang, Qujing, Hechi, Kunming, Zunyi, Laibin, Guiyang, Liupanshui, Zhaotong, Chongzuo, Lhasa, Qinzhou, Tongren, Lijiang, Anshun, Yuxi, Bijie, Xingyi, Kaili	841
Northeastern		0

Note: The table "Yuxi (1)" said Yuxi's karst collapse is one.

Table 3. Distribution population of Karst Collapse in cities of different-sizes.

City size	Cities with karst collapse	Total number
Large	Nannig, Guangzhou, Hangzhou, Kunmig, Changsha, Jinan, Guiyang, Nanjing, Lhasa	147
Medium	Guilin, Hezhou, Yulin, Yongzhou, Liuzhou, Tai'an, Guigang, Foshan, Tangshan, Chenzhou, Xiangtan, Qujing, Jiujiang, Yichun, Huangshi, Huainan, Liyi, Shien, Ji'an, Zunyi, Laibin, Anqing, Liupanshui, Zhuzhou, Xianning, Xuzhou, Zaozhuang, Zhaotong, Ezhou, Pingxiang, Shaoxing, Huaibei, Qinzhou, Jinhua, Yueyang, Xinyu, Anshun, Huzhou, Wuxi, Suzhou, Bijie, Xingyi, Wuhu, Shaoyang, Zhenjiang	1069
Small	Longyan, Yunfu, Tongling, Hechi, Loudi, Zhangjiajie, Jingdezhen, Shangrao, Zongzuo, Tongren, Lijiang, Yuxi, Huaihua, Kaili, Sanming	116

Note: The table "Yuxi (1)" said Yuxi's karst collapse is one.

3 THE FORMATION CHARACTERISTICS OF KARST COLLAPSE

3.1 *Formation conditions of karst collapse*

Three conditions are indispensable for the formation of karst collapse, first is lower terrain, the lower strata with the presence of karst caves, which is the channels for the drain and transportation for groundwater and overlying soil, and accumulation places after collapse; second is certain thickness of the loose cap layer at the quaternary soil layer of the bed rock, not too thick but loose, which is the material source of collapse; third is the karst groundwater whose hydrodynamic conditions are easy to change, which is also the driving force and prerequisites of collapse.

3.2 *The main influencing factors of karst collapse*

3.2.1 *Developmental degree of Karst*

The developmental karst of the underlying bedrock is a necessary condition for the formation of karst collapse. The more the karst develops, the more favorable it is to form the karst collapse (Huang X F,1998). The upward-opening cave fissure is the window to accept collapse materials. So, the opening degree of the karst cave is another important factor for occurrence of collapse (Zhang L F et al., 2007).

3.2.2 *Quaternary soil characteristics*

The characteristics includes three aspects: soil structure, soil type and soil thickness (Huang X F, 1998).

Soil layer can be composed of clay, sand, gravel. From the type of soul, homogeneous sandy soil is the most prone to collapse, sand and gravel-like non-homogeneous soil followed. The higher the clay content is, the less favorable the collapse is caused under the same conditions, i.e. fine-grained soils are relatively difficult to collapse, sandy soil and gravel soil are easier. The thicker the soil, the longer it takes to form collapse and the larger the size of the collapse pit.

3.2.3 *Topographic features*

Different types of landforms and different geomorphological locations have a significant impact on the collapse, in general, negative terrain areas, peak cluster depression, peak forest plains are prone to collapse (Zhang L F et al., 2007).

3.2.4 *Hydrodynamic conditions*

Groundwater activities are mainly manifested by water level movements and changes in flow rates, flow volume and hydraulic gradients.

The presence of surface water is also conducive to the occurrence of subsidence, especially when the underlying bedrock karst is developed near the surface water, and the karst water level is maintained below the bedrock surface in the long time run. With the long-term infiltration of surface water and atmospheric rainfall, ground collapse will be formed (Tan N J & Wang N SH, 2002).

3.2.5 *Human activities*

There are two kinds of human activities that can induce or trigger karst collapse, one is to change the groundwater power conditions, the other is to change the existing soil hole stress conditions. Human activities that may affect the karst collapse include the extraction of karst water, mine water inrush, reservoir water storage, irrigation and surface load blasting. According to statistics, 67.87% of karst collapse in China is related to human activities.

4 SUMMARY

1. The karst collapse occurred in low-lying areas, and occurred along the fault-affected zone and anticline, diagonal axis and wings.
2. From the distribution of karst collapse in different geographical areas in China, the karst collapse occurred more in the southern districts where karst development exists, and in view of quantity, more collapse occurs in the medium-sized cities of the western region.
3. The developmental degree of karst collapse is closely related to the lithology, thickness and karst development degree of overburden layer. Karst collapse occurs more in the karst fissure where the overburden layer is loose and less thick.
4. The formation of karst collapse is influenced both by natural factors and human factors. Human activities are dominant, and groundwater activity is the most active factor in the formation of karst collapse. Therefore, in the dangerous area of karst collapse, good monitoring and forecasting is necessary, significant reduction in groundwater levels caused by underground engineering and the construction of basic engineering should be avoided.

REFERENCES

Huang X F. (1998) Basic features and development conditions of karst collapse in Yulin, Guangxi. Carsologica Sinica, 17(2), 111–118.

Liu B C, Song Q SH, Chen X J, et al. (2000) Analysis on the features of the karst collapse in western region of Guilin city. Journal of Geological Hazards and Environment Preservation, 11(3), 200–203.

Lu Y R. (2010) Karst in China-marvelous peak and isopropyl cave. Beijing, Higher Education Press.

Qian J P. (2007) Basic characteristics of karstic collapse in Guilin city and its prevention countermeasure. *Mineral Resources and Geology,* 21(2), 200–204.
Tan N J, Wang N SH. (2002) Relations between natural factor pumping factor and karst collapse. Guangxi Geology, 15(2), 65–68.
Zhang L F, Zeng X SH, Yao Y SH, et al. (2007) Review on karst collapse in China. The Chinese Journalof Geological Hazard and Control, 18(3), 126–130.
Zhu J. (2002) Characteristics and countermeasures of karst collapse in Guangxi province. Hydrogeology & Engineering Geology, 3, 75–76.
Zhu SH Z, Zhou J H, Chen X J. (2000) Analysis of forming conditions and main influential factors of karst collapse in west urban district Guilin city. *Journal of Guilin Institute of Technology*, 20(2), 100–105.

Land Reclamation in Ecological Fragile Areas – Hu (Ed.)
© 2017 Taylor & Francis Group, London, ISBN 978-1-138-05103-4

Reclaimed soil PSD and thermal property under machinery compaction

X.Y. Min, X.J. Li & Q.C. Li
College of Resources and Environment, Shandong Agricultural University, Taian, Shangdong, China

ABSTRACT: Soil Particle Size Distribution (PSD) and soil thermal property were observed under different machineries and compaction times, in order to find a change law of soil PSD and soil thermal property during a process of reclamation. Researches has shown that, 1) During compacting, a small range increase of pressure refined the soil particle size, but a greater pressure fluctuated it. 2) When simulating compaction of a crawler dozer, fractal dimensions increased and then decreased with the increasing of compaction time, while simulating compaction of a dump truck, the fractal dimensions fluctuated with the increasing of compaction time. 3) Both compaction machine and compaction time affected soil thermal conductivity, the compaction of machineries could affect soil thermal property through affecting PSD. Results indicated that compared with the dump truck, the soil PSD was more manageable when using the crawler dozer to compact, and the best compaction time was 3 times.

1 INTRODUCTION

In the process of reclamation, topsoil stripping has attracted a broad attention (Yang et al., 2010. Goosen, 2015. Lamb et al., 2015), however, an effect from the construction machinery to reclaimed soil is still a serious issue (Liu et al., 2014. Lu et al., 2015). Recently, experts and scholars at home and abroad have done a lot of studies on the reclaimed soil quality, Wang et al. (2014) studied a multi-fractal characteristics of reconstructed soil particle in opencast coal mine dump in loess area. Hang et al. (2014) studied characteristics of soil particles fractal dimension under different reclamation years in coal mining subsidence. Béasse et al. (2015) studied soil microbial communities identify organic amendments for use during oil sands reclamation. Lee et al. (2014) studied a influence of amendments and aided phytostabilization on metal availability and mobility in Pb/Zn mine tailings. However, related researches on change law of reclaimed soil quality in the process of compaction are still lacking. This research did a change law of soil texture during compacting, not only could help find the compaction mechanism, but also had guiding significance to reconstruct soil, it also did a study on thermal conductivity, which lays a foundation on heat conduction and moisture migration of reclaimed soil.

2 MATERIALS AND METHODS

2.1 *Experiment design*

Experimental zone was located in the experimental station of Shandong agricultural university. Digging 10 soil-bins separated by cardboards, whose volume are all 1 m^3, and filling 60 cm coal gangue at the bottom and covering 40 cm new-soil from the topsoil of the experimental field at the top. The basic physical properties of new-soil follows, the soil bulk density is 1.29 g/cm^3, the moisture content is 14.7%, besides, the volume fraction of clay is 0.28%, the silt is 41.18%, the very fine sand is 15.31%, the fine sand is 40.86% and the medium sand is 2.27%.

This experiment adopted a hammer, 25 kg of quality and 40 cm in diameter, to simulate the compacting effect from machinery to reclaimed soil. Two machineries simulated were the dump truck and the crawler dozer. Pressures from the hammer to the ground are different when it falling from different heights, so the heights could be calculated based on the pressures from different machineries. Thus if letting the hammer fall freely in design heights, a compacting effect from machineries could be simulated favorably. Besides, the simulated compaction times were 1, 3, 5, 7, 9. The experiment design is seen in Figure 1.

2.2 *Analytical method*

Collecting soil samples after 2 years' natural aging, and fishing the measurement of soil particle size distribution and soil thermal conductivity in lab.

Figure 1. Experiment design.
Note: T represent the plots that simulating the dump truck to compact. D represent the plots that simulating the crawler dozer to compact.

The soil samples were 0–20 cm topsoil, shifting them with 2 mm sieve after natural drying, putting. 0.5 g soil through using a electronic balance into 20 ml volumetric flask, adding 5 ml sodium oxalate of 0.25 mol/L as the dispersing agent, and diluting with water to 20 ml and shaking up, then adding the liquid into the groove of winner 2000 laser particle size analyzer and analyzing, the measuring range of this instrument is 0~0.3 mm. The analyzing instrument of soil thermal conductivity and temperature are is PC-2R soil thermal properties recorder, repeating 3 times of each processing.

2.3 *Soil single fractal dimension*

Adopting a model of soil particle size distribution to calculate soil single fractal dimension D (Tyler & Wheatcraft, 1992), the formula is:

$$\frac{V_i}{V_0} = \left(\frac{\bar{d}i}{\bar{d}\max}\right)^{3-D} \quad (1)$$

Type: $\bar{d}_{max}$—The biggest size of soil particle size grading,
$\bar{d}_i$—A certain size,
V_i—The total volume of soil particles whose sizes are less than R,
V_0—The volume of the soil particles,
D—The volume fractal dimension of soil particle.

Thus, establishing a coordinate system with $\lg(V_i/V_0)$ and $\lg(\bar{d}_i/\bar{d}_{max})$ for vertical and horizontal respectively and doing linear fitting, and getting D through the straight slope.

3 RESULT AND ANALYSIS

3.1 *PSD of reclaimed soil under simulated compacting*

According to the U.S. agriculture department soil texture classification standards (Grant, 1982), control soil and test soil were all belong to loam, which are easy to cultivate. When under the compaction of simulating the crawler Dozer (D), the content of silt was increasing (36.37%~60.10%) while the content of sand was decreasing (63.36%~39.78%) with the increase of the compaction time, and it showed that a small range increase of pressure refined the soil particle size, and the soil permeability was worse with the soil texture changed from sandy loam to silty loam; When under the compaction of simulating the dump Truck (T), the content of silt and sand fluctuated with the increase of the compaction time, the soil textures were changed among sandy loam, loam and silty loam (Table 1), and it showed that a bigger pressure to the ground had a serious and inhomogenous effect on soil particle and texture, which is a problem noticeably in the actual construction process.

Table 1. Soil texture under different processing.

Compaction time	Particle size distribution (%)			Soil texture
	Clay	Silt	Sandy	
D1	0.27	36.37	63.36	Sandy loam
D3	0.13	69.33	30.54	Silty loam
D5	0.07	67.47	32.46	Silty loam
D7	0.10	64.59	35.31	Silty loam
D9	0.12	60.10	39.78	Silty loam
T1	0.24	71.46	28.30	Silty loam
T3	0.09	24.82	75.09	Sandy loam
T5	0.20	47.47	52.33	Loam
T7	0.06	62.21	37.73	Silty loam
T9	0.25	24.98	74.77	Sandy loam
CK	0.28	41.18	58.54	Sandy loam

3.2 *Reclaimed soil fractal dimension of PSD*

Soil fractal dimension of PSD is a parameter that used to describe the effectiveness of soil particles' space. When under the compaction of simulating the crawler Dozer (D), the soil fractal dimensions reduced after increased first, followed by 2.757, 2.917, 2.914, 2.901 and 2.880 with the increasing of compaction time, and it reached the peak at 3 times' compaction, besides, there was a sharp fluctuation of the soil fractal dimension when the compaction times increased from 1 to 3; When under the compaction of simulating the dump Truck (T), the soil fractal dimension appeared a wavy fluctuation, followed by 2.928, 2.635, 2.805, 2.893 and 2.647 with increasing of compaction time, and it reached the peak at 1 times' compaction (Figure 2).

3.3 *Reclaimed soil thermal conditions under simulated compaction*

As seen on statistical results from multiple factors analysis of variance (Table 2), Compaction machine, compaction time and their interactions could cause significant differences to the soil thermal conductivity ($P < 0.01$). And from partial Eta square statistics, the partial Eta squares of compaction machine, compaction time and their interactions were 0.794, 0.617 and 0.814, which showed that compaction machine (79.4%), compaction time (61.7%) and compaction machine × compaction time (81.4%) were all main sources of the variation of soil thermal conductivity.

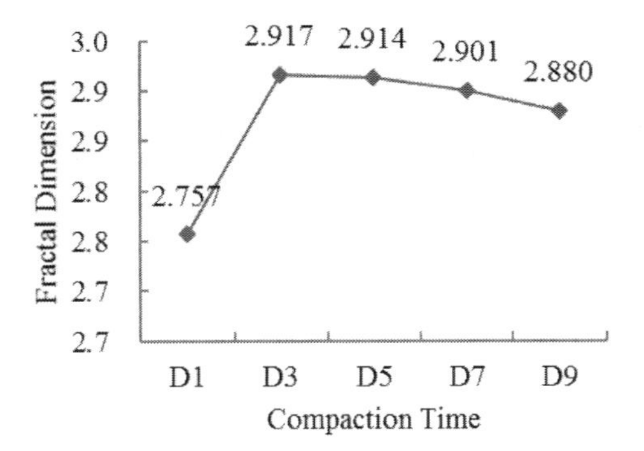

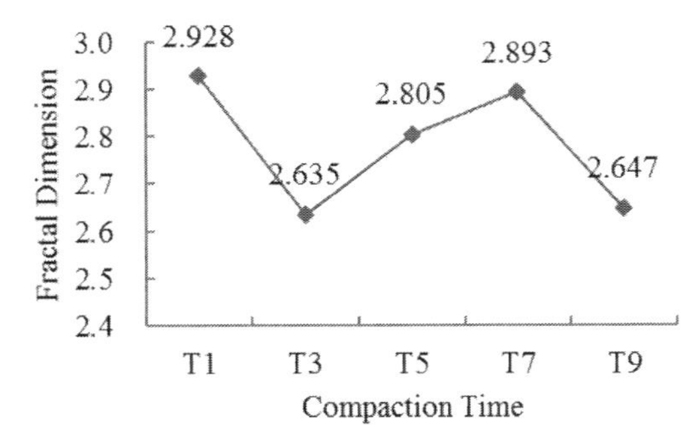

Figure 2. Soil fractal dimension under different compaction times.

Table 2. Effect factor analysis of soil thermal conductivity.

Source	df	Mean-square	F	Sig.	Partial Eta square
Compaction machine	1	0.411	76.894**	0.000	0.794
Compaction time	4	0.043	8.060**	0.000	0.617
Compaction machine × Compaction time	4	0.117	21.867**	0.000	0.814

Note: $R^2 = 0.908$ (Adjusted $R^2 = 0.866$); **represent $P < 0.01$.

When simulating the crawler Dozer (D), the soil thermal conductivities were 1.077,.1.366, 1.271, 1.134 and 1.048 W/m·k with the increasing of compaction times, and there was a highest soil thermal conductivity at 3 times' simulated compaction. When simulating the dump Truck (T), the soil thermal conductivities were 1.188, 0.717, 0.985, 1.032 and 0.804 W/m·k with the increasing of compaction times, and there was a highest soil thermal conductivity at first times' compaction. The comparison of soil thermal conductivity was

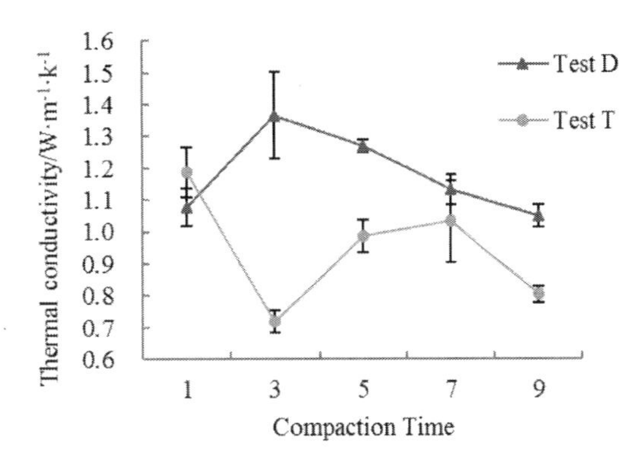

Figure 3. Test soil thermal conductivity.

the crawler dozer > the dump truck (Figure 3), which showed that soil still had a good thermal conductivity under small pressure compaction but the larger one would destroy this condition.

4 CONCLUSION

In a process of compaction, a smaller pressure on the ground could refine the sizes of soil particles, but the larger one would fluctuate them; When simulating the crawler dozer, soil fractal dimensions increased and then decreased with the increasing of compaction time, and the highest one, which was 2.917, appeared on 3 times' compaction; When simulating the dump truck, soil fractal dimensions fluctuated with the increasing of compaction time, and the highest one was 2.928 and appeared on first times' compaction; The comparison of soil thermal conductivity was the crawler dozer > the dump truck, which showed that soil still had a good thermal conductivity under small pressure compaction but the larger one would destroy this condition; Compared with the dump truck, the soil PSD was more manageable and the soil thermal property was better when using the crawler dozer to compact, and the best compaction time was 3 times.

REFERENCES

Béasse, M.L. Quideau, S.A. & Oh, S.W. (2015) Soil microbial communities identify organic amendments for use during oil sands reclamation. *Ecological Engineering*, 75, 199–207.

Deng, T.F. Liu, Y. Yan, Q.X. He, T.B. & Gao, A.Q. (2014) Mechanical composition and soil nutrient characteristics and their relationships in typical lonicera cinfusa soil of Guizhou. *Journal of Soil and Water Conservation*, 28 (05), 209–214.

Goosen, J. (2015) Topsoil Stripping and Management for Mine Rehabilitation. *Mining and the*

Environment, [Online] Available from: doi: 10.13140/RG.2.1.2432.0484 [Accessed 15th September 2015].
Grant, K. (1982) Applications of Soil Physics. *Journal of Applied Ecology*, 19 (1), 70–71.
Guan, X.Y. Yang, P.L. & Lv, Y. (2011) Relationships between soil patrical size distribution and soil physical properties based on multifractal. Transactions of the *Chinese Society for Agricultural Machinery*, 2 (3), 44–50.
He, S.Q. Gong, Y.B. Zheng, Z.C. & Kong, X.D. (2013) Changes and influences of soil anti-erodibility under different vegetation types. *Journal of Soil and Water Conservation*, 27 (05), 17–22.
Huang, X.N. Li, X.J. Liu, N & Min, X.Y. (2014) Characteristics of soil particles fractal dimension under different reclamation years in coal mining subsidence. *Journal of China Coal Society*, 39 (6), 1140–1146.
Lamb, D. Erskine, P.D. & Fletcher, A. (2015) Widening gap between expectations and practice in Australian minesite rehabilitation. *Ecological Management & Restoration*, 16 (3), 186–195.
Lee, S.H. Ji, W.H. Lee, W.S. Koo, N. Koh, I.H. Kim, M.S. & Park, J.S. (2014) Influence of amendments and aided phytostabilization on metal availability and mobility in Pb/Zn mine tailings. *Journal of Environmental Management*, 139, 15–21.
Liu, N. Li, X.J. Guo, B. & Min, X.Y. (2014) Simulation analysis on influencing factors of reclamation soil compation in mechanical compaction process. *Transactions of the Chinese Society of Agricultural Engineering*, 30 (1), 183–190.
Lu, Q.Q. Wang, N.H. & Chen, X.W. (2015) Effect of mechanical compaction on soil micro-aggregate composition and stability of black soil. *Transactions of the Chinese Society of Agricultural Engineering*, 31 (11), 54–59.
Tyler, S.W. & Wheatcraft, S.W. (1992) Fractal Scaling of Soil Particle-Size Distributions: Analysis and Limitations. *Soil Science Society of America Journal*, 56 (2), 362–369.
Wang, J.M. Zhang, M. Bai, Z.K. & Yang, R.X. (2014) Multi-fractal characteristics of reconstructed soil particle in opencast coal mine dump in loess area. *Transactions of the Chinese Society of Agricultural Engineering*, 30 (4), 230–238.
Yang, H.C. Shang, T. Zhou, W, Kun, X.U. & Zhang, G.Y. (2010) Integrating operation study of soil stripping and land reclamation in Zhungeer mine area. *Coal Technology*, 29 (09), 67–69.

Land Reclamation in Ecological Fragile Areas – Hu (Ed.)
 ISBN 978-1-138-05103-4

Bacterial community structure in reclaimed soil filled with coal wastes in different reclamation years

H.P. Hou, C. Wang, J.R. Li, Z.Y. Ding, S.L. Zhang, L. Huang, J. Dong,
J. Ma & Y.J. Yang
School of Environment Science and Spatial Informatics, China University of Mining and Technology, Xuzhou, China

ABSTRACT: Soil microbe is the core component of soil ecosystem and the soil microbial community structure plays a key role in the formation of soil fertility. Therefore, it is important to study soil microbial community structure and monitor the quality level of reclaimed soil, and it has practical guiding significance to improve the soil quality through microbial measures in ecological restoration. In this study, the soil samples (RS) from the reclaimed land filled with coal gangue with different reclamation years in Xuzhou mining area were compared with soil samples (CS) from normal farmland which were not affected by coal mining subsidence. Illumina PE250 sequencing method was used to determine the sequence numbers of six levels of bacteria phylum, class, order, family, genus and species bacterial and analyzed the vertical structure and time structure of bacterial community. (1) Compared with CS samples, the number of bacteria species and the bacterial diversity in RS samples decreased. (2) RS belonged to Firmicutes and Proteobacteria at the phylum level. After the disturbance of the reclamation project, Firmicutes increased and transferred from 20–40 cm to 0–20 cm soil layer. Bacilli had an absolute advantage at the class level. Quantity of Bacilli in the 0–20 cm soil layer was higher in RS than in CS samples, and with the increase of the reclamation years the quantity of Bacilli in 20–40 cm soil layers of RS samples decreased while in 0–20 cm soil layer had no changing trend. RS belonged to Lactobacillales and Bacillales at the order level. Desulfuromonadales played an important role in the remediation of heavy metal pollution and the quantity in 0–20 cm of RS was 74.81% –99.59% less than that in CS samples. (3) In RS samples, Bacillus, Enterococcus and Streptococcaceae were the dominant family, while Bacillus, Enterococcus and Lactococcus were the dominant genus, and Bacillus sp. JH7, Enterococcus faecium and Lactococcus piscium were the dominant species. All of these in RS samples were less in number than CS samples especially in 0–20 cm soil layer. The quantity of these in 20–40 cm soil layer of RS decreased with the increase of the reclamation years.

Keywords: Bacterial community structure; Coal gangue filling reclamation; Reclamation years; High-throughput sequencing

1 INTRODUCTION

Coal is the most important energy in China, and its large-scale exploitation inevitably brings a series of ecological and environmental problems. The damage of land resources in mining area is one of the most prominent problems (Bian et al., 2012), which has been highly concerned by scholars and governments at home and abroad. The Land reclamation and landscape reconstruction are the effective ways to control the land damage of coal mining (Bradshaw,1997; Bian 2000). Soil microbes are an important part of soil ecosystem. The structure and function of soil microbes play a key role in the decomposition and transformation of soil organic matter, nutrient cycling and utilization, soil fertility formation. The characteristics and changes of soil bacterial community structure can reflect the diversity and ecological function of soil microbial community in reclaimed soil (Luna et al., 2016; Dangi et al., 2012). The success of soil reclamation is the restoration of soil microbial community structure, in the natural state 5 to 14 years after the soil microbial community can be basically restored (Meng et al., 2016). Therefore, it is of great practical significance to study the changes of soil microbial community structure in reclaimed land and control land and its different reclamation years, and to improve the quality of reclaimed soil by using biological measures. In this study, we compared the coal gangue reclamation sites with different reclamation years in Xuzhou. The Illumina PE250 sequencing method was used to determine the sequence number of six levels of bacteria. The structure of the bacterial community in the reclaimed soil provide the theoretical basis for the bioremediation technology of the coal gangue filling reclaimed soil, and provide the basis for the improvement of soil quality in different reclamation years.

2 MATERIALS AND METHODS

2.1 *Overview of the study area*

The study area is located in Xuzhou City, Jiangsu Province. Three coal gangue filling and reclamation demonstration sites were selected, the first sample S was 1 year of reclamation, and the control sample was SC. The second reclamation sample was reclamation period of 6 years, and control sample is MC. The third reclamation sample L was reclamation period of 15 years, and the control is LC. The soil thickness of the coal gangue reclamation land was 50 cm, the ground crops were mainly cultivated with maize and soybean for two crops one year, and the soil was composed of the Yellow River alluvial. The soil type was ordinary cinnamon soil, soil PH was 6–7, the conductivity was 65.20–66.30.

2.2 *Soil sample collection*

Soil sampling was performed in June, 2016, coal gangue filling reclamation land, numbered S, M and L, the control sample was near the test site without coal mining collapse of the normal farmland, numbered SC, MC and LC.

2.3 *Sequencing of 16S rRNA gene of soil bacteria*

2.3.1 *DNA extraction and PCR amplification*

12 soil sample DNAs were extracted using the E.Z.N.A. Soil DNA kit (Omega Bio-tek, Norcross, GA, U.S.). The V4-V5 region of 16S rRNA for bacteria was subjected to PCR amplification. The parenting primers used were: 515F 5′-barcode-GTGCCAGCMGCCGCGG-3′ and 907R 5′-CCGTCAATTCMTTTRAGTTT-3′. PCR amplification procedure is as follows: an initial denaturation step of 95°C for 2 min, followed by denaturation at 95°C for 30 s, annealing at 55°C for 30 s, and elongation at 72°C for 30 s with a final extension step at 72°C for 5 min, a total of 25 cycles. The PCR reaction was carried out in triplicate in 20 μL of a mixture containing 4 μL of 5 × FastPfu buffer, 2 μL of 2.5 mM dNTP, 0.8 μL of each primer (5 μM), 0.4 μL of FastPfu polymerase and 10 ng of template DNA. The amplicons were extracted from 2% agarose gels and purified using AxyPrep DNA Gel Extraction Kit (Axygen Biosciences, Union City, CA, U.S. S.) and quantified using QuantiFluor TM-ST (Promega, U.S.A.).

2.3.2 *Library construction and sequencing*

The purified PCR product was quantified by 3.0 (Life Invitrogen) and mixed every 24 amplicons of different sequences. The pooled DNA product was used to construct the Illumina Pair-End library after the Illumina genomic DNA library preparation procedure. Pairwise sequencing of the amplicon library (2 × 250) was then performed on the Illumina MiSeq platform (Shanghai BIOZERON Co., Ltd.) according to the standard protocol. The original reading was stored in the NCBI Sequence Read Archive (SRA) database.

3 RESULTS AND ANALYSIS

3.1 *Contrastive analysis of bacterial community types in reclamation and control soil*

Bacterial species were found in 12 samples at the level of phylum, class, order, family, genus and species, as shown in Table 1. The results showed that the number of bacterial species in the reclaimed soils at each classification level was reduced compared to the normal farmland unaffected by coal mining collapse. In the sample with 1 and 6 years of reclamation period, the bacterial species of 0–20 cm soil layer was larger than 20–40 cm soil layer, while in the sample with 15 years of reclamation period, the bacterial species of 0–20 cm soil layer was less than 20–40 cm soil layer. And the results of control sample were compared with those of the reclaimed land. In the sample with 1 and 6 years of reclamation period, the bacterial species of 0–20 cm soil layer was less than 20–40 cm soil layer, while in the sample with 15 years of reclamation period, the bacterial species of 0–20 cm soil layer was large than 20–40 cm soil layer. So, the soil bacterial community structure changed after reclamation.

3.2 *Comparison of dominant bacterial community in reclaimed and control soil*

3.2.1 *Comparison of different soil layers in reclamation and control*

1 Phylum

Firmicutes and Proteobacteria is the dominant species at the phylum in the soil samples.

Table 1. Number of bacterial species in sample.

Classification level	S-1	S-2	SC-1	SC-2	M-1	M-2	MC-1	MC-2	L-1	L-2	LC-1	LC-2
Phylum	25	22	30	32	26	27	32	30	18	29	31	29
Class	58	45	65	70	60	56	68	65	38	63	69	66
Order	116	86	138	140	123	117	139	134	72	127	139	135
Family	203	138	244	243	213	204	244	232	117	223	239	233
Genus	442	216	618	603	469	416	589	551	178	488	548	518

Note: -1 represents 0–20 cm soil layer, -2 represents 20–40 cm soil layer.

Table 2. Number of sequence of the main bacteria in soil sample.

	S-1	S-2	SC-1	SC-2	M-1	M-2	MC-1	MC-2	L-1	L-2	LC-1	LC-2
Firmicutes	25129	32864	9496	9617	20961	26091	8189	15201	34487	24295	4830	23348
Proteobacteria	6571	2499	11384	11641	7203	3613	11428	7848	2156	7111	14903	9258
Bacilli	23967	31378	8235	8438	19883	24746	7224	14140	32881	22838	3806	21916
Lactobacillales	12575	16261	4230	3441	10410	12949	3400	7284	17344	11929	1872	11449
Bacillales	11393	15117	4005	4997	9474	11798	3824	6856	15537	10909	1934	10467
Bacillaceae	10132	13207	3356	4107	8249	10128	3174	5941	13547	9525	1592	9128
Streptococcaceae	6160	7590	2094	1533	4926	6052	1654	3517	8178	5955	864	5590
Enterococcaceae	5894	7870	1951	1739	4991	6273	1534	3391	8319	5412	912	5301
Bacillus	10058	13119	3306	3866	8184	10058	2997	5798	13450	9462	1574	9071
Enterococcus	5894	7870	1951	1739	4991	6273	1534	3391	8319	5412	912	5301
Lactococcus	5830	7128	1973	1426	4663	5692	1554	3317	7711	5641	816	5255
Bacillus sp. JH7	9223	12047	2908	2489	7490	9230	2518	5192	12371	8726	1430	8342
Enterococcus faecium	5894	7870	1951	1739	4991	6273	1534	3391	8319	5412	912	5301
Lactococcus piscium	5520	6719	1855	1348	4394	5371	1464	3117	7278	5361	770	4965

The relative abundance of them were shown in Table 2. The results showed that the advantage of Firmicutes in the reclaimed soil was slightly larger than that in the soil Proteobacteria, the relative abundance of the two soils in the normal soils varied with the soil nutrient conditions, and the difference between the two was not significant. The relative abundance of Firmicutes in the 0–20 cm soil layer of the reclaimed soil less than 20–40 cm soil layer in the reclamation period of 1 year and 6 years is, and the relative abundance of Firmicutes in the 0–20 cm soil layer was greater than 20–40 cm soil layer in the reclamation period is 15 years. Proteobacteria is gram-negative bacteria, is the largest number of bacteria, can be cultivated for the most, the highest metabolic diversity of the phylum. Agricultural significance of the bacteria are derived from the strain of bacteria, such as nitrogen and bacteria and legumes symbiotic formation of nodules. Mycobacterium played an important role in maintaining soil ecosystem functions and helped the cycle of nitrogen and energy throughout the system. According to the results of the experiment, the proportion of Proteobacteria in 0–20 cm soil layer or 20–40 cm soil layer was lower in the reclaimed soil in this study area, and the recovery of soil microbial community structure could be achieved by applying the bacteria fertilizer.

2 Class

At the level of the class, Bacilli in the reclaimed land samples is an absolute dominant species (Table 2). The number of Bacilli in the reclaimed soil was significantly higher than that in the normal farmland. Bacilli has the characteristics of fast metabolism, fast propagation, strong acid resistance, strong alkali resistance, high temperature resistance and high volume. Compared with other bacteria classes, Bacilli possesses the conditions for survival in the barren soil environment such as reclaimed soil, so Bacilli occupies an absolute advantage in the reclaimed soil, and Bacilli's dominance is not significant in normal farmland. In the samples with 1 year and 6 years of reclamation, the number of Bacilli in 0–20 cm soil layer is less than 20–40 cm soil layer. In the sample with 15 years of reclamation, the number of Bacilli in 0–20 cm soil layer is greater than 20–40 cm soil layer. The number of Bacilli in the soil at the vertical soils of 0–20 cm and 20–40 cm of the soil was different depending on the nutritional conditions of the study sample area.

3 Order, family, genus, species
Lactobacillales and Bacillales are dominant species at the order (Table 2). At the level of family, Bacillus, Bacillaceae, Streptococcaceae and Enterococcaceae were the dominant species. In the genus, Bacillus, Enterococcus and Lactococcus were dominant species. In the species level, Bacillus sp. JH7, Enterococcus faecium, Lactococcus piscium were dominant species. In the reclaimed soil, the dominance of Lactobacillales and Bacillales was equal, the dominance of bacteria at the family was Bacillaceae> Streptococcaceae> Enterococcaceae, the dominance of bacteria at the genus was Bacillus> Enterococcus> Lactococcus, the dominance of bacteria at the species was Bacillus sp. JH7> Enterococcus Faecium> Lactococcus piscium. And the relative abundance of the dominant species of order, family, genus, and species level was independent of 0–20 cm soil layer or 20–40 cm soil layer. In the normal farmland, the relative abundance of the dominant species in the 0–20 cm soil layer and 20–40 cm soil layer changed with the different nutritional conditions of the sampling sites. Bacillus is a special kind of physiological structure, which can resist the extreme environment, and it is a kind of micro organism with high heat, low temperature, radiation and other harsh environments. Thus, this physiological characteristic can explain the fact that Bacillus is the largest proportion of reclaimed soil, and Bacillus subtilis—JH7 inherits the characteristics of Bacillus, which occupies a great advantage in the level of bacteria. However, in the agricultural soil improvement, bacterial order, family, genus and species have little research and application. The relative abundance of the dominant species had nothing to do with the vertical change of soil, because the difference of soil nutrient condition at 0–20 cm and 20–40 cm at the vertical level was not obvious, or the difference was not enough to cause the above several advantages of the relative abundance of bacteria changes.

3.2.2 *Comparison of different reclamation years*

By analyzing the composition of different samples of OTU (97% similarity), the Principal Component Analysis (PCA) was carried out on the soil microbial community data of 6 plots. Samples from different environments show a decentralized or aggregated distribution, with the results shown in Fig. 1. The credibility of the PC1 axis is 97.12%, which explains most of the differences in all variables. PC2 axis of the credibility of 1.01%, the cumulative contribution rate of 98.13%. The results of principal component analysis showed that the difference between the soils of the reclaimed soil and the soil was significant, which

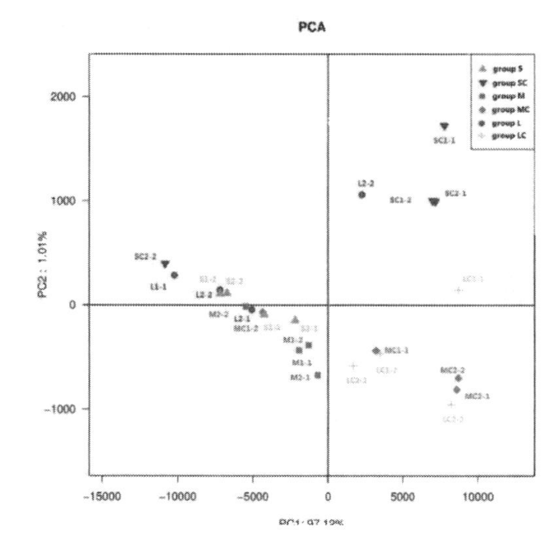

Figure 1. PCA of soil samples.

indicated that the effect of different reclamation years on soil microbes was better reflected in PC1. The results were consistent with the results of cluster analysis between different samples. After the mining of coal resources, land subsidence, coal gangue filling reclamation project and other serious disturbance process, the soil bacterial community structure changed greatly. At the same time, on the PC1 axis, the A, F and D samples of the reclaimed land were close to the value of 0, and the reclamation period of L, M and S reclamation land was 15 years, 6 years, 1 Year, indicating that over time, the microbial community structure of reclaimed soils will gradually return to normal soil levels. The results showed that with the increase of the reclamation period, the number of bacterial dominant bacteria in the reclaimed soil was the same, and the proportion of the bacteria was slightly changed.

3.3 *Contrastive analysis of soil bacterial community diversity in reclamation and control*

The sequence of microbial gene in all soil samples was determined by Illiumina high throughput sequencing technique. The diversity and abundance of soil microbial community were reflected by Shannon index and Chao index, as shown in Fig. 2. The results showed that the diversity and abundance of bacterial communities in the reclaimed soil decreased by 33.39%–66.94% and 34.75%–85.32% in the 0–20 cm soil layer, and decreased by 13.41%–65.6 2% and 12.43%–84.80% in the 20–40 cm soil layer.

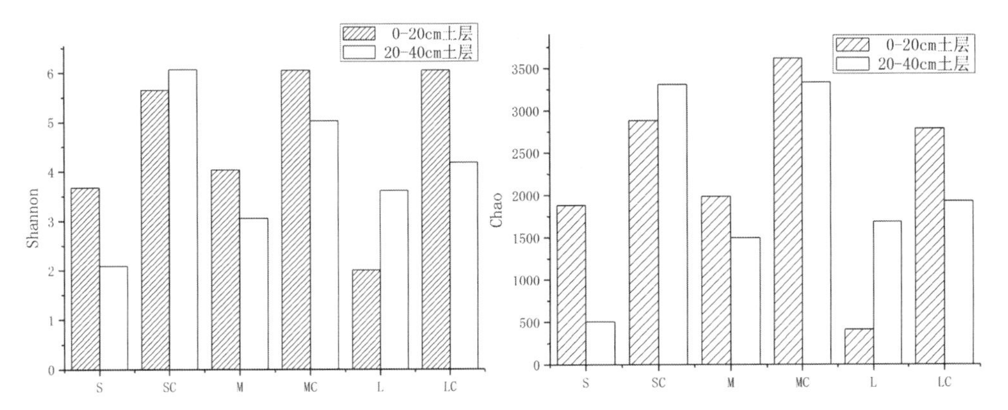

Figure 2. Shannon and Chao index of bacterial community in each vertical layer under 97% similarity level.

4 CONCLUSION AND DISCUSSION

This paper took the coal gangue filling and reclamation area in northern Xuzhou as the research object and analyzed the correlation between soil biota community structure and dehydrogenase activity in different reclamation years. The conclusions were as follows:

1. Compared with the normal soil in the farmland which was not affected by coal mining subsidence, the number of bacteria in the reclaimed soils decreased and the diversity of the communities decreased.
2. Firmicutes and Proteobacteria were the dominant bacteria group of the soil at the phylum in study area; Bacilli occupied the absolute advantage at the class; Lactobacillales and Bacillales were the dominant community at the order; Bacillaceae, Enterococcaceae and Streptococcaceae were the dominant community at the family; Bacillus, Enterococcus and Lactococcus were the dominant community at the genus; Bacillus sp. JH7, Enterococcus faecium and Lactococcus piscium were the dominant community at the species.
3. The improvement of soil ecological environment should pay attention to soil microbial diversity and increase the quantity of organic matter and beneficial bacteria in the soil by increasing the use of bio organic fertilizer or specific microbial fertilizer, so as to improve the quality of the soil. By comparing the characteristics of bacterial community in reclaimed soil and normal farmland soil, it was found that the reclamation of unstable soil ecosystem in the lack of certain bacteria and rate of soil bacterial community succession and the stability of soil ecosystem can be accelerated by the way of sampling artificially adding bacteria. It is of great significance to guide soil ecological restoration.

ACKNOWLEDGMENTS

This research was funded This study was supported by the National Natural Science Foundation of China (No. 51474214). The authors are thankful to Shanghai BIOZERON Co. Ltd. for test data.

REFERENCES

Bian Z, Miao X, Lei S, et al. The challenges of reusing mining and mineral-processing wastes[J]. Science, 2012, 337(6095): 702–703.

Bian Zhengfu. A review of research on land reclamation at home and abroad [J]. China Land Science, 2000, 14(1): 6–11. (in Chinese with English abstract)

Bradshaw A. Restoration of mined lands—using natural processes [J]. Ecological Engineering, 1997, 8(4): 255–269.

Dangi S R, Stahl P D, Wick A F, et al. Soil microbial community recovery in reclaimed soils on a surface coal mine site[J]. Soil Science Society of America Journal, 2012, 76(3): 915–924 Dong.

Luna L, Pastorelli R, Bastida F, et al. The combination of quarry restoration strategies in semiarid climate induces different responses in biochemical and microbiological soil properties [J]. Applied Soil Ecology, 2016, 107: 33–47.

Meng Huisheng, Hong Jianping, Yang Yi, et al. Effect of applying phosphorus bacteria fertilizer on bacteria diversity and phosphorus availability in reclaimed soil[J]. Chinese Journal of Applied Ecology, 2016, 27(9): 3016–3022. (in Chinese with English abstract).

Land Reclamation in Ecological Fragile Areas – Hu (Ed.)
© 2017 Taylor & Francis Group, London, ISBN 978-1-138-05103-4

Research on caving deep ore mining without surface subsidence

Y. Wang & W.B. Sun
School of Energy Engineering, Xi'an University of Science and Technology, Xi'an, Shanxi, China

ABSTRACT: For the large iron mines buried more than 1500 meters deep, the sublevel caving method is the most economical and reasonable, but the caving mining will cause the surface subsidence, brings a series of environmental problems. According to the char-acteristics of deep buried ore, put forward a kind of caving mining, cut shallow protective layer and monitoring collapse area, utilization of tailings and waste rock filling protective layer and protect the environment from damage. Both the use of caving method has the advantages of high strength, small mining cross engineering quantity and avoided the caving method of surface subsidence, at the same time, shallow filling not only uses the waste rock and tailings, but also simplifies the filling process, short-en the filling distance, reduce the cost of filling.

1 INTRODUCTION

In recent years, More and more 1 kilometers of iron mines have been found, such as Da Taigou iron mine, Si Shanling iron mine, ore bodies are in the maximum depth of 2000 meters. In the future, more than 1 kilometers of deep mining iron mine will be more and more. At the same time, deep mining has also increased the cost of mining, the need to adopt the most efficient economic mining methods. Caving method is an efficient method for metal mine, it is more suitable for iron and manganese metal mines than filling method. But use the caving method, surface subsidence must be allowed, with the mining, goaf surrounding rock failure, will appear in the surface collapse pit (Li et al., 2009). If the deep mining caused subsidence, affected areas will be huge, so we adopted the caving method in deep mining, we need to take the necessary measures to control the goaf surrounding rock damage, eliminate the surface subsidence situation (Li et al., 2007). At present, many achievements have been made in the study of the collapse area (Quan, 2008) (Hu et al., 2014) (Li et al., 2012), but there is little research on how to eliminate the surface subsidence.

1.1 Stability analysis of goaf

Caving mining caused overburden caving is a complicated problem of space and time (Wang et al., 2015). Generally, there are 3 kinds of stability state of the surrounding rock in the goaf, stable, unstable and metastable states (Jiang et al., 2016). In a stable state, the formation of a balanced arch in the surrounding rock, the stability of the surrounding rock support goaf, the surrounding rock is not moving in a static state. When the state is unstable, the spatial span of goaf exceeds the stability of surrounding rock, the surrounding rock cannot be self-stabilized for a long time. The metastable state is between stable state and unstable state, and the movement of the surrounding rock is small, which does not affect the continuous use of the underground engineering and surface facilities, and does not destroy the surface environment. The formation of the caving mining goaf, stability of surrounding rock is changing in the three condition. If the state of the granular in the goaf is changed, and the surrounding rock is fully supported by the granular to keep the stable or metastable state of the goaf, it will control surface without collapse.

In goaf, the granular contine development with mining, became the formation witch loose at ends and dense in middle. In order to maintain the stability of the surrounding rock in the goaf, it is necessary to support the surrounding rock under the premise that the granular have good mobility in the process of moving down. The supporting effect of the particles on the surrounding rock can be attributed to the lateral pressure of the surrounding rock and the pressure can be reflected by the lateral pressure coefficient.

After the continuous treatment of granular, in the limit equilibrium state, the active stress state of the solid particles can be expressed as:

$$\sigma_z = \gamma z \tag{1}$$

$$\sigma_x = \gamma z \frac{1 \mp \sin\varphi}{1 \pm \sin\varphi} \tag{2}$$

Table 1. Micro-parameters used to represent rock mass.

Type	Meso parameters	Number
Granular	Size ratio Rmax/Rmin	3.5
	Density (kg/m^3)	1550
	E(Pa)	6.0e7
	Normal/tangential stiffness ratio	0.45
	Friction coefficient	0.7
	Normal strength/ standard deviation (Pa)	4.5e1/0.1e1
	Tangential strength/ standard deviation (Pa)	5.0e1/0.1e1

According to the measured results of a certain iron mine, at loose state, it is concluded that the natural repose angle of pranular is about 37 degrees. The upper and lower operational symbols are used to characterize the active and passive stresses respectively. The internal friction angle should be close to the natural repose angle of about 37 degrees, and the maximum value of the passive compressive stress is about 17.5 times of the active compressive stress, the passive compressive stress coefficient:

$$\eta_1 = \frac{1+\sin\varphi}{1-\sin\varphi} = \frac{1+\sin 37^\circ}{1-\sin 37^\circ} = 4.03 \tag{3}$$

The iron ore rock broken expand coefficient is 1.45, for $4.03/1.45 = 2.78 > 1$. Obviously, the largest passive pressure of granular is much greater than the maximum principal with rock thickness entity gravity field stress. According to this, it is concluded that the passive compressive stress of the granular in the goaf should be the main force to prevent the surrounding rock from expanding and limiting the falling of the surrounding rock.

In order to observe the comparison of the passive lateral pressure and the lateral pressure of the surrounding rock, the numerical simulation of the stress change of the surrounding rock is carried out. According to the characteristics of the goaf, the calculation parameters of the granular are determined as shown in Table 1.

The active compressive stress of the granular can not prevent the falling of the surrounding rock, and the surrounding rock will produce passive compressive stress when the chip is falling. A numerical model of high 30 m, wide 15 m, dip angle of 70 degrees, 75 degrees, 80 degrees, 85 degrees and 90 degrees is established by using PFC discrete element software. As shown in Figure 1, the grain size of the filled particles is consistent with the field measurement. In the initial equilibrium model, a monitoring and recording of xforce wall on the side, the force monitoring on the horizontal direction, the force changes to the stability of surrounding rock and rock observation falling film. After set the record points, to 500 iterative steps, the setting plate wall can move to the granular direction at a certain speed, simulation of surrounding rock fall, to 500 iterative steps. Can get on side wall of active and passive state under x direction stress curve, as shown in Figure 1 shows.

The results appeared by the stress map of each angle, dispersion balance after the plate wall stress in x direction remains unchanged, this stage is

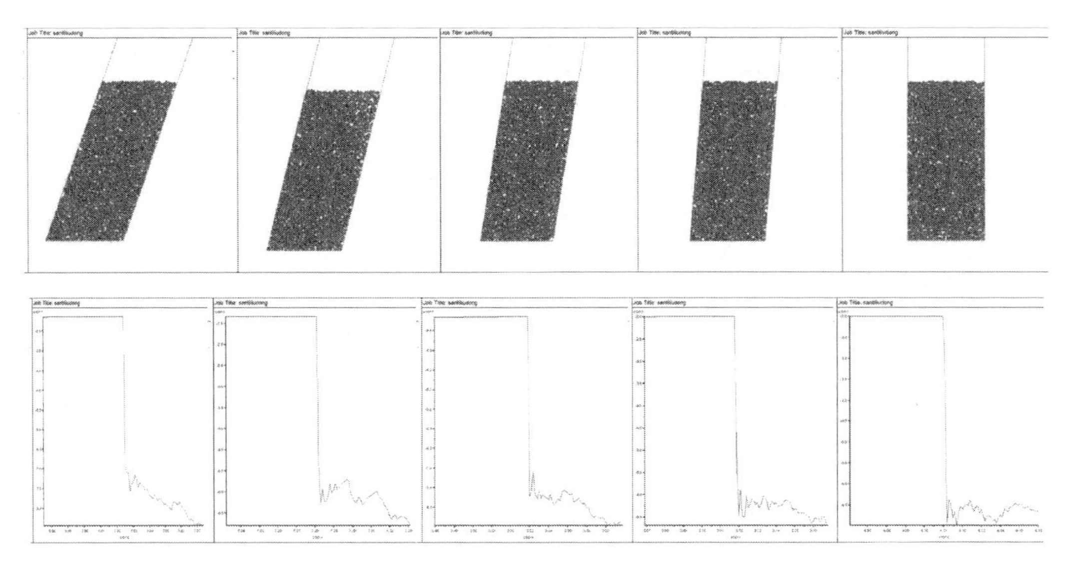

Figure 1. Side wall stress condition.

the active pressure stage; when the side wall of a moment, dispersion is compressed, than the active stress much passive lateral stress.

From the numerical simulation process, when the active lateral dispersion to the goaf surrounding rock pressure is not enough to hinder the rock fall, rock produced greater pressure on the passive side to prevent falling, active lateral pressure and passive lateral pressure together prevent the falling of the rock. At the same time, it shows that the inclination angle is big and small, and when the surrounding rock is falling, the passive side pressure is very large, which plays a leading role.

2 GOAF TREATMENT SCHEME

According to the above analysis, the use of caving mining in deep mining, after the formation of covering layer, granular filling chamber is arranged in the roadway and the sides of the covering layer, covering layer in a small space as a protective layer, a sensor on the inside, the underground waste rock dump, waste rock and other solid dispersions as granular fill in the top cover layer, which layer height is basically unchanged, which can guarantee the goaf surrounding rock has always been dispersion of effective support, in a stable or metastable state of goaf surrounding rock does not fall, and he surface will not collapse.

3 CONCLUSION

1. The active side pressure which the granular in goaf is not enough to hinder the rock fall, when a rock falling trend, the passive side pressure dispersion to prevent fall, active and passive side pressure prevent the piece of rock block together.
2. When the surrounding rock is falling, the passive side pressure is very large, and it plays a leading role in preventing the falling of the surrounding rock, little relationship with ore body dip.
3. Through supply the granular to the above of the caving mining covering layer, maintain the space of the protective layer above the granular, can effectively control the goaf surrounding rock stability, eliminate surface subsidence.

REFERENCES

Hu jingyun & Li Shulin & Lin Feng. (2014) Research on disaster monitoring of overburden ground pressure and surface subsidence in extra-large mined-out area. *Rock and Soil Mechanics*. (4), 1117–1122.

Jiang Renyi & Zhang Mingfeng & Wu Peng. (2016) Study on β caving of large scale goaf in underground mine. Journal of Safety Science and Technology. (S1), 13–17.

Li Haiying & Ren Fengyu & Chen Xiaoyun. (2012) The method for predicting and controlling the range of surface subsidence during deep ore body mining. *Journal of Northeastern University*. 33(11), 1624–1627.

Li Wenxiu & Dai Lanfang & Hou Xiaobing. (2007) Fuzzy genetic programming method for analysis of ground movements due to underground mining. *International Journal of Rock Mechanics and Mining Sciences*. 44(6), 954–961.

Li Wenxiu & Guo Yugui & Hou Xiaobing. (2009) A Visco-elastic Model for Analysis of Ground Subsidence Due to Underground Mining by Pillarless Sublevel Caving Method. *Engineering Mechanics*. (07), 227–231.

Quan Guiguang. (2008) Mine mined-out area of surface subsidence area forecast. *Non-ferrous Mining and Metallurgy*. 24 (5), 11–12.

Wang Yunmin & Lu Yugen & Sun Guoquan. (2015) Study on the Law of Rock Movement and Surface Subsidence by Deep Mining with Sublevel Caving. *Metal Mine*. (06), 6–9.

Land Reclamation in Ecological Fragile Areas – Hu (Ed.)
 ISBN 978-1-138-05103-4

Aggregate stability and organic carbon distribution in reconstruction soil filled with gangue from coal mining areas

Z.Y. Hu
School of Earth and Environment, Anhui University of Science and Technology, Huainan, China

X.Y. Chen
School of Earth and Environment, Anhui University of Science and Technology, Huainan, China
Engineering and Technology Research Center on Coal Exploration of Anhui, Suzhou, China
School of Earth and Space Sciences, University of Science and Technology of China, Hefei, China

M. Chen, Y.Z. Zhou & F. Wang
School of Earth and Environment, Anhui University of Science and Technology, Huainan, China

ABSTRACT: Experimental site were selected from reclamation soil filled with coal gangue of two typical coal mining subsidence areas in Huainan, China. To discuss mechanical stability aggregate, water stable aggregates size distribution and soil organic carbon content in different coversoil thickness. Results showed that the composition proportion of the mechanical stable aggregate was in shape of "W" with decreasing size fraction, the percentage of size fraction >3.2 mm accounted the largest proportion and existed significant difference from that of other size fraction ($P < 0.05$), the water stable aggregate is similar to the mechanical stable aggregate. Mean weight diameter of soil water stable aggregate, the content of total soil organic carbon and each size fraction of Water-stable aggregate and aggregate-associated C in ecological remediation areas were all less than that of the control area. The thickness of 60–80 cm would promotes soil aggregate formation, the accumulation and stability of organic carbon.

Keywords: Soil aggregate, Soil organic carbon, Reconstruction soil, Coal mining subsidence areas, Covering soil thickness

1 INTRODUCTION

The surface soil resources is seriously damaged, including the destruction of soil structure, soil environment changes, water-soil loss, and so on (Zou et al., 2014, Cheng et al., 2014, and Chen et al., 2016). Coal gangue as base matrix are often used from local government departments or mining enterprises to repair ecological environment and restore the mining area land productivity by reconstruct soil profile in China (Hu et al., 2005). However, Forming naturally stable soil structure of the reconstructing soil profile need long time due to the special properties of the filling matrix and human disturbance behavior (Chen et al., 2016, and Li et al., 2015), thus, we should design science soil profile reconstruction technology to guarantee the vegetation and ecological system of the area repair fast and stable. Soil aggregate structure and its stability play an important role, which it can effectively effect the soil carbon storage and organic carbon conversion (Wei et al., 2011).

Soil Aggregate (SA) is a secondary particles formed by a number of relatively stable soil mineral particles under the action of biological and non-biological cementing material, which not only is the main carrier of soil material and energy exchange, but also the main active site of soil micro organisms (Liao et al., 2015). The quantity and quality of SA can affect on the soil water thermal characteristics, soil fertility and soil sustainability (Wei et al., 2011, and Wei et al., 1990). The diameter of the SA is generally between 10 and 0.25 mm. Soil organic carbon content was positively correlated with large aggregate, where negatively correlated with micro aggregate in the forest plantations (Yu et al., 2013). The soil water stable aggregate composition change is closely related to the soil environment, and the long term no-till infiltration conditions is more advantageous to the formation of water stable aggregates (Wei et al., 2011). Fine SA structure can be satisfied the plant growth process due to suitable moisture, nutrients, temperature and gas conditions (Huang et al., 2012). At present, SA and its associated carbon from reclamation soil filled with coal gangue in mining subsidence areas are less reported, and the reconstruction soil aggregate and its organic carbon content are different with natural soil according vegetation types and cover soil thickness in coal mining subsidence areas.

2 MATERIALS AND METHODS

2.1 *Site description and sampling*

The experiments were carried out at two typical ecological restoration area of mining subsidence area filling with coal gangue in Huainan, Dongchen ecological park (A) and Panyi mine ecological restoration area (B) respectively. According to the thickness of overlying soil, the two sites are divided into different sampling areas, 20 cm, 20~40 cm, 40~60 cm, 60~80 cm, >80 cm respectively. However, the overlying soil of Dongchen ecological park is higher than 20 cm, so that we set up 2 sampling areas where the thickness of overlying soil is higher than 80 cm. In addition, selecting a natural soil as control area (CK) at nearby Dongchen ecological park. Every sampling area sets three sampling points, which 0~20 cm depths surface soil were collected. The soil samples were air-dried and sieved through a 0.2 mm mesh to measure the content of soil organic carbon, which determined by redox reaction with potassium dichromate titration (Lu., 2000).

2.2 *Measurements*

The machinery-stable aggregate was measured by dry sieving method, while the water-stable aggregate was measured by wet sieving methodthen weigh the weight after drying at 60°C for 48 h. Water-Stable Aggregate Rate (WSAR), machinery-stable aggregate average weight diameter (DMWD) (Van., 1950), machinery-stable aggregate geometric mean diameter (DGWD) (Li et al., 2015), water-stable aggregates average weight diameter (WMWD), water-stable aggregate geometric mean diameter (WGWD) soil structural failure rate (PAD) (Chen et al., 2012) and soil aggregate index (Huang et al., 2012) (ELT) were used to measure the stability of SA, the equations are as follows:

$$WSAR = \frac{WSA}{A} \times 100\% \tag{1}$$

where WSA is the quality of water-stable aggregates that the diameter is >0.25 mm, A is the quality of soil aggregates that the diameter is >0.25 mm.

$$MWD = \sum_{i=1}^{n} \bar{x}_i w_i \tag{2}$$

$$GWD = \exp\left[\frac{\sum_{i=1}^{n} w_i \ln \bar{x}_i}{\sum_{i=1}^{n} w_i}\right] \tag{3}$$

$$PAD = 100\left[\frac{(DA_{0.25} - WA_{0.25})}{DA_{0.25}}\right] \tag{4}$$

where MWD is aggregate average weight diameter, GWD is aggregate geometric mean diameter, X_i is average diameter of aggregates at any particle size, W_i is percentage of aggregate weight corresponding to X_i, $DA_{0.25}$ is content of machinery-stable aggregate that the diameter is >0.25 mm, $WA_{0.25}$ is content of water-stable aggregate aggregates that the diameter is >0.25 mm.

$$E_{LT} = \frac{WT - W_{0.25}}{WT} 100\% \tag{5}$$

where WT is total soil sample weight, $W_{0.25}$ is water-stable aggregate quality.

The contribution rate of soil organic carbon to soil organic carbon at every level (AOCCR) is calculated by the following formula:

$$\text{ACOOR} = \frac{\text{organic carbon content} \cdot \text{aggregate content}}{\text{Soil organic carbon content}}$$

3 RESULTS AND DISCUSSION

3.1 *Effect of cover soil thickness on soil aggregate of reconstruction soil*

With the increase of the soil thickness, the change trend of soil mechanical-stable approximately is increasing firstly and then decreasing. The mechanical-stable aggregate content of sampling area of cover soil thickness at 40 to 60 cm sampling plot that the content of large-size (>3.2 mm) is decreasing and small-size (<0.25 mm) significantly increasing (Table 1) is different with other sampling area, while there is no obvious regularity in water-stable aggregate change (Table 2). Compared with the CK, the contents of water-stable aggregate is higher at two study area except the diameter >2 mm and <0.053 mm. In general, soil aggregate mainly exists in the form of large aggregate.

3.2 *The distribution of organic carbon of water-stable aggregate in reconstituted soil*

The total soil organic carbon in A and B was both smaller than CK, but compared with B,A have more total organic carbon content and organic carbon content in water stable aggregates (Table 3). It indicated that vegetation play a role in soil organic carbon content in reconstituted soil, woodland is more beneficial to organic carbon accumulate than

Table 1. Machinery-stable aggregate composition of reconstruction soil under different cover soil thickness.

Covering soil thickness/cm	Each grade of soil aggregate/%						
	>3.2 mm	2–3.2 mm	1–2 mm	0.5–1 mm	0.25–0.5 mm	<0.25 mm	>0.25 mm
0–20	66.96	10.71	9.95	5.49	2.05	4.85	95.16
20–40	68.62	8.72	9.53	6.03	2.3	4.81	95.20
40–60	55.97	11.27	13.17	9.79	3.81	5.98	94.01
60–80	80.26	6.07	5.69	3.64	2.01	2.87	97.67
>80	71.93	7.56	8.73	6.09	2.42	3.28	96.73

Table 2. Water-stable aggregate composition of reconstruction soil under different covering soil thickness.

Covering soil thickness/cm	Each grade of SA/%							
	>2 mm	1–2 mm	0.5–1 mm	0.25–0.5 mm	0.1–0.25 mm	0.053–0.1 mm	<0.053 mm	>0.25 mm
0–20	24.50	12.88	15.98	13.80	9.37	3.29	20.18	67.16
20–40	24.00	4.23	11.63	13.36	13.03	4.69	29.06	53.22
40–60	26.66	8.16	16.29	14.03	9.19	3.05	22.62	65.14
60–80	25.63	10.49	12.75	6.88	5.83	3.06	35.36	55.76
>80	23.13	8.93	16.89	12.75	7.36	3.63	27.31	61.70
CK	36.03	4.44	4.90	4.45	3.69	1.85	44.64	49.82

Table 3. Effect of vegetable on organic carbon content in water-stable aggregate.

Sampling area	Total organic carbon	The organic carbon content in water-stable aggregates/g · kg^{-1}					
		>2 mm	2–1 mm	1–0.5 mm	0.5–0.25 mm	0.1 0.25 mm	0.053–0.1 mm
A	6.37	7.07	5.21	4.29	5.72	5.38	5.78
B	5.51	4.34	4.24	3.73	4.52	4.45	4.41
CK	8.49	7.31	8.41	10.60	11.73	11.46	14.23

grassland. In addition, the decomposition rate of soil organic carbon was faster than natural soil in reconstituted soil. Soil organic carbon content were positively correlated with clay and sand content and soil organic carbon is mainly distributed in the small-size aggregate (Christensen, 1986).

The overlying soil thickness has a significant effect on the organic carbon content in water-stable aggregate. Figure 1 shows that the organic carbon content in water-stable aggregate are higher than other overlying soil thickness at 40–60 cm and 60–80 cm overlying soil thickness. When the overlying soil thickness at 0–20 cm, the organic carbon content is the lowest in 1–0.5 cm SA, while the highest in 0.25–0.1 cm. With particle size of soil aggregate decreasing, the organic carbon content decreased firstly and then increased in aggregates. When the overlying soil thickness at 20–40 cm, the organic carbon content are lower in >2 cm and 1–0.5 cm soil aggregate for 3.62 g/kg

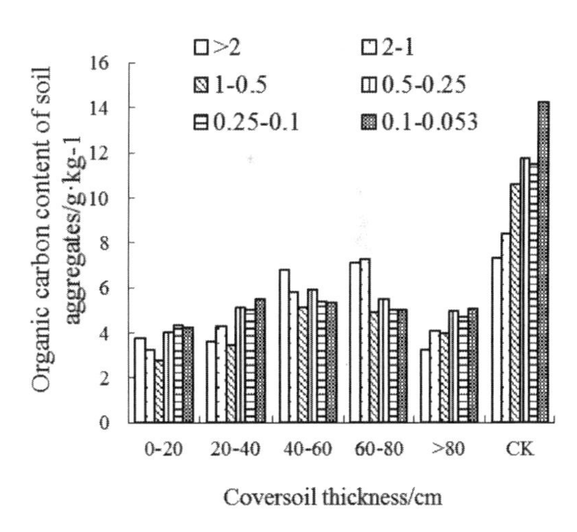

Figure 1. Effect of covering soil thickness on organic carbon content in water stable aggregates.

Table 4. Effect of vegetation types on soil aggregate stability parameters.

Vegetation type	DMWD	DGWD	WMWD	WGWD	WSAR/%	PAD/%
A	2.69	2.34	0.82	0.38	63.43	36.54
B	2.54	2.13	0.67	0.36	62.09	33.61
CK	2.94	2.75	0.87	0.41	70.66	29.34

Table 5. Effect of covering soil thickness on soil aggregate stability parameters.

Covering soil thickness/cm	DMWD	DGWD	WMWD	WGWD	WSAR/%	PAD/%
0–20	2.63	2.27	0.70	0.38	68.50	31.52
20–40	2.63	2.26	0.67	0.30	56.99	43.04
40–60	2.38	1.94	0.63	0.33	60.82	35.68
60–80	2.85	2.57	1.21	0.69	77.98	22.01
>80	2.69	2.36	0.80	0.37	63.79	36.21
CK	2.94	2.75	0.87	0.40	70.66	29.34

and 3.46 g/kg respectively. With particle size of soil aggregate decreasing, organic carbon content is mainly increase in the aggregate.

3.3 *Stability of reconstituted soil aggregates*

SA stability is directly related to the soil erosion resistance and soil organic carbon stability. DMWD, DGWD, WMWD, WGWD, WSAR, PAD and ELT are reliable indicators to assess the SA stability and an important parameter to evaluate the soil structure (Pulido et al., 2015). The aggregate stability parameters were all larger than study area in the CK, which indicated that the aggregate stability and soil structure of reconstruction soil are worse than natural soil. However, the aggregate stability parameters of A is great than B (Table 4).

Covering 60–80 cm thickness soil, the DMWD and DGWD were both higher than other covering thickness, especially higher than the covering thickness at 40–60 cm ($P < 0.05$), and close to the CK (Table 5). In addition, the larger WSAR and smaller PAD value are, the better the soil structure have (He et al., 2014). The value range of WSAR is from 56.99% to 77.98%. Covering 60–80 cm soil, the WSAR value is the largest, while the smallest value at covering 20–40 cm thickness soil. Moreover, the PAD values of A and B were both higher than CK, which explained that the process of soil reconstruction destroyed the soil structure of ecological restoration.

4 CONCLUSIONS

The mechanical-stable aggregate and water-stable aggregate content are firstly decreasing and then increasing, then decreasing and then increasing with the particle size of soil aggregate decreasing in reconstruction soil filled with gangue from coal mining areas, which present a shape of "W". It is more complex than the study results by Changting Shi (2010), they found that the aggregate content change trend presented a "Λ" type with the particle size of soil aggregate decreasing. The mechanical-stable and water-stable aggregate mainly exist in a form of large aggregate. The soil aggregate stability parameters of CK were larger than study area. In addition, vegetation types exert a big influence on soil aggregates stability. The total soil organic carbon in A and B was both smaller than CK, but compared with B, A have higher total organic carbon and organic carbon content in water stable aggregates. Covering soil thickness play a role in SA composition and organic carbon distribution, because the special pore structure and material composition of coal gangue will affect the microorganisms activity in the overlying soil (Li et al., 2013).

ACKNOWLEDGMENTS

This research was supported by the projects of the Natural Science Foundation of China, Topsoil respiration response to gas/heat gradient of reconstruction soil profiles filled with coal gangue and its environmental purpose (NO. 41572333) and carbon dioxide transport mechanism in reclaimed Soil filled with coal gangue under the conditions of pyrite oxidation heat release (NO. 51274013).

REFERENCES

Bavel, C. H. M. V. (1950) Mean weight-diameter of soil aggregates as a statistical index of aggregation. Soil Science Society of America Journal,14(C), 20–23.

Chen, S., Yang, F., Lin, S., Liu, S. R., Tang, S. R., Cai, C.F., & Hu, R. G. (2012) Impact of Land Use Patterns on Stability of Soil Aggregates in Red Soil Region of South China. Journal of water and soil conservation, 2012, 26(5): 211–216.

Chen, X. Y., M. (2016) Reconstruction of soil water movement and quality evolution in coal mining area. Hefei, Anhui science and Technology Press.

Chen, X. Y., Wang, F., Yan, J. P., Liu, Y., Tang, H. Y., & Zhou. Y. Z. (2016) Effect of soil cover thickness on diurnal variation of soil respiration in mining area. Journal of China University of Mining and Technology, 2016, 45 (1): 163–169.

Cheng, J. X., Nie, X. J., & Liu, C. H. (2014) Spatial variation of soil organic carbon in coal-mining subsidence areas. Journal of the China Coal Society, 39(12), 2495–2500.

Christensen, B. T. (2006) Straw incorporation and soil organic matter in macro-aggregates and particle size separates. European Journal of Soil Science, 37(1), 125–135.

Dongmei He, H. R. (2014) Long term effect of land reclamation from lake on chemical composition of soil organic matter and its mineralization. Plos One, 9(6), e99251.

Hu, Z. Q., (2005). Concept of and methods for soil reconstruction in mined land reclamation. Soils.

Huang, D. D., Liu, S. X., Zhang, X. P., Xu, J. P., Wu, L. J., & Lou, Y. J. (2012) Constitute and organic carbon distribution of soil aggregates under conservation tillage. Journal of Agro-Environment Science.

Hui, Z., Yinli, B. I., Zhu, C., Tao, D. U., & Bo, H. (2014) Effect of mining subsidence on soil moisture dynamic changes of sandy land. Journal of China University of Mining & Technology, 43(3), 496–501.

Li, H. T., Yu, J., Fang, F. M., & Wang, F. (2015) Soil Particle Size Distribution, Fractal Dimension, Soil Aggregate Feature and Change after Land Leveling Reclamation. Environmental science and technology, 36(8): 998.

Liao, H. K., Juan, L. I., Long, J., Liu, L. F., Yang, H., & Zhang, W. J. (2015) The study of relationship between land use and aggregate associated SOC based on between-within classes PCA in the karst region. Journal of Natural Resources.

Lu, R. K., M. (2000) Soil agricultural chemical analysis methods. Beijing, China agricultural science and technologypublishing house.

Na, L. I., Han, X., You, M., & Xu, Y. Z. (2013) Research review on soil aggregates and microbes. Ecology & Environmental Sciences.

PulidoMoncada, M., Gabriels, D., Cornelis, W., & Lobo, D. (2015) Comparing aggregate stability tests for soil physical quality indicators. *Land Degradation & Development*, 26(8), 843–852.

Shi, C. T., Wang, E. H., & Chen, X. W. (2010) Organic carbon in soil aggregates and its dynamic changes of larch plantations in black soil region. Journal of Soil & Water Conservation, 24(5), 208–212.

Wei, C., Gao, M., Che, F., & Deng, C. (1990) Study on soil aggregate and soil moisture-thermal regime in paddy field under no-tillage and ridge culture. Acta Pedologica Sinica.

Wei, Y. W., Su, Y. R., Chen, X. B., He, X. Y., Qin, W. G., & Wei, G. F. (2011) Effects of human disturbance on soil aggregates content and their organic c stability in karst regions. Ying Yong Sheng Tai XueBao, 22(4), 971–978.

Yuefeng, Y. U., Lingxiao, L. U., Hu, D. U., Peng, W., Zeng, F., & Song, T. (2013). Distribution of organic carbon and aggregation of calcareous soil in different forest types. Acta Botanica Boreali-Occidentalia Sinica, 33(5), 1011–1019.

Land Reclamation in Ecological Fragile Areas – Hu (Ed.)
 ISBN 978-1-138-05103-4

Protection and tourism development of ancient villages in resource-exhausted mining areas—a case study of Mentougou district

C. Yang & M.C. Fu
School of Land Science and Technology, China University of Geosciences, Beijing, China

ABSTRACT: In recent years, lots of mines have been closed constantly, facing various problems of industrial transformation, ecological environmental management etc. Giving full play to the advantages of ancient villages, exploiting the sustainable development of resource-depleted mining areas have become the keys in solving the problems that impede the development of green industry. By investigating ancient villages and consulting literature reviews, this paper analyzed the resource situation and distribution of mining wasteland, spatial distribution and protection of ancient villages. The paper adopting case study method, considering the orientation towards "Western Capital City Ecological Conservation Area" and tourism value of mining industry site. Results showed that, combining the tourism resources in mining sites and ancient villages into the green industrial chain not only can protect the ancient villages, but also can form the eco-tourism brand of "scenery-ancient village-mining site", which can effectively guarantee the regional sustainable development.

1 INTRODUCTION

With the urgent need of economic transformation in resource-exhausted mining area, researches on tourism development of abandoned mining area has become a hot spot in recent years. Serving as western protection barrier in Beijing, Mentougou district contains rich mineral resources and many ancient villages which have crucial scientific, historical and cultural value. At present, academic researches mainly focused on protection of cultural heritage, the environmental spatial structure of the ancient village, and the development of tourism resources of the ancient village, etc. The long-term mining exploitation in the region has caused lots of damages to the ancient village and the natural environment. The destruction of the ecological environment in the abandoned mining area has become an important problem that restricts the economic and ecological sustainable development of the region which needs be solved urgently.

2 BASIC PROFILE OF THE STUDY AREA

Mentougou area is located in the west of Beijing where is rich in coal resources, with coal storage area of nearly 700 km^2. It has a long history which can be traced back to the Liao Dynasty. There are 126 abandoned coal mines, 41 of them can be used as land for building reconstruction (230.25 hm^2). Abandoned coal mines are mainly distributed in Zhaitang Town, Wangping Town and Longquan Town. In recent years, due to the gradual depletion of resources, Beijing Government transforms Mentougou's function from" Western Mining Area" to "Western Capital City Ecological Conservation Area". Lots of coalmines are gradually closed in resource depletion areas, while the research and practice of ecological restoration and industrial transformations of the mine areas are gradually carried out.

3 DISTRIBUTION AND PROTECTION OF THE ANCIENT VILLAGES

3.1 *Distribution of ancient villages*

According to statistics, the existed 112 villages in Mentougou district, which were developed during the Qing Dynasty. There are about 30 villages are still preserved well, located in 8 towns of the region (Figure 1). For a long time, the region takes coal as the main industry, villages were built due to coal mines and the villagers are engaged in coal mining for a living.

3.2 *Protection and development of ancient villages*

With the closure of resource-exhausted coal mines in recent years, ancient villages in the area get the corresponding protection and development. However, affected by factors such as resources, location

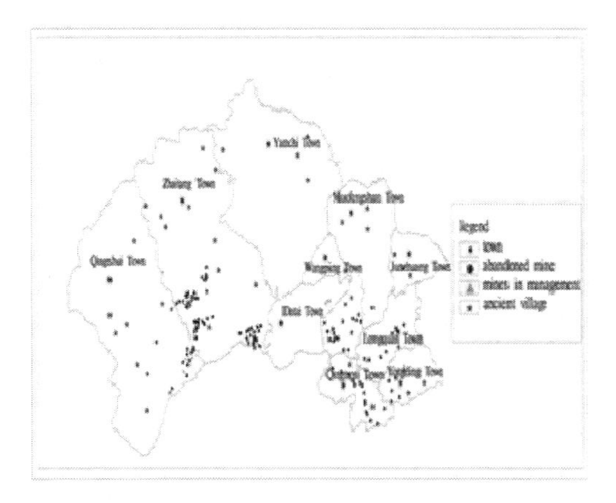

Figure 1. Distribution of coal mine and ancient villages in Mentougou district.

conditions, the present situation of protection and development of ancient villages are uneven. High-profile ancient villages such as Cuandixia Village, Lingshui Village, and Liuliqu Village are well preserved, the development of tourism brings certain income and have been rated as "Chinese Historical and Cultural Village". However, in those ancient villages which are relatively closed such as Meijian Village in Qingshui Town, economy in these places are relatively backward, the following problems are about to be solved: first, the villagers are not fully aware of protecting ancient villages, which leads to lots of ancient buildings randomly dismantled and its original style and features are destroyed; second, tourism products are quite monotone, which are mainly depending on catering accommodation that lack of creation; third, the natural environment has been seriously affected by coal mining before, the ecological construction is absent.

4 CASE ANALYSIS ABSENT ITS ORIGINAL STYLE AND TOWN JIUYUAN VILLAGE

4.1 *Jiuyuan village development orientation*

Jiuyuan Village is located in Wangping Town, Mentougou district. The area of village is about 3.18 km². In recent years, as the native resources are gradually depleted, the State Council issued a policy which focused on closing mines and cutting down its production. About 20 coal mines in Wangping Town got shut down, and the decline of the mines have a certain negative impact on Jiuyuan Village economic development. But with the adjustment of local economic structure, the village takes the advantages of itself, setting the tourism as the new direction for economic growth. In this way, the village not only develops its ecological and cultural tourism villages, which guarantees sustainable development.

4.2 *Measures for tourism development*

1. To change the traditional and single mode of tourism development

Jiuyuan Village changed its original single tourism mode which was entirely based on the farmhouse-tours, and came up with the idea of integrating its unique tourism resources altogether, exploring the new exploitation model, such as developing Jinyuan business street of folk culture and natural landscape of karst cave. Making full use of the surrounding mine sites, building Magezhuang Coal Mine Museum, and constantly developing other mining tourist projects, which shows a fully integration of various tourism resources.

2. Develop eco-tourism to promote coordinated development of environment and economy.

A nice ecological environment is the prerequisite to attract tourists. Jiuyuan village has mass vegetation, the coverage rate of forest reached more than 90%. In recent years, adhering to the principle of sustainable development, villagers have vigorously developed its featured fruit-picking tourism parks. Now, there are 100 mu of walnut garden and 30 mu of apricot garden. At the same time, the village is also focusing on construction of agricultural gardens, demonstration base of circular agricultural and other eco-tourism projects, for instance exploring Yongding River valley wetland. In this way, the ecological environment are improved, and he economic income has obtain a striking rise as well.

3. Fully integrate tourism resources to achieve multi-line development

Jiuyuan village develops its tourism resources by conducting joints development with resources from peripheral areas. For example, combining with the Ju-ren culture tour from Lingshui Village and coloured glaze industrial tour from Liuliqu Village, which forms the "Cultural Model of Ancient Villages" tourism line. Meanwhile, it combines natural scenery with mining sites from other areas, for instance, it unites the development of Jinyuan residential area—18 ponds in the west of Beijing Mining ecological restoration base. With the help of typical industrial characteristics in mining wasteland, visitors would achieve the purpose of learning and leisure.

4.3 *Effects of Jiuyuan village tourism development*

Through exploration of new tourism development mode, Jiuyuan Village promotes the transformation and upgrading of the industry, which not only conserve the original ancient villages, roads and the ancient sites, but also make the income increase. Through the development of eco-tourism, the environment of the resource-exhausted mining areas has been improved. Jiuyuan Village was named as the "Beijing municipal folk tourism village", "the most beautiful village in Beijing" in 2010.

5 SUMMARY

Based on the investigation of the protection and exploitation of the ancient villages in the resource-exhausted mining Areas of Mentougou District, this paper explored the development mode of the tourism resources of the abandoned land in Wangping Town through case study. Results have showed that changing traditional tourism model and developing the eco-tourism, creating a multi-line combination of tourism development model, can create ancient village groups tourism brand in Mentougou District. The problem that coordinating the development of ancient village and its original style, and how to facilitates harmony of old and new buildings are worthy of our follow-up exploring. This paper would provide some reference for the ecological construction and economic development of the Mentougou area to achieve the coordinated development of social economy and ecological environment.

ACKNOWLEDGEMENTS

Supported by the National Natural Science Foundation of China (41641008).

REFERENCES

Akcil A, Koldas S. (2006) Acid Mine Drainage (AMD): causes, treatment and case studies. Journal of Cleaner Production 14(2):15–18.

Long Jinghua, Hu Zhenqi, Xiao Wu. (2015) Field Investigation and Reutilization Evaluation of Abandoned Mine Land in Mentougou District of Beijing. Coal Engineering 47(9):120–123.

Shi Shaohua, Huang Fengqing. (2015) On Utilization and Protection of Folk-Custom Tourism Resources in Beijing Ancient Village—Taking Beijing Mentougou District as an Example. *Journal of Wuhan Business University* 29(2):8–11.

Sun Keqin. (2009) Study on Protection and Exploitation of Heritage Resources of Ancient Villages in Mentougou District of Beijing. Areal Research and Development, 28(4):72–76.

Wang Yuqin, Wang Linlin, Li Xiaojing. (2010) Strategy of Spatial Redistribution of Ecological Tourism Development in Abandoned Mine Land. Progress in Geography 29(7):811–817.

Ye Zhen Spatial Distribution Evolution of West Beijing Traditional Villages in China. Doctoral Dissertation, Beijing. Beijing Forest University, 2014.

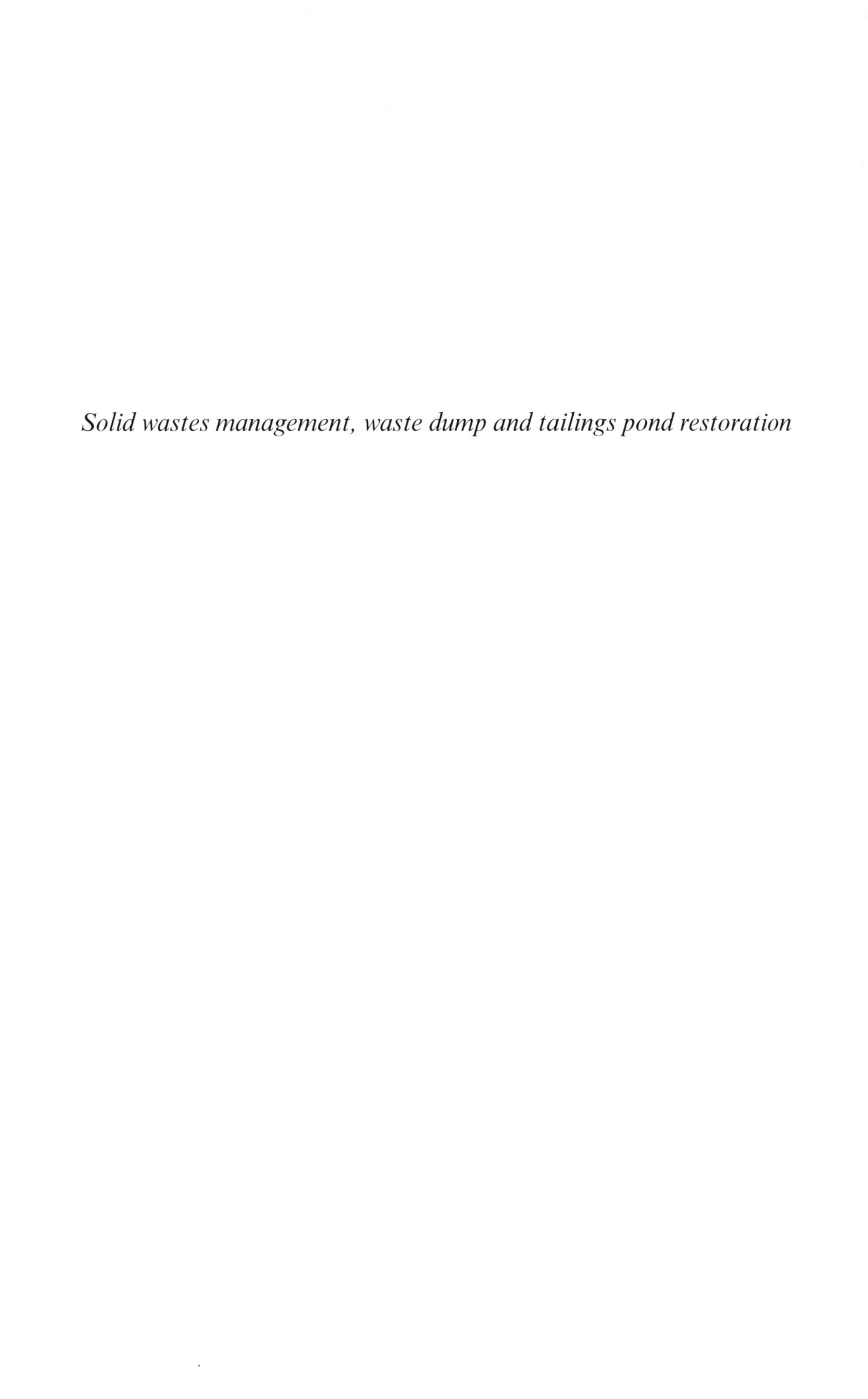

Solid wastes management, waste dump and tailings pond restoration

Land Reclamation in Ecological Fragile Areas – Hu (Ed.)
 ISBN 978-1-138-05103-4

Paste technology integrated solutions for effective management of sulfidic mine tailings

E. Yilmaz
Cayeli Bakir Isletmeleri A.S., First Quantum Minerals Ltd., Cayeli, Rize, Turkey
Department of Mining Engineering, Karadeniz Technical University, Trabzon, Turkey

ABSTRACT: This paper presents an overview of paste technology integrated means for effective management of sulfidic tailings by laboratory and field investigations. The properties of tailings, such as grain size, water content, porosity, specific surface and degree of saturation were determined experimentally. The effect of cement content, type and curing on properties of paste backfill was examined by using the laboratory setups which can well replicate the backfill's field conditions. Effect of placement and curing regimes on environmental and geotechnical behavior of surface paste disposal was examined by using the field experimental cells. Results indicate that underground backfill and surface disposal can be utilized concurrently without creating environmental damages (e.g., tailings dam failure, acid formation) over the mine's life.

1 INTRODUCTION

The mining industry has a large potential to affect industrialized and developing economies by providing employment, income, and opportunities for economic growth and diversification. However, it inevitably generates the huge volumes of environmentally harmful mine tailings that need to be managed properly to diminish or eradicate the environmental impact. Sulfidic tailings contain pyrite and iron minerals which can react with air and water in order to form sulfuric acid and dissolved iron (Aubertin et al., 2002). Waste rocks are defined as the coarse-sized material resulting from the decomposition of unwanted rocks by weathering during mining without being processed while mine tailings are defined as the fine-sized material left over after the process of removing the valuable constituents from the uneconomic fraction of the ore (Gutiérrez et al., 2016; Schoenberger, 2016). Tailings management poses many challenges on the mining industry, especially concerning the impact on the environment. To prevent the environmental pollution and accomplish sustainable mining operations, there are a number of different tailings management ways (Lottermoser 2010).

One practical solution is to use mining tailings as paste for underground mines and as paste disposal for surface facilities. Using paste integrated solutions as re-filling and storage has become a popular operating system in recent years because of increased tailings usage, reduced rehabilitation costs and environmental impact, and improved ore extraction and safety conditions (Yilmaz and Fall, 2017). In parallel, the exploration and extraction of the ores at modern mines need to be maximized while minimizing their costs (Adiansyah et al., 2015). The need for a low cost and efficient tailings management technique in mines is driving mine operators to consider using the tailings as a raw material source for hard rock mines worldwide. In recent times, paste backfill was used for underground mines as ground support and waste reduction element (Benzaazoua et al., 2008; Yilmaz 2015a; Khaldoun et al., 2016). An innovative way of tailings management has also been proposed recently by mines to replace the usual methods. The principle of this way is to use dewatered tailings (with 70–75% solids) and drop them surface in paste form (Yilmaz 2015b). Figure 1 shows a schematic view of sustainable waste management. Open pit or underground mining creates indispensably vast amounts of wastes which are hugely reduced using paste technology in terms of waste volume and environmental impact.

Indeed, paste technology offers a number of technical, operational, financial and environmental benefits for the secure management of destructive mining wastes (Boger 2012; Bussière 2007; Johnson et al., 2013; Yilmaz and Fall, 2017). The key benefits of the use of paste technology for underground backfilling include: reduced binder amount to reach equal strength, decreased quantity of the backfill linked with water, reduced labor necessities, faster backfill placement and mining cycle, reduced maintenance, and improved mining activities.

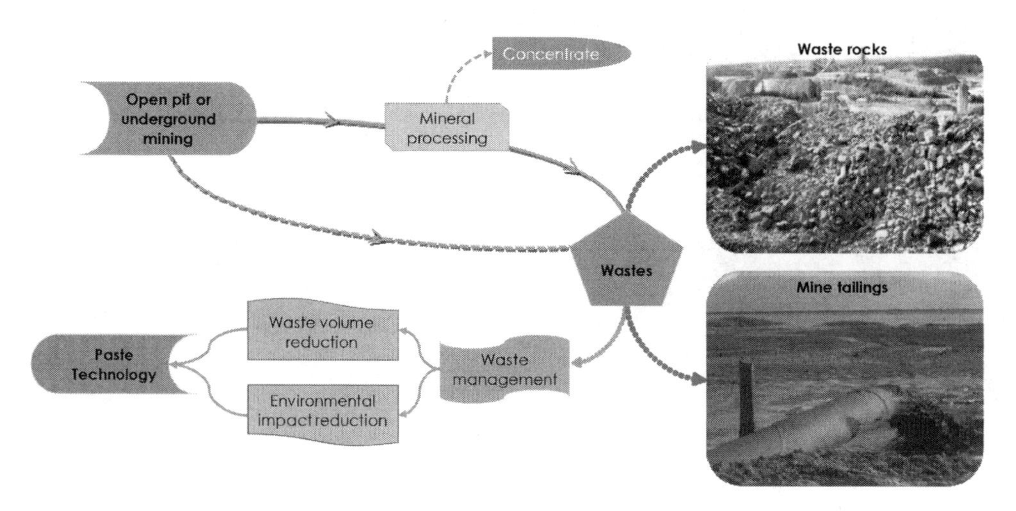

Figure 1. Schematic view of sustainable mining waste management with paste technology.

In addition, the key benefits of the use of paste technology for surface disposal application include: reduced storage footprint, decreased water consumption, reduced rehabilitation and closure costs, and decreased operational risks. The paste is usually defined as a non-settling and flowable material of solids and water which shows negligible water bleed when sitting freely. It has a low yield stress of 200 kPa and a fair slump consistency of up to 10 inches, measured by a slump cone and quickly monitored by operators (Wijewickreme et al., 2005; James et al., 2011; Simon and Grabinsky, 2013). Founded on its underground or surface application, cement and sand can be added to the paste material without having any adverse effect on its homogeneous nature and flow characteristics (Tariq and Yanful, 2013; Li et al., 2016).

Cemented Paste Backfill (CPB) is a high density mixture that consists usually of total or deslimed tailings (75–85 wt% solids), hydraulic binders (2–7 wt%) and enough water (6–10 inches) to ease the paste's transport to underground stopes by gravity or pump (Kesimal et al., 2005; Belem and Benzaazoua, 2008). CPB are used for backfilling stopes in underground mines. The CPB's quality and behavior are not only controlled by internal material characteristics, but also their external field conditions. Various works have been so far done on both laboratory-made paste backfills (Bertrand et al., 2000; Fall and Benzaazoua, 2005; Nehdi and Tariq, 2007; Ouellet et al., 2007; Ercikdi et al., 2010; Cihangir et al., 2012; Aldhafeeri et al., 2016; Koohestani et al., 2016; Zheng et al., 2016) and field paste backfills (Hassani et al., 2004; Yumlu and Guresci, 2007; Li and Aubertin, 2010; Veenstra et al., 2011; Karaoglu et al., 2013; El Mkadmi et al., 2014; Emad et al., 2015). Due to the fact that field testing are a daunting task and risky, different experimental setups or methods were used recently in order to mimic CPB's field conditions at laboratory (Fahey et al., 2011; Yilmaz 2015a; Yilmaz and Ercikdi, 2016). Therefore, measuring CPB's performance is vital for a capable mine backfill design. Surface Paste Disposal (SPD) has lately been proposed in the mining industry as an option to conventional mine tailings disposal methods for above ground storage (Cincilla et al., 1997; Oxenford and Lord, 2006; Deschamps et al., 2008; Fisseha et al., 2010; Simms et al., 2010; Bascetin et al., 2013; Daliri et al., 2016). Derived from the CPB technique being used for underground mines, it can be implemented with or without cement addition. It is worth mentioning that the SPD system has been used less often by mines worldwide due to a pretty new technique for tailings management. The Bulyanhulu mine (Tanzania) was the first gold mine to adopt this system (Shuttleworth et al., 2005; Martin et al., 2006). The non-cemented paste tailings have 73 wt% solids concentration which corresponds to a slump of 250 mm. The dewatered tailings are placed serially in 30 cm layers (each layer builds up every 5 days) which are allowed to dry out before the next layer is put in place. It is believed that using paste technology as both disposal and backfill purposes will bring a new light to the sustainable tailings management for mines.

2 MATERIALS AND METHODS

2.1 *Laboratory and field test systems*

In this study, the two experimental lab set-ups were used to study both CPB and SPD applications. Figure 2 shows photos of laboratory apparatus

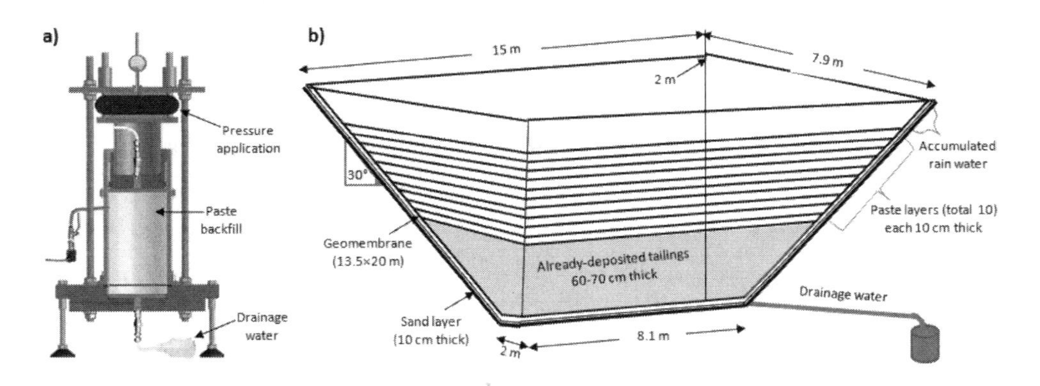

Figure 2. Schematic view of laboratory apparatus and field experimental cells used in this study.

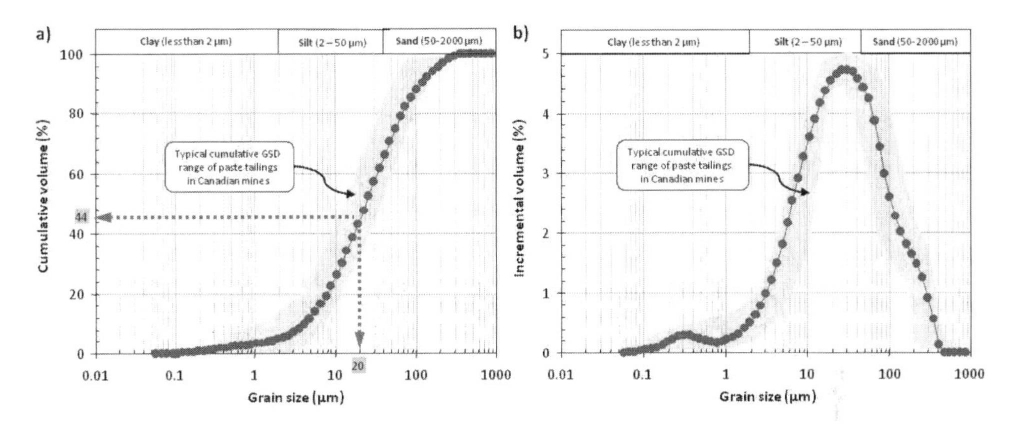

Figure 3. Grain size distribution of paste tailings: a) cumulative and b) incremental curves.

used to imitate the properties of underground paste backfills and field cells used to duplicate the application of surface paste disposal. Laboratory apparatus (Figure 2a) is consisted of three parts: i) a top loading platen to apply a vertical total stress of up to 800 kPa, ii) a transparent sample container ($D \times H = 102 \times 204$ mm), and iii) a bottom drainage hole to collect water drained from samples. Further information on this apparatus can be found in Yilmaz (2015a). The field conditions of paste backfills can be well simulated by this apparatus. After each test, CPB specimen is first extracted and then subjected to compressive strength testing. In addition, the two experimental field cells were used to study the SPD's application. Cell one is designed by filling partly with tailings having a low cement rate of 2 wt% and termed as Cemented Cell (CC) while cell two is designed by filling completely with tailings without cement and termed as Uncemented Cell (UC). Field cells (Figure 2b) were designed as an inverted pyramid, with a basal area (8.1×2 m) and a top surface area (15×7.9 m). Tailings are prepared in surface paste plant and sent to the site using a ready-mix truck. Tailings are deposited in thin layers (10 cm).

2.2 *Characterization of paste backfill ingredients*

Paste tailings were sampled as representative of the tailings streams from the paste plant of a polymetallic mine. Samples were received in sealed plastic containers to avoid any oxidation and prepared to characterize physical, chemical and mineralogical properties. A Malvern Mastersizer laser diffraction-type particle size analyzer was employed to measure the grain size distributions of paste tailings (Figure 3).

The laboratory analyses indicate that the tailings have an average water content w of 24.5 wt%, a specific gravity G_s of 3.3, a specific surface S_s of 2.5 m²/g, a total porosity n_{tot} of 48.7%, an optimum water content w_{opt} of 9.8 wt%, a maximum dry unit weight γ_{dmax} of 23.6 kN/m³, a relative compaction R_c of 93 wt%, a liquid limit w_L of 22.8 wt%, a plastic limit w_P of 17.4 wt%, a liquidity index *LI* of 1.3 wt% and a plastic index *PI* of 4.8 wt%. From the Atterberg limit results, the tailings can be designated as CL-ML, silty clay. Most of the grain size falls into silts to fine sand range, having a clay-sized grain of 9.5%. Tailings can be also classified as medium tailings (amount of grains finer than 20 μm ~44%) according to the fines content.

The mineralogical composition of tailings was determined using both a Bruker AXS D8 Advance X-ray diffractometer XRD and a Hitachi® 3500-N scanning electron microscopy coupled with energy dispersive X-ray spectroscopy SEM-EDS (Figure 4a). The XRD results revealed the dominance of quartz (53.5 wt%) and pyrite (27.5 wt%) phases within tailings. To better identify the mineral phases' chemical composition, the EDS analysis was also performed and the results obtained are shown in Figure 4b. SEM-EDS results confirm that sulfur (S = 29.2%) and iron (Fe = 22.3%) leading to acid and sulfate production were the two main elements within the tailings. Silicon (Si = 13.5%) and aluminum (Al = 7.3%) precipitate in tailings as stable hydroxides while a small part of calcium (Ca = 0.4%) precipitates as expansive gypsum.

As well, the remaining CPB ingredients such as hydraulic binders and mixing waters were characterized as follows. A blended cement consisting of normal Portland Cement type 10 (PCI) and ground granulated blast furnace slag (Slag) in a ratio of 20:80% was used for CPB applications while only PCI was used in part for SPD applications. The cement contents used were respectively 4.5 wt% and 2 wt% for both backfill and disposal purposes. The chemical characteristics of both binders and waters were listed in Table 1.

The two different waters (tailings pore water and municipal tap water) were used as mixing water. ICP-AES (Perkin Elmer Optima 3100RL) was used to analyze the chemical composition of mixing waters and that of the water collected during the paste deposition. pH, redox potential Eh and electrical conductivity EC were measured using a Benchtop pH/ISE Meter Orion Model

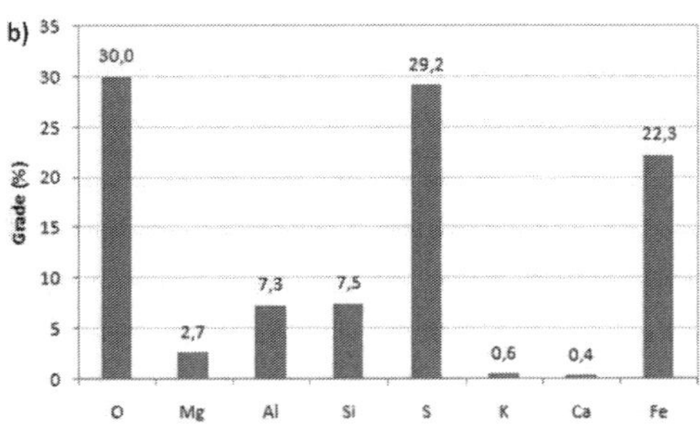

Figure 4. Mineralogical and chemical compositions of paste tailings: a) SEM and b) EDS analyses.

Table 1. Chemical characteristics of hydraulic binders and mixing waters used in this study.

Element	Detection limit (mg/L)	PCI 100% (mg/L)	Slag 100% (mg/L)	Blended cement (mg/L)	Tailings pore water (mg/L)	Tap water (mg/L)
Al	0.010	2.57	5.42	4.44	0.21	0.01
Ba	0.001	0.01	0.15	0.03	0.05	0.01
Ca	0.030	47.0	22.45	30.6	559	40.9
Fe	0.006	1.71	0.38	0.45	0.01	0.06
K	0.003	0.69	0.42	0.46	0.29	0.84
Mg	0.001	1.33	6.81	3.73	1.83	2.27
Mn	0.002	0.03	0.75	0.14	0.01	0.01
Na	0.004	1.57	1.49	1.51	0.05	0.01
S	0.033	1.47	1.31	1.34	1630	46.0
Si	0.020	9.12	16.9	14.5	0.89	0.90
Ti	0.002	0.15	0.37	0.29	0.01	0.02
Zn	0.005	0.03	0.01	0.01	0.11	0.10

920A coupled with a Thermo Orion Triode combination electrode (Pt-Ag-AgCI). The measured pH, Eh and EC were 9.3, 152 mV and 7.6 mS/cm for pore waters, and 8.1, 425 mV and 0.3 mS/cm for tap waters, respectively.

2.3 *Sample preparation, placement and curing conditions*

Numerous CPB samples were prepared at fixed 4.5 wt% binder rate to study the effect of curing and stress conditions. The tailings, binder and water were mixed and thoroughly homogenized using a heavy-duty food Hobart mixer for about 12 minutes. After mixing, the mixture's slump was measured and set to ~18 cm by adding mixing water. The resulting CPB samples were first poured into molds (both transparent molds and conventional plastic molds) and then placed in a fog room at a relative humidity of 80% and a temperature of 24°C (Figure 5a). Note that a pressure sequence of up to 400 kPa was applied to CPBs during curing. The applied pressures exhibit a paste filled stope height of 18 m which was calculated from vertically-applied pressures (p_v) and knowing the CPB's wet unit weight γ (23.1 kN/m^3).

For SPD application, tailings are first prepared at the paste plant and then delivered by using a ready-mix truck to two mini construction cells (Figure 5b). The tailings that contain a slump ranging from 200 to 290 mm were disposed in these cells with or without cement. Each disposal design (cemented and uncemented) was made of 10 paste layers, each about 10 cm.

To get a flat layer of tailings, a *Topcon® RL-H3C* Rotating Laser was used across the field cells. Only first layer is kept cemented to show the effect of cement addition as a barrier layer on mitigation and/or prevention of acid mine drainage or tailings oxidation. In case of cemented cell, the binder rate used was only 2% by dry weight of tailings. The selected binding agent was normal Portland cement.

2.4 *Unconfined compression tests*

The unconfined compression tests were conducted only on CPB samples by using a stiff universal testing machine (MTS Sintech 10/GL load frame) having 50 kN load capacity and 1 mm/min deformation rate. The three tests were undertaken on samples to reach an average strength value. A small portion of samples were oven-dried for about 3 days at 45°C temperature until mass stabilization to measure their final water content. To evaluate the influence of curing time the strength testing were done on 7-, 14-, and 28-day cured paste backfill samples.

Note that no mechanical strength testing was conducted for SPD application.

2.5 *Geotechnical and microstructural tests*

The geotechnical index tests were done on representative samples broken after strength testing. The water content and specific gravity of CPB samples were initially determined. Using these two data (e.g., w and G_s), the remaining parameters (e.g., void ratio e and degree of saturation S_r) can be calculated. The microstructural properties of CPB samples were determined using a Micromeritics Autopore III 9420 mercury intrusion porosimeter which allows applying a pressure of up to 414 MPa and measuring throat pore diameter of up to 0.003 μm. MIP was analyzed according to ASTM D 4404 [48] standard. After strength testing, small cylindrical samples ($D \times H = 12 \times 24$ mm) were taken from the middle point of samples. For a given CPB mix, two MIP tests were done for reaching an average value.

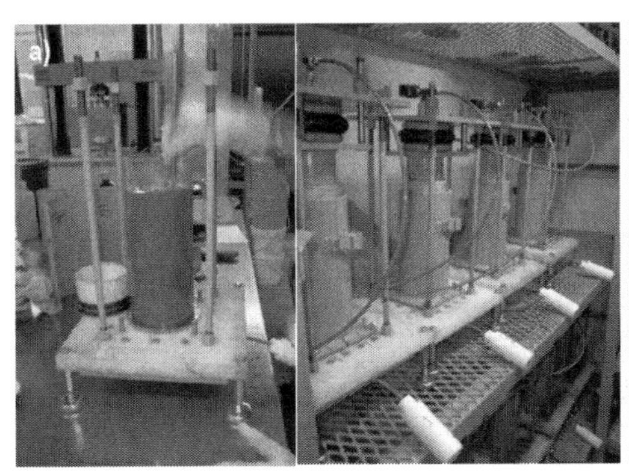

Figure 5. The tailings sample preparation and placement conditions: a) CPB and b) SPD applications.

3 RESULTS AND DISCUSSION

3.1 *Effect of curing and stress conditions on strength properties of paste backfills*

Figure 6a exhibits the influence of curing conditions on strength development of CPB samples containing 4.5% cement as a function of curing time. The different paste backfill samples (consolidated, drained and undrained ones) were compared with the curing times of 7, 14 and 28 days. Generally, consolidated backfills produced always the best strengths among others, regardless of the curing time. For example, the strength of consolidated backfills cured at 28 days was 2.9 MPa while the strength of drained backfills was 1.5 MPa, indicating a strength increase of 50%. Observation can make that the variation of strength reduces somewhat with increasing curing time due to reduced void spaces in backfill after the formation of cement hydration products. Another reason why consolidated backfills present higher strengths than other types of backfilling (drained or non-drained) may be linked with drainage water and consolidation loadings. The drainage of free waters available in fresh backfills causes a reduction in the overall porosity of filling and hence the higher strength. The drainage of free waters also plays a role on hydration, leading to more dense materials.

Figure 6b demonstrates the change in strength of stress application to 4.5 wt% cemented paste backfills under during the early stages of curing. The equivalent strength of CPB samples increases with incremental pressure. At a curing time of 28 days, the paste backfill strength were respectively 2.1 MPa, 2.5 MPa, and 2.7 MPa for a vertical stress application of 400 kPa, 600 kPa and 600 kPa, when comparing with a strength gain of 1.6 MPa of non-drained backfills cured under 0 kPa. An increase of 50% in pressure (vertical stress is increased from 400 kPa to 800 kPa) produced a 29%, 25%, and 21% increase of strength for a curing time of 7, 14 and 28 days, respectively. Increasing the pressure leads to an increase of drainage water, which may have favored better hydration process due to over-saturation of dissolved ions. Application of stress during the early stages of curing could increase the corresponding CPB's strength. Overall, if the stress is applied right before the start of curing which contributes to paste's hardening, as is being practiced in this study, the rate of strength increase, final backfill strength become higher, compared with the application of stress after the completion of hydration.

3.2 *Effect of stope depth on strength properties of field paste backfills*

This distinction is important in terms of groundwater flow, which has very different characteristics in fractures and conduits compared to the bulk of the rock. Figure 7 shows the evaluation of strength gain with curing time for 4.5 wt% binder paste backfills placed in an underground stope, obtained at different depths from the top of paste-filled stope. The strength increases with increasing stope depth for a given curing age, following identical trend as overburden pressure ($\sigma_v = \gamma * h$). At 28 days, a strength gain of 1.6 MPa, 2.1 MPa, 2.5 MPa, 2.6 MPa, 2,7 MPa and 2.9 MPa were obtained from CPB samples at a stope depth of 0 m, 4 m, 8 m, 12 m, 16 m and 20 m, respectively. This proves that the strength of paste-filled stopes increases from the top to the bottom due to the consolidation effect. For a curing time of 7, 14 and 28 days, the strength of backfills located at the bottom of the stope was respectively 81%, 64% and 51% higher than that of backfills located at the top of the stope.

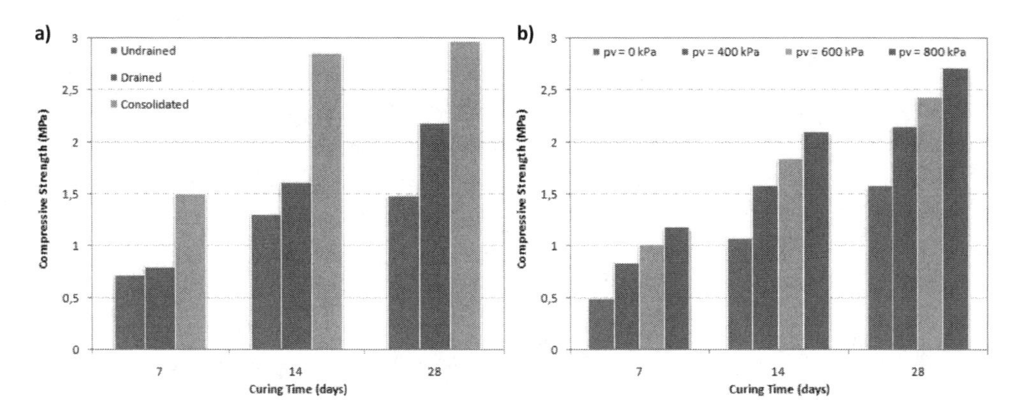

Figure 6. Change in the strength of different backfills in terms of a) curing and b) stresses conditions.

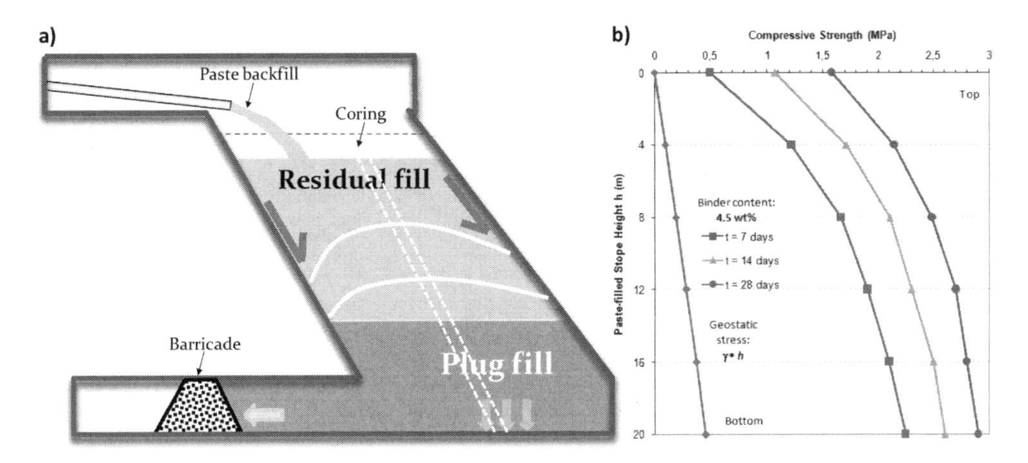

Figure 7. Assessment of stope depth effect on paste backfill: a) schematic view and b) test results.

The relative density of consolidated cemented paste backfills increases with increasing stope depth, thus decreasing void ratios as a result of effective stress during curing. This is because the lower the void ratio, the closer the cemented paste backfill particle packing, and hence, the lower the friction angle that develops and the stronger the paste backfill mass. Experiences show a relatively lower porosity and higher strength near the bottom of paste backfilled stope and a higher porosity and lower strength near the top of paste backfilled stope at modern mines. For an efficient paste backfill design, it is vital to evaluate the stress distribution in cemented paste backfill, which often evolves over placement conditions.

3.3 *Effect of paste layer addition on SPD's water content and degree of saturation*

Figure 8a shows the evaluation of volumetric water content VWC measured in paste layers 1, 3, 5, 8 and 10 for field cells with or without cement. For the first layer, VWC of cemented cell CC is the lowest (0.46) when comparing with the one gained from uncemented cell UC (0.48). The remaining paste layers always exhibit the same trend where UC specifies lower water contents than CC because of full water saturation of cemented paste layer which is assumed to be crucial for the maximal reduction of oxygen transport through the cover. The volumetric water content in paste layer 3, 5, 8 and 10 increases because of re-saturation by free water drained after the layer of the paste placed is fully settled. The main part of water flowed through desiccation crack network reaches and stay in cemented (2 wt%) layer, which acts as a barrier cover.

Figure 8b presents the variation in degree of saturation of paste tailings in both CC and UC with time. As expected, the bottom layers (Layers 1 and 3) are more saturated than top layers due to the consolidation settlements which may decrease void spaces among solid particles. For fine-sized tailings, thinly-placed layers give faster settlement than thickly-placed layers. The degree of saturation of tailings varies between 94% and 100% and reaches a plateau following a slight increase after paste deposition is completed. Layers 1, 3 and 8, respectively provide a degree of saturation close to 100%. Due to the fact that the first layer is cemented (2 wt%) in CC, and acts as a barrier cover, the layers in CC keeps more water within the matrix than the ones of UC, resulting in higher saturation.

3.4 *Effect of stress and drainage conditions on geotechnical properties of paste backfills*

Figure 9 shows change in water content and degree of saturation of different paste backfills containing a cement content of 4.5 wt% (i.e. consolidated, drained and undrained paste backfills) as a function of curing time. The lowest values were gained from consolidated paste backfills at different curing times, irrespective of curing time.

For a 4.5 wt% cement, water content decreased from 18.7 to 13.2% for consolidated cemented paste backfills, from 20.8 to 18.5% for drained cemented paste backfills, and from 25.1 to 24.2% for undrained cemented paste backfills (used as control sample), when the equivalent curing time increased from 7 to 28 days (Figure 9a). Water content is governed by the final water-to-cement ratios of samples and the quantity of excess water needed for hydration products. In general,

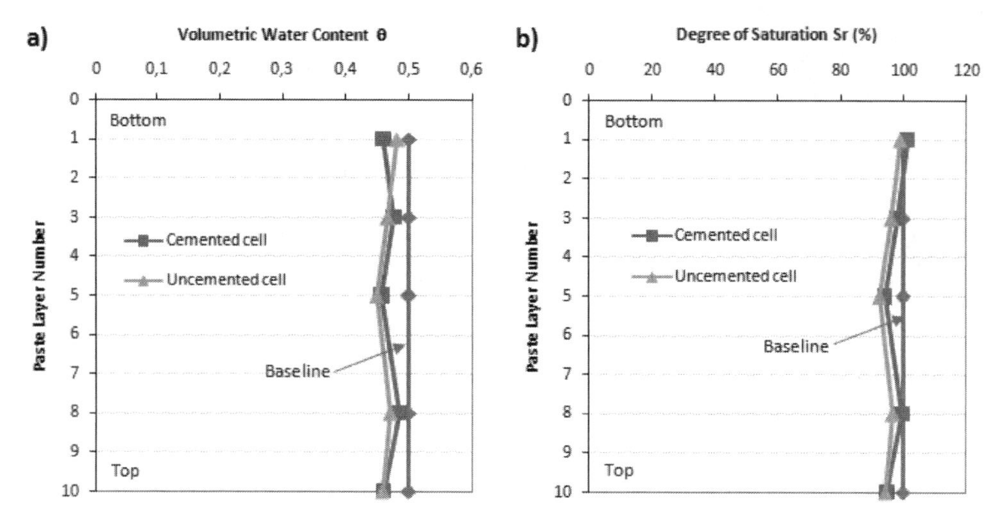

Figure 8. Change in the water content (a) and degree of saturation (b) with paste layer addition.

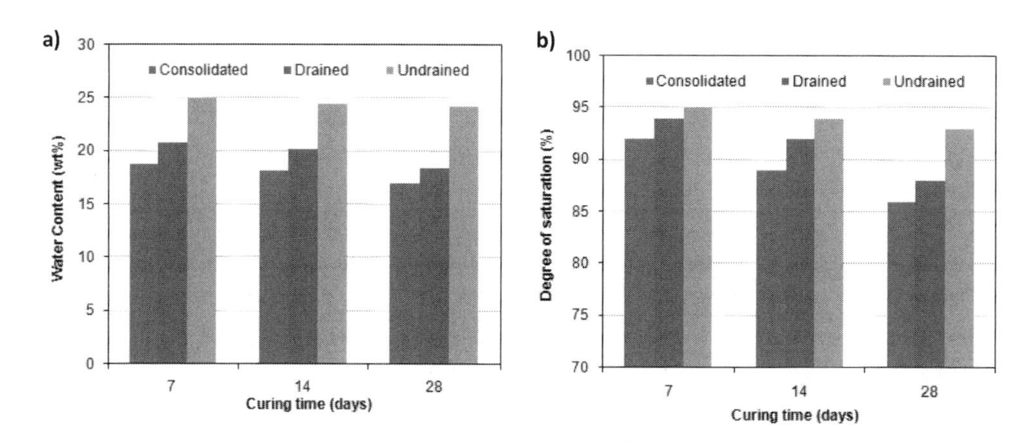

Figure 9. Change in the water content (a) and degree of saturation (b) with different paste backfills.

consolidated cemented paste backfills where stress is incrementally applied during the early stages of curing contain less water in their own structures, when comparing with both drained and undrained cemented paste backfills where no stress is applied to samples. Figure 9b shows evolution of degree of saturation with curing time for different cemented paste backfills. The degree of saturation reduces with time, irrespective of the paste backfill samples tested. The highest reduction in the degree of saturation was gained from consolidated cemented paste backfills. At 28 days, the degree of saturation was 86% for consolidated cemented paste backfills, 88% for drained cemented paste backfills, and 93% for undrained cemented paste backfills. It is noteworthy that variation of the degree of saturation within undrained cemented paste backfills was limited (on average varies 95–93%) for a given curing time.

3.5 *Effect of binder content and drainage conditions on pore structure of paste backfills*

The two different Mercury Intrusion Porosimetry (MIP) tests were undertaken on both Cemented Paste Backfill (CPB) and Surface Paste Disposal (SPD) samples and the obtained results were presented in Figure 10a. For the SPD applications, the MIP results indicate that the total porosity n_{tot} was 48.6% for uncemented paste tailings and 45.4% for paste tailings containing 2 wt% normal Portland cement (referred to as paste backfill).

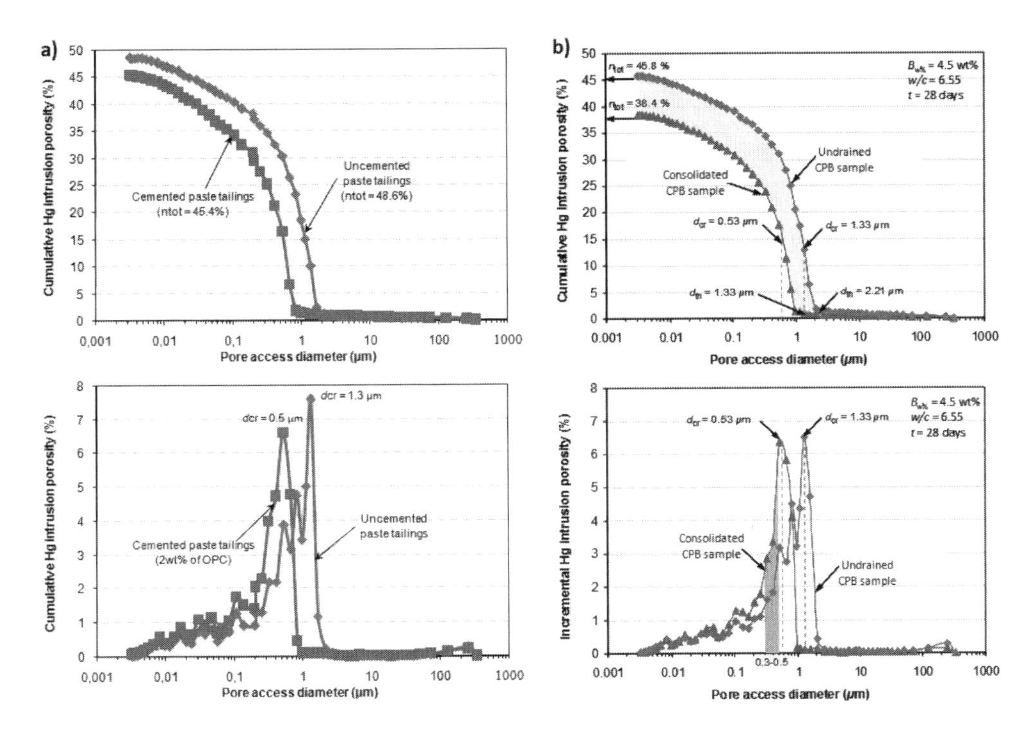

Figure 10. The cumulative and incremental pore size curves: a) SPD and b) CPB applications.

The critical pore diameter d_{cr} was 0.5 µm and 1.3 µm for uncemented paste tailings and cemented paste tailings backfill, respectively. The observed difference between paste tailings without cement and 2 wt% cemented paste backfill justifies an evolution of texture and pore structure as a function of the binder type and amount used in the paste backfill mixes and its cement hydration products (as the pores are filled completely by the hydrates or precipitates).

Figure 10b shows the change in pore size distribution curves for 28-day cured CPB samples with a binder content of 4.5 wt%. The MIP total porosity n_{tot} corresponds to the last recorded point (or the highest intrusion pressure and the smallest equivalent pore throat size). The consolidated paste backfills constantly show lower total porosity n_{tot} (38.4%) than that of undrained ones (45.8%). The curing conditions (with and without pressure during curing) play a role in total porosity n_{tot} of CPB samples for a given binder content. The effective curing conditions (the combined effects of applied pressure and drainage on the consolidated paste backfills) appear to accelerate the cementitious processes through the formation of increasing amounts of cementitious products in a mechanically reduced pore space. However, hydration/precipitation appears to occur to a lesser extent in the undrained paste backfills that are not allowed to drain. As a result, the porosity of the undrained paste backfills is greater than that of the consolidated paste backfills. In addition, the pore size distribution curves are dependent on water-to-cement (*w*/*c*) ratio. The *w*/*c* ratio of consolidated paste backfills decreases with the removal of excess water from samples due to the applied pressure during curing. The reduction in *w*/*c* ratio is proportionally accompanied by a reduction in n_{tot}, which could be explained in part by a reduction in pore size as pores are partially filled by hydrates and/or precipitates.

4 CONCLUSIONS

This paper presents the results of the works intended for both underground backfill and surface disposal applications. Several lab and field testing were done on paste tailings received from a Canadian mine. Lab investigation included mechanical testing, geotechnical and micro-structural testing on consolidated, drained and undrained backfills with 4.5 wt% cement at curing of 7, 14 and 28 days while field investigation included volumetric water content and degree of saturation of tailings disposed within two experimental cells with and without cement. Only 2 wt% cement was used for paste tailings placed in Layer 1 of cemented cell. Based on the different lab and field test results, the following conclusions can be drawn:

- The best ideal CPB mixtures were obtained from the consolidated paste backfills when comparing with the others: both drained and undrained paste backfills.
- Application of consolidation stress to lab-made paste backfills during curing imitates CPB's placement and field conditions which must be considered for an efficient backfill design for mines.
- The cement addition to paste tailings layers implemented in field cells plays an important role on quality and behavior of surface paste disposal.
- Surface paste disposal method can be well practiced over conventional tailings impoundments in order to reduce waste management and future reclamation costs.

These results advice that consolidated backfills done using a new lab tool can be favorably used for cost and performance effective paste backfills instead of both drained and undrained backfills done using conventional moulds. Besides, the suggested new approach for the secure surface disposal of hazardous paste tailings allows operators to reduce waste treatment- and management-related costs. Finally, additional work is in progress to validate the benefits of underground backfill and surface disposal test results.

REFERENCES

Adiansyah, J.F., Rosano, M., Vink, S. & Keir, G. (2015) A framework for a sustainable approach to mine tailings management: disposal strategies. *Journal of Cleaner Production,* 108, 1050–1062.

Aldhafeeri, Z., Fall, M., Pokharel, M. & Pouramini, Z. (2016) Temperature dependence of the reactivity of cemented paste backfill. *Applied Geochemistry,* 72, 10–19.

Aubertin, M., Bussière, B. & Bernier, L. (2002) *Environnementet Gestion des Résidus Miniers,* Presses Internationales Polytechnique, 210 p.

Bascetin, A. & Tuylu, S. (2017) Application of Pb-Zn tailings for surface paste disposal: Geotechnical and geological observations. *International Journal of Mining, Reclamation and Environment,* In-press, http://dx.doi.org/10.1080/17480930.2017.1282411.

Belem, T. & Benzaazoua, M. (2008) Predictive models for pre-feasibility cemented paste backfill mix design. *2nd International Symposium on Post-Mining,* Nancy, France, 1–13.

Benzaazoua, M., Bussière, B., Demers, I. Aubertin, M., Fried, E. & Blier, A. (2008). Integrated tailings management by combining environmental desulphurization and cemented paste backfill: Application to mine Doyon, Quebec, Canada. *Minerals Engineering,* 21, 330–340.

Bertrand, V.J., Monroy, M.G. & Lawrence, R.W. (2000) Weathering characteristics of cemented paste backfill: mineralogy and solid phase chemistry. *5th International Conference on Acid Rock Drainage,* Colorado, United States, 863–876.

Boger, D. (2012) Personal perspective on paste and thickened tailings: A decade on. *IMMM Transactions. Section A: Mining Technology,* 121, 29–36.

Bussière, B. (2007) Hydro-geotechnical properties of hard rock tailings from metal mines and emerging geoenvironmental disposal approaches. *Canadian Geotechnical Journal,* 44, 1019–1052.

Cihangir, F., Ercikdi, B., Kesimal, A., Turan, A. & Deveci, H. (2012) Utilisation of alkali-activated blast furnace slag in paste backfill of high-sulphide mill tailings: Effect of binder type and dosage. *Minerals Engineering,* 30, 33–43.

Cincilla, W.A., Landriault, D. & Verburg, R. (1997) Application of paste technology to surface tailings disposal of mineral wastes. *4th International Conference on Tailings and Mine Waste,* Colorado, United States, 343–356.

Cooper, R.A. & Smith, M.E. (2011) Case study: operation of three paste disposal facilities. *14th International Conference on Paste and Thickened Tailings,* Perth, Australia, 261–270.

Daliri, F., Simms, P. & Sivathayalan, S. (2016) Shear and dewatering behaviour of densified gold tailings in a laboratory simulation of multi-layer deposition. *Canadian Geotechnical Journal,* 53, 1246–1257.

Deschamps, T., Benzaazoua, M., Bussière, B., Aubertin, M. & Belem, T. (2008) Microstructural and geochemical evolution of paste tailings in surface storage. *Minerals Engineering,* 21, 341–353.

El Mkadmi, N., Aubertin, M. & Li, L. (2014) Effect of drainage and sequential filling on behaviour of paste backfill in mine stopes. *Canadian Geotechnical Journal,* 51, 1–15.

Emad, M.Z., Mitri, H. & Kelly, C. (2015) State-of-the-art review of backfill practices for sublevel stoping system. *International Journal of Mining, Reclamation and Environment,* 29, 544–556.

Ercikdi, B., Cihangir, F., Kesimal, A., Deveci, H. & Alp, I. (2010) Effect of natural puzzolans as mineral admixture on the performance of cemented paste backfill of sulphide-rich tailings. *Waste Management Resources,* 28, 430–435.

Fahey, M., Helinski, M. & Fourie, A. (2011) Development of specimen curing procedures that account for the effect of effective stress during curing on the strength of cemented mine backfill. *Geotechnical and Geological Engineering,* 29, 709–723.

Fall, M. & Benzaazoua, M. (2005) Modeling the effect of sulphate on strength development of paste backfill and binder mixture optimization. *Cement and Concrete Research,* 35, 301–314.

Fisseha, B., Bryan, R. & Simms, P. (2010) Evaporation, unsaturated flow, and salt accumulation in multilayer deposits of paste tailings. *Geotechnical and Geoenvironmental Engineering* 136, 1703–1712.

Gutiérrez, M., Mickus, K. & Camacho, L.M. (2016). Abandoned Pb-Zn mining wastes and their mobility as proxy to toxicity: A review. *Science of The Total Environment,* 565, 392–400.

Hassani, F., Ouellet, J., Zhu, Z. & Roy, A. (2004) Paste backfill behaviour in a narrow vein mine: in situ stress and strain monitoring. *8th Int. Symposium on Mining with Backfill,* Beijing, China, 257–267.

James, M., Aubertin, M., Wijewickreme, D. & Wilson, G.W. (2011) A laboratory investigation of the dynamic properties of tailings. *Canadian Geotechnical Journal,* 8, 1587–1600.

Johnson, D., Moghaddam, R., Ahmed, I. & Laroche, C. (2013) Comparative evaluation of surface disposal of thickened versus slurry tailings. *16th International Seminar on Paste and Thickened Tailings,* Belo Horizonte, Brazil, 291–304.

Kam, S., Girard, J., Hmidi, N., Mao, Y. & Longo, S. (2011) Thickened tailings disposal at Musselwhite Mine. *14th International Seminar on Paste and Thickened Tailings,* Perth, Australia, 225–236.

Karaoglu, K., Kucukates, K. & Thompson, B. (2013) Paste backfill pressure monitoring at Inmet's Cayeli underground copper and zinc mine. *23rd World Mining Congress,* Montreal, Canada, 15–28.

Kesimal, A., Yilmaz, E., Ercikdi, B., Alp, I. & Deveci, H. (2005) Effect of properties of tailings and binder on the short and long terms strength and stability performance of cemented paste backfill. *Materials Letters,* 59, 3703–3709.

Khaldoun, A., Ouadif, L., Baba, K. & Bahi, L. (2016) Valorization of mining waste and tailings through paste filling solution, Imiter operation, Morocco. *Mining Science and Technology,* 26, 511–516.

Koohestani, B., Belem, T., Koubaa, A. & Bussière, B. 2016. Experimental investigation into the compressive strength development of cemented paste backfill containing nano-silica. *Cement and Concrete Composites,* 72, 180–189.

Li, L. & Aubertin, M. (2010) An analytical solution for the nonlinear distribution of effective and total stresses in vertical backfilled stopes. *Int. Journal of Geomechanics and Geoengineering,* 5, 237–245.

Li, W., Hou, Y. & Han, S. (2016) GGBFS effect on relationship between uniaxial compressive strength and elastic modulus of backfill. *Electronic Journal of Geotechnical Engineering,* 21, 1897–1907.

Lottermoser, B. (2010) *Mine Wastes: Characterization, Treatment and Environmental Impacts*, Springer-Verlag Berlin Heidelberg, 400 p.

Martin, V., Aubertin, M. & McMullen, J. (2006) Surface disposal of paste tailings. *5th ICEG Environmental Geotechnics: Opportunities, Challenges and Responsibilities for Environmental Geotechnics,* Cardiff, United Kingdom, 1471–1478.

Nehdi, M. & Tariq, A. (2007) Developing durable paste backfill from sulphidic tailings. *Waste Resources Management,* 160, 155–166.

Ouellet, S., Bussière, B., Aubertin, M. & Benzaazoua, M. (2007). Microstructure evolution of paste backfill: Mercury intrusion porosimetry test results. *Cement and Concrete Research,* 37, 1654–1665.

Oxenford, J & Lord E. (2006). Canadian experience in the application of paste and thickened tailings for surface disposal. *6th Int. Seminar on Paste and Thickened Tailings,* Limerick, Ireland, 93–105.

Schoenberger, E. (2016). Environmentally sustainable mining: the case of tailings storage facilities. *Resources Policy,* 49, 119–128.

Simms, P., Dunmola, A., Fisseha, B. & Bryan, R. (2010) Generic modelling of desiccation for cyclic deposition of thickened tailings to maximize density and to minimize oxidation. *13th International Seminar on Paste and Thickened Tailings,* Ontario, Canada, 293–302.

Simon, D. & Grabinsky, M.W. (2013) Apparent yield stress measurement in cemented paste backfill. *International Journal of Mining, Reclamation and Environment,* 27, 231–256.

Shuttleworth, J., Thomson, B. & Wates, J. (2005). Surface paste disposal at Bulyanhulu – practical lessons learned. *8th International Seminar on Paste and Thickened Tailings,* Santiago, Chili, 207–218.

Tariq, A. & Yanful, E.K. (2013). A review of binders used in cemented paste tailings for underground and surface disposal practices. *Journal of Environmental Management,* 131, 138–149.

Veenstra, R.L. Bawden, W.F., Grabinsky, M. & Thompson, B. 2011. Matching stope scale numerical modelling results of early age cemented paste backfill to in-situ instrumentation results. *64th Canadian Geotechnical Conference,* Ontario, Canada, 1–8.

Wijewickreme, D., Sanin, M. & Greenaway, G. 2005. Cyclic shear response of fine-grained mine tailings. *Canadian Geotechnical Journal,* 42, 1408–1421.

Yilmaz, E. & Fall, M. (2017) *Paste Tailings Management,* Springer International Publishing, 303 p.

Yilmaz, E. (2015a) *Geotechnical Characterization of Cemented Paste Backfill,* LAP Lambert Academic Publishing, 379 p.

Yilmaz, E. (2015b) *Environmental Characterization of Surface Paste Disposal,* LAP Lambert Academic Publishing, 152 p.

Yilmaz, T. & Ercikdi, B. (2016) Predicting the uniaxial compressive strength of cemented paste backfill from ultrasonic pulse velocity test. *Nondestructive Testing and Evaluation,* 31, 247–266.

Yumlu, M. & Guresci, M. (2007) Paste backfill bulkhead monitoring: A case study from Inmet's Cayeli Mine, Turkey. *CIM Bulletin,* 100, 1–10.

Zheng, J., Zhu, Y. & Zho, Z. (2016) Utilization of limestone powder and water-reducing admixture in paste backfill of coarse copper mine tailings. *Construction and Building Materials,* 124, 31–36.

Land Reclamation in Ecological Fragile Areas – Hu (Ed.)
© 2017 Taylor & Francis Group, London, ISBN 978-1-138-05103-4

An investigation of surface paste disposal for lead-zinc underground mine tailings

A. Bascetin, S. Tuylu & D. Adiguzel
Department of Mining Engineering, Istanbul University, Avcilar, Istanbul, Turkey

ABSTRACT: Removing and storing the high amounts of mine tailings properly are the main problems of mining industry in the economic and environmental aspects, recently. Various studies have been done to solve this problem. Therefore, surface and underground tailing storage methods have been developed using paste technology. In this study, the storage possibilities of process tailings of a Pb-Zn underground mine were investigated in detail. In this context, the physical properties, mineralogical, and elemental composition of the tailings was revealed in the aspect of convenience for the material. Additionally, physical changes of the paste material in the laboratory environment was measured. The results obtained from this study showed that the surface paste disposal method has some advantages compared to the alternative tailing disposal methods in terms of environmental risk and storage of more tailings than conventional methods.

1 INTRODUCTION

In the world, disposal methods of process tailings have become controversial due to tailings dam accidents in recent years. For this reason, recently, studies about disposal of tailings on the surface with a material with lower water content have gained momentum in the leading countries such as Canada, Australia, and the USA. In this method, effect of climatic conditions, consideration of the rehabilitation conditions, maintaining of physical and geochemical stability, and finally topographic condition's fluidity of material are the most important factors. Additionally, environmentally friendly mining activities are alternative to accidents of tailing dams.

Environmental damages occurred during the deposition of mine process tailings at surface using conventional methods (tailings dam, etc.) can be reduced and/or eliminated with the use of new technologies and methods to an important extent. Nowadays, it is investigated to safely deposit metal mine wastes generally discharged into surface tailings dams, seas, rivers, creeks, etc. using Paste Tailing Technology, which gains the growing importance in recent years. Meanwhile, Paste Tailings Technology has two applications in industry as an underground paste backfill and surface paste deposition.

In recent years, paste tailings technology which is developed as an alternative to surface at deposition methods is defined as pumpeable fine grained material, which is originated from dewatering of mineral processing tailings with filtration or thickening process, and water addition at certain ratio and binder if necessary (Newman et al., 2001; Verburg, 2001).

One of the most important distinguishing properties of paste technology compared to conventional deposition methods is the particle size distribution of the mix material. Paste material must contain fine grained material in order to keep enough amount of water. According to studies in literature and experience, the particle size of 15% w/w of a typical paste material must be less than 20 μm in size. Therefore, it is possible to produce typical paste flow behavior materials due to the ability of keeping water of fine particles. If the paste is left motionless for a long time, it can release a little leachate. Furthermore, it is stable for long term, and doesn't show any risk such as the pipeline blockage (Benzaazoua et al., 2004; Yilmaz et al., 2014; Bascetin et al., 2015).

This study will also lead the disposal of all mining tailings on the surface as a paste. In addition, since tailings can be stored for longer years without any change on tailings surfaces with paste disposal method on surface, valuable metals in disposed tailings will be concentrated with new technological developments.

2 MATERIAL AND METHODS

In this study, a Pb-Zn underground mine located in the western part of Turkey was determined as a working site. Inadequacy of the land in terms of storage areas according to the amount of tailing produced, environmental risks that may result from metal mine tailings, and the rainy climate of the region were taken into account in the selection of the land in question. The region has a climate between temperate and cold in which winter months are rainy and summer months are hot and dry. There are four tailing dam sites, which belong to Pb-Zn underground mine.

This study consists of three basic stages, including, respectively, determination of the properties of tailings, preparation of paste material and the storage of paste material. At first, physical, chemical, geochemical and mineralogical properties of tailing sample taken from the filtration output were determined. The determination of these properties is extremely important in identifying the applicability of tailing with paste technology. After the determination of the properties of the tailing, preparation of the paste material was initiated and the optimum solid/pulp ratio value of the paste material was determined in terms of its storability. Finally, paste material mixture was formed according to this value determined and the storage of the tailings layer-by-layer in a laboratory scale cabin was ensured. The matric suction, volumetric water content, oxygen consumption, crack formations, cutting strength and the geochemical seepage analyze of the paste material stored in the cabin were performed, and the storing of the mentioned tailing as a surface paste material was evaluated in terms of environmental and stability according to the results obtained.

3 RESULTS AND DISCUSSION

3.1 *Properties of tailings*

In this study, physical, chemical, mineralogical and geochemical analyzes were performed to reveal the characterization of tailing material, and its physical properties are given in Table 1.

When analyzed the particle size distribution which is directly proportional to water holding ability of the material, it was observed that the particle ratio could be at least 15% thinner than 20 μm required for pumpability in the particle size distribution of paste materials is ensured by the value of 42%. It was observed that 8% of the particle size distribution of the tailing material was clay, 72% of it was silt and the remaining portion of 20% was fine sand sizes according to BS 1377-2 and ASTM: D 2487 standards.

The tailing material contains risk in terms of both its physical and geotechnical properties and Acid Mine Drainage (AMD) it can create and the heavy-metal mobilization. Therefore, it is very important to examine the mineralogical components and elemental contents constituting the tailing material. In this context, the tailing material in question was examined by ICP-MS and XRD analysis, and the chemical analysis results are given in Table 2.

It was seen that the data obtained from ICP-MS and XRD analysis results supported each other. It was observed that there was Quartz mineral in the tailing at most by 36%. Besides, CaO (23%) which was ranked as the second was observed to be one of the highest values that were contained in solid tailing because the wall rock of the ore zone was limestone, and that lime was used in schist and pyrite

Table 1. Physical properties of tailing.

Parameters	Values
Density	3.15 g/cm^3
Specific surface area	3.698 m^2/g
Porosity	53%
D_{10}	2.83 μm
D_{30}	11.48 μm
D_{50}	33.17 μm
D_{60}	45.71 μm
D_{90}	170 μm
Coefficient of Uniformity, Cu	16.15
Coefficient of Curvature, Cc	1.02

Table 2. Chemical analysis results of the tailing material.

Basic Oxides (%)					
SiO_2	Al_2O_3	Fe_2O_3	MgO	CaO	Na_2O
36.19	8.08	13.58	2.54	23.26	0.20
K_2O	TiO_2	P_2O_5	MnO	Cr_2O_3	
2.45	0.29	0.08	0.38	0.010	
Base and Precious Metals (ppm)					
Mo	Cu	Pb	Zn	Ni	As
5.7	217.2	1500.1	1548	29.4	601.8
Cd	Sb	Bi	Ag	Au	Ba
8.2	23.6	26.5	3.7	0.0558	404
Co	Rb	Sn	Sr	V	W
15.7	76.4	11	191.3	62	44.6
Zr	Hg				
76.8	0.37				
Rare Earth Elements (ppm)					
La	Ce	Pr	Nd	Sm	Eu
30.4	57.2	6.00	20.9	3.86	0.99
Dy	Ho	Er	Tm	Yb	Lu
2.87	0.55	1.62	0.27	1.58	0.23
Gd	Tb	Y	Sc		
3.37	0.54	17.2	7		
Radioactive Elements (ppm) and Other Values (%)					
Th	U	TOT/C	TOT/S	LOI	Sum
9.0	6.4	3.16	7.49	12.4	99.48

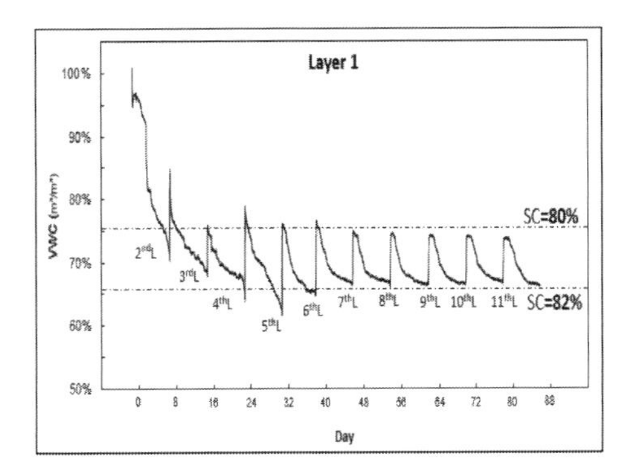

Figure 1. The volumetric water content and solid/pulp ratio (SC) values of Layer 1.

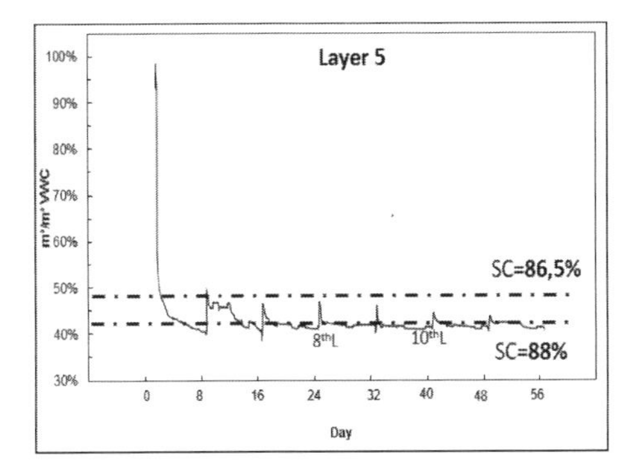

Figure 2. The volumetric water content and solid/pulp ratio (SC) values of Layer 5.

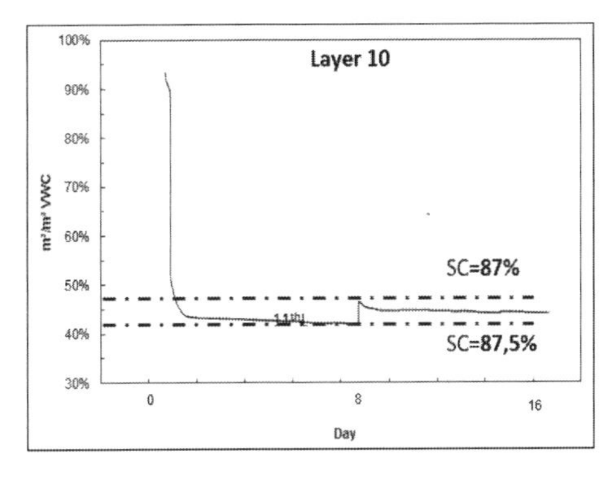

Figure 3. The volumetric water content and solid/pulp ratio (SC) values of Layer 10.

settling in the flotation process and in adjusting the pH value of the pulp. It can be expressed iron is mainly consisted of Pyrite, and aluminum consists of Albite. When looking at the loss on ignition values (Loi), it was determined that organic substances and volatile compounds constituted about 12% of the tailing material. These heavy-metal contents in Table 2 can significantly produce toxicity on soil and water from an environmental point (Simsek et al., 2012).

3.2 *Solid/water ratios of paste material*

The average solid/pulp ratio values of the layers were ensured to be 71.48%. Thus, all layers were ensured to be about 25 cm slump value during the preparation of paste material, and solid/pulp ratio values, which are one of the criteria of paste material was ensured to be between 65–75%.

3.3 *Measurements of sensors*

The volumetric water content of the paste material in the storage process was measured by the sensors placed on 1st, 5th and 10th layers, and the changes of the solid/pulp ratio values calculated accordingly were given in Figure 1 for the 1st layer, in Figure 2 for the 5th layer and in Figure 3 for the 10th layer.

Layer 1 was consolidated and quickly lost its water, and 82% of solid/pulp ratio value were reached at the end of 24 days in Figure 1. In the next layer casting process, it was determined that layer 1 was not affected by the effective stresses created by the 7 layers which were stored one on the top of the other after layer 4 and by the seepages, and it remained constant for the next 64 days in 82% solid/pulp ratio value. Also, according to seepage analyses performed, the seepage of the layer 1 was determined to be in the value of 7.4 and 2.9 mS/cm bearing no risk in terms of pH and EC values AMD.

As seen in Figure 2, it is seen that layer 5 remained constant in the range of approximately 86.5–88% after first casting. Also, it was measured that the volumetric water content of the layer 5 which was 100% after a short time after being cast rapidly reduced to 50%. After the casting of layer 6 which was an upper layer, the volumetric water content of the layer 5 increased by approximately 10%.

As seen in Figures 2 and 3, the solid/pulp ratio value during casting was 71.5 as in the layer 1. However, unlike the layer 1, by the effect of matric suction forces in the material layers, it became stable with the ~%88 solid/pulp ratio value, which was quite close to optimum water content value (10.07%). This also showed that the consolidation of the material was quickly completed. According to the seepage analyzes performed, pH and Ec values of the layers 5 and 10 were observed to be at similar values in the Layer 1.

4 CONCLUSIONS

In this study, the storability of the Pb-Zn mine tailings which are in environmentally hazardous waste class as a paste material aboveground was revealed. As a result of all analyses obtained, it is seen that the storage of the process wastes aboveground with the paste technology is possible so as to minimize the environmental risks. Also, it will be possible to monitor the behavior of the stored material under different climatic conditions and to keep its possible environmental impacts under control by monitoring in terms of physical, chemical and mechanical parameters after wastes are stored.

REFERENCES

ASTM D2487-11, Standard Practice for Classification of Soils for Engineering Purposes (Unified Soil Classification System).

Bascetin, A., Tuylu, S., Adiguzel, D. & Akkaya U.G (2015). The study of optimum tailing storing conditions for surface paste disposal method. *J. SME Annual Meeting & Exhibit*, 15–18.

Benzaazoua, M., Perez, P., Belem, T., Fall & M (2004). A laboratory study of the behaviour of surface paste disposal. Proceedings of the 8th International Symposium on Mining with Backfill, Beijing, China, September 19–21, pp. 180–192.

BS 1377-2:1990. Methods of test for soils for civil engineering purposes. Classification tests.

Newman, P., Cadden, A., White & R (2001). Paste—The Future of Tailings Disposal? Securing the Future. *J. International Conference on Mining and the Environment*. June 25 - July 1, Skelleftea, Sweden, pp. 594–603.

Simsek, C., Gunduz, O.& Elci, A (2012). Assessment of Environmental Quality Effects of Heavy Metal and Natural Radioactivity from Abandoned Balya (BALIKESIR) Pb-Zn Mine Waste. *J. Journal of Engineering Science and Design*, Vol:2 No:1 pp. 43–55.

Verburg, R.B.M. 2001. Use of Paste Technology for Tailings Disposal: Potential Environmental Benefits and Requirements for Geochemical Characterization. *J. International Mine Water Association (IMWA) Symposium*, Belo Horizonte, Brazil, p. 1–13.

Yilmaz, E., Benzaazoua, M., Bussière, B. & Pouliot, S (2014). Influence of disposal configurations on hydrogeological behaviour of sulphidic paste tailings: a field experimental study. *J. Int. J. of Miner. Process.*, 131, 12–25.

Land Reclamation in Ecological Fragile Areas – Hu (Ed.)

Ecological restoration of coal mine degraded lands using a grass-legume mixture—a case study from India

Subodh Kumar Maiti & Jitendra Ahirwal
Department of Environmental Science and Engineering, Centre of Mining Environment, Indian Institute of Technology (Indian School of Mines), Dhanbad, Jharkhand, India

ABSTRACT: In India, coal deposit is locked under forest cover and more than 92% is extracted by surface mining that causes massive destruction of land and vegetation, alterations in landscape and aesthetics. Minesoils act as continuous source of pollution until they are reclaimed. Since last five decades, reclamation by tree plantation was the sole aim by the mining companies. From last decade only, it has been realized that degraded areas have to ecologically restore by developing three-tier canopy cover. In this paper, ecological restoration of coal mine overburden was under taken by using grass-legume seeds – as an initial coloniser and observed that after 3 years overall increment of soil organic carbon was 80%. Available N was increased from 70 mg kg^{-1} to 225 mg kg^{-1} after 3 years. Available P and potassium also increased substantially. Cost of ecological restoration by using grass-legume mixture an initial coloniser and plantation of tree sapling (2500 nos/ha and maintenance for 3 years) comes to USD 11300 ha^{-1}.

1 INTRODUCTION

Globally, India is the 3rd largest producer of coal and produces 639 Million Tonnes (MT) of coal in year 2015–16, against 609 MT in 2014–15 showing an increase of 4.93% over the previous year and meet 60% of total energy demands. Out of total coal production of 639 MT, 92.74% production from opencast mines (593 MT) and rest 7.26% from underground mines (46 MT). In the year 2015–16, overall stripping ratio was 2.67. Government of India (GOI) targeted the production of coal up to 1.5 BT by 2022. To meet this target, Coal India Limited (CIL) will substantially enhance production to 1 BT and rest 500 MT by private sector. Therefore, CIL has to almost double their current level of production of 538.75 MT in 2015–16. As per annual plan document of Ministry of Coal (MOC), all India target of coal production for 2016–17 was fixed at 724.71 MT. Ministry of coal, emphasized that the focus of the government is to increase the domestic production of coal which includes efforts to expedite environment clearances & forest clearances, pursuing with state government for assistance in land acquisition and coordinated effort with railways for movement of coal. This includes capacity addition from new projects, use of mass production technologies and identification of existing on-going projects with growth potential. The minister added that steps have also been taken by CIL and its subsidiaries to improve the production of coal by adopting latest available technologies such as continuous miners, selective mining, surface miners and clean coal technologies, coal sizing and sampling technologies.

In India, majority of new and unmined mineral and coal deposits is under forest cover; thus, complete degradation of land, destruction of forest ecosystem and fragmentation of habitat are inevitable, and magnitude is so massive that entire landscape is changed (Photo 1–4). Currently, to meet the target of production, a number of large capacity opencast project ranging from 40–50 MT are envisaged. Fortunately, unlike other industries, mining is very temporary user of land, and proper scientific ecorestoration can restore the functioning of ecosystem and may bring better landscape. There is a growing concern to make the land useful yet again. Several organisations are concerned about these aspects and are actively involved to carry out restoration work. But in some places, in small scale, alternative land use were developed (eco-park, fruit orchards, etc.).

Reclamation, Rehabilitation and Restoration are commonly used term in the field of restoration ecology to describe the state of an ecosystem. In mining context the term reclamation describes the general process whereby the land surface returned to some form of beneficial use (Maiti, 2012). According to Society for Ecological Restoration, reclamation encompasses the stabilization of the terrain, assurance of public safety and overall aesthetic, whereas rehabilitation emphasizes the reparation of ecosystem processes, productivity and service. Ecological restoration is the process of assisting the recovery of an ecosystem that has been damaged or destroyed by the anthropogenic intervention (SER, 2004).

1.1 *Land reclamation practices in India*

Restoration to pre-disturbed condition is often impossible, although several efforts have been taken to restore ecosystem services mostly through development of forest as end -use. In India, technical reclamation of mine degraded sites is largely carried out by the mining companies/owners. Second most task "biological reclamation" is carried out by the State forest Department/other State Department on contractual basis by planting tree saplings in plantation pits with a density of 25000 trees/ha. Most of the reported reclamation studies were invested in these reclaimed lands covering—accretion of nutrients in mine soil and

Photo 1. Distance view of opencast coal mines showing magnitude of land disturbance.

Photo 2. View of barren overburden dumps. Nearby OB is reclaimed with tree plantation.

Photo 3. View of reclaimed dump.

Photo 4. View of an eco-park developed on reclaimed dump.

mine reclamation (Srivastava et al., 1989); spoil character and vegetation development, influence of microsites on redevelopment of vegetation on coalmine spoils (Jha and Singh, 1992); importance of VAM fungi in coalmine overburden reclamation (Mukhopadhyay and Maiti, 2011); biological reclamation of coal mine spoils without topsoil (Maiti and Saxena, 1998); litter fall, litter decomposition and nutrient release pattern (Singh et al., 1999) are few of them. Earlier the concept of reclamation only aims of plantation to improve the scenic and aesthetic values of the environment. Type of plant species commonly used for restoration in the coalmines of Coal India Limited, India is shown in Table 1.

Plantation play constructive role in the reclamation of mine spoil because it modifies the soil characteristics (Dutta and Agrawal, 2002) nevertheless, it the oldest technology used for the rehabilitation of the degraded land. Moreover, in the new age

Table 1. Type of plant species commonly used for reclamation of coal mine overburden dumps in Indian (Ahirwal et al., 2016).

Trees	Ramagundan SCCL	Jharia Coalfields	KD Heslong, CCL	SECL
Acacia auriculiformis	√	√	√	√
Acacia mangium	–	–	√	–
Acacia catechu	–	–	√	–
Ailanthus excelsa	√	√	√	–
Albizia labbeak	√	–	√	–
Alstonia scholaris	–	√	–	–
Azadirachta indica	–	√	–	–
Cassia fistula	–	√	–	–
Cassia siamea	√	√	√	√
Casuarina equsetifolia	–	–	–	–
Dalbergia sissoo	√	√	√	√
Dendrocalamus sp	√	√	√	–
Delonix regia	–	√	√	–
Eucalyptus spp.	√	√	√	–
Gmelina arborea	√	√	√	√
Gravellia robusta	–	–	–	√
Heterophragma adenophylum	–	–	√	–
Inga dulcis		–	–	√
Leucaenea leucocephala	–	√	√	√
Melia azedarach	√	√	√	√
Peltophorum pterocarpum			√	
Pongamia pinnata		√	√	√
Prosopis juliflora	√	√	√	–
Terminalia arjuna	–	–	–	√
Tectona grandis	–	–	–	√

of ecological restoration, several studies like comparative performance and restoration potential of species planted on mine spoil (Singh et al., 2004), bioreclamation of coalmine overburden dumps and metal accumulation in tree species (Maiti, 2006; Maiti, 2007), ecological restoration of mined-out areas by planting trees (Tripathi and Singh, 2008), strategies for development of sustainable ecosystem on mine spoil dumps by adopting Microbe Assisted Green Technology (Juwarkar et al., 2009), use of reclaimed mine soil index for screening of tree species for reclamation of coal mine degraded land (Mukhopadhyay et al., 2013), mine spoil acts as a sink of carbon dioxide (Tripathi et al., 2014), soil development in reclaimed coal mine spoils (Kumar et al., 2015) carbon sequestration on reclaimed lands (Das and Maiti, 2016a, 2016b; Tripathi et al., 2014); Carbon sequestration and CO2 flux (Ahirwal et al., 2017) were conducted to demonstrate the potential of the restoration practices and to determine the role of that practice in the development of the ecosystem services.

Restoration to pre-disturbed condition is often impossible, although several efforts have been taken to restore ecosystem services. Very few studies concerning the nutrients in mine soil and mine reclamation; spoil character and vegetation development, influence of microsites on redevelopment of vegetation on coalmine spoils; coal based environmental problems, importance of VAM fungi in coalmine overburden reclamation (Maiti, 2012); biological reclamation of coal mine spoils (Ahirwal and Maiti, 2016); litter fall and nutrient release; tree canopy development; carbon sequestration in restored sites, use of coir-mat (Maiti and Maiti, 2015) can be found in the research literature.

The restoration process can generally include the following seven successive key steps – (i) identify the processes those causes damage to the environment, (ii) suggest restoration measures that minimize degradation, (iii) suggest sensible objectives for the restoration, (iv) design parameters that asses the recovery process, (iv) design concrete restoration measures, (v) application of these measures to the project and its practical implementation, and (vi) monitoring of restored sites.

1.2 *Essential activities of forestry reclamation approach*

1. Forestry Reclamation Approach (FRA) applicable to Indian conditions are simplified as follows:
2. Ensure that all objectionable/toxic strata are buried carefully while during construction and backfilling waste dumps;
3. Backfill and regrade the waste materials to the surface configuration or contour.
4. Spread and regrade the material from B- horizon (subsoil), over the graded waste materials in a planned thickness to form the sub-surface layer of the reconstructed site (at least 2 feet).
5. Spread and regrade *A* horizon (topsoil) over the sub surface horizon in a uniform thickness (0.5–0.6 m or 2 feet);
6. Design proper drainage (top 3% and 28–30% at slope) to control erosion and stability of dumps. Design of Sediment retention basin and design of sediment trap;
7. Sampling and analysis of the finally reclaimed mine soils for addition further amendments like, pH modification, organic and inorganic fertilization for creating favourable conditions for seeds germination and plant growth;

Ecorestoration in mine degraded site seeks to stimulate natural succession processes leads to forest. All vegetation types are established at the initial stage (grasses and legumes) of reclamation. As time passes, grasses and legume covers enhance yield of organic matter and nitrogen for the fast-growing trees and desire Multipurpose Trees (MPT), which are gradually mature and develop forest.

1.3 *Development of revegetation programme during ecorestoration*

Successful tree seedling establishment on drastically disturbed lands is contingent on seven major variables (Miller, 1998): (i) selection of proper native species; (ii) purchase of the best quality planting stock; (iii) correct handling of planting stock; (iv) correct planting techniques; (v) effective control of competing vegetation; (vi) proper soil conditions and preparation; and (vii) weather. While determining appropriate revegetation programme for a site, the following aspects should be considered (Maiti, 2012)

1. Future land use or land cover on derelict sites intend to develop (as per closure plan report);
2. Climatic conditions including mean daily temperature, the growing season, the duration of critical moisture deficits, and precipitation,
3. Size of the revegetation area in order to assess material requirements (e.g. planting stock, seed and soil amendment, geotextile)
4. Contouring of the area to mimic local topography and blend into surrounding landscape consistent with future land use or land cover
5. Creating of water bodies in the low lying area (if possible) or other special considerations both pre and post mining
6. Availability of stockpiled materials for revegetation
7. Success of natural revegetation and species present
8. Contouring to ensure proper drainage or re-establish previous drainage (if possible),
9. Identify erosion prone areas, and the necessity for erosion control work, including the use of bio-engineering techniques (geotextile, coir-mat)
10. Mine soil characteristics including texture, pH, moisture regime, soluble salts and content of nutrients and organic matter and required amendments that may affect revegetation success
11. Use of original or native species present on the site
12. Re-use of soils on the site that were shifted during mining activities,
13. Timing of seeding to coincide with optimal germination times (depends on local climatic conditions)

1.4 *Use of a grass-legume mixture and tree species*

Ecorestoration in mine degraded site seeks to stimulate natural succession processes leads to forest. All vegetation types are established at the initial stage (grasses and legumes seed are sown in the interspacing of tree rows) of reclamation. As time passes, grasses and legume covers enhance yield of organic matter and nitrogen for the fast-growing trees and desire multipurpose trees (MPT), which are gradually mature and develop forest (Figure 1).

Grass and legume mixture is primarily used for erosion control and minimization of run-off volume. It is a technique for quickly covering the surface of a disturbed or degraded site. Native revegetation is different from temporary reseeding, which a practice is used to provide short term cover for a site scheduled for future disturbance (Maiti, 2012). It performs roles of mulch and hold loose materials and ameliorate habitat by (i) enhancing infiltration; (ii) improve drainage, because it increases local hydraulic conductivities; (iii) reduce soil erosion; (iv) enhance slope stability and reinforcement; and (v) amelioration of site conditions for vegetation establishment and growth. Important grass-legume mixture proved to be enormous success for Indian conditions are - *Stylosanthes humilis, S. hamata* (Caribbean Stylo),

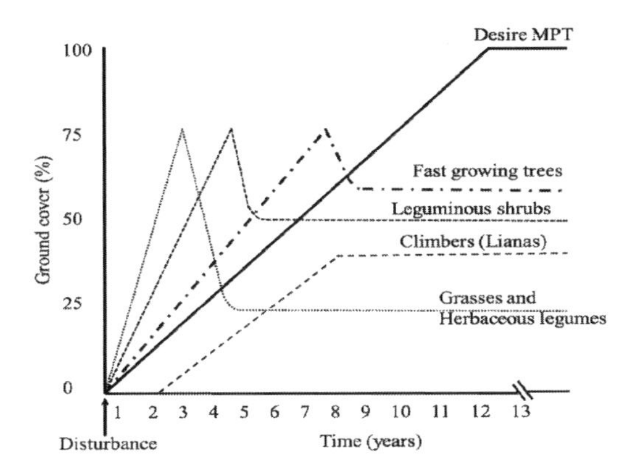

Figure 1. Schematic vegetation model of ecological restoration of coalmine degraded site (Ahirwal et al., 2016).

Sesbania sesban, S. cannabinis (Sesban); *Crotalaria spectabilis, C. juncea; Vetiver zizaniodes* (Vetiver grass); *Cybopogon citratus* (Lemon grass); *Pennisetum pedicellatus* (annual grass) and *Hibiscus sabdariffa.*

This mix plantation creates a nitrogen balance in the soil, and decomposition of dry plant parts creates nitrogen rich litter and mulch for soil. Grasses have extensive fibrous root systems which can reduce erosion by holding the loose soil particles, can tolerate adverse soil conditions and form mulches after drying (Maiti, 2012). Generally perennial forage type legumes are used but native species show greater improvement in soil fertility. Nitrogen fixed by forage legumes are dependent on legume growth and persistence and ranged from 0.3 to 40 kg N ha^{-1} 12 $weeks^{-1}$ during west season and found highest with *Stylosanthes* (Thomas et al., 1997). Apart from N, P is the main limiting nutrients for establishment of legumes. Phosphorus tolerance to legumes varies greatly amongst the species. The tropical legume *Stylosanthes humilis* is very efficient in P uptake from low-P soils and may have a relatively low P requirement for maximum growth (Coppin and Bradshaw, 1982). It can fix atmospheric N, enhance soil fertility and encourage biomass production. Application of P @ 10 kg ha^{-1} increases N fixation from 5 kg ha^{-1} to > 50 kg ha^{-1} but also and biomass productivity from 26% to 40% (Mapfumo et al., 2005).

Legume litter has a high N concentration and thus has high early decomposition rates which underscores its distinctive role in ecosystems and nutrient cycling. Its presence also improves litter quality (Spehn et al., 2002), decomposition (Scherer-Lorenzen, 2008; Milcu et al., 2008) and certain legume species assist nitrogen transfer to co-occurring non-leguminous plants (Spehn et al., 2002). In nutrient-poor soils, legumes have a competitive advantage over the other species in the community. Study confirms a higher decomposition rate of legumes over grasses (Agbenin and Adeniyi, 2005). Thus, during the hydroseeding of grass-legume mixture, a higher percentage of legumes seeds are used, which will accelerate nutrient cycling, improve the quality of soils, biodiversity and sustainability of the plant community.

The aim of this study are (1) to investigate the changes in physicochemical and nutritional properties of waste dump surface due to growth of grass-legume mixture and (2) effectiveness of use of grass legume seeds for stabilisation of steep slope and amelioration of habitat.

2 MATERIAL AND METHODS

2.1 *Site description*

Revegetated coal mine overburden dumps of Jharia Coalfields (Bharat Coking Coal Limited, BCCL), Jharkhand, India was selected to study the impact of grass-legume mixture on soil fertility. The Jharia coalfields falls between latitudes 23°39′–23°48′N and longitudes 86°11′–86°27′E covering an area of 450 km^2. The climate of the study area is dry tropical and received an average 1306 mm of rainfall annually out of which 80% of the rainfall received during the monsoon season. The mining is carried out by shovel–dumper combination. The average height of dumps was 40–50 m with a quarry depth of approximately 60–70 m. The rock type consist of sandstone, carbonaceous shale, sandstone and coarse coal particles.

2.2 *Engineering reclamation of the study site*

Top surface of the dump was levelled by dozer and a 3° slope towards the central drainage system was provided. The slope was seeded with grass-legume mixture before the onset of monsoon. Again seeds of grass-legume mixed with soil and Farm Yard Manure (FYM) (soil: FYM; 1:1) were re-spread in bare areas where germination of grass-legume were not observed. About 2000 nos of tillers of *Cymbopogon citratus* were planted along the brim of the dump in 0.5 m × 0.5 m dimensions, and towards inner side one row of *Azadirachta indica* was developed by seeding. The vegetated slope was irrigation in the summer months. Plant growth monitoring and soil sampling was done after seven months. Reinforced concrete drainage was provided to drain runoff from top-surface of the dump, while seepage from the slope was drain-out through earthen channel.

2.3 *Soil and vegetation sampling*

Replicate soil samples were collected from underneath the canopy of *Pennisetum pedicellatum* and *Stylosanthes hamata* growing on the revegetated coal mine dumps. A total 5 replicates mine soil samples were randomly collected by laying 10 m × 10 m quadrate and using the soil corer at 0–15 cm depth. Sample were packed in air tight polyethylene bags and brought to laboratory for physicochemical analysis. Five replicates of each species were collected using quadrates of size 1 m × 1 m at the each corners and centre point of the 10 m × 10 m quadrate from where soil samples were collected. Initial growth measurement of species *P. pedicellatum* and *S. hamata* was measured by measuring steel tape. Species were uprooted carefully to measure the shoot and root length and then taken to laboratory for measurement of fresh biomass. Fresh biomass of each species was dried in a drying hot air oven at 65°C for 48 h to measure the dry weights of the species.

2.4 *Sample analysis*

All the collected soil samples were air-dried and lightly crushed with mortar and pestle then passed through a 2-mm sieve for physicochemical analysis. Fine earth fraction (< 2 mm size) was determined by the sieving method. Soil pH and Electrical Conductivity (EC) were determined in soil: water suspensions (1:2.5, w/v) with a pH meter and conductivity meter. Soil Organic Carbon (SOC) was determined by the rapid dichromate oxidation method (Nelson and Sommers, 1996). To determine the available nitrogen (N), samples were distilled with alkaline potassium permanganate and titrated against sulfuric acid (Subbiah and Asija, 1956). Soil total nitrogen was quantified by the dry combustion methods using CHNS analyser (Model: Eurovector EA 3000). Available phosphorus was extracted by Bray's method and quantified with a spectrophotometer (Bray and Kurtz, 1966). Exchangeable potassium (K) were extracted by 1 N ammonium acetate solution and quantified by flame photometer determination (Jackson, 1973).

3 RESULTS AND DISCUSSION

3.1 *Germination of grass-legume seeds*

Most of grasses and legume seeds (herbaceous and forage categories) have little seed-dormancy and do not require treatment to increase germination. The required quantity of seed is determined by weight. Seed germination generally expressed as percentage, measure the number of seeds in a lot that can be expected to germinate and grow to became a health plant. Seed test can be performed by placing some seeds on moist paper in a covered petri dishes. Leguminous seeds have a low dormancy period and within 6-days, 60% of the seeds are germinated and rest germinated in 10 days. Maximum germination of grass seeds occurred within a week (Figure. 2).

3.2 *Vegetation growth*

Entire slope was covered, stabilized and erosion was controlled during the monsoon with lash green lash green growth of *Stylosanthes* legume. Distance view of restored dump is shown in **Photos 5** and **6**. During field survey in monsoon, it was observed that, creeper nature of *Stylosanthes* formed 30–40 cm carpet above the coir-mat, and underneath decomposition of leaves (black colour) contributes SOM and initiate formation of humus. Natural colonization of vegetation on slope surface due to accretion of habitat by *Stylosanthes* legume was observed. For example, colonization of *Evolvulus alsiniodes*, which was not sown, rather colonized in the later stage, and Co-existed with *Stylosanthes*. At the toe of restored dump regermination and luxurious growth of *S. sesban* in the monsoon season was observed. *S. sesban* is tall woody annual shrubs, excellent nitrogen fixer, stabilizes toe part of dump, control erosion and die

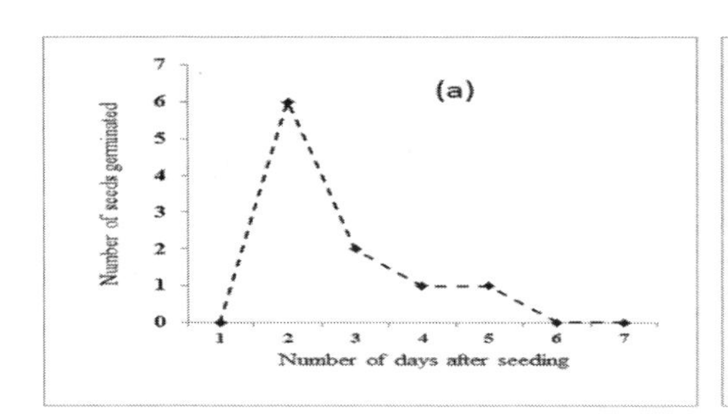

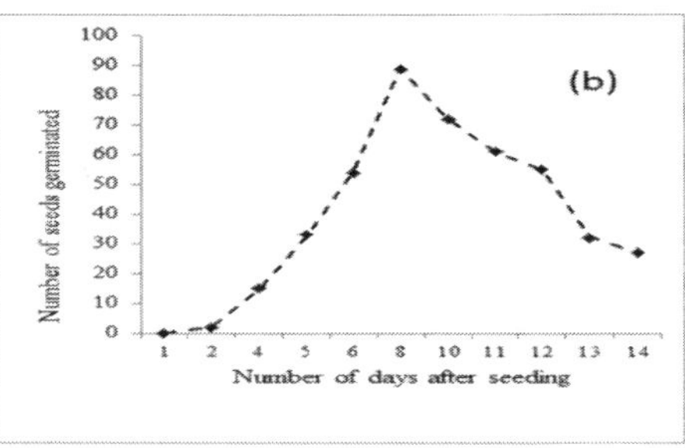

Figure 2. Laboratory Seed germination of (a) *Stylosanthes hamata,* and (b) *Pennisetum pedicellatusm.*

Photo 5. Growth of grass-legume mixture in the slope.

Photo 6. Close view of stylosenthus legume.

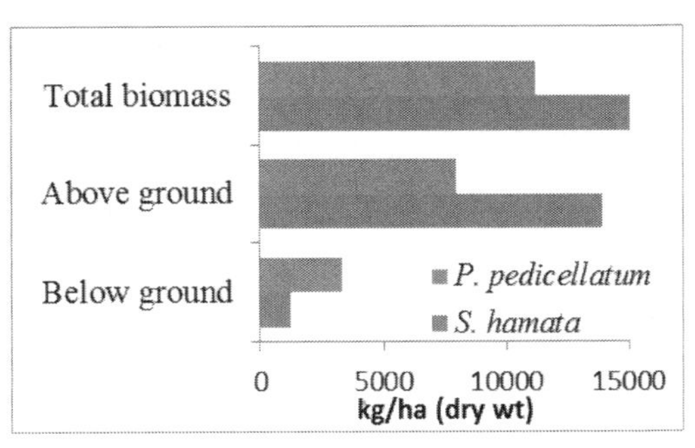

Figure 3. Biomass production by grass-legume mixture sown at the reclaimed site.

Photo 7. View of dry mulch in the reclaimed site.

after monsoon. Dominance of *Stylosanthes* on the slope was observed, and with passage of time (2nd yr onwards), *Stylosanthes* substantially reduced abundance of Dinnath Grass on the slope. Highest total biomass production was observed under *S. hamata* and then *P. pedicellatum*. Highest leaf litter production was observed under *S. hamata*, while highest amount roots were produced by grass species, which actually help the stabilisation of surface.

3.3 *Shoot and root biomass*

Growth and development of biomass by *S. hamata* and *P. pedicellatum* is shown in Figure 3. Maximum biomass is contributed by legumes *(S. hamata)* in the order of 15000 kg/ha (dry wt basis) out of which 90% is contributed by above-ground biomass. *S. hamata* legume is a perennial legume, survived for 3–4 years and during dry season, its above ground biomass formed a dry mulch that ameliorate spoil surface temperature (Photo 7). In case of grass, contribution of above ground biomass is lesser than legumes. *P. pedicellatum* being a annual grass, they also contribute mulch formation at the dump surface. Comparatively higher root biomass than legumes helps to anchor loose soil particles which provides stability of dump.

Table 2. Comparison of accumulation of dry mulch in ecologically restored dump with litter fall amount in reference site (forest area).

Study site	Vegetation cover	Mulch /litter biomass (t ha^{-1})
Reclaimed dump	*Stylosanthe hamata*	9.03–9.28
Reclaimed dump	*Pennisetum pedicellatum*	9.21–9.54
Forest site	*Butea monosperma*	6.13–6.81

3.4 *Amelioration of soil temperature due to vegetation development*

Soil temperature is an important parameter controls biochemical processes and growth of vegetation which helps to restored nutrient cycling in the derelict site (Maiti and Maiti, 2015). Mulch properties like quantity and architecture also affect soil microenvironment. Mulching makes soil less prone to erosion, increase infiltration, increase biomass, seedling establishment, ameliorates soil surface temperature and different types of mulches modify soil temperatures in different ways (van Donk and Tollner, 2000; Ji and Unger, 2001; Mollard et al., 2014). Accumulation of dry mulch contributed by grass-legumes was observed 9–9.5 t ha^{-1} on the waste dump, which was higher than the reference forest site (6.65 t ha^{-1}). Comparison of accumulation of dry mulch in ecologically restored dump with the litter fall amount on the reference site (forest area) during the spring season is given in Table 2. The plant survival rate could decrease due to extreme temperatures. High temperature could impede the germination of seeds or kill seedlings before their root systems grow deep and reach a more suitable soil environment (Peper and McPherson, 2003; Fehmi and Kong, 2012). Therefore, mulch-covered surface soil could provide the appropriate soil temperature condition for seed germination, seedling survival and plant growth (Mollard et al., 2014).

Higher accumulation of dry mulch was due to the massive growth of the leguminous species. Their accumulation also had a significant influence in the reduction of rhizospheric temperature. A significant reduction in the rhizospheric temperature of 32% was observed under mulch cover, while it found higher in the unreclaimed dump. The nearby reference forest site also showed 11% reduction in soil temperature under litter. Wang et al. (2017) studied the effect of mulches and legumes for the restoration of urban abandoned land and reported decrease in temperature from 6% to 18%. This indicates dry mulches on the reclaimed site not only enhance SOC content but also their dry parts ameliorate surface temperature during summer and help to conserve moisture (Figure 4).

3.5 *Soil properties*

Physicochemical characteristic of the revegetated mine soils and reference site are summarized in Table 3. Soil pH after was decreased (7.8 to 6.8) after three 3 years of the establishment of grass and legume mixtures and found suitable for plant growth. The symbiotic association between the N fixing bacteria and plants root was decreased at low pH (Bordeleau and Prevost, 1994). Generally, the soil pH of 5.0 or higher is found suitable for nitrogen fixation because bacteria will not penetrate many of the legumes if the soil pH is less than 5.0. The change in electrical conductivity from 0. 73 dS m^{-1} to 0.30 dS m^{-1} was observed after three years of revegetation. In reference soil, EC was ranged from 0.15 to 0.16 dS m^{-1} which is lower than that of revegetated mine soils. In both the revegetated and reference sites values were optimal for plant growth.

Fine earth fraction (< 2 mm) in the revegetated soils increased with the age of reclamation and showed the increment of 77% after three years of revegetation. Low fine earth fraction in RMS can be explain by the nature of mining, characteristics of parent rocks materials, and magnitude of technical reclamation. Over the years, growth of the grass-legume mixture along with the extensive fibrous root system trapped the loose mine spoils and helps in paedogenesis process. In Indian mining context, very low fine earth fraction up to

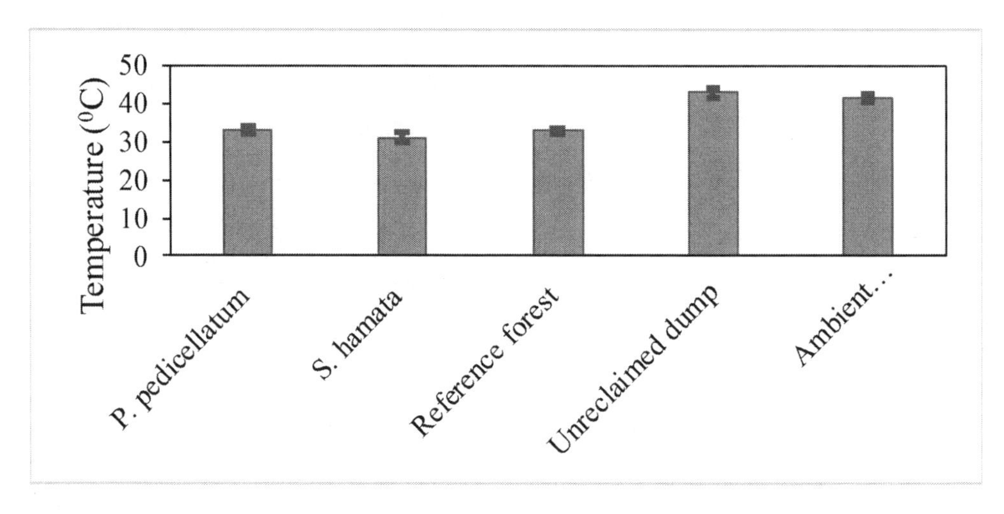

Figure 4. Variation in the temperature amelioration under different vegetation cover and land use.

Table 3. Soil physicochemical characteristic of the revegetated coal mine dumps and reference site.

	Revegetated site			
Soil properties	1 year	2 years	3 years	Reference site
pH (1:2.5; soils: water)	6.4 (5.7–6.6)	7.5 (7.2–7.6)	6.8 (6.7–7.0)	6.6 (6.5–6.8)
EC (dS m^{-1})	0.81 (0.76–0.83)	0.73 (0.70–0.75)	0.30 (0.28–0.33)	0.16 (0.15–0.16)
Fine earth fraction%	27 (24–29)	32 (28–34)	48 (45–52)	46 (43–54)
Soil organic carbon%	0.43 (0.38–0.50)	1.05 (0.90–1.15)	1.90 (1.6–2.0)	1.10 (0.63–1.40)
Av. nitrogen (mg kg^{-1})	25 (18–29)	70 (65–73)	225 (215–238)	175.2 (130–195)
Total nitrogen%	0.008 (0.007–0.008)	0.017 (0.15–0.020)	0.164 (0.154–0.182)	0.150 (0.118–0.162)
Av. phosphorus (mg kg^{-1})	0.80 (0.73–0.91)	1.60 (1.25–2.0)	4.75 (4.0–5.5)	1.8 (0.7–2.8)
Ex. potassium (mg kg^{-1})	52.1 (48.2–55.3)	58.2 (55.5–60.0)	132.66 (120–140)	115 (105–140)

15–20% was reported (Maiti and Saxena, 1998). Roberts et al. (1988) studied the mine soil genesis and reported 65 to 71% of coarse fragments in unreclaimed mine soil of Virginia, USA. Higher coarse fraction in the mine soils may restrict the vegetation growth by limiting the root proliferation, water-holding capacity and long-term nutrient availability of plants (Maiti, 2012).

Initially, surface mining activities remover all the topsoil and subsoils materials resulted in loss of SOC content. In the present study, initially organic carbon concentration in the mine was 0.43% that can be explain by lack of fresh C input in to the soils and removal of topsoil for the mining. With passage of time, the level of SOC was increased from 0.43 and reached to 1.9% at the end of three years. The increment of SOC was 144% and 81% for second and third year, respectively. In the control plot, SOC concentration was found 1.10%. The increase in the SOC concentration may be attributed to the legume growth and decomposition of the grasses and legumes on the spoil surface. Vegetation restoration also can improve the soil's microorganism activity that enhanced the decomposition of organic matter. Soil organic matter is closely associated with various physical, biological, and chemical properties of soil and thus plays a significant role in most soil processes and functioning (Maiti, 2012; Yang et al., 2016). The restored urban green area accumulation of organic matter was primarily the result of legume growth and the decomposition of leaf litter. However, the degradation of natural mulch may also have an important effect on the soil organic matter.

It has been well established that nitrogen accumulation and building up of nitrogen cycle is the most crucial factor in soil development in mined out areas. In present study, initially Av-N concentration was very low (25 mg kg^{-1}). After two years accumulation available N was increased up to 70 mg kg^{-1} and it increased to 225 mg kg^{-1} at the end of three years. Accumulation of available N by legumes was compared with the control plot and found the increment of 600%. Higher percentage of mineralizable N in restored dump is due to higher microbial activity and nitrogen fixation by legumes (Maiti and Maiti, 2015). Available N-stock recovery of 61% was observed after 8 years of revegetation of *Prosopis juliflora* plantation in coal mine overburden dumps and found suitable in try tropical climate (Ahirwal et al., 2017). Nitrogen accumulation is controlled by the organic carbon input and nitrogen fixation (Fehmi and Kong, 2012). The higher percentage of total nitrogen in a restored urban green area is caused by the biological nitrogen fixation of legumes (Andrews et al., 1998). It has been reported that most legumes can fix 50–150 kg N $year^{-1}$ (Coppin and Bradshaw, 1982). Li and Daniels (1994) studied the N accumulation in the young mine soils of Appalachian region of USA and reported 250–475 mg N kg^{-1} in the > 10 years of mine soil, held on micas and nonhydrolyzable organic N bound in coal. Both the N fractions were geologic N, stable in chemical character and unavailable to plants. With time, N accumulation occurred primarily in the surface 0 to 5 cm, associated with A- horizon development. The main source of nitrogen accumulation is the association of the symbiotic nitrogen fixation bacteria with the leguminous tree species growing on the reclaimed sites.

The soil C:N ratio after the three years of revegetation on mine soil was found 11.6 which is higher than that of the reference site (7.33). The improvement in C: N ratio can be explain by the higher accumulation of organic C by mineralisation process. The C: N ratio is important to assess the soil fertility status because it has a direct impact on SOM decomposition and nitrogen cycling. Singh et al. (2007) studied the impact of forest degradation on soil properties and reported that undisturbed soil showed the C: N of 10.47 and degraded soils showed relatively higher C: N ratio of 17.15.

With the passage of time, soil P and K contents increased in the revegetated mine soils from

0.8 mg kg^{-1} to 4.75 mg kg^{-1} and 58 mg kg^{-1} to 115.0 mg kg^{-1}. Reference soil showed the lower concentrations of P and K than that of three years old revegetated site. The main source of available P in mine spoil is the breakdown of spoil materials. The low level of P may be attributed to the low release of P from shale and fixation of a considerable portion of P by shale (Maiti, 2007). The P concentration below 20 mg kg^{-1} indicates the low fertilizer of the soil. In soils 100 mg K kg^{-1} is sufficient for plant growth, 50–100 mg kg^{-1} indicates moderate deficiency range and 0–50 mg kg^{-1} indicates high deficiency. Concentration of K in the revegetated mine soils found satisfactory for the plant growth. Soil N, P and K are the major limiting nutrients returned through litter fall and play an important role in soil fertility and tree growth (Das and Mondal, 2016). With the age of revegetation, increase in the plant biomass and litter fall increases the plant available nutrients through decomposition processes.

3.6 *Cost of ecological restoration*

The cost of restoration is location specific and varies between countries as well as within the country (Maiti and Maiti, 2015). The major factors that influence the restoration cost are: nature of waste, height and slope of waste dump, designated end land use, quantity and quality of soil (availability and distance from the site), source of water, quality of seeds, site maintenance (roads, infrastructure, power line, fences, security guard), labour, regulatory compliance and climatic conditions. Topsoil is regarded as a strategic resource for the success of any ecological restoration project. Nicholas and McGinnies (1982) reported that herbage yields were seven times greater and root yields six times greater when the grasses were grown in topsoil (25 cm) than when grown in spoil without topsoil. Even though, for the present work, soil was available free of cost, but the cost of topsoil was considered @ USD 0.62 m^3 for restoration work. Running cost of Heavy Earth Moving Machinery (HEMM) accounts 8–9% of the total cost, which again depends on efficiency of use of HEMM and supervision. Cost of seeding material comes close to 8%, and irrigation cost comes to 33% assuming there is no close sources on water, and underground boring of 150 feet (current cost of boring is Rs 1500/ feet, which includes cost of pipe, labour charges and machinery used).

Table 4. Cost of ecological restoration of waste dump blanketing with soil and seeding with grass–legume mixture for initial development of vegetation cover (per hectare basis).

Sl. no.	Activity	Cost in Indian Rs ha^{-1} (US$ ha^{-1})
	Technical or Engineering restoration	
1.	Running cost of dozers, loader, pockland for grading of dump, salvaging of topsoil, transportation, topsoil application on slopes, compact soil on the top surface and provision of drainage	35,000 (540)
2.	Excavation and transportation of soil for blanketing the flat surface of waste dump (50–60 cm thick) and in slope (20–30 cm thick), 5,00 m^3 (approx) @ Rs 50 m^3	25,000 (385)
3.	Cost of soil materials up to a depth of 100 cm with moderate fertility from forest area 500 m^3 @ Rs 40 m^3	20,000 (310)
	Biological restoration	
4.	Grass–legume seeds (*S. hamata* and *P. pedicellatum*) – 300 kg @ Rs 180 kg^{-1}	54,000 (830)
5.	*Development of irrigation facility*: Construction of boring with 15.24 cm GI pipe, depth of 45.72 m; 5HP pump (7.5 kwh, 3 stages) – submersible (lump sum Rs 1500 per feet) – one time	225,000 (3500)*
6.	(a) Aftercare and maintenance of site for 3 years which includes, watering arrangement during lean seasons, provision of watch guard, day to day unskilled labor charges, running cost of pump for 2 h per day etc. (lump sum @ Rs 1,00,000 per year/ for 10 ha) (b) Cost of tree plantation (indigenous species) by pit plantation method (pit size: 45 cm × 45 cm × 30 cm) @ 2500 trees/ha (cost of sapling @ Rs 30–45/ sapling based on types of tree species)- Total cost comes to Rs 75,000 to 112,500/- which includes 3 yr maintenance and payment will be made after actual counting of tree species survived after 3 years).	3,75,000 (5769)
	Total	734,000/ha (USD = 11300/ha)

Conversion unit: 1USD = Rs 65* One time investment only. The cost may be varied depending on the sources supply of water. In the present cost estimate, assuming that no nearby river/stream is available, and such situation construction of boring is essential.

Aftercare and maintenance for 3 years constitutes 51% of the cost (higher cost estimates comes due to lesser area (1 ha), however one watch-guard can look after close to 10 ha area). It is always advisable to test the Real Value (RV) or Pure Live Seed (PLS) count for seed lot (Maiti, 2012). The knowledge of seed dormancy and methods of seed treatment to overcome dormancy is essential to leguminous seeds. Seeds of *A. indica* do not have longer dormancy periods, therefore, for better results, seeds were sown immediately after collection. The quantity of seeds to be ordered depends on the types of seeds, seed viability, methods of sowing and time of broadcasting. It is advisable that, rather than single broadcasting operation, it is always worth wise to sown the seeds 3 times, in an interval of 7–10 days. Average cost ecorestoration worked out as US$ 11,300 ha^{-1} (Rs 7.34 lakhs) (Table 4). Yang et al (2016) carried out ecological restoration of highway slope by covering with straw-mat and seeding with grass–legume mixture reported unit cost of US$ 6 m^{-2}, and according to them this cost is suitable for developing countries. As a thumb rule, at present, the mine closure cost will cover the following activities for which a corpus escrow account @ Rs. 6.0 lakhs per ha for OCP and @ Rs.1.0 lakh per ha for UG mine of the property leasehold shall be opened with the Coal Controller Organization. In case of mines having acid mine drainage, post closure acid mine drainage management cost shall also be included in the total closure cost (Debnath et al, 2011). However, the additional amount beyond the escrow account will be provided by the mine operator after estimating the final mine closure cost five years prior to mine closure (as per the mine closure guideline).

4 CONCLUSIONS

Success of ecorestoration project depends on availability resources and designated end use of the area. However, cost of restoration varies from place to place due to constraint in the availability of resources. The important variables like, topsoil (quality and quantity), earth work (excavation, hauling, placement and regrading), seed mix, watering facility, drainage design, labour and aftercare and maintenance of site. In the present ecorestoration project, cost of soil materials considered @US$ 0.50 m^{-3}, is too less in comparison to developed countries. Likewise, labour, seed mixture, expert advice, aftercare and maintenance cost etc., will be less in developing countries. Therefore, derivation of unit cost of ecorestoration @ US$ 11300 ha^{-1} may be applicable to developing countries. Grass–legume mixture (*Stylosanthes–Pennisetum*) can be used as an initial colonizers for stabilization of a very steep slope, after blanking with topsoil. Addition of fast growing, annual, high biomass producing yielding species is essential to increase the soil organic matter and moisture. Fast growing species can form massive green cover in very short time and plays an important role to reduce erosion and conserve moisture. Initial year, both *Stylosanthes* and *Pennisetum* colonize together, but after 2nd year onwards, *Stylosanthes* cover the entire slope surface and eradicate the *Pennisetum*. High biomass yielding herbaceous species with short life cycle can play a role of green manure for the soil as it would add organic carbon and nitrogen to the soil after drying. Aftercare and maintenance of the ecorestored site, particularly watering and protection from cattle is essential.

REFERENCES

Agbenin, J.O. & Adeniyi, T. (2005) The microbial biomass properties of a savanna soil under improved grass and legume pastures in northern Nigeria. *Agriculture, Ecosystems & Environment*, 109 (3), 245–254.

Ahirwal, J., Maiti, S.K. & Reddy, M.S. (2017) Development of carbon, nitrogen and phosphate stocks of reclaimed coal mine soil within 8 years after forestation with *Prosopis juliflora* (Sw.) Dc. *Catena*, 156, 42–50.

Ahirwal, J. & Maiti, S.K. (2016) Assessment of soil properties of different land uses generated due to surface coal mining activities in tropical Sal (*Shorea robusta*) forest, India. *Catena,* 140, 155–163.

Ahirwal, J., Maiti, S.K. & Singh, A.K. (2016). Ecological restoration of coal mine degraded lands in dry tropical climate: What has been done and what needs to be done? *Environmental Quality Management*, 26(1), 25–36.

Ahirwal, J., Maiti, S.K. & Singh, A.K. (2017) Changes in ecosystem carbon pool and soil CO2 flux following post-mine reclamation in dry tropical environment, India. *Science of the Total Environment*, 583, 153–162.

Andrews, J.A., Johnson, J.E., Torbert, J.L., Burger, J.A. & Kelting, D.L. (1998) Minesoil and site properties associated with early height growth of eastern white pine. *Journal of Environmental Quality*, 27, 192–199.

Bordeleau, L.M. & Prevost, D. (1994) Nodulation and nitrogen-fixation in extreme environments. *Plant & Soil,* 161, 115–125.

Bray, R. & Kurtz, L.T. (1966) Determination of total, organic and available forms of phosphorus in soil. *Soil Science*, 59, 39–45

Coppin, N.J, Bradshaw, A.D. (1982) *Quarry reclamation*. Mining Journal Books, London, pp. 8–17

Das, C. & Mondal, N.K. (2016) Litterfall, decomposition and nutrient release of *Shorea robusta* and *Tectona grandis* in a sub-tropical forest of West Bengal, Eastern India. *Journal of Forestry Research*, 27, 1055–1065.

Das, R. & Maiti, S.K. (2016a). Estimation of carbon sequestration in reclaimed coalmine degraded land

dominated by *Albizia lebbeck, Dalbergia sissoo* and *Bambusa arundinacea* plantation: a case study from Jharia Coalfields, India. *International Journal of Coal Science & Technology*, 3(2), 246–266.
Das, R. & Maiti, S.K. (2016b) Importance of carbon fractionation for the estimation of carbon sequestration in reclaimed coalmine soils—A case study from Jharia coalfields, Jharkhand, India. *Ecological Engineering*, 90, 135–140.
Debnath, A.K., Shekhar, S., & Ranjan, R. (2011) Mine closure- World Bank Appraoch via-a-vis Indian context. *Journal of Mines Metals Fuels*, 59(9), 274–278.
Dutta, R.K. & Agrawal, M. (2002) Effect of tree plantation on the soil characteristics and microbial activity of coal mine spoil land. *Tropical Ecology*, 43, 315–324.
Fehmi, J.S. & Kong, T.M. (2012) Effects of soil type, rainfall, straw mulch, and fertilizer on semi-arid vegetation establishment, growth and diversity. *Ecological Engineering*, 44, 70–77.
Jackson, M.L. (1973). *Soil Chemical Analysis*. New Delhi, PHI Pvt. Ltd.
Jha, A.K. & Singh, J.S. (1992) Influence of microsites on redevelopment of vegetation on coalmine spoils in a dry tropical environment. *Journal of Environmental Management*, 36, 295–116.
Ji, S. & Unger, P.W. (2001) Soil water accumulation under different precipitation, potential evaporation, and straw mulch conditions. *Soil Science Society of America Journal*, 65(2), 442–448.
Juwarkar, A.A., Yadav, S.K., Thawale, P.R., Kumar, P., Singh, S.K. & Chakrabarti, T. (2009) Developmental strategies for sustainable ecosystem on mine spoil dumps: a case of study. *Environmental Monitoring & Assessment*, 157(1), 471–481.
Kumar, S., Maiti, S.K. & Chaudhuri, S. (2015) Soil development in 2–21 years old coalmine reclaimed spoil with trees: A case study from Sonepur-Bazari opencast project, Raniganj Coalfield, India. *Ecological Engineering*, 84, 311–324.
Li, R.S. & Daniels, W.L. (1994) Nitrogen accumulation and form over time in young mine soils. *Journal of Environmental Quality*, 23(1), 166–172.
Maiti, S.K. (2006). Properties of mine soil and its effects on bioaccumulation of metals in tree species: Case study from a large opencast coalmining project. *International Journal of Mining, Reclamation & Environment*, 20 (2), 96–110.
Maiti, S.K. & Maiti, D. (2015) Ecological restoration of waste dumps by topsoil blanketing, coir-matting and seeding with grass-legume mixture. *Ecological Engineering*, 77, 74–84.
Maiti, S.K. & Saxena, N.C. (1998) Biological reclamation of coalmine spoils without topsoil: An amendment study with domestic raw sewage and grass-legume mixture. *International Journal of Surface Mining, Reclamation & Environment*, 12, 87–90.
Maiti, S.K. (2007) Bioreclamation of coalmine overburden dumps—with Special empasis on micronutrients and heavy metals accumulation in tree species. *Environmental Monitoring & Assessment*, 125, 111–122.
Maiti, S.K. (2012) *Ecorestoration of the coalmine degraded lands*. Springer Science & Business Media.
Mapfumo, P., Mtambanengwe, F., Giller, K.E. & Mpepereki, S. (2005) Tapping indigenous herbaceous legumes for soil fertility management by resource-poor farmers in Zimbabwe. *Agriculture, Ecosystems & Environment*, 109(3), 221–233.
Milcu, A., Partsch, S., Scherber, C., Weisser, W.W. & Scheu, S. (2008) Earthworms and legumes control litter decomposition in a plant diversity gradient. *Ecology*, 89(7), 872–1882.
Miller, S. (1998) Successful tree planting techniques for drastically disturbed lands: A case study of the propagation planting of container-grown oak and nut trees in Missouri. *American Society for Surface Mining and Reclamation*, Princeton, WV (United States).
Mollard, P.O.F., Naeth, A.M. & Cohen-Fernandez, A. (2014) Impacts of mulch on prairie seedling establishment: facilitative to inhibitory effects. *Ecological Engineering*, 64, 377–38.
Mukhopadhyay, S. & Maiti, S.K. (2011) Trace metal accumulation and natural mycorrhizal colonisation in an afforested coalmine overburden dump: a case study from India. *International Journal of Mining, Reclamation & Environment*, *25*(2), 187–207.
Mukhopadhyay, S., Maiti, S.K. & Masto, R.E. (2013) Use of Reclaimed Mine Soil Index (RMSI) for screening of tree species for reclamation of coal mine degraded land. *Ecological Engineering*, 57, 133–142.
Nelson, D.W. & Sommers, L.E. (1996) Total carbon, organic carbon, and organic matter. In: *Methods of soil analysis*, Part, 3(3), 961–1010.
Nicholas, P.J. & McGinnies, W.J. (1982) An evaluation of 17 grasses and 2 legumes for revegetation of soil and spoil on a coal strip mine. *Journal of Range Management*, 288–293.
Peper, P.J. & McPherson, E.G. (2003) Evaluation of four methods for estimating leaf area of isolated trees. *Urban Forestry & Urban Greening*, 2(1), 19–29.
Roberts, J.A., Daniels, W.L., Burger, J.A. & Bell, J.C. (1988). Early stages of mine soil genesis as affected by topsoiling and organic amendments. *Soil Science Society of America Journal*, 52, 730–738.
Scherer-Lorenzen, M. (2008). Functional diversity affects decomposition processes in experimental grasslands. *Functional Ecology*, 22(3), 547–555.
Singh, A.N., Raghuvanshi, A.S. & Singh, J.S. (2004) Comparative performance and restoration potential of two albizia species planted on coal mine soil in a dry tropical region, India. *Ecological Engineering*, 22, 123–140.
Singh, K.P., Singh, P.K. & Tripathi, S.K. (1999) Litterfall, litter decomposition and nutrient release patterns in four native tree species raised on coal mine spoil at Singrauli, India. *Biology & Fertility of Soils*, 29, 371–378.
Singh, R.S. Tripathi, N. & Singh, S.K. (2007) Impact of degradation on nitrogen transformation in a forest ecosystem of India. *Environmental Monitoring & Assessment*, 125(1–3), 165–173.
Society for Ecological Restoration (SER) (2004) *SER International Primer on Ecological Restoration*. Society for Ecological Restoration International Science & Policy Working Group, Version 2, October, 2004.

Spehn, E.M., Scherer-Lorenzen, M., Schmid, B., Hector, A., Caldeira, M.C., Dimitrakopoulos, P.G., Finn, J.A., Jumpponen, A., O'donnovan, G., Pereira, J.S. & Schulze, E.D. (2002) The role of legumes as a component of biodiversity in a cross European study of grassland biomass nitrogen. *Oikos*, 98(2), 205–218.

Srivastava, S.C., Jha, A.K. & Singh. J.S. (1989) Changes with time in soil biomass C, N and P of mine spoils in a dry tropical environment. *Canadian Journal of Soil Science*, 6, 849–855.

Subbiah, B.V. & Asija, G.L. (1956). A rapid procedure for the determination of available nitrogen in soils. *Current Science,* 25, 259–260.

Thomas, R.J., Asakawa, N.M., Rondon, M.A. & Alarcon, H.F. (1997) Nitrogen fixation by three tropical forage legumes in an acid-soil savanna of Colombia. *Soil Biology & Biochemistry*, 29(5–6), 801–808.

Tripathi, N. & Singh, R.S. (2008) Ecological restoration of mined-out areas of dry tropical environment India, *Environmental Monitoring & Assessment*, 146, 325–337.

Tripathi, N., Singh, R.S. & Nathanail, C.P. (2014) Mine spoil acts as a sink of carbon dioxide in Indian dry tropical environment. *Science of the Total Environment*, 468, 1162–1171.

Van Donk, S.J. & Tollner, E.W. (2000) Apparent thermal conductivity of mulch materials exposed to forced convection. *Transactions-American Society of Agricultural Engineers*, 43(5), 1117–1128.

Wang, J., Liu, H., Wu, X., Li, C. & Wang, X. (2017) Effects of different types of mulches and legumes for the restoration of urban abandoned land in semi-arid northern China. *Ecological Engineering*, 102, 55–63.

Yang, Y., Yang, J., Zhao, T., Huang, X. & Zhao, P. (2016) Ecological restoration of highway slope by covering with straw-mat and seeding with grass–legume mixture. *Ecological Engineering*, 90, 68–76.

Land Reclamation in Ecological Fragile Areas – Hu (Ed.)
 ISBN 978-1-138-05103-4

Study on plant selection of a mine park based on ecological restoration theory

J.Y. Yang & Z.Y. Qi
Architecture and Art College, Central South University, Changsha, Hunan, China

ABSTRACT: The objective of ecological restoration in mining area is to restore the ecological system to its initial state. The establishment of the mine park aims to improve the ecological environment of the mine with plants, which also has a certain landscape effect. Based on numerous previous studies on the plants of mine park, two suggestions can be made to determine the mine park plant collocation: 1. To establish three-dimensional space ecological system with arbor, grass and shrub; 2. To build an integrated system, dominated by hyperaccumulator, native plants, nitrogen fixing plants and all kinds of ornamental plants.

1 INTRODUCTION

With the rapid development of the landscape design, urban planning and other disciplines, more and more designers were joined to mine renovation project, which brought a lot of new point of view. In China, the number of mine has more than 150000, 20% of which are in the growth stage, 58% of which are in mature period, 12% of which are in terminal decline. A multitude of mining faced the transformation and sustainable development of the resource utilization. (Qiguo & Bohong, 2013) A main problem about mining repair is the disposal of heavy metals. According to the difference of heavy metal pollution and soil category, there are several methods to solve those problems, such as physical remediation, chemical remediation and biological remediation.(Yao et al., 2012) Physical remediation including soil replacement and thermal desorption, which could not purified pollutants completely, just affect the topsoil with contaminants remain in the soil. (Yildirim & Sasmaz, 2016) Chemical remediation such as chemical leaching, fixation and electrokinetic remediation are cost intensive and not a bonafide way to environment (Epelde et al., 2016), just a way to evasion problems temporarily. (Del Río et al., 2002) Phytoremediation as one of the biological remediation is the cheapest and ecologically sustainable method. (Stanley et al., 2013, Rahman & Hasegawa, 2011, Yoon et al., 2006) In this method, plants rely on its physiological and anatomical characteristic to absorb the heavy mental into shoots from the polluted area. The aim of this work was to identify the plant in mine park and derived indexes in mine soil remediation. This was accomplished by: 1) analyzing the problems that built a mine park may face; 2) determining the standard of mine park that need to reach; 3) according to the successful case of mine park to determine the choice of plants.

2 PROBLEMS IN REPAIR OF MINE PARK

2.1 *Heavy metal pollution*

Mining activities, especially opencast mining, can play havoc with slope in soil structure, cause surface collapse and soil fissure. It caused the decrease of the bearing capacity of soil, the increase of vegetation destruction and soil erosion. (Carvalho et al., 2014) Simultaneously, Heavy metal as one of the main pollutant in mines, which effected both terrestrial and aquatic environments, has emerged as a global problem. The most common metal pollutants are Fe, Cu, Zn, Cd, Ag, Cr, Pb, As, Ni, and Hg, (Yurkevich et al., 2017) which are non-biodegradable and can enter the soil surface then accumulate in living animals and plants. (Yildirim & Sasmaz, 2016)

2.2 *Lack of biodiversity*

Vegetation clearing, soil degradation, soil erosion, water pollution and so on have destroyed the habitats of wildlife. Coal mine soil structure is poor and lack of organic matter that plants benefit to grow, which caused a sharp decline in biodiversity. (Burges et al., 2016, Carvalho et al., 2014)

2.3 *Atmospheric pollution*

The process of production in mines produced large amounts of dust and poisonous gases. The dust produced in the strip mine and the poisonous gas produced by spontaneous combustion of coal gangue can generate mine dust storms which not only polluted atmosphere but also menaced the biocenosis. In china, there are about 1500 hillock with the increase of 0.2*104 ha per year. Even in the positive governance, there are about 100 in the spontaneous combustion state. (Ruxun, 2014)

3 THE MINE PARK PLANT COLLOCATION

Phytoremediation is a new way to remediate contaminated sites by using plant to improve soil and groundwater quality, and made the ecological functions revitalize. (Willscher et al., 2016) it is crucial to choice the plant as its the main part of phytoremediation in mine park. Plant select of mine park as a landscape elements play an important role in mining wasteland. One the one hand, plants can be used as the pioneer of soil improvement and environmental restoration. On the other hand, plants have the function of softening hard coordination landscape and space landscape.

3.1 *Hyperaccumulator*

Hyperaccumulators have been found to absorb metals at extraordinarily high levels (>1%) in contrast to normal plants. (Del Río et al., 2002) (see Table 1) So far, plentiful hyperaccumulators have been identified. (Baker, 1995). The native species in Guadiamar river area in Aznalcollar mine such as *Malva nicaeensis*, *Datura stramonium*, *Portulaca oleracea*, *Chenopodium album* and *Amaranthus blitoides* have showed predominant ability in mine area.

Table 1. Several plant species are reported as different metal-accumulators.

Species	Contaminants
Alyssum wulfenianum	Ni
Arabidopsis thaliana	Zn, Cd
Azolla pinnata	Cu, Cr
Brassica juncea	Cu, Ni
Brassica napus	Cd
Carthamus oxyacantha	Cd, Fe
Caladium	CN, Hg
Digitaria radicosa	Hg
Epilobium fragilis	As
Lemna minor	Cu, Cr
Limnocharis flava	CN
Lindernia crustacea	Hg
Mikania cordata	CN
Mucuna pruriens	CN
Paspalum conjugatum	Hg, CN
Porippa globosa	Cd
Pteris vitata	Cu, Ni, Zn
Verbascum speciosum	Cu
Zingiber	Hg

3.2 *Local plant*

Perfect adaptability is the advantage of native plants. (Fernández et al., 2017) In Mediterranean region, *hemicryptophyte* is a common perennial herbaceous plant, it multiplied quickly and adjusted to the polluted environment for bioremediation purposes. (Buscaroli et al., 2016) In Indonesia, *Myrtaceae* is a dominant family with a good adjustment in lack condition. (Oktavia et al., 2015) In Papua New Guinea, the Namie Mine survived several potential plant species such as *Piper anduncum*, *Branchiariareptans* and *Phragiteskarka*, which are indigenous species can adopt the polluted mine. (Stanley et al., 2013) In central South Africa, indigenous *Eucalyptus* species growth fastest and showed excellent ability in phytoremediation in the Witwatersrand Basin Goldfields. (Dye et al., 2017) Native plants have a high growth rates to adapt to extreme environment and toxic soil. (Buscaroli et al., 2016) In china, plant such as *Eremochloa ophiuroides*, *Festuca arundinacea*, *Miscanthus sinensis*, *Eragrostis curvula*, *Cynodon dactylon*, *Paspalum natatum*, *Vetiveria zizanioiaes*, *Pennisetum purpureum*, *Arthraxon hispidus*, *Equisetum ramosissimum* and *Dysophylla yatabeana* can make the bare land covered by plants quickly, then turn to hassock community, finally it can be covered by arbors. Local plants and hyperaccumulator play a pioneering role in mining area.

3.3 *Nitrogen-fixing plants*

Nitrogen fixing plant with high ecological value. The research showed that nitrogen-fixing plants can nitrogen fixation 50–150 kg each year for per hectare. Nitrogen-fixing plants mainly includes the following: *Robinia*, *Albizia*, *Amorpha*, *Caragana*, *Acacia*, *Lespedeza*, *Glycine*, *Pisum*, *Phaseolus*, *Medicago*, *Myrica*, *Hippophae*, *Elaeagnus*, *Alnus*, *Coriaria*, *Casuarina*, *Leguminous* plant and *Cycas*. Nitrogen-fixing plants play an important role in repair diggings. In Bangka Island, the technique by using legume cover crop was effective to land revegetation. (Oktavia et al., 2015)

3.4 *Other plants*

Aromatic plant is a kind of extremely high ornamental. In mine area, it can make environment more health and more humanization. Design it as fragrance avenue, fragrant plant area and ecological sightseeing park.

Ornamental grass has a strong adaptability with a low management requirements. In Italy, Butchart Garden was built on a quarry pit. The garden had been divided into four parts, with many kinds of plants, especially ornamental grass. Until now, it became a blossoms park.

3.5 *Arbor—shrub – grass three-dimensional space plant framework*

Steady ecological system institutions include arbor, shrub and herb. In Xishan mine park, shrub and grass had been cultivated by rows and zonal respectively. With same number of arbor and shrub. In Leiyang mine, *clover*, *erect milkvetch* and *wheatgrass* etc. herbaceous plant had been used for stabilized sands. The shrub including *sea-buckthorn*, *crape myrtle* and *ostryopsis davidina* etc. *Saliz matsudana*, *elm*, *masson pine*, *Chinese pine* were used for protect water and soil. The construction of three-dimensional plant not only has the landscape effect can also build perfect biological systems.

4 SUMMARY AND DISCUSSION

The purpose of phytoremediation was using plant to revitalize the ecological functions. Plant useful for mining rehabilitation should be metal tolerance and adapt to the local environment with high growth rate. Native plants and hyperaccumulator have an advantage for this. Three-dimensional space ecological system with grass and shrub were beneficial to mine restoration. However, The choice of plant varieties is only one part of the repair work, how to establish a self-regulating ecosystem is more important.

REFERENCES

Burges, A., L. Epelde, F. Blanco, J.M. Becerril & C. Garbisu (2016) Ecosystem services and plant physiological status during endophyte-assisted phytoremediation of metal contaminated soil. *Science of The Total Environment*.

Buscaroli, A., D. Zannoni, M. Menichetti & E. Dinelli (2016) Assessment of metal accumulation capacity of Dittrichia viscosa (L.) Greuter in two different Italian mine areas for contaminated soils remediation. *Journal of Geochemical Exploration*.

Carvalho, F.P., J.M. Oliveira & M. Malta (2014) Radioactivity in Soils and Vegetables from Uranium Mining Regions. *Procedia Earth and Planetary Science,* 8, 38–42.

Del Río, M., R. Font, C. Almela, D. Vélez, R. Montoro & A. De Haro Bailón (2002) Heavy metals and arsenic uptake by wild vegetation in the Guadiamar river area after the toxic spill of the Aznalcóllar mine. *Journal of Biotechnology,* 98, 125–137.

Dye, P., V. Naiken, A. Clulow, E. Prinsloo, M. Crichton & I. Weiersbye (2017) Sap flow in Searsia pendulina and Searsia lancea trees established on gold mining sites in central South Africa. *South African Journal of Botany,* 109, 81–89.

Epelde, L., O. Muñiz & C. Garbisu (2016) Microbial properties for the derivation of critical risk limits in cadmium contaminated soil. *Applied Soil Ecology,* 99, 19–28.

Fernández, S., C. Poschenrieder, C. Marcenò, J.R. Gallego, D. Jiménez-Gámez, A. Bueno & E. Afif (2017) Phytoremediation capability of native plant species living on Pb-Zn and Hg-As mining wastes in the Cantabrian range, north of Spain. *Journal of Geochemical Exploration,* 174, 10–20.

Oktavia, D., Y. Setiadi & I. Hilwan (2015) The Comparison of Soil Properties in Heath Forest and Post-tin Mined Land: Basic for Ecosystem Restoration. *Procedia Environmental Sciences,* 28, 124–131.

Qiguo, L. & Z. Bohong (2013) Context integration, functional composite, and marker conjunction: strategic thinking about hunan Baoshan mine state park planning. *Industrial Construction*, 29–32.

Rahman, M.A. & H. Hasegawa (2011) Aquatic arsenic: Phytoremediation using floating macrophytes. *Chemosphere,* 83, 633–646.

Ruxun, W. 2014. Study on Landscape Ecological Environment Restoration and Sustainable Utilization of Coal Mining Wasteland–Taking Xi Shan Coal Minng Wastelan in Tai Yuan City for Example. Dalian University of Technology.

Stanley, R., G. Arpa, H. Sakulas, A. Harakuwe & D. Timi (2013) Phytoremediation—An Eco-friendly and Sustainable Method of Heavy Metal Removal from Closed Mine Environments in Papua New Guinea. *Procedia Earth and Planetary Science,* 6, 269–277.

Willscher, S., L. Jablonski, Z. Fona, R. Rahmi & J. Wittig (2016) Phytoremediation experiments with Helianthus tuberosus under different pH and heavy metal soil concentrations. *Hydrometallurgy*.

Yao, Z., J. Li, H. Xie & C. Yu (2012) Review on Remediation Technologies of Soil Contaminated by Heavy Metals. *Procedia Environmental Sciences,* 16, 722–729.

Yildirim, D. & A. Sasmaz (2016) Phytoremediation of As, Ag, and Pb in contaminated soils using terrestrial plants grown on Gumuskoy mining area (Kutahya Turkey). *Journal of Geochemical Exploration*.

Yoon, J., X. Cao, Q. Zhou & L.Q. Ma (2006) Accumulation of Pb, Cu, and Zn in native plants growing on a contaminated Florida site. *Science of the Total Environment,* 368, 456–464.

Yurkevich, N., S. Bortnikova, V. Olenchenko, N. Abrosimova, O. Saeva & Y. Karin (2017) Study of Water-rock Interaction in Sulfide Mining Tailings using Geochemical and Geoelectrical Methods. *Procedia Earth and Planetary Science,* 17, 112–115.

Land Reclamation in Ecological Fragile Areas – Hu (Ed.)
 ISBN 978-1-138-05103-4

Phytoaccumulation of trace elements by *Grevillea pteridifolia* Knight growing on iron ore tailings: An implication of phytoremediation

Nilima Chaturvedi
The Energy and Resources Institute (TERI), Darbari Seth Block, IHC Complex, New Delhi, India

ABSTRACT: Pot experiments were conducted to investigate the effect of Iron Ore Tailings (IOT) both individually as well as in combination with soil (at different proportions) on growth, photosynthetic pigments, antioxidant enzymes and accumulation heavy metals (Fe, Cu, Zn, Ni, Cr, and Pb) from Iron ore tailings by *Grevillea pteridifolia*. Results suggested that the plants grown on tailings showed an increased growth, chlorophyll content, as well as metal accumulation with increasing proportion of tailings in the soil. Furthermore, an increase in antioxidant activities in plants grown on tailings as compared to control suggests plant efficiency to overcome stress generated due to excess accumulation of heavy metals. The order of accumulation of various heavy metals in the plant parts was observed to be Fe > Zn > Cr > Cu > Pb > Ni. Overall, *Grevillea pteridifolia* was found to be well adapted in iron ore tailings and it may be recommended for phytoremediation of most of the studied metals.

Keywords: Iron Ore Tailings (IOT), Heavy Metals, Phytoremediation, Accumulation, Translocation, Heavy Metals

1 INTRODUCTION

Mining operations generate considerable amount of waste materials and tailings, which are either deposited on the surface as mine spoil dumps or stored in large size ponds called tailing ponds. Removal of fertile topsoil, formation of unstable slopes prone to sliding and erosion, and siltation of water bodies due to wash off of mineral overburden dumps are some other negative effects of mining. The metals released from mining, smelting, forging, and other sources would accumulate in the soil, altering its chemistry (Khan et al., 2009, Kumar, 2013). Mine contaminated soils represent a very harsh environment for crop production with low pH, nutrients and limited topsoil availability. Thus, reclamation of mine dumps and abandoned mine lands is a complex multi-step process involving improvement of physical and chemical nature of the site (ameliorative) and careful selection of species, cultivars, or ecotypes (adaptive), both to be used in juxtaposition with one another (Johnson et al., 1994).

In recognition of the role of trees to improve soil fertility (Nair et al., 2010), agroforestry systems (growing trees and crops in an integrated manner) are believed to have a great potential to reclaim the mine contaminated sites. This conjecture is based on the notion that tree incorporation would result in greater export of pollutants, improve site fertility, and render the sites productive (Kumar, 2013). The present study aimed at studying the phytoremediation potential of *Grevillea pteridifolia*, which is widely used in afforestration owing to its fast growing nature and pleasant appearance. Because of thick and fleshy leaves with petioles flexible and capacity to withstand vibration *Grevillea pteridifolia* is also used for noise control at the industry sites (Kumar et al., 2013). In spite of its wide spread use in agroforestry programmes and multiple uses, phytoremediation potential of *Grevillea pteridifolia* remains unexplored. Thus, the objectives of the study were to examine the (1) growth of *Grevillea pteridifolia* on iron ore tailings and (2) accumulation and translocation of various heavy metals within the plant body under different treatments.

2 METHODOLOGY

2.1 *Tailings substrata analysis*

pH, EC and WHC (%) of IOT as well as soil samples were determined according Chaturvedi et al. (2013). Exchangeable Ca, Na, K and Mg were extracted by 1(N) ammonium acetate solution (Gupta, 2000). Organic carbon content was determined by rapid dichromate oxidation technique (Walkey and Black 1934) and CEC by 1(N) ammonium acetate extraction method (Jackson 1973). Diethylene Triamine Penta Acetic acid (DTPA) extractable (plant available) metals were determined using 0.005 M DTPA solution (Lopez—Sanchez et al. 2000) while Available N and P by alkaline permanganate (Subbiah and Asija 1956) and ammonium fluoride extraction (Bray and Kurtz 1945) methods, respectively.

2.2 *Biochemical parameters*

2.2.1 *Photosynthetic pigments*

Photosynthetic pigments like chlorophyll a, b and total was quantified spectrophotometrically following the method of Porra et al. (1989).

2.2.2 *Antioxidant enzyme assay*

Activities of CAT, POD and SOD were measured following the method of Chance and Maehly (1955), Singh et al. (2006) and Misra and Fridovich (1972) respectively. The activity of these enzymes was expressed as specific activity (U^{-1} mg protein).

2.3 *Heavy metal analysis from soil and plant samples*

Rhizospheric soil samples were obtained following Yanai et al. (2003). Oven dried (60°C) soil tailing and plant samples were ground using a mortar and pestle and digested in aquaregia (HNO3/HCl, 1:3), and thereby, concentration of heavy metals were determined using the AA-6300 SHIMADZU Atomic Absorption Spectrophotometer after adjustment of required dilution factor. All the reagents and reference standards were of analytical grade from Merks (Darmstandt, Germany) and Suprapure hydrochloric and nitric acids (Merks, Darmstandt, Germany) were used for sample digestion and preparation of standards.

3 RESULTS AND DISCUSSION

3.1 *Physico-chemical parameters*

The Physico-chemical characteristics of soil and tailings samples are presented in Table 1. The chemical analysis of the iron ore tailings revealed about 59.70% Fe_2O_3, 18.1% Al_2O_3 and 1.77%, SiO_2 and 9.7% LOI. Tailings were comparatively acidic than the garden soil with pH 5.5 for tailings and 6.2 for garden soil. The percentage of WHC was found maximum for the garden soil (38.7%) and minimum for IOT (22.7%). Scanning Electron images revealing the morphology of Soil and tailing samples has been presented in Fig. 1.

The soil and tailing samples showed large differences between their nitrogen content, but little differences were observed between the potassium

Table 1. Showing the various physico-chemical characteristics of soil and tailing samples (n = 4, Mean ± SE).

Treatment	pH	EC (μS/cm)	OC (%)	CEC c mol (+) kg^{-1}	WHC (%)
T0	6.2 ± 0.025*	103 ± 0.017*	3.3 ± 0.007*	0.464 ± 0.033*	38.7 ± 0.009*
T1	6.1 ± 0.26*	108 ± 0.021*	2.4 ± 0.024*	0.533 ± 0.043*	33.4 ± 0.012*
T2	5. 9± 0.026*	112 ± 0.015*	1.4 ± 0.009*	0.599 ± 0.025*	29.8 ± 0.021*
T3	5.7 ± 0.017*	115 ± 0.022*	0.98 ± 0.013*	0.654 ± 0.042*	25.9 ± 0.020*
T4	5.5 ± 0.014*	119 ± 0.029*	0 ± 0.002*	0.731 ± 0.030*	22.7 ± 0.015*

Figure 1. Scanning electron image of Garden soil and Iron Ore Tailings (IOT).

Table 2. Showing the nitrogen, phosphorus and exchangeable cations of soil and tailing samples (n = 4, Mean ± SE).

	Available N	Available P	Exchangeable cations [c (+) mol/kg]			
Treatment	(mg/kg)	(mg/kg)	Ca	Na	K	Mg
T0	159 ± 0.003*	7.6 ± 0.035*	0.34 ± 0.015*	1.36 ± 0.013*	0.33 ± 0.009*	2.8714 ± 0.013*
T1	116.9 ± 0. 019*	6.1 ± 0.203*	2.94 ± 0.002*	1.97 ± 0.003*	0.27 ± 0.012*	2.5876 ± 0.005*
T2	86.12 ± 0.120*	4.2 ± 0.045*	3.83 ± 0.009*	1.22 ± 0.005*	0.20 ± 0.035*	1.8192 ± 0.017*
T3	44.67 ± 0.035*	2.3 ± 0.111*	5.18 ± 0.016*	0.86 ± 0.029*	0.14 ± 0.014*	1.1856 ± 0.003*
T4	28.00 ± 0.009*	1.2 ± 0.169*	8.13 ± 0.005*	0.58 ± 0.111*	0.10 ± 0.005*	0.3142 ± 0.009*

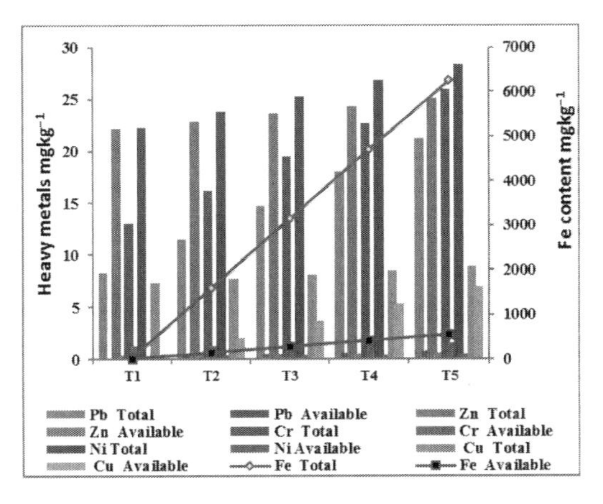

Figure 2. Environmentaly and plant available metal contents (mgkg^{-1}) in control and various treatments.

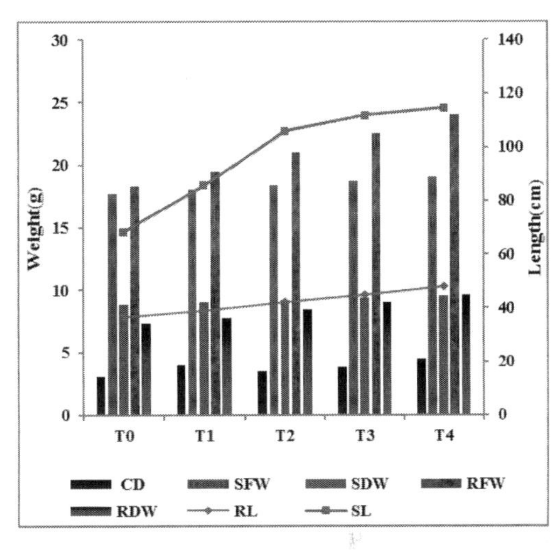

Figure 3. Effect of iron ore tailings on growth of *G. pteridifolia.*

and phosphorus content of the two (Table 2). The concentration of environmentally and plant available metals increased or decreased proportionately with the increasing or decreasing proportion of tailings in the soil for most of the metals (Fig. 2).

3.2 *Growth and photosynthetic pigments*

A significant increase in plant height was observed in all the treatments in *Grevillea pteridifolia* Fig. 3. Significant positive correlations ($p < 0.01$) between root length, shoot length, root fresh weight, shoot fresh weight, root dry weight as well as shoot dry weight and the concentration of various heavy metals (Fe, Cu, Pb, Zn, Cr and Ni) in root and shoot parts of *Grevillea pteridifolia* further confirms synergistic effect of these metals/ IOT on growth of the plant Regarding the effect of heavy metals on photosynthetic pigments significant positive correlations between Zn, Ni and chlorophyll content was observed which suggests excellent tolerance mechanism of *G. pteridifolia* towards these toxic metals. Moreover, Chl b synthesis was much lower than Chl a in all the treatments as well as in control. This change in the ratio of Chl a/b suggests differential effect of metals on light-harvesting complexes like LHC2 of PS2, (Aravind and Prasad, 2004).

Furthermore Carotenoids, which are regarded as non-enzymatic antioxidants, serve as an accessory pigment for photosynthesis and protect the chlorophyll pigment under stress conditions by quenching the photodynamic reactions, replacing peroxidation and collapsing membrane in chloroplasts (Kenneth et al., 2000). And an increase in the concentration of carotenoids with increase in metal uptake confirms the same (Prakash et al., 2007).

3.3 *Antioxidant activity*

Exposure of plants to tailings (both with and without additives) led to an increase in the activities of CAT, POD and SOD Fig. 4. The activity of enzymes increased with increase in doses and duration of exposure of tailings. Significant positive correlations ($p < 0.01$) were observed between the

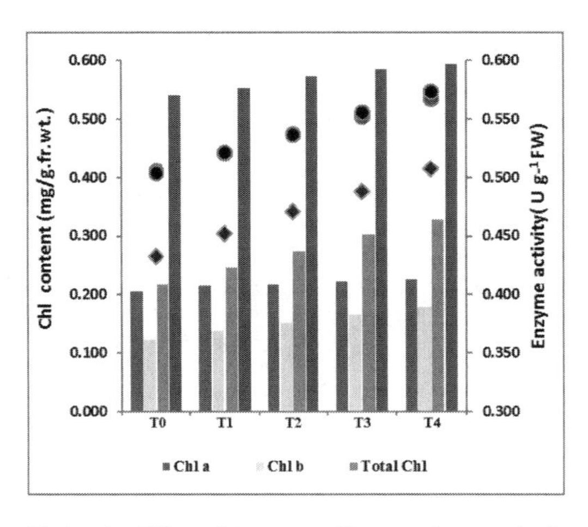

Figure 4. Effect of iron ore tailing on photosynthetic pigments and antioxidant enzymes of the selected plants.

shoot heavy metals (Fe, Cu, Zn, Ni, Cr and Pb) and activity of CAT, POD and SOD in both control and the treatments. The higher oxidative enzymes activity is possibly a result of gradual shift of reductive metabolism to oxidative metabolism. These results suggest that heavy metals (Fe, Cu, Zn, Ni, Cd and Pb) present in the tailings induced oxidative stress in the plants and that elevated activity of antioxidant enzymes could play an important role in mitigating oxidative injury.

3.4 *Metal accumulation pattern*

The comparative accumulation of different heavy metals by *Grevillea pteridifolia* subjected to various treatments is shown in Fig. 5. The metal concentrations in IOT was found to be significantly ($p < 0.01$) higher than the control. The mean metal concentration in the plants increased with increasing IOT (and hence metal) conc. in the soil. Also, the maximum and minimum values of each metal were found to be comparatively higher in treated plants than control. Accumulation of Pb, Cr, Ni and Zn was maximum in the root while Fe and Cu in the shoot. The overall order of accumulation of various heavy metals by *Grevillea pteridifolia* was Fe > Zn > Cr > Cu > Pb > Ni. Though there is severe dearth of literature on accumulation of heavy metals by *Grevillea pteridifolia,* a different species of *Grevillea* namely *Grevillea exul* has been reported to accumulate substantial amount of manganese content in the epidermal tissues. Similarly, Léon et al. (2005) and Rabier et al. (2008) reported accumulation of Ni in different parts of *Grevillea exul.* like seed coat and phloem of basal

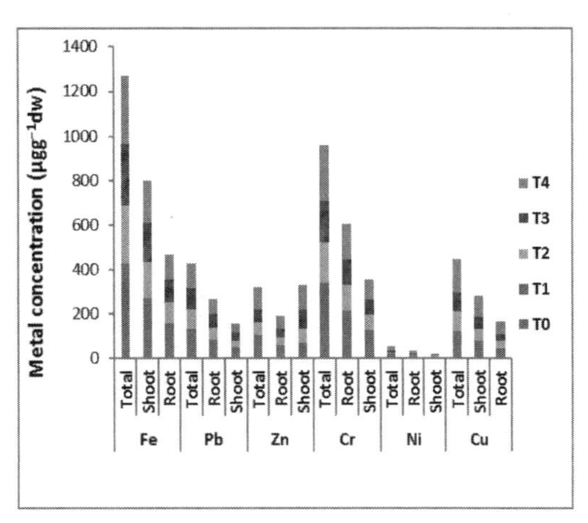

Figure 5. Metal concentration (μgg^{-1} dw) in different parts of *G. pteridifolia.*

stem and roots respectively.al. (2008) and Rabier et al. (2008) reported accumulation of Ni in different parts of *Grevillea exul.* like seed coat and phloem of basal stem and roots respectively.

4 CONCLUSION

Thus, the present study which is to the best of our knowledge is the first detailed report on assessment of heavy metal accumulation potential of *G. pteridifolia* on iron ore tailings clearly suggests that, the plant has not only the potential to survive on metallic wastes like Iron ore tailings but can also accumulate substantial amount of heavy metals. And hence, can be used as a potential tool for remediation of industrial wastes. Considering its multipurpose uses and beautiful appearance *G. pteridifolia* must be given a serious try in remediation and revegetation of industrial wastes and mining zones.

REFERENCES

Aravind P. and Prasad M.N.V. (2004). Zn protects chloroplasts and associated photochemical functions in cadmium exposed *Ceratophyllum demersum* L., a freshwater macrophyte, Plant Sci. 166:1321–1327.

Bray R.H., Kurtz L.T. (1945). Determination of total, organic, and available forms of phosphorus in soils. Soil Sci 59:39–45.

Chance B., Maehly A.C. (1955). Assay of catalase and peroxidase. Methods Enzymol 2:764–775.

Gupta, P.K. (2000). Chemical methods in environmental Analysis: Water, Soil and Air. India, Agrobios, 240–241.

Jackson, M.L. (1973). Soil chemical analysis. New Delhi, Prentice Hall Pvt. Ltd.

Johnson, M.S., Cooke, J.A. and Stevenson, J.K. (1994). Revegetation of metalliferous wastes and land after metal mining. In: Mining and its Environmental Impact. Hester R.E. and Harrison R.M. (eds). Issues in Environmental Science and Technology, Royal Society of Chemistry, Letchworth, England, pp. 31–48.

Kenneth E., Pallet K.E., and Young J., (2000). Carotenoids, in Antioxidants in Higher Plants, R.G. Alscher and J.L. Hess, eds., CRC Press, Boca Raton, FL, 60–81.

Khan, M.S., Zaidi, A., Wani, P.A. and Oves, M. (2009). Role of plant growth promoting rhizobacteria in the remediation of metal contaminated soils. Environmental Chemistry Letters 7:1–19.

Kumar B. M. (2013). Mining waste contaminated lands: an uphill battle for improving crop productivity. Journal of Degraded and Mining Lands Management. 1 (1): 43–50.

Léon V., Rabier J., Notonier R., Barthelémy R., Moreau X., Bouraïma-Madjèbi S., Viano J. and Pineau R. (2005). Effects of three nickel salts on germinating seeds of *Grevillea exul* var. rubiginosa, an endemic serpentine Proteaceae. *Ann Bot.* 2005 Mar; 95(4):609–18.

Lopez-Sanchez J.F., Sahuquilo A., Rauret G., Lachica M., Barahona E., Gomez A., Ure A.M., Muntau H., Quevauviller P.H. (2000). Extraction procedures for soil analysis. In: Quevauviller PH (ed) Methodologies in Soil and Sediment Fraction Studies: Single and Sequential Extraction Procedures. Royal Society of Chemistry, Cambridge, UK, p 28–65.

Misra H.P., Fridovich I. (1972). The generation of superoxide radical during auto oxidation. J Biol Chem 247:6960–6966.

Nair, P.K.R., Nair, V.D. Kumar, B.M. and Showalter, J.M. (2010). Carbon sequestration in agroforestry systems. Advance Agronomy108:237–307.

Porra R.J., Thompson W.A., Kriedmann P.E. (1989). Determination of accurate extinction coefficients and simultaneous equations for assaying chlorophylls a and b extracted with four different solvents: verification of the concentration of chlorophyll standards by atomic absorption spectroscopy. Biochim Biophys Acta 975:384–394.

Prakash D., Suri S., Upadhyay G. and Singh B.N. (2007). Total phenols, Antioxidant and Free radical scavenging activities of some medicinal plants, Int. J. Food Sci. Nutr. 58:18–28.

Rabier J., Laffont-Schwob I., Notonier R., Fogliani B., Bouraïma-Madjèbi S. (2008). Anatomical element localization by EDXS in Grevillea exul var. exul under nickel stress. *Environ Pollut.* 2008 Dec; 156(3):1156–63. doi: 10.1016/j.envpol.2008.04.001.

Singh S., Melo J.S., Eapen S., D'Souza S.F. (2006). Phenol removal by hairy roots; role of inherent peroxidise and HO. J Biotechnol 123:43–49.

Subbiah B.V., Asija G.L. (1956) A rapid procedure for the determination of available nitrogen in soils. Curr Sci 25:259–260.

Walkey A., Black I.A. (1934). An examination of the Degtjareff method for determining soil organic matter and a proposed modification of the chromic acid titration method. Soil Sci 37:29–38.

Yanai R.D., Majdi H., Park B.B. (2003). Measured and modelled differences in nutrient concentrations between rhizosphere and bulk soil in a Norway spruce stand. Plant Soil 257:133–142.

Land Reclamation in Ecological Fragile Areas – Hu (Ed.)
 ISBN 978-1-138-05103-4

Characteristics of capillary water rise in coal-fired cinders by lab test

X.Y. Zhou, C.L. Jiang & P.F. Liu
School of Resource and Environmental Engineering, Anhui University, Hefei, Anhui, China

ABSTRACT: This study applied the indoor standpipe method to test the capillary rise in coal-fired cinders (fly ash, bottom ash and their mixture at ratio of 4:1) and found that the correlation between the height of the capillary rise and time in coal-fired cinders follows a power function. Based on the relationship, the capillary rise process is divided in to three stages, including the rapid rise stage, the slow rise stage and the roughly stable stage. The effect of coarse particles in fly ash on capillary water movement has a dual nature: the negative capillary pressure was greater in the early stage, and macro pores formed by coarse particles facilitate rapid capillary rise; whereas at the late stage, coarse particles prolongs the length of capillary water movement path, making capillary water movement more difficult.

1 INTRODUCTION

Coal-fired power plants produce a large quantity of coal-fired cinders, including fly ash, bottom ash and their mixtures, which have attracted global attention as precious recyclable waste resources (Komonweeraket *et al.*, 2015). The use of coal-fired cinders to fill mining-induced subsidence or low-lying fields for land reclamation not only properly disposes industrial solid wastes, but also greatly reclaim wastelands, thus it is a good approach which has been widely applied in recent years (Chen *et al.*, 2013). Capillary water plays an important role in soil moisture migration, salinity change and soil engineering properties, etc (Miao *et al.*, 2011; Liu *et al.*, 2013). The height of capillary rise can be predicted using Hazen empirical formula (Yuan *et al.*, 2014), SIMWASER simulation (Stenitzer *et al.*, 2007), soil physical parameters (Dong *et al.*, 2008) and the like. Lu (2004) *et al.* investigated the process of capillary water movement and proposed the analytical solution of the capillary rise speed over time and the fitting formula of the capillary rise height over time. High salinity not only changes the porous structure of soils but also impacts capillary rise process (Li *et al.*, 2011); Dullien (1990) and Fredllund (1993) studied in detail the mechanism of capillary rise based on the shrink film equilibrium theory.

The above-mentioned studies were performed mainly targeting the problems with their focuses on capillary rise in sand and soil media. However, coal-fired cinders aren't soil in the strict sense; the physicochemical properties of filled or reclaimed soils are also obviously different (Hu *et al.*, 2002). Therefore, this study focuses on the characteristics of capillary water movement in coal-fired cinders.

2 EXPERIMENT DESIGN

2.1 *Materials and methods*

The coal-fired cinders used in the experiment were fly ash bottom ash, and their mixture at ratio of 4:1 from the cinders dumpsite in an industrial park, Wanbei City, Anhui Province, China. The particle size of bottom ash and fly ash was analyzed using the sieving method and the laser particle-size analyzer. The direct observation method at the positive hydraulic head was utilized to test the capillary rise in coal-fired cinders so as to analyze the height and speed of them. The test process is listed as follows. 1) At laboratory temperature (10°C~30°C), densely and uniformly pack fly ash, bottom ash and their mixture into three transparent silicone tubes with inner diameter of 1.6 cm respectively, and mark them with different labels; 2) Seal the bottoms of the tubes with geotextile to prevent ash leaking; 3) Fill an appropriate amount of water into the water sink and lay the rock with good water permeability on the sink bottom, making the rock surface level as high as the water surface; 4) Begin recording time when the bottom of the sample tube contacts the rock surface; 5) Measure and record the height of capillary rise with the accuracy of 1 mm according to the lifting, wetting front position in the sample tube with the rock surface as a reference. Because the speed of capillary rise at the first several minute is very fast, the observation points were more intense at the beginning and the situation of water lifting in the first 30 min was recorded using a high-speed digital camera to facilitate recording.

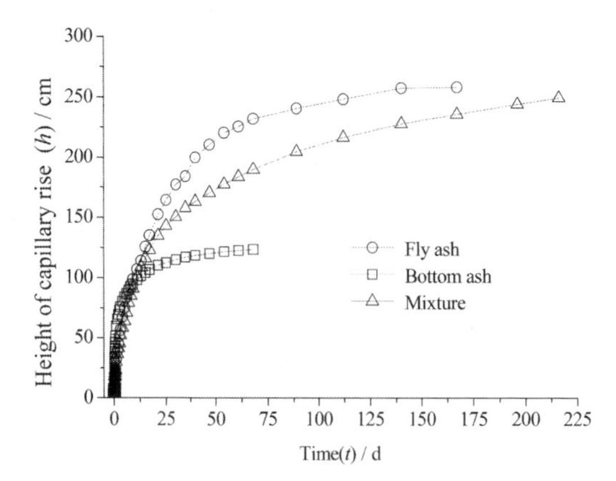

Figure 1. Relationship between the height of capillary rise and time of different coal-fired cinders.

2.2 *Experimental results*

The capillary rise was stabilized in the bottom ash at the maximum lifting height of 123.5 cm at 68th day, fly ash of 258.2 cm at 167th day, and their mixture of 249.4 cm at 216th day of the experiment. Fig. 1 shows the curves of capillary rise height over time in the three samples. It is clear from Fig. 1 that the capillary rises of the three samples behave consistently; very rapid at first, then slow with time advancing, which means more time is needed for the capillary water to lift the same height; and the time needed for capillary water reaching the maximum height is different.

3 ANALYSIS AND DISCUSSION

3.1 *Relationship between capillary rise and time*

Regression analysis showed that the relationship between the height of capillary rise and time can be better described by a power function,

$$h = At^{B} \quad (1)$$

where h is the height of capillary rise, cm; t is the time for which the capillary water reaches h, day; A and B are fitting parameters, dimensionless.

Table 1 lists the corresponding regression coefficients of different samples. From Table 1, it is evident that the fitting equations are highly accurate and can better describe the process and estimate the height of capillary rise in different samples over time.

Comprehensive analyses indicate that the capillary water lifts rapidly at the early stage, reaching a great height. It can be seen that the height of capillary rise is the greatest in bottom ash during the first 7 days of experiment, mixture on the 7th day, fly ash on the 11th day, and the height is always smaller in fly ash than in the mixture within the first 11 days.

Table. 1. Fitting parameters of the height of capillary rise with time..

	Regression coefficients			
Specimens	A	B	R^2	Remarks
Fly ash	42.385	0.386	0.980	$t \leq 167$
Bottom ash	50.429	0.243	0.969	$t \leq 68$
Mixture	38.950	0.363	0.988	$t \leq 216$

According to the test procedure and the related capillary rise theory, the reasons for the above trend of capillary rise might be: 1) surface tension much greater than the gravity of the rising water column at the initial stage leads to a faster capillary rise speed; and 2) with the capillary water column heightening, its gravity gradually increases and gets close to the surface tension, resulting in a gradually slowing in capillary rise speed (Kong, 2010). Further analysis finds that at the initial stage, the total amount of lifted capillary water is comparatively small, thus its related viscosity resistance and own gravity are relatively small and subsequently the capillary pressure and the inertial forces to be dominant then, resulting in a very great capillary rise speed. With the height of capillary rise increasing, both the viscosity resistance and its own gravity can counteract most of the capillary rise force, the contents of capillary water in the coal-fired cinders become very small and the counteraction becomes very weak. Therefore, at the final stage, the speed of capillary rise becomes very slow and long-lasting. Comprehensive analysis of the entire process indicates that the driving force for capillary rise shows a significantly damping trend as manifested followed by slow capillary rise in the coal-fired cinders.

3.2 *Difference in capillary rise among different coal-fires cinders*

The main difference of coal-fired cinders is their particle composition. In this experiment, the bottom ash is mainly composed of particles with diameter > 0.1 mm and $d_{10} \approx 0.09$ mm; while the diameter of 90% fly ash particles is < 0.045 mm and the diameter of 78.9% mixed cinder is < 0.075 mm. The impact of different particle composition on the process of capillary rise is mainly reflected in the maximum capillary rise height and the rise process.

In this paper, considering the overall situation, capillary rise with speed $v > 5$ cm/d, $0.5 < v \leq 5$ cm/d and $v \leq 0.5$ cm/d was divided into rapid rise stage, slow rise stage and basically stable stage, respectively. Currently, there is no standard criterion on how to divide them.

Fig. 2 shows the average speed of different samples at the first two stages. In the rapid rise stage, the bottom ash has the fastest average rise speed, up to 25.3 cm/d, fly ash of 8.4 cm/d and the mixed cinder of 10.1 cm/d. This phenomenon is probably due to that the fly ash and the mixed cinder had significantly smaller average particle size. In the slow rise stage, all the capillary rise speeds show a monotonically decreasing trend.

Fig. 3 shows the duration of rapid and slow rise stages of different samples according to the above dividing criterion for the capillary rise. It is seen from Fig. 3 that the bottom ash had the shortest rapid and slow rise duration, which were 3 and 27 days, respectively. In other words, the bottom ash reached the maximum rise height in the shortest time. In addition, the fly ash had longer rapid rise

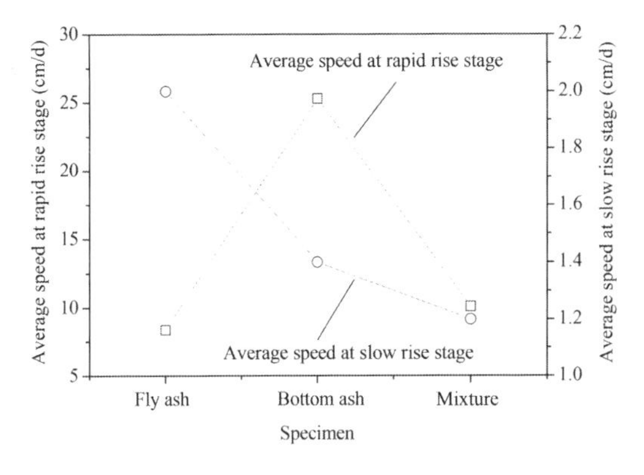

Figure 2. Average speed of capillary rise.

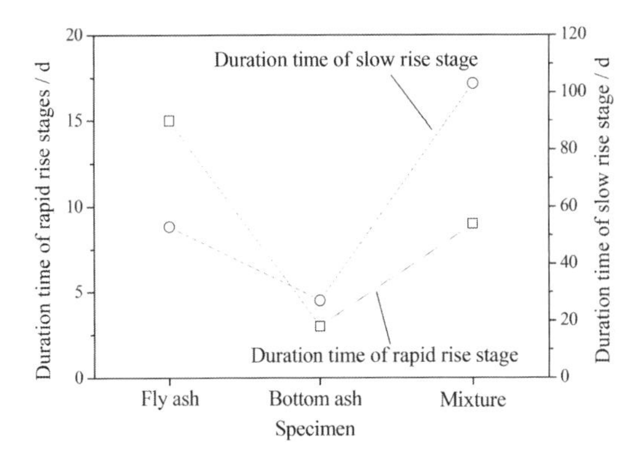

Figure 3. Duration time of capillary rise.

duration and shorter slow rise duration than the bottom ash, which were 15 and 53 days respectively; while the mixed cinder had the longer rapid rise duration and shorter slow rise duration, which were 9 days and 103 days, respectively.

From Fig. 2 and 3, it can be seen that the maximum capillary rise height in fly ash and mixed cinder is similar, but the capillary rise process is obviously different. The phenomenon might be explained by the followings:

1. The particle size of main components of fly ash and mixed cinder is similar, the former one's d_{10}, d_{60} and uniformity coefficient C_u were 0.0014 mm, 0.013 mm and 9.29 and 0.0018 mm, 0.017 mm and 9.44 for the latter. Thus, their maximum capillary rise height is not obviously different (Yuan *et al.*, 2014);
2. The mixed cinder is composed of 80% of fine particles (fly ash) wrapped around 20% coarse particles (mixed). Due to the presence of coarse particles and the subsequent high content of macro pores in the mixed cinder, capillary water, which has larger negative pressure and movement driving force in the early stage, rises rapidly (Dullien *et al.*, 1990). Decreasing with time, capillary negative pressure gradually reaches a balance with gravity, thus, the advantageous movement paths of capillary water formed by macro pores disappeared. Therefore, the coarse particles become an obstacle factor of capillary water movement (Kong, 2010). Thus, the effect of coarse particles on capillary water movement has a dual nature.

4 SUMMARY

1. Capillary rise in fly ash, bottom ash and mixed cinder shows a trend of rapid increase followed by a slow increase and could be well described using three different power functions of water height change with time. The calculation formulas of the height of capillary rise in the three samples were obtained. (2) According to the changes in trend of capillary rise speed, the process of capillary rise is divided into rapid rise stage, slow rise stage and stable stage. Samples of different grain composition have different average capillary rise speeds and duration time at each stage. (3) The effect of coarse particles on capillary water movement has a dual nature. In early stage when the negative capillary pressure is large, the advantageous macropores path formed by the coarse particles facilitates rapid rise of capillary water; while in the late stage coarse particles increase the length of capillary rise path, which obstructs capillary water movement.

ACKNOWLEDGMENTS

This work was financial supported by the National Natural Scientific Foundation of China (41602310), the Natural Science Foundation of Anhui Province (1408085QD70), and the Natural Science Foundation of the Department of Education of Anhui Province (KJ2014A018).

REFERENCES

Chen X (2013) Water-salt variation properties and crop response of reconstruction soil filled with fly ash. Anhui *University of Science and Technology doctoral dissertation*, Huainan, China, pp. 16–28.

Dong B, Zhang X, Li X, *et al.* (2008) Comprehensive tests on rising height of capillary water. *Chinese Journal of Geotechnical Engineering* 30 (10):1569–1574 (in Chinese).

Dullien F (1990) *Porous media: fluid transport and pore structure.* Petroleum Industry Press, Beijing, China, pp. 46–85.

Fredlund DG. & Rahrdio H (1993) *Soil mechanics for unsaturated soils.* Wiley & Sons Inc, New York, pp. 65–114.

Hu Z, Qi J. & Si J (2002) Physical and chemical properties of reclaimed soil filled with fly ash. *Journal of China Coal Society* 27(6):639–643.

Komonweeraket K, Cetin B, Aydilek A. & Edil TB (2015) Geochemical analysis of leached elements from fly ash stabilized soils. *J Geotech Geoenviron* 141(5): 687–695.

Kong X (2010) *Advanced seepage mechanics.* 2nd edition. University of Science & Technology China Press, Hefei, China, pp. 123–165.

Li X, Zhou J, Zhao Y, *et al.* (2011) Effects of high-TDS on capillary rise of phreatic water in sand soil. *Transactions of the CSAE* 27(8): 84–89.

Liu X, Zhang Z, Li Y. & Pan W (2013) Study on a method of determining the coefficient *C* of capillary in tailings sand of a certain tailings dam. *Journal of University of South China* 27(2):23–27.

Lu N. & Likos WJ (2004) Rate of capillary rise in soil. *J Geotech Geoenviron* 130(6): 646–650.

Miao Q, Chen Z, Tian Q, Qian N. & Yao Z (2011) Experimental study of capillary rise of unsaturated clayey sand. *Rock and Soil Mechanics* 32(S1):327–333.

Stenitzer E, Diestel H, Zenker T. & Schwartengraber R (2007) Assessment of capillary rise from shallow groundwater by the simulation model SIMWASER using either estimated pedotransfer functions or measured hydraulic parameters. *Water Resour Manag* 21(9): 1567–1584.

Yuan J, Qian J, Zhang H. & Liang F (2014) *Soil tectonics and soil mechanics (Fourth Edition).* China Communications Press, Beijing, China, pp. 37–40.

Land Reclamation in Ecological Fragile Areas – Hu (Ed.)
© 2017 Taylor & Francis Group, London, ISBN 978-1-138-05103-4

Ecology analysis of invasive plants after 12 years of natural restoration in karst desertification area

Z.Q. Yang
College of Tourism and Sport Health, Hezhou University, Hezhou, China

ABSTRACT: After 12 years natural restoration, research of invasive plants in karst desertification area of north Guilin was done by field investigation and statistical analysis. The results shows that, vegetation coverage of karst desertification area has been greatly changes, in addition to the artificial plants, a large number of natural plants has invaded, which including 21 types of plants and belong to 7 family and 13 category, most of them are herbaceous plants, it accounts for 52%, 4 arbors, it accounts for 20%, 3 lianes, it accounts for 14%. Most of them are heliophobes, it accounts for 92%, and heliophytes accounts 8%. According to ecology analysis, different mixed trees model can be adopted in different slopes for ecological reclamation of karst desertification area, which also have the prevention function of soil and water loss, disease and insect pest, forest fire.

1 INTRODUCTION

Karst environment is a type of ecosystem which developed from soluble carbonate rocks. Karst rock Hill area accounts for about 12% world land, karst is most widely distributed across the country with an area of 344000.30 km^2 in China, exposed area is 91000.79 km^2, mainly located in Yunnan, Guizhou, Sichuan, Hunan, Hubei, Guangdong and the northern Shanxi, Shandong, Henan, Hebei area. For a long time, because of intensify man-made interference such as logging, grazing, producing fire, which resulting in repeated damage to vegetation, frequent natural disasters, serious soil erosion. Karst revegetation is a world-wide problem, serious soil erosion leading to rock-desertification, and local habitat becomes more harsh, all that bringing about more and more bare land with typical characters of difficult to reclamation (Cai Yunlong 1999). In the harsh land of karst area, large number of artificial forest have been cultivated, however, which have poor re-restoration capability after destruction, because of low survival rate, low growth rate, slow reclamation. Many karst hills are bare in a long time, showing of fragile ecological environment and small environmental capacity and low threshold of anti-interference system to maintain stability.

2 MATERIAL AND METHOD

2.1 *Experiment site*

Experiment site is along the highway of Guiyang, location is 110°12'58" ~ 110°29'66"'25°07'54" ~ 25°22'72". Broken terrain with a steep slope and 71% of that is more than 25 °. Soil is dark or brown lime soil, exposed bedrock of low-lying land area accounted for 20%, soil thickness is 15–120 cm, exposed bedrock area of sloping land is over 42%, soil thickness is thin which generally in 10–40 cm. The average annual temperature is 18.8°C, annual sunshine is 1553.09 hours, and annual rainfall is 1894 mm, the annual frost-free period of about 309 days. April to August is rainy season, rainfall accounted for 80% of the whole year, September to March of following year is dry season. Which located in the subtropical monsoon climate zone, it belongs to typical karst peak cluster depression in the northwest of Guangxi.

2.2 *Study method*

Setting up randomly sample area in the standard site, investigating arbor and shrub in the 2 m × 2 m sampling then survey its height and crown width and diameter at breast height, investigating herb in the 1 m × 1 m sampling then survey its types and frequency and coverage, moreover, investigating ecological conditions and human interference of sampling site. After investigation, ecological index of each invasive plant was carried out, which including frequency, density, coverage, biomass and important value (Wang Delu 2005).

3 RESULT AND ANALYSIS

3.1 *Composition of invasive plants in karst desertification area*

After natural restoration, vegetation coverage was changed greatly in the karst desertification area, canopy density reached 0.95, however, the vegetation coverage of bare karst desertification hill is only 0.26. There are 21 types plants, belongs to pteridophyte 36 families and 85 genus. According to the ecotype, its vegetation mainly composed of 11 herbaceous species and accounting for 52%, among them, there are 7 species 1–2 year-old herbs; there are only 4 arbors and accounting for 20%; there are 3 vines and accounting for 14%. From the ecological habits, there are more heliophytes and accounting for 92%, less shade plants and only accounting 8%, all that shows that the ecological environment is more drought in the karst desertification area. Among the invasive plants, most of them are Gramineae and Compositae species, there are 4 Gramineae species and accounting 20%, 6 Compositae species and accounting 28%, legume plant is also accounting 28%. The remaining family is always have one specie, and some species such as *Broussonetia papyrifera* and *Wisteria sinensis (Sims) Sweet* only appeared occasionally. Roegneria serotina Keng. and Plantago asiatica L. have comprehensive advantages, in addition, *Mimosa pudica Linn.* and *Plantago asiatica L.* appeared intensively in some site, while the *Prunus persica* only distribute occasional, which may be related to interference of animal and human beings.

3.2 *Ecology analysis of invasive plants*

3.2.1 *Invasive vegetation of Sapium rotundifolium stand*

In the *Sapium rotundifolium* stand, there are a large number of shade-tolerant plants due to the better growth and lush foliage of *Sapium rotundifolium*, such as *Oplismenus compositus (L.) P. Beauv.* and its relative coverage reached 51%, then it is *Plantag2o asiatica L.* its relative coverage accounted for 34%, furthermore, it also scattered distribution of some plants such as *Artemisia princeps* and *Compositae Bellis perennis Linn* (Table 1). It shows that Sapium rotundifolium grows well in karst desertification area, and nutrient contest is not obvious between *Sapium rotundifolium* and other pioneer plants such as drought-resistant and poor soil-resistant species.

3.2.2 *Invasive vegetation of Celtis tetrandrassp stand*

In the Celtis tetrandrassp stand, due to fully use of moisture and nutrients, Celtis tetrandrassp grows well and have large crown and more branches, so the invasive plants are relatively less with 12% total coverage and 235 g biomass. In addition to some area where sunshine is easy to reach, there are some heliophytes species of Roegneria serotina Keng. with 75% relative coverage, the rest are scattered shade-tolerant plants such as Plantago asiatica L. and Oplismenus compositus (L.) P. Beauv. with 20% relative total biomass (Table 2). All that have created good conditions for the growth of Celtis tetrandrassp.

3.2.3 *Invasive vegetation of Ficus sublanceolata stand*

In the Ficus sublanceolata stand, there are more invasive plants because of smaller crown and less branches of Ficus sublanceolata, and most of them are heliophytes such as Roegneria serotina Keng., the total coverage reach to 50% because the lower layer plants can take full advantage of sunshine. Roegneria serotina Keng. has a obvious advantage with a 60% relative coverage and 45.32% relative biomass. Broussonetia papyrifera is also scattered distribution here (Table 3). Due to species competition, Ficus sublanceolata grows is secondary to Celtis tetrandrassp.

3.2.4 *Invasive vegetation of Ulmus parvifolia stand*

In the Ulmus parvifolia stand, it has the same characteristics of Ficus sublanceolata because of its small crown and less branches, the total coverage reach to 60% of all invasive plants with 1250 g biomass, due to rich sunshine, Roegneria serotina

Table 1. Vegetation growth and distribution in sample area of Sapium rotundifolium.

Species	TC	D/(number/m^2)	AH/m	TB/(g/m^2)	RF	RC	RB	RD	SDR4
Oplismenus compositus (L.) P. Beauv.	35	42	0.20	450	45.20	51	56.79	57.20	52.50
Plantago asiatica L.					33.24	34	25.46	27.66	30.09
Artemisia princeps					12.36	12	9.58	7.48	10.36
Compositae Bellis perennis Linn					4.70	4	4.00	3.82	4.13
Roegneria serotina Keng.					4.50	4	4.17	3.78	4.11

TC-total coverage, D-density, AH-average height, TB-total biomass.

Table 2. Vegetation growth and distribution in sample area of *Celtis tetrandrassp.*

Species	TC	D/(number/m²)	AH/m	TB/(g/m²)	RF	RC	RB	RD	SDR4
Roegneria serotina Keng.	12	21	0.15	235	62.24	75	80.25	81.76	74.75
Oplismenus compositus (L.) P. Beauv.					17.31	12	7.2	12.33	12.38
Plantago asiatica L.					16.57	10	6.37	5.65	9.76
Herba Taraxaci					3.88	3	6.18	0.26	3.11

TC-total coverage, D-density, AH-average height, TB-total biomass.

Table 3. Vegetation growth and distribution in sample area of *Ficus sublanceolata.*

Species	TC	D/(number/m²)	AH/m	TB/(g/m²)	RF	RC	RB	RD	SDR4
Roegneria serotina Keng.	50	75	0.34	870	47.63	60	45.32	72.31	56.31
Bidens pilosa Linn.					22.18	21	16.61	13.92	18.42
Oplismenus compositus (L.) P. Beauv.					12.35	10	8.43	10.6	10.34
Plantago asiatica L.					15.84	5	2.25	2.87	6.49
Broussonetia papyrifera					1.00	2	14.83	0.15	4.49
Wisteria sinensis (Sims) Sweet					1.00	2	12.56	0.15	3.92

TC-total coverage, D-density, AH-average height, TB-total biomass.

Table 4. Vegetation growth and distribution in sample area of Ulmus parvifolia.

Species	TC	D/(number/m²)	AH/m	TB/(g/m²)	RF	RC	RB	RD	SDR4
Roegneria serotina Keng.	60	160	0.40	1250	44.52	53	66.73	82.57	61.71
Oplismenus compositus (L.) P. Beauv.					27.61	21	23.65	13.61	21.47
Miscanthus floridulu (Labnll.)Warb					18.42	16	6.76	3.54	11.18
Parthenocissus tricuspidata					9.45	10	2.86	0.28	5.65

TC-total coverage, D-density, AH-average height, TB-total biomass.

Keng. grows vigorous with 53% relative coverage and 82.57 relative density, it is the largest invasive species in In the Ulmus parvifolia stand (Table 4). It is precisely because of the competitiveness of these plants, Ulmus parvifolia is need to care management to grows well.

3.2.5 *Invasive vegetation of Rhus chinens stand*

In the Rhus chinens stand, there are more invasive heliophytes species (such as Roegneria serotina Keng., Artemisia princeps, Pueraria lobata (Willdenow) Ohwi, plant total coverage reach to 40% and biomass up to 1120 g. In all invasive plants, Roegneria serotina Keng. still showed significant compttitive advantage with a 55% relative coverage and 67.73% relative biomass, then, it is Oplismenus compositus (L.) P. Beauv. and Artemisia princeps, the relative biomass were 15.66% and 13.78 individually (Table 5). Rhus chinens is also a pioneer species with characters of drought-resistant and poor soil-resistant, it will grows better if there are proper tending management.

3.2.6 *Invasive vegetation of Pyracantha fortuneana stand*

In the Pyracantha fortuneana stand, Pyracantha fortuneana belongs to shrub with small branches and some even degraded into thorn, so there are much more invasive plants and almost surrounded Pyracantha fortuneana, total coverage reaches to 65% and with 0.50 m average height, that shown a clear competitive advantage. Since the invasion plant have shown their own advantages, there are only three species under the stand, namely

Table 5. Vegetation growth and distribution in sample area of *Rhus chinens.*

Species	TC	D/(number/m²)	AH/m	TB/(g/m²)	RF	RC	RB	RD	SDR4
Roegneria serotina Keng.	40	100	0.40	1120	60.13	55	67.73	76.42	64.82
Oplismenus compositus (L.) P. Beauv.					23.25	23	15.66	18.55	20.12
Artemisia princeps					11.72	12	13.78	4.13	10.4
Vicia L.					2.45	5	1.75	0.45	2.41
Pueraria lobata (Willdenow) Ohwi					2.45	5	1.08	0.45	2.25

TC-total coverage, D-density, AH-average height, TB-total biomass.

Table 6. Vegetation growth and distribution in sample area of *Pyracantha fortuneana.*

Species	TC	D/(number/m²)	AH/m	TB/(g/m²)	RF	RC	RB	RD	SDR4
Roegneria serotina Keng.	65	140	0.50	940	74.83	75	78.5	81.21	77.38
Oplismenus compositus (L.) P. Beauv.					22.93	23	18.4	16.58	20.23
Plantago asiatica L.					2.24	2	3.1	2.21	2.39

TC-total coverage, D-density, AH-average height, TB-total biomass.

Roegneria serotina Keng. Oplismenus compositus (L.) P. and Beauv. Plantago asiatica L., Roegneria serotina Keng. is the majority with 75% relative coverage, while Plantago asiatica L. only scattered in the bottom (Table 6).

3.2.7 *Comparative analysis of invasive plants in different stands*

Basing on the comprehensive analysis of coverage, height, density and biomass of invasive plants, then vegetation difference in different stands was carried out.

From the above figure, in the 6 different stands, total coverage changes of invasive plants as followings: *Celtis tetrandrassp* < *Sapium rotundifolium* < *Rhus chinens* < *Ficus sublanceolata* < *Ulmus parvifolia* < *Pyracantha fortuneana*; height changes are: *Sapium rotundifolium* < *Celtis tetrandrassp* < *Ficus sublanceolata* < *Ulmus parvifolia* < *Rhus chinens* < *Pyracantha fortuneana*; density changes are: *Celtis tetrandrassp* < *Sapium rotundifolium* < *Ficus sublanceolata* < *Rhus chinens* < *Pyracantha fortuneana* < *Ulmus parvifolia*; total biomass changes are: *Celtis tetrandrassp* < *Sapium rotundifolium* < *Ficus sublanceolata* < *Pyracantha fortuneana* < *Rhus chinens* < *Ulmus parvifolia*. There are less undergrowth vegetation in the *Celtis tetrandrassp* stand, while there are much more undergrowth vegetation in *Ulmus parvifolia* and *Pyracantha fortuneana* stands, that shows *Celtis tetrandrassp* can adjust to the special environment in the shortest possible by fully use of sunshine and nutrients and moisture, and let other plants difficult to disturb. At the same time, another ecological problem was found, the erosion of slope by heavy rain would be more obvious in the *Celtis tetrandrassp* stands with scare undergrowth vegetation, and such problem does not exist in stand with more undergrowth vegetation. As a result, it can afford suggestion for the restoration of karst desertification area, in a steep slope, it is suitable to cultivate such species as *Pyracantha fortuneana*, and in a more moderate slope for mixture cultivation is more appropriate, such as *Celtis tetrandrassp+ Ficus sublanceolata+ Pyracantha fortuneana*, *Celtis tetrandrassp+ Rhus chinens+ Ulmus parvifolia*, *Celtis tetrandrassp+ Sapium rotundifolium + Ulmus parvifolia*, Which not only play a role in the prevention of soil erosion, but also to prevent pests and disease sand forest fires.

4 CONCLUSIONS AND DISCUSSION

1. After natural restoration, vegetation coverage was changed greatly in the karst desertification area, There are 21 types plants, belongs to pteridophyte 36 families and 85 genus. According to the ecotype, its vegetation mainly composed of 11 herbaceous species and accounting for 52%, From the ecological habits, there are more heliophytes and accounting for 92%, less shade plants and only accounting 8%, all that shows that the ecological environment is more drought in the karst desertification area.
2. Growth and distribution of invasive plants are very different among different stands: In the *Sapium rotundifolium* stand, there are a large

number of shade-tolerant plants due to the better growth and lush foliage of *Sapium rotundifolium*; In the *Celtis tetrandrassp* stand, due to fully use of moisture and nutrients, *Celtis tetrandrassp* grows well and have large crown and more branches, so the invasive plants are relatively less with 12% total coverage; In the *Ficus sublanceolata*, *Ulmus parvifolia*, *Rhus chinens* stand, there are more invasive plants because of smaller crown and less branches of *Ficus sublanceolata*, and most of them are heliophytes, the total coverage reach to 50% and 60% and 40%; In the *Pyracantha fortuneana* stand, *Pyracantha fortuneana* belongs to shrub with small branches and some even degraded into thorn, so there are much more invasive plants and almost surrounded *Pyracantha fortuneana*, total coverage reaches to 65%. It will grows better if there are proper tending management in the high coverage area.

3. *Pyracantha fortuneana* stand have largest total coverage and average height of invasive plants, that is 65% and 0.50 m. *Ulmus parvifolia* stand have largest average density and total biomass, that is 160/m^2 and 1250 g/m^2. There are less undergrowth vegetation in the *Celtis tetrandrassp* stand, while there are much more undergrowth vegetation in *Ulmus parvifolia* and *Pyracantha fortuneana* stands, that shows *Celtis tetrandrassp* can adjust to the special environment in the shortest possible by fully use of sunshine and nutrients and moisture, and let other plants difficult to disturb. At the same time, different mixture cultivated mode should be taken in different slope for restoration of karst desertification area, which not only play a role in the prevention of soil erosion, but also to prevent pests and disease sand forest fires.

ACKNOWLEDGEMENTS

Supported by Ministry of Education humanities social sciences research project (15YJCZH145).

Supported by the Doctor's scientific research foundation of Hezhou University (HZUBS201406).

REFERENCES

Cai Yunlong. (1999) Ecological rehabilitation and development of agriculture forestry and animal husbandry in karst mountain areas of southwest china. J. *Resource Science, 21(5)*:37~41.

Wang Delu. (2005) Preliminary study on types and quantitative assessment of Karst rocky desertification in Guizhou Province. J. *Acta Ecologica Sinica, 25(5)*:1057~1063.

Land Reclamation in Ecological Fragile Areas – Hu (Ed.)
© 2017 Taylor & Francis Group, London, ISBN 978-1-138-05103-4

Old mine dumps recovery: An environmental and techno-economical challenge

V. Dentoni, B. Grosso, P.P. Manca & G. Massacci
DICAAR—Department of Civil and Environmental Engineering and Architecture, University of Cagliari, Italy

ABSTRACT: The paper discusses from a techno-economical point of view the possibilities offered by the combined mineralurgical and metallurgical treatment of mining residues as a methodology for the reclamation of abandoned mine areas. The proposed methodology permits to achieve better results in the case of: original minerals high residual concentration grades; possible recovery of the rare earth minerals often associated with the main ore; presence of a homogeneous mining domain along with a metallurgical district in which the production cycle could be carried out locally. Last but not least, other favorable circumstances are represented by the fact that tailings were originated in the last century during which mining technologies were not well developed, leaving a certain amount of useful minerals still to be recovered. The proper setting of the mineralurgical processes aimed at obtaining both a commercial concentrate and a non-polluting residue proved to be of fundamental importance in the remediation processes.

1 INTRODUCTION

The paper discusses the applicability of the Circular Economy Approach to those wastes originated from past Pb and Zn mixed sulfides ore processing activities, very common across the mining landscape of Sardinia. Mining and mineral-processing wastes are considered one of the world's largest problem. Their reuse should be included in future sustainable development plans, even though their potential impacts on a number of environmental processes are highly variable and should be thoroughly assessed. The chemical composition and geotechnical properties of the waste determine which uses are most appropriate, and whether their reuse is economically feasible. When properly evaluated, mining and mineral-processing wastes can be reused for mineral-extraction, providing additional fuel for power plants, construction materials, and for remediating any abandoned mining activities (Bian et al., 2012).

Mine wastes as solid and liquid materials found at or near mine sites are uneconomic as well. Moreover, they constitute one of the world's largest waste accident, containing high concentration of elements and compounds that effect both ecosystems and humans.

Multidisciplinary research on mine wastes focuses on their character, stability, impact, remediation and reuse, according to the general trend to exploit low-grade ore deposits of large dimension. According to Lèbre, et al. (2017) three critical aspects of the mining waste should be taken into consideration: 1) residual metal grade; 2) the environmental strategy; and 3) its economic context.

2 THE CIRCULAR ECONOMY STRATEGY APPLIED TO THE REUSE OF MINING AND MINERAL-PROCESSING RESIDUES

The Mining Circular Economy Strategy should use the 3R principle as follows: Reducing, Reusing and Recycling (Zhao et al., 2012). In particular, the 3R principle applied to mining and mineral processing residues and their associated minerals can reduce pollution converting low-grade wastes into valuables, saving crushing and grinding costs (around 50% out of the total cost of the mineral processing, in traditional mining). On the other hand, new substances could be produced, and used as building materials (Manca et al., 2015), raw materials, concrete, ceramics, glass or these wastes could be used for filling up large areas affected by voids and subsidence (Manca et al., 2014).

The origin of wastes in the metal production chain is represented in the block diagram in Figure 1. Three types of solid residues having higher metal concentration are produced: waste rocks, tailings, and slag. Moreover, the water circulating in the mining voids and surface embankments can be polluted by transportation and/or leaching when different types of wastes are involved. It should be said, however, that pollution might propagate from the soil (solid waste) to the air, due to wind action.

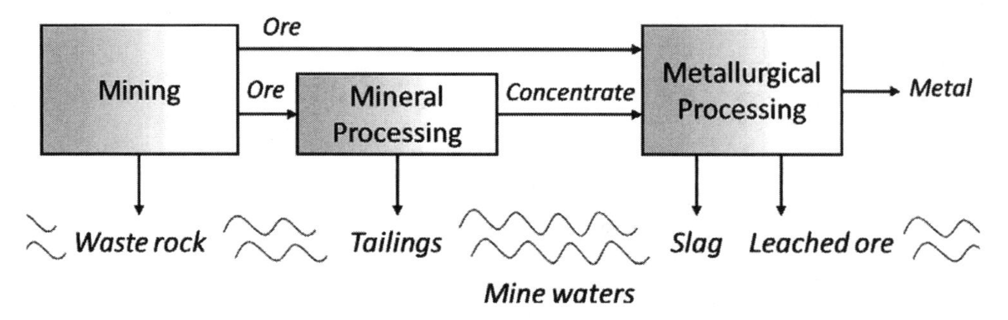

Figure 1. Origin of wastes in the metal production chain (Lèbre et al., 2017).

It can be argued that the solid mining residues, along with the parts of unexploited ore body exposed to leaching, represent the primary source of pollution originating from the abandoned mining areas. Therefore, their removal should be the most effective remediation activity, especially for those sites for which no securing works or insulation techniques have been carried out. Indeed, this option excludes the need of insulation or any other treatments that would require unlimited duration or continuous control. The complete waste removal, however, requires that the useful minerals still present in the waste are recovered using appropriate processing methods in the attempt of obtaining a residue that could be reused according to the circular economy principles.

For a first evaluation of the applicability of the Circular Economy Strategy, in the present work the simplest case considering some lead and zinc sulfide wastes from which extractable concentrates having economic value, are discussed. The following pages explain the process, the economic evaluation of the results, and experimental investigations needed to check the economic feasibility of the process.

3 MINERAL PROCESSING OF WASTE DUMPS: ECONOMIC EVALUATION

3.1 *The process*

The operations producing a change in the metal grade, and the steps preceding the waste metallurgical treatment are represented in Figure 2. Here the part including the flotation and the inerting processes will be discussed in Figure 3.

The gravity separation represents a pre-concentration stage useful in both the thermal processes (Waelz klin) and the physico-mechanical treatments (flotation), the later one been even directly fed. The tails are considered as a polluting waste to be inerted and disposed.

The section including the gravity separation and thermal processes concerns the oxidized Zn compounds (calamines), whereas the flotation is applied to the sulfides. A scavenger phase (dotted line) is required in the special case in which the waste characteristics do not comply with the current regulation related to their classification as an inert. Thus, the flotation concentrates (having a grade unacceptable for the metallurgy) should be disposed after inertization.

Figure 3 shows a detail of the flotation process in which four products are present: two Pb and Zn concentrates, a final inert residue and a mixed product (after scavenging) to be inertized.

3.2 *Economic evaluation of the results*

The economic outputs of the relevant technological choices linked to an ore dressing process are expressed through some indexes based on the material balance, as shown in Table 1.

From Table 1, the following physical quantities can be defined:

Percent of weight recovered:

$$X_W = 100\frac{M_c}{M_f} = 100\frac{f-t}{c-t} \tag{1}$$

Enrichment ratio:

$$E = \frac{c}{f} \tag{2}$$

The following relationship is deduced:

$$t = (M_f \cdot f - M_c \cdot c)/T \tag{3}$$

The goal to achieve in the mining economy practices is that to maximize the net profit:

NP = (value of concentrate—mining cost—processing cost).

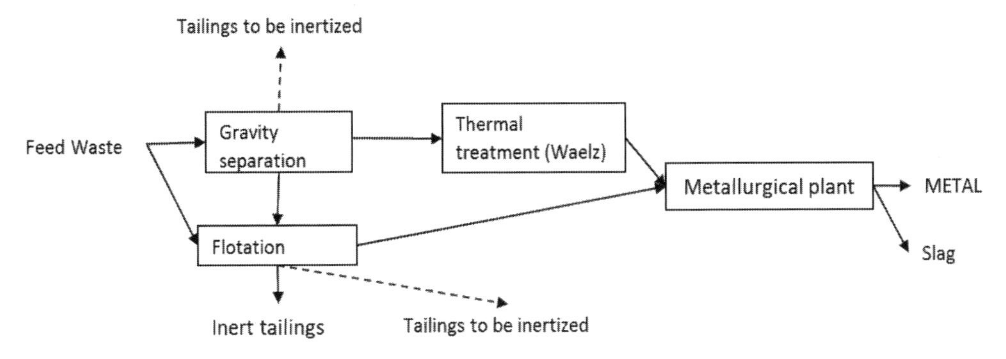

Figure 2. Waste treatment steps preceding the metallurgical process.

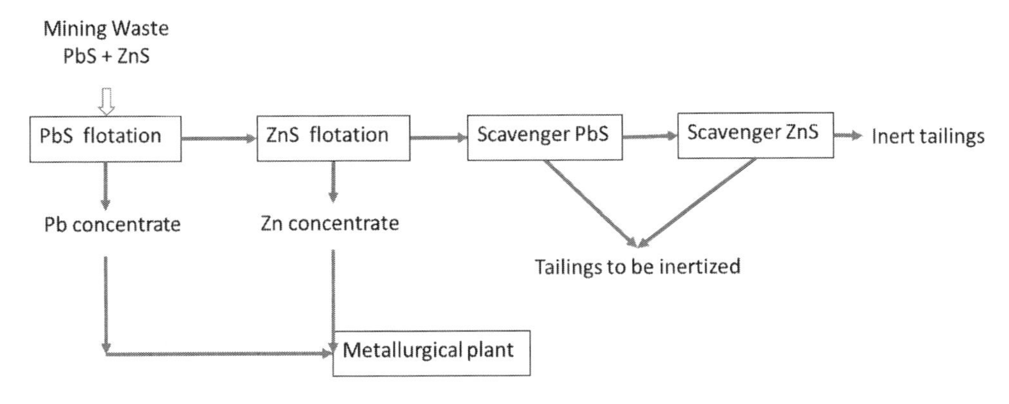

Figure 3. Flotation.

Table 1. Material balance in mineral processing.

Products	Mass	Assay	Metal mass	% metal recovery
Feed	M_f	f	$M_f \cdot f$	—
Concentrate	M_c	c	$M_c \cdot c$	$\rho_c = 100\,(M_c \cdot c)/(M_f \cdot f)$
Tailings	M_t	t	$M_t \cdot t$	$\rho_t = 100\,(M_t \cdot t)/(M_f \cdot f)$

Note: MT = MF—MC

In applying this process to an environmental remediation two main differences should be considered:

1. there is no mining cost just because the material was already extracted and dumped there; 2) after having been mineral processed, the waste can have an economic value. Indeed, the final tailing grade should reach a limit value permitting its classification as an inert. Lower values are to be avoided due to their economic costs.

The values of c and of t are therefore to be imposed; the amount of M_c and of M_t must be determined in order to maximize the net profit NP = (value of concentrate—processing cost + value of final tailings), according to the equation:

$$NP = M_c \cdot \left(c \cdot \alpha \cdot LME - C_{met}\right) - M_f \cdot C_{proc} + M_t \cdot V_t \quad (4)$$

where: NP = net profit (\$); M_c = mass of concentrate (t); c = metal assay in the concentrate; α = metallurgical yield; LME = London Metal Exchange metal price (\$/t); C_{met} = unit metallurgical expenses (\$/t); C_{proc} = unit mineral-processing cost (\$/t), inclusive of tailings management cost; V_{tail} = unit value of final tailings (\$/t).

In the case of a negative outcome ($NP < 0$), the condition to be attained is NP lower than the cost of an alternative remediation operation which, in turn, should have an equal environmental benefit.

3.3 *Experimental aspects*

Once defined the objectives to achieve and the relationships between the physical quantities expressing the results of a mineral processing operation, the values to be determined through experimental tests or known *a priori* should be specified.

It should be noted that the proposed application is here used in a non-conventional industrial way, as its main objective is not to reach a high metal recovery but the declassification of the original waste into an inert material. The commercial value of the concentrate compensates or attenuates the processing costs.

The experimental investigations should allow to determine the following quantities:

- Feed characteristics. The amount of waste to be treated and its grade of useful metals are assumed to be known; the chemical composition, including the minor compounds and the valuable accessories elements, has to be determined, as well as the size and mineralogical composition and the possible alterations due to weathering.
- Refuse and concentrate grades (t and c). The obtainable values are defined by the experimental relationships between them and the feed grade. In addition, the value of t to be obtained must be defined based on various conditions according to the environmental legislation, among which there are: the land use of the site, the leachability of the refuse obtained and the conditions of the natural background. The grade c is established by the metallurgy according to international formulas or agreements between producer and user.
- The cost of the treatment process. The cost of flotation is related to c and t values to be achieved and to the consequent plant complexity. In addition, in the case in which the experimental tests would demonstrate the impossibility to obtain concentrations below the required limits in the final rejection, it would be necessary to complete the treatment process with a scavenger section, in which the value of t is imposed (according to the environmental regulations); M_c and its grade c are those obtainable from the formulas. In this case, however, M_c is no longer a commercial concentrate but a mixture to be disposed after inertization. The values of M_c and c must be chosen in order to minimize the additional inertizing cost, that must be included in the comprehensive economic evaluation.

For some given values of M_f and f the following aspects must require an experimental assessment:

1. The relationships between f, c and t, and the consequent values of metal recovery ρ_c and percent of weight recovered X_W.
2. The cost of treatment, which increases for a higher complexity of the flotation circuit, including the scavenger section too.
3. The grade of the Pb-Zn mixture produced by the scavenger phase, according to the need of minimizing the inertizing cost (e.g. the percent amount of Portland Cement to be added)

A first evaluation of the above-mentioned relationships is proposed, based on the available scientific and industrial data as discussed in the following paragraphs.

3.3.1 *Relationship between feed grade and tailings grade*

For a given value of c (metal assay in the concentrate), the minimum value of grade t of flotation tailings depends on the feed grade f. For the case considered in this paper (Pb and Zn mixed sulfides), t increases at increasing f according to the curves shown in Figures 4 and 5 (respectively for Pb and for Zn); c values in the range 20% ÷ 60% for the Pb circuit and 20% ÷ 50% for the Zn circuit were considered.

3.3.2 *Flotation costs*

The unit cost of flotation depends on the complexity of the flotation circuit, the grades of the feed and the final concentrate, and on the plant capacity. In the example considered (Pb and Zn mixed sulfides) the relationship between the flotation unit cost and the feed grade may be considered linear, on first approximation, as shown in Figure 6. For a

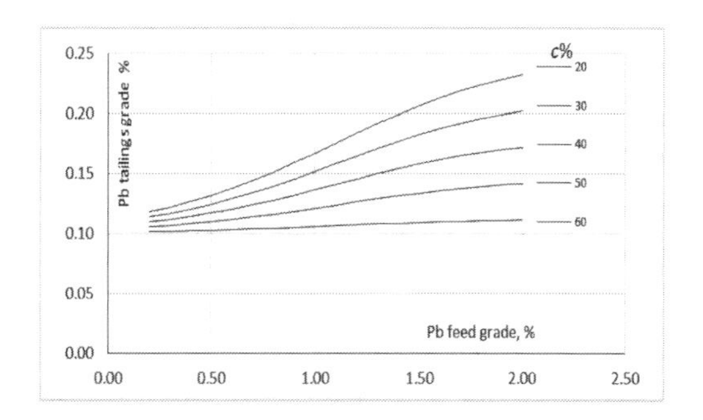

Figure 4. Pb tailings grade as a function of Pb feed grade.

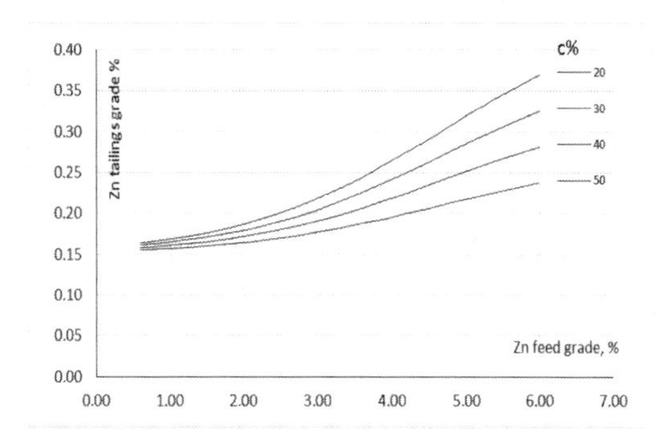

Figure 5. Zn tailings grade as a function of Zn feed grade.

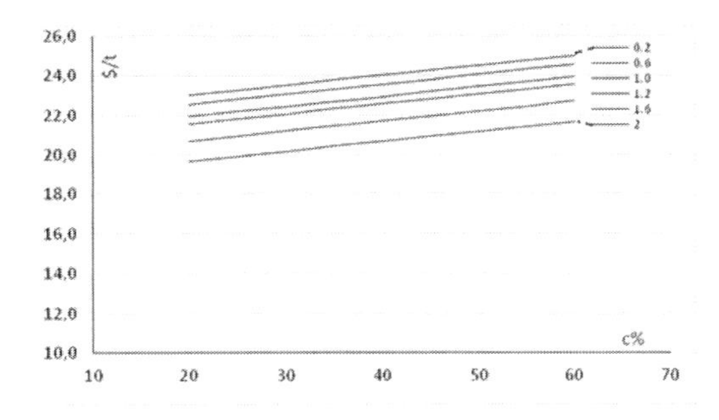

Figure 6. Unit cost of flotation as a function of the concentrate grade.

given value of the concentrate grade, the unit cost of flotation is lower the richer is the feed.

The reported case refers, as already mentioned, to some mixed Pb-Zn sulfides in which the lead grade range varies between 0.2% and 2.0%, on the assumption that the Pb/Zn ratio is 1 to 3. The range of variation of the concentrate grade includes low values lower than 50%, unusual in metallurgy.

3.3.3 *Choice of grades in scavenger circuit*

The scavenger cycle is used in flotation plants as roughing phase, in the first part of the process. However, in this special case, it is used in the last phase with a different function, which is, that of providing two different products: the first one having a lower grade of metals, must accomplish to the national regulations on the solid residues classification. The second one at higher grade must be disposed of as hazardous waste. In order to keep the inerting operation costs as low as possible some specific tests (for stabilization/solidification, Desogus et al. 2012 and for leachability, Desogus et al. 2013) must be carried out, taking into account the optimal ratio between mass and content. However, the resulting costs should not be higher than the costs of the non-hazardous declassified material landfilling, amounting to about \$ 75/t.

4 SENSITIVITY OF THE ECONOMIC OUTCOME IN CASE OF LEAD AND ZINC MIXED SULFIDES

4.1 *Assumptions*

As a first example, a simple case even though realistic concerning the grade materials and their relationships was calculated.

It considers a waste of Pb and Zn mixed sulfides, with no oxidized forms. The grade range and Pb/Zn ratio taken into consideration were those of the 26 mine dumps of Campo Pisano (Iglesias, Italy) reported in the histogram in Figure 7. The average ratio between the concentrations of Pb and Zn is about 1/3 and the concentrations of Pb variation range is about 0.2–2.0, Zn about 0.6–6.0.

In the calculations were considered to following elements:

- The Pb and Zn limit content in the final tail declassified to inert corresponding to the respective Concentration of Contamination Threshold (CCT_{Pb} = 1000 mg/kg, CCT_{Zn} = 1500 mg/kg).
- The c grade variation field of the concentrate for the metallurgy is equal to 20%–60% for Pb and 20%–50% for Zn.
- The flotation costs versus grade of feed f and grade of concentrate c according to the functions represented graphically in Figure 6.
- The value of final tail, if reusable (as building material) should be equal to 5 \$/t.
- The Pb and Zn concentration in the scavenger cycle should be equal to 5%, after evaluation of the cost of the subsequent inerting.
- The inerting cost should be equal to 75 \$/t, corresponding to the landfill costs of delivering non-hazardous wastes.
- The metallurgical yield $\alpha_{Pb} = 0.95$ and $\alpha_{Zn} = 0.85$; the metal prices LME_{Pb} = 1875 \$/t and LME_{Zn} = 2195 \$/t; the unit metallurgical expenses C_{met} = 200 \$/t.

For the obtained result traceability, the sale formula (see in the theoretical part) does not consider the penalties related to metal grade lower than those usually accepted in the market did.

4.2 *Final economic outcomes*

Taking into account the concentrated products commercial value, the reusable tailing value and the cost of flotation (principal and scavenger cycles), the final economic result calculated for ranges of feed grade (see legend) can be summarized as indicated in the Figure 8.

The histograms in Figure 9 show the weight changes between the different products: Pb and Zn concentrated; final tailings; and scavengers to be inertized. The final grade levels at the extremes of the range variation (c = 20% and c = 60%) and the two extreme conditions of the feed grade (Pb = 0.2%, Zn = 0.6% and Pb = 2.0%, Zn = 6.0%) were chosen for the calculations.

The observation of the figures and their comparison allows pinpointing the following:

- Higher-grade levels feed determines reduction of the final tailing and increasing of both the Pb and Zn concentrates and the scavenger.
- This effect is lowered as the grade of the final concentrate c increases.

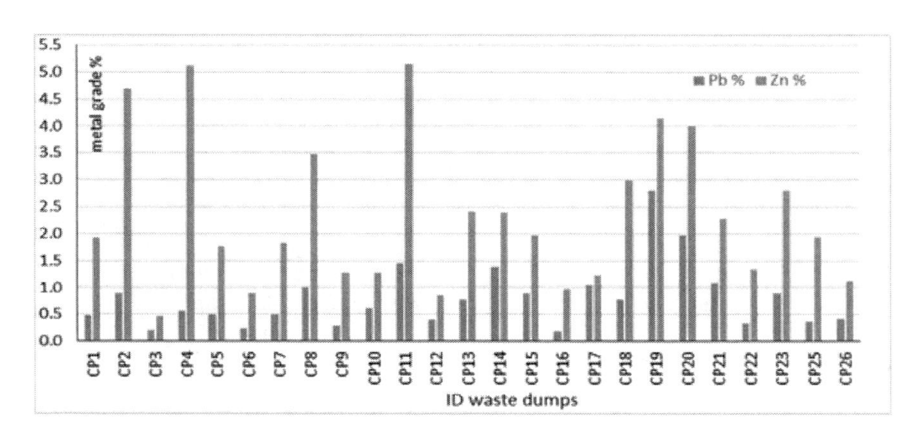

Figure 7. Pb and Zn grade of Campo Pisano (Iglesias, Italy) Mine dumps.

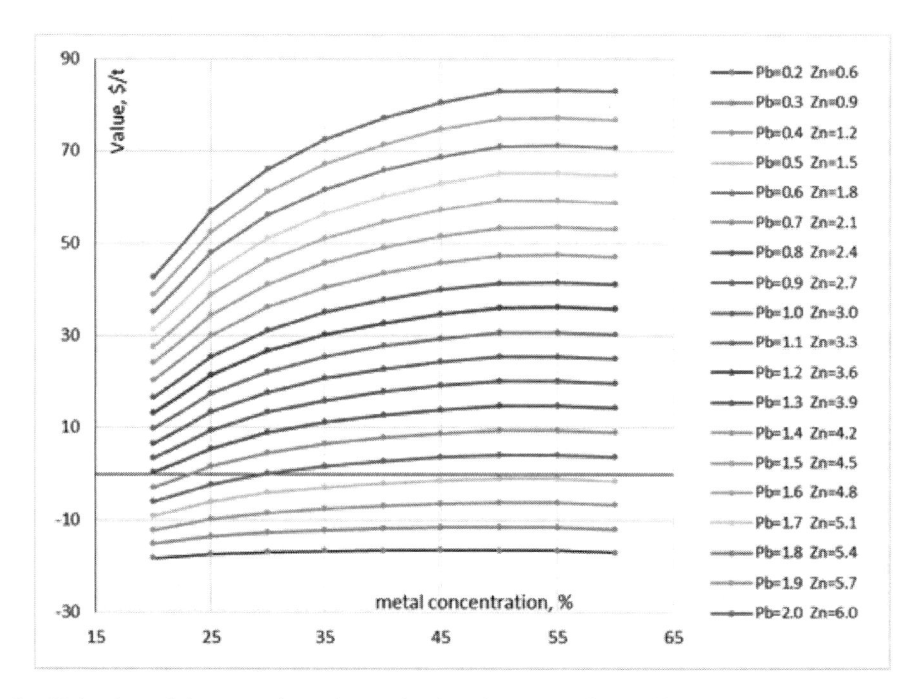

Figure 8. Unit values of the economic results as a funtion of concentration and feed grade.

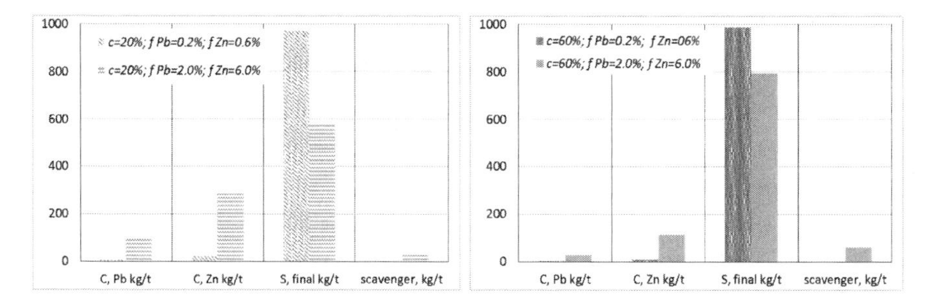

Figure 9. Comparison of the products weight distribution in extreme conditions.

5 CONCLUSIONS

The results obtained allowed the following considerations:

- The convenience to increase the flotation towards the *c* highest levels is greater the higher the feed grade. When the unit flotation cost for the lower feed grade is higher than the richest one, the *NP* curve reaches a maximum beyond which it is not convenient to push the process further (in the example considered 50%);
- An increase in the concentrate grade *c* above 50% is irrelevant to the operation of the economic budget because it affects lead only and not zinc; the latter (according to the considered sale formula) compared to lead has a metallurgical performance lower than 10%, but a higher market value of about 15%;
- The lower feed grade considered, Pb = 0.2% and Zn = 0.6%, proved to have a cost of around $ 15/t, not very sensitive to the increase of content of the concentrate. This cost, however, is almost corresponding to that of a new landfill construction. Using the proposed method, better enviromental performances are obtainable because the original pollutin agents are removed for good.
- The theoretical treatment and the numerical application developed demonstrated that (although it is a simple and for some aspects optimistic case due to absence of oxidized metal compounds) the reprocessing of mining landfills could solve the environmental problem with significant net profits.

The process here used is the traditional flotation method, applied in a non-traditional way for both the low-grade contents and the plant use. The technical and economic objective is the optimal combination between the useful metals maximum recovery and the reusable waste production. In same special cases, however, the better solution could be linked to a high-grade concentration, while for others the better solution could be to reach a high-classified inert M_t, calibrated on specific values for reuse on site. Minimizing the residues to treat and the final tailings to reuse transport costs proved fundamental to the process economy. In particular, situations non-traditional metallurgical treatments, such as a Plasma torch (Rao et al., 2013), for example, preceded by selective mineralurgical processes could be used to extracting peculiar useful minerals.

REFERENCES

Bian, Z., Miao, X., Lei, S., Chen, S., Wang, W. & Struthers, S. (2012) The Challenges of Reusing Mining and Mineral-Processing Wastes. *Science*, 337, 702–703.

Desogus, P., Manca, P.P. & Orrù, G. (2013) Heavy Metal Leaching Contaminated Soils From A Metallurgical Plant. *International Journal of Mining, Reclamation and Environment*, 27, 202–214.

Lèbre, É., Corder, D.G. & Golev, A. (2017) Sustainable practices in the management of mining waste: A focus on the mineral resource. *Minerals Engineering*, 107, 34–42.

Manca, P.P., Desogus, P., & Orrù, G. (2014) The reuse of abandoned Acquaresi mine voids for storage of the Masua flotation tailings. *International Journal of Coal Science & Technology*, 1, 213–220.

Manca, P.P., Orrù, G. & Desogus, P. (2015) Recycling of sludge from ornamental stone processing as resource in civil construction. *International Journal of Mining, Reclamation and Environment*, 29, 141–155.

Rao, L., Rivard, F. & Carabin, P. (2013) Thermal plasma torches for metallurgical applications. *Proc. of 4th International Symposium on High Temperature Metallurgical Processing*, TMS, 57–66.

Zhao, Y., Zanga, L., Li, Z. & Qin, J. (2012) Discussion on the Model of Mining Circular Economy. *Energy Procedia* 16 (2012) 438–443.

Land Reclamation in Ecological Fragile Areas – Hu (Ed.)

Quantitative study on the soil reconstruction of a root system in the coal resource-concentrated distribution of Shenfu mineral regions

Q. Li
Yulin University, Shaanxi Yulin, P.R. China
Shaanxi Key Laboratory of Ecological Restoration in Shanbei Mining Area, Shaanxi Yulin, P.R. China

G.B. Liu
State Key Laboratory of Soil Erosion and Dryland Farming on the Loess Plateau, Institute of Soil and Water Conservation, Northwest A&F University, Yangling, P.R. China

F.R. Kang
Yulin University, Shaanxi Yulin, P.R. China
Shaanxi Key Laboratory of Ecological Restoration in Shanbei Mining Area, Shaanxi Yulin, P.R. China

W.L. Wang
State Key Laboratory of Soil Erosion and Dryland Farming on the Loess Plateau, Institute of Soil and Water Conservation, Northwest A&F University, Yangling, P.R. China
Shaanxi Key Laboratory of Ecological Restoration in Shanbei Mining Area, Shaanxi Yulin, P.R. China

Y.J. Bo
Yulin University, Shaanxi Yulin, P.R. China
Shaanxi Key Laboratory of Ecological Restoration in Shanbei Mining Area, Shaanxi Yulin, P.R. China

G.P. Zhao
Shaanxi Key Laboratory of Ecological Restoration in Shanbei Mining Area, Shaanxi Yulin, P.R. China

ABSTRACT: The exploitation of coal mineral resources brings convenience and wealth, as well as the damage of ecological environment. This study selected Sandy loessin in the mine-concentrated region, using soil samples with roots, pure soil, and designed roots for soil anti-scouribility experiment. The primary goal of this study was to evaluate the relative contribution of root net-link, soil-root bond, and biochemistry functions, to create soil configuration in soil anti-scouribility. The results showed that root physical consolidation effect is the governing factor, contributing 69.5% in soil reconstruction process of new structural soils. Compared with alfalfa density of 90 plants/m^2, physical consolidation effect in the treatment of 360 plants/m^2 increased by 12.4%. In addition, the root link function is the key man-ner in a proportion of 79.6% in the physical consolidation effect. Exponential function can well express the relationship between root physical consolidation effect and root surface area density.

1 INTRODUCTION

Coal plays a leading role in the energy structure of China and contributes greatly to China's economic development. The exploitation of coal mineral resources brings convenience and wealth to human (Luca *et al.*, 2015). However, the construction and production of mine follows by destruction of the land and the damage of ecological environment attracted much attention from governors and scientists, especially after 'The Belt and Road Initiative' implemented in 2013, which puts forward new requirements for the mining exploration and ecological environment protection (Liu *et al.*, 2015).

Plant roots significantly affect soil erosion process of overland flow. In general, Root Total Effect (RTE, %) on soil erosion resistance can be divided into Root Physical Enlacing (RPE, %), including net-link and soil-root bond functions (NF and BF, %), and Root Biochemistry Effect (RBE, %) (Liu *et al.*, 1998; Li, *et al.*, 2016, 2017). For example, Gyssels *et al.*, (2003, 2005) pointed out that plant (Beet, Maize and Endive) roots could reduce sediment loss by 20%, averagely, in silt loam soils, Zhou & Shangguan (2005) found that soil loss could be cut as much as 96% in the Ryegrass root-penetrated soils, and such effect can be well explained by the root features like root biomass, root surface area density, etc. (De Baets *et al.*, 2012; Burylo *et al.*, 2012). Even though, the relative contribution of root physical enlacing and biochemistry effect to soil reinforcement is still not clear recently in the mine-concentrated distribution of flow-induced erosion regions.

This study selected soil samples from dumping site in the Shenfu mine-concentrated distribution, and was performed to investigate the relative contribution of root physical enlacing and biochemistry effect to soil erosion resistance. Such kind of study could strengthen the mechanism of roots to soil erosion resistance in the root-penetrated soil.

2 MATERIALS AND METHODS

2.1 *Soil sample preparing*

This experiment was conducted at Institute of Soil and Water Conservation, Northwest A&F University in China. The Loess soil samples were collected from a slope land, with slope gradient of 16° in the coal resource-concentrated distribution of Shenfu mineral regions (37°16′56″N, 110°4′28″E). In the present study, soil samples were obtained from two parts, one part (CK1) is Loess parent material, sampled from 2.5–3 m-depth, representing soil samples without RBE. The other soil was tillage sandy soils, representing soil samples with RBE, but without BF. The part1 and part2 soil were independently passed through a 5-mm sieve and air-dried after the removal of root fragments. *Purple alfalfa* seeds were sown at 10 cm row spacing with plant density of 90 (R1), 180 (R2), 270 (R3) and 360 (R4) stems/m^2 with 4 replicates in a metal box, with dimensions of 200 cm length × 28 cm width on 6 May 2013 (Figure 1). Similar root-texture cotton thread in a diameter of 0.4 mm was chosen as the designed roots, and was gently worn buried in CK2 soil sample at a horizontal angle of 10° using stainless steel needle. Thus, the designed roots in the soil have NF and BF, but no RBE, to soil erosion resistance.

Figure 1. Basic experimental set-up and process.
(A. Experimental treatment diagram; B. Soil samples with designed roots in various density; C. Scouring process in soil samples with designed roots; ①. CK1; ②. CK2).

2.2 *Soil sample collection for flume experiment*

Laboratory-simulated flow experiments were conducted nine weeks after *Purple alfalfa* was planted. The above-ground biomass was chipped to be level to the soil surface, and the residues were cleared. Four special rectangular sampling metal boxes with dimensions of 20 cm length × 10 cm width × 10 cm depth were collected in each plot. Prior to scouring experiment, the samples were placed in a container with a constant water level of 5 cm below the soil surface to allow 12 h for slow capillary rise, and they were taken out of the water, drained for 8 h to obtain the same soil moisture. A concentrated flow experiment was conducted with a hydrological flume (length = 2 m, width = 0.10 m). Simulated runoff flux of 4 L min^{-1} and flume slope of 15° was used in the present study. The scouring time (15 min) referred to the maximum time frequency for rainstorms in the research region. During flow experiment, samples of runoff and detached soil were collected every 1 min during the first 3 min and every 2 min thereafter using 10 L buckets. After the suspended particles had settled, the clear water was drained off, and the sediments were sampled and oven-dried at 105°C.

2.3 *Soil indicator determination*

Roots were separated from the soil samples by hand washing on a sieve after each flow experiment. Each root segment was dried with filter paper and then was scanned to obtain the root images for calculating root surface area density (RSAD, cm^2 root surface-area cm^{-3} soil). Soil indicators were determined by traditional methods (Nelson and Sommers, 1982). Soil Erosion Resistance (SER, L g^{-1}) was expressed as soil erodibility:

$$RTE = \frac{y_{CK1} - y_i}{y_{CK1}} \times 100\% \tag{1}$$

where f is the flow rate (L min^{-1}), t is the abrasion time (min), and W is the weight of oven-dried sediment (g).

Root Total Effect (RTE, %) was calculated as equation (2).

$$RTE = \frac{y_{CK1} - y_i}{y_{CK1}} \times 100\% \tag{2}$$

where y_{CK1} is the soil loss in CK1, y_i is the soil loss in different root density treatments (i = 1, 2, 3, 4, the same follows).

RPE denotes the soil loss in the soil samples with and without root enlacing, and was calculated by following equation:

$$RPE = \frac{y_{CK2} - y_i}{y_{CK2}} \times 100\% \tag{3}$$

where y_{CK2} is soil loss in CK2.

NF denotes that roots like a net fixing soil in root-penetrated soils, and it is obtained by the soil loss between designed roots and soil parent material treatments.

$$NF = \frac{y_{CK1} - y_i'}{y_{CK1}} \times 100\% \quad (4)$$

where y_i' is the soil loss in designed roots treatment (i = 1, the same follows).

The RTE was assumed as 100% to soil erosion resistance in the present study, and BF (root-soil bonding effect on erosion resistance) was calculated by RPE excluding NF as follows:

$$BF = RPE - NF \quad (5)$$

RBE denotes the functions of root exudates to enhance soil erosion resistance, and was calculated by RTE excluding RPE as follows:

$$RBE = RTE - RPE \quad (6)$$

3 RESULTS AND DISCUSSION

3.1 *Soil properties in different root density treatments*

Soil structural stability could be enforced by plant roots through root physical enlacing and biochemistrical exudates, and thus the erosion resistance of soil was improved. Table 1 indicates that soil bulk density decreased slightly in the root-penetrated soil as compared with CK2. 1.8–2.6 times were found in the aggregate content with increment of root density treatments. Compared with CK2, the soil organic matter content, total nitrogen content and available phosphorus in the root-penetrated soil were increased significantly, which demonstrates that roots can ameliorate soil structural stability, and increase soil nutrient content.

3.2 *Relative contribution of roots to soil erosion resistance*

As depicted in Table 2, a gently increasing trend was found in root physical enlacing as the root density increases from R1 to R4 treatment. This result agreed with the conclusion of previous studies that root mass density had a significant impact on soil erosion resistance in the flow-induced erosion regions (Zhang *et al.*, 2013; Wang *et al.*, 2015). As the roots increased in quantity and size, a 6.7% percentage increase in RPE was observed, whereas the relative contribution of RBE in the root total effect reduced slightly. This result indicated that root physical enlacing is the main reason, accounting for 64.7–77.2% of root total effect in strengthening soil erosion resistance. For comparison, the root physical enlacing in fibrous root-penetrated soils may play a more important role in reducing soil loss due to its larger root mass density and root surface area density (Knapen *et al.*, 2008).

3.3 *Relationship between RPE and RSAD*

There is a remarkable relationship between soil erosion resistance and plant roots. Fig. 2 showed that an exponential function of $y = 72.11(1-\exp(-0.025x))$, $R^2 = 0.866^*$ could fit well the relationship between RPE effect and RSAD. This result is in accordance with the findings of Zhou & Shangguan (2005) that RSAD is an effective indicator in evaluating soil erosion resistance. Therefore, the RSAD could effectively reflect the effect of root physical enlacing, and thus could forecast the changes in soil erosion resistance.

Table 1. Soil properties in different root-penetrated treatments.

Soil properties	Treatment					
	CK1	CK2	R1	R2	R3	R4
Bulk density/(g cm^{-3})	1.32 ± 0.01b	1.33 ± 0.02b	1.36 ± 0.03a	1.32 ± 0.01b	1.32 ± 0.01b	1.29 ± 0.01b
Aggregate content/ (g kg^{-1})	39.4 ± 1.05e	57.6 ± 2.12d	107.9 ± 3.12c	137.0 ± 1.19b	136.4 ± 3.21b	148.2 ± 3.83a
Soil organic matter/ (g kg^{-1})	1.83 ± 0.11c	2.05 ± 0.13c	2.18 ± 0.05c	2.55 ± 0.12b	2.92 ± 0.08b	4.17 ± 0.14a
Total nitrogen/(g kg^{-1})	0.11 ± 0.00c	0.09 ± 0.00c	v0.13 ± 0.02b	0.14 ± 0.01b	0.14 ± 0.00b	0.27 ± 0.00a
Available phosphorus/ (mg kg^{-1})	1.53 ± 0.02f	2.09 ± 0.03e	2.88 ± 0.03d	3.74 ± 0.02c	3.95 ± 0.07b	4.45 ± 0.06a

Note: CK1, CK2 and R1 to R4 denote treatments of parent soil, tillage soil and root density level 1 to 4, the same follows. Different small letter in the same row means significant at P ≤ 0.05.

Table 2. Relative contribution of roots to soil erosion resistance (%).

Root effect (%)		Treatment R1		R2		R3		R4		Mean	
RTE		100.00		100.00		100.00		100.00		100.00	
RPE	NF	64.77	50.27	67.14	51.37	69.05	53.41	77.21	66.26	69.54	55.32
	BF		14.5		15.77		15.64		10.95		14.22
RBE		35.23		32.86		30.95		22.79		30.46	

Note: R1 to R4 denote treatments of root density level 1 to 4, the same follows.

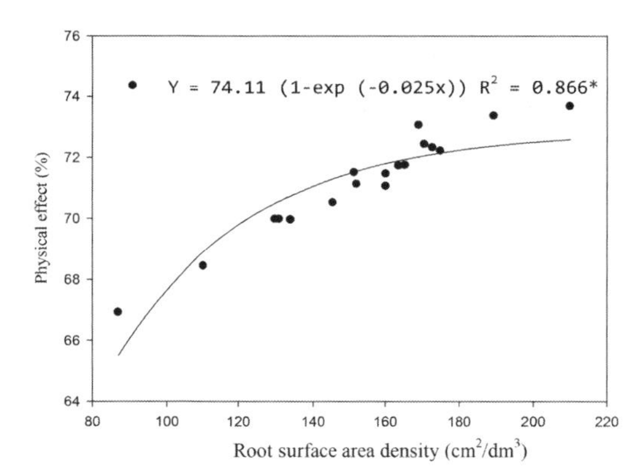

Figure 2. Regression between root physical effect and root surface area density.

4 SUMMARY

In conclusion, our study demonstrated that root physical enlacing is the main form, accounting for 64.7–77.2%, in total root effect strengthening soil erosion resistance. Exponential function of $y = 72.11 (1–\exp(–0.025x))$, $R^2 = 0.866*$ could well express the relationship between root physical enlacing effect and root surface area density to soil erosion resistance. Root surface area density could be regarded as the key indicator in forecasting the changes in soil erosion resistance in in the mine-concentrated distribution of flow-induced erosion regions.

ACKNOWLEDGMENTS

Financial assistance was provided by the projects of the National Natural Science Foundation of China (41661101, 41641013), Doctoral research foundation of Yulin University (16GK19).

REFERENCES

Burylo, M., Rey, F., Bochet, E., Dutoit, T. (2012) Plant functional traits and species ability for sediment retention during concentrated flow erosion. *Plant Soil*, 353, 135–144.

De Baets, S., Poesen, J., Reubens, J., Wemans, K., De Baerdemaeker, J., Muys, B. (2008). Root tensile strength and root distribution of typical Mediterranean plant species and their contribution to soil shear strength. *Plant Soil*, 305, 207–226.

Gyssels, G., Poesen, J. (2003) The importance of plant root characteristics in controlling concentrated flow erosion rates. *Earth Surface Processes and Landforms*, 28, 371–384.

Gyssels, G., Poesen, J., Bochet, E., Li, Y. (2005) Impact of plant roots on the resistance of soils to erosion by water: A review. *Progress in Physical Geography*, 29(2), 189–217.

Knapen, A., Poesen J., Govers, G. (2007) Resistance of soils to concentrated flow erosion: A review. *Earth Science Review*, 80(1), 75–109.

Li, Q., Liu, G.B., Zhang, Z., Tuo, D.F. (2017) Relative contribution of root physical enlacing and biochemistrical exudates to soil erosion resistance in the Loess soil. *Catena*, 153, 61–65.

Li, Q., Liu, G.B., Zhang, Z., Tuo, D.F., Xu, M.X. (2015) Effect of root architecture on the structural stability and erodibility of topsoils during concentrated flow in the hilly Loess Plateau. *Chinese Geography Science*, 25 (6), 757–764.

Liu, G.B. (2012) Study on soil anti-scouribility and its mechanism. *J. Soil Water Conser*. 4, 93–96.

Liu H, Yeerken W Z T, Wang C L. (2015) Impacts of the Belt and Road Initiative on the spatial pattern of territory development in China. *Progress in Geography*, 34(5): 545–553.

Luca, M., Mohamed, B., Victor, C. (2015) Status of the world's soil resources. 2015 International Year of soils.

Nelson, D.W., Sommers, L.E. (1982) Methods of soil analsis, Part 2, Chemical and microbial properties. *American Society of Agronomy*, 9, 539–552.

Wang, B., Zhang, G.H., Shi, Y.Y., Shan, Z.J. (2015) Effects of near soil surface characteristics on the soil detachment process in a chronological series of vegetation restoration. *Soil Science Society of America Journal*, 79, 1213–1222.

Zhang, G.H., Tang, K.M., Ren, Z.P. (2013) Impact of grass root mass density on soil detachment capacity by concentrated flow on steep slopes. *Transactions of the Asabe*, 56(3), 927–934.

Zhou, Z.C., Shangguan, Z.P. (2005) Soil anti-scouribility enhanced by plant roots. *Journal of Integrative Plant Biology*, 47, 676–682.

Land Reclamation in Ecological Fragile Areas – Hu (Ed.)
 ISBN 978-1-138-05103-4

Alders (*Alnus sp.*) as a potenial biological stabilizer on fly ash disposal sites

M. Pietrzykowski, B. Woś & M. Pająk
Department of Forest Ecology and Reclamation, Institute of Ecology and Silviculture, Faculty of Forestry, University of Agriculture in Krakow, Krakow, Poland

T. Wanic
Department of Soil Science, Institute of Ecology and Silviculture, Faculty of Forestry, University of Agriculture in Krakow, Krakow, Poland

W. Krzaklewski
Department of Forest Ecology and Reclamation, Institute of Ecology and Silviculture, Faculty of Forestry, University of Agriculture in Krakow, Krakow, Poland

M. Chodak
Department of Management Protection of Environment, AGH University of Science Technology, Krakow, Poland

ABSTRACT: Paper presents an assessment of growth of black and grey alder introduced on the fly ash disposal sites resulting from lignite combustion in central Poland. The research was conducted at 3 variants: control with pure fly ash (CFA), with addition of lignite culm (CFA+L) and acidic sands from overburden of lignite mine (CFA+MS). Before tree planting whole surface was covered by biosoilds (4 Mg ha^{-1}), mixed with grass seeds and NPK fertilization and applied by hydroseeding. Black alder displayed better growth parameters than the grey alder, however, the demonstrated differences may arise from the specific ecology of each species. Positive impact of lignite amendment was observed in higher density index (SDI) and tree height (H), whereas the addition of acidic sand had no impact on the growth of alders. According to our results an intrduction of black alder at CFA+L variant is an optimal solution for fly ash disposal site reclamation.

1 INTRODUCTION

Fly ash landfills adversely affect the environment in adjacent areas through wind erosion and dust pollution as well as alkaline leachate and saline water frequently containing heavy metals and leading to ground water contamination (Haynes 2009). Moreover, ash from landfills remains suspended in the air for a long time and thus becomes a source of pollution. This negatively affects the health of the local population, causing irritation of the upper respiratory tract and a number of adverse health effects, including even lung cancer (Dellantonio et al. 2009; Pandey et al. 2009). However, fly ash may also have a positive impact on the properties of the poor acidic soils (Frąc et al. 2016). Dust dry deposition raising forest soil fertility status through an increase in nutrient content, especially calcium and magnesium, as well as stimulation of microbial activity (Frąc et al. 2016).

An effective manner of limiting wind erosion and other adverse effects of fly ash landfills is biological containment by planting shrubs and trees on the slopes and tops of such landfills (Pandey & Singh 2012; Pandey 2015; Żołnierz et al. 2016). Introduction of vegetation on combustions waste, however, is difficult due to unfavourable habitat conditions resulting from a number of negative physicochemical properties of FA technosols including: high susceptibility to compaction, poor air and water ratios, excessively alkaline pH, high EC variability, lack of nitrogen and available phosphorus, and in some cases, high content of heavy metals (Adriano et al. 1980; Haynes 2009; Krzaklewski et al. 2012; Pietrzykowski et al. 2015). So far, in Central Europe the following tree species have been introduced on combustion waste landfills with various results: Scots pine (*Pinus sylvestris* L.), European larch (*Larix decidua* Mill.), Silver Birch (*Betula pendula* Roth), black locust (*Robinia pseudoaccacia* L.), red oak (*Quercus rubra* L.), common oak (*Quercus robur* L.), box elder maple (*Acer negundo* L.), sycamore maple (*Acer platanoides* L.), green ash (*Fraxinus pennsylvanica* Marsh.), black alder (*Alnus glutinosa* (L.) Gaertn.), poplar (*Populus sp.*) and willow (*Salix* sp.) (Wysocki 1979; Čermák 2008; Pietrzykowski et al. 2010; Żołnierz et al. 2016). Most of these experiments were carried out on ashes covered with a layer of soil mineral (Čermák 2008; Pietrzykowski et al. 2010). However, in addition to substantial cost, this practice also entails the risk of root system deformation due to the fact that it develops primarily in the surface horizons containing mineral soil (Čermák 2008). This is important for afforestation and the stability of introduced tree stands in later phases of development. As mentioned above, this technology is very expensive and stocks of more fertile soil are limited. Therefore, continued research on the introduction of trees directly on to the ash is still needed (Krzaklewski et al. 2012).

As demonstrated in previous studies, alders (*Alnus* sp.) (Wysocki 1979; Čermák 2008; Krzaklewski et al. 2012; Pietrzykowski et al. 2015) are one of the most promising species of trees for land reclamation and biological restoration of combustion waste landfills. In Central Europe, two woody species and a shrubby species of alder occur naturally; they are black and grey alder and green alder, respectively. Alders are species associated with damp and fertile natural habitats. In Central Europe the black alder occurs mainly in river lowlands which are periodically flooded and in depressed areas with water flow (Pancer–Kotejowa & Zarzycki 1980). The grey alder occurs both in the northern Boreal and in southern mountains (Boratyński 1980), growing mainly on mineral soils, often on skeletal alluvial mountain soils (Pancer–Kotejowa & Zarzycki 1980). In Central Europe, the green alder is a mountain species mainly occurring above the upper limit of forests (Ellenberg 1988).

Alders are a pioneering species and, therefore, they often occur for instance in secondary succession on former agricultural lands (Vacek et al. 2016). They are capable of coexisting with papillary bacteria and of binding atmospheric nitrogen which makes them a valuable phytoremediation species used in the restoration of degraded habitats for many years. Easily degradable organic precipitation with a narrow C: N ratio associated with high nitrogen content also makes alders important in the process of reclamation (Krzaklewski et al. 2012). Due to organic precipitation and intensive rooting, alders increase the amount of Soil Organic Matter (SOM) and the bioavailability of nutrients, acting as a natural "fertiliser" in former industrial sites (Šourková et al. 2005). The black and grey alder were used in the reclamation of former industrial sites in a wide range of conditions and soil substrates, including but not limited to alkaline Neogene clays, quaternary sands and gravel from the over burden of former lignite mines in Sokolov Lignite District (Frouz et al. 2001; Šourková et al. 2005). Black alder was introduced on Neogene alkaline clays of the overburden in opencast sulphur mines and on quaternary sand–clay deposits of sand mine opencasts in southern Poland (Józefowska et al. 2016) as well as on strongly alkaline and skeletal deposits in opencast oil shale mines (Lõhmus et al. 2007; Kuznetsova et al. 2011). The green alder was introduced on quaternary deposits of lignite mine overburden and on sand mine opencasts (Krzaklewski et al. 2003).

In Czech studies (Čermák 2008), the black alder was considered a species that adapts well to the conditions of post–mining areas with good growth parameters in combustion waste sedimentation tanks. However, it should be added that the experiment was carried out on ashes covered with 0.4–0.5 layer of soil mineral. Pioneering experiments on the introduction of the black and grey alder on combustion waste sedimentation tanks without using the insulating layer were carried out in the 1970s on a sedimentation tank for ashes following bituminous coal combustion in Halemba power plant (southern Poland) and in lignite power plant in Konin (central Poland). The cited study showed, however, low survival rate and poor growth of the introduced alder species (Wysocki 1979). However, recent studies (Krzaklewski et al. 2012) suggest that these two species, i.e. the black and grey alder, can be successfully used in the biological restoration of lignite combustion waste landfills. In other studies, it was pointed out that the green alder may also be a prospective species for biological reclaim and anti–erosion protection (Pietrzykowski et al. 2015). In view of the studies conducted so far and of preliminary results, the following issues remain important: i) survival, growth rate and size of the biomass of individual alder species in the context of their phytoremediation role and nutrient enrichment of ecosystems emerging on fly ash techno soils; ii) determining the optimal length for alders as a forecrop with a phytoremediation function and possible time for their replacement with target species.

The aim of the work was continuation of alder (*A. glutinosa* (L.) GAERTN. And *Alnus incana* (L.) MOENCH) growth parameter and adaptation assessment within 10 years of their introduction on a landfill containing fly ash generated by lignite combustion. In a permanent long–term experiment, substrate amendments were applied (lignite culm and Miocene acidic sands) available in the immediate vicinity of the site and a variant in which trees were introduced on to the fly ash with no insulating layer was also included. The results were compared with data obtained during the initial period, 5 years after the trees were planted (Krzaklewski et al. 2012).

2 MATERIALS AND METHODS

2.1 *Study site*

Bełchatów power plant and Lubień combustion waste disposal sites which belongs to it, are located in Central Poland (N 51 27; E 19 27), in a temperate climate zone with precipitation ranging from 550 to 600 mm annually and an average annual temperature of around 7.6–8.0°C. The vegetation period lasts from 210 to 218 days (Woś 1999). The Lubień landfill has been in operation since 1980 and currently takes up approximately 440 ha of land. Combustion waste containing about 85% ash and 15% slag is deposited there with the use of hydro–transport. The main component of combustion waste are thermally processed aluminosilicates. The average content of Al_2O_3 and SiO_2 compounds is from about 60 to 70%, and calcium oxide CaO about 20%. The content of trace elements generally does not exceed the average reported for natural soils. In the case of the landfill Lubień, adverse environmental impact is caused mainly by leaching sulphate, chloride and calcium, which in turn affects the growth of concentrations of these ions, increased mineralisation and increased overall hardness and alkalinisation of ground water (Krzaklewski et al. 2012).

2.2 *Description of the experiment*

The experiment started in September 2005 in a part of a sedimentation tank flat shelf set up between 2003 and 2004. Before the start of the experiment and the planting of trees on the entire surface, it was first subject to hydro–seeding with biosolids (sewage sludge of 4 Mg dry mass ha^{-1}) mixed with the seeds (200 kg ha^{-1}) of Cock's–foot grass (*Dactylis glomerata* L.) and Italian ryegrass (*Lolium multiflorum* Lam.). Next, NPK start up mineral fertilising was applied with N – 60, P – 36 and K – 36 kg ha^{-1}. Afterwards in this experiment, 24 plots for grey and black alder introduction, measuring 6 m × 13 m were laid out. In the described fly ash landfill and experimental sites, the green alder was also planted, but with different replication and combination of substrates, and not included in presented study (Pietrzykowski et al. 2015). The plots were separated with a 2–m–wide buffer strip. 50 seedlings of black alder or grey alder (6410 trees ha^{-1}) were planted on the plots in holes of 40 cm × 40 cm × 40 cm in 3 soil treatment variants (with 4 replications for each variant): lignite addition (CFA + L), Miocene acidic sand addition (CFA + MS) and control without any soil amendment (CFA) in planting holes (Krzaklewski et al. 2012).

2.3 *Soil sampling*

Soil samples were collected in 2015 from 0 to 40 cm horizon at five points—four of them were located at the corners of the experimental site and the one was taken from the interior of a sapling site 24 mixed samples of technosols (1.0 kg mass of fresh sample) representative of individual plots were selected to determine basic soil properties in the lab.

2.4 *Assesment of tree growth parameters*

In each experimental plot, the survival of trees as stand density (D—number of trees per ha) was assayed, Diameter at Breast Height (DBH) with an accuracy of 0.1 cm and Height (H) of all trees with an accuracy of 0.01 m were measured. Tree measurements were conducted in the autumn in 2015 after 10 years of growth, later based on these measurements and results from previous measurements in spring 2011 (Krzaklewski et al. 2012). Based on these results, the current annual growth was calculated for tree height (ΔH) and change of stand density (ΔD) from spring 2011 to autumn 2015. In addition, the upper height (H100) was determined, i.e. the average height of 100 tallest trees per hectare, and Reineke's Stand Density Index (SDI) from the equation (Zeide 2005):

$$SDI = D \times (DBH/25)^{1.6} \quad (1)$$

where: SDI = stand density index; D = density (number of trees per hectare); DBH = diameter at breast height.

2.5 *Lab tests*

Soil samples were dried and sieved through a 2.0 mm sieve. The basic soil parameters were determined in the soil samples: particle size using Fritsch GmbH Laser Particle Sizer ANALYSETTE 22; soil pH was determined in 1 M KCl at a 1:2.5 soil:solution ratio; Electrical Conductivity (EC) by conductometric methods at a 1:5 soil:solution ratio at 21°C; Carbon (C), Nitrogen (N) and Sulphur (S) content using the thermal conductivity method with the LECO TruMac® CNS analyser.

2.6 *Statistical analyses*

Two–way analysis of variance (ANOVA) was used to test the effect of the soil treatment and tree species on alder growth parameters (D, ΔD, SDI, DBH, H, ΔH, H100). Prior to ANOVA, the data sets were tested for normality using the Kolmogorov–Smirnov test and for variance homogeneity using Leven's test. The Tukey's Honestly Significant Differences (HSD) test for unequal sample sizes was run if any significant differences were found ($p < 0.05$). The statistical analyses applied STATISTICA version 12 software (StatSoft Inc. 2013).

3 RESULTS

3.1 *Basic soil properties*

Fly ash technosols typically displayed low silt (7–8%) and clay (1%) fraction content, pH was neutral or slightly alkaline (7.3–7.4) (Soil Survey Division Staff 1993). The determined EC was from 185 to 206 μS cm^{-1}. Nitrogen content was from 0.029 to 0.043%, carbon content from 3.14 to 4.73%, and sulphur content from 0.067 to 0.74%. There were no significant differences in the mean values of the analysed soil parameters between the studied soil treatment variants (CFA, CFA+L, CFA+MS) and technosols under various alder species (black alder and grey alder) (Table 1).

3.2 *Alders growth parameters*

Similarly to the initial period after 5 years of growth (Krzaklewski et al. 2012), after 10 years of growth, the black alder typically displayed better growth parameters compared to grey alder growth. D density ranged from 4423 to 4583 trees ha^{-1}, and changes in density ΔD in the period 2011–2015 ranged from −81 to −141 trees ha^{-1} yr^{-1} respectively for the black alder and the grey alder. Density index

Table 1. Effects of used treatment (CFA, CFA+L, CFA+MS) and tree species soil properties.

Factor	Level	Silt	Clay	$CaCO_3$	pH_{KCl}	EC[3]	N	C	S
		%				μS cm^{-1}	%		
		N.S.[2]	N.S.	N.S.	N.S.	N.S.	N.S.	N.S.	N.S.
Tree Species (Sp)	Black alder	8 ± 0.9[1]	1 ± 0.1	3.36 ± 0.41	7.4 ± 0.04	185 ± 15	0.032 ± 0.005	3.37 ± 0.48	0.069 ± 0.008
	Grey alder	7 ± 0.9	1 ± 0.1	3.24 ± 0.39	7.3 ± 0.03	196 ± 14	0.044 ± 0.005	3.91 ± 0.45	0.074 ± 0.008
		N.S.	N.S.	N.S.	N.S.	N.S.	N.S.	N.S.	N.S.
Soil treatment (Tr)	CFA[4]	7 ± 1.0	1 ± 0.2	2.83 ± 0.49	7.4 ± 0.04	186 ± 17	0.029 ± 0.006	3.14 ± 0.56	0.067 ± 0.010
	CFA + L	8 ± 1.0	1 ± 0.2	4.15 ± 0.49	7.3 ± 0.04	178 ± 17	0.043 ± 0.006	4.73 ± 0.56	0.072 ± 0.010
	CFA + MS	8 ± 1.1	1 ± 0.2	2.93 ± 0.41	7.4 ± 0.05	206 ± 18	0.041 ± 0.006	3.74 ± 0.60	0.076 ± 0.011
Interaction	Tr*Sp	N.S.	N.S.	N.S.	N.S.	N.S.	N.S.	N.S.	N.S.

Note: [1] – mean ± SE; [2] – *, $p < 0.05$; N.S. – not significant; [3] – EC – electrical conductivity; [4] – CFA – control variant (fly ash) with any soil amendment in planting holes; CFA+L – fly ash with lignite addition in planting holes, CFA+MS – Miocene acidic sand addition in planting holes.

Table 2. Effects of treatment (CFA, CFA+L, CFA+MS) and tree species on alder growth parameters on the fly ash landfill.

Factor	Level	D[4]	ΔD	SDI	DBH	H	ΔH[1]	H100
		trees ha^{-1}	trees ha^{-1} yr^{-1}		cm	m	m yr^{-1}	m
		N.S.	N.S.	*	*	*	*	N.S.
Tree Species (Sp)	Black alder	4423 ± 714[1]	−81 ± 102	507.6 ± 109.8[b] [3]	6.4 ± 0.2[b]	6.3 ± 0.2[b]	0.65 ± 0.11[b]	9.0 ± 1.0
	Grey alder	4583 ± 798	−141 ± 80	234.5 ± 50.9[a]	3.9 ± 0.2[a]	3.9 ± 0.2[a]	0.29 ± 0.11[a]	8.5 ± 1.3
		N.S.	N.S.	*	N.S.	*	N.S.	N.S.
Soil treatment (Tr)	CFA[5]	4407 ± 699	−147 ± 110	321.3 ± 157.6[a]	4.8 ± 0.2	4.9 ± 0.3[a]	0.43 ± 0.20	8.2 ± 1.0
	CFA+L	4888 ± 835	−112 ± 63	424.5 ± 179.01[b]	5.4 ± 0.2	5.6 ± 0.2[b]	0.51 ± 0.21	9.1 ± 1.4
	CFA+MS	4214 ± 603	−73 ± 101	367.4 ± 154.8[ab]	5.4 ± 0.2	5.0 ± 0.2[a]	0.46 ± 0.25	9.0 ± 1.2
Interaction	Tr × Sp	N.S.	N.S.	N.S.	N.S.	N.S.	N.S.	N.S.

Note: [1] – mean ± SE; [2] – *, $p < 0.05$; N.S. – not significant; [3] – Different letters indicate significant ($p < 0.05$) differences between the studied variants; [4] – D – stand density, ΔD – change of stand density between values in 2011 and 2015, SDI – Reineke's stand density index, DBH – diameter at breast height, H – trees height, ΔH – change of trees height between values in 2011 and 2015, H100 – top height; [5] – CFA – control variant (fly ash) with any soil amendment in planting holes; CFA+L – fly ash with lignite addition in planting holes, CFA+MS – Miocene acidic sand addition in planting holes.

SDI ranged from 234.5 to 507.6, Diameter at Breast Height (DBH) from 3.9 to 6.4 cm, height H from 3.9 to 6.3 m, increase ΔH in the period from 2011 to 2015 from 0.29 to 0.65 yr^{-1}, and upper height from 8.5 to 9.0 m, respectively for grey alder and black alder.

Significantly higher values of the average black alder growth parameters compared to the grey alder parameters were found in SDI, DBH, H and ΔH (Table 2).

There were no significant interactions between the tree species and the type of soil substrate (Tr × Sp) (Table 2). Soil treatment impact on tree parameters was displayed in the case of SDI and H. Significantly higher SDI index was found in CFA+L (424.5) compared to CFA (321.3) and H in CFA+ L (5.6 m) compared to CFA (4.9 m) and CFA+MS (5.0 m) (Table 2).

In each experimental plot, the survival of trees as stand density (D—number of trees per ha) was assayed, Diameter at Breast Height (DBH) with an accuracy of 0.1 cm and Height (H) of all trees with an accuracy of 0.01 m were measured. Tree measurements were conducted in the autumn in 2015 after 10 years of growth, later based on these measurements and results from previous measurements in spring 2011 (Krzaklewski et al. 2012). Based on these results, the current annual growth was calculated for tree height (ΔH) and change of stand density (ΔD) from spring 2011 to autumn 2015. In addition, the upper height (H100) was determined, i.e. the average height of 100 tallest trees per hectare, and Reineke's Stand Density Index (SDI) from the equation (Zeide, 2005):

4 DISCUSSION

The impact and growth parameters of tree species are associated with their adaptation to the harsh habitat conditions of anthropogenically transformed sites (Kuznetsova et al. 2011; Pietrzykowski & Socha 2011). Similarly to the results obtained after 5 years of growth following their introduction on FA (Krzaklewski et al. 2012), after 10 years the black alder typically displayed better growth parameters than the grey alder. Poorer growth parameters of the grey alder may be due to its ecological properties. In Central European conditions the grey alder is smaller in size compared to the black alder (Boratyński 1980). However, when compared to their size in natural habitats, it may be assumed that both species display good growth parameters 10 years after their introduction, which shows their good adaptability. In optimal forest habitats in southern Poland (Niepołomice Forest), the black alder aged from 5–9 years displayed D – 3900–5200 trees ha^{-1}, DBH – 2.7–4.9 cm, HL – 4.0–8.0 m (Orzeł et al. 2005). Grey alder tree stands in the Baltic states (Estonia) displayed the following parameters aged 10–12: D – 6266–15900 trees ha^{-1}, DBH – 3.7–4.6 cm, H – 5.9–8.3 m (Uri et al. 2014). In harsh habitat conditions occurring in fly ash landfills, maximising the growth of the introduced alder species is not the most important goal. The main task of alders is anti–erosion protection of landfills and the species phytoremediation role connected with their ability to produce large amounts of easily degradable organic matter with high nutrient content, their ability to bind atmospheric nitrogen using symbiotic bacteria of the *Frankia* genus, and the resulting quicker rate of initial soil formation and preparation of the habitat for long–living target species. In the harsh habitat conditions of FA landfills, the process of spontaneous regression of alder, similar to that on reclaimed mining sites, may occur in the future. Observations from afforested post–mining areas where alders were introduced as a phytoremediation addition indicate their regression within approximately 15–20 years, after they have fulfilled an important role in shaping the biotope (Pietrzykowski 2015). The process of alder regression is due to both habitat conditions which are far from ecological optimum (rainfall–retention water management), as well as from competition with target species of trees whose growth rate accelerates (Pietrzykowski 2015).

The obtained results indicate that an addition of lignite to planting holes after 10 years of growth had an impact on the value of the SDI index and height H. No significant differences were found between the mean values of other growth parameters depending on the variant used (CFA, CFA + L, CFA + MS). On this basis, it may be said that to initiate the growth of alder plantings it would be sufficient to apply NPK mineral fertilisation (60–36–36 kg ha^{-1}) at the onset and stabilisation by using of hydro–seeding with biosolids (sewage sludge 4 Mg ha^{-1}) and a mixture of appropriate grass seeds (200 kg ha^{-1} *Dactylis glomerata* and *Lolium multiflorum*). For best results, however, the use of lignite in the planting holes is recommended. This has an impact on the value of Height (H), and SDI index. Height in forestry is widely used as an indicator of habitat fertility (Socha 2008), while the SDI index informs us about the potential number of trees growing in a given area and with a certain thickness of 25.4 cm (Zeide 2005). In earlier studies (Krzaklewski et al. 2012) after 5 years of growth it was also pointed out that the best way to improve the properties of the ashes for alder growth and biomass production, was to insert lignite into the planting holes. Similarly to the situation after 5 years of growth (Krzaklewski et al. 2012), the addition of acidic sand had no effect on any of the analysed growth parameters. This confirms earlier observations that the addition of sand is not enough to improve the properties of fly ash for the growth of alders (Krzaklewski et al. 2012). An addition of acidic sand was used to lower the strongly alkaline pH of fly ash. However, the black and grey alder were successfully introduced in the course of reclaim treatment on post–mining deposits typically displaying strongly alkaline pH similar to the analysed fly ash: tertiary clays on lignite overburden (Frouz et al. 2001; Šourková et al. 2005), sulphur mine overburden (Józefowska et al. 2016) and strongly alkaline deposits accompanying oil shale opencast mines (Lõhmus et al. 2007; Kuznetsova et al. 2011). For these reasons, it may be considered that the strongly alkaline soil pH is not a limiting factor for the growth and adaptation of the analysed alder species. Pioneering studies on the introduction of the grey and black alder on fly ash landfills without the use of mineral topsoil from bituminous coal (Halemba, southern Poland) and lignite (Konin, central Poland) combustion carried out in the 1970s did not produce positive results due on the low survival rate and poor tree growth after 3 years. Site preparation for planting was different than in our experiments and included various fertilisation combinations, addition of fertile soil, tertiary acidic sands, bentonite and peat (treatment of ash pits – dug in disposal area). In order to protect from erosion, the site was sprayed with latex emulsions and a mix of herbaceous plants was sown. Lack of success might have resulted however also from the reported severe weather conditions (drought and low temperatures) and the impact of industrial pollution in those years, in particular pollution by sulphur compounds and the resulting acid rain (Wysocki 1979).

5 CONCLUSION

The study confirms considerable usefulness of the black and grey alder as a biostabilizers with good growth parameters. The alders play important role in soil and habitat preparation for next step of afforestation and the possible introduction of the target (climax) species at a later date. After 10 years of growth, the introduced species typically displayed good growth parameters (density, SDI index, average diameter at breast height and height, increase in height), comparable with alders of similar age growing on natural habitats. The black alder, however, displayed better growth parameters than the grey alder. The observed differences between the species growth may be associated with their ecological characteristics and different growth dynamic in young age. The impact of lignite addition on alder growth parameters was observed in the value of higher density index (SDI) and tree height (H), whereas the addition of acidic sand had no significant impact on the growth of alders. Prior to afforestation, pre–treatment of the landfill with initial NPK mineral fertilisation and stabilisation using hydro–seeding with biosolids (sewage sludge 4 Mg ha^{-1}) are recommended along with the insertion of lignite as a substrate into planting holes to improve the properties of fly ash.

ACKNOWLEDGEMENTS

The study were financed by The National Science Centre, Poland, grant No. 2015/17/B/ST10/02712. We thanks to Iwona Skowrońska MSc. from Laboratory of Geochemistry and Reclamation, Dept. of Forest Ecology and Reclamation AUC for laboratory tests and her kind collaboration. Special thanks to PGE GiEK SA Elektrownia Bełchatów power plant for field working support and access to experimental area of Lubień fly ash disposal sites.

REFERENCES

Adriano, D.C., Page, A.L., Elseewi, A.A., Chang, A.C. & Straughan, I. (1980). Utilization and disposal of fly ash and other coal residues in terrestrial ecosystems: a review. *J. Environ. Qual.* 9 (3), 333–344.

Boratyński, A. (1980). Systematics and geographic distribution of alders. In: Białobok, S. (eds.). *Alders Alnus Mill. Our forest trees. Popular science monographs.* Polish Academy of Sciences, Institute of Dendrology, Warszawa – Poznań, pp. 35–71 (in Polish, English summary).

Dellantonio, A., Fitz, W.J., Repmann, F. & Wenzel, W.W. (2009). Disposal of coal combustion residues in terrestrial systems: contamination and risk management. *J. Environ. Qual.* 39 (3), 761–775.

Ellenberg, H. (1988). *Vegetation Ecology of Central Europe.* Fourth Edition. Cambridge University Press, Cambridge.

Frouz, J., Keplin, B., Pižl, V., Tajovský, K., Starý, J., Lukešová, A., Nováková, A., Balík, V., Háněl, L., Materna, J., Düker, C., Chalupský, J., Rusek, J. & Heinkele, T. (2001). Soil biota and upper soil layer development in two contrasting post–mining chronosequences. *Ecol. Eng.* 17 (2–3), 275–284.

Frąc, M., Weber, J., Gryta, A., Dębicka, M., Kocowicz, A., Jamroz, E., Oszust, K. & Żołnierz, L. (2016). Microbial functional diversity in Podzolectohumus horizons affected by alkaline fly ash in the vicinity of electric power plant. *Geomicrobiol. J.* DOI: 10.1080/01490451.2016.1220651

Haynes, R.J. (2009). Reclamation and revegetation of fly ash disposal sites – challenges and research needs. *J. Environ. Manage.* 90, 43–53.

Józefowska, A., Woś, B. & Pietrzykowski, M. (2016). Tree species and soil substrate effects on soil biota during early soil forming stages at afforested mine sites. *Appl. Soil Ecol.* 102, 70–79.

Krzaklewski, W., Pająk, M., Pietrzykowski, M. & Strutyński, M. (2003). Possible applications of green alder (*Alnus viridis* (Charix) DC. And In Lam. & DC.) in the reclamation of post–mining sites. *Advances of Agricultural Sciences Problem Issues* 493 (3), 905–912 (In Polish, English summary).

Krzaklewski, W., Pietrzykowski, M. & Woś, B. (2012). Survival and growth of alders (*Alnus glutinosa* (L.) Gaertn.and *Alnus incana* (L.) Moench) on fly ash technosols at different substrate improvement. *Ecol. Eng.* 49, 35–40.

Kuznetsova, T., Lukjanova, A., Mandre, M. & Lõhmus, K. (2011). Aboveground biomass and nutrient accumulation dynamics in young black alder, silver birch and Scots pine plantations on reclaimed oil shale mining areas in Estonia. *Forest Ecol. Manag.* 262 (2), 56–64.

Lõhmus, K., Kull, A., Truu, J., Truu, M., Kaar, E., Ostonen, I., Meel, S., Kuznetsova, T., Rosenvald, K., Uri, V., Kurvits, V. & Mander, Ü. (2007). The reclamation of the North Estonian oil shale mining area. In: Mander, Ü., Wiggering, H. & Helming K. (eds.). *Multifunctional Land Use. Meeting Future Demands for Landscape Goods and Services.* Springer, pp. 387–401.

Orzeł, S., Forgiel, M., Socha, J. & Ochał, W. (2005). Biomass and annual production of common alder stand of the Niepołomice Forest. *EJPAU* 8 (1): #25. Available Online: http://www.ejpau.media.pl/volume8/issue1/art-25.html

Pancer–Kotejowa, E. & Zarzycki, K. (1980). Outline of ecology. In: Białobok, S. (eds.). *Alders Alnus Mill. Our forest trees. Popular science monographs.* Polish Academy of Sciences, Institute of Dendrology, Warszawa – Poznań, pp. 229–257 (in Polish, English summary).

Pandey, V.C. (2015). Assisted phytoremediation of fly ash dumps through naturally colonized plants. *Ecol. Eng.* 82, 1–5.

Pandey, V.C., Abhilash, P.C. & Singh, N. (2009). The Indian perspective of utilizing fly ash in phytoremediation, phytomanagement and biomass production. *J. Environ. Manage.* 90, 2943–2958.

Pandey V.C. & Singh, B. (2012). Rehabilitation of coal fly ash basins: Current need to use ecological engineering. *Ecol. Eng.* 49, 190–192.

Pietrzykowski, M. & Socha, J. (2011). An estimation of Scots pine (*Pinus sylvestris* L.) ecosystem productivity on reclaimed post–mining sites in Poland (central Europe) using of allometric equations. *Ecol. Eng.* 37 (2), 381–386.

Pietrzykowski, M. (2015). Tree response to soil reconstruction on reclaimed post–mining sites—a key issue in forest ecosystem restoration. Chapter 1. In: Orzeł, S. (eds.). *Problems of forestry in the mountains and industrial region.* Agricultural University of Krakow Publishing, Krakow, Poland, pp. 15–39 (in Polish, English summary).

Pietrzykowski, M., Krzaklewski, W. & Gaik, G. (2010). Assessment of forest growth with plantings dominated by Scots pine (*Pinus sylvestris* L.) on experimental plots on a fly ash disposal site at the Bełchatów power plant. University of Zielona Góra Scientifical Reports. *Environ. Eng.* 137 (17), 65–74 (in Polish, English summary).

Pietrzykowski, M., Krzaklewski, W. & Woś, B. (2015). Preliminary assessment of growth and survival of green alder (*Alnus viridis*), a potential biological stabilizer on fly ash disposal sites. *Journal of Forestry Research*, 26 (1), 131–136.

Socha, J. (2008). Effect of topography and geology on the site index of *Picea abies in the West Carpathian*, Poland. *Scand. J. Forest Res.* 23, 203–213.

Soil Survey Division Staff (1993). *Soil survey manual. Chapter 3. Examination and Description of Soils.* Soil Conservation Service. U.S. Department of Agriculture Handbook 18.

StatSoft Inc. (2013). STATISTICA (data analysis software system), version 12.

Uri, V., Aosaar, J., Varik, M., Becker, H., Ligi, K., Padari, A., Kanal, A. & Lõhmus, K. (2014). The dynamics of biomass production, carbon and nitrogen accumulation in grey alder (*Alnus incana* (L.) Moench) chronosequence stands in Estonia. *Forest Ecol. Manag.* 327 (1), 106–117.

Vacek, Z., Vacek, S., Podrázský, V., Král, J., Bulušek, D., Putalová, T., Baláš, M., Kalousková, I. & Schwarz, O. (2016). Structural diversity and production of alder stands on former agricultural land at high altitudes. *Dendrobiology* 75, 31–44.

Woś, A. (1999). *Polish Climate.* PWN, Warsaw, pp. 1–301 (in Polish).

Wysocki, W. (1979). *Reclamation of alkaline ash piles and protection of their environment against dusting.* Industrial Environmental Laboratory Office of Research and Development U.S. Environmental Protection Agency Cincinnati, Ohio.

Zeide, B. (2005). How to measure stand density. *Trees* 19, 1–14.

Čermák, P. (2008). Forest reclamation of dumpsites of coal combustion by–products (CCB). *J. For. Sci.* 54 (6), 273–280.

Šourková, M., Frouz, J. & Šantrùčková, H. (2005). Accumulation of carbon, nitrogen and phosphorus during soil formation on alder spoil heaps after brown–coal mining, near Sokolov (Czech Republic). *Geoderma* 124 (1–2), 203–214.

Żołnierz, L., Weber, J., Gilewska, M., Strączyńska, S. & Pruchniewicz, D. (2016). The spontaneous development of understory vegetation on reclaimed and afforested post–mine excavation filled with fly ash. *Catena* 136, 84–90.

Land Reclamation in Ecological Fragile Areas – Hu (Ed.)
© 2017 Taylor & Francis Group, London, ISBN 978-1-138-05103-4

Management of coal processing wastes: Studies on an alternate technology for control of sulfate and chloride discharge

Paul T. Behum
Office of Surface Mining and Reclamation Enforcement, Alton, Illinois, USA

Yoginder P. Chugh
Southern Illinois University Carbondale, Carbondale, Illinois, USA

Liliana Lefticariu
Department of Geology, Southern Illinois University Carbondale, Illinois, USA

ABSTRACT: Management of coal mining and coal processing wastes, particularly of high sulfur coals, can generate excessive amounts of Sulfate (SO_4^{2-}) and Chloride (Cl^-) in mine drainage that are known to negatively impact quality of both surface and ground water. The U.S. Environmental Protection Agency (USEPA) provides guidance on allowable SO_4^{2-} and chloride Cl^- discharges from mine sites. This research evaluates the hypothesis that co-disposal of CCPW and FCPW with appropriate compaction can result in improved geochemical and geotechnical environments that will minimize Acid Mine Drainage (AMD) formation and SO_4^{2-} and Cl^- discharges. Addition of ground limestone (ag-lime) to the mix was also evaluated as a drying agent and for improvement in overall geochemistry by buffering higher pH values within the coal waste. These objectives were to develop and implement innovative concepts for engineered co-management of CCPW and FCPW at coal mining sites. The authors performed long-term field column leaching studies to analyze improvement in SO_4^{2-} and Cl^- in water quality. Requirements for stricter standards in some states led to the need for development of potentially improved environmental practices. This paper presents the overall encouraging results of the field kinetic studies.

1 INTRODUCTION

A typical underground coal mining complex in the mid-continental USA consists of a combination of mine shafts and slopes to access the coal seam/s along with coal preparation and refuse disposal facilities located on the surface as shown in Figure 1. The term "coal refuse" in this paper refers to reject rock from crushers prior to Run-of-Mine (ROM) coal entering the processing plant and other waste rock derived from the coal cleaning processes. The reject rock from the crushers is typically a small portion of the coal refuse. The ROM coal is generally wet processed to remove mineral matter and sulfur and to improve the quality of coal shipped to power plants. The marketable coal recovery typically varies between 50 and 60% with the remaining fraction disposed as waste on-site. Coal refuse typically consists of two size fractions: 1) Coarse Coal Processing Waste (CCPW), which is generally larger than 150 microns (100 mesh) in size and in some cases, is larger than 3.2 mm (1/8-inch) in size; and 2) Fine Coal Processing Waste (FCPW), which is generally a slurry (+/– 15% solids content) with solids being less than 150 microns (100 mesh) in size. CCPW refers to reject from heavy media vessels, jigs, cyclones, and spirals, while FCPW refers to reject from flotation columns and cells and effluent from filter presses, screen-bowl centrifuges, and de-sliming cyclones, all of which is typically concentrated in a thickener. The FCPW constitutes about 5–10% of the ROM coal processed from the Interior Basin coal mines in the USA.

FCPW is typically disposed as: 1) a slurry contained within an embankment constructed of CCPW (conventional practice) or incised ponds, 2) dewatered FCPW (50–70% solids), contained within CCPW embankments as illustrated in Figure 1, or potentially 3) a mono-fill constructed of a blend of CCPW and dewatered FCPW. The sizes of coal refuse disposal areas have been increasing in the Interior Basin due to the recent trends of mining thinner coal seams with larger equipment capable of cutting roof and floor rock. This has increased the percentage of the Out-of-Seam dilution (OSD) material to 50% or more of the ROM coal. Figure 1 also shows that underground coal mine complexes within the region typically operate

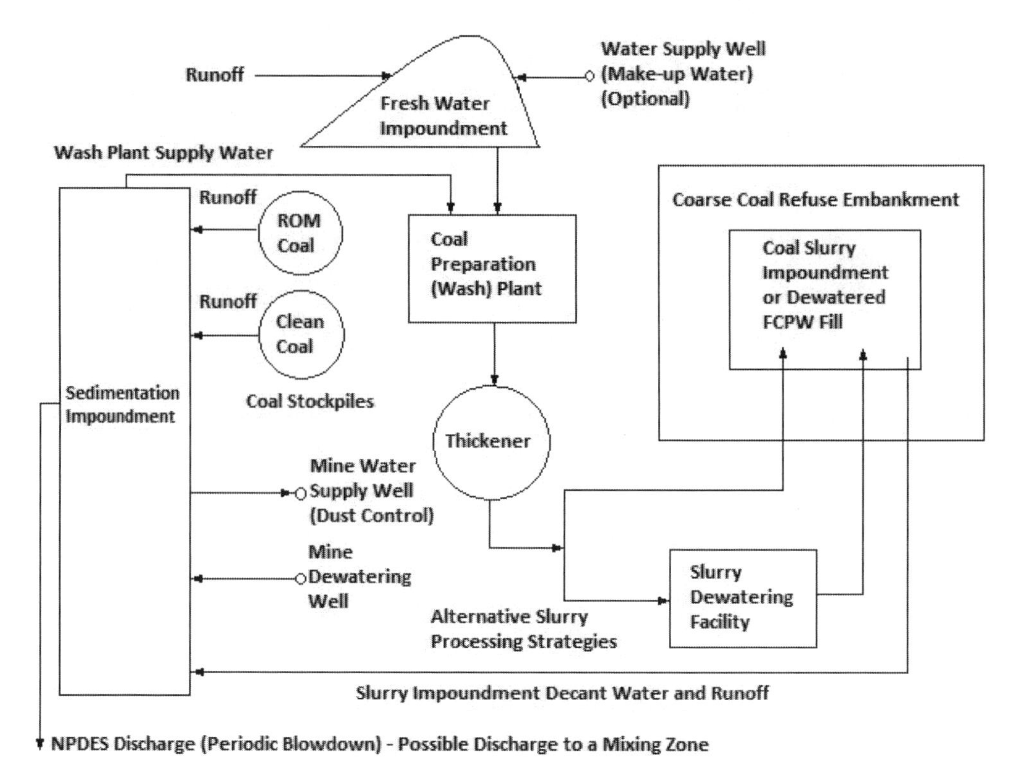

Figure 1. Schematic of coal processing and water handling operations on the surface at a typical underground coal mine complex.

a closed loop water-handling system where onsite water resources are continuously recycled and any regulated surface water discharges are from sedimentation basins, but typically only during larger precipitation events when either less stringent regulations is applied or there is a considerable amount of dilution water available.

2 PROBLEM STATEMENT

Coal waste disposal facilities have been identified as a principal source of elevated SO_4^{2-} discharges by both state and federal environmental regulatory agencies (The Advent Group, 2005; Chugh *et al.*, 2007a; Chugh *et al.*, 2007b). Infiltration of oxygen- and ferric iron-bearing water into coal waste stockpiles contributes to the oxidation of pyrite and other sulfide minerals with the subsequent release of SO_4^{2-} and metals into mine drainage (Evangelou, 1995; Moses *et al.*, 1987; Moses and Herman, 1991). Development of Good Management Practices (GMPs) for refuse disposal and water management to reduce pyrite oxidation and SO_4^{2-} and chloride levels in these discharges are necessary to meet compliance levels for regulated constituents in mine drainage. Such practices should: 1) reduce water treatment costs during mining and subsequent reclamation of the facility, 2) reduce liability associated with maintaining and abandoning a large permanent impoundment structure, and 3) allow release of reclamation bonds in a timely fashion.

Weathering of pyrite can be controlled if water rich in oxygen and ferric ions is prevented from contacting sulfide minerals inherent in the waste rock (Evangelou, 1995; Nordstrom, 1982). Dissolved oxygen concentration, temperature, pH of water, reactive surface area of pyrites, form of pyrite, catalytic agents, flushing frequencies, and time allowed for contact are some of the variables that control the rate of pyrite oxidation and the associated SO_4^{2-} discharge. The inherent pyrite content within the coal and the associated strata is an important factor in determining the amount of sulfate and metals in mine drainage. Although abiotic chemical conversion of Fe (II) to Fe (III) is an extremely slow reaction, the presence of iron-oxidizing bacteria increases reaction rate by an order of magnitude and, as a result, the production of SO_4^{2-} during the weathering process (Singer and Stumm, 1970; Kleinmann and Crerar, 1979). Within the Interior Basin, several researchers have provided noteworthy contributions to the control of Acid Mine Drainage (AMD), including Nawrot and Gray, 2000; Naumann and Wiram, 1995; Wiram, 1984, which has emphasized the advantages of adding alkalinity producing materials during coal mine reclamation.

3 RESEARCH HYPOTHESIS AND OVERALL RESEARCH

This research has evaluated the hypothesis that co-disposal of CCPW and FCPW with or without addition of alkaline materials, and with appropriate compaction can improve geochemical and geotechnical parameters of coal waste that will minimize AMD formation and SO_4^{2-} discharges. This hypothesis was developed based on data available in the literature and also collected as part of a previous study (Chugh *et al.*, 2007a) that concluded that the different-size CCPW fractions have a higher acid-forming potential (Potential Acidity or PA) and lower Neutralization Potential (NP) as compared to FCPW. In current practice, CCPW is typically disposed separately as a structural fill in surface disposal areas, while FCPW is disposed as a dilute slurry (about 15% solids) in impoundments that provide an overall lower disposal cost (NRC, 2002). Previous research also indicates that, in coal, the minerals which provide NP (e.g., calcite) weather faster than minerals that are acid-producing such as pyrite (e.g. Hossner and Doolittle, 2003; Nawrot and Gray, 2000). Therefore, drainage from coal waste, which is initially alkaline, remains alkaline for a certain period of time and then tends to become acidic. This period can be extended by addition of alkaline materials which is a widely accepted mine reclamation practice (Dreher et al., 1994; Nawrot and Gray, 2000; Stewart et al., 2001). Chugh et al. (2007a, 2007b) and others have suggested that differential AP and NP rates could be used advantageously in waste management operations to allow for improved placement and compaction practices, and reduction of SO_4 discharges. There is a potential improvement in the overall NP and a reduction in the PA in a blend of CCPW and FCPW because the FCPW has a net neutralization potential (NNP; NNP = NP-PA) that in some cases is near neutral or, at least, a level slightly above (-) 5 tons of $CaCO_3$ equivalent per 1,000 tons of material. Most permitting agencies consider a NNP less than (-) 5 tons of $CaCO_3$ equivalent per 1,000 tons of material (in this case coal waste) potentially acid-forming and requiring management practices to control problematic mine drainage. Co-disposal of CCPW and FCPW offers an interesting alternative concept and is discussed in the next section.

Most Illinois mine operators would consider co-disposal if FCPW could be dewatered inexpensively to about 60–65% solids and the long-term stability of a co-disposal fills can be ensured. As it is a relatively few number of U. S. mining operations currently practice refuse co-disposal. However, there remains a need to minimize disposal area space requirements and to eliminate the need for slurry disposal areas, which are becoming increasingly difficult and more expensive to permit. None of the operators interviewed in the preceding studies (Chugh *et al.*, 2007a and 2007b), have considered the potential improvements in geochemical and geotechnical environments as the result of co-disposal. A judicious addition of limestone or other alkaline materials to the co-disposal blend can provide additional advantages by buffering drainage pH, reducing high levels of total dissolved solids, and minimizing SO_4^{2-} discharge (Nawrot and Gray, 2000). Both hypotheses presented above were evaluated in a field setting by constructing six large-size leaching columns – two each of 100% CCPW (the control sample); 90% (by volume) CCPW and 10% FCPW; and 84% CCPW, 8% FCPW, and 8% limestone. Material in each column was compacted to approximately 50% of the maximum Proctor density. All columns were equipped with a lysimetric and gravity drainage ports for collecting water samples.

This research was motivated by the Illinois Environmental Protection Agency (IEPA) policy to implement lower standards for SO_4^{2-} and Cl^- discharge from coal mining and processing operations. Historically, Illinois allowed compliance level was 3,500 mg/L for SO_4^{2-} discharges and 1,000 mg/L for Cl^- discharges, whereas a "General Use" standard applicable to other industries had been set at 500 mg/L for both of these anions (Chugh *et al.*, 2007a and 2007b). The IEPA and the USEPA agreed that this difference was not legally defensible and implemented more restrictive SO_4^{2-} levels (USEPA, 2009). These IEPA regulations also established a SO_4^{2-} discharge levels in a range of 500 to 2,000 mg/L based on the hardness and Cl^- levels in the receiving stream; Cl^- discharge levels were also set at 500 mg/L.

3.1 *Overall research*

The overall goal of this study was to investigate the rate of SO_4^{2-} and Cl^- leaching from simulated kinetic tests in the field to demonstrate commercial potential for co-disposal technology. Co-disposal should enhance geotechnical stability through improved cohesion and compaction, while improving the geochemical environment. As a result, the overall refuse facility liabilities should be lowered. Addition of ground limestone (ag-lime) was also evaluated for use as a drying agent and as measure to improve in overall geochemistry by buffering higher pH values (<8.0) inherent in fresh refuse from the basin. The study would aid mine operators in developing and implementing innovative concepts for engineered waste disposal. Although field testing was conducted at an Illinois mining operation, the concepts involved should be applicable throughout the Interior Basin and, with minor modifications, at other coal mines in the USA.

4 RESEARCH STUDIES

These included: 1) physical and chemical characterization of CCPW, FCPW, and limestone materials; 2) characterization of geotechnical properties for identified waste disposal options; and 3) field-scale kinetic tests and data analysis.

4.1 *Physical properties of coal waste*

Both FCPW and CCPW were collected from a large coal preparation plant cleaning ROM coal from two commonly mined seams: Springfield (No. 5) and Herrin (No. 6). Most of this coal was produced from two underground mines. An approximately 18.9-liter (L) (5-gallon) sample of CCPW was collected and dried at low temperature (< 38°C or < 100°F), and then screened for particle size distribution. The sampled materials were crushed to less than 10 mm (3/8-inch) in size. Proctor compaction tests were conducted in 15 cm (6-inch) diameter molds to develop moisture-density data for optimizing compaction. The mold samples were also tested for uniaxial compressive strength for an undrained condition (Chugh *et al.*, 2007a; Chugh *et al.*, 2007b).

A similar sized sample of FCPW was also collected from the thickener underflow discharge. FCPW was decanted and air-dried before drying in a low temperature furnace. A 3.79-L (1-gallon) sample of agricultural ground limestone was also dried at low temperature and analyzed for particle size distribution.

4.1.1 *Results*

Table 1 shows average particle size distributions for CCPW and limestone samples. Since FCPW is uniformly less than 200 mesh (0.074 mm or 0.0029 inches), additional particle size analyses were not conducted on this fraction. A majority of CCPW (64.52%) is in the 4.75 to 50.8 mm (0.187 to 2.0 inch) size range. Agricultural ground limestone was a finer-grained material with a majority (72.97%) in the 1.70 to 19.05 mm (0.066 to 0.75 inch) size range. The initial moisture content of CCPW was 11.5%, whereas dewatered FCPW was 31.9% (68.1% solids content). In contrast, the initial moisture content of agricultural limestone was only 0.1%. Therefore, by blending limestone the moisture content could be significantly reduced.

Table 1. Particle size distribution for CCPW and ground limestone.

	Percent finer	
Sieve opening (mm)	CCPW	Ground limestone
50.8	94.31	100
19.05	62.20	100
4.75	29.79	78.09
1.7	15.09	27.03
0.425	4.44	13.17
0.075	0.53	4.58
< 0.075	0.52	3.58

4.2 *Chemical properties of coal waste*

The geochemical evaluation of CCPW, FCPW, and agricultural limestone included: 1) acid-base accounting, 2) determination of major and key trace elements in High Temperature Ash (HTA) product, and 3) X-Ray Fluorescence (XRF) analysis of all materials. Following SIUC laboratory determination of paste pH, the U.S. Geological Survey Energy Resources Program (USGS-ERP), the State of Illinois Office of Mines and Minerals (ILDNR-OMM), and a commercial testing laboratory determined the total and pyritic sulfur content along with an elemental analysis of all materials. Sulfur values were used to calculate Potential Acidity (PA) for CCPW and FCPW with respective values of 120.37 and 63.91 tons $CaCO_3$ per ton of coal waste.

Neutralization Potential (NP) was also determined for all materials to determine the Net Neutralization Potential (NNP). One CCPW and one FCPW coal waste sample was subjected to HTA analysis following heating in a 750°C air-recirculating chamber furnace using ICP/MS and ICP-OES. Limestone elemental analysis was conducted using a Varian 725 ICP-OES for major elements (except SO_3, which was determined using a LECO sulfur analyzer) and presented in oxide form. A hand-held XRF analyzer was used to determine a large suite of elemental concentrations in whole dried and ground CCPW, FCPW and limestone samples.

4.2.1 *Results*

Acid-base accounting analyses results are presented in Table 2. The "mine permit" application provided representative data, while the remaining data was compiled from analyses performed in this study. Chugh *et al.* (2007a) presented additional acid-base accounting data for CCPW and FCPW samples collected from another similar preparation plant and found that: 1) the FCPW represented 10–15% of the coal waste, and 2) the sulfur content is typically lower for FCPW as compared to CCPW. Elemental data and chemical analyses of HTA are presented in Tables 3 and 4. The second row of limestone data in Table 3 was provided by the quarry operator.

Table 2. Geochemical properties of coal waste and ground limestone samples.*

Coal waste fraction	Mean sulfur content (%) Total	Pyritic	Median paste pH	Tons $CaCO_3$ per 1000 tons of waste Maximum Potential Acidity (PA)	Neutralization Potential (NP)	Net Neutralization potential (NNP)
Mine Permit (Coarse)	5.70 (n = 2)	3.41 (n = 47)	7.12 (n = 47)	106.4 (n = 47)	23.8 (n = 47)	–84.5 (n = 47)
Coarse	4.84 (n = 2)	3.90 (n = 1)	5.92 (n = 1)	121.88 (n = 1)	1.51 (n = 1)	–120.37 (n = 1)
Fine	2.56 (n = 2)	2.13 (n = 1)	6.31 (n = 1)	66.53 (n = 1)	2.65 (n = 1)	–63.91 (n = 1)
Limestone	NT	0.17 (n = 1)	8.35 (n = 1)	5.31 (n = 1)	58.17 (n = 1)	52.86 (n = 1)

*n = no. of samples; NT = not tested.

Table 3. Comparison of major elements in coal waste HTA and ground limestone.

Sample ID	Al_2O_3 (%)	CaO (%)	Fe_2O_3 (%)	K_2O (%)	MgO (%)	Na_2O (%)	P_2O_5 (%)	SiO_2 (%)	SO_3 (%)	TiO_2 (%)
Coarse	17.9	3.24	10.7	3.59	1.40	0.817	0.249	59.3	2.33	0.699
Fine	18.2	4.30	8.12	3.50	1.35	0.732	0.135	58.0	2.39	0.721
Limestone	0.22	49.67	0.31	0.06	1.63	0.050	0.001	10.1	0.38	0.040
Limestone	0.50	54.52	0.30	0.05	0.64	0.030	0.043	1.07	0.11	0.0005

Table 4. Comparison of trace elements in coal waste HTA and ground limestone.

Sample ID	Ba (ppm)	Cr (ppm)	Cu (ppm)	Li (ppm)	Mn (ppm)	Mo (ppm)	Ni (ppm)	Pb (ppm)	Sr (ppm)	V (ppm)	Zn (ppm)
Coarse	480	132	59.5	46.1	264	43.6	80.0	114	113	279	293
Fine	561	125	54.2	61.1	282	21.6	54.3	80.1	109	205	213
Limestone	294	25.8	10.9	NT	171	2.5	11.5	ND	532	ND	34.5

Note: ND = not detected; NT = not tested.

4.3 *Geotechnical characteristics of different coal waste disposal options*

Three waste disposal options were considered: 1) CCPW disposal in embankments or fills (the control) - Disposal Practice 1 or DP 1; 2) Co-disposal of blended CCPW (90%) and FCPW (10%) – Disposal Practice 2 or DP 2; and 3) Co-disposal of blended CCPW (84%) and FCPW (8%) with limestone addition (8%) - Disposal Practice 3 or DP 3. Moisture-density relationships were developed using standard proctor tests for all disposal options. All tests were performed in a 15-cm (6-inch) mold using ASTM D698-12e1 Method C; ASTM, 2013) prior to column construction (Behum *et al.*, 2010). The mold size limited the top size of the material to about 20 mm (¾-inch).

4.3.1 *Results*

Particle size distribution for the materials used in Proctor tests are summarized in Table 5. These area based on three tests for DP 1 materials, two for DP 2 materials and two for DP 3 materials. The data is based on percent retained rather than percent passing.

Table 5. Particle size distribution of materials used in Proctor tests.

Sieve opening (mm)	Percent retained DP 1	DP 2	DP 3
4.76	63	58	54
1.68	20	18	20
0.42	11	10	11
0.75	5	4	4
< 0.075	1	10	11

The Proctor test results for all disposal options and tests are summarized in Figure 2. For DP 1, the testing moisture contents varied between 2% and 17%. The field columns had maximum dry unit weight = 18.5 kN/m^3 or 118 pcf. Similar data for DP 2 and DP 3 Proctor tests were 2% to 9.5%, and 18.0 kN/m^3 or 115 pcf; 2.5% to 11.5% and 18.52 kN/m^3 or 118 pcf. Moisture contents for

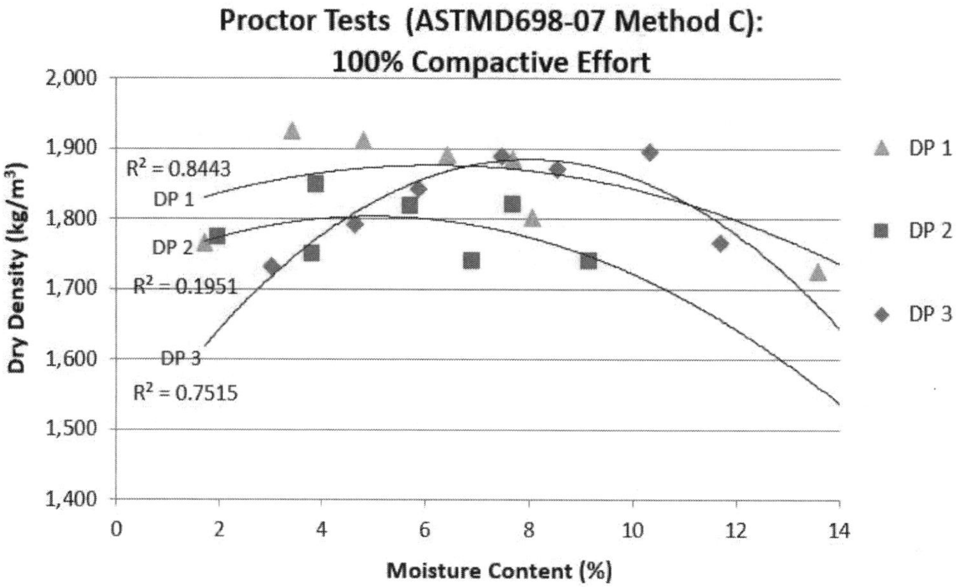

Figure 2. Composite proctor test results for DP 1, DP 2 and DP 3 options.

Figure 3. Field kinetic test columns (left); photograph of field test site (right).

maximum Proctor density values for DP 1, DP 2, and DP 3 mixes were 6.0%, 5.5%, and 7.5%, respectively. The addition of ground limestone advantageously increased the moisture content at which maximum density was achieved by about 2%.

4.4 *Field kinetic testing*

Field-scale kinetic tests were conducted at an active coal mining complex. During the first phase of field-scale testing, three sets of duplicate test columns (FC-1 through FC-6) were constructed using 200-L (55-gallon) barrels. These columns were monitored for 19 months. Unfortunately, a tornado damaged some columns. A new set of field-scale kinetic-leaching columns were assembled from relatively inexpensive and readily available components; this second field experiment is the subject of this study. The revised field test columns were enlarged by using about 380–L (100-gallon), polycarbonate oval-tanks intended for livestock activities. These stock tanks measured 113.4 cm (52.5 inches) long by 94 cm (37 inches) wide by 53.3 cm (21 inches) tall (Figure 3). Two test columns received 100% CCPW; two received a blend of 93.3% CCPW and 6.7% dewatered FCPW by volume; and two received a blend of 86.7% CCPW, 6.7% dewatered FCPW, and 6.7% ground limestone by volume. Assuming a normal porosity of 16%, approximately 300 kg (650 lb.) of CCPW was packed into each column. Sand cone tests were performed on four columns to determine dry unit weight and moisture content achieved during construction. Proctor dry unit weights of 21.55 kN/m^3 (137.3 pcf) and 14.27 kN/m^3 (90.92 pcf) were achieved for DP 1 and DP 3 mix materials, respectively. The kinetic test columns were leached in free-draining mode for 19.3 months (16 leach cycles; Figure 3).

4.4.1 *Leachate sampling and analysis*
Pore-water and leachate samples were collected at about 38 days intervals using a 500 ml Nalgene® bottle. Field measurements of pH, temperature, and Dissolved Oxygen (DO) were conducted on each pore water and leachate sample with a professional-grade multi-parameter meter. In some cases, there was insufficient pore water sample volume to measure DO. Temperature, pH, and conductivity (Specific Conductance – SC) of pore water and leachate samples were measured immediately upon return to the laboratory. Alkalinity and ferrous iron were determined within 24 hours after sample collection. All water sample quality analyses were statistically analyzed using spreadsheet-based models.

4.4.2 *Anion analyses*
Concentrations of SO_4^{2-}, Cl^-, F^-, and nitrate (NO_3^{2-}) anions were determined using an Ion Chromatograph (IC). For comparison, SO_4 concentration was also determined using colorimetric analysis (Hach Method 8051; Hach Company, 2004).

4.4.3 *Major and trace metal analyses*
Major metals and a suite of trace metals were determined by either an Atomic Absorption Mass Spectrometer (AA-MS) or by an Ion-Coupled Plasma Optical Emission Spectrometer (ICP-OES). Additional metals analyses were performed by a commercial laboratory. These were supplemented by colorimetric tests for key mine drainage components: Iron (Fe), Manganese (Mn), and Aluminum (Al) using the Hach FerroVer Method (Method 8008), Periodate Oxidation Method (Method 8034), and the Aluminon Method (Method 8012), respectively (Hach Company, 2004).

Geochemical modeling was performed to predict reaction pathways and evaluate reaction kinetics. Models were used to better understand geochemical conditions such as pore water and leachate composition in column materials. Geochemical and modeling studies results are described in detail elsewhere (Behum, *et al.*, 2009; Behum *et al.*, 2010; and Lefticariu, *et al.*, 2014).

4.4.4 *Kinetic studies results*
4.4.4.1 Leachate chemistry
Because this paper is focused more on assessing the overall geochemical advantages of the proposed DP 2 and DP 3 co-disposal practices over DP 1, only broad-based geochemical observations are made here. Table 6 shows the median pH of leachate collected from all columns during the initial leach period (≤7 months) ranged from 7.83 to 8.32, which was well above the carbonate buffer pH level of 6.37. However, there was a sharp decline in pH after that time for all DP 1 columns, and for three of the four blended refuse columns (DP 2 and DP 3). The median values of pH for DP 1, DP 2, and DP 3 during months 8 to 19 were 2.50, 3.56 and 5.82, respectively.

Table 6 also shows results for major anions monitored, including Cl^-, HCO_3^- and SO_4^{2-}. Of these, the HCO_3^- and Cl^- were the most easily leached anions. HCO_3^- is desired because it provides alkalinity needed to buffer the pH, but it declined after 7 months of testing. Similarly, Cl^- declined from an average of 198.0 mg/L in the initial 7-month period to an average of 10.8 mg/L during the remaining test period for DP 1. Similarly, DP 2 and DP 3 showed declines in Cl^- from 197.2 mg/L to 12.7 mg/L, and 126.2 mg/L to 11.4 mg/L, respectively. Thus, the discharge of the relatively easily leached and mobile anion Cl^- was relatively unaffected by the alternate disposal practices.

Conversely, the average SO_4^{2-} concentrations were overall greater and rose significantly after seven months of leaching from an average of 3,437 mg/L to 10,445 mg/L in DP 1, 2,490 mg/L to 4,502 mg/L in DP 2, and 3,093 mg/L to 3,456 mg/L in DP 3. Thus, the SO_4 concentrations in leachate from the blended columns were considerably lower especially in DP 3. Similar results were also obtained for major and trace cation (Ca^{2+}, Mg^{2+}, Na^+, K^+, Fe^{2+}, Mn^{2+}, Ni^{2+}, Zn^{2+}, Cd^{2+}, Sr^{2+} and Pb^{2+}) concentrations (Lefticariu, *et al.*, 2014). The principle cations during early leaching

Table 6. Average concentration of selected mine drainage parameters in leachate.*

Refuse type	Interval	pH	ORP	SO_4	Cl	TDS	Alkalinity	Acidity	Fe	Mn	Al
CCPW-Only	≤ 7 months	8.02	0.132	3,437	198.0	3,865	236.6	9.0	0.76	0.89	1.22
	> 7 months	2.50	0.769	5,449	10.8	10,445	0.0	4,909	1,258	78.8	140.6
CCPW/FCPW Blend	≤ 7 months	8.32	0.077	2,490	197.2	2,968	266.2	2.7	0.23	0.62	0.66
	> 7 months	3.56	0.621	4,502	12.7	5,253	18.2	1,008	204.6	23.1	44.00
CCPW/FCPW/ Limestone Blend	≤ 7 months	7.83	0.133	3,093	126.2	3,698	203.7	1.5	0.08	0.66	0.01
	> 7 months	5.82	0.454	3,456	11.4	2,549	35.9	100.2	12.43	7.10	5.14

*Data in mg/L except pH (median value), ORP (volts), acidity/alkalinity (mg/L CCE).

(< 7 months) were alkali metal ions sodium (Na^+) and potassium (K^+), and alkali earth elements calcium (Ca^{2+}) and magnesium (Mg^{2+}), which is consistent with the earlier laboratory leaching tests (Behum et al., 2009; Behum et al., 2010). Initial Na^+ concentrations were elevated in the initial 7-month period averaging 912.4, 839.1, and 809.1 mg/L for CCPW, CCPW/FCPW, and limestone blend leachate, respectively. Subsequently, Na^+ concentrations declined rapidly to relatively low levels in the final (>7 months) leach period (203.5, 375.7, and 179.1 mg/L for DP 1, DP 2, and DP 3, respectively). Ca^{2+} and Mg^{2+} concentrations both increased during the study; however, Mg^{2+} increased at a much higher rate (from 30.8 mg/L to 253.4 mg/L in DP 1; similar, but less significant, increases in Ca^{2+} and Mg^{2+} concentrations were observed in DP 2 and DP 3 columns. Metals commonly associated with pyrite weathering (Fe, Mn, Al, Ni, Zn, Cu, and Cd) increased in the later leaching period (>7 months), but were again lower in the DP 2 and DP 3 columns. These results are presented in detail in a research report by Lefticariu, *et al.* (2014).

4.4.4.2 Pore-water chemistry

Overall, the volume of pore water collected was much lower than the leachate volume. After seven months of kinetic testing, all but two of the lysimeter ceramic sample cups ceased producing pore water samples because of Fe-rich precipitates. The remaining two lysimeters ceased producing pore water samples after nine months of testing. This limited analysis of pore water chemistry during later leaching periods (>7 months). During the kinetic testing, the average pore water sample collected was only 30.8 ml. The average SO_4^{2-} levels in pore water samples from DP 1, DP 2, and DP 3 columns were 2,844, 3,022 and 1,995 mg/L, respectively, as compared to 3,437, 2,490, and 3,093 mg/L, respectively, in leachate samples. Similar to the leachate sample observations, higher Cl^- concentrations were observed in pore water samples collected during the initial 7-month leach testing period. Again, additional data and discussion is provided in Lefticariu, *et al.* (2014).

Elemental extraction in column leachate: Data for most major and trace constituents were converted to a mass (loading) basis by multiplying concentration values and leachate volume. This conversion allowed determination of the cumulative elemental extraction by calculating the percentage of leachate mass in each cycle, then comparing this to the original mass of the element within the column. Mass data was plotted as a function of time, which is represented by the leach cycle. The complete kinetic testing program consisted of 16 leach cycles completed in 568 days (19 months) with an average of 19,813 ml of leachate collected for each cycle from all of the refuse columns. Leachate volume was then compared to the estimated pore volume of 54,501 ml to yield an initial liquid-to-solids (L/S) ratio of 0.19. As a result, the average rate of pore volume flushing is approximately 0.36 volumes per

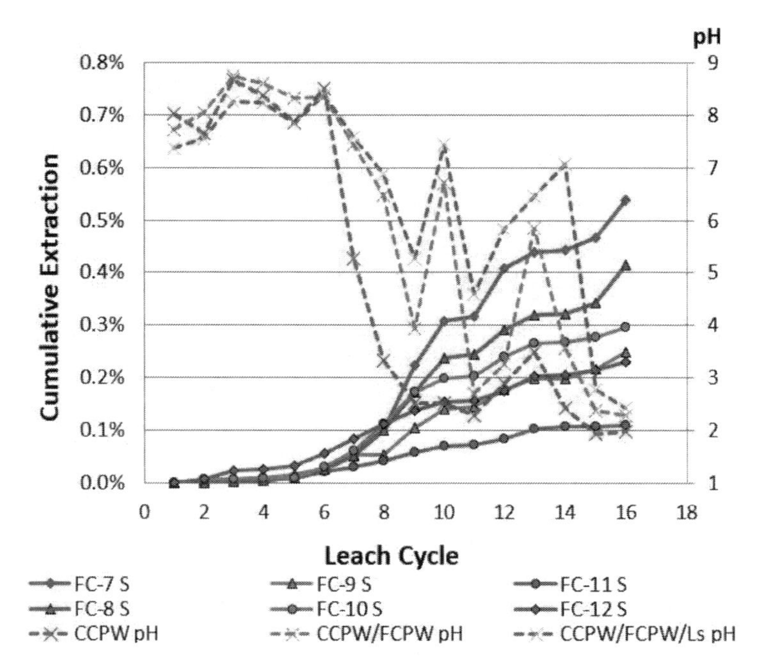

Figure 4. Cumulative extraction of sulfur and pH (dashed lines) in different leach cycles.

leach cycle, which yields an approximate of 5.82 pore volumes that were leached over the course of the 19 month/16-cycle study. The cumulative extraction of S and Cl versus leach cycle is shown in Figures 4 and 5. Although the extraction percentage was greater for Cl than for S during early leach cycles (<7 months), overall amount of S extracted was higher, especially for CCPW (DP 1) columns. After seven months of testing, S extraction increases in all leachate. However, only S extraction from DP 1 column leachate exceeded Cl extraction during this period.

Figure 6 plots the percentage extraction of major and selected trace elements during the 19-month kinetic test. S and Cl are the major anions in high Total Dissolved Solids (TDS) discharges and Na and K are the major cations. Ca extraction was somewhat greater than Mg and Sr extraction; Mg extraction was elevated in the CCPW-only columns (DP 1). Mobility of Ca, Mg, and

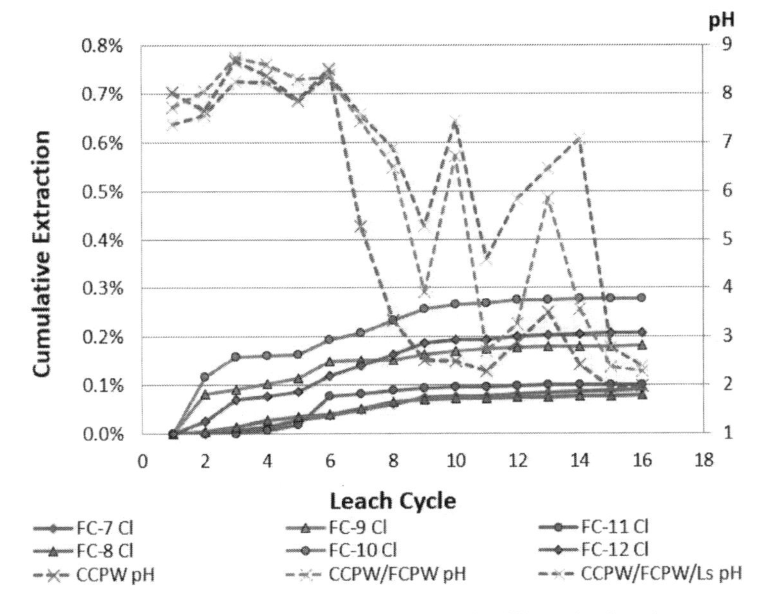

Figure 5. Cumulative extraction of chloride and pH (dashed lines) in different leach cycles.

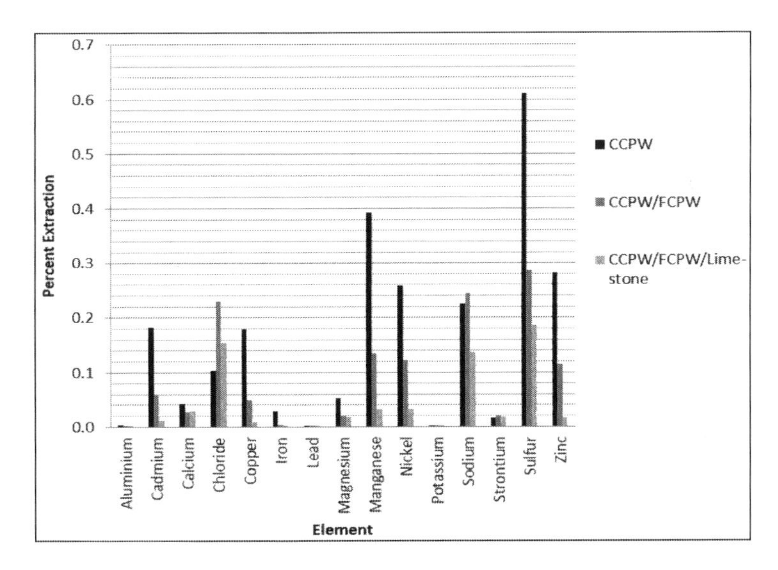

Figure 6. Elemental extraction during 19-month kinetic test.

Sr elements commonly associated with calcareous elements were relatively low during the 19-month kinetic test compared to Mn, Ni, and Zn trace elements that are commonly associated with pyrite (Figure 6). Fe extraction from the DP 1 columns, and to a much later extent from CCPW/FCPW blend columns (DP 2), increased as pH dropped in later leaching periods (>7 months). Mn, Ni, and Zn also had higher extraction rates then Fe suggesting considerable precipitation of Fe phases within the test columns. Only a very small amount of Fe extraction occurred in the limestone blend leachate (DP 3), which also maintained an overall higher pH (Lefticariu *et al.*, 2015).

5 CONCLUSIONS AND RECOMMENDATIONS

5.1 *Conclusions*

1. The co-disposal practices of CCPW and FCPW with or without limestone addition are a significant improvement over the current practice of disposing CCPW alone. The limestone addition practice further improves SO_4^{2-} release.
2. The proposed practice had minimal effect on release of chloride (Cl^-). Most of the discharge occurs soon after placement and must consider using water management practices of retention ponds, dilution, and discharge after large precipitation events.
3. The decline in HCO_3^- concentrations was much greater than sulfate (SO_4^{2-}) concentrations in alternate disposal practices. This suggests that either carbonate weathering rates are faster than pyrite weathering in disposed waste a mine waste pile environment or that the alkalinity-producing minerals are being coated with mineral precipitates that are limiting dissolution.
4. Mobility of SO_4^{2-} was significantly lower in alternate disposal technology especially with limestone addition. The higher extraction of S compared to Ca suggests that the formation of calcium SO_4^{2-} such as gypsum or anhydrite was relatively small.
5. Elements associated with alkalinity-producing minerals such as calcite and dolomite (Ca, Mg, and Sr) were leached to a greater extent than heavy metals typically associated with pyrite (Mn, N, and Zn) and lithophile elements (Al and K).
6. For the duration of experiments, the percentage of S extracted was smaller than Cl.
7. There was at least a 7-month period before additional fresh refuse or the final reclamation cover must be placed in order to avoid excessive release of SO_4^{2-}.

5.2 *Recommendations*

The encouraging results obtained in this study allow us to disseminate this alternate disposal concept to mine operators globally and seek input on how best to pursue commercialization of technology. Variables that might affect its success in the field need to be identified in cooperation with industry prior to wide-scale implementation of this technology. The authors have identified several areas for further studies that would aid commercialization: 1) a larger scale leaching study should employ commercially dewatered fine coal processing waste; 2) laboratory- and field scale leaching studies on DP 2 and DP 3 (co-disposal) alternatives should consider alternative alkaline amendment materials such as Cement Kiln Dust (CKD), acetylene production waste lime, and steel slag, which may be available at a lower cost than agriculture ground limestone; 3) filed demonstration of commercial dewatering technologies to dry coal waste to the necessary solids content of about 65% is needed to develop an economic analysis; and 4) a comprehensive economic analysis of the alternative disposal technologies should be completed. The research team believes that the concepts presented here should be considered for developing other innovative concepts for management of coal waste materials to improve both the economics and environmental impact of disposal.

REFERENCES

American Society for Testing and Materials (ASTM). (2013) "ASTM D698-12e1: Standard Test Methods for Laboratory Compaction Characteristics of Soil Using Standard Effort (12400 ft-lbf/ft3(600 kN-m/m3))." Annual Book of ASTM Standards, Vol. 04.08, West Conshohocken PA, website: http://www.astm.org/Standards/D698.htm accessed on January 8, 2015.

Behum P.T., Chugh, Y.P., Teklehaimanot, Y., and Lefticariu, L. (2009) "Geochemistry of Coal Mine Drainage in Response to Improved Material Handling: Preliminary Results of Kinetic Testing." Paper No. 28-2, Geological Society of America, North-Central Section, 43rd Annual Meeting, Rockford IL, April 2–3.

Behum P.T., Lefticariu, L., and Chugh, Y.P. (2010) "Results of Simulated Weathering of Coal Stockpiles and Coal Refuse Disposal Areas." *Geochimica et Cosmochimica Acta*, Vol. 74, No. 12, Supplement 71.

Chugh, Y.P., Behum P.T., Nawrot, J., and Mohanty, S. (2007b) "Identification of Good Management Practices in Illinois Mining Operations to Minimize Sulfate and Chloride Discharges: An Overview." Proceedings, 16th International Symposium on Mine Planning and Equipment Selection (MPES) and 10th International Symposium on Environmental Issues and Waste Management in Energy and Mineral

Production (SWEMP), Singhal, R.K., Fytas, K., Jongsiri, S., and Ge, H. (Eds.), Bangkok, Thailand, December 11–13.
Chugh, Y.P., Mohanty, S., Behum P.T., Nawrot, J., Bell, W, and Chen, S. (2007a) "Best Management Practices to Minimize Sulfate and Chloride Discharges." Final Technical Report on Illinois Clean Coal Institute Project Number DEV05-8.
Dreher, G.B., Roy, W.B. and Steele, J.D. (1994) "Laboratory studies of fluidized bed combustion residues in mixtures with coal slurry solids." In: Proc. of Management of High Sulfur Coal Combustion Residues: Issues and Practices, Southern Illinois University at Carbondale, Springfield, Illinois, 91–101.
Evangelou, V.P. (1995) Pyrite Oxidation and Its Control, CRC Press, Boca Raton FL.
Hach Company. (2004). "DR/890 Colorimeter procedures manual." Loveland, CO, June 2004, 616 p.
Hossner, L.R. and Doolittle, J.J. (2003). "Iron sulfide oxidation as influenced by calcium carbonate application." Jour. of Environ. Qual., 32, 773–780.
Kleinmann, R.L.P. and Crerar, D.A. (1979) "*Thiobacillus Ferroxidans* and the Formation of Acidity in Simulated Coal Environments." *Geomicrobiology Journal,* Vol. 1, No. 4, pp. 373–387.
Lefticariu, L., Chugh, Y.P., Behum P.T. (2014) Field demonstration of alternative coal processing waste disposal technology, Final technical report, Illinois Clean Coal Institute, 28 p.
Moses, C.O., Nordstrom, D.K., Herman, J.S., and Mills, A.L. (1987) "Aqueous Pyrite Oxidation by Dissolved Oxygen and by Ferric Iron." *Geochimica et Cosmochimica Acta*, Vol. 51, Issue 6, pp. 1561–1571.
Moses, C.O. and Herman, J.S. (1991) "Pyrite Oxidation at Circumneutral pH." *Geochimica et Cosmochimica Acta*, Vol. 55, Issue 2, pp. 471–482.
National Research Council (NRC). (2002). Coal Waste Impoundments, Risks, Responses and Alternatives. National Academy Press, Washington, D.C., 658–667.
Naumann, H.S. and Wiram, V.P. (1995) "Alkaline Additions to the Backfill – A Key Mining/Reclamation Component to Acid Mine Drainage Prevention." Proceedings, 16th Annual West Virginia Surface Mine Drainage Task Force Symposium, Morgantown WV, April 4–5.
Nawrot, J.R. and Gray, B. (2000) "Principles and Practices of Tailings Reclamation: Coal Refuse." *Reclamation of Drastically Disturbed Lands*, Chapter 18, Barnhisel, R.I, Darmody, R.G., and Daniels, W.L. (Eds.), American Society of Agronomy, Crop Science Society of America, Soil Science Society of America, Madison WI, pp. 451–488.
Nordstrom, D.K. (1982) "Aqueous Pyrite Oxidation and the Consequent Formation of Secondary Iron Minerals." *Acid Sulfate Weathering*, Chapter 3, Kittrick, J.A., Fanning, D.S., and Hossner, L.R. (Eds.), Soil Science Society of America, Madison WI, pp. 37–56.
Singer, P.C. and Stumm, W. (1970) "Acidic Mine Drainage: The Rate Redetermining Step." *Science*, Vol. 167, No. 3921, pp. 1121–1123.
Stewart, B.R., Daniels, W.L., Zelazny, W.L., and M.L. Jackson. (2001) "Evaluation of leachates from coal refuse blended with fly ash at different rates." Jour. of Environ. Qual., 30, 1382–1391.
The Advent Group, Inc. (2005) "Changing the Water Quality Standard for Sulfate in Discharges from Coal Mines: The Economic Impact on the Coal Industry." Final Technical Report on Illinois Clean Coal Institute Project Number 03-1/US-4.
US Environmental Protection Agency. (2009) "EPA's Rationale for Approval of Amendments to the Existing Illinois Pollution Control Board Regulation." 35 Illinois Administrative Code 302.102, 302.208, 309.103, 405.109, 406.100, 406.203, and 407. Revised Water Quality Criteria for Sulfate and Total Dissolved Solids (WQSTS IL2008-274), March 18.
Wiram, V.P. (1984) "Challenges to Mining Coal in the Midwest." Proceedings, 5th Annual West Virginia Surface Mine Drainage Task Force Symposium, Morgantown WV, March 21–22.

Land Reclamation in Ecological Fragile Areas – Hu (Ed.)
 ISBN 978-1-138-05103-4

Mine water management as an important part of mining and transition to post-mining in hard coal mining areas in Quang Ninh, Vietnam

K. Broemme, H. Stolpe, C. Jolk, Q.V. Trinh, F. Bilek & A. Ulbricht
Environmental Engineering and Ecology, University of Bochum, Bochum, NRW, Germany

ABSTRACT: As the hardcoal mining areas in Quang Ninh, Vietnam, are located close to the coast of Ha Long Bay mine water management is essential for the protection of coastal water qualities and a sustainable mine operation. Ha Long Bay is a unique seascape of limestone pillars which was recognized as World Natural Heritage Site by UNESCO. In recent years many mine water treatment facilities were implemented and the overall environmental situation of Ha Long Bay has improved. But like in many other areas mine water management is not static due to continued changes in morphology and technology.

When open pit mines are closed down or move to underground mining the mine water management has to be adapted, too. A newly established research and development project funded by the German Federal Ministry of Education and Research (BMBF) will analyze the mine water management from a material flow perspective in order to further optimize the use of water, materials and energy not only for the current situation but also for future mining and post-mining situations. Especially the potential for reuse of treated mine water is investigated as the area regularly faces water shortages during dry seasons. The paper introduces the project approach.

1 INTRODUCTION

1.1 *Project history*

With emerging environmental issues and a rising awareness in the public and in governmental authorities in 2005 Vietnam National Coal – Mineral Industries Holding Corporation Limited (VINACOMIN, former VINACOAL) started a research cooperation with the German Research Association Mining and Environment (RAME) in order to solve some of the most pressing environmental issues in Quang Ninh hardcoal mining areas. RAME was funded by BMBF (German Federal Ministry of Education and Research) from 2005 until 2015 and is coordinated by the Institute of EE+E Environmental Engineering+Ecology, Ruhr-University of Bochum (Prof. Dr. Harro Stolpe, Dr. Katrin Brömme).

The joint research project RAME consisted of six subprojects:

- Environmental management, environmental information system, capacity development (subproject I, 2007–2011),
- Stabilization of waste rock dumps (subproject II, 2008–2013),
- Mine water treatment (subproject III, 2007–2012),
- Dust Mitigation and Monitoring (subproject IV, 2009–2013),
- Plant based methods for waste rock dump recultivation and mine water treatment with constructed wetlands (subproject V, 2008–2011),
- Methods for Post-Mining Land Use Planning (subproject VI, 2011–2015).

In subproject III an efficient mine water treatment concept was developed and implemented in Vang Danh underground coal mine in Uong Bi area (Bilek et al., 2011). In the period 2009–2015 VINACOMIN invested into 14 new mine water treatment plants in Hon Gai area which improved the overall environmental situation of the water resources in the area.

Subproject VI developed the first concept for post-mining land use (Brömme et al., 2014). The huge morphological changes due to open pit mining induce strong affects on the water resources. The boundaries of water catchments change, too. One important question for the future water management of the area is the integration of post-mining lakes which could take over the function of clean water storage for the area.

Based on the subprojects III and VI in 2016 members of RAME developed a new research project on mine water reuse and recycling which develops technical solutions for mine water reuse for different time periods starting from the current situation with 3 large open pit mines still ongoing until the period where all open pit mining is completed but underground mining is ongoing and even further extended. The project is called WaterMiner.

1.2 *Project area*

Hon Gai region is located in the province Quang Ninh in northeastern Vietnam (Fig. 1). It is a part of the most important hard coal mining area in Vietnam, the Quang Ninh coal basin. The region is also famous for Ha Long Bay which is listed as a World Natural Heritage Site by the United Nations Educational, Scientific and Cultural Organization (UNESCO) since 1992.

The Hon Gai region is a part of Ha Long City approximately 170 km north-east of Hanoi. It is a peninsula with a total area of about 12,400 ha. The development of the city is closely linked with the development of the coal mines within its administrative boundaries.

The largest inland surface water source of Hon Gai peninsula is the Dien Vong River in the North where also the local urban water supply facility is located. Other inland rivers are short and small, some of them fall dry during the dry season. The mines mostly discharge their treated mine water into these small rivers.

The main water users in the area are domestic water supply, industry, tourism and services, agriculture, aquaculture and environment. The current water demand of Ha Long City (including Hon Gai and Bai Chay area) amounts to 50 million m^3 per year and is expected to increase up to 109 million m^3 per year in the year 2030. While the total balance between available water resources and water demand shows still a surplus the city already now lacks water in the dry season, especially in the months November to February with a total amount of 10 million m^3 per year (DONRE Quang Ninh, 2016).

2 PROJECT WATERMINER

2.1 *Project objectives*

Treated mine water is a readily available resource in the mining areas in Quang Ninh. The current annual volume is about 90 million m^3 (2016). Water is needed in the mines for sanitary and for industrial purposes. The reuse of treated mine water for these purposes saves other water resources and the expenses to buy clean water from the local water supply services.

Available water volumes strongly vary with the seasons and in their spatial distribution. The water supply for coal production and preparation could be enough in the rainy season but there are

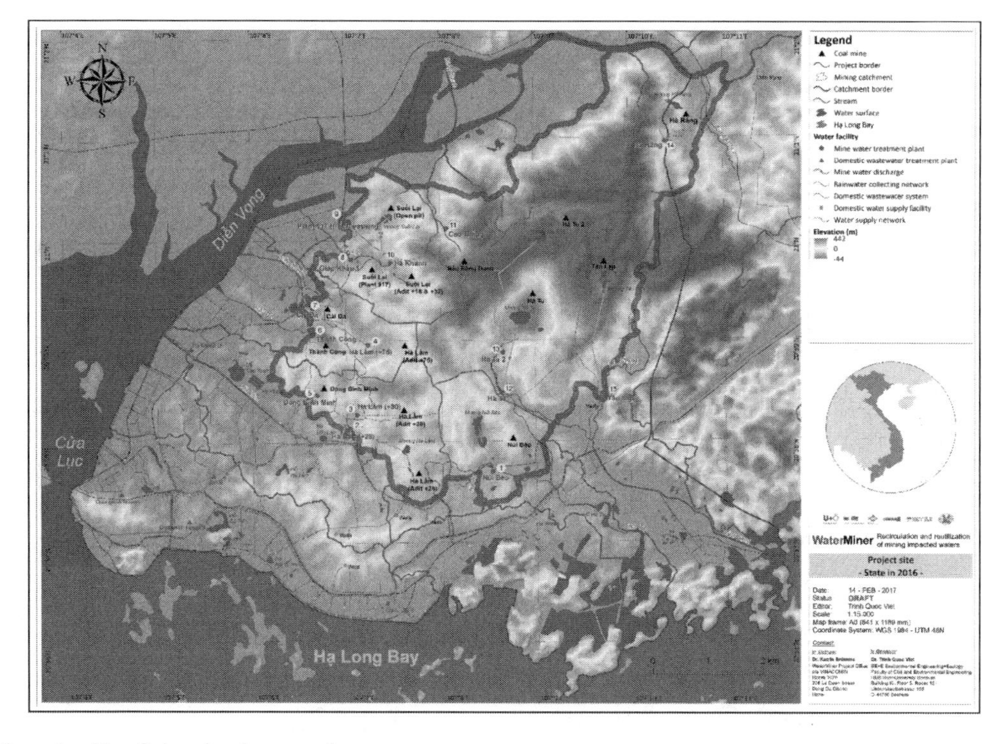

Figure 1. Hon Gai peninsula, research area.

water deficits in the dry season. Mine-internal reuse of water then will in some cases not be sufficient any more to meet the demands. Thus an integrated approach for mine water reuse in the whole Hon Gai area is necessary in order to achieve a spatial and temporal redistribution according to the water demand in the mines. Outside the mining areas there are potential external water users.

In summary there is a need for a spatial and temporal redistribution of water resources between the water sources and the water users in Hon Gai area as there are already water deficits occurring which are expected to increase in the future with higher demands and due to climate change.

The coal dust particles in the surface runoff water represent another problem. These surface runoff waters are discharged most of the time untreated into rivers and canals. The coal dust load represents an economically usable fuel source. However instead of reusing it, the suspended coal is polluting the downstream. The detention of the coal particles and the treatment of the polluted river waters is necessary in order to further reduce the environmental pollution due to coal mining and to reclaim the coal. Although at the moment there are no fees for surface runoff from rainwater but the Law on Environmental Protection (2014) already requires the separate drainage of wastewater and rainwater.

The solution approaches and measures developed in the project later can be transferred to other mining areas in Quang Ninh, for instance Dong Trieu-Uong Bi and Cam Pha where similar issues need to be dealt with.

The main project results are integrated technical concepts for mine water treatment and re-use for the mining area Hon Gai for three different specific dates including the current situation and 2 future scenarios. The technical concepts are based on a data management system for mine water, a corresponding material flow model, an exemplary online monitoring system for mine water and an economic efficiency analysis.

Based on this, the project will develop a detailed technical concept for one selected drainage unit including measures for the collection and treatment of surface runoff water. A pilot im-plementation of the detailed technical concept is intended.

For the regional technical concept within the scope of this project, the term "mine water" includes mine drainage from open pit and underground mines, waste water from all industrial processes related to coal production, sanitary waste water from kitchens, bathrooms and laundries in the mines. Part of the detailed technical concept are the surface runoff and the coal dust reclamation.

2.2 *Project approach*

The project is divided into 7 work packages (see Fig. 2) where WP 1–3 establish the necessary data management structures. WP 4 uses the data to set up a regional water-related material flow management model and scenarios. The WP 5 and 6 support the optimization of scenarios related to treatment technologies and economic efficiency. WP 7 supports the exemplary implementation of developed concepts in a selected area.

For the analysis the research area is divided into drainage units. A drainage unit is very similar to a catchment area but for the mining area the drainage direction depends not only on the natural morphology but also on the mine affiliation and the mine water management.

For each drainage unit the processes related to mine water treatment (see Fig. 3) and sanitary

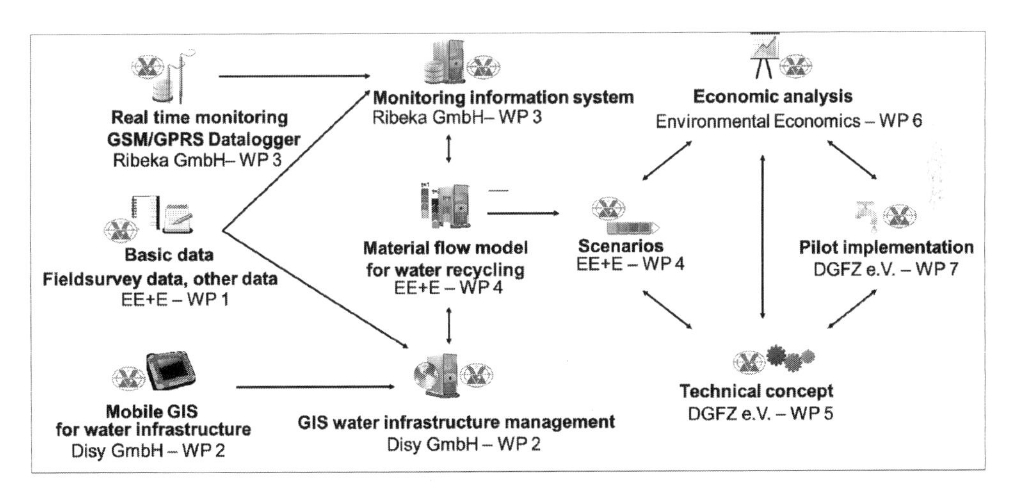

Figure 2. WaterMiner project structure (WP = work package).

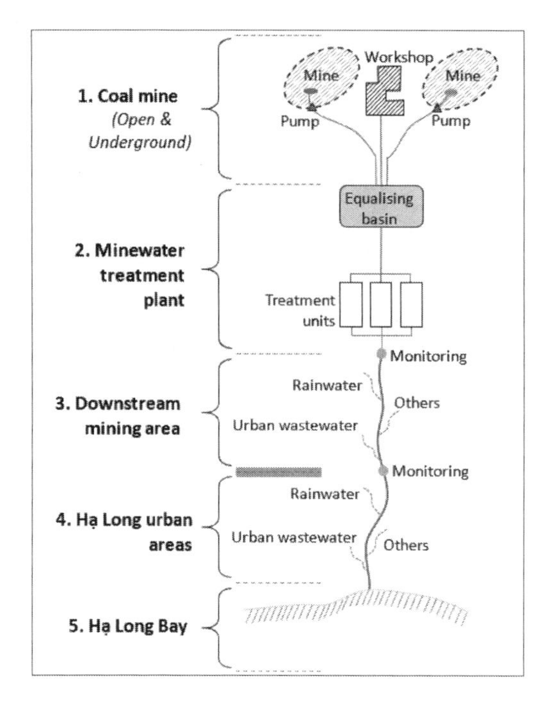

Figure 3. Flowchart of mine water treatment.

water treatment are analysed in detail. For further data processing all necessary information are divided into fixed and variable data and a corresponding database structure will be designed.

The project will work out mine water re-use options for three different time periods. The current situation includes 3 large open pit mines in Nui Beo, Ha Tu and Suoi Lai. The next period representing the year 2020 will include only 2 large open pit mines as Nui Beo then has completed its open pit activities and is using underground mining technology. The last period represents the year 2030 where Suoi Lai and Ha Tu open pit mines are completed as well.

2.3 *First investigation results*

By 2017, 14 mine water treatment plants with a total capacity of 6,250 m^3 per hour have been constructed. The largest plant with a capacity of 4,000 m^3 per hour is scheduled to come into operation at the end of 2017 to increase mine water treatment capacity in the region to 10,250 m^3 per hour. Other plants are planned to be constructed by 2020 at open pits of Suoi Lai (1,000 m^3/h) and Ha Rang (400 m^3/h), Lang Khanh Coal Port (100 m^3/h), Hon Gai Coal Preparation Plant (50 m^3/h) and Cuoc Be Underground Mine (1,000 m^3/h).

A small part of treated mine water is already re-used for dust mitigation. In mines like Ha Lam and Nui Beo treated mine water is also used for sanitary purposes like bathing and laundry. Further potential internal users are: mechanical workshops, industrial cleaning, truck washing, coal screening, coal preparation, cooling system inside underground mines, drinking water.

Besides internal users the project also will involve potential external users like other industrial users, seafood processing, ice making, urban road cleaning, irrigation for urban parks and gardens, irrigation for golf courses, agriculture, water for laundries of hotels etc.

3 OUTLOOK

A further important aspect of the project investigation is the surface runoff from the different surfaces in the mines like roads, workshops, waste rock dumps, coal stores etc. The runoff is mostly not collected and runs untreated into the streams. During a heavy rainfall the volumes of surface runoff are very high. The project will check potential measures to catch and treat surface runoff before it is discharged into the streams.

If it is possible to catch and to treat the main part of the surface runoff the ecological function of the water in the streams is increased again. Creating a more regular flow in the small streams of Ha Long City also could be a type of water re-use after treatment. For the economic assessment of this eco-waters the increase of quality of life as well as the economic benefits for the local residents have to be evaluated.

As the inhomogeneous distribution of water resources between the seasons is one of the main driving factors for water re-use the project also will look deeply into the pros and cons of the establishment of artificial water reservoirs in former open pit mines.

REFERENCES

Bilek, F.; Brömme, K.; Stolpe, H. (2011) Active Treatment of Fe-, Mn- and Coaldust Contaminated Mine Water as Part of the RAME-Project in Vietnam. In: *Mine water—Managing the challenges: Proceedings of 11th International Mine Water Association Congress, September 4th-11th 2011, Aachen, Germany*, pp. 281–286.

Brömme, K.; Stolpe, H.; Jolk, C.; Greassidis, S.; Borgmann, A.; Zindler, B.; Tran, M. (2014) Development of methods for post-mining land use planning for coal mines in urban areas in Quang Ninh, Vietnam. *In: Hu, Zhenqi (ed.): Legislation, Technology and Practice of Mine Land Reclamation: Proceedings of Beijing International Symposium Land Reclamation and Ecological Restoration, LRER*. Leiden: CRC Press/ Balkema. pp. 517–526.

DONRE Quang Ninh (2016) *Water resources planning of the province Quang Ninh until 2020 with an orientation to 2030*. Department of Natural Resources and Environment Quang Ninh.

Land Reclamation in Ecological Fragile Areas – Hu (Ed.)

Utilization of submarinetailings disposal as alternative tailings management system

E. Yilmaz
Mining Engineering Department, Karadeniz Technical University, Trabzon, Turkey
Cayeli Bakir Isletmeleri A.S., Madenli Beldesi, Rize, Turkey

ABSTRACT: The mining industry creates huge quantities of sulphidic or non-sulphidic tailings which may pose a number of environmental problems (e.g., acid mine drainage formation and tailings dam failures) mainly due to their sheer volumes and toxicities. To manage properly these hazardous tailings without causing any environmental and geotechnical risks, the industry is always searching for new, sustainable and efficient waste management options. This paper demonstrates sustainable use and application of a Submarine Tailings Disposal (STD) system which has been currently operating in a Cu-Zn mine located in north-eastern Turkey. A complete tailings characterization and discharge monitoring results are also presented. Results indicate clearly that the SPD system runs well as a combined tailings management technique in the mine. Results indicate that underground backfill and surface disposal can be utilized concurrently without creating environmental damages (e.g., tailings dam failure, acid formation) over the mine's life.

1 INTRODUCTION

Even though the mining activities greatly contribute to the development of modern societies, they also lead inevitably to the formation of large volumes of unwanted wastes (Polling 1995). This will continue increasingly in the future since ore grade decreases as mines go deeper and deeper for sustainable operations. This massive decrease in ore grades will also cause increases in the waste-to-metal production ratio (Dold 2008; Edinger 2012). If wastes are not properly managed, due to their sheer quantities and toxicities, they may lead to several environmental (air, water and land) pollutions such as acid mine drainage and release of heavy metals (Kwong 1993; Asmund et al., 1994; Akcil and Koldas, 2006; Walker et al., 2009; Jamieson 2011). There are some methods of tailings management which can remove the environmental impacts (Franks et al., 2011; Adiansyah et al., 2015; Yilmaz and Fall, 2017). Tailings are usually managed with three main options: i) disposal into the surface tailings dams or impoundments, ii) discharge into the deep sea zones or lakes, iii) backfill into underground mined-out voids.

One can generally say that many modern mines worldwide employ land-based techniques for sustainable tailings management. However, there is a feasible alternative to land-based tailings disposal methods which is also known as Submarine Tailings Disposal (STD, Ellis and Ellis 1994; Ellis et al. 1995a, b; Vogt 2012; Cornwall 2013; Dold 2014; Ma et al., 2017). This system is considerably less costly than tailings dams or impoundments, and is often used in coastal areas where topographic (e.g., high seismic activity and landslides) and climatic restrains (e.g., heavy rains and less sunny) are dominantly taken place (Ellis 2008; Angel et al., 2013; Ramirez-Llodra et al., 2015). The STD system has been well used as a sustainable tailings management practice by most mines worldwide (Garnett and Ellis, 1995; Jones and Ellis, 1995; Poling et al., 2002; Berkun 2005; Blackwood and Edinger, 2007; Lin et al., 2009; Sahami et al., 2011; Søndergaard et al., 2011; Reichelt-Brushett 2012; Rusdinar et al., 2013). Figure 1 shows a schematic diagram of the STD system. It consists of a head tank, which collects tailings/water at the site, overland tailings/water pipeline, a near-shore mixing tank, a submarine outfall pipeline, which discharges tailings from the mix tank to sea, and seawater intake pipeline.

Storage of tailings via the STD system is indeed considered as a more secure way since sulfide-rich minerals such as pyrite are geochemically stable under reduced conditions in the submarine environments (Blanchette et al., 2001; Dold et al., 2009; Matthies et al., 2011; Rzepka et al., 2013; Pedersen et al., 2017). This paper explains the aspects and application of the STD system and presents results of the experiments needed for assessing the technique's efficiency.

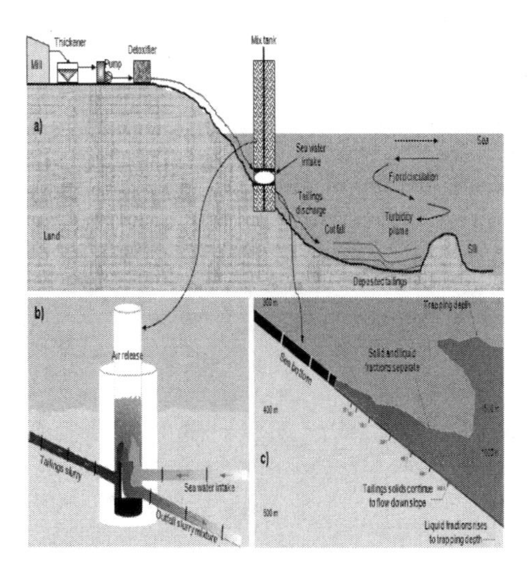

Figure 1. Schematic view of (a) submarine tailings disposal, (b) mix tank and (c) submarine pipeline.

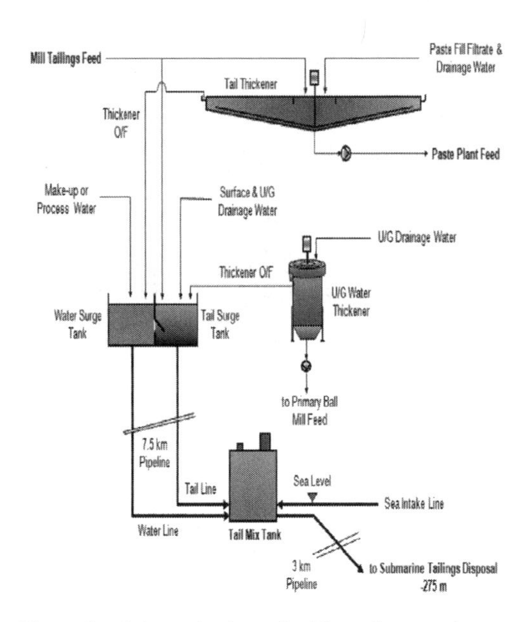

Figure 2. Schematic view of tailings slurry and waste water management at the mine studied.

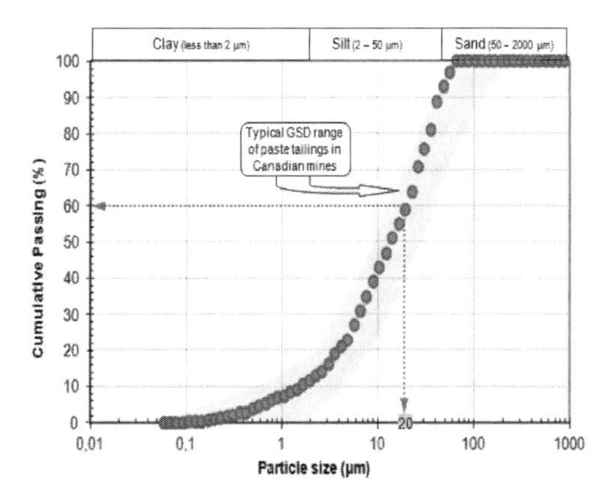

Figure 3. Grain size distribution of tailings, comparing to typical Canadian mine tailings.

2 MATERIALS AND METHODS

2.1 *Mine site description*

The mine studied is located in the black sea coast of northeastern Turkey. The mine produced its first concentrate in 1994 and extracts about 1.2 million tons of ore per year to produce about 220,000 tons of copper and zinc concentrates. Yearly, the mean ore feed head grades were 3.2% Cu and 6.3% Zn. Concentrate grade averaged 20% for Cu at a 76% recovery and Zn concentrate grade averaged 49% at a 71% recovery. In the last two decades, the average recoveries of zinc and copper have been 72% and 81%, respectively.

Large amounts of mill tailings with a high acid generating potential is produced during the mineral processing. Nearly 55% of the tailings generated are pumped underground as paste backfill while the remaining tailings (45%) are discharged to the Black Sea at a depth of 350 m at approximately 3.5 km offshore. There are two 7.5 km long overland pipelines: one for tailings slurry and the other for excess water from thickener overflows. Mine wastewater and sewage plant effluent are also added water side. Both tail line and water line are inter-connected at the de-aeration tank located Black Sea coast. Tailings gravitate down to 275 m sea level via an outfall pipeline by the density differences of tailings and sea water. Figure 2 shows a flow sheet of tailings and waste water management implemented at the mine.

2.2 *Tailings slurry characterization*

The tailings slurry was sized by using a Mastersizer S Ver. 2.15 (Malvern Instruments Ltd, UK) particle size analyzer and the results are shown on Figure 3. Tailings had approximately 60 wt% finer than 20 μm and classified as a medium size tailings material. The specific gravity of the tailings was also measured using picnometer. Results indicated that as-received tailings had a specific gravity of 3.5 and a specific surface area of 2.1 m^2/g and a pH of 10%.

The rheological properties of the as-received tailings slurry were measured by a rotational viscometer Rheomat 15T at 23°C and investigated at 30, 35, 40, and 45% dry solids by weight at a pH of 10, as mixed (Table 1). In addition, settling tests were performed at solids concentrations of 15 and 30% dry solids by weight at pH of 10, as mixed (Table 2).

Table 1. Rheological properties of tailings slurry used in the experiments.

Solid content Cw, wt%	Saturated volume fraction, ϕsat	Saturated volume ratio	Viscosity of slurry/ Viscosity of water	Yield stress (Pascal)
45	0.19	0.23	6.9	3.1
40	0.16	0.19	5.2	1.6
35	0.13	0.15	4.3	1.0
30	0.11	0.12	3.4	0.5

Table 2. Settling properties of tailings slurry used in the experiments.

Cw% - initial	Cw% - maximum	ϕsat- initial	ϕsat- maximum	Maximum settling rate, cm/hour
30	58	0.11	0.28	8.15
15	53	0.05	0.25	17.0

Note that a 7 mm diameter HDPE rod, positioned with its lower end at the settled solids surface, fell hurriedly through the settled solids layer under its own weight, striking the bottom of the cylinder audibly—indicating that the solids were very softly settled. Very little resistance was felt to lateral (sideways) movement of the rod.

The chemical composition of the tailings is determined by atomic absorption spectrometry, spectrophotometer (e.g., K_2O and Na_2O), and wet chemical analysis (SO_3). Tailings are also dominated by iron oxide, Fe_2O_3 (43.7%) and minor quantities of silicon dioxide, SiO_2 (10.9%) and aluminum oxide, Al_2O_3 (3.9%), together with trace amounts of magnesium, calcium, potassium, sodium, titanium, chromium, manganese and phosphorous oxides (all less than 2%). A loss-on-ignition value of 27.7% for the tailings slurry is indicative of loss of sulfur, as pyrite is burned off to reveal the high iron oxide reading.

The polished section was also prepared and displays a sharp contact between two completely different types of ore, as shown clearly in Figure 4. Note that gg stands for gangue minerals.

The first type (I) possesses a clastic texture with pyrite (py) in the form of clasts, crystals, layery aggregates, and dissemination. Pyrite is often replaced by chalcopyrite (cp). Sphalerite (sl) and chalcopyrite occur as clasts and irregular patches as well as a matrix for finely disseminated pyrite. The second type (II) is characterized by a rather massive matrix of sphalerite associated with galena (gn) in the form of irregular inclusions in dimensions ranging from less than 10 microns to a few hundred microns. Sphalerite also bears minute inclusions of chalcopyrite without really being "diseased".

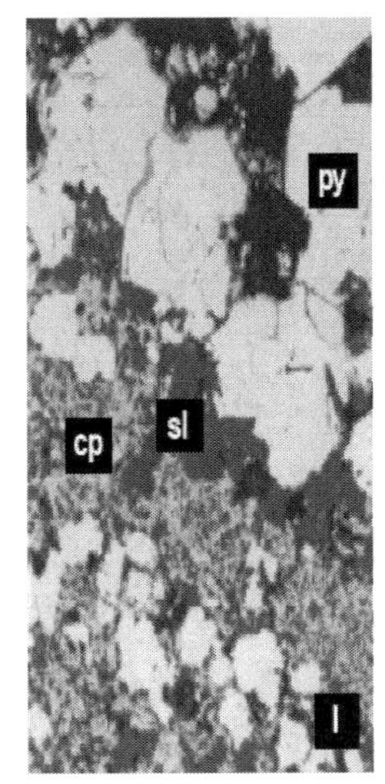

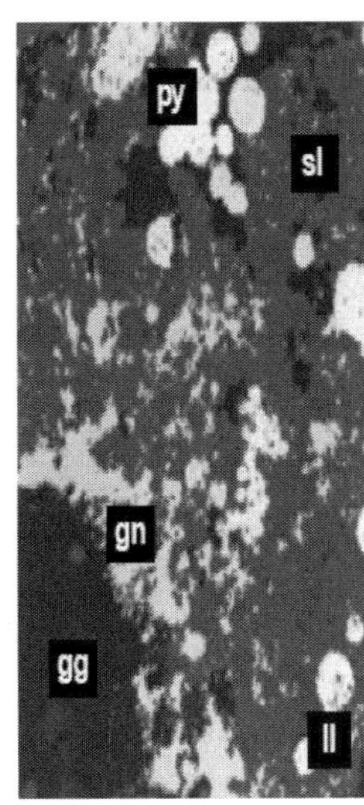

Figure 4. The polished sections of two types of ore samples: type I (left); type II (right).

2.3 *Submarine tailings disposal system*

The studied mine's STD system includes: i) head tank, which collects tailings and waste water at the mine site; ii) overland tailings pipeline from the head tank at the mine site to the shore of the Black Sea (~7.25 km); iii) overland waste water pipeline from the head tank at the mine site to the shore of the Black Sea (~7.25 km); iv) near-shore mixing tank; v) submarine outfall pipeline, which discharges tailings from the mix tank to 298 meters below sea level in the Black Sea (~2.85 km); and vi) seawater intake pipeline.

To prevent the mix tank from overflowing and the subsea pipeline from running in slack flow, the mix tank level needs to stay between 2 to 10 meters. With the seawater intake pipeline, the mix tank level stays at 4.3 m. Given the set point for the mix tank is 4.3 m, the highest throughput available for the submarine tailings pipeline is 180 tph at 12.9–13.7%. The limiting factor for tailing throughput is the maximum flow rate and slurry concentration of the overland tailings pipeline. Typical operating slurry concentrations for the overland tailings pipeline are from 25 to 30%. The maximum solid throughput for the overland tailings pipeline at a slurry concentration of 30% is 122 tph at a flow rate of 320 m^3/h. If the flow rate is greater than 320 m^3/h for a throughput of 122 tph then head tank overflow would occur. In order for the STD system to operate at a throughput greater than 122 tph, a higher slurry concentration would be needed. Table 3 lists the maximum flow rates at the operating slurry concentration of 30% for overland and submarine pipelines without risk of head and mix tank overflow.

2.4 *Pipeline characteristics and STD system criteria*

Table 4 summarizes the pipeline system characteristics for the overland tailings, water, and subsea tailings pipeline.

Operating Velocity: Deposition velocity is defined as the point at which the heaviest particles are no longer fully suspended and begin to accumulate near the bottom of the pipe. Slurry pipelines are operated in a turbulent flow regime to prevent solids from building up on the bottom of the pipe.

Mix Tank Level: Tailings and waste water flow from the overland pipelines into the mix tank. From the mix tank, the pressure difference from the outlet of the submarine tailings pipeline and the mix tank dictate the submarine tailings flow rate. At high solid concentrations, the mix tank level will drop below 3 meters, which causes seawater to flow into the mix tank via the seawater intake pipeline to keep the mix tank level at 3 meters.

Head Tank Level: The overland tailings and waste water pipelines also operate with a gravity line. Tailings and waste water flow from the outlet of the head tanks at an elevation of 97 meters to the inlet of the mix tank at an elevation of 3 meters below sea level. To prevent overflow on the water side of the head tank there exists a flap valve, which allows water to flow into the tailings side of the head tank if the level on the water side is higher.

Slack Flow: Slack flow occurs when the hydraulic gradient line is below the line profile. When this occurs, the pipeline would run partially full for the areas below the line profile. This causes accelerated erosion to the pipeline. To prevent slack flow for the system, additional process water is added in to the head tank for the overland tailings pipeline and/or seawater is added into the mix tank via the seawater intake pipeline for the submarine tailings pipeline.

Table 3. Maximum flow rate of both overland and submarine pipelines.

Parameter	Overland water pipeline	Overland tailings pipeline	Submarine tailings pipeline
Max volumetric flow rate, m^3/h	270	320	1043
Maximum solid throughput, tph	0	122	122
Slurry content at max flow rate, wt%	0	30	11

Table 4. Overland water, overland tailings, and submarine tailings pipeline system characteristics.

Parameter	Overland water pipeline	Overland tailings pipeline	Submarine tailings pipeline
Design solid throughput, tph	180	180	180
Pipeline nominal diameter, inch	14	12	22
Pipeline standard dimensional ratio	17	7	9
Pipeline outer diameter, mm	356	324	559
Pipeline wall thickness, mm	21	46	62
Pipeline inner diameter, mm	314	231	435

2.5 *Atomic absorption spectrometry and pH measurement*

There are two types of atomic spectroscopy, namely Atomic Absorption Spectroscopy (AAS) and Atomic Emission Spectroscopy (AES). In this study, A Varian AA 240 FS flame atomic absorption spectrometry AAS (based on the absorption of ultraviolet or visible radiation by free atoms in the gaseous state) was used for the determination of Cu, Zn, and Pb within tailings slurry samples. Slurry samples were digested using a $HF\text{-}HClO_4\text{-}HNO_3$ total dissolution technique. In addition, pH of the tailings slurry samples was measured using a Benchtop pH/ISE Meter Orion Model 920A coupled with a Thermo Orion Triode combination electrode (Pt-Ag-AgCI).

3 RESULTS AND DISCUSSION

3.1 *Submarine tailings pipeline hydraulic design results*

Hydraulic models for the overland tailings/waste water pipelines and the submarine tailings pipeline are built to predict hydraulic performance of the system. Assuming the maximum allowable tank height for the mix tank is 10 meters, the maximum flow rate for the subsea pipeline at 180 tph is 1264 m^3/h at 12.9%. Given the seawater intake pipeline, the mix tank height is typically 4.3 meters with a low of 2 meters. Table 5 depicts the submarine tailings pipeline operating range based on the mix tank height of 4.3 to 10 meters. At a mix tank height of 10 meters, the mix tank is at risk of overflow. At a mix tank height of 4.3 meters, the submarine outfall pipeline is operating without the need of seawater intake.

Figure 5 depicts the operating range for the submarine outfall tailings pipeline. Using the operating range, it is possible to determine the operating data for the submarine tailings pipeline at the ideal mix tank heights and solid throughputs. For example, in order for the submarine tailings pipeline to operate at a mix tank height of 4.3 meters and a solids throughput of 107 tph, the submarine tailings pipeline would need 1,000 m³/h of slurry at a 10% weight concentration. Any flow rates (m³/h) and throughputs (tph) above the mix tank height 10 m line would cause the mix tank to overflow. Any flow rates and throughputs below the mix tank height 2 m line imply the need for seawater dilution to prevent the mix tank from emptying. The section in between the two lines represents the operating range where the submarine tailings pipeline can function without risk of mix tank overflow or require additional seawater.

Table 5. Operating range for submarine outfall tailings pipeline.

Solid throughput, tph	Submarine slurry content, %	Flow rate, m³/h	Flow velocity, m/s	Mix tank height, m
180	12.9	1,264	2.37	10
180	13.7	1,182	2.21	4.3
120	9.9	1,131	2.12	10
120	10.7	1,037	1.94	4.3
60	6.1	943	1.77	10
60	6.9	824	1.54	4.3
0	0	556	1.04	10
0	0	189	0.35	4.3

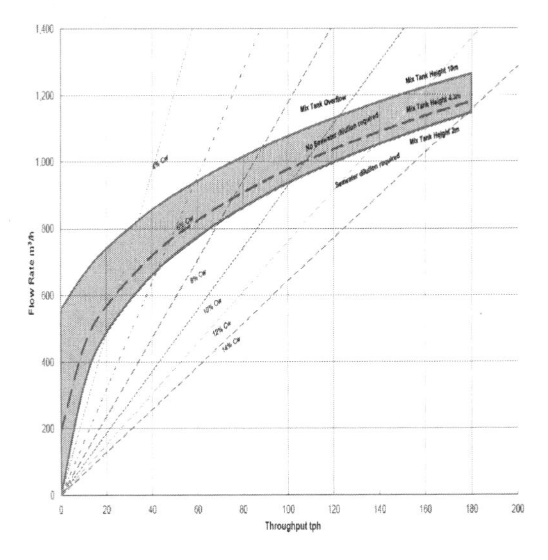

Figure 5. Operating range diagram for submarine tailings pipeline.

3.2 *Overland tailings pipeline hydraulic design results*

It was assumed that the level of the mix tank would range from 4.3 to 10 meters. In addition, it is assumed that the head tank for the overland tailings pipeline is approximately 4–5 meters in height, giving an additional head of 1 to 3 meters to the gravity line. While the overland tailings pipeline can operate at a lower head tank level, it is assumed that waste water from the waste water side of the head tank would flow into the overland tailings head tank side in order to prevent slack flow. Table 6 depicts the overland tailings pipeline operating range.

Figure 6 depicts the operating range for the overland tailings pipeline. Using the operating range, it is possible to determine the operating data for the overland tailings pipeline at the maximum and minimum flow rates. For example, given the solid throughput of 180 tph at the maximum flow rate of 302 m³/h, the slurry content needs to be 41.8%. Any flow rates and throughputs above the maximum flow rate line would cause the head tank to overflow. Any flow rates and throughputs below the minimum flow rate line imply the risk of slack flow in the overland tailings pipeline. The section in between the two lines represents the operating range where the overland tailings pipeline can function without risk of head tank overflow or slack flow.

3.3 *Overland waste water pipeline hydraulic design results*

Figure 7 depicts the operating range for the overland waste water pipeline. Any flow rate above the maximum flow rate line would cause the head tank to overflow. Any flow rate below the minimum flow rate line would result in the head tank being emptied by open channel flow for the first section of the overland waste water pipeline. The section in between the two lines represents the operating range where the head tank is not at risk of overflow and the overland waste water pipeline is operating with packed flow.

Table 6. Operating range for overland tailings pipeline.

Solid throughput, tph	Overland slurry content, %	Flow rate, m³/h	Flow velocity, m/s	Head tank height, m	Mix tank height, m
180	43.4	286.0	1.89	1	10
180	41.8	302.3	2.00	3	4.3
120	30.8	304:3	2.01	1	10
120	29.6	319.5	2.11	3	4.3
60	16.7	315.6	2.09	1	10
60	16.1	330.5	2.18	3	4.3
0	0	326.6	2.16	1	10
0	0	341.4	2.26	3	4.3

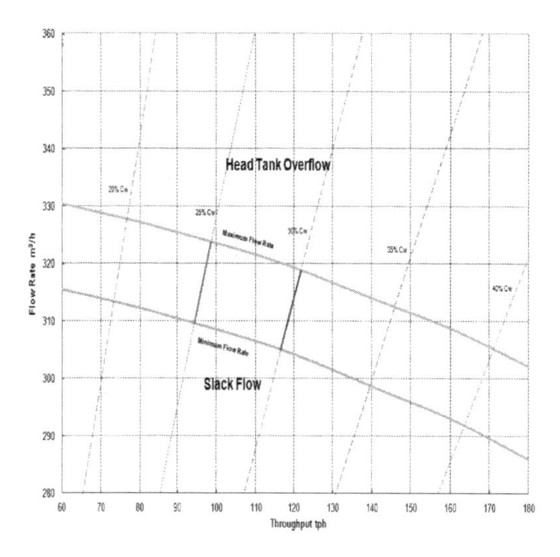

Figure 6. Operating range diagram for overland tailings pipeline.

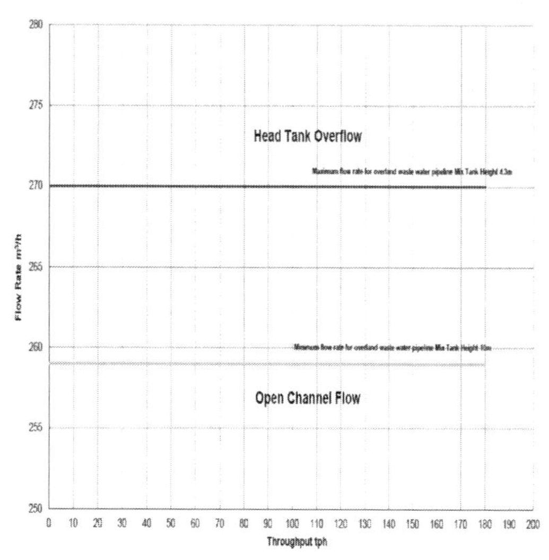

Figure 7. Operating range diagram for overland waste water pipeline.

3.4 *Disposal of mine tailings into the deep sea zones*

Figure 8 shows the usage amounts of the created process tailings as Submarine Tailings Disposal (STD) technique. Note that a mean process tailings of 85,677 m^3 has been produced from the concentrator plant of the mine studied. Approximately 49% of the tailings generated are delivered using the overland tailings and submarine outfall pipelines to undersea as tailings disposal. Due to the fact that there is no enough space vicinity the mine for constructing a surface tailings dams or impoundments and there are too much rainfalls in the region, the mine uses the STD system as a unique tailings management option as well as cemented paste backfill in which the generated tailings are mixed with cements first and then delivered to underground mined-out stopes or openings for either backfill or disposal purposes.

It is good to stress that there has been no mechanical or operational problem of using the STD system. Between 2010 and 2017, a total of more than 7 million tons process tailings are produced and almost 3.5 millions tones of these tailings are delivered to undersea for disposal.

3.5 *Tailings discharge monitoring results*

Samples are collected from the mix tank right before the generated tailings are delivered to undersea as disposal. These samples are brought to assay laboratory for pH and some heavy metal analyses. Figure 9a shows the change in pH of waste water samples collected between 2010 and 2017. Results

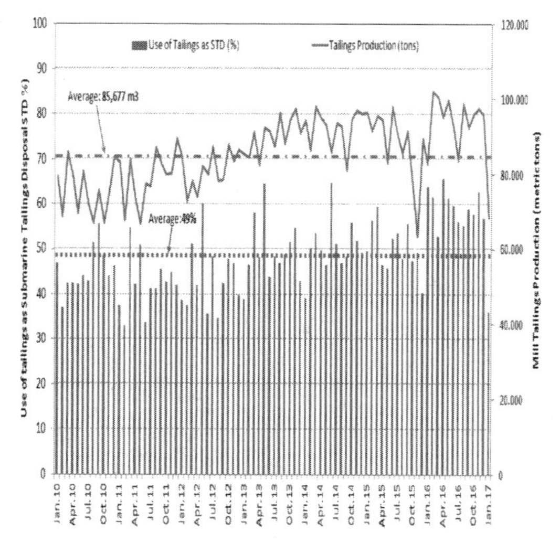

Figure 8. Disposal of mine tailings to undersea using submarine tailings placement system.

indicate clearly that pH of all samples are higher than 5 which reflects allowable limit. A mean pH value is 10, showing an effect of the lime added to copper-zinc circuits in the studied mine as pH modifier. In similar manner, the concentration of lead Pb was assessed for the same time period. Results indicate that samples' Pb values were under an allowable limit of 0.05 mg/L (Figure 9b). In addition, copper Cu concentrations were under the allowable limit of 0.05 mg/L (Figure 9c). Zinc Zn concentrations are found to be scattered in the

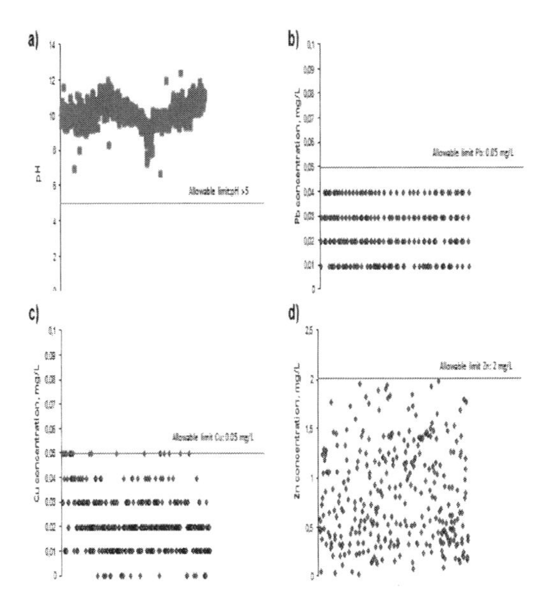

Figure 9. Change in pH (a), lead (b), copper (c), and zinc (d) concentrations over time.

collected samples and were under an allowable limit of 2 mg/L (Figure 9d). One can overall say that sulphidic minerals, such as copper Cu, zinc Zn, and lead Pb are not as stable in undersea reducing environments. The liberation of heavy metals from sulphidic ores under reducing conditions has been observed in both laboratory—and field-scale testing. There are some indications that high chloride concentration might increase the corrosion of sulfides.

A field investigation undertaken by Berkun (2005) focused on the potential for tailings plume separation and upwelling above the disposal depth. The researcher used oceanographic sampling combined with laboratory experimentation in a water tank in order to determine plume dispersal characteristics. The experimental findings showed a worst-case potential for plume separation and upward movement of fine particles up to 89 m above the 350 m discharge depth. This was taken into account to be acceptable as this region was within the deep anoxic zone of the Black Sea and still deeper than the permanent pynocline existing at ~150 m.

4 CONCLUSIONS

This paper presents the utilization of submarine tailings disposal as alternative tailings management system. This system was particularly used in the regions where topographic and climatic restraints are two main obstacles of constructing a proper tailings impoundment or dam on the surface. From the stability and economy points of view, the STD system provided a unique technique of safe disposal of the process tailings generated into the deep sea zones or lakes. The limiting factor for tailings throughput is the maximum flow rate and slurry concentration of the overland tailings pipeline. At slurry concentration of 30%, the overland tailings pipeline can operate at a maximum throughput of 122 tph at 318.8 m^3/h without risk of head tank overflow. The main restriction on the STD system is the total flow rate of the overland pipelines without head tank overflow. To prevent head tank overflow, the overland tailings pipeline flow rate must run between 302.3 to 341.4 m^3/h. It must be emphasized that for the overland tailings pipeline to process high solid throughput, high slurry concentrations are required. This may be an issue since the waste water side of the head tank has a flap valve, which enables excess waste water to flow into the overland tailings side of the head tank. While this prevents the water side of the head tank from overflowing, this also dilutes the slurry concentration of the overland tailings pipeline. A lower solid throughput or less waste water disposal is thus needed for optimal STD system operation. As a result, the STD system can be advantageously used as alternative tailings management in most modern mines.

REFERENCES

Adiansyah, J.S., Rosano, M., Vink, S. & Keir, G. (2015) A framework for a sustainable approach to mine tailings management: disposal strategies. *Journal of Cleaner Production,* 108, 1050–1062.

Akcil, A. & and Koldas, S. (2006). Acid Mine Drainage (AMD): Causes, treatment and case studies. *Journal of Cleaner Production,* 14 (12–13), 1139–1145.

Angel, B.M., Simpson, S.L., Jarolimek, C.V., Jung, R., Waworuntu, J. & Batterham, G. (2013) Trace metals associated with deep-sea tailings placement at the Batu Hijau copper-gold mine, Sumbawa, Indonesia. *Marine Pollution Bulletin,* 73 (1), 306–313.

Asmund, G., Broman, P.G. & Lindgren, G. (1994) Managing the environment at the Black Angel Mine, Greenland. *International Journal of Surface Mining, Reclamation and Environment,* 8, 37–40.

Berkun, M. (2005) Submarine tailings placement by a copper mine in the deep anoxic zone of the Black Sea. *Water Research,* 39 (20), 5005–5016.

Blackwood, G.M. & Edinger, E.N. (2007) Mineralogy and heavy metal speciation patterns of shallow marine sediments affected by submarine tailings disposal and artisanal gold mining, Buyat-Ratototok district, North Sulawesi, Indonesia. *Environmental Geology,* 52, 803–818.

Blanchette, M.C., Hynes, T.P., Kwong, Y.T.J., Anderson, M., Veinott, G., Payne, J.F., Stirling, C. & Sylvester, P.J. (2001) A chemical and eco-toxicological assessment of the impact of marine tailings disposal. In: The 8th

International Conference on Tailings and Mine Waste, Vail, Fort Collins, Colorado, USA, 16–19 January, 323–331.

Cornwall, N. (2013) Submarine tailings disposal in Norway's fjords. *M.Sc. Thesis*, Lund Univ., 41 p.

Dold, B. (2008) Sustainability in metal mining: From exploration, over processing to mine waste management. *Rev. Environ. Sci. Biotechnol.*, 7, 275–285.

Dold, B. (2014) Submarine Tailings Disposal (STD) – A review. *Minerals*, 4, 642–666.

Dold, B., Wade, C. & Fontbote, L. (2009) Water management for acid mine drainage control at the poly-metallic Zn-Pb-(Ag-Bi-Cu) deposit of Cerro de Pasco, Peru. *Journal of Geochemical Exploration*, 100, 133–141.

Edinger, E. (2012) Gold mining and submarine tailings disposal: Review and case study. *Oceanography*, 25 (2), 184–199.

Ellis, D. & Ellis, K. (1994) Very deep STD. *Marine Pollution Bulletin*, 28 (8), 472–476.

Ellis, D.V. (2008) The role of deep Submarine Tailings Placement (STP) in the mitigation of marine pollution for coastal and island mines. *Marine Pollution: New Research* (Editor: T.N. Hofer), Nova Science Publishers, 23–51.

Ellis, D.V., Pedersen, T.F., Poling, G.W., Pelletier, C. & Horne, I. (1995) Review of 23 years of STD: Island copper mine, Canada. *Marine Georesources & Geotechnology*, 13 (1–2), 59–99.

Ellis, D.V., Poling, G.W. & Baer, R.L. (1995) Submarine Tailings Disposal (STD) for mines: An introduction. *Marine Georesources & Geotechnology*, 13 (1–2), 3–18.

Franks, D.M., Boger, D.V., Côte, C.M. & Mulligan, D.R. (2011) Sustainable development principles for the disposal of mining and mineral processing wastes. *Resources Policy*, 36 (2), 114–122.

Garnett, R.H.T. & Ellis, D.V. (1995) Tailings disposal at a marine placer mining operation by West Gold, Alaska. *Marine Georesources & Geotechnology*, 13 (1–2), 41–57.

Jamieson, H.E. (2011) Geochemistry and mineralogy of solid mine waste: Essential knowledge for predicting environmental impact. *Elements*, 7, 381–386.

Jones, S.G. & Ellis, D.V. (1995) Deep water STD at the Misima gold and silver mine, Papua, New Guinea. *Marine Georesources & Geotechnology*, 13 (1–2), 183–200.

Kwong, Y.T.J. (1993) Prediction and prevention of acid rock drainage from a geological and mineralogical perspective. *Canada Centre for Mineral and Energy Technology*, Ottawa, Ontario, 47 p.

Lin, H.K., Walsh, D.E., Chen, X. & Oleson, J.L. (2009) Release of heavy metals from sulphide flotation tailings under deep water discharge environments. *Miner. Metall. Process.*, 26, 174–178.

Ma, W., Schott, D. & Lodewijks, G. (2017) A new procedure for deep sea mining tailings disposal. Minerals, 7 (4), 1–14.

Matthies, R., Bowell, R.J. & Williams, K.P. (2011) Geochemical assessment of gold mine tailings proposed for marine tailings disposal. *Geochemistry: Exploration, Environment, Analysis*, 11, 41–50.

Pedersen, K.B., Jensen, P.E., Ottosen, L.M., Evenset, A., Christensen, G.N. & Frantzen, M. (2017) Metal speciation of historic and new copper mine tailings from Repparfjorden, Northern Norway, before and after acid, base and electrodialytic extraction. *Minerals Engineering*, 107, 100–111.

Poling, G., Ellis, D.J., Murray, J.W., Parsons, T. & Pelletier, C. (2002) Underwater tailings placement at Island Copper Mine: A success story. *Society for Mining, Metallurgy, and Exploration*, Englewood, Colorado, USA, 1–15.

Poling, G.W. (1995) Mining/milling processes and tailings generation. *Marine Georesources & Geotechnology*, 13 (1–2), 19–31.

Ramirez-Llodra, E., Trannum, H.C., Evenset, A., Levin, L.A., Andersson, M., Finne, T.E., Hilario, A., Flem, B., Christensen, G., Schaanning, M. & Vanreusel, A. (2015) Submarine and deep-sea mine tailing placements: A review of current practices, environmental issues, natural analogs and knowledge gaps in Norway and internationally. *Marine Pollution Bulletin*, 97 (1–2), 13–35.

Reichelt-Brushett, A. (2012) Risk assessment and ecotoxicology limitations and recommendations for ocean disposal of mine waste in the Coral Triangle. *Oceanography*, 25, 40–51.

Rusdinar, Y., Edraki, M., Baumgatl, T., Mulligan, D. & Miller, S. (2013) Long term performance of geochemical riverine mine tailings deposition at Freeport Indonesia. *Mine Water Environ.*, 32, 56–70.

Rzepka, P., Walder, I., Bozecki, P. & Rzepa, G. (2013) Subsea tailings deposition leach modeling. *Mineralogical Magazine*, 77, 2107.

Sahami, A., Waworuntu, J. & Fawcett, S. (2011) Geochemical reactivity of submarine tailings from the Batu Hijau mine. *Mineralogical Magazine*, Vol. 75, pp. 1777.

Søndergaard, J., Asmund, G., Johansen, P. & Rigit, F. (2011) Long-term response of an arctic fiord system to lead-zinc mining and submarine disposal of mine waste (Maarmorilik, West Greenland). *Marine Environmental Research*, 71, 331–341.

Vogt, C. (2012) International assessment of marine and riverine disposal of mine tailings. In: *The 34th Meeting of the London Convention and the 7th Meeting of the London Protocol*, London, UK, 1 November, 1–50.

Walker, S.R., Parsons, M.B., Jamieson, H.E. & Lanzirotti, A. (2009) Arsenic mineralogy of near-surface tailings and soils: Influences on arsenic mobility and bio-accessibility in the Nova Scotia gold mining districts. *Canadian Mineralogist*, 47, 533–556.

Yilmaz, E. & Fall, M. (2017) *Paste Tailings Management*, Springer International Publishing, 303 p.

Contaminated land remediation

Land Reclamation in Ecological Fragile Areas – Hu (Ed.)
© 2017 Taylor & Francis Group, London, ISBN 978-1-138-05103-4

Treatment of metals and metalloids in neutral mine effluents using modified materials

I.L. Calugaru
Technology Center for Industrial Waste (Centre Technologique des Résidus Industriels—CTRI), Rouyn-Noranda, Québec, Canada

C.M. Neculita
Research Institute on Mines and Environment (RIME), University of Québec in Abitibi-Témiscamingue (UQAT), Rouyn-Noranda, Québec, Canada

T. Genty
Agnico Eagle Mines Limited, Rouyn-Noranda, Québec, Canada

G.J. Zagury
RIME, Department of Civil, Geological, and Mining Engineering, Polytechnique Montréal, Montréal, Québec, Canada

ABSTRACT: Neutral mine effluents can contain above regulated concentrations of hazardous metals and metalloids that could adversely impact the quality of receiving water bodies. To prevent, limit, and treat water pollution, extensive research has been conducted in the last two decades. Among the proposed options, the use of modified natural and residual materials seems effective and inexpensive. In the present review paper, the modified materials performance and metal removal mechanisms, their use in active and passive treatment, and their efficiency for the treatment of synthetic and real mine effluents are critically discussed. The influence factors on performance improvement following the modification are also emphasized. In summary, metal recovery and sorbent re-use could potentially diminish water treatment costs and entail responsible management of resources. Moreover, waste valorization eliminates disposal concerns while metals recovery eliminates leaching risks. However, further research is necessary to evaluate field scale performance and to complete costs evaluation.

1 INTRODUCTION

Mine effluents represent a particular category of contaminated waters due to their mechanism of generation, contamination level and flows, as well as environmental impacts that may persist hundreds to thousands of years after mine closure (Lee et al., 2005; Parviainen et al., 2012; Sapsford, 2013; Kučerová et al., 2014; Frau et al., 2015). Mine effluents can be acidic (AMD), neutral (CND), basic, dilute, mineralized or saline (Nordstrom et al., 2015). Neutral mine effluents (pH 6–8) are generated whenever one or more of the following conditions are present: (1) neutralization capacity of the mine waste or of the bedrock (related to carbonate mineral content) overcomes acid generation (associated with sulfide content) (Lindsay et al., 2009; Kučerová et al., 2014; Bright and Sandys, 2015; Lindsay et al., 2015); (2) presence of certain sulfides (e.g., sphalerite, pentlandite, chalcopyrite, galena), which dissolve without acid generation (Stantec, 2004; Mayes et al., 2009); (3) partially efficient methods applied to control acid drainage generation from mine waste (Stantec, 2004). The main contaminants in neutral mine effluents are metals (e.g., Cd, Cr, Co, Cu, Fe, Hg, Mn, Mo, Ni, U, Zn), metalloids (e.g., As, Sb, Se) as well as SO_4^{2-} (Stantec, 2004; Bright and Sandys, 2015; Lindsay et al., 2015). Contaminant concentrations, which are variable (from relatively high, to trace), are related to the mineralogy, geology, hydrology, and climatic conditions and events (temperature, rainfalls, and snowmelt) on the mine site (Frau et al., 2015; Haakensen et al., 2015; Nordstrom et al., 2015). Climatic conditions can cause round year variations in flow and charge of contaminants, whereas a long-term evolution of the quality of mining effluents is to be expected, with respect to mineralogy (Frau et al., 2015; Haakensen et al., 2015). However, not all contaminants (e.g. Fe, Mn, Cd, As) follow the general pattern of decrease in concentration with time (Frau et al., 2015; Genty et al., 2016). Metal and metalloids dissolved in water can present significant risks to humans and natural ecosystems because of their toxicity, persistence and accumulation potential (Naser, 2013). Thereupon, several types of disorders affecting humans are reported in literature (Table 1).

To prevent, limit, and treat water pollution by metal and metalloids originating from mine effluents, extensive research has been conducted in the last two decades. The remedial options are designed for confining the contaminants in the waste (Seidel et al., 2005; Bhattacharyya et al., 2009; Hwang and Neculita, 2013; Chen et al., 2016) or collecting and treating contaminated effluents. With regard to treatment, both active and passive options can be considered to treat the effluents of an active mine, prior to their discharge into environment, whereas in the case of an abandoned mine, passive treatment is preferable (Rakotonimaro et al., 2016; Skousen et al., 2016).

Table 1. Toxic effects of metals and metalloids.

Metal / Metalloid	Toxic effect	References
As	Skin lesions, cancer (lung, uterus, skin, bladder, liver, genito-urinary tract)	Chowdhury et al., 2000
Cd	Cancer (lung, kidney, prostate), kidney dysfunctions, skeletal disorders, bone fractures	Candeias et al., 2015
Cr	Nausea, diarrhea, liver and kidney damage, dermatitis, asthma, internal hemorrhaging, irritation and ulceration of the nasal septum, eye damage	Mohan et al., 2011
Cu	Vomiting, cramps, convulsions	Fu & Wang, 2011
Hg	Neuropsychological dysfunctions, gastrointestinal and genito-urinary troubles, frequent cough, pain (musculoskeletal, abdominal, head, teeth), skin lesions, blurring vision, easy fatigability	Cortes-Maramba et al., 2006
Mn	Manganism (weakness, anorexia, muscle pain, apathy, rigidity, postural difficulties, tremor, decreased mental status)	García-Mnedieta et al., 2009
Ni	Skin allergies, lung fibrosis, cancer of the respiratory tract, asthma, conjunctivitis, immunotoxicity, hepatic toxicity, nephrotoxicity	Cempel & Nikel, 2006
Pb	Damage to central nervous system, kidney, liver, reproductive system, brain; anemia, insomnia, weakness, hallucinations	Fu & Wang, 2011
Zn	Stomach cramps, skin irritation, vomiting, nausea, anemia	Fu & Wang, 2011

Treatment methods are based on one or several of the following processes: chemical precipitation and coprecipitation, coagulation-flocculation, flotation, membrane filtration, solvent extraction, electrochemical treatment, ion exchange, complexation/sequestration, evaporation, distillation, crystallization, oxidation, reduction (chemical, enzymatic, microbial, bacterial, and photo reduction), phytoremediation, and adsorption (Peters et al., 1985; O'Connell et al., 2008; Sandy and DiSante, 2010; Haakensen et al., 2015; Skousen et al., 2016). Conventional water treatment technologies drawbacks include high initial capital, operational and/or maintenance costs, waste management issues (e.g. quantities, physical and chemical stability of generated sludge) and chemical products consumption (transport, storage, cost and safety concerns) (O'Connell et al., 2008; Lim & Aris, 2014). In contrast, locally easily available materials (naturals or residuals) might suppress some of the above mentioned cost and safety related limits. Moreover, the modification of such materials facilitate the reutilization and stabilization of residues (Miyake et al., 2002; Vadapalli et al., 2010), offers flexibility in terms of treated pollutants (type, concentration, mixed solutions) and flow and charge variations, and allows metal recuperation (O'Connell et al., 2008; Wang et al., 2009) and sorbent regeneration (O'Connell et al., 2008; Wang et al., 2009) etc.

2 TREATMENT PROCEDURES AND EFFICIENCY OF MODIFIED MATERIALS

Most natural and residual materials can be modified by means of relatively simple procedures to effectively treat toxic metals and metalloids (Wan Ngah & Hanafiah, 2008; Lim & Aris, 2014). For instance, charring can be successfully performed on dolomite, sludge and sea shells (Table 2), at temperatures between 600 to 900°C, and durations between 1 to 8 h, under atmospheric conditions. Charring modifies chemical and mineralogical composition, enhances specific surface and porosity, and increases basicity compared to the original material (Jha et al. 2006; Alidoust et al., 2015; Salameh et al., 2015; Calugaru et al., 2016).

Table 2. Type of treatment procedures for materials modification and metal/metalloid treated.

Material	Treatment	Metal/Metalloid	References
Dolomite	Charring	As Co Mn Ni, Zn	Salameh et al., 2015 Ivanets et al., 2014 Mamchenko et al., 2008; Calugaru et al., 2016
Goethite	Alginate coating	As	Lee et al., 2015
Perlite	Chitosan coating	Co, Cu, Ni	Swayampakula et al., 2009
Sand	Iron oxide coating Aluminum oxide coating	As Se	Thirunavukkarasu et al., 2003, Kuan et al., 1998
Peat	Fe(III) grafting	As	Ansone et al., 2013
Moss	Fe(III) grafting	As	Ansone et al., 2013
Grass	Hydrothermal pyrolysis	Cd, Cu	Regmi et al., 2012
Ash	Urea grafting Alkaline fusion and hydrothermal treatment Hydrothermal treatment	Cu, Hg Co, Cu, Ni Ni, Zn Cu, Zn Cu, Mn, Pb, Zn Cu, Ni	Orlando et al., 2002 Mishra and Tiwari, 2006 Calugaru et al., 2017 Wang et al., 2009 Nascimento et al., 2009 Shah et al., 2013
Sludge	Pelleting Pyrolysis Charring Hydrothermal treatment Alginate coating	Zn As, Cr, Pb Ni Al, Cu, Fe, Mn, Ni, Zn As	Mayes et al., 2009 Zhang et al., 2015 Jha et al., 2006 Vadapalli et al., 2010 Lee et al., 2015
Wood residues	Activation with HCl Pyrolysis Pyrolysis Acrylic acid grafting Amidoxime grafting Sulfonic acid grafting	Cr, Cu, Ni, Cr Cu, Zn Cd, Cu, Ni Cd, Cr, Cu, Ni Cd	Argun et al., 2007 Mohan et al., 2011 Chen et al., 2011 Geay et al., 2000 Saliba et al., 2005 Shin & Rowell et al., 2005
Agriculture residues	Pyrolysis Pyrolysis Pyrolysis Pyrolysis Fe(III) grafting	Cr Cu, Zn Cd, Cu, Ni, Pb Cd, Cu, Pb, Zn As	Dong et al., 2011 Chen et al., 2011 Inyang et al., 2012 Kołodyńska et al., 2012 Ansone et al., 2013
Textile industry waste	Acrylamide grafting Xanthate grafting	Hg Cd, Cu, Pb	Biçak et al., 1999 Bediako et al., 2015
Food industry waste	Fe(III) grafting Fe(III) / Zr (IV) grafting Activation with H_2SO_4 Activation with H_2SO_4 Pyrolysis	As Sb Se Cd, Cu, Ni, Zn Cu	Murugesan et al., 2006 Biswas et al., 1999 El-Shafey, 2007 Kumar et al., 2012 Baccar et al., 2009
Seafood waste	Charring Activation with H_2SO_4 Activation with NaOH	Cd Cu Cr	Alidoust et al., 2015 Liu et al., 2009 Dima et al., 2015
Biochar	Activation with H_2O_2 Activation with KOH Fe(III) grafting	Pb Cd, Cu As, Cr	Xue et al., 2012; Regmi et al., 2012; Agrafioti et al., 2014

In the case of dolomite [$CaMg(CO_3)_2$], a thermal decomposition in two steps, shown by the following equations, can be conducted:

$$CaMg(CO_3)_2 \rightarrow CaCO_3 + MgO + CO_2 \quad (1)$$

$$CaCO_3 \rightarrow CaO + CO_2 \quad (2)$$

If the decomposition process is stopped after the first step (~800°C), a solid rigid product consisting of calcite ($CaCO_3$) and periclase (MgO) is obtained. The second step of thermal decomposition of dolomite (≈900°C) adds CaO to the MgO generated within the first step, and the resulting product is even more alkaline (paste pH 12.9) compared to the one obtained after the first charring step (paste pH 11.4, Ivanets et al., 2014). One step charring of dolomite was reported to double its removal capacity for Ni (from 2 to 4.7 mg/L), to triple its removal capacity for the As (V) (from 0.65 to 2.16 mg/L), and to enhance about sevenfold its removal capacity for Mn and Zn (from 0.23

to 1.79 mg/L and from 1.7 to 12.2 mg/L respectively) (Mamchenko et al., 2008; Salameh et al., 2015; Calugaru et al., 2016). In addition, complete charring of dolomite was reported to enhance Co removal from 2.84 to 520 mg/L (Ivanets et al., 2014). Calcination of oyster shells (2h, 900°C) enhanced their sorption capacity for Cd from 29.5 mg/g up to 1.7 g/g, due to complete decomposition of $CaCO_3$ into CaO (Alidoust et al., 2015).

Chemical activation with acids (HCl, H_2SO_4), bases (NaOH, KOH) or oxidants (H_2O_2) performed on ash, wood residues, biochar, food industry waste and seafood waste frees/generates chemical functions effective in metals and metalloids withdrawal, enhances Cation Exchange Capacity (CEC) and specific surface (available for binding contaminants). For instance, peanut hull biochar oxidation with H_2O_2 increased the number of its surface oxygen-containing functional groups (particularly carboxyl group). This resulted in 25 time increase of the sorption capacity for Pb, from 0.88 up to 22.82 mg/g (Xue et al., 2012). Moreover, grafting of functional groups on the backbone of natural polymers, such as cellulose, originates from wood residues or textile industry wastes, enhances metals and metalloids removal capacity, in addition to stabilize the biopolymer (Biçak et al., 1999; O'Connell et al., 2008). For instance, acrylic acid grafting on sawdust (*Picea abies* mixed with *Fagus sylvatica*) increased 50 fold its removal capacity for Cu (from 2 to 104 mg/g), 34 fold its removal capacity for Cd (from 4.9 to 168 mg/g), and 36 fold its removal capacity for Ni (from 2.7 to 97.3 mg/g) (Geay et al., 2000). Metals [e.g. Fe (III) and Zr (IV)] can also be grafted in order to treat anionic species of metals and metalloids. For instance, Fe (III) was grafted on peat, moss, agriculture residues (straw, cane), food industry waste (orange waste, waste tea fungal biomass), or biochar, mainly to treat As (15 mg/g) but also Sb (up to 145 mg/g) and Cr (0.75 mg/g) (Table 2).

Activated carbon can be prepared by pyrolysis of wood residues, agriculture residues and food industry waste (Baccar et al., 2009; Yahya et al., 2015), whereas biochar can be prepared by pyrolysis of municipal solid waste, waste tires, plastic, bones, newspaper scrap, sludge, grass, compost, wood residues, agriculture residues, food waste and food industry waste (Mohan et al., 2014; Inyang et al., 2016). Activated carbon is often seen as a universal adsorbent in water treatment (Mohan et al., 2014), being widely used for the removal of metals and metalloids due to its high surface area and large pore volumes (Fu and Wang, 2011). Initially, it was produced from coal. However, depletion of the raw material increased its price and, therefore, conversion of diverse low-cost carbonaceous precursors into activated carbon was successfully investigated and performed (Baccar et al., 2009; Fu and Wang, 2011). Even though the activated carbon is still the most common sorbent for removing contaminants from water, it is evaluated as more costly to produce than the biochar (Mohan et al., 2014).

Coating of a rigid and inert mineral support (e.g. goethite, perlite, sand, sludge) can be performed in order to enhance the mechanical stability and the availability of the active surface, and to improve hydraulic properties of reactive materials (e.g. chitosan, alginate, iron oxide, aluminum oxide) (Table 2).

Originated from coal, municipal waste and agricultural waste combustion, ash is extensively treated by hydrothermal treatment (NaOH solution, 100–150°C, up to 72 h) or by alkaline fusion (solid NaOH or KOH, 600–750°C, 1 to 2 h) followed by hydrothermal treatment (100–150°C, 6 to 24 h). The resulting materials differ from the original ash based on chemical and mineralogical composition. Moreover, specific surface area and porosity are enhanced, basicity and CEC are increased, as well as the stability of the modified material and of the sludge resulting from wastewater treatment (Miyake et al., 2002; Vadapalli et al., 2010). Satisfactory results in metal remediation from wastewaters were achieved for Cd (185 mg/g; Izidoro et al., 2013), Co (110 mg/g; Mishra and Tiwari, 2006), Cu (148 mg/g; Mishra and Tiwari, 2006), Mn (60 mg/g; Nascimento et al., 2009), Ni (92 mg/g; Mishra and Tiwari, 2006), Pb (549 mg/g; Jha et al., 2008), Zn (220 mg/g; Izidoro et al., 2013).

The uptake kinetic exhibited by a particular sorbent depends on the contaminant and its concentration. Charred dolomite adsorbed 98%–99% of the Co (initial concentration of 100 mg/L) within 15–20 minutes (Ivanets et al., 2014). On the other hand, a solution containing 0.05 mg/L As (III) required 24 h, and 48 h for the same concentration of As (V), to reach equilibrium. Higher initial concentrations of 0.5, 1 and 2 mg/L As (III) and As (V) required 120 h to reach equilibrium (Salameh et al., 2015). Oyster shells calcined (2 h at 900°C) exhibited a Cd removal capacity of about 1.7 mg/g, which was acquired mostly in the first hour of the kinetics batch testing (Alidoust et al., 2015). Thermally activated paper sludge ash acquired about 70% of the sorption capacity of 266 mg/g Ni, in the first hour of the kinetics batch testing (Jha et al., 2006).

Numerous materials grafted to complex metals and metalloid exhibit rapid kinetics. Cellulose grafted with acrylamide (Biçak et al., 1999) attained the equilibrium state within 8 minutes, in contact with a 50 mg/L Hg contaminated solution.

However, 24 h of reaction time to reach equilibrium were also reported. Although, this could be associated with the high concentration of contaminant in solution, of 1120 mg/L Cr, 580 mg/L Ni, 630 mg/L Cu or 520 mg/L Cr (Saliba et al., 2005).

Biochars can reach sorption equilibrium instantaneously in aqueous solutions containing 63–190 mg/L Cu, whereas 30 minutes to 2h were necessary in the case of Pb at concentration of 207 mg/L, and 414–621 mg/L, respectively (Kolodynska et al. 2012). The uptake of Se by aluminium-oxide-coated sand can also be considered rapid, as equilibrium state has been reached in less than 60 minutes (Kuan et al., 1998).

A zeolite prepared from coal fly ash by alkaline fusion followed by hydrothermal treatment showed a rapid and quantitatively predominant sorption (within 30 minutes) of Ni, Cd, Cu (500 mg/L) and Pb (1000 mg/L), followed by a slower, quantitatively insignificant step, during which the equilibrium was attained. This was related to the abundance of active sites available in the beginning of the kinetics testing, whereas their progressive occupation led to a slower sorption and therefore lesser removal efficiency (Jha et al., 2008). Bagasse fly ash zeolite removed at least 70% of Ni and Cu within the first 30 minutes, whereas it took 5 to 6 h to reach equilibrium (Shah et al., 2013). Nonetheless, synthetic cancrinite showed faster kinetics for Pb (60 min needed to reach the equilibrium) whereas for Co, Cu, Ni and Zn solutions, 3 h were necessary to reach the equilibrium state (Qiu and Zheng, 2009).

A higher ratio sorbent: solution might reduce the time needed to reach equilibrium condition. For instance, to remove 10 mg/g Pb from an initial solution of 113 mg/L, a ratio of 10 g biochar/L solution required 20 h equilibrium time whereas doubling the ratio, the previous equilibrium time was decreased to its half (Mohan et al., 2007).

For various sorbents and contaminants concentrations, 24 h and 48 h contact time needed to reach equilibrium were also reported. However, in many cases, 6 h to 12 h were sufficient to achieve 90% removal efficiency (Zhang et al., 2008; Sasaki et al., 2014).

In summary, the modification of natural and residual materials substantially enhances their efficiency in metal and metalloid removal. Raw material can be subject to several types of treatment, whereas new (modified) material was found effective for the treatment of several metallic contaminants.

3 INFLUENCE FACTORS ON EFFICIENCY IMPROVEMENT AND METAL/METALLOID REMOVAL MECHANISM

Commonly, the following factors support the acquired performance: enhanced specific surface, grain-size distribution and porosity, surface charge, surface functionalization, oxidation-reduction potential, neutralizing capacity and alkalinity generation (Table 3).

The mechanism of metal removal by modified products is usually complex and can involve one or several of the following: ion exchange, chemisorption by precipitation, chemisorption by complexation, oxydo-reduction, precipitation, co-precipitation, physical adsorption (Lu et al., 2012; Wang and Chen, 2014). Moreover, the mechanism might differ with reference to the same material, but a different contaminant (Robinson-Lora and Brennan, 2011), or if the speciation of the contaminant changes (Lee et al., 2015).

Metalloids are presents mostly in theirs anionic form in neutral mine effluents, under oxic conditions (Atlas of Eh- pH diagrams, 2005). Their removal by physical adsorption, involving the electrostatic attraction between anionic species of the contaminant and protonated sorption sites, is greatly sensible to pH and even to ionic strength variations. The pH affects the speciation of the contaminant and the percentage of protonated adsorption sites, whereas the ionic strength affects the interfacial potential, the activity of the ions, and the adsorption process, entailing a competition with the electrolyte ions for the available sorption sites (Niu et al., 2007).

Metals in superior states of oxidation are present in neutral mine effluents under oxic conditions as anions (Cr, Mn, Mo, V), neutral species (Mn, V), and cationic species (V), whereas metals in inferior states of oxidation are present only as hydrated cations or hydrolyzed cationic species (Atlas of Eh- pH diagrams, 2005). They compete with protons at lower pH, whereas higher pH is more favorable. However, pH increase may also entail their precipitation.

Ion exchange is recognized as the main mechanism for the uptake of metallic cations by zeolites (Jha et al., 2008; Qiu and Zheng, 2009), the exchangeable cations being mainly Na^+ and K^+. Adsorption also contributes to metal removal process by surface reactions of metal cations with terminal hydroxyl groups on zeolite and by electrostatic attraction between positive metallic species and negative charge on zeolite surface (Hui et al., 2005; Qiu and Zheng, 2009). The release of exchanged cations (Na^+, K^+, Ca^{2+} or Mg^{2+}) increases the pH in the aqueous solution, and therefore, promoting the dissociation of Si-OH and Al-OH zeolite groups, which increase the number of anionic surface sites of adsorption (Hui et al., 2005). The pH increase also entails precipitation of metals as oxides and hydroxides (Hui et al., 2005; Jha et al., 2008).

Ion exchange was found as dominating mechanism for removal of Pb and Cd by wood and bark originated biochars at initial pH of 5 (Mohan et al., 2007). Precipitation as carbonates and phosphates was found as main mechanism for Pb, Cu, Zn, and Cd removal by manure originate biochar (Xu

Table 3. Type of treatment procedures and resulting modifications.

Treatment	Material	Modification/improvement
Charring	Dolomite, oyster shells	Chemical and mineralogical composition modification, enhanced specific surface and porosity, increased neutralizing and alkalinity generation capacity
	Sludge	Chemical and mineralogical composition modification, enhanced specific surface
Pyrolysis	Sludge, wood and agriculture residues, grass, food industry waste	Chemical composition modification, enhanced specific surface and porosity, increased number of oxygen-containing and aromatic functional groups
Alkaline fusion/hydrothermal treatment	Ash, sludge	Chemical and mineralogical composition modification, enhanced specific surface and porosity, increased basicity and CEC
Chemical activation with oxidants, acids or bases	Wood residues, biochar, food industry waste	Chemical composition modification, increased number of active functional groups (e.g. oxygen-containing groups: carboxyl, carbonyl, hydroxyl; amino), increased CEC, favorable surface shape; increased neutralizing and alkalinity generation capacity
Grafting of metals or organic functions	Peat, moss, ash, wood and agriculture residues, biochar, textile industry and food industry waste	Chemical composition modification, clustering of functional groups [e.g. carboxyl, carbonyl, amino, mercapto, hydroxyl, amidoxime, sulfonic, xanthate, acrylamide, Fe(III), Zr(IV)]
Coating	Sludge, goethite, perlite, sand	Enhanced mechanical stability and availability of the active surface, improved hydraulic properties
Pelleting	Sludge	Enhanced surface area and porosity, mechanical stability, improved hydraulic properties

et al., 2013), whereas the same metals were removed mainly by complexation with phenolic –OH by rice husk originate biochar (Xu et al., 2013). Obviously, chemical composition of biochar determines the mechanisms of metal removal. Biochars remove Cu, Zn, Cd, and Pb by forming inner-sphere complexes and surface complexes involving carboxyl and hydroxyl functional groups, mineral oxides and free hydroxyl of mineral oxides, and by surface precipitation and co-precipitation (Lu et al., 2012; Kołodyńska et al., 2012). Metal removal is enhanced in neutral solutions as the competition of protons for binding sites decreases, metal ions hydrolyze, and precipitation occurs (Kołodyńska et al., 2012).

Metals can be removed from aqueous solution by mean of chemical bonds formed with different ligands (electrons donors) positioned at the sorbent surface. Some sorbents can be very selective for a specific metal, i.e. polyacrylamide grafted onto cellulose was reported selective for Hg (Biçak et al., 1999). Lignocellulosic biomass, hydrolyzed to enhance the number of sites for cations adsorption, combine complexation (by mean of phenolic and/or carboxylic sites) and ion exchange (accompanied by H_3O^+ release and pH decrease) to recover metals from aqueous solutions. Moreover, chemical functions, as carboxyl groups (–COOH) grafted on the biomass sorbent, further enhance the number of adsorption sites. Nevertheless, the mechanism of metal uptake remains a combination of complexation and ion exchange (Geay et al., 2000), or complexation solely (Saliba et al., 2005).

Iron-based sorbents remove arsenic species by formation of Fe-O-As bonds (Ansone et al., 2013). At pH 6–8, the As (III) and As (V) species are: $HAsO_2$, $As(OH)_3$, $H_2\,AsO_4^-$ and $HAsO_4^{2-}$ (Atlas of Eh- pH diagrams, 2005). The oxidation of the As (III) species in the presence of loaded Fe (III) was already reported, therefore only the anionic species of As (V) were sorbed (Mondal et al., 2007). AMD sludge alginate beads were effective to remove As (III) in the pH interval 2 to 10 and the As (V) in the interval 2 to 9. However, As (III) was oxidized to As (V) by MnO_2 and MnOOH originated from the AMD sludge, which enhanced its sorption (Lee et al., 2015). Sorption of Se (IV) by sulfuric acid treated peanut shells is increased at low pH values (1.5). Se (IV) is adsorbed on the surface and further reduced to elemental Se, process that requires protons (El-Shafey, 2007). A 3-steps mechanism of Cr (VI) removal by sugar beet tailing biochar was suggested by Dong et al. (2011). This mechanism involves: (1) electrostatic attraction of $HCrO_4^-$ (pH 1 to 6) to positively charged biochar's surface; (2) reduction of Cr (VI), due to its elevated redox potential (+1.3V) to Cr (III) in the presence of organic matter, under acidic conditions; (3) complexation of Cr (III) by carboxyl and hydroxyl functions of biochar. However, reduction is not always involved in the removal mechanism of highly oxidized metals as in the study of Dima et al. (2015), Cr (VI) was adsorbed by chitosan particles without previous reduction.

When bivalve mollusk shells are acid treated, they withdraw metals (Cu, Fe, Zn, Cd) by ion exchange with Ca^{2+} and by adsorption employing functions of negative charge (–OH, –NH, C = O and S = O). The main mechanism should be chemisorption by complexation, as the regeneration of the sorbent requires a solution of EDTA (Liu et al., 2009). Basic and strongly basic sorbents, such as calcined dolomite and calcined bivalve mollusk shells, increase the pH of the contaminated effluent, and entail precipitation of metals in the form of carbonates, oxides, and hydroxides (Ivanets et al., 2014; Alidoust et al., 2015). In addition to precipitation, ion exchange with Ca^{2+} enhances the metal uptake (Alidoust et al., 2015).

The effectiveness of a specific modified material towards several metallic contaminants is related to the multiple mechanisms available for their removal, as well as an improved accessibility and reactivity of the active surface.

4 UTILIZATION AND COST OPPORTUNITIES

Most published studies dealing with modification of natural and residual materials and the evaluation of their performance for the removal of metals in contaminated waters only involve synthetic effluents. The findings usually show better efficiency of modified materials. However, the modified materials showed in some cases comparable performance upon testing with real effluents. Nonetheless, in the case of a real effluent, all the soluble species in presence should be quantitatively determined prior to the treatment trial. For instance, synthetic zeolite NaP1 (Álvarez-Ayuso et al., 2003) showed almost similar sorption capacities towards Zn, Cr and Ni, in synthetic solution versus real electroplating wastewater. However, in the case of a cyanide zinc wastewater (730 mg/L of CN^-), the zeolite (cation exchanger) was not effective for Zn removal due to complexation of Zn cation in the anionic form $Zn(CN)_4^{2-}$. Chromium and bronze electroplating wastewater had to be chemically treated for cyanide destruction, prior to metal removal by activated alumina (Mazumder et al., 2011). Acid-pretreated bivalve mollusk shells were found effective (Liu et al., 2009) for the treatment of an electroplating wastewater as 99.98% Fe, 99.4% Zn and 92% Cu were removed from a sample containing 233 mg/L Fe, 75 mg/L Zn and 9.4 mg/L Cu. The CND (pH 6.2) was successfully treated with synthetic zeolite-P (Vadapalli et al., 2010). The zeolite was prepared from solid residues resulting from the treatment of real AMD with coal fly ash. Hydrous ferric oxide pellets were prepared from the waste resulting from coal mine water treatment, to treat Zn from DNC (pH 7.8). The designed pilot-scale treatment system was found substantially more effective for Zn removal, than several alternative passive units (e.g. pilot-scale aerobic wetland, pilot-scale anoxic limestone drain, subsurface flow wetland, etc.) (Mayes et al., 2009). A ground water sample collected from Ram Nagar, Kolkata (West Bengal, India) was treated with tea fungal biomass impregnated of $FeCl_3$. The As (III) was completely removed, from an initial concentration of 1.3 mg/L, whereas 77% of the initial 0.9 mg/L As (V), was treated (Murugesan et al., 2006).

Being indispensable prior to designing a full scale treatment system, laboratory batch testing should be followed by pilot testing. While many published studies stop after batch testing, several evaluations continued with minicolumns testing, using fly ash zeolites (Shah et al., 2013; Belviso et al., 2014), thermally activated dolomite (Mamchenko et al., 2008; Ivanets et al., 2014), acid hydrolyzed bark (Gaballah and Kilbertus, 1998), polyacrylic acid grafted sawdust (Geay et al., 2000), chitosan-coated activated alumina (Boddu et al., 2008), H_2O_2-treated biochar (Xue et al., 2012), and ferric activated carbon (Zhang et al., 2007). Shaken batch reservoirs or percolation through columns were also employed at semi-pilot plant scale (500 L) to evaluate modified bark (Gaballah and Kilbertus, 1998), while pilot plant scale trials (1000 L) were successfully conducted with modified bark (Gaballah and Kilbertus, 1998), or pelletised hydrous ferric oxide recovered from coal mine water treatment (Mayes et al., 2009).

However, in the case of a passive treatment system, variable flow and charge of contaminants engendered by climatic conditions (temperature, rain, and snowmelt) might be challenging for its consistent performance (Mayes et al., 2009). Despite this, treatment systems such as permeable reactive barriers, inorganic media passive systems, passive biochemical reactors, treatment tanks, are promising for the passive treatment of contaminated mine effluents, and for in situ groundwater remediation, allowing at the same time the re-use of waste materials (Mayes et al., 2009; Warrender et al., 2011; Song et al., 2012). The United States Environmental Protection Agency (USEPA) includes biochemical reactors and Permeable Reactive Barriers (PRB) in its «Reference guide to treatment technologies for mining-influenced water», of 2014. Zeolites and activated carbon are mentioned among reactive materials used in PRBs. Several contaminants can be treated, i.e. Cd, Co, Cr, Cu, Fe, Mn, Ni, Pb, Se, Tc, U, Zn, and a period of 5 to 10 years is indicated for operation with no or extremely low cost as routine compliance and monitoring.

Although thermal and chemical modification might seem by far the most expensive side of using such processed natural and residual materials, transport might also have an important part within the total cost of water treatment (Hengen et al., 2014). As a consequence, sorbent regeneration and reuse, metals recuperation, resulting process water and sludge reuse, are opportunities to reduce treatment cost.

Since metals and metalloids removal mechanisms influence the stability/lixiviation of contaminants, they will determine the possibility of sorbent regeneration and reuse, and metal recuperation.

Elution allows the recovery of contaminants in concentrated solutions as well as the reuse of sorbents. Depending on the metal removal mechanism, an acid, a saline, a basic or a chelating agent solution needs to be employed. Spent activated bark (Gaballah and Kilbertus, 1998), graft copolymer of acrylamide with cellulose cotton (Biçak et al., 1999), cellulose xanthate (Bediako et al., 2015), sawdust bearing polyacrylic chains (Geay et al., 2000), Fe (III) and Zr (IV)-loaded orange waste (Biswas et al., 2009), require acidic solutions to desorb metals and metalloids. Several sorbents acting mainly through complex formation with metals might require chelating agents for regeneration (Saliba et al., 2005; Liu et al., 2009). Recovery and sorbent regeneration can be achieved with strong acid solutions (Aryal et al., 2010) or strong alkaline solutions (Zhang et al., 2007; Boddu et al., 2008). Metal recovery and synthetic zeolite regeneration are generally performed with saline solutions (5% NaCl), acidic solutions (0.1M HCl, 0.1M HNO_3) or chelating agent solutions (0.1M EDTA) (Wang et al., 2009; Shah et al., 2013). In some cases, solely the contaminant can be recovered, by acidic rinse or incineration of the spent sorbent, as rinsing might solubilize the chemical functions involved in the uptake of the contaminant (Orlando et al., 2002). For instance, a basic solution (0.1–0.5 N NaOH) was employed by Murugesan et al. (2006) to desorb the As from the saturated sorbent (Fe-modified waste tea fungal biomass) which became then, easily biodegradable. Adsorbed metals can be also recovered by burning the spent biomass employed as sorbent and further processing the resulting ash, rich in metallic oxides (Gaballah and Kilbertus, 1998; Geay et al., 2000; Argun et al., 2007).

For inorganic sorbents that cannot be (suitably) regenerated, recovery and reuse of sludge lower water treatment cost (up to 25%) and reduce sludge management issues. Several metals (i.e. Cu, Fe, Ni, Zn) and compounds (e.g. iron oxides, barium hydroxide, magnesium hydroxide, calcium carbonate) can be recovered for further applications. Sludge was also found effective in wastewater treatment (removal of P, As, dye, AMD and CND treatment), soil remediation, agriculture and horticulture, construction materials, mine drainage prevention, and CO_2 sequestration (Rakotonimaro et al., 2017).

Further studies should focus on real effluents, field scale trials, metal (or sludge) and water recovery, and sorbent regeneration/reuse.

5 CONCLUSION AND RESEARCH NEEDS

Natural and residual materials can be transformed by mean of simple procedures into effective sorbents for metals and metalloids, with promising results for the treatment of contaminated waters. In the context of raw materials depletion and with the aim to decrease the environmental footprint of the society, both raw materials and transport issues should be reconsidered. Metal recovery and sorbent re-use decrease water treatment costs. Moreover, waste valorization diminishes disposal concerns and metals recovery eliminates the risks associated with leaching, in addition to insuring a responsible management of resources.

Unfortunately, many studies available in the literature are solely conducted at the laboratory scale. They focus on parameters optimization to enhance sorbent's yield, quality and performances, and metal and metalloid removal performance are investigated in batch testing. In many cases, column tests do not follow the batch testing while pilot scale production and testing of the new sorbents are scarce. Metal recovery, and sorbent and treated water reuse, are not approached in most of the studies, although they should be part of the process. Moreover, synthetic solutions are used in many studies to evaluate the performance of the developed sorbents.

Since most published studies were performed at a laboratory scale, further research is needed to transfer to treatment technologies, at the field scale. New (sealable) products could be developed based on both sorbent and recovered metal/metalloid. The results of cost evaluation together with the life cycle analysis should supply powerful arguments in favor of the new (emergent) treatment technologies. Locally available materials (natural and residual) still need to be evaluated for the treatment of specific wastewater, like neutral mine effluents. However, speciation of contaminants, ionic competition and strength, and possible mechanisms of metals removal should guide any attempt of material modification and evaluation of treatment performances. Finally, metal/sludge recovery, sorbent and water reuse should be part of the treatment technology.

REFERENCES

Agrafioti, E., Kalderis, D. & Diamadopoulos, E. (2014) Ca and Fe modified biochars as adsorbents of arsenic and chromium in aqueous solutions. *Journal of Environmental Management*, 146, 444–450.

Alidoust, D., Kawahigashi, M., Yoshizawa, S., Sumida, H. & Watanabe, M. (2015) Mechanism of cadmium biosorption from aqueous solutions using calcined oyster shells. *Journal of Environmental Management*, 150, 103–110.

Álvarez-Ayuso, E., García-Sánchez, A. & Querol, X. (2003) Purification of metal electroplating waste waters using zeolites. *Water Research*, 37 (20), 4855–4862.

Ansone, L., Klavins, M. & Viksna, A. (2013) Arsenic removal using natural biomaterial-based sorbents. *Environmental Geochemistry and Health*, 35 (5), 633–642.

Argun, M.E., Dursun, S., Ozdemir, C. & Karatas, M. (2007) Heavy metal adsorption by modified oak sawdust: Thermodinamics and Kinetics. *Journal of Hazardous Materials*, 141 (1), 77–85.

Aryal, M., Ziagova, M. & Liakopoulou-Kyriakides, M. (2010) Study on arsenic biosorption using Fe(III)-treated biomass of *Staphylococcus xylosus*. *Chemical Engineering Journal*, 162 (1), 178–185.

Atlas of Eh- pH diagrams (2005) National Institute of Advanced Industrial Science and Technology.

Baccar, R., Bouzid, J., Feki, M. & Montiel, A. (2009) Preparation of activated carbon from Tunisian olive-waste cakes and its application for adsorption of heavy metal ions. *Journal of Hazardous Materials*, 162 (2–3), 1522–1529.

Bediako, J.K., Wei, W., Kim, S. & Yun, Y.-S. (2015) Removal of heavy metals from aqueous phases using chemically modified waste Lyocell fiber. *Journal of Hazardous Materials*, 299, 550–561.

Belviso, C., Cavalcante, F., Di Gennaro, S., Lettino, A., Palma, A., Ragone, P. & Fiore, S. (2014) Removal of Mn from aqueous solution using fly ash and its hydrothermal synthetic zeolite. *Journal of Environmental Management*, 137, 16–22.

Bhattacharyya, S., Donahoe, R.J. & Patel, D. (2009) Experimental study of chemical treatment of coal fly ash to reduce the mobility of priority trace elements. *Fuel*, 88 (7), 1173–1184.

Biçak, N., Sherrington, D.C. & Senkal, B.F. (1999) Graft copolymer of acrylamide onto cellulose as mercury selective sorbent. *Reactive & Functional Polymers*, 41 (1–3), 69–76.

Biswas, B.K., Inoue, J.-I., Kawakita, H., Ohto, K. & Inoue, K. (2009) Effective removal and recovery of antimony using metal-leaded saponified orange waste. *Journal of Hazardous Materials*, 172 (2–3), 721–728.

Boddu, V.M., Abburi, K., Talbott, J.L., Smith, E.D. & Haasch, R. (2008) Removal of arsenic (III) and arsenic (V) from aqueous medium using chitosan-coated biosorbent. *Water Research*, 42 (3), 633–642.

Bright, D.A. & Sandys, N. (2015) Beyond ML/ARD: the many faces of neutral mine drainage in the context of mine closure. In: Fourie, A.B., Tibbett, M., Sawatsky, L. and van Zyl D. (eds) *Mine Closure 2015, 1–3 June 2015, Vancouver, Canada*, 10p.

Calugaru, I.L., Neculita, C.M., Genty, T., Bussière, B. & Potvin, R. (2016) Performance of thermally activated dolomite for the treatment of Ni and Zn in contaminated neutral drainage. *Journal of Hazardous Materials*, 310, 48–55.

Calugaru, I.L., Neculita, C.M., Genty, T., Bussière, B. & Potvin, R. (2017) Removal of Ni and Zn in contaminated neutral drainage by raw and modified wood ash. *Journal of Environmental Science and Health Part A – Toxic / Hazardous Substances & Environmental Engineering*, 52 (2), 117–126.

Candeias, C., Ávila, P.F., Da Silva, E.F., Ferreira, A., Durães, N. & Teixeira, J.P. (2015) Water-rock interaction and geochemical processes in surface waters influenced by tailings impoundments: impact and threats to the ecosystems and human health in rural communities (Panasqueira Mine, Central Portugal). *Water, Air, and Soil Pollution*, 226:23.

Cempel, M. & Nikel, G. (2006) Nickel: A review of its sources and environmental toxicity. *Polish Journal of Environmental Studies*, 15 (3), 375–382.

Chen, X., Chen, G., Chen, L., Chen, Y., Lehmann, J., McBride, M.B. & Hay, A.G. (2011) Adsorption of copper and zinc by biochars produced from pyrolysis of hardwood and corn straw in aqueous solution. *Bioresource Technology*, 102 (19), 8877–8884.

Chen, Z., Pan, X., Chen, H., Guan, X. & Lin, Z. (2016) Biomineralization of Pb(II) into Pb-hydroxyapatite induced by *Bacillus cereus* 12-2 isolated from Lead-Zinc mine tailings. *Journal of Hazardous Materials*, 301, 531–537.

Chowdhury, U.K., Biswas, B.K., Chowdhury, T.R., Samanta, G., Mandal, B.K., Basu, G.C, Chanda, C.R., Lodh, D., Saha, K.C., Mukherjee, S.K., Roy, S., Kabir, S., Quamruzzaman, Q. & Chakraborti, D. (2000) Groundwater arsenic contamination in Bangladesh and West Bengal, India. *Environmental Health Perspectives*, 108 (5), 393–397.

Cortes-Maramba, N., Reyes, J.P., Francisco-Rivera, A.T., Akagi, H., Sunio, R. & Panganiban, L.C. (2006) Health and environmental assessment of mercury exposure in a gold mining community in Western Mindanao Philippines. *Journal of Environmental Management*, 81 (2), 126–134.

Dima, J.B., Sequeiros, C. & Zaritzky, N.E. (2015) Hexavalent chromium removal in contaminated water using reticulated chitosan micro/nanoparticles from seafood processing wastes. *Chemosphere*, 141, 100–111.

Dong, X., Ma, L.Q. & Li, Y. (2011) Characteristics and mechanisms of hexavalent chromium removal by biochar from sugar beet tailing. *Journal of Hazardous Materials*, 190 (1–3), 909–915.

El-Shafey, E.I. (2007) Removal of Se(IV) from aqueous solution using sulfuric acid-treated peanut shell. *Journal of Environmental Management,* 84 (4), 620–627.

Frau, F., Medas, D., Da Pelo, S., Wanty, R.B. & Cidu, R. (2015) Environmental effects on the aquatic system and metal discharge to the Mediterranean Sea from a near-neutral zinc-ferrous sulfate mine drainage. *Water, Air, and Soil Pollution,* 226:55.

Fu, F. & Wang, Q. (2011) Removal of heavy metal ions from wastewaters: A review. *Journal of Environmental Management,* 92 (3), 407–418.

Gaballah, I. & Kilbertus, G. (1998) Recovery of heavy metal ions through decontamination of synthetic solutions and industrial effluents using modified barks. *Journal of Geochemical Exploration*, 62 (1–3), 241–286.

García-Mendieta, A., Solache-Ríos, M. & Olguín, M.T. (2009) Evaluation of the sorption properties of a Mexican clinoptilolite-rich tuff for iron, manganese and iron-manganese systems. *Microporous and Mesoporous Materials*, 118 (1–3), 489–495.

Geay, M., Marchetti, V., Clément, A., Loubinoux, B. & Gérardin, P. (2000) Decontamination of synthetic solutions containing heavy metals using chemically modified sawdust bearing polyacrylic acid chains. *Journal of Wood Science*, 46 (4), 331–333.

Genty, T., Bussière, B., Paradie, M. & Neculita C.M. (2016) Passive biochemical treatment of ferriferous mine drainage: Lorraine mine site, Northern Quebec, Canada. In: Drebensteds, C. and Paul, M. (eds.) *IMWA 2016: Mining Meets Water – Conflicts and Solutions, IMWA 2016, 11–15 July 2016, Leipzig, Germany*. Freiberg, Technische Universität Bergakademie Freiberg. pp. 790–795.

Haakensen, M., Pittet, V., Spacil, M.M., Castle J.W. & Rodgers Jr., J.H. (2015) Key aspects for successful design and implementation of passive water treatment systems. *Journal of Environmental Solutions for Oil, Gas, and Mining,* 1 (1), 59–81.

Hengen, T.J., Squillace, M.K., O'Sullivan, A.D. & Stone, J.J. (2014) Life cycle assessment analysis of active and passive acid mine drainage treatment technologies. *Resources, Conservation and Recycling*, 86, 160–167.

Hui, K.S., Chao, C.Y.H. & Kot, S.C. (2005) Removal of mixed heavy metal ions in wastewater by zeolite 4A and residual products from recycled coal fly ash. *Journal of Hazardous Materials*, 127 (1–3), 89–101.

Hwang, T. & Neculita, C.M. (2013) In situ immobilization of heavy metals in severely weathered tailings amended with food waste-based compost and zeolite. *Water, Air, and Soil Pollution,* 224:1388.

Inyang, M., Gao, B., Yao, Y., Xue, Y., Zimmerman, A.R., Pullammanapallil, P. & Cao, X. (2012) Removal of heavy metals from aqueous solution by biochars derived from anaerobically digested biomass. *Bioresource Technology*, 110, 50–56.

Inyang, M.I., Gao, B., Yao, Y., Xue, Y., Zimmerman, A., Mosa, A., Pullammanapallil, P., Ok, Y.S. & Cao, X. (2016) A review of biochar as a low-cost adsorbent for aqueous heavy metal removal. *Critical Reviews in Environmental Science and Technology*, 46 (4), 406–433.

Ivanets, A.I., Shashkova, I.L., Kitikova, N.V. & Drozdova, N.V. (2014) Extraction of Co(II) ions from aqueous solutions with thermally activated dolomite. *Russian Journal of Applied Chemistry*, 87 (3), 270–275.

Izidoro, J., Fungaro, D.A., Abbott, J. & Wang, S. (2013) Synthesis of zeolites X and A from fly ashes for cadmium and zinc removal from aqueous solutions in single and binary ion systems. *Fuel*, 103, 827–834.

Jha, V.K., Kameshima, Y., Nakajima, A., Okada, K. & MacKenzie, K.J.D. (2006) Effect of grinding and heating on Ni^{2+} uptake properties of waste paper sludge. *Journal of Environmental Management*, 80 (4), 363–371.

Jha, V.K., Matsuda, M. & Miyake, M. (2008) Sorption properties of the activated carbon-zeolite composite prepared from coal fly ash for Ni^{2+}, Cu^{2+}, Cd^{2+} and Pb^{2+}. *Journal of Hazardous Materials*, 160 (1), 148–153.

Kołodyńska, D., Wnętrzak, R., Leahy, J.J., Hayes, M.H.B., Kwapiński, W. & Hubicki, Z. (2012) Kinetic and adsorptive characterization of biochar in metal ions removal. *Chemical Engineering Journal*, 197, 295–305.

Kuan, W.-H., Lo, S.-L., Wang, M.K. & Lin, C.-F. (1998) Removal of Se(IV) and Se(VI) from water by aluminium-oxide-coated sand. *Water Research*, 32 (3), 915–923.

Kučerová, G., Majzlan, J., Lalinská-Voleková, B., Radková, A., Bačík, P., Michňová, J., Šottník, P., Jurkovič, L., Klimko, T., Steininger, R. & Göttlicher, J. (2014) Mineralogy of neutral mine drainage in the tailings of siderite-Cu ores in Eastern Slovakia. *The Canadian Mineralogist*, 54 (2), 779–798.

Kumar, P.S., Ramalingam, S., Abhinaya, R.V., Kirupha, S.D., Murugesan, A. & Sivanesan, S. (2012) Adsorption of metal ions onto the chemically modified agricultural waste. *Clean-Soil, Air, Water*, 40 (2), 188–197.

Lee, H., Kim, D., Kim, J., Ji, M.-K., Han, Y.-S., Park, Y.-T., Yun, H.-S. & Choi, J. (2015) As(III) and As(V) removal from the aqueous phase via adsorption onto acid mine drainage sludge (AMDS) alginate beads and goethite alginate beads. *Journal of Hazardous Materials*, 292, 146–154.

Lee, J.Y., Choi, J.C., Yi, M.J., Kim, J.W., Cheon, J.Y., Choi, Y.K., Choi, M.J. & Lee, K.K. (2005) Potential groundwater contamination with toxic metals in and around an abandoned Zn mine, Korea. *Water, Air, and Soil Pollution,* 165 (1), 167–185.

Lim A.P. & Aris, A.Z. (2014) A review on economically adsorbents on heavy metals removal in water and wastewater. *Reviews in Environmental Science and Biotechnology*, 13 (2), 163–181.

Lindsay, M.B.J., Condon, P.D., Jambor, J.L., Lear, K.G., Blowes, D.W. & Ptacek, C.J. (2009) Mineralogical, geochemical, and microbial investigation of a sulfide-rich tailings deposit characterized by neutral drainage. *Applied Geochemistry*, 24 (12), 2212–2221.

Lindsay, M.B.J., Moncur, M.C., Bain, J.G., Jambor, J.L., Ptacek, C.J. & Blowes, D.W. (2015) Geochemical and mineralogical aspects of sulfide mine tailings. *Applied Geochemistry*, 57, 157–177.

Liu, Y., Sun, C., Xu, J. & Li., Y. (2009) The use of raw and acid-pretreated bivalve mollusk shells to remove metals from aqueous solutions. *Journal of Hazardous Materials*, 168 (1), 156–162.

Lu, H., Zhang, W., Yang, Y., Huang, X., Wang, S. & Qiu, R. (2012) Relative distribution of Pb^{2+} sorption mechanisms by sludge-derived biochar. *Water Research*, 46 (3), 854–862.
Mamchenko, A.V., Kiy, N.N., Chernova, L.G. & Misochka, I.V. (2008) The investigation of the impact of the modification methods of natural dolomite on water demanganesation. *Journal of Water Chemistry and Technology*, 30 (4), 191–196.
Mayes, W.M., Potter, H.A.B. & Jarvis, A.P. (2009) Novel approach to zinc removal from circum-neutral mine waters using pelletised recovered hydrous ferric oxide. *Journal of Hazardous Materials*, 162 (1), 512–520.
Mazumder, D., Ghosh, D. & Bandyopadhyay, P. (2011) Treatment of electroplating wastewaters by adsorption technique. *International Journal of Civil and Environmental Engineering*, 3:2, 101–110.
Mishra, T. & Tiwari, S.K. (2006) Studies on sorption properties of zeolite derived from Indian fly ash. *Journal of Hazardous Materials*, 137 (1), 299–303.
Miyake, M., Tamura, C. & Matsuda, M. (2002) Resource Recovery of Waste incineration Fly Ash: Synthesis of Zeolites A and P. *Journal of the American Ceramic Society*, 85 (7), 1873–1875.
Mohan, D., Pittman Jr., C.U., Bricka, M., Smith, F., Yancey, B., Mohammad, J., Steele, P.H., Alexandre-Franco, M.F., Gómez-Serrano, V. & Gong, H. (2007) Sorption of arsenic, cadmium, and lead by chars produced from fast pyrolysis of wood and bark during bio-oil production. *Journal of Colloid and Interface Science*, 310 (1), 57–73.
Mohan, D., Rajput, S., Singh, V.K., Steele, P.H. & Pittman Jr., C.U. (2011) Modeling and evaluation of chromium remediation from water using low cost bio-char, a green adsorbent. *Journal of Hazardous Materials*, 188 (1–3), 319–333.
Mohan, D., Sarswat, A., Ok, Y.S. & Pittman Jr., C.U. (2014) Organic and inorganic contaminants removal from water with biochar, a renewable, low cost and sustainable adsorbent – A critical review. *Bioresource Technology*, 160, 191–202.
Murugesan, G.S., Sathishkumar, M. & Swaminathan, K. (2006) Arsenic removal from groundwater by pretreated waste tea fungal biomass. *Bioresource Technology*, 97 (3), 483–487.
Nascimento, M., Soares, P.S.M. & Souza, V.P. (2009) Adsorption of heavy metal cations using coal fly ash modified by hydrothermal method. *Fuel*, 88 (9), 1714–1719.
Naser, H.A. (2013) Assessment and management of heavy metal pollution in the marine environment of the Arabian Gulf: A review. *Marine Pollution Bulletin*, 72 (1), 6–13.
Niu, C.H., Volesky, B. & Cleiman, D. (2007) Biosorption of arsenic (V) with acid-washed crab shells. *Water Research*, 41 (11), 2473–2478.
Nordstrom, K., Blowes, D.W. & Ptacek, C.J. (2015) Hydrogeochemistry and microbiology of mine drainage: An update. *Applied Geochemistry*, 57, 3–16.
O'Connell, D.W., Birkinshaw, C. & O'Dwyer, T.F. (2008) Heavy metal adsorbents prepared from the modification of cellulose: A review. *Bioresource Technology*, 99 (15), 6709–6724.
Orlando, U.S., Baes, A.U., Nishijima, W. & Okada, M. (2002) Preparation of chelating agents from sugarcane bagasse by microwave radiation as an alternative ecologically benign procedure. *Green Chemistry*, 4, 555–557.
Parviainen, A., Lindsay, M.B.J., Pérez-López, R., Gibson, B.D., Ptacek, C.J., Blowes, D.W. & Loukola-Ruskeeniemi, K. (2012) Arsenic attenuation in tailings at a former Cu-W-As mine, SW Finland. *Applied Geochemistry*, 27 (12), 2289–2299.
Peters, R.W., Ku, Y. & Bhattacharyya, D. (1985) Evaluation of recent treatment techniques for removal of heavy metals from industrial wastewaters. *AICHE Symposium Series, Separation of Heavy Metals*, 243, 165–203.
Qiu, W. & Zheng, Y. (2009) Removal of lead, copper, nickel, cobalt and zinc from water by a cancrinite-type zeolite synthesized from fly ash. *Chemical Engineering Journal*, 145 (3), 483–488.
Regmi, P., Garcia Moscoso, J.L., Kumar, S., Cao, X., Mao, J. & Schafran, G. (2012) Removal of copper and cadmium from aqueous solution using switchgrass biochar produced via hydrothermal carbonization process. *Journal of Environmental Management*, 109, 61–69.
Robinson-Lora, M.A. & Brennan, R.A. (2011) Anaerobic precipitation of manganese and co-existing metals in mine impacted water treated with crab shell-associated minerals. *Applied Geochemistry*, 26 (5), 853–862.
Rakotonimaro, T., Neculita, C.M., Bussiere, B., Benzaazoua, M. & Zagury, G.J. (2017) Recovery and reuse of sludge from active and passive treatment of mine drainage-impacted waters: a review. *Environmental Science and Pollution Research*, 24 (1), 73–91.
Rakotonimaro, T.V., Neculita, C.M., Bussiere, B. & Zagury, G.J. (2016) Effectiveness of various dispersed alkaline substrates for the pretreatment of ferriferous acid mine drainage. *Applied Geochemistry*, 73, 13–23.
Salameh, Y., Albadarin, A.B., Allen, S.J., Walker, G. & Ahmad, M.N.M. (2015) Arsenic (III, V) adsorption onto charred dolomite: Charring optimisation and batch studies. *Chemical Engineering Journal*, 259, 663–671.
Saliba, R., Gauthier, H. & Gauthier, R. (2005) Adsorption of heavy metal ions on virgin and chemically-modified lignocellulosic materials. *Adsorption Science & Technology*, 23, 313–322.
Sandy T. & DiSante, C. (2010) *Review of available technologies for the removal of selenium from water*, prepared for North American Metals Council.
Sapsford, D.J. (2013) New perspectives on the passive treatment of ferruginous circumneutral mine waters in the UK. *Environmental Science and Pollution Research*, 20 (11), 7827–7836.
Sasaki, T., Iizuka, A., Watanabe, M., Hongo, T. & Yamasaki, A. (2014) Preparation and performance of arsenate (V) adsorbents derived from concrete wastes. *Waste Management*, 34 (10), 1829–1835.
Seidel, H., Görsch, K., Amstätter, K. & Mattusch, J. (2005) Immobilization of arsenic in a tailings material by ferrous iron treatment. *Water Research*, 39 (17), 4073–4082.
Shah, B.A., Mistry, C.B. & Shah, A.V. (2013) Sequestration of Cu(II) and Ni(II) from wastewater by synthesized zeolitic materials: Equilibrium, kinetic and column dynamics. *Chemical Engineering Journal*, 220, 172–184.
Shin, E.W. & Rowell, R.M. (2005) Cadmium ion sorption onto lignocellulosic biosorbent modified by sulfonation: the origin of sorption capacity improvement. *Chemosphere*, 60 (8), 1054–1061.

Skousen, J., Zipper, C.E., Rose, A., Ziemkiewicz, P.F., Nairn, R., McDonald, L.M. & Kleinmann, R.L. (2016) Review of passive systems for acid mine drainage treatment. *Mine Water and the Environment,* doi:10.1007/s10230-016-0417-1.

Song, H., Yim, G.-J., Ji, S.-W., Neculita, C.M. & Hwang, T. (2012) Pilot-scale passive bioreactors for the treatment of acid mine drainage: Efficiency of mushroom compost vs. mixed substrates for metal removal. *Journal of Environmental Management*, 111, 150–158.

Stantec Consulting Ltd. (2004) *Review of water quality issues in neutral pH drainage: examples and emerging priorities for the mining industry in Canada*. Report prepared for the MEND initiative.

Swayampakula, K., Boddu, V.M., Nadavala, S.K. & Abburi, K. (2009) Competitive adsorption of Cu(II), Co(II) and Ni(II) from their binary and tertiary aqueous solutions using chitosan-coated perlite beads as biosorbent. *Journal of Hazardous Materials*, 170 (2–3), 680–689.

Thirunavukkarasu, O.S., Viraraghavan, T & Subramanian, K.S. (2003) Arsenic removal from drinking water using iron oxide-coated sand. *Water, Air, and Soil Pollution*, 142 (1), 95–111.

USEPA (United States Environmental Protection Agency) (2014) *Reference guide to treatment technologies for mining-influenced water*. EPA 542-R-14-001, 94p.

Vadapalli, V.R.K., Gitari, W.M., Ellendt, A., Petrik, L.F. & Balfour, G. (2010) Synthesis of zeolite-P from coal fly ash derivative and its utilization in mine-water remediation. *The South African Journal of Science*, 106, 7p.

Wang, C., Li, J., Sun, X., Wang, L. & Sun, X. (2009) Evaluation of zeolites synthetized from fly ash as potential adsorbents for wastewater containing heavy metals. *Journal of Environmental Sciences,* 21 (1), 127–136.

Wang, J. & Chen, C. (2014) Chitosan-based biosorbents: Modification and application for biosorption of heavy metals and radionuclides. *Bioresource Technology*, 160, 129–141.

Wan Ngah, W.S. & Hanafiah, M.A.K.M. (2008) Removal of metal ions from wastewater by chemically modified plant wastes as adsorbents: A review. *Bioresource Technology*, 99 (10), 3935–3948.

Warrender, R., Pearce, N.J.G., Perkins, W.T., Florence, K.M., Brown, A.R., Sapsford, D.J., Bowell, R.J. & Dey, M. (2011) Field trials of low-cost reactive media for the passive treatment of circum-neutral metal mine drainage in Mid-Wales, UK. *Mine Water and the Environment*, 30 (2), 82–89.

Xu, X., Cao, X. & Zhao, L. (2013) Comparison of rice husk—and dairy manure—derived biochars for simultaneously removing heavy metals from aqueous solutions: Role of mineral components in biochars. *Chemosphere,* 92 (8), 955–961.

Xue, Y., Gao, B., Yao, Y., Inyang, M., Zhang, M. & Zimmerman, A.R. (2012) Hydrogen peroxide modification enhances the ability of biochar (hydrochar) produced from hydrothermal carbonization of peanut hull to remove aqueous heavy metals: Batch and column test. *Chemical Engineering Journal*, 200–202, 673–680.

Yahya, M.A., Al-Qodah, Z. & Ngah, C.W.Z. (2015) Agricultural bio-waste materials as potential sustainable precursors used for activated carbon production: A review. *Renewable and Sustainable Energy Reviews*, 46, 218–235.

Zhang, N., Lin, L.-S. & Gang, D. (2008) Adsorptive selenite removal from water using iron-coated GAC adsorbents. *Water Research*, 42 (14), 3809–3816.

Zhang, Q.L., Lin, Y.C., Chen, X. & Gao, N.Y. (2007) A method for preparing ferric activated carbon composites adsorbents to remove arsenic from drinking water. *Journal of Hazardous Materials*, 148 (3), 671–678.

Zhang, W., Zheng, J., Zheng, P., Tsang, D.C.W. & Qiu, R. (2015) Sludge-derived biochars for arsenic (III) immobilization: effects of solution chemistry on sorption behavior. *Journal of Environmental Quality*, 44 (4), 1119–1126.

Land Reclamation in Ecological Fragile Areas – Hu (Ed.)
 ISBN 978-1-138-05103-4

Carbon and mineral feed additive produced from rice husk

S.V. Yefremova & Yu.I. Sukharnikov
National Center on Complex Processing of Mineral Raw Materials of the Republic of Kazakhstan, Almaty, Kazakhstan

N.B. Sarsembayeva
Kazakh National Agrarian University, Almaty, Kazakhstan

A.A. Kablanbekov
National Center on Complex Processing of Mineral Raw Materials of the Republic of Kazakhstan, Almaty, Kazakhstan

N.I. Bogdanovich
Northern (Arctic) Federal University, Arhangelsk, Russia

D.B. Murtazayeva & A.A. Zharmenov
National Center on Complex Processing of Mineral Raw Materials of the Republic of Kazakhstan, Almaty, Kazakhstan

ABSTRACT: Current work deals with investigation of carbonized rice husk as a feed additive. Different dose of this product was tested in the diet for the laying hens as a feed additive and substitute of a basal feed part. It was determined that using of carbonized rice husk as a carbon and mineral feed additive in the concentration of 2–3% is optimum amount to improve physiological functions of the hens.

1 INTRODUCTION

It is well known that waste of industry and agriculture can be used as raw material in different processes. Rice husk is one of them. It is raw material to produce activated carbon (Farajzaden & Vardast 2003, Lattuada *et al.* 2014), silicon carbide (Qadri 2012), silica, rubber filler, furfural and others (Efremova 2012). The problem of rice husk processing is that known technological processes are not profitable (Gorzkowski 2013) because only one finished product is produced. In this connection we developed the complex technology of thermal recycling of rice husk and created the equipment to realize this process. The major unit of this equipment is the rotating reactor into which rice husk are continuously fed through the feeding bunkers, and the prepared solid product is unloaded from the opposite end. Pyrolysis of rice husk is performed in the flue gases atmosphere at 550–650°C for 30 min. Yield of carbonized product was about 35% from the rice husk. This product consists of nanoparticles of carbon and silicon dioxide. Another product with the same yield represents a mixture of various organic compounds. Earlier we reported that the composition, structure and properties of both products enable their multifunctional use (Yefremova *et al.* 2016).

The object of this work is to investigate of carbonized rice husk as a carbon and mineral feed additive in the diet for the laying hens.

2 SCHEME OF LAYING HENS FEEDING

Under vivarium conditions an experiment was set up to study the effect of carbon and mineral feed additive produced from rice husk (CMFA) on physiological functions of laying hens and egg's quality. Egg-laying flock in a number of 63 birds in control and experimental groups was used in the experiment. The experiment was twice repeated. CMFA feeding was started from 150-day of age. The dose of carbon and mineral feed additive produced from rice husk in all-mash for the laying hens of experimental groups was 2 and 5% (Table 1). All-in-one feed for the birds was crumbled (Table 2).

For the whole period of raising the laying hens, composition and food value of all-mash receipts was not changed.

Table 1. Scheme of the laying hens feeding.

Hens age	The additives added in feed mix	CMFA weight ratio, %	Basal feed weight ratio, %
Control group			
150–180	–	–	100
181–210	–	–	100
211–240	–	–	100
240–270	–	–	100
271–300	–	–	100
I experimental group			
150–180	CMFA	2	98
181–210	CMFA	2	98
211–240	CMFA	2	98
240–270	CMFA	2	98
271–300	CMFA	2	98
II experimental group			
150–180	CMFA	5	95
181–210	CMFA	5	95
211–240	CMFA	5	95
240–270	CMFA	5	95
271–300	CMFA	5	95

Table 2. Program and conditions of the laying hens feeding.

Components, %	Control group	Experimental group	
		I	II
All-mash	87,2	85,2	82,2
Meat and bone meal	5,0	5,0	5,0
Grass meal	2,0	2,0	2,0
Limestone flour	5,0	5,0	5,0
Grit	0,4	0,4	0,4
Fine salt	0,4	0,4	0,4
CMFA	–	2,0	5,0
Total	100,0	100,0	100,0

3 RESULTS AND DISCUSSION

3.1 *Effect of CMFA on physiological functions of laying hens*

The results of research showed high resiliency of experimental laying hens: according to liveability the hens from experimental group left behind the chickens from control group to 3.2%. The inclusion of the studied CMFA additive to the composition of all-in-one feed of 150-day of age laying hens generally had a positive effect on intensity of body weight gain. The weight gain for the whole feeding period at experimental laying hens after change of 2% all- mash on CMFA was higher to 10.5%, and after change of 5% - to 2.3% (Table 3). The results of hematologic studies (Table 4) also show a positive effect of a new carbon and mineral additive on the bird growth.

Table 3. Effect of CMFA on the growth of the laying hens of 150-day of age.

Groups	Body weight of 1 bird at the		Absolute gain, g	Weight gain, %
	Beginning of experiment	End of experiment		
Control	2420,4	2721,4	301,0	100
I experimental	2468,9	2801,6	332,7	110,5
II experimental	2431,42	2739,16	307,7	102,3

Table 4. Hematological indices of laying hens in feeding with different doses of CMFA.

Indice	Unit measure	Control group	Experimental group	
			I	II
Hematocrit	%	31 ± 0,18	39 ± 0,13	37 ± 0,34
Hematog-lobulin	g/L	52,0 ± 1,8	54,0 ± 1,7	63,1 ± 2,0
Leucocytes	10^9/L	42,6 ± 0,12	41,2 ± 0,80	43,2 ± 0,49
Red blood counts	10^{12}/L	3,1 ± 0,18	3,1 ± 0,19	3,6 ± 0,14
Bactericidal activity	%	44,2 ± 0,91	58,1 ± 0,81	59,0 ± 0,73
Lysozyme activity	%	39,1 ± 0,93	38,7 ± 0,71	40,8 ± 0,20
Crude protein	g/L	34,5 ± 0,41	35,1 ± 0,18	37,8 ± 0,30
Calcium	mmol/L	2,68 ± 0,13	3,10 ± 0,14	3,10 ± 0,16
Inorganic phosphorus	mmol/L	1,5 ± 0,03	1,4 ± 0,08	1,49 ± 0,12

Table 5. Levels and intervals of variability factors.

Variables	Plan characteristics					
		Factor levels				
	Variability step, λ	−1,682 (−α)	−1	0	1	1,682 (+α)
X_1, CMFA additive, %	1,5	0	0,5	2	3,5	5
X_2, feeding period, month	1	1	2	3	4	5
X_3, all-mash spending, g/bird· month	50	2470	2490	2540	2590	2620

3.2 *Recommendations on optimal doses of CMFA*

Experimental data, received in different feeding regimes, were elaborated by using the method of three-factor experiment of second order with central compositional rotatable design (Bogdanovich *et al.* 2010) in order to determine optimal doses of carbon and mineral additive for hens feeding. A proportion of CMFA additive (X_1), feeding period (X_2) and all-mash spending (X_3) were chosen as the variables and body weight gain (Y) was chosen as an output parameter because it is the indicator of health and healthy growth of poultry.

Table 6. Planning matrix and experiment results by adding CMFA to the basal diet of the laying hens.

Experiment No.	X_1	X_2	X_3	Y
1	–1	–1	–1	270,5
2	1	–1	–1	315,3
3	–1	1	–1	296,5
4	1	1	–1	300
5	–1	–1	1	275
6	1	–1	1	305
7	–1	1	1	300
8	1	1	1	302,1
9	–1,682	0	0	300
10	1,682	0	0	306,6
11	0	–1,682	0	322,5
12	0	1,682	0	370
13	0	0	–1,682	346,8
14	0	0	1,682	395,2
15	0	0	0	351,3
16	0	0	0	393
17	0	0	0	380,3
18	0	0	0	391
19	0	0	0	381,3
20	0	0	0	364,4

The values and intervals of variability factors are presented in Tables 5–7. The received experimental data were used for calculation of the coefficients of regression equation of second order (1), which describe the process the laying hens feeding in the regime of replacement of basal feed part with carbon and mineral feed additive produced from rice husk (Bogdanovich *et al.* 2010):

$$y = b_0 + b_1x_1 + b_2x_2 + b_3x_3 + b_{11}x_1^2 + b_{22}x_2^2 + b_{33}x_3^2 + b_{12}x_1x_2 + b_{13}x_1x_3 + b_{23}x_2x_3. \quad (1)$$

The calculated values of coefficients of regression equation and F-ratio test are presented in Table 8.

After exclusion from Equation (1) the effects with insignificant coefficients, the final equation for assessing the model adequacy is as follows:

$$y = 377{,}8 + 8{,}25 \cdot X_2 + 5{,}94 \cdot X_3 - 34{,}85 \cdot X_1^2 - 19{,}66 \cdot X_2^2 - 6{,}74 \cdot X_3^2. \quad (2)$$

At α = 0,05 of significance point, F-ratio test F_{table} = 4,007. Taking into account, that the estimated value of F-ratio test (3,706) for equation does not exceed the table one, one can consider that received equation quite describes correctly experimental data.

On the basis of created mathematical model the yield surfaces are built (Figure 1), which characterize the effect of adding different doses of CMFA on body weight gain of the laying hens.

Yield surface shown on Figure 1a describes the dependence of body weight gain of laying hens on the dose added in the basal diet of CMFA (from 0 to 5%)

Table 7. Plan of experiments performing.

Day	Exp. No.	Code	X_1, CMFA additive, %		X_2, feeding period, month		X_3, all-mash spending, g/bird·month		Y, Gain, g
1	1	CMFA-3	–1	0,5	1	4	–1	2490	296,5
	2	CMFA-15	0	2	0	3	0	2540	351,3
	3	CMFA-9	–1,682	0	0	3	0	2540	300
	4	CMFA-13	0	2	0	3	–1,682	2470	346,8
	5	CMFA-1	–1	0,5	–1	2	–1	2490	270,5
	6	CMFA-17	0	2	0	3	0	2540	380,3
	7	CMFA-8	1	3,5	1	4	1	2590	302,1
2	8	CMFA-2	1	3,5	–1	2	–1	2490	315,3
	9	CMFA-16	0	2	0	3	0	2540	393
	10	CMFA-14	0	2	0	3	1,682	2620	395,2
	11	CMFA-7	–1	0,5	1	4	1	2590	300
	12	CMFA-4	1	3,5	1	4	–1	2490	300
	13	CMFA-20	0	2	0	3	0	2540	364,4
	14	CMFA-11	0	2	–1,682	1	0	2540	322,5
3	15	CMFA-18	0	2	0	3	0	2540	391
	16	CMFA-5	–1	0,5	–1	2	1	2590	275
	17	CMFA-10	1,682	5	0	3	0	2540	306,6
	18	CMFA-6	1	3,5	–1	2	1	2590	305
	19	CMFA-19	0	2	0	3	0	2540	381,3
	20	CMFA-12	0	2	1,682	5	0	2540	370

and feeding period in the limit till 5 months and it is almost symmetric about the axes. In this condition, the value of body weight gain varies in the interval of 220–340 g, and achieves maximum value (340 g) by adding carbon and mineral feed additive in concentration of 2–3% for 3–5 months. The data of Figure 1b characterize the dependence of body weight gain of laying hens on the dose added in the basal diet of CMFA (from 0 to 5%) and concentration of basal diet (2456–2624 g/bird·month). It is symmetric about CMFA axis and has maximum in the center of the plan. Maximum gain (340 g) is observed by adding 2–3% of CMFA at maximum spending of the feed, which indicates a preferential variant of the use of CMFA as the additive, but not in the replacing regime. In case of absence of CMFA (Figure 1c) in the diet, body weight gain in the studied conditions does not exceed 250 g.

Thus, the best result in laying hens feeding can be achieved by adding CMFA in concentration of 2–3% as the additive to the basal diet.

3.3 *Assessment of laying hens' egg quality in feeding with CMFA*

An experiment on using CMFA as the ingredient of feed mix in concentration of 2% to total diet mass was tested on 6th and 7th month of age of laying hens. Control kill of pullets was done to study the condition and organogeny of ovogenesis. The data of Table 9 show more active development of reproductive system at pullets from experimental group against their higher mass by the beginning of reproductive period in comparison with the control pullets.

Biometric parameters show a stable increase of eggs mass in experimental samples in all periods of the egg-laying by a mean of more than 2 g (Table 10). 6-months laying hens, keeping to a diet with CMFA, laid the eggs with a mass at 47,6 ± 1,03 g, which is accurately higher ($P < 0{,}05$) than the masses of control samples of the eggs (45,2 ± 0,71 g).

Measuring of shell thickness and inner shell thickness does not show significant difference between control and experimental samples. However, index analysis, which indirectly characterize shell thickness (Elastic Strain – ES), accurately shows high physical quality of the eggs' shell, received from the layers of experimental groups. Thus, in the control ES was from 24,4 to 24,6 mcM, and in the experiment this index was in the limit from 23,7 to 26,9 mcM ($P < 0{,}05$).

At 6-month of age of the layers it is registered accurate increase ($P < 0{,}05$) of yolk mass in the eggs, laid by experimental hens, white index has a tendency to increase by 0,6% in experimental samples of the eggs, laid by hens at 7-month of age. As this index has a positive connection with Haugh units, the analogical changes of the last one is observed.

Peculiar attention should be paid to the data reflecting vitamin sufficiency of the hen eggs in feeding

Table 8. The values of coefficients of regression equation of second order, which describes the process of laying hens feeding by adding CMFA, and F-ratio test.

Coefficient	Full equation	Equation with significant coefficients	Equation with added coefficients
b_0 =	377,80	377,80	377,80
b_1 =	6,70	0,00	0,00
b_2 =	8,25	0,00	8,25
b_3 =	5,94	0,00	5,94
b_{12} =	–8,65	0,00	0,00
b_{13} =	–2,03	0,00	0,00
b_{23} =	1,43	0,00	0,00
b_{11} =	–34,85	–34,85	–34,85
b_{22} =	–19,66	–19,66	–19,66
b_{33} =	–10,91	0,00	–6,74
f_1 =		12	9
F-ratio test =		4,007	3,706

Table 9. Biometric parameters of ovogenesis organs at pullets of 150-day of age, 2% CMFA.

Indices	Control group	Experimental group
Egg tube mass, g	36,8 ± 1,6	43,7 ± 1,8
Ovarium mass, g	14,6 ± 0,9	16,31 ± 1,2
Egg tube length, cm	46,4 ± 2,3	52,1 ± 2,7
Follicle counts, pcs		
Large	3,0 ± 0,01	6,0 ± 0,13
Small	3,0 ± 0,11	7,0 ± 0,09

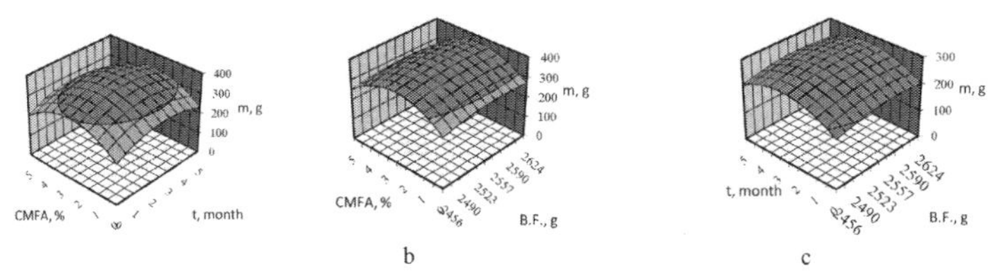

a – CMFA dose and feeding period; b – CMFA dose and basal diet; c – feeding period and basal diet

Figure 1. Effect of feeding regime on body weight gain of the hens.

Table 10. Dynamics of biometric and biochemical indices of the quality of hen eggs, 2% CMFA.

Indices	Hens of 6-month of age		Hens of 7-month of age	
	Control	Experiment	Control	Experiment
Egg mass, g	45,2 ± 0,71	47,6 ± 1,03	52,0 ± 0,40	54,1 ± 1,41
Index form, %	73,5 ± 0,63	76,6 ± 0,52	75,0 ± 0,46	76,6 ± 0,54
Crack and internal cracks, %	26,6 ± 0,37	16,6 ± 0,18	18,9 ± 0,42	15,4 ± 0,09
Elastic strain, mcM	24,6 ± 0,79	23,7 ± 0,62	24,4 ± 1,60	26,9 ± 2,20
Yolk height, mm	–	–	175,3 ± 2,51	178,1 ± 2,73
Yolk diameter, mm	34,4 ± 0,9	34,8 ± 1,1	38,7 ± 1,2	41,1 ± 0,6
Yolk mass, g	8,7 ± 0,37	10,9 ± 0,28	14,8 ± 0,20	15,3 ± 0,20
White diameter, mm	65,9 ± 1,18	68,3 ± 0,93	77,5 ± 1,23	74,0 ± 0,89
White index, %	7,9 ± 0,03	8,0 ± 0,09	7,8 ± 0,07	8,4 ± 0,10
Haugh unit	76,8 ± 1,03	76,4 ± 1,20	76,0 ± 0,69	78,3 ± 0,95
Carotinoids, μgr/g	14,5 ± 0,16	13,7 ± 0,16	8,2 ± 0,05	13,4 ± 0,21
MU	24,7	23,3	13,9	22,8
Vitamin A, μgr/g	7,1 ± 0,23	6,9 ± 0,19	7,7 ± 0,28	8,1 ± 0,13
MU	23,4	22,8	25,4	26,7
Sum of carotinoids and vitamin A, MU	48,1	46,1	39,3	49,5
Shell thickness, mcM	332 ± 6,85	337 ± 4,67	341 ± 0,08	335 ± 0,08
Inner shell thickness, mcM	32,0 ± 0,04	35,2 ± 0,12	34,1 ± 0,08	35,6 ± 0,08

Table 11. Chemical and amino acid composition of hen eggs received from layers, which are kept by 2% CMFA diet.

Indices	Control sample			Test samples of the eggs					
	Yolk	White	Total	Yolk	% to control	White	% to control	Total	% to control
Moistness, %	55,0	87,4	142,4	54,1	98,4	87,4	100,0	141,5	99,4
Dry solid, %	45,0	12,6	57,6	45,9	102,0	12,6	100,0	58,5	101,6
Protein, %	15,21	9,96	25,17	14,52	95,5	10,57	106,1	25,09	99,7
Lipids, %	56,4	0,03	56,43	55,9	99,1	0,02	66,7	55,92	99,1
Carbohydrates %	1,6	0,20	1,80	1,8	112,5	0,40	200,0	2,2	122,2
Amino acids, % DI									
Lysin	3,686	2,971	6,657	3,964	107,5	2,338	78,7	6,302	94,7
Methionine	1,078	1,707	2,785	1,066	98,9	1,370	80,3	2,434	87,5
Arginine	3,557	2,637	6,194	3,520	99,0	2,091	79,3	5,611	90,6
Histidine	2,240	2,551	4,791	2,118	94,6	2,113	82,8	4,231	88,3
Threonine	2,253	2,250	4,503	2,306	102,4	1,846	82,0	4,152	92,2
Leucine	3,615	3,539	7,154	3,466	95,9	3,006	84,9	6,472	90,5
Isoleucine	1,780	1,896	3,676	1,782	100,1	1,609	84,9	3,391	92,2
Phenylalanine	2,493	4,058	6,551	2,665	106,9	3,319	81,8	5,984	91,3
Valine	1,870	2,409	4,279	2,050	109,6	1,828	75,9	3,878	90,6
Glycine	1,413	1,775	3,188	1,419	100,4	1,757	99,0	3,176	99,6
Asparaginic	4,212	4,820	9,032	4,201	99,7	4,085	84,8	8,286	91,7
Serine	2,975	2,905	5,880	3,036	102,7	2,637	90,8	5,673	96,5
Glutaminic	5,573	6,827	12,400	6,076	109,0	5,385	78,9	11,461	92,4
Alanine	2,067	2,711	4,778	2,208	106,8	2,610	96,3	4,818	100,8
Proline	2,355	3,118	5,473	2,737	116,2	3,192	102,4	5,929	108,3
Tyrosine	2,711	2,073	4,784	2,575	95,0	1,787	86,2	4,362	91,2
Sum of amino acids	43,878	48,247	92,125	45,189	103,0	41,573	86,2	85,762	93,1
Including nonreplacable	16,775	18,830	35,605	17,299	103,1	15,316	81,4	32,615	91,6

with CMFA diet. The detailed analysis showed insignificant decrease of carotinoids in experimental samples of the eggs, received from the hens at the age of 6 months, by 0,8 μgr/g, and increase of this index by 5,2 μgr/g in the eggs of 7-month hens.

Vitamin A content was lesser by 0,2 μgr/g in the experimental eggs of 6-month hens and higher by 0,4 μgr/g in the experimental eggs of 7-month hens. A sum level of carotinoids and vitamin A in control sample was by 2 MU higher, than in experimental one at six-month of age. Then, it changed and it was observed accurate increase of carotinoids in experimental samples of the eggs ($P < 0,05$) and vitamin A, that in total showed the difference at 10,2 MU in favor of experimental samples of eggs, laid by the hens at the age of 7 months.

Despite the indicated differences and instability of indices it was constantly registered the decrease of inner and external defects of shells in the experimental samples of the eggs by 10,0% and 3,5% at 6 and 7 -month of age, respectively. Quite possibly, this is due to the fact that observed tendency of increase of inner shell thickness by 3,2 mcM and 1,5 mcM, caused by inclusion CMFA to metabolic processes, which happen in the organism of the layer.

The results of chemical research of the eggs are given in Table 11. The data from Table 11 show that dry solids weight ratio in experimental samples increases by 1,6% against control one. Besides, the content of protein in the white is more than 6,1%, total sum of carbonhydrates – by 22,2%. However, protein-lipid composition of experimental eggs has a tendency to decrease. Total amino acid content was by 6,9% lesser in experimental samples, than in control one, and respectively was 85,8 and 92,1% in air-dry substance. The level of nonreplaceable amino acids in experimental samples of the yolk was higher by 3,1%, and in the white – lower by 18,6%, on the average the content of nonreplaceable amino acids in the egg was lower by 8,4% against control one.

4 SUMMARY

1. It is defined that inclusion of studied CMFA additive to the all-in-one feed of 150-day of age laying hens generally had a positive effect on intensity of body weight gain: it was observed body weight gain for the whole period of feeding by replacing of 2% (10,5%) and in lesser degree (2,3%) – by replacing of 5% all-mash to CMFA.
2. It is defined that hematological indices of laying hens by feeding with different doses of carbon and mineral feed additive produced from rice husk corresponds to the physiological norms.
3. It is educed more active development of reproductive system of pullets from experimental groups as a result of their higher mass at the beginning of reproductive period in comparison with the controlled pullets. There was not educed accurate difference between biometric values, reflecting the quality of laid eggs in control and experimental variants. It is not observed significant dependence in the dynamics of biometric and biochemical indices of hen eggs quality depending on CMFA dose and feeding period of the bird. However, it is defined stable increase of eggs mass in experimental samples in all periods of egg laying and decrease of a number of defects and shell cracks in experimental samples of the eggs.
4. On the basis of research conducted within studied quantity range of adding CMFA in laying hens feeding, 2–3% CMFA to the basal diet can be recommended as optimal one.

ACKNOWLEDGEMENT

We acknowledge the Ministry of Education and Science of the Republic of Kazakhstan (Project No. 2254/GF4) for the financial support of the current work.

REFERENCES

Bogdanovich, N.I., Kuznetcova, L.N., Tretyakov, S.I. & Zhabin, V.I. (2010) *Examples and calculations of experiment planning.* Arhangelsk, Northern (Arctic) Federal University.

Efremova, S. (2012) Rice hull as a renewable raw material and its processing routes. *Russian journal of General Chemistry*, 5, 999–1005.

Farajzaden, M.A. & Vardast, M.R. (2003) Rice bran as an excellent sorbent for heavy metals from aqueous media. 1. Optimizations of conditions. *Journal of Chinese Chemical Society,* 50 (2), 245–250.

Gorzkowski, E., Qadri, S., Rath, B., Goswami, R. & Caldwell, J. (2013). Formation of Nanodimensional 3C-SiC Structures from Rice Husk. *J. Electronic Materials*, 5, 799–804.

Lattuada, R.M., Peralba, M.C.R., Dos Santos, J.H.Z. & Fisch, A.G. (2014) Peat, Rice Husk and Rice Husk Carbon as Low-Cost Adsorbents for Metals from Acidic Aqueous Solutions. *Separation Science and Technology,* 49 (1), 101–111.

Qadri, S., Imam, M., Fliflet, A., Rath, B., Goswami, R. & Caldwell, J. (2012) Microwave-induced transformation of rice husk to SiC. *J. Applied Physics*, 111, 073523-1-073523-5.

Yefremova, S., Sukharnikov, Yu., Bunchuk, L., Kablanbekov, A., Anarbekov, K., Yesengarayev, Ye., Sarsembayeva, N. & Zharmenov, A. (2016) Alternative Materials for Metallurgy and Agriculture. In: Bascetin, A.; Kursun, I.; Ozdemir, O. (eds.) *16th International Symposium on Environmental Issues and Waste Management in Energy and Mineral Production (SWEMP)/ International Symposium on Computer Applications (CAMI): Proceeding of 16th International symposium on environmental issues and waste management in energy and mineral production, SWEMP 2016, 5-7 October 2016, Istanbul, Turkey.* Istanbul, AGRO ARGE Danismanlik San. ve Tic. A.S.

Land Reclamation in Ecological Fragile Areas – Hu (Ed.)

Enrichment and analysis of soil heavy metals in different turfs on the golf course

S.X. Tang, P.H. Qiu, D. Wu, Z.Z. Zhao, H.P. Wu & Y. Chen
College of Geography and Environmental Sciences, Hainan Normal University, Haikou, Hainan, China

ABSTRACT: The contents of As and Pb in the turf soil on the golf course do not exceed the national standard limit but are higher than the background values. The Cd and Hg contents and the Cr contents in the fairway plow pan and plow layer of the putting green exceed both the national standard limit and the background values, thereby presenting as erious enrichment phenomenon. The vertical distribution of soil heavy metals in the turf soil on the golf course is mainly influenced by the applied quantity of chemical fertilizers and rainfall capacity, where as the soil heavy metal contents of different turfs on the golf course are determined by the quality of applied fertilizers.

1 INTRODUCTION

By the end of 2010, the number of golf courses in China has reached nearly 600[1]. Grass land is the core part of a golf course. Fertilizers, pesticides, bactericides, and herbicides must be applied during the planting and maintenance of the turf of the golf course to obtain a high quality turf and a good striking turf surface. The quantity of the applied pesticide must be three to eight times that of the crops[2]. However, researchers have noted adverse effects of fertilizers and pesticides on the soil environment. Existing studies on the influence of turf soil have mainly concentrated onthe leaching and enrichment phenomena after the application of nitrogen, phosphorus, and organochlorine and organo-phophorus pesticides with water[3–5]. These works have concentrated mostly on fields, vegetable lands, and orchards[6]. Analysis of the differences among turfs in terms of soil heavy metal contents and their enrichment degree is of great theoretical significance and realistic value.

2 MATERIALS AND METHODS

2.1 *General situation of the research region*

The studied golf course is located in Haikou City, Hainan Province (19°32′N–20°05′N, 110°10′E–110°41′E). It is a tropical humid region at the northern edge with atropical sea island monsoon climate zone featured by warm spring with small rainfall, high-temperature summer with large rainfall and strong eluviation, autumn with plenty of typhoon rain, and winter with small rainfall and weak eluviation. The golf course was built in a sandy laterite area near the mountain and by the river on the coastal platform. The construction was completed and the operation was started in 1996. The golf course has an area of 124 hm^2 and an overall fairway length of 6, 446.5 m, and it can be classified as an 18-hole and 72-shot international standard golf course. The grass variety of the tee ground is Bermuda 419 (thick leaves); the grass variety of the putting green is Bermuda 328 (fine leaves); and the grass variety of the fairway is local Cynodon dactylon. The main fertilizers applied on the turf mainly included 18-6-18 compound fertilizer, 18-3-18 compound fertilizer, high-nitrogen fertilizer, and high-potash fertilizer. The annual average application quantity of fertilizers is 3.08 kg/m^2.

2.2 *Sample collection*

Based on the differences in the maintenance of the turf types and the differences in soil eluviation between rainy and dry seasons, a professional geotome was used to collect soil samples in the 0–20 cm plow layer and the 21–40 cm plow pan in the fairway, the tee ground, and the putting green. Two points were randomly selected in three areas, and sampling was conducted for five times in 2015, for a total of 60 samples. Approximately 200 g of soil was collected at each sampling site. After the soils at various layers were adequately mixed, the mixture was sealed in polyethylene plastic bags and brought back to the lab, where it was processed (natural airing → crushing → impurities removal → blending → milling → running through 100-mesh screen) for standby application. HNO3-HF dispelling method recommended by USEPA was used to treat soil samples before analysis, and ICP-MS (Agi-lent 7700 × model) was used to measure heavy metal contents in the solution to be measured. All reagents used in the analysis were guaranteed reagents. The utilized water was ultrapure water. The national standard soil references (GSS-1 and GSS-4) were added for quality control. The analysis results were consistent with the quality control requirements.

2.3 *Data processing*

Excel2007, SPSS.18, and Matlab software were used for data statistics and difference analysis. Origin 9.0 software was used for drawing.

2.4 *Evaluation method*

Single pollution index method (calculation formula: Pi = Ci/Si) was used as the evaluation method for soil heavy metal enrichment. In the equation, Pi is single heavy metal pollution index of soil, Ci is the measured value of heavy metal "i" in the soil sample, and Si is evaluation criterion of "i". When Pi ≤ 1, no "i" accumulation (pollution) is present in the soil; when Pi > 1.0, "i" accumulation (pollution) is present in the soil. A greater Pi value corresponds to a more serious accumulation (pollution).[7]

3 RESULTS

3.1 *Differences in soil heavy metal contents among different turf types*

Table 1 presents a comprehensive descriptive statistics of heavy metals (Cr, Ni, As, Cd, Pb, and Zn) in 60 soil plow layers and plow pans of three types of turfs. As indicated by the variation coefficients of the contents of the seven heavy metals in the plow layers and the plow pans of the fairway, the tee ground, and the putting green, the variation coefficients of heavy metals in all soil types are large, where Cd (36.78%–109.69%) and Cr (22.82%–95.22%) have the maximum variation coefficients, followed by Pb (20.91%–83.39%) and As (26.69%–56.07%). As shown in Table 1, the heavy metal contents in the plow pan are obviously greater than those in the plow layer in the following cases: Cr, Ni, Cd, Pb in the fairway; As, Cd, Hg, Pb in the tee ground; and Cr, Ni, As, Pb in the putting green. The As and Hg contents in the fairway and the Cd contents in the putting green in the plow layer are obviously greater than those in the plow pan. Such distribution indicates that the contents of most heavy metals in the plow pan are greater than those in the plow layer. A comparison was conducted among the plow layers and the plow pans of the fairway, the tee ground, and the putting green in heavy metal contents. No significant differences were observed except for the following cases: Cr and As contents in the plow layers of the putting green and the tee ground; Cr contents in the plow pans of the fairway and the tee ground; and Pb contents in the plow pans of the fairway and the putting green. However, the heavy metal contents in the plow layer and the plow pan were averaged, as shown in Table 2. The average contents of Cd, Pb satisfied fairway > tee ground > putting ground; those of Hg and Hi satisfied tee ground > fairway > putting green; those of Cr satisfied putting green > fairway > tee ground; those of As satisfied fairway > putting green > tee ground.

3.2 *Evaluation of soil heavy metal enrichment under different turf types*

The stan-dard limiting value[8] of the national soil environmental quality grade II and the background values of the natural grassland soil in the lowest wind frequency zone in the NWW direction in the peripheral golf course were taken as the calculation basis. Single-factor pollution index method was employed as the evaluation method for soil heavy metal enrichment and pollution. As shown by the results in Table 3, four heavy metals (Ni, As, and Pb) in the plow layers and the plow pans of the fairway, the tee ground, and the putting green satisfy Pi (Ci/SiG) <1; Cd and Hg in the plow layers and the plow pans of the fairway, the tee ground, and the putting green satisfy Pi (Ci/SiG) >1; and Cr in the plow pan of the fairway and in the plow layer of the putting green satisfy Pi (Ci/SiG) >1. As, Cd, Hg, and Pb satisfy Pi (Ci/SiD) >1; Cr satisfies Pi(Ci/SiD) >1, exceptfor the plow pan of the tee ground.

Thus, through the 20-year operation of the golf course, although the As and Pb contents do not exceed the national standard limiting values, they exceed the background values of local soil. The Cd and Hg contents exceed both the national standard limiting values and the background values of local soil, exhibiting obvious enrichment. In particular, the Hg content exceeds the national standard limiting value by more than twice, and the Hg content in the plow pan of the tee ground even exceedsthe national standard limiting value by more than 18 times. Moreover, the Cr contents in the plow pan of the fairway and the plow layer of the putting green exceeds both the national standard limiting value and the background value. The Ni contents exceed the local background values but not the national standard limiting value.

Table 1. Statistic values of total elemental contents of heavy metals in the soil under different turf types of the golf (mg/kg).

Elements	Types	Symbol	Range of content	Average value	Standard deviation	CV(%)
Cr	Fairway (n = 20)	d-1	17.25–123.04	82.69	30.66	37.08
		d-2	125.87–446.72	253.07 A	109.74	43.36
	Tee ground (n = 20)	f-1	19.95–222.81	84.86 B	80.81	95.22
		f-2	10.93–159.86	64.60 A	60.82	94.15
	Putting green (n = 20)	g-1	153.81–378.07	277.59 B	63.35	22.82
		g-2	34.19–361.40	182.96	132.09	72.20
Ni	Fairway (n = 20)	d-1	7.8–35.84	22.37	10.07	45.01
		d-2	14.47–73.42	36.22	22.16	61.19
	Tee ground (n = 20)	f-1	4.2–51.15	26.27	19.23	73.19
		f-2	2.81–55.76	24.60	20.02	81.36
	Putting green (n = 20)	g-1	4.83–28.02	9.02	6.91	76.60
		g-2	7.9–91.35	30.23	28.09	92.92
As	Fairway (n = 20)	d-1	7.48–12.63	6.47	3.32	51.28
		d-2	0.82–7.48	4.98	2.64	52.98
	Tee ground (n = 20)	f-1	2.04–6.13	4.19 C	1.26	30.12
		f-2	3.37–8.03	5.71	1.71	29.95
	Putting green (n = 20)	g-1	3.47–8.24	5.83 C	1.56	26.69
		g-2	0.48–7.75	5.08	2.85	56.07
Cd	Fairway (n = 20)	d-1	0.11–1.60	0.59	0.56	94.29
		d-2	0.35–1.13	0.70	0.26	36.78
	Tee ground (n = 20)	f-1	0.06–2.16	0.79	0.80	101.00
		f-2	0.11–2.03	0.82	0.77	93.44
	Putting green (n = 20)	g-1	0.09–2.43	0.95	1.04	109.19
		g-2	0.09–2.25	0.83	0.91	109.69
Hg	Fairway (n = 20)	d-1	0.58–9.88	3.91	3.01	77.06
		d-2	0.26–4.52	1.55	1.44	93.11
	Tee ground (n = 20)	f-1	0.32–14.13	4.10	4.32	105.23
		f-2	0.65–20.40	5.89	6.13	104.11
	Putting green (n = 20)	g-1	0.32–2.43	0.99	0.85	85.77
		g-2	0.16–10.06	2.52	3.34	132.69
Pb	Fairway (n = 20)	d-1	3.96–14.37	8.52	3.39	39.84
		d-2	10.21–18.99	13.79 D	2.88	20.91
	Tee ground (n = 20)	f-1	3.04–12.43	6.65	3.33	50.14
		f-2	4.64–17.32	8.57	5.12	59.69
	Putting green (n = 20)	g-1	2.243–14.71	5.98	4.98	83.39
		g-2	3.19–14.91	6.97 D	4.34	62.30
		g-2	3.54–11.21	6.70	2.74	40.86

Note: 1 stand plow layer; 2 stand plow pan; d: Fairway; f: Tee ground; g: Putting green, Different capital letter means significant difference at $P < 0.05$.

Table 2. Average content of heavy metals in the soil of the pitch (mg/kg).

	Cr	Ni	As	Cd	Hg	Pb
Fairway	200.087	38.121	7.016	0.460	2.014	9.773
Tee ground	29.922	39.256	5.061	0.423	7.128	5.280
Putting green	260.659	11.497	5.793	0.404	0.392	3.653

3.3 *Correlations between soil heavy metals with TN and TP in golf turf*

A large quantity of organic fertilizers and compound fertilizers have been applied during maintenance process of golf turf, through correlation analysis between soil heavy metal contents with TN and TP, among Cr, Ni, As, Cd, Hg, Pb, except that correlations between Cd with TN and TP and between Hg and TP satisfy $P < 0.05$ with strong correlation, contents of other heavy metals have no obvious correlation with TN or TP. The accumulation trend of the Cd content with TN is $y^{(-0.5)} = 1.3722 + 0.0115x$, $R^2 = 0.4204$, SSE = 3.2954, whereas that of the Cd content with TP is $y = 0.0610 + 0.0143x$, $R^2 = 0.2537$, SSE = 1.8938, and that of the Hg content with TP is $y = 8.0734 - 0.1661x$, $R^2 = 0.1706$, SSE = 28.377.

Table 3. Index of soil pollution by heavy (pi) on the surface and deep layers of the soil for different turf types.

Site location		Cr		Ni		As		Cd		Hg		Pb	
		Ci/SiG	Ci/SiD	Ci/SiG	Ci/SiD	Ci/SiG	Ci/SiD	Ci/SiG	Ci/SiD	Ci/SiG	Ci/SiD	CiSiG	Ci/SiD
Fairway	d-1	0.33	1.44	0.56	1.83	0.16	2.35	1.97	1.04	13.03	12.22	0.03	1.94
	d-2	1.01	3.62	0.91	0.93	0.12	1.22	2.33	1.21	5.17	5.17	0.06	3.46
Tee ground	f-1	0.34	1.48	0.66	2.14	0.10	1.52	2.63	1.39	13.67	12.81	0.03	1.51
	f-2	0.26	0.92	0.62	0.63	0.14	1.40	2.73	1.41	19.63	19.63	0.03	2.15
Putting green	g-1	1.11	4.84	0.23	0.74	0.15	2.12	3.17	1.67	3.30	3.09	0.02	1.36
	g-2	0.73	2.62	0.76	0.78	0.13	1.25	2.77	1.43	8.40	8.40	0.03	1.75

Note: Ci: Measured value of heavy metal i in soils at present, SiG: Limit value of heavy metal i of Grade f Grade e value of heavy metal i of Grade; Cr:250; As; 30; Ni:40; Cd:0.3; Hg:0.3; Cu:50; Pb:250, unit: mg/kg, SiD: Background values of heavy metal i in soils in Dongshan Golf Course: Cr:69.84; As;4.07; Ni:38.91; Cd:0.58; Hg:0.3; Pb:3.98, unit: mg/kg.

4 CONCLUSION

1. Heavy metal contents in all area of golf course satisfy plow pan > plow layer, except for As and Hg in fairway soil and Cd in putting green soil. From the average content of heavy metals, The average contents of Cd, Pb satisfy fairway > tee ground > putting green; those of Hg and Ni satisfy tee ground > fairway > putting green; that of Cr satisfies putting green > fairway > tee ground; and that of As satisfies fairway > putting green > tee ground.
2. The As and Pb contents in the soil of golf turf do not exceed the national standard limiting values, but they exceed the background values of local soil. The Cd and Hg contents exceed both the national standard limiting values and the background values of the local soil, thereby presenting obvious enrichment. In particular, the Hg content exceeds the national standard limiting value by more than twice, and the Hg content in the plow pan of the tee ground even exceeds the national standard limiting value by more than 18 times. Moreover, the Cr contents in the plow pan of the fairway and the plow layer of the putting green exceed both the national standard limiting value and the background value. The Ni contents locally exceed the background values even if they do not exceed the national standard limiting value.
3. The vertical distribution of the heavy metals in the golf turf soil is mainly influenced by the quantity of applied fertilizer and rainfall. The soil heavy metal contents under different turf types on the golf course are related to quality of the applied fertilizers. Furthermore, a non-polluted irrigation water source has no influence on the enrichment of soil heavy metals of the golf soil.

ACKNOWLEDGMENT

Fund program: NSFC (Natural Science Foundation of China) fund program (41361108, 41361090), special program of provincial applied technology research and demonstrative promotion (ZDXM2015019), Hainan key subject (geography).

REFERENCES

Bai Ling-yu, Zeng Xi-bai, Li Lian-fang, Pen Chang. Li Shu-hui (2010). Effects of Land Use on Heavy Metal Accumulation in Soils and Source Analysis. *J. Scientia. J. A gricultura Sinica.* 43(1): 96–104. (in Chinese).

Chang Z I, Han L B. (2002). Research Progress about the Impact of Fertilizer and Pesticide on the Environment of Golf Course. *J. Grassland and Turf*, 4 (99): 6–9.

Ji C D, Zhou Y Y, Gu Y. (2015). Determination of organic pesticide residues in turf soil environment of Qingshuihu golf course. *J. Grassland and Turf*, (3): 55–61.

Li G F. (1994). Environmental Quality Evaluation. *J. Beijing: Chinese Environmental Science Press.*

Liu F Z. (2001). A Guide to Agricultural Environment Quality Monitoring. *J. Beijing: China Standard Press*, 609–610.

Reichman R, Rolston D E, Yates S R, Skaggs T H. (2011). Diurnal Variation of Diazinon Volatilization: Soil Moisture Effects. *J. Environ. Sci. Technol.* 45(6): 2144–2149.

Schueler T R, Holland H K. (2000). The Practice of watershed protection . Center for Watershed Protection. *J. Ellicott City, M D*, 673–675.

Yang G Y. (2011). China's Golf Course under the ban. N. People's daily, 06, 20.

Land Reclamation in Ecological Fragile Areas – Hu (Ed.)

Experimental study of the production of typical pollutants from underground coal fires

Y. Pu & Q. Zeng
Institute for Arid Ecology and Environment, Xinjiang University, Urumqi, China
School of Resource and Environment Sciences, Xinjiang University, Urumqi, China

Z.M. Cao
School of Metallurgical and Ecological Engineering, Beijing University of Science and Technology, Beijing, China

S.K. Jin
Xinjiang Coal Research Institute, Urumqi, China

L.H. Zhao
Institute for Arid Ecology and Environment, Xinjiang University, Urumqi, China
School of Resource and Environment Sciences, Xinjiang University, Urumqi, China

ABSTRACT: Underground Coal Fire (UCF) is a natural disaster associated with coal mining activities around the world, including in the Xinjiang region of China. It is difficult to quantify the impacts of UCFs on the environment. The present study focused on characterizing the production of typical pollutants from UCFs, which is needed for assessing the impacts of UCFs on the environment. A programmed heating–oxidation experiment was used to study coal oxidation and combustion. A thermogravimetric analysis was also performed to quantify the kinetics of coal oxidation, such as the characteristic temperatures, the reaction activation energy and the frequency factor. In addition, a chemistry simulation tool, FACTSage, was used to study the production of typical pollutants from coal oxidation and combustion under different conditions. Results showed that the main gaseous pollutants were carbon monoxide, carbon dioxide, and some trace CnHm gases. Carbon monoxide, carbon dioxide, and methane were the main greenhouse gases produced. The results also showed that changes in the reaction rate constant with temperature followed a similar pattern to changes in the oxygen consumption rate and the production rate of carbon monoxide. The reaction temperature affected the distribution of products from coal oxidation to some extent.

1 INTRODUCTION

Underground Coal Fire (UCF) is a natural disaster associated with coal mining. Globally, UCFs are widely reported in all major coal-producing countries (Alfred, 2004; Claudia, 2007; Glenn, 2004, 2011; J. Denis, 2007; Melissa, 2004; R.S. Chatterjee, 2006; Robert B., 2004; Zeng, 2012), causing large economic losses and severe ecological impacts. In an UCF event, coal–oxygen reactions produce heat and generate reaction products. The resultant products are ultimately responsible for the environmental impacts of UCFs and, thus, have been the targets of quantitative studies. Hower et al. (2011, 2009) monitored pollutant concentrations (e.g., CO, CO_2, Hg, H_2S) in gases released from UCFs. O'Keefe et al. (2011, 2010) measured CO, CO_2, and Hg emissions from UCFs in the field. Mark et al. (2013, 2012, 2011) measured greenhouse gas emissions from UCFs by ground-based and airborne surveying techniques. Yaci Liang et al. (2014) analyzed Hg concentrations in fumes released from an UCF in Inner Mongolia (China). Recently, Zeng et al. (2016, 2014) and Nie et al. (2017) investigated the distribution of heavy metals in soils at two UCF zones and the migration of As (a typical pollutant associated with UCFs) in those regions for the first time. Fu Gang (2002) characterized the emission of polycyclic aromatic hydrocarbons during coal combustion. Feng (2004) studied the thermodynamic and kinetic behaviors of typical trace pollutant elements in coal (i.e., Hg, Cd, As, Pb, Cr, Ni) during combustion. Xu et al. (2005) compared the effects of experimental conditions on the combustion characteristics of coal samples and found heating rate to be the dominant influencing factor. Kong (2005) studied the emission of NO and SO_2 from coal combustion in a circulating fluidized bed. However, no study has reported the patterns of product formation during coal–oxygen reactions under conditions relevant to UCFs.

The present study characterized the production of typical pollutants under conditions similar to those experienced during UCFs, by combining programmed heating–oxidation, Thermogravimetric Analyses (TGA), and chemistry thermodynamic simulations (FACTSage). The goal of this study was to quantify the production of typical pollutants during UCFs, which will aid assessments of UCF impacts on local environments.

2 MATERIALS AND METHODS

2.1 *Sample sites*

Coal samples were collected from the Beitashan UCF zone, in the East Junggar coalfield (Xinjiang, China). The sampling site (90°56′26″ E, 44°50′59″ N) was 178 km northeast of Qitai, Changji, Xinjiang. In the Beitashan UCF zone, seam B_6 (thickness: 27.97 m, dip: 7°) has been combusting at a depth of 17–18 m over an area of 150,888 m^2, causing up to 80,000 tons loss of coal annually. Figure 1 shows the topography and temperature map of the UCF zone. Table 1 lists the physicochemical characteristics of the samples collected from the zone.

2.2 *Methods*

2.2.1 *Programmed heating–oxidation*

Coal samples were treated by programmed heating–oxidation using a temperature controller, an adiabatic heating furnace, a Gas Chromatography (GC) station, gas pipes, a temperature monitoring system, and required accessories (Fig. 2). Each sample (100 g) was loaded into the furnace; dry air was set to flow through the furnace at 80 mL min^{-1} and it was heated to 500°C in a progressive manner (room temperature to 50°C in 15 min, 50°C to 75°C at 1.67°C min^{-1}, 75°C to 100°C at 1.67°C min^{-1}, and 100°C to 500°C at 1°C min^{-1}). After reaching each temperature point (50°C, 75°C, 100°C and 500°C), the temperature was maintained for 5 min and then the gas was collected from the outlet of the oven. GC analyses were performed to characterize the gases (e.g., O_2, CO, C_nH_m) present. The temperatures of the furnace and coal samples were continuously monitored until a coal sample's temperature exceeded that of the furnace, indicating coal burning, defined as the critical temperature.

2.2.2 *Thermogravimetric analysis*

Coal samples were analyzed by TGA to characterize their mass loss with the increases in temperature. Samples were loaded in a Hitachi STA 7300 TGA station and heated from 30°C to 1000°C in a dry air atmosphere (flow rate: 300 mL min^{-1}). TG and differential TG curves were recorded.

Figure 1. Topography and temperature map of the Beitashan UCF zone.

Table 1. Characteristics of coal seam B6 in the underground Beitashan coal-fire zone.

Seam	Properties (%)			Element (%)							
	Fixed carbon	Moisture	Volatile matter	Ash	C	H	O	N	S	Calorific value (MJ/Kg)	Coal type
B6	59.30	9.98	28.30	2.42	69.47	3.75	13.59	0.67	0.12	26.68	Non-caking

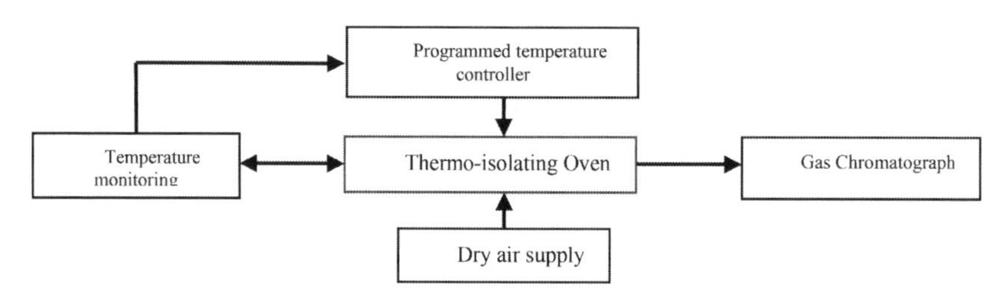

Figure 2. Schematic of the experimental apparatus.

2.2.3 *Thermodynamic simulations*

Coal–oxygen reactions were simulated with FactSage (Thermfact, CRCT, Montreal, QC, Canada). In modeling, minimization of Gibbs free energy was used as the criterion for equilibration and the composition of the reaction system at equilibrium was determined by the Lagrange multiplier method.

3 RESULTS AND DISCUSSION

3.1 *Programmed heating–oxidation*

The critical temperature and the concentrations of O_2, CO, and C_nH_m during low-temperature oxidation were monitored and analyzed by programmed heating–oxidation. Table 2 summarizes the results.

3.2 *Thermogravimetric analysis*

TG and differential TG curves of the samples (Fig. 3) showed a series of features during heating: moisture loss–oxygen absorption (T_1), thermal decomposition (T_2), ignition (T_3), maximum combustion rate (T_4), and burning-out (T_5). Table 3 lists the characteristic temperatures for these events.

Coal combustion in a coal-fire zone is a complicated process. Combustion reaction kinetic parameters are key to understanding coal combustion and the basis for investigating pollutant formation at the microscopic level. According to theory, coal combustion may involve the following coal–oxygen reactions:

a. $C + O_2 \rightarrow CO_2 + q_a$
b. $C + 1/2O_2 \rightarrow CO + q_b$
c. $C + CO_2 \rightarrow 2CO + q_c$
d. $CO + 1/2O_2 \rightarrow CO_2 + q_d$

Of these, reactions (a), (b), and (c) are solid phase exothermic reactions at the coal surface and reaction (d) is a gas-phase endothermic reaction; the letter q (with subscripts) represents the heat of the reaction. During coal combustion, reactions (a) and (b) are the leading reactions.

It has been established that the rate constant for a coal combustion reaction follows the Arrhenius equation:

$$k = A\exp(-E / RT) \tag{1}$$

where k is the rate constant; A is the frequency factor; E (KJ mol^{-1}) is the activation energy for the reaction; T (K) is the absolute temperature; and R is the universal gas constant (8.314 J K^{-1} mol^{-1}).

The activation Energy (E) is a critical factor that determines the rate of a reaction and how easily it occurs. It is also an important thermal kinetic parameter that characterizes the coal–oxygen reaction process.

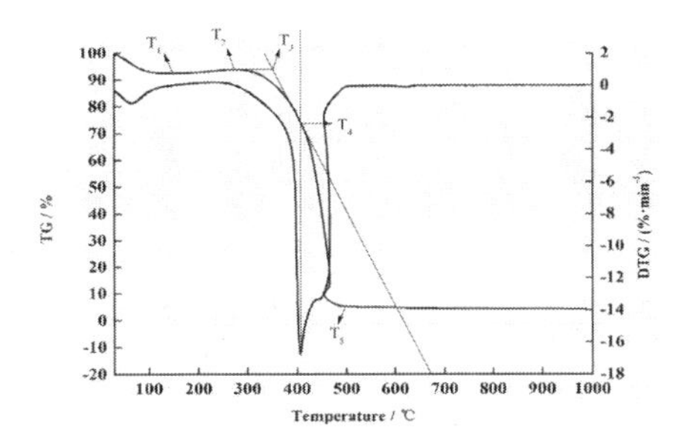

Figure 3. TG and differential TG curves of the coal samples.

Table 3. Characteristic thermogravimetric temperatures of the coal samples.

T1 (°C)	T2 (°C)	T3 (°C)	T4 (°C)	T5 (°C)
146.97	274.69	351.10	406.54	505.32

Table 2. Results of the programmed heating–oxidation of the coal samples.

	Sample temperature (°C)	Furnace temperature (°C)	O_2 (%)	CO (ppm)	CH_4 (ppm)	CO_2 (ppm)	C_2H_4 (ppm)	C_2H_6 (ppm)	C_2H_2 (ppm)	C_3H_8 (ppm)
B_6	11.2	12.0	21.09	0.00	0.00	617.21	0.00	0.00	0.00	0.00
	36.8	43.9	20.92	0.00	0.00	804.70	0.00	0.00	0.00	0.00
	60.2	69.7	20.84	0.00	0.00	873.68	0.00	0.00	0.00	0.00
	85.1	95.2	20.73	94.67	0.00	1240.70	0.00	0.00	0.00	0.00
	107.4	117.0	20.31	442.84	0.32	1957.30	0.00	0.00	0.00	0.00
	123.5	128.4	20.13	472.40	0.34	2554.11	0.00	0.00	0.00	0.00
	141.0	143.5	19.53	1236.88	0.61	3775.32	0.00	0.00	0.00	0.00
	157.2	156.5	18.79	5016.91	5.61	7398.13	0.00	0.00	0.00	0.00
	173.5	168.6	16.44	17283.33	6.80	16556.05	9.53	3.41	4.50	10.19
	246.1	175.2	3.16	65550.80	3377.60	42475.28	214.28	286.76	7.19	200.30
	226.5	158.0	3.97	249836.58	2695.64	174504.92	168.99	224.34	5.01	139.59
	216.6	165.5	3.44	190575.21	1673.44	185998.94	145.26	167.05	11.12	104.69

This parameter can be calculated using multiple methods, such as the frequently used Freeman–Carroll, Flynn–Wall–Ozawa, and Coats–Redfern methods. In the present study, the activation energies for the coal–oxygen reactions were calculated by the Coats–Redfern method. In the calculation, coal combustion was treated as a first-order reaction and the conversion efficiency of coal was expressed as:

$$\alpha = \frac{m_0 - m_t}{m_0 - m_\infty} \tag{2}$$

where α is the conversion efficiency of the coal sample (i.e., the proportion of mass converted to products); m_0 is the mass of the sample at the start of the experiment; m_t is the mass at time t; and m_∞ is the mass of the sample at the end of the experiment.

The reaction rate was expressed as:

$$\frac{d\alpha}{dt} = k(1-\alpha) \tag{3}$$

where k is the rate constant and t represents time.

Thermal analysis kinetics indicates that, for reactions of different orders (n = 1 or n ≠ 1), the Coats–Redfern equation is written as:

$$n=1,\ \ln\left[\frac{-\ln(1-\alpha)}{T^2}\right] = \ln\left[\frac{AR}{\beta E}\left(1-\frac{2RT}{E}\right) - \frac{E}{RT}\right] \tag{4}$$

$$n\neq 1,\ \ln\left[\frac{1-(1-\alpha)^{1-n}}{T^2(1-n)}\right] = \ln\left[\frac{AR}{\beta E}\left(1-\frac{2RT}{E}\right)\right] - \frac{E}{RT} \tag{5}$$

Generally, because $2RT/E$ is close to 0, $1 - 2RT/E$ approaches 1. Under this condition, Eqs (4) can be written as:

$$\ln\left[\frac{-\ln(1-\alpha)}{T^2}\right] = -\frac{E}{RT} \tag{6}$$

Equations (2), (3), and (6) were used to calculate kinetic parameters for the coal–oxygen reactions in this study as follows: $\ln[-\ln(1-\alpha)/T^2]$ was plotted against $1/T$ and the slope of the regression line was the activation energy. Table 4 summarizes the results.

3.3 *FACTSage modeling*

According to the analyses of the coal samples and the data from the programmed heating–oxidation experiments, parameters were selected for the FactSage simulation (Table 5). Four reaction

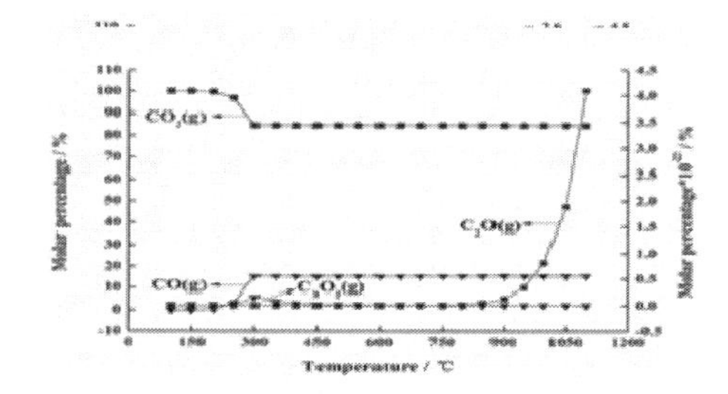

Figure 4. Composition for the system C + O.

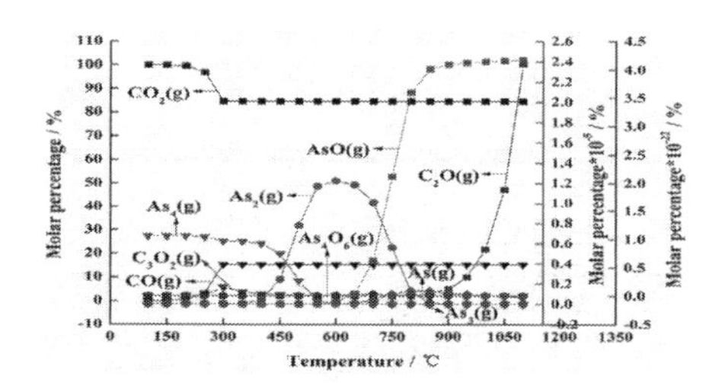

Figure 5. Composition for the system C + O + As.

Table 4. Kinetic parameters for the coal–oxygen reactions of the coal samples.

Sample \ Parameter	Temperature range (°C)	Regression equation	Activation energy (KJ/mol)	Frequency factor	Coefficient of correlation
B6	30–146.97	y = –4975.01x + 2.2618	41.36	4.78×10^2	–0.93
	147.05–274.69	y = –11517.06x + 10.4560	95.75	4.00×10^6	–0.95
	274.77–505.32	y = –16294.80x + 9.8924	135.47	3.22×10^6	–0.96

Table 5. Reactant configurations selected for FactSage modelling.

	C (mol)	O_2 (mol)	H (mol)	S (mol)	N (mol)	As (mol)	Hg (mol)	Cr (mol)	Cd (mol)	Ni (mol)	Si (mol)
B_6	26.84	24.77	17.70	0.09	0.30	6.67×10^{-6}	3.42×10^{-6}	4.13×10^{-4}	8.90×10^{-7}	1.70×10^{-5}	0.01

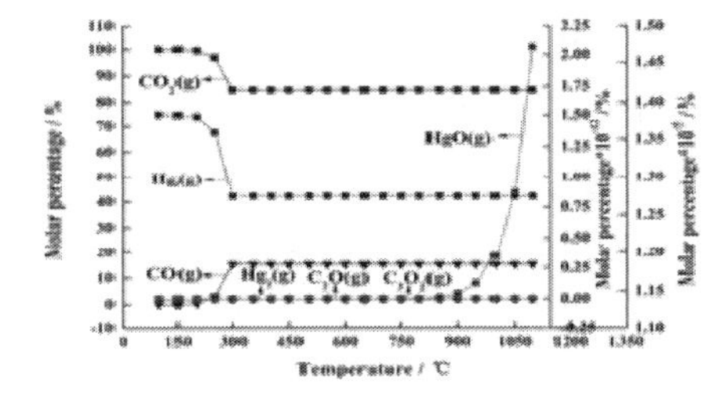

Figure 6. Composition for the system C + O + Hg.

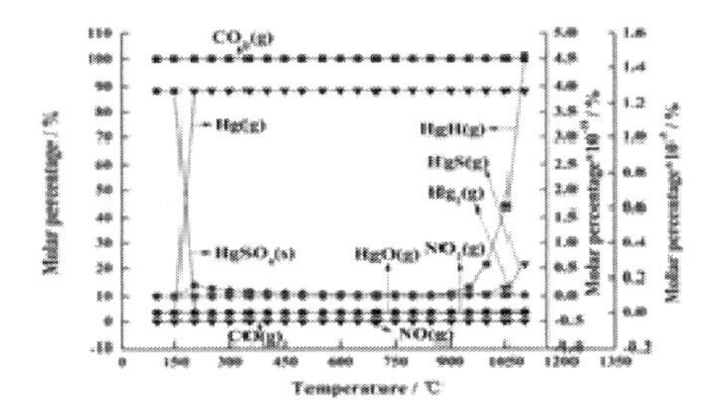

Figure 7. Composition for the system C + O + H + S + N + Hg.

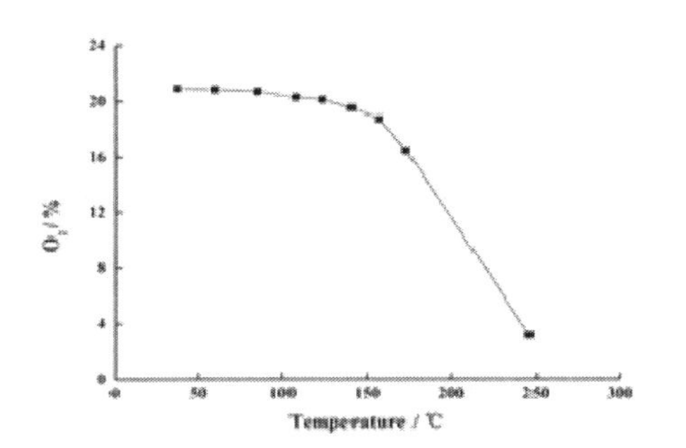

Figure 8. O_2 concentration change with temperature.

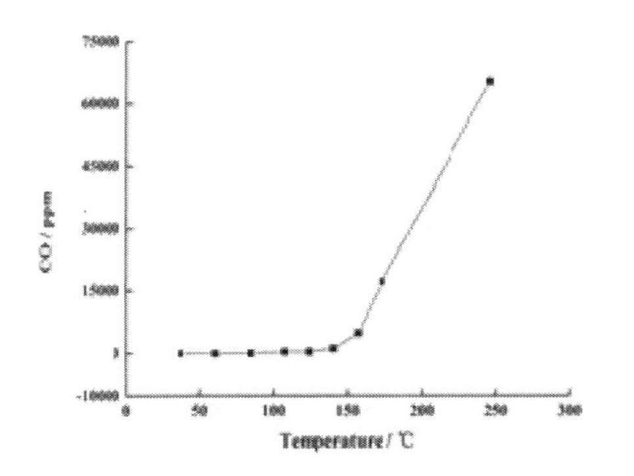

Figure 9. CO concentration change with temperature.

systems (C + O; C + O + As; C + O + Hg; C + O + H + S + N + Hg) were constructed and the results are depicted in Figs. 4–7.

3.4 *Analyses*

3.4.1 *Programmed heating–oxidation*

Figures 8 and 9 depict the changes in O_2 and CO concentrations with temperature during programmed heating–oxidation; with increasing temperature, the O_2 concentrations decreased gradually. When the temperature reached approximately 150°C, the O_2 concentrations started to decrease rapidly. The CO concentrations followed a similar but reverse pattern, of gradually increasing initially and then quickly increasing when the temperatures were above 150°C. These O_2 and CO concentration patterns were closely related to the progression of the coal–oxygen reactions.

3.4.2 *Thermogravimetric analysis*

In TGA (Table 3), the samples exhibited a series of features: moisture loss–oxygen absorption (T_1: 146.97°C), thermal decomposition (T_2: 274.69°C), ignition (T_3: 351.10°C), maximum combustion rate (T_4: 406.54°C), and final burning-out (T_5: 505.32°C). Thermodynamic calculations (Table 4) revealed that the activation energy for the temperature range from ambient temperature to T_1 was 41.36 KJ mol^{-1}; for T1 –T2 the activation energy was 95.75 KJ mol^{-1} and for T2 –T5 it was 135.47 KJ mol^{-1}. Based on the activation energy and

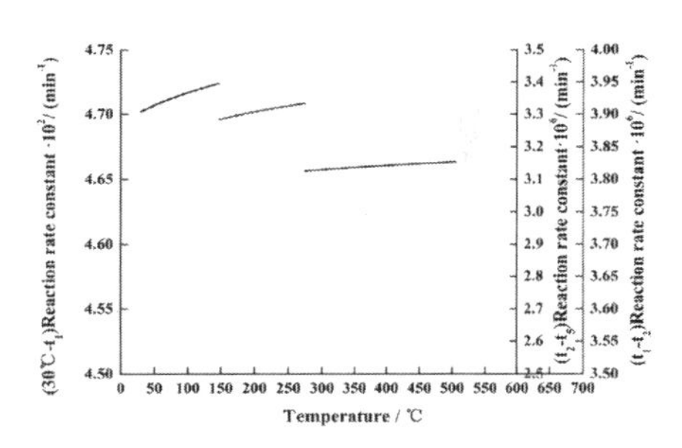

Figure 10. Reaction rate constants for temperatures.

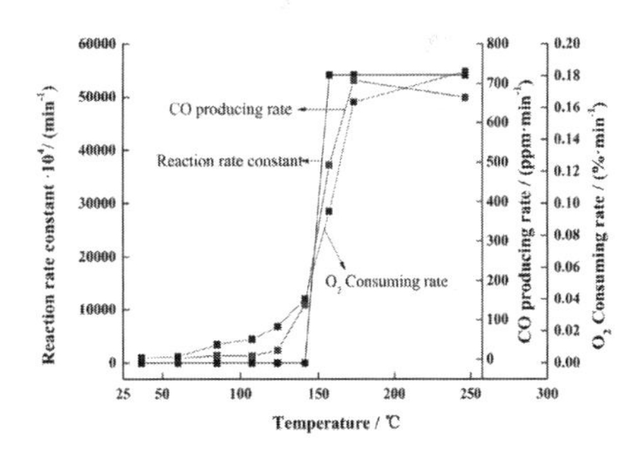

Figure 11. Changes in reaction rate constant, O_2 consumption, and CO producing with temperature.

frequency factor for each stage, the rate constants for the three ranges were expressed by Eqs (7)–(9):

Figure 10 depicts the rate constants calculated from these equations. Additionally, from the data in Table 2 and Eqs (7)–(9), the changes in the rate constant, oxygen consumption rate, and CO formation rate with temperature were calculated (Fig. 11) and all of these varied in generally similar patterns.

$$k = 4.78 \times 10^2 \exp\left[-\frac{41.36}{8.314 \times (t + 273.15)}\right] \quad (7)$$

$$k = 4.00 \times 10^6 \exp\left[-\frac{95.75}{8.314 \times (t + 273.15)}\right] \quad (8)$$

$$k = 3.22 \times 10^6 \exp\left[-\frac{135.47}{8.314 \times (t + 273.15)}\right] \quad (9)$$

3.4.3 *FactSage modeling*

The FactSage simulation for the system C + O (Fig. 4) predicted that the products consisted primarily of CO_2 and CO. At ≥ 800°C, trace C_2O (g) also formed. For the system C + O + As (Fig. 5), the products also primarily consisted of CO_2 and CO, with the formation of trace As_4 starting at 100°C and trace As_2 starting at 350°C. The formation of As_4 and As_2 fluctuated with increasing temperature, showing a valley and a peak respectively. For the system C + O + Hg (Fig. 6), CO_2 and CO remained the major products, and Hg (g) decreased with increasing temperature, reaching a steady concentration at 300°C. Starting from 800°C, trace HgO (g) formed and it proportionally increased with further increases in temperature. The system C + O + H + S + N + Hg (Fig. 7) produced predominantly CO_2, in addition to trace Hg (g) and $HgSO_4$ (s). From 950°C, trace HgS (g) and HgH (g) started to form and these increased with further increases in temperature.

4 CONCLUSIONS

The production of typical pollutants from an UCF was studied by programmed heating–oxidation, TGA, and a FACTSage simulation for the first time. The gaseous pollutants predominantly comprised CO and CO_2, in addition to trace C_nH_m (g). Moreover, the products included two major greenhouse gases: CO_2 and CH_4. With increasing temperatures, the coal–oxygen reactions accelerated, yielding products at higher rates. The reaction constant of the coal–oxygen reactions, oxygen consumption rates, and CO formation rates varied in similar patterns. The FACTSage simulations suggested that temperature affected the composition of the products. The simulated products were primarily CO_2 and CO, and the concentrations of other trace products varied with temperature. Further studies are required to characterize the formation of solid products in coal–oxygen reactions and their accumulation characteristics under UCF conditions.

REFERENCES

Alfred E. Whitehouse, Asep A.S. Mulyana. (2004) Coal fires in Indonesia. *International Journal of Coal Geology*, 59:91–97.

Claudia Kuenzer, Jianzhong Zhang, Anke Tetzlaff, et al. (2007) Uncontrolled coal fires and their environmental impacts: Investigating two arid mining regions in north-central China. *Applied Geography*, 27:42–62.

Fu Gang. (2002) Study on the characteristics of organic PAHs pollutants emission from coal combustion (Thesis). *Zhejiang University*. (Chinese)

Feng Rong. (2004) Study and comparison of chemical thermodynamic and kinetic calculation of representative trace elements in coal combustion process (Thesis). *Huazhong University of Science and Technology.* (Chinese)

Glenn B. Stracher, Tammy P. Taylor. (2004) Coal fires burning out of control around the world: thermo dynamic recipe for environmental catastrophe. *International Journal of Coal Geology*, 59:7–17.

Glenn B. Stracher, Anupma Prakash, Ellina V. Sokol. (2011) Coal and Peat Fires: A Global Perspective. Elsevier, Oxford.

James C. Hower, Jennifer M.K. O'keefe, Kevin R. Henke, et al. (2011) Time series analysis of CO concentrations from an Eastern Kentucky coal fire. *International Journal of Coal Geology*, 88:227–231.

James C. Hower, Kevin Henke, Jennifer M.K. O'keefe, et al. (2009) The Tiptop coal-mine fire, Kentucky: Preliminary investigation of the measurement of mercury and other hazardous gases from coal-fire gas vents. *International Journal of Coal Geology*, 80:63–67.

Jennifer M.K. O'keefe, Erika R. Neace, et al. (2011) Old Smokey coal fire, Floyd County, Kentucky: Estimates of gaseous emission rates. *International Journal of Coal Geology*, 87:150–156.

Jennifer M.K. O'keefe, Kevin R. Henke, James C. Hower, et al. (2010) CO_2, CO, and Hg emissions from Truman Shepherd and Ruth Mullins coal fires, eastern Kentucky, USA. *Science of the Total Environment*, 408:1628–1633.

J. Denis N. Pone, Kim A.A. Hein, Glenn B. Stratcher, et al. (2007) The spontaneous combustion of coal and its by—products in the Witbank and Sasolburg coalfields of South Africa. *Interntional Journal of Coal Geology*, 72:124–140.

Kong Dehao. (2005) Study on the emission of pollutants from coal combustion of circulating fluidized bed (Thesis). *Shandong University*. (Chinese)

Mark A. Engle, Lawrence F. Radke, Edward L. Heffern, et al. (2012) Gas emissions, minerals, and tars associated with three coal fires, Powder River Basin, USA. *Science of the Total Environment*, 420:146–159.

Mark A. Engle, Lawrence F. Radke, Edward L. Heffern, et al. (2011) Quantifying greenhouse gas emissions from coal fires using airborne and ground-based methods. *International Journal of Coal Geology*, 88:147–151.

Mark A. Engle, Ricardo A. Olea, Jennider M.K. O'keefe, et al. (2013) Direct estimation of diffuse gaseous emissions from coal fires: Current methods and future directions. *International Journal of Coal Geology*, 112:164–172.

Melissa A. Nolter, Daniel H. Vice. (2004) Looking back at the Centralia coal fire—a synopsis of its present status. *International Journal of Coal Geology*, 59:99–106.

Nie Jing, Zeng Qiang, Pu Yan, Jin Shi-kui. (2017)Migration of the typical heavy Arsenic within the soil of underground coal fire. *Journal of the China Coal Society*, 42(2):527–537. (Chinese)

R.S. Chatterjee. (2006) Coal fire mapping from satellite thermal IR data—a case example in Jharia Coalfield, Jharkhand. *ISPRS Journal of Photogrammetry and Remote Sensing*, 60:113–128.

Robert B. Finkelman. (2004) Potential health impacts of burning coal beds and waste banks. *International Journal of Coal Geology*, 59: 9–24.

Xu Chaofen, Zhang Penyu, Xia Ming, Sun Xuexin. (2005) Effect of the experiment condition on performance of coal combustion. *Journal of Huazhong University of Science and Technology(Nature science)*, 33(5):73–75. (Chinese)

Yanci Liang, Handong Liang, Shuquan Zhu. (2014) Mercury emission from coal seam fire at Wuda, Inner Mongolia, China. *Atmospheric Environment*, 83:176–184.

Ze-yang Song, Claudia Kuenzer. (2014) Coal fires in China over the last decade: A comprehensive review. *International Journal of Coal Geology*, 133:72–99.

Zeng Qiang. (2012) Study on the thermal dynamic characteristics of combustion system for coal fires in Xinjiang region (Thesis). *China University of Mining and Technology*, (Chinese)

Zeng Qiang, Tashpolat·Tiyip, Manfred W. Wuttke, et al. (2014) Characteristics of the heavy metal distribution in the soils of underground coal fire and the evaluation of its impact on environment. *Journal of China University of Mining and Technology*, 43(4):695–700. (Chinese)

Zeng Qiang, Nie Jing, Pu Yan. (2016) Characteristics of the distribution of typical heavy metals in the soils of underground coal fire. *Journal of the China Coal Society*, 41(8):1989–1996. (Chinese)

Land Reclamation in Ecological Fragile Areas – Hu (Ed.)
© 2017 Taylor & Francis Group, London, ISBN 978-1-138-05103-4

Isolation of halotolerant bacteria for degrading petroleum hydrocarbons and the optimum conditions for their activity

X.Q. Zhou & L. Yang
College of Agriculture, Shihezi University, Shihezi Xinjiang, China

ABSTRACT: The objectives of this experiment were to isolate halotolerant bacteria capable of degrading petroleum hydrocarbons and to determine the optimum conditions for their oil-degrading activity. Bacterial strain DM-1 could utilize oil as a sole carbon and energy source and efficiently degrade petroleum hydrocarbon. Based on its morphological and physiochemical characteristics as well as on Biolog analysis, strain DM-1 was identified as *Bacillus subtilis*. Strain DM-1 grew well at NaCl concentrations ranging from 5 to 70 g/L, and thus was classified as moderately halotolerant. Environmental conditions significantly influenced the oil-degrading ability of DM-1. A single factor experiment showed that the optimum degradation conditions for strain DM-1 were an initial pH of 7.5, a temperature of 30°C, a salt concentration of 5 g/L, and an inoculation amount of 10%. Under these conditions, strain DM-1 could degrade 66.94% of the petroleum hydrocarbon in 7 days.

1 INTRODUCTION

Soil is an important natural resource for sustainable human development. However, accidental discharge and leakage of petroleum during the process of exploitation, refining, transportation and use have caused significant decline in soil quality and function (Kumari et al., 2016; Rahman et al., 2002). Petroleum is complex mixture, including alkanes, aromatics, nitrogen-oxygen-sulfur compounds and asphaltenes. These substances may pose a risk to both humans and the ecological environment due to the high toxicity and persistence (Tahseen et al., 2016). Bioremediation is widely considered as a promising techniques for treating petroleum contaminants in the environment because of its low cost, high effectiveness and environmental friendliness (Gojgic-Cvijovic et al., 2012). In this process, the microorganisms are the crucial to degradation of crude oil as bioremediation agents (Hazaimeh et al., 2014).

Xinjiang is an important production base of petroleum in China. As a consequence of crude oil accompanied with saline formation water (Dastgheib et al., 2011), there is a large area of oil-contaminated saline soil. Therefore, the halotolerant strains with capable of degrading hydrocarbons obtained from local oil field is very crucial. To this aim, this research was carried out to isolate and characterize halotolerant strain DM-1 capable of degrading hydrocarbons, and to determine the optimal conditions for the degradation of crude oil. It could provide technical support and theoretical basis for bioremediation of oil contaminated soil in Xinjiang.

2 MATERIALS AND METHODS

2.1 *Sample collection and material*

In this study, samples of crude oil-contaminated soil was used to enrich degrading bacteria from the Kelamayi Oilfield. The crude oil was derived from the production well of Kelamayi Oilfield.

Mineral Salt Medium (MSM) was used to enrich and screen the degrading bacteria, which contained (g/l): 1.0 $K_2HPO_4.3H_20$, 1.0 KH_2PO_4, 0.5 $MgSO_4{\cdot}7H_2O$, 1.0 NH_4NO_3, 10 NaCl, 0.02 $CaCl_2$, pH 7.0.

2.2 *Screening and identification of oil-degrading strain*

Five-gram soil sample was added into 250 ml Erlenmeyer flasks with 100 ml MSM, containing 0.3 g of crude oil as the sole carbon source. After 3 days of cultivation at 30°C with shaking at 150 rpm, 10 ml cultures were transferred to the flasks having the same fresh MSM added with the crude oil as the first cultivation. After four repeated transformations, the active inoculums from the flask were streaked onto MSM agar plates with crude oil, and the different colonies were purified on nutrient agar medium respectively.

Among the isolates, strain DM-1 was identified by bacterial identification manual (Dong & Cai,

2001) and Biolog microbial identification system (Biolog Inc., Hayward, CA, USA).

2.3 *Growth and crude oil removal*

The growth curves of the isolated strain was measured through culture densities, using a visible spectrophotometer at 600 nm. The optimum fermentation conditions were determined by investigating the culture densities of strain DM-1 under different initial pH, temperature and NaCl concentration.

The effects of environmental and biological factors on DM-1 degradation of petroleum were studied by single factor experiment, and the degradation rate of crude oil was measured by a weight method.

3 RESULTS AND DISCUSSION

3.1 *Isolation and identification of oil-degrading strain*

Fourteen strains were isolated from oil-contaminated soils, including 13 strains of bacteria and 1 strain of fungus. Only one of these isolates could grow well on crude oil in mineral medium, named DM-1, and it could degrade 59.02% of the petroleum hydrocarbon in 7 days. Therefore, strain DM-1 was selected for further investigation. After observing colony characteristics and microscopic examination, the strain DM-1 was identified as Bacillus subtilis by using Biolog microbial identification system (Table 1).

3.2 *Growth characterization of strain DM-1*

Cell age was important for petroleum degradation by bacteria. The optimum cell age is general at late growth phase, which could adopt easily to new environment and enhance the resistance. (Le et al., 2008). Research showed that after lag phase of 4 h, strain DM-1 reached early stationary phase at 28 h. Therefore, the seed culture was incubated for 24 h.

To determine growth range of the strain DM-1, the effects of pH, temperature and salinity on the growth were studied. The strain DM-1 was adapted to alkaline environment and could grow over a range of pH from 6.0 to 9.0, with the optimum growth at pH 7.5. This strain grew well in the range of 15°C to 45°C, and the optimal temperature for the growth was 30°C, while below or above the range of 15–45°C, the growth of strain DM-1 was severely inhibited. The strain DM-1 had a range of growth salinity from 0 to 7% of NaCl (w/v), with the highest growth at 0.5%, and is a moderate halotolerant bacterium (Le et al., 2008).

Table 1. Identification of DM-1 by Biolog microbial identification system.

SIM	DIS	Org Type	Family	Genus	Match details
0.768	4.590	GP-Rod-SB	*Bacillaceae*	*Bacillus*	*Bacillus subtilis*

3.3 *Optimization of degradation condition*

3.3.1 *Effect of pH on degradation*

As shown in Fig. 1, the degradation efficiency of crude oil by strain DM-1 increased with increasing initial pH from 4.0 to 7.5, but percentage of degradation decreased when pH continued increase. The reason may be that the activity of the enzyme is affected by the pH (Silva et al., 2009). Hence, the optimum pH for hydrocarbons degradation was 7.5 for this strain, and the degradation rate reached 65.13%.

3.3.2 *Effect of temperature on degradation*

Temperature changes the physical and chemical properties of petroleum pollutants, which also affects the growth and the enzymes activity of microorganisms (Rodríguezblanco et al., 2010), thereby affecting the rate of microbial metabolism of petroleum hydrocarbons. As can be seen in the picture (Fig. 2), the optimum degradation temperature is 30°C, which is consistent with the optimum growth temperature. At 20°C and 40°C, the degradation rate of DM-1 was 32.33% and 35.61%, respectively. Degradation of petroleum contaminants will be inhibited if the temperature is lowered or elevated. The reason may be that the viscosity and the water solubility of the petroleum hydrocarbon decreased at low temperature, delaying the time of biodegradation. While above

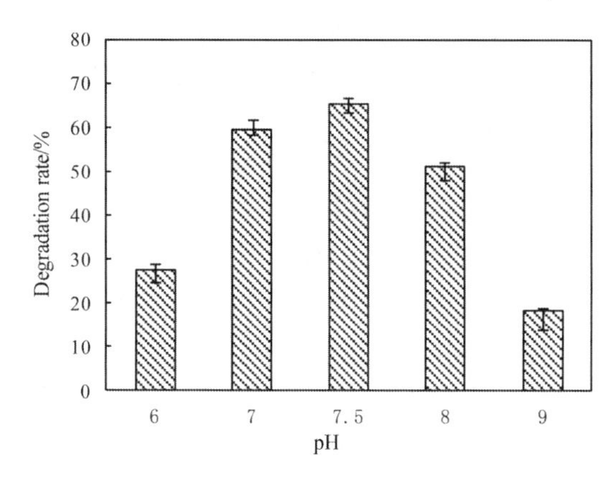

Figure 1. Effect of pH on degradation of crude oil by strain DM-1.

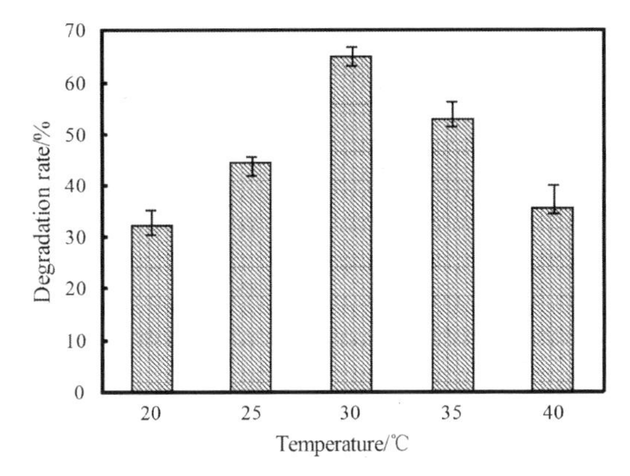

Figure 2. Effect of temperature on degradation of crude oil by strain DM-1.

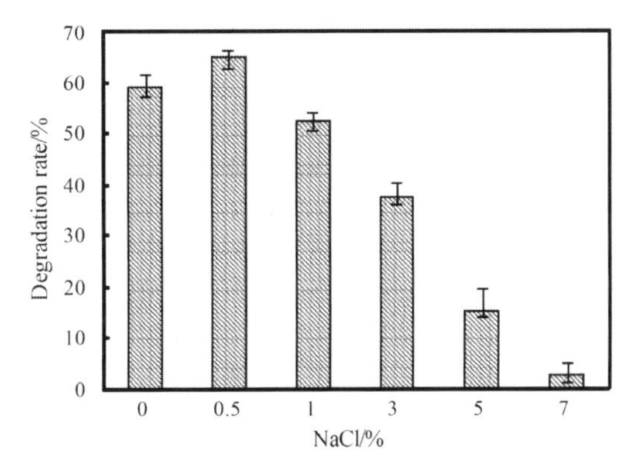

Figure 3. Effect of NaCl concentration on degradation of crude oil by strain DM-1.

the 40°C, the growth of the micro organisms was severely inhibited due to the membrane toxicity of petroleum hydrocarbon increased (Zeman et al., 2014).

3.3.3 *Effect of salinity on degradation*

As shown in Fig. 3, the biodegradation was enhanced with these additional NaCl in certain range. The optimal salinity for the degradation of strain DM-1 was determined to be 0.5% (w/v), suggesting that the strain was a suitable bioremediation agents under the salt environments. When the salinity increased from 3% to 7% (w/v), there was significant decreased in degradation efficiency. With the increase of salt concentration, the solubility of petroleum hydrocarbon and the concentration of dissolved oxygen decreased, and the microbial activity was inhibited, influencing the microbial biodegradation of petroleum (Dastgheib et al., 2011).

3.3.4 *Effect of crude oil concentration on degradation*

As can be seen in the picture (Fig. 4), crude oil degradation by strain DM-1 showed decrease tendency with the concentration of oil. It may be toxic to micro organisms at high concentrations. When crude oil concentrations increased from 1% to 5% (w/v), the degradation rate decreased significantly from 65.13% to 10.65%. Therefore, the high concentration of oil pollutants in the soil is a key problem of microbial remediation technology.

3.3.5 *Effect of biological factor on degradation*

The percentages of degradation by strain DM-1 at different incubation amounts was shown in Fig. 5. It is seen from the results that there was a rise in the percentage degradation of petroleum hydrocarbon, along with a rise in the incubation amount from 1% to 20%. However, no significant rise in biodegradation was detected at 20% of inoculum amounts. The increase in degradation of petroleum hydrocarbon by high amounts of inoculums may

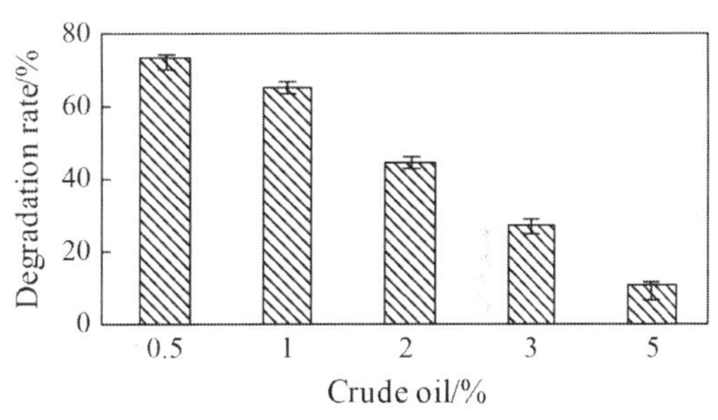

Figure 4. Effect of oil concentration on degradation of crude oil by strain DM-1.

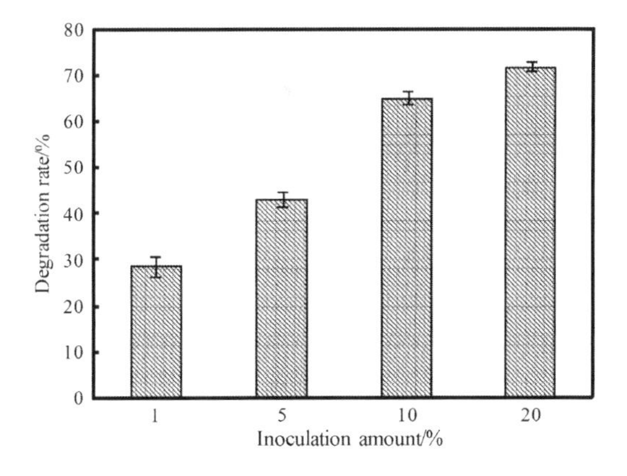

Figure 5. Effect of inoculation on degradation of crude oil by strain DM-1.

be due to the increasing metabolism of availability substrate, while above 10%, inoculum concentration might have not much influence on degradation as a result of the capability of microbial metabolism and the competition of available nutrients in the system (Cai et al., 2013).

Under the optimized conditions, strain DM-1 could degrade 66.94% of the petroleum hydrocarbon in 7 days.

4 CONCLUSIONS

The strain DM-1 with crude oil as sole carbon and energy source was isolated and identified as *Bacillus subtilis*. *B. subtilis* DM-1 could grow at salinities up to 7% of NaCl in nutrient media, which would be useful in bioremediation of saline sites contaminated with petroleum hydrocarbon. In addition, the ability of *B. subtilis* DM-1 to biodegrade petroleum hydrocarbon was influenced by changes at pH, temperature, salinities, substrate concentration, and addition of inoculum amount, and the maximum percentage of degradation could be achieved under the optimised conditions.

ACKNOWLEDGMENTS

This work was supported by financial support from the National Natural Science Foundation of China (Grant No. 41401349).

Yang Le and Zhou Xiao-qin contributed equally to this work.

REFERENCES

Cai, H., Yin, H. & Ye, J.S. et al.(2013) Isolation of an effective benzo [a] pyrene degrading strain and its degradation characteristics. *J. Environmental science*, 34(5):1937–1944.

Dastgheib, S.M.M, Amoozegar, M.A & Khajeh, K. et al. (2011) A halotolerant Alcanivorax sp. strain with potential application in saline soil remediation. *J. Applied Microbiology and Biotechnology*, 90(1): 305–312.

Dong, X.Z & Cai, M.Y. (2001) Handbook of common bacterial system identification. Beijing, Science Press.

Gojgic-Cvijovic, G.D., Milic, J.S., Solevic, T.M., Beskoski, V.P., Ilic, M.V., & Djokic, L.S., et al. (2012). Biodegradation of petroleum sludge and petroleum polluted soil by a bacterial consortium: a laboratory study. *J. Biodegradation*, 23(1), 1–14.

Hazaimeh, M., Mutalib, S.A., Abdullah, P.S., Kee, W.K., & Surif, S. (2014). Enhanced crude oil hydrocarbon degradation by self-immobilized bacterial consortium culture on sawdust and oil palm empty fruit bunch. *J. Annals of Microbiology*, 64(4), 1769–1777.

Kumari, B., Singh, S.N., & Singh, D.P. (2016). Induced degradation of crude oil mediated by microbial augmentation and bulking agents. *J. International Journal of Environmental Science and Technology,* 13(4), 1029–1042.

Le, B.S., Paniagua, D., & Vazquez-Duhalt, R. (2008). Biodegradation of organic pollutants by halophilic bacteria and archaea. *J. Journal of Molecular Microbiology & Biotechnology*, 15(2–3), 74–92.

Rahman, K.S., Thahira-Rahman, J., Lakshmanaperumalsamy, P., & Banat, I.M. (2002). Towards efficient crude oil degradation by a mixed bacterial consortium. *J. Bioresour Technol*, 85(3), 257–261.

Rodríguezblanco, A., Antoine, V., Pelletier, E., Delille, D., & Ghiglione, J.F. (2010). Effects of temperature and fertilization on total vs. active bacterial communities exposed to crude and diesel oil pollution in nw mediterranean sea. *J. Environmental Pollution*, 158(3), 663–73.

Silva, I.S., Grossman, M., & Durrant, L.R. (2009). Degradation of polycyclic aromatic hydrocarbons (2–7 rings) under microaerobic and very-low-oxygen conditions by soil fungi. *J. International Biodeterioration & Biodegradation*, 63(2), 224–229.

Tahseen, R., Afzal, M., Iqbal, S., Shabir, G., Khan, Q.M., & Khalid, Z.M., et al. (2016). Rhamnolipids and nutrients boost remediation of crude oil—contaminated soil by enhancing bacterial colonization and metabolic activities. *J. International Biodeterioration & Biodegradation*, 115, 192–198.

Zeman, N.R., Renno, M.I., Olson, M.R., Wilson, L.P., Sale, T.C., & Long, S.K.D. (2014). Temperature impacts on anaerobic biotransformation of lnapl and concurrent shifts in microbial community structure. *J. Biodegradation*, 25(4), 569–585.

Land Reclamation in Ecological Fragile Areas – Hu (Ed.)
© 2017 Taylor & Francis Group, London, ISBN 978-1-138-05103-4

Case study of the evaluation of the soil pollution situation in urban industrial wasteland

P. Shi
College of Jangho Architecture, Northeastern University, Shenyang, Liaoning, China

G.X. Zhang
Environmental Protection Bureau of Benxi County, Benxi, Liaoning, China

Y.H. Fu
College of Jangho Architecture, Northeastern University, Shenyang, Liaoning, China

S. Guo
Texas A&M University, College Station, Texas, USA

ABSTRACT: Urban industrial wasteland might have been affected by toxic and hazardous substances emitted by the original production activities. Consequently, pollution survey analysis and environmental risk assessment should have been conducted before redevelopment. In the paper, by surveying soil sample from waste dump, tailing pond and surrounding area in Lead-Zinc Mining area, fuzzy mathematics method was adopted to establish Fuzzy Comprehensive Evaluation model as to analyze conditions of soil pollution by heavy metal and soil nutrient. It would provide important scientific basis for pollution control and ecological restoration in the mining area and thus carry out comprehensive ecological restoration in Lead-Zinc Mine. Fuzzy Comprehensive Evaluation results indicated that soil pollution by heavy metal were severe, in which could be given priority to develop forestry. Most deficient soil nutrient elements were available P and N. Due to the lack of nitrogen and phosphorus, it was necessary to improve the soil before phytoremediation.

1 INTRODUCTION

Waste involves those unused and useless materials that are discarded (wastes), excessive expenditure of economically useful materials (to waste), and damage to natural capital (e.g. industrial wasteland) (JH Willison & RP Côté, 2009). Urban industrial wasteland refers to the abandoned land which was used for the industrial production or the related industrial production in the past, but now is partially or completely becoming abandoned or facing the function readjustment because of the resource exhaustion, the economic recession, industrial transformation or planning changes in the process of urban development and urbanization(J Chang et al., 2012), mainly including abandoned mines, quarries, factories, railway stations, docks, municipal waste disposal sites and so on. Those abandoned area might have been affected by toxic and hazardous substances emitted by the original production activities. Direct use of those abandoned land would be of ecological risk and original ecosystems would have been seriously damaged. Especially in abandoned mining area, in the process of exploiting and mineral processing, a mass of waste water, tailing, dust and noise have had serious impact on the ecological environment in the surrounding area, greatly changing the material cycling and energy flow of the ecosystem in the mining area. Consequently pollution control in the mining area as well as habitat restoration in the surrounding area became an urgent issue.

In the paper, taking Bajiazi Lead-Zinc Mine (Liaoning Province) as an example, a lot of investigation on soil environment quality had been conducted on the waste dump, tailing pond and surrounding area in the study area. Fuzzy comprehensive—weighted average model and fuzzy comprehensive index had been adopted to evaluate soil quality in the area, aimed at understanding conditions of soil pollution by heavy metal and soil nutrient as to provide important scientific basis for pollution control and ecological restoration in urban mining wasteland.

Located in Jianchang County, Bajiazi Lead-Zinc Mine covered an area of 66 km^2 (25.48 $mile^2$) with a total of 8 zones. The mining area was located in low mountains and hills with relatively small elevation. Climate condition was continental monsoon with large temperature difference. Annual average rainfall was 440.6 mm, mostly concentrated in the summer and autumn, and most cases multi rainstorm following. Frozen period was from

December to February with average 1.10 meter-thickness of the permafrost. Cinnamon was dominated in the area. Plant resources was abundant, mainly occupied by broadleaf forests as Quercus.

A mass of waste rocks were piled up on waste dumps in Bajiazi Lead-Zinc Mine, occupying large area of land. Surrounded by mountains on three sides, tailing pond covered an area of more than 2 km^2 (0.77 $mile^2$). With loose tailing, serious erosion, severe land degradation, bare ground, such scene could be seen everywhere. Not even a blade of grass grew, just little plant growing at the edge of the tailing pond. Our study was designed to improve the efficiency of phytoremediation through understanding the content of soil heavy mental and nutrition (S Rau et al., 2007) together.

2 SAMPLE COLLECTION AND ANALYSIS

2.1 *Sample collection*

Samples were collected from three sorts of typical areas, such as waste dump, tailing pond and surrounding areas, representing a wide range of abandoned mining area types (CL Lauber et al., 2009). Considering sample uniformity in the regional distribution, GPS was employed to determine the sample location. Briefly, all of soils were collected near the root of the plant growing area from nonagricultural to agricultural land that were seriously affected by mining activity for the majority of the year. At every site, soil from the top 20 cm of mineral soil was collected from 4 to 8 randomly selected locations in S-shaped line within an area of 200 m^2. And then each sample was mixed four points apart, weighed 1 kg, according to quartile which would be later placed in the sample bag and then brought back to the laboratory following the procedures of registering sample number, air-drying, grinding and grading, and mixing, bottling backup (Bao Shidan, 2005). Total of 48 soil samples were eventually collected with waste dump, tailing pond and surrounding area each accounting for 16.

2.2 *Sample analysis and determination*

Cu, Zn, Pb, Cd content in sample was determined by Atomic Absorption Spectrophotometry (AAS). The content of Hg, As was detected via Atomic Fluorescence Spectrometry (AFS). The content level of Cr was measured by DPCI Spectrometry. Available N content was measured by Alkaline Hydrolysis Diffusion Method. Total P content was measured by Sulfuric Acid and Per chloric Acid Digestion—Mo-Sb Colorimetry Method. Available P content was measured by leaching with sodium bicarbonate extraction—Mo-Sb Colorimetry Method. Available K content was measured by Ammonium Acetate Method. Analytical results are shown in Table (Table 1).

3 SOIL QUALITY EVALUATION AND ANALYTICAL RESULTS

3.1 *Determination of evaluation factor and criteria*

3.1.1 *Evaluation criteria of soil pollution by heavy metal*

Evaluation factors were set as U = {Cu, Zn, Pb, Cr, Cd, Hg, As}. Comprehensively considering background value of soil environment in Jianchang County, Liaoning Province (Wu Yanyu, 1994), critical content, National Soil Environmental Quality Standard (GB15168–1995) and assessment criteria of soil pollution by heavy mental was divided into five levels (Table 2), evaluation set was V = {I, II, III, IV, V}.

3.1.2 *Evaluation criteria of soil nutrients*

Evaluation factors were set as U = {Organic Matter, Total N, Available N, Total P, Available P, Total K, Available K}. Referring to Soil Nutrient Grading Standard in 2nd National Soil Survey, assessment *Cij* = criteria of soil nutrient was divided into five levels (Table 3), the evaluation set was V = {I, II, III, IV, V}.

Table 1. Average content of heavy metal and nutrient in sample ($mg \cdot kg^{-1}$).

Sample position	pH	Cu	Zn	Pb	Cr	Cd	Hg	As	Organic Matter	Total Nitrogen	Available Nitrogen	Total Phosphorus	Available Phosphorus	Total Potassium	Available Potassium
Waste dump	8.03	1283.2	2160.4	1458.1	32.1	0.75	0.048	14.30	1.51	0.035	66.50	0.012	0.72	0.432	121.03
Tail pond	7.93	1035.9	2499.3	1104.8	28.3	17.70	0.260	13.20	2.00	0.006	7.00	0.018	2.73	0.449	16.85
Surrounding	7.97	235.4	1473.8	534.1	53.7	12.90	0.078	10.38	6.14	0.214	246.40	0.043	12.60	1.042	156.74

N: Nitrogen; P: Phosphorus; K: Potassium; Organic Matter (%); Total Nitrogen/Phosphorus/Potassium (%).

Table 2. Assessment criteria of soil pollution by heavy metal (mg·kg^{-1}).

Evaluation factor	Cleanliness	Passably cleanliness	Light pollution	Mediate pollution	Severe pollution
Cu	21.52	37.04	100	200	400
Zn	88.28	129.99	200	300	500
Pb	24.38	39.97	200	350	500
Cr	61.70	102.08	150	250	300
Cd	0.07	0.28	0.4	0.6	1.0
Hg	0.04	0.11	0.5	1.0	1.5
As	9.28	20.83	25	30	40

Cleanliness (I); Passably Cleanliness (II); Light Pollution (III); Mediate Pollution (IV); Severe Pollution (V)

Table 3. Assessment criteria of soil nutrient (mg·kg^{-1}).

Soil nutrient	Extreme lacking	Lacking	Moderation	Richness	Abundant
Organic Matter	0.6	1	2	3	4
Total Nitrogen	0.05	0.075	0.1	0.15	0.2
Available Nitrogen	30	60	90	120	150
Total Phosphorus	0.04	0.075	0.1	0.15	0.2
Available Phosphorus	3	5	10	20	40
Total Potassium	0.5	1	1.5	2	3
Available Potassium	30	50	100	150	200

Organic Matter (%); Total N/P/K (%); Extreme Lacking (I); Lacking (II); Moderation (III); Richness (IV); Abundant (V)

3.2 *Calculation of membership degree*

Half-falling trapezoidal distribution membership function expression (DUAN Yong-hui, 2004):

When quality level $j = 1$,

$$r_{ij} = \begin{cases} 1 & C_i \le S_{i1} \\ (S_{i2} - C_i)/(S_{i2} - S_{i1}) & S_{i1} < C_i < S_{i2} \\ 0 & C_i \ge S_{i2} \end{cases} \tag{1}$$

When quality level $1 < j < 5$,

$$r_{ij} = \begin{cases} 0 & C_i \le S_{i(j-1)}, C_i \ge S_{i(j+1)} \\ (C_i - S_{i(j-1)})/(S_{ij} - S_{i(j-1)}) & S_{i(j-1)} < C_i < S_{ij} \\ \left(S_{i(j+1)} - C_i\right)/(S_{i(j+1)} - S_{ij}) & S_{ij} < C_i < S_{i(j+1)} \end{cases} \tag{2}$$

When quality level $j = 5$,

$$r_{ij} = \begin{cases} 0 & C_i \le S_{i4} \\ (C_i - S_{i4})/(S_{i5} - S_{i4}) & S_{i4} < C_i < S_{i5} \\ 1 & C_i \ge S_{i5} \end{cases} \tag{3}$$

where, r_{ij} was Evaluation Factor i belonging to membership degree of Level j, C_i was the measured value of Factor i, S_{ij} was the standard value of Factor i and Level j. Membership degree of each evaluation factor could be calculated by the formula above and six fuzzy relation matrix R could be established after integration. Taking assessment criteria of soil pollution by heavy metal in the waste dump as an example, fuzzy relation matrix was as follows

$$R_{\text{Heavy Mental in Waste Dump}} = \begin{vmatrix} 0 & 0 & 0 & 0 & 1 & \text{Cu} \\ 0 & 0 & 0 & 0 & 1 & \text{Zn} \\ 0 & 0 & 0 & 0 & 1 & \text{Pb} \\ 1 & 0 & 0 & 0 & 0 & \text{Cr} \\ 0 & 0 & 0 & 0.625 & 0.375 & \text{Cd} \\ 0.893 & 0.107 & 0 & 0 & 0 & \text{Hg} \\ 0.565 & 0.435 & 0 & 0 & 0 & \text{As} \end{vmatrix} \tag{4}$$

3.3 *Calculation of weight*

Due to various evaluation factors affecting differently on the soil environmental quality in the study area, indicators of level varied correspondingly. Consequently it was necessary to give weight to each evaluation factor. Weight formula of each pollution factor by Pollution Value Method was calculated as: $a_i = (Ci/Si)/(\Sigma_{i=1}^{n}(Ci/Si))$, where a_i was the weight of Pollution Factor i; C_i was the measured value of Pollution Factor i; S_i was the arithmetic mean of standard value of soil environmental quality as Pollution Factor i; C_i/S_i was the ratio of the measured value of Pollution Factor i and the arithmetic mean mentioned above (ZHOU Lin-fei, 2005). Taking assessment criteria of soil pollution by heavy metal in the waste dump as an example, individual weight was calculated and weight is normalized. Weight of seven evaluation factor constituted weight vector of three evaluation areas. After careful treatment, normalized result was $A_{\text{Heavy Mental in Waste Dump}}$ = (0.32Cu, 0.34Zn, 0.25Pb, 0.007Cr, 0.06Cd, 0.003 Hg 0.02 As) (Table 4).

3.4 *Fuzzy comprehensive evaluation*

It was primary to select fuzzy composite operator according to comprehensive assessment based on the results of fuzzy matrix. General method was to select big or small algorithm (ie, M "$\wedge \vee$" operator), however its weight effect was not obvious as well as easily losing information. Weighted average (ie M · + operator) algorithm was then adopted to avoid losing information in the fuzzy matrix R caused by selecting small weight. Calculation formula was $b_j = \Sigma_{i=1}^{n}(ai \cdot rij)$ (Song Xiaoqiu, 2004), where b_j was belonging to membership degree of Level j. Consequently, fuzzy comprehensive evaluation result vector of soil environmental quality in various parts could be drawn from above (Table 5).

Based on fuzzy comprehensive index principle, fuzzy comprehensive index F was calculated by Fuzzy Comprehensive Evaluation result vector and standard vector (Table 6) adopting Fuzzy Comprehensive Evaluation model F = Y·S (Li Ping, 2003), where F was Fuzzy Comprehensive Evaluation index, Y was Fuzzy Comprehensive Evaluation result vector, S was standard vector. Standard vector S = {35 50 65 80 95}. Taking waste dump as an example, Fuzzy Comprehensive Evaluation score of soil pollution by heavy metal was Y_1 = (0.0840, 0.0088, 0.0000, 0.0375, 0.9325); $F_1 = Y_1 \cdot S = 94.9675$. Other parts see Table 7 (Table•7).

3.5 *Result of analysis*

In the process of Fuzzy Comprehensive Evaluation, fuzzy characteristic of soil quality evaluation was taken into consideration as well as combined effects of both the weight of each evaluation factor and the interaction between evaluation factors for soil quality. Through site investigation, analysis validation, evaluation results fully met the actual situation of evaluation areas. Waste dump and tailing pond suffered the most serious soil pollution by heavy metal, in contrast soil nutrient in the surrounding area was in desirable condition. In addition, according to Table 5, conditions of soil pollution by heavy metal in waste dump, tailing pond and surrounding area were severe pollution. Condition of soil nutrient in waste dump was from moderation to lacking, condition of soil nutrient in tailing pond was from moderation to extreme lacking and condition of soil nutrient in surrounding area was abundant. According to Table 7, soil pollution

Table 4. Weight of evaluation factor on soil heavy metal and soil nutrient in the study area.

Sample position	Cu	Zn	Pb	Cr	Cd	Hg	As	Organic Matter	Total Nitrogen	Available Nitrogen	Total Phosphorus	Available Phosphorus	Total Potassium	Available Potassium
Waste dump	0.32	0.34	0.25	0.007	0.06	0.003	0.02	0.21	0.09	0.23	0.03	0.015	0.075	0.35
Tail pond	0.11	0.17	0.08	0.02	0.62	0.006	0.009	0.51	0.03	0.04	0.08	0.09	0.16	0.08
Surrounding	0.04	0.16	0.063	0.008	0.717	0.003	0.011	0.27	0.17	0.25	0.04	0.07	0.06	0.14

Table 5. Fuzzy comprehensive evaluation result of soil pollution by heavy metal and soil nutrient.

Sample position		I	II	III	IV	V
Heavy Mental	Waste dump	0.0840	0.0088	0.0000	0.0375	0.9325
	Tail pond	0.0259	0.0068	0.0023	0.0000	0.9800
	Surrounding	0.0180	0.0025	0.0015	0.0329	0.9471
Soil Nutrient	Waste dump	0.2100	0.2830	0.3597	0.1474	0.0000
	Tail pond	0.4800	0.0000	0.5100	0.0000	0.0000
	Surrounding	0.0366	0.0584	0.0568	0.0371	0.8111

Table 6. Cross reference list of quality level and fuzzy comprehensive evaluation score.

Quality level	1	2	3	4	5
F Range	<35	35–50	50–65	65–80	>80

Table 7. Fuzzy comprehensive evaluation score of soil pollution by heavy metal and soil nutrient.

Evaluation factor	Waste dump	Tail pond	Surrounding area
Soil Pollution	94.9675	94.4960	93.4590
Soil Nutrient	56.6425	49.9500	87.9155

by heavy mental in waste dump, tailing pond and surrounding area was Level IV—severe pollution. Soil nutrient in waste dump was Level III—moderation, soil nutrient in tailing pond was Level II—lacking, and soil nutrient in surrounding area was Level IV—abundant. Two evaluation results which were consistent, showed that Fuzzy Comprehensive Evaluation method was indeed able to objectively reflect real situation of soil quality in the study area. It fully indicated that Fuzzy Comprehensive Evaluation method was entirely feasible to evaluate the soil quality condition in such study area.

4 CONCLUSIONS

Fuzzy Comprehensive Evaluation results indicated that soil pollution by heavy metal in the study area was severe pollution, which in waste dump pollution factor of heavy metal was Zn, Cu, Pb, tailing pond was Cd, Zn, Cu and surrounding area were Cd, Zn. Two heavy metal Cd, Zn were dominate in the study area combined with Cu, Pb, which led to soil pollution by heavy metal. Consequently, it could be inferred that soil condition in the study area had already suffered serious heavy metal pollution, belonging to the third type of soil, which called for the third standard level implementation of the National Soil Environmental Quality. In order to ensure the safety of agricultural production, it was not suitable to plant crops. Instead it could be given priority to develop forestry and transplant plant varieties which were resistant to Cd, Zn and other heavy metals. After improvement of soil environmental quality, planting crops could be taken into consideration.

In the study area, most deficient soil nutrient elements in waste dump were total P and available P, in tailing pond were total N and available N, and in surrounding area were total P. Due to the lack of nitrogen and phosphorus, it was necessary to improve the soil before phytoremediation by means of applying composting city sludge to the contaminated soil, or applying reasonable amount of nitrogen and phosphate fertilizer to increase soil nutrient content in order to improve the poor physical structure and reduce the bioavailability of heavy metal, eventually improving soil quality. Plant communities tolerant of barren and unproductive and plant varieties capable of sand-fixing nitrogen-fixing would be gradually recovered to improve ecological system in the mining area as to achieve sustainable development of area.

ACKNOWLEDGEMENTS

This work was supported by the National Natural Science Foundation of China under grant 51504066 and the Fundamental Research Funds for the Central Universities under grant N161104002.

REFERENCES

Bao, S.D. (2005) Soil and Agricultural Chemistry Analysis. China, China Agriculture Press. Chang, J. (2009) Case study on the redevelopment of industrial wasteland in resource—exhausted mining area. *J. Procedia Earth and Planetary Science,* (1): 1140–1146.

Chen, L. & Yang, B.S. (2012) Analysis of Landscape Reformation on Urban Derelict Land. *J. Journal of Green Science and Technology*,(7):45–47.

Duan, Y.H. (2004) Application of fuzzy comprehensive assessment on soil environmental pollution evaluation. *J. System Sciences and Comprehensive Studies in Agriculture*, 20(4): 303–305.

Lauber, C.L. (2009) Pyrosequencing-based assessment of soil pH as a predictor of soil bacterial community structure at the continental scale. *J. Applied and environmental microbiology*, (15): 5111–5120.

Li, P. (2003) Fuzzy Overall Evaluation of Nutrients in Farming Brown Forest Soil in Southeastern Tibet. *J. Soils*, 35(5): 435–437.

Rau, S. (2007) Biochemical responses of the aquatic moss *Fontinalis antipyretica* to Cd, Cu, Pb and Zn determined by chlorophyll fluorescence and protein levels. *J. Environmental and Experimental Botany*, (3): 299–306.

Song, X.Q. (2004) The principle and method of fuzzy mathematics. China, China University of Mining and Tecnology Press.

Willison, J.H. & Raymond, P.C. (2009) Counting biodiversity waste in industrial eco-efficiency: fisheries case study. *J.* Journal of *Cleaner Production*, (3): 348–353.

Wu, Y.Y. (1994) Geochemical Background Values for the Soil in Liaoning Province. China, China Environmental Science Press.

Zhou, L.f. (2005) Research on application of fuzzy mathematics to water quality evaluation of wetland. *J. Water Resources and Hydropower Engineering*, 36(1):35–38.

Land Reclamation in Ecological Fragile Areas – Hu (Ed.)
 ISBN 978-1-138-05103-4

Effects of industrial by-products on the food chain

B. Tóth, Cs. Bojtor, G. Hankovszky, D. Kaczur & Á. Illés
Institute of Nutrition, University of Debrecen, Debrecen, Hungary

B. Kovács
Institute of Food Science, University of Debrecen, Debrecen, Hungary

ABSTRACT: Application of large amount of nitrogen fertilizer, has been reported to decrease soil pH, nitrate leaches into the groundwater, surface runoff water from the field transports phosphorus, ammonia and nitrate to water reservoirs and rivers. During the different industrial and production procedures and probably during everyday use some by-products and wastes are generated which have high micro- and macro element content and they do not endanger the environment. They should not be handled as wastes but rather as nutrient amendments and use in the plant nutrition with the reduction of chemical fertilizers. We have come to the conclusion that all of the examined by-products can be used in the plant nutrition but it is essential to determine the concentrations of application accurately for field use.

1 INTRODUCTION

The increasing use of fertilizer in agricultural production has been associated with a substantial increase in agricultural productivity. This increase in fertilizer use has been driven by a variety of economic forces including changes in the output price and changing relative factory prices. Associated with the increase in the use of fertilizer there are adverse environmental consequences that are not reflected in the costs and returns of agricultural production.

The growth of agricultural chemical use in an integral part of the technological revolution in agriculture has generated major changes in production techniques, shifts in input use, and growth in output and productivity. The mechanization revolution of the 1930s ans 1940s has been augmented since 1945 by a biological revolution in terms of fertilizer (Carlson and Castle, 1972). That biological revolution continues up to the present. A consequence of these changes is that farmers use more fertilizer, and less labor.

Between 1948 and 1981, for example fertilizer use increased at an annual rate of 2.6%, but was approximately unchanged between 1982 and 1993. Agricultural productivity expanded by 1.8% annually over this period (Ball and Nehring, 1996; Ball et al., 1997).

From the European settlement of the United States until the 19th century, increased agricultural production came almost entirely from expanding the cropland base. When there was nowhere to expand, maure and other farm refuse increase was productivity. Manure and other farm refuse were applied to the soil. Later, applications of manure were supplemented with fish, seaweed, leaves, straw, leached ashes and Peruvian guano, materials that contained a higher percentage of nitrogen, phosphate and potash than manure did (Wines, 1985). Crop rotations were also used to supply needed nutrients.

Commercial fertilizers provide relatively low-cost nutrients needed to fully realize the yield potential of new high-yielding varietes of crops (Ibach and Williams, 1971).

An adequate supply of nutrients is essential to crop growth. Ideally, soil nutrients should be available in the proper amounts at the time the plant can use them. Relatively heavy use of nitrogen and some other fertilizers, however, can lead to soil acidification, other changes in soil properties, and offsite environmental problems. Fertilizers are sometimes overapplied. When this occurs, the total amount of plant nutrients available to growing crops not only exceeds the need or ability of the plant to absorb them but exceeds the economics optimum as well. Estimates of crop absorption of applied nitrogen range from 25 to 70% and generally vary as a function of plant growth anf health and the method and timing of nitrogen application (National Research Council, 1989). Crops are much more likely to fully utilize nitrogen that is properly timed. Unused nitrogen can be immobilized, denitrified, washed into surface water or leached into groundwater (Huang et al., 1992).

As alluded to previously, the greatest problem associated with the increase in the use of fertilizer is the potential for environmental harm. Water pollution is probably the most damaging and widespread environmental effect of agricultural production. Phipps and Crosson (1986) and Nielson and Lee (1987) estimate that between 50 and 70% of all fertilizer reaching surface waters, principally nitrogen and phosphorous, originate from land in the form of fertilizer. Nitrate moves with water. Thus, nitrogen movement into surface waters more fully reflects the effects of agricultural activity than phosphorous movement does. Phosphorous moves as a passenger bound with sediment, much of which erodes from fields but is not deposited in surface waters (Smith et al., 1987). Increased nutrient levels stimulate algal growth, which can accelerate the natural process of eutrophication. In its later stages, the algal growth stimulated by nutrients will die and decay, which can significantly deplete available oxygen and reduce higher order aquatic plant and animal production (Kahn and Kemp, 1995).

2 MATERIALS AND METHODS

The experimental plant was maize (*Zea mays* cvs. Norma). The seeds were soaked in 10 mM $CaSO_4$ for 4 hours after sterilization and then germinated on moistened filter paper at 22°C. The seedlings were transferred to continuously aerated nutrient solution of the following composition: 2.0 mM $Ca(NO_3)_2$, 0.7 mM K_2SO_4, 0.5 mM $MgSO_4$, 0.1 mM KH_2PO_4, 0.1 mM KCl, 1 μM H_3BO_3, 1 μM $MnSO_4$, 10 μM $ZnSO_4$, 0.25 μM $CuSO_4$, 0.01 μM $(NH_4)_6Mo_7O_{24}$. Iron was added to the nutrient solution as Fe(III)-EDTA at a concentration of 100 μM. The native materials were used in the quantity of 2 g dm^{-3}.
The seedlings were grown under controlled environmental conditions (light/dark regime 10/14 h at 24/20°C, relative humidity of 65–70% and a photosynthetic photon flux of 300 μmol m^{-2} s^{-1}).

The relative chlorophyll content was measured on the second and third leaves. The relative chlorophyll content was measured with SPAD-502 (MINOLTA, Japan) Chlorophyll Meter.

To measure the total elements concentration, whole roots and shoots were blotted with filter paper, dried at 85°C and then digested. Ten ml HNO_3 (65v/v%) was added to each gram of the samples for overnight incubation. Then, the samples were pre-digested for 30 min at 60°C. Finally, 3 ml H_2O_2 (30 m/m%) was added for 90 min. boiling at 120°C. The solution were filled up to 50 ml, homogenised and filtered through MN 640 W filter paper.

The absolute chlorophyll a, b and carotene contents were measured by Meterek SP 80 Spektrometer according to Moran and Porath (1980). The plant samples were dried at 85°C for weight permanence and were measured with analytical scale (OHAUS) after cooling back them to room temperature.

All of the data were analyzed statistically using Microsoft Office Excel 2003 (Microsoft) for calculating mean and SE and Sigma Plot 8.0 version for t-test.

3 RESULTS

The effect of any nutrients can be considered directly, when the plants take it up, and the element gets into the living cell. Then the elements will be transferred to the different part of plant. Among the elements there are plenty of harmful ones, like heavy metals. The tolerant plants exclude the harmful elements from their metabolism, through the excretion of the elements to the vacuole. The sensitive plants are not able to transfer these elements from the roots, therefore these accumulate, causing the changing of membrane functions, and also affect on osmotic potential. The concentrations of examined elements in the shoots and roots of corn, treated with the examined side-products, can be seen in Table 1.

The root to shoot transport of K was extremely intensive at the control, and at the treatment by sewage sludge compost. The observation was the same in case of P. The "together" movements of Fe and Al are well seen in this case as well, but it is not so expressed as it was in the other cases. The high concentration of P and K in the shoots let us make a conclusion of a more intensive carbohydrate metabolism.

The concentration of all examined elements decreased in the shoot of corn when extruded poppy head treatment was applied. This decrease happened in spite of the fact that the concentrations of measured elements were extremely high in the roots.

Table 1. Elements content in the shoots and roots of maize effected by different treatments (sewage sludge compost, extruded poppy-heads, grinding sludge) mg kg^{-1} n = 3 ± S.E.

Elements concentration in the shoots of maize (mg kg^{-1})

	Contr.	SScomp	Extr.poppy	Grind.sludge
Al	67.60	18.50	10.05	7.95
Ca	8006	6479	4656	5567
Fe	92.70	137	76.15	284.50
K	80071	67856	37820	71057
Mg	2003	1755	1218	2022
P	13987	11769	7500	5899
S	3507	3219	2198	3735
Zn	33.40	28.85	29.80	46.15

Elements concentration in the shoots of maize (mg kg^{-1})

	Contr.	SScomp	Extr.poppy	Grind.sludge
Al	50	92	470	74
Ca	7773	5311	8142	4423
Fe	117	284	595	50507
K	51065	41212	48986	34085
Mg	4715	2948	4404	2554
P	5413	4020	6453	6690
S	10470	8891	10790	8234
Zn	55	48	73	119

This observation draws our attention to the importance of transport processes. Intensive root to shoot transport of K was observed at the control plants.

The concentrations of examined elements decreased in the shoots of maize, except the Fe, Mg and S, where slight increases were observed, when we applied grinding sludge treatments. The decrease of Al concentration in shoots was about 8.5 fold due to the grinding sludge treatments. The concentrations of Fe were 432 fold higher in the roots of treated plants in comparison to the control. It can be explained with the extremely high Fe concentration of grinding sludge (683,000 mg kg^{-1}).

The SPAD-unit is increased on all measuring days in the 2nd leaves of maize at the extruded poppy-heads. The increase was significant on the 11th day in the 3rd leaves. We suppose that the synthesis of photosynthetic pigments is a slower process than the growth of leaf. Thus, the total pigment concentration has not finished at the total leaf size developing. The synthesis of photosynthetic pigments needs energy and materials. So, we suppose that one of the highest quantities of the produced energy during the leaf-differentiation turns to the structure formation and after the finishing the synthesis of photosynthetic pigments (Table 2).

The SPAD-unit is significantly increased in the second and third leaves of maize on all measuring days at the grinding sludge treatment.

Since the SPAD-values gives only relative readings, the volumes of photosynthetic pigments

Table 2. The relative chlorophyll content (SPAD-unit) in second and third leaves of maize on the 6th, 9th and 11st days of the experiment. n = 60 ± S.E. Significant differences compared to the control: $^{**}p < 0.01$; $^{***}p < 0.001$.

Relative chlorophyll content in the 2nd leaves of maize			
Treatments	6th day	9th day	11th day
Control	36.77 ± 3.35	37.32 ± 3.92	38.38 ± 2.31
SScomp	39.42 ± 3.25	41.95 ± 3.83**	41.62 ± 2.70**
Extr.poppy	39.35 ± 3.56	42.30 ± 4.61**	39.49 ± 2.67
Grind. sludge	43.10 ± 3.09***	48.68 ± 2.85***	46.35 ± 2.09***
Relative chlorophyll content in the 2nd leaves of maize			
Treatments	6th day	9th day	11th day
Control	29.74 ± 3.46	28.61 ± 4.99	24.94 ± 2.20
SScomp	31.30 ± 1.95	32.11 ± 4.54	34.42 ± 3.09***
Extr poppy	28.73 ± 2.66	27.94 ± 6.09	31.48 ± 4.21***
Grind sludge	35.39 ± 3.99**	43.85 ± 3.07***	45.63 ± 2.01***

(chlorophyll-*a*, chlorophyll-*b*, carotenoids) were also measured (Table 3).

The increases in SPAD-values under the influence of the compost treatment are supported by the values showing absolute concentrations of photosynthetic pigments.

The absolute volumes of photosynthetic pigments decreased in the case of leaves of any age when extruded poppy-heads treatment was applied. In the third, the younger leaf, this decrease is significant, which supports the earlier explanation, namely that the time curves of leaf growth and syntheses of photosynthetic pigments do not coincide. The decrease in the pigment levels in older leaves in relation to the control may be a cause of the more moderate organic matter accumulation in the plants treated (Table 4).

Table 3. Chlorophyll -*a* and *b* carotenoid contents (mg g^{-1}) in the 2nd and 3rd leaves of maize under the effects of different treatments (sewage sludge compost, extruded poppy-heads, grinding sludge) n = 3 ± S.E. Significant differences compared to the control: $^{*}p < 0.05$, $^{**}p < 0.01$; $^{***}p < 0.001$.

Chlorophyllcontents in the 2nd leaves of maize			
Treatments	Chl-a	Chl-b	Carotenoids
Control	12.44 ± 0.61	3.51 ± 0.41	8.53 ± 0.89
SScomp	13.18 ± 1.55	3.89 ± 0.75	8.54 ± 0.86
Extr.poppy	11.22 ± 0.52	2.97 ± 0.59	6.95 ± 0.76
Grind. sludge	16.75 ± 0.98*	5.57 ± 0.37*	10.86 ± 0.37
Chlorophyll contents in the 3rd leaves of maize			
Treatments	Chl-a	Chl-b	Carotenoids
Control	8.79 ± 0.38	3.48 ± 0.99	5.92 ± 0.69
SScomp	13.45 ± 0.44**	3.54 ± 0.98	8.63 ± 0.02*
Extr.poppy	7.55 ± 0.43*	1.79 ± 0.34*	5.24 ± 0.14
Grind. sludge	16.64 ± 0.61***	6.17 ± 0.63*	11.77 ± 0.65***

Table 4. Effect of by-products (sewage sludge compost, extruded poppy-heads and grinding sludge) on the dry matter of maize (g $plant^{1}$) n = 9 ± S.E. Sigcant differences compared to the control: $^{***}p < 0.001$.

Dry weight of shoots and roots of maize (mg kg^{-1})		
Treatments	Shoots	Roots
Control	0.199 ± 0.02	0.064 ± 0.01
Sewage sludge Compost	0.254 ± 0.03***	0.064 ± 0.01
Extruded poppy-heads	0.180 ± 0.04	0.052 ± 0.01
Grinding sludge	0.282 ± 0.03***	0.066 ± 0.01

4 CONCLUSIONS

The use of these by-products in agriculture offers us the possibility to moderate the quantities of expensive mineral fertilizers to be used in agriculture. As a side effect, the total CO_2 emission could be decreased, as well. The criteria od their application include that they should not exercise any toxic effect on plants or pollute the environment, but can be used in replacement of the expensively manufactures chemicals.

All of the by-products involved into the experiment are generated in huge volumes; therefore, they can be suitable for solving nutrient deficiency problems in large agricultural fields. It has been confirmed by the experiments focusing on the nutritional elements on the by-products that all the by-products contain several essential elements in large quantities. Our examinations concentrated on the elements that occur in all the by-products. From among them, we have selected the eight elements sevring as the basis of comparison: Al, Ca, Fe, K, Mg, P, S and Zn.

On the base of experiments, we have come to the conculsion that all the examined by-products can be used in the nutrition supply of plants, but it is essential to determine the concentarations of applicarion accurately for field use. This statement is particularly true for grinding sludge.

In summary of the results, we come to the conculsion that the by-products that were involved to our experiment can be applied for full or partial replacement of the costly made chemicals as the case may be, it is necessary the continuous of the by-products, e.g. seage sludge compost, may change, it is necessary the continuous monitoring of the by-products offered for agricultural use.

ACKNOWLEDGEMENT

This research was supported by the European Union and the State of Hungary, co-financed by the European Social Fund in the framework of TÁMOP 4.2.4.A/2-11-1-2012-0001 'National Excellence Program' and Hungarian State Post Doctoral Fellowship.

REFERENCES

Ball E., Bureau J.C., Nehring R., Somwaru A. (1997). Agricultural productivity revisited. Am. J. Agric. Econ. Forthcoming.

Ball E., Nehring P. (1996). Productivity: Agriculture's Growth Engine, Agricultural Outlook. U.S. Department of Agriculutre: Economic Research Service, May.

Carlson G.A., Castle E.N. (1972). Economics of pest control: pest control strategies for the future. Washington: National Academy of Sciences.

Huang W., Hansen L., Uri N.D. (1992). The effects of the timing of nitrogen fertilizer application and irrigation on yield and nitrogen loss in cotton production. Environ Plann A 24, 1449–1462.

Ibach D.B., Williams M.S. (1971). Economics of fertilizer use. In: (eds. Olson R.A., Army T.J., Hanway J.J., Kilmer V.J.) Fertilizer technology and use. Madison, WI: Soil Science Society of America.

Kahn J., Kemp W. (1995). Economics losses associated with the degradation of an ecosystem: the case of submerged aquatic vegetation in the Chesapeake Bay. J. Environ Econ Manage 16, 246–263.

Moran R., Porath D. (1980). Chlorophyll determination in intact tissues using N, N-dimethylformamide. Plant Physiol 65(3), 478–479.

National Research Council (1989). Alternative Agriculture. Washington DC: National Academy Press.

Nielsen E., Lee L. (1987). Magnitude and costs of groundwater contamination from agricultural chemicals, Economic Research Service. Washington DC. Department of Agriculture.

Phipps T., Crosson P. (1986). Agriculture and the Environment. Washington DC: The National Center for Food and Agriculture Policy.

Sanchez P.A. and Salinas G. (1981). Low input technology for managing Oxisols and Ultisols in tropical America. Adv. Argon 34, 280–406.

Wines R.A. (1985). Fertilizer in America: from waste recycling to resource exploitation. Philadelphia, PA: Temple University Press.

Land Reclamation in Ecological Fragile Areas – Hu (Ed.)
© 2017 Taylor & Francis Group, London, ISBN 978-1-138-05103-4

A review of land reclamation technologies of saline-alkali soil in China

L. Zhang & J.Y. Zhang
Institute of Land Reclamation and Ecological Restoration, China University of Mining and Technology, Beijing, China

L.H. Gao & L.F. Lv
National Center for Quality Supervision and Testing for Fertilizers, Shandong, Linyi, China

Y.S. Liang & Y. Zhao
Institute of Land Reclamation and Ecological Restoration, China University of Mining and Technology, Beijing, China

ABSTRACT: Saline-alkali soil land was a great environmental problem in China, which should be reclaimed to promote agricultural sustainable development. This article analyzed the formation of saline-alkali soil, discussed mechanism of reclamation technologies, and introduced the practice of saline-alkali land reclamation in China. More high-efficient technologies and the integrated technologies of saline-alkali soil land reclamation would be developed. The saline-alkali soil problem would be solved successfully in the future.

1 INTRODUCTION

There were more than 99 million hm^2 saline-alkali soil land in China, which was a great environmental problem (Zhang, 2008; Mei *et al.*, 2013). In the face of ever-increasing population, decreasing arable land, the development of saline-alkali soil reclamation technology was of great significance (Wu, 2013). There were three major technologies, including physical technology, chemical technology and biotechnology, which were practiced in saline-alkali soil reclamation in China. It would improve the utilization of saline-alkali soil resources and ecological environment, and promote agricultural sustainable development by reclamation.

2 FORMATION OF SALINE-ALKALI SOIL

Saline-alkali soil mainly occurs in low-lying areas of arid, semi-arid, semi-humid or coastal plain (Liu, 2008). Salt and alkali gradually accumulated in the topsoil because of redistribution of a variety of soluble salts. In arid and semi-arid regions, salt and alkali, which were in the under earth or underground water, were carried to the surface of the land because of soil capillary action. Salt and alkali were kept in the topsoil after water evaporation, which formed the saline-alkali soil.

3 MECHANISM OF RECLAMATION TECHNOLOGIES

3.1 *Physical technologies*

3.1.1 *Salt washing*

The method of salt washing could leach the salt by irrigation water. Two requirements should be met: the first one was fresh water source, and the second one was the drainage system. This method could reduce the salinity of the soil to meet the salt concentration requirements of plants growth, but could not drastically remove the salt in the soil (Liu, 2008).

3.1.2 *Deep plough*

Deep plough could loosen the topsoil and enhance the permeability to prevent salt water rise before the rainy season. Deep plough could also prevent the soil compaction, improve soil aggregate structure, keep moisture and fertilizer, and reduce the damage of salt.

3.1.3 *Mulching*

Plastic film and biologic material mulching could reduce the evaporation of soil moisture, prevent the salt accumulation in topsoil. Zhao *et al.* (2001) used plastic film mulching method in the west of Jilin province, which had successfully reduced soil salinity and increased crop yields. Straw stubble mulching was used in Canadian prairies area, which could increase soil organic matter, and reduce the evaporation of soil moisture (Wang & Klein, 1996) Mulching could control salinity and keep it in the deep soil. However, this method could not radically exclude the salinity, which also had the potential risk of re-salinization.

3.2 *Chemical technologies*

Application of gypsum could increase soluble calcium (Ca^{2+}) and replace soluble sodium (Na^{+}) in the soil. This method was generally used in the high salinity saline-alkali soil to reduce soluble sodium (Na^{+}), which would reduce the harm of plants (Qadir & Oster, 2004).

3.3 *Biotechnologies*

3.3.1 *Organic fertilizer*

Organic materials, such as livestock manure, fallen leaves and litter, could increase soil organic matter, buffer performance, adjusting the pH, reducing its salinity, and reduce salt and alkali injury by decomposing (Wang *et al.*, 2010). The mathematical model could be established to determine optimum fertilization by relationship study between fertilization and crop production (Ruan *et al*, 2006; Wang *et al*, 2008), which would reclaim saline-alkali soil and obtain high and stable yield.

3.3.2 *Saline-alkali-tolerant plants*

Saline-alkali-tolerant plants could be selected to reclaim the saline-alkali soil, which had been proved to be a successful method. These plants generally had large biomass, resistance ability of salt and alkali. Some plants had strong ability to resist salt and alkali, such as Lemus chinensis, Sesbania cannabina, Medicago spp., Populus spp., *et al.* (Sun *et al.*, 2004; Cui *et al.*, 2008; Yu *et al.*, 2007; Cui *et al.*, 2008).

3.3.3 *Bacteria and fungi*

Soil bacteria, such as silicate bacteria (Li *et al.*, 2006), organic phosphorus bacteri and photosynthetic bacteria (Liu *et al.*, 2002), had an important effect on improving and utilizing of the saline-alkali soil. Fungi had been studied to reclaim saline-alkali soil. Cellulolytic fungi was inoculated on the wheat straw, which creased nodulation and growth of fenugreek (ABD & OMAR, 1998). The AM fungi in salt condition could promote plants to absorb mineral nutrients to improve salt tolerance and increase growth (Wang *et al.*, 2004).

4 THE PRACTICE OF SALINE-ALKALI LAND RECLAMATION

4.1 *Physical technologies reclamation practice*

Saline-alkali washing and deep plough technology had been applied mainly in the east of China, where had plenty of water source. In 2000, Shengli oilfield introduced the technique of "pipe alkali-modified" from Netherlands, and implemented it in the project. More than 6,700 hm^2 saline-alkali soil was reclaimed in north of the Yellow River estuary in Dongying city, Shandong province (Li *et al.*, 2010). Mulching technology was applied arid regions in the west of China because there was not enough water for salt washing. However, physical technology could only be a temporary solution. It just mechanically moved and eventually discharged the saline and alkali of soil into rivers, which lead to water pollution and saline-alkali hazard could not be fundamentally solved.

4.2 *Chemical technologies reclamation practice*

Lime was the main chemical element applied in the chemical technology reclamation project of saline soil. Considering the economic factors, desulfurization waste of flue gas was used as calcium ion source. It had been applied and proved to be an effective chemical technology. But it had no effect on lowering the underground water level. If groundwater rise and fall over and over again, calcium ion would be replaced depleted soon, which would lead to soil saline again (Li *et al.*, 2010). Because of the high cost of long distance transportation, chemical technologies were restrictive in the reclamation practice.

4.3 *Biotechnologies reclamation practice*

In recent years, great progress had been achieved on the issues of biotechnology reclamation practice in China. Organic fertilizer, bacteria and fungi were not widely applied because of economic factors, while saline-alkali-tolerant plant was popular in 21th century. Medlar could adapt to cold, drought and salt environment in the northwest of China. Medlar had been planted more than 20 hm^2 in the area of saline-alkali soil in Tongxin County nearby Yellow River, which had harvested every year (Yuan, 2007). A salt tolerance gene was successfully extracted from a common plant by Saline-alkali Land Bio-resources and Environment Research Center of Northeast forestry University (Yuan, 2012). This gene would enhance saline-alkali resistance ability of the plant after gene transformation. Zhongtian Poplar (Ba Li Zhuang Poplar) was successfully bred and application in the field in Dongying City in Shandong Province in China, which was Genetically Modified (GM) (Zhang, 2014). With molecular biology development and biotechnology innovation, more saline-alkali resistance transgenic crops would be used in biotechnology reclamation practice (Chen & He, 2010). However, it is difficult to reclaim saline-alkali land only by biotechnology reclamation.

5 SUMMARY

Saline-alkali soil land was a great environmental problem in China, which should be reclaimed to promote sustainable agricultural development. There were three major technologies for saline-alkali soil land reclamation. Firstly, physical technologies could mechanically move saline and alkali and were easy operated, but saline-alkali hazard could not be fundamentally solved. Secondly, chemical technologies could replace soluble sodium (Na^+) in the soil to reduce the harm of plants. However, chemical technologies were restrictive in the reclamation practice, because of the high cost of long distance transportation. Thirdly, biotechnologies were the best-known method for saline-alkali soil land reclamation. Organic fertilizer could not only reduce salt and alkali injury, but also increase soil organic matter. Saline-alkali-tolerant plants had a strong ability to resist salt and alkali. Bacteria and fungi could crease plants growth.

The saline-alkali soil problem would be solved successfully by technologies and methods progress of land reclamation. More high-efficient technologies of saline-alkali soil land reclamation would be developed in the future. The integrated technologies would become research focus, which would be more economic, maneuverable, less resource consumption and environmental friendly.

REFERENCES

ABD-ALLA M.H. & OMAR S. A. (1998) Wheat straw and cellulolytic fungi application increase nodulation, nodule efficiency and growth of fenugreek (Rigonella foenum—graceum L.) grown in saline soil. *Biology and Fertility of Soils*, 26(1), 58–65.

Chen, L.P. & He, D.Y. (2010) Research Advance on Drought and Salt Resistant Genes in Transgenic Plants. *Review and Progress*, 29(3), 542–549.

Cui, B.S., He, Q., Zhao, X.S. (2008) Ecological thresholds of Suaeda salsa to the environmental gradients of water table depth and soil salinity. *Acta Ecologica Sinica*, 28(4), 1408–1418.

Cui, X.Y., Wang, Y., Guo, J.X. (2008) Osmotic regulation of betaine in Leymus chinensis under saline-alkali stress and cloning and expression of Betaine Aldehyde Dehydrogenase (BADH) gene. *Chemical Research in Chinese Universities,* 24(2) 204–209.

Li, Y., Li, D.Y., Huang & M.Y., *et al.* (2006) Study on biological characteristics of several silicate bacteria isolated from saline soil. *Chinese Journal of Soil Science*, 37(1), 206–208.

Li, Y., Yi, H.F. & Zhao, B., *et al.* (2010) Study on improving Xinjiang sodic soils amelioration with desulfurized gypsum. *Ecology and Environmental Sciences*, 19(7), 1682–1685.

Liu, F.J., Hu, W.Y. & Li, Q.Y. (2002) Phytosynthetic Bacteria (PSB) as a water quality improvement mechanism in saline-alkali wetland ponds. *Journal of Environmental Sciences*, 14(3), 339–344.

Liu, J.H. (2008) Research progress of development and control of saline land. *Journal of Shanxi Agricultural Sciences,* 36(12), 51–53.

Mei, D.X., Zhan, D.S. & Wand, Z. (2013) The Status and Trends of Ecological Restoration of Saline-alkali Land in Coastal Region. *Chinese Agricultural Science Bulletin*, 29(5), 167–171.

Qadir, M. & Oster, J.D.. (2004) Crop and irrigation management strategies for saline-sodic soils and waters aimed at environmentally sustainable agriculture. *Science of the Total Environment*, 323, 1–19.

Ruan, C.C., Li, Y. & Zhang J.J., *et al.* (2006) Effect of Zn fertilizer on rice yield and yield characters in salinized soil. *Journal of Jilin Agricultural University*, 28(6), 91–593.

Sun, Z.Y. (2004) *Saline—alkaline soil greening technology*. Beijing, China Forestry Publishing House.

Wang, G.J., Zhang, L.H. & Zhao J.M., *et al.* (2004) The AM fungi on salt condition and the effect of AM fungi on plants. *Journal of Changchun Normal University*, 23(4), 64–68.

Wang, L.C., Xie, J.G. & Qin, Y.B., *et al.* (2008) The methods of soil testing and fertilizer recommendation. *Chinese Journal of Soil Science*, 39(4), 865–870.

Wang, L.M., Chen, J.L. & Liang, Z.H, *et al.* (2010) Research progress of improvement and utilization of saline and alkali soil. *Journal of Zhejiang Forestry College,* 27(1): 143–148.

Wang, X.B. & Klein, K.K. (1996) Management of stubble cover in Canadian steppe region. *Soil and Fertilizer Sciences in China,* 2, 34–38.

Wu, Y.P. (2013) *Effects of earthworms and organic matter on coastal saline—alkaline soils.* Beijing, China Agricultural University.

Yu, W.W., Cao, B.H. Cao F.L. (2007) Growth ionic absorption and ionic distribution re sponses to soil drought and drought-salt stresses in two Robinia pseudoacacia clones. *Journal of Zhejiang A & F University,* 24(3), 290–296.

Yuan, H.M. & Wu, H.X. Zhao, X.M. (2007) Efficient use of water in central ningxia drought area. *Modern agricultural science and technology*. 2004. 3B: 26–27.

Yuan, H.M. (2012) Investigation and Reflection on the Improvement and Utilization of Saline—Alkali Land in Australia, *World Agriculture*, 3, 57–59.

Zhang, J.F. (2008) Discussion on Ecological Rehabilitation of Salt-affected Soils. *Research of Soil and Water Conservation,* 15(4), 74–78.

Zhang, Y.S. (2014) The first new varieties of transgenic salt-resistant poplar poplar—Zhongtian Yang let salty land is no longer barren, *Forestry of China*, 3B, 26–27.

Zhao, L.P., Wang, Y. & Ma, J., *et al.* (2001) Study on Improvement of Soda Saline-Alkaline Soil in Western Jilin Province. *Chinese Journal of Soil Science*, 32, 91–96.

Land Reclamation in Ecological Fragile Areas – Hu (Ed.)
© 2017 Taylor & Francis Group, London, ISBN 978-1-138-05103-4

Does biochar promote reclaimed soil qualities?

Z. Cai
School of Environment Science and Spatial Informatics, China University of Mining and Technology, Xuzhou, China

T. Zheng
Institute of Soil Science, Chinese Academy of Science, Nanjing, China

M. Ahsan, H. Wang & Z.F. Bian
School of Environment Science and Spatial Informatics, China University of Mining and Technology, Xuzhou, China

ABSTRACT: Soil, taken from lake bottom and river bed, the selective sources of top soil construction in land reclamation, need to be improved with their properties in general. Biochar can be used to improve some soil quality. In this research, produced five different kinds of straw biochar at corresponding pyrolysis temperatures respectively and used 3 added levels into 2 kinds of soil followed seeding ryegrass to check alteration of soil quality. After the application of biochar, it decreases soil chroma with 1–3 units in sandy soil and 1–4 units in clay soil. Electrical Conductivity (EC) of soil is increased significantly and pH of soil is moderated after 30 days plantation of ryegrass. In the addition of biochar, MSA increases significantly but doesn't change WSA significantly. Furthermore, ryegrass planted soils was promoted significantly in height. It conclude that the production of different types biochars must be relating to enhancement of soil properties through specific pyrolysis conditions and soil types.

Keywords: Straw biochar; Soil properties; Soil quality; Ryegrass

1 INTRODUCTION

Biochar is a charcoal made of raw materials from thermal cracking of plants or waste residues (Suliman *et al.*, 2016). It has the following characteristics: fine particle, rich carbon (Woolf *et al.*, 2010) and difficult to be degraded. Incorporation of biochar into soil may have a significant effect on soil's physical and chemical properties. From the related studies, biochar is featured not only with large pore size, but also with more porosities and specific surface areas. Therefore, if it is added to the sandy soil (Case *et al.*, 2012), it can significantly reduce the soil's bulk density, and increase soil's porosity, thereby enhancing water capacity of soil (Dugan *et al.*, 2010), and improving soil's physical properties. Partial biochar could improve the pH value, EC value and nutrient level of soil (Abujabhah *et al.*, 2016). These characteristics of biochar microbial population distribution also played a certain degree of regulation (Grossman *et al.*, 2010). Biochar can promote crop growth by changing the physical and chemical properties of the soil. With stable structure, biochar can improve the performance of carbon negative soil via absorbing the carbon dioxide in the atmosphere and then transferring carbon into highly resistant soil (Case *et al.*, 2012; Troy *et al.*, 2013; Xie *et al.*, 2010). The emergence of biochar is a promising solution to the current ecological problems, and a positive catalyst for agricultural development (Atkinson *et al.*, 2012).

However, a large number of studies have shown that biochar cannot produce a positive effect on the improvement of soil properties. For example, its effect on water binding capacity of loam and clay is not obvious. The surface area of biochar prepared at high temperature is large, but they have less surface charges, and the ability to keep nutrient ions is not strong (Rehrah *et al.*, 2014). The biochars under different treatment can produce different effects on organic carbon contained in the soil (Abujabhah *et al.*, 2016).

This paper studied straw biochar under different temperatures with the treatment of mixing different amounts of reconstructed plough horizon soil for understanding the biochar's role in improving the quality of reconstructed soil (Major et al., 2010), and exploring the possibility of using biochar for reconstructing high-quality soil.

2 MATERIALS AND METHODS

2.1 *Experiment materials*

In this experiment, 5 kinds of biochar: C3, C4, C5, C6 and C7 were used. The raw materials used in this experiment were tobacco farm, and 19 winter wheat straws under different temperatures (Demirbas, 2004; Ahmad et al., 2012; Domene et al., 2015): 300°C, 400°C, 500°C, 600°C and 700°C. It took 30 minutes to reach the targeted temperature, and the temperature was kept stable for 2 hours in the muffle furnace and then was cooled down to room temperature (Atkinson et al., 2014).

Two kinds of soil textures from the Xuzhou City Peixian section of the Yellow River terraces and Xuzhou City Peixian Datun mining area for coal mining subsidence were filled in Reclamation of Lake Weishan Lake Sediment (They are the main sources of Xuzhou: remediated and reconstructed soils). The annual ryegrass was used in the experiment.

2.2 *Experimental method*

Soil was grinded through 1 mm sieve. Set 0.5%, 1% and 2% (mass ratio of carbon to soil) as three kinds of biochar addition levels. Two control groups (CK), a total of 32 treatment, and 3 replicates were made for each treatment, with a total number of 96 samples.

Soil's pH was measured via a pH of 2.5: 1 soil solution and then a pH meter (pHS3E, Shanghai Youke). The Electrical Conductivity (EC) value of the soil solution's conductivity was determined by setting the soil solution with soil-to-water ratio of 5: 1 and then by using the conductivity meter (SX713 type, Shanghai Sanxin Instrument Factory). The soil color was first qualitatively observed and then the color was graded via the Munsell color card. The soil aggregates were determined by dry sieve method and wet sieve method. The biomass of perennial ryegrass with electronic scales weight (0.01 g) was weighed to measure Fresh weight.

3 RESULTS AND DISCUSSION

3.1 *Effects of biochar on soil's pH value*

The pH value of biochar is related to prepared materials and corresponding conditions. The biochar prepared by wheat straw is generally alkaline and its alkalinity increases with the increase of temperature (Rehrah et al., 2014). In this experiment, the pH of prepared wheat straw biochar are 7.2, 8.1, 10.2, 10.5 and 10.3 respectively.

For sand, biochar generally reduces its pH value. There is a big difference on the effect of biochar on sand's pH of different added levels and different kinds of biochar. Figure 1 reflects the effect of biochar on sand's pH. The effect of biochar on lake mud's pH value is almost the same as that on sand, and all of them play a significant role in reducing the effect. Figure 2 reflects the effect of biochar on

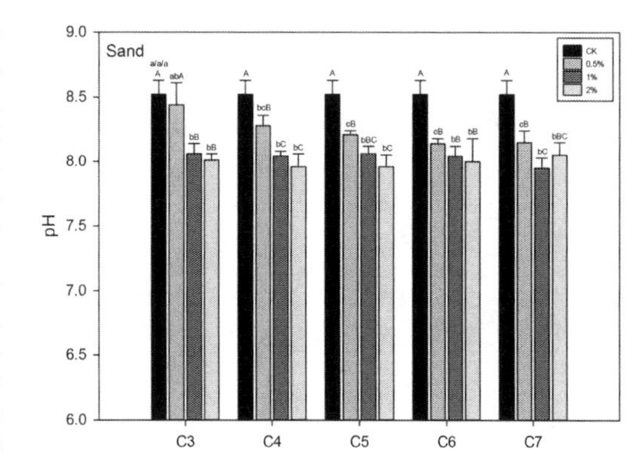

Figure 1. The effect of biochar on sand's pH. In this figure, different lowercase letters represent the significant level achieved by the difference between the different types of carbon treatment with the same carbon content (P<0.05); different capital letters refer to the significant level achieved by the difference between the different contents of carbon treatment with the same carbon species (P<0.05), the same below.

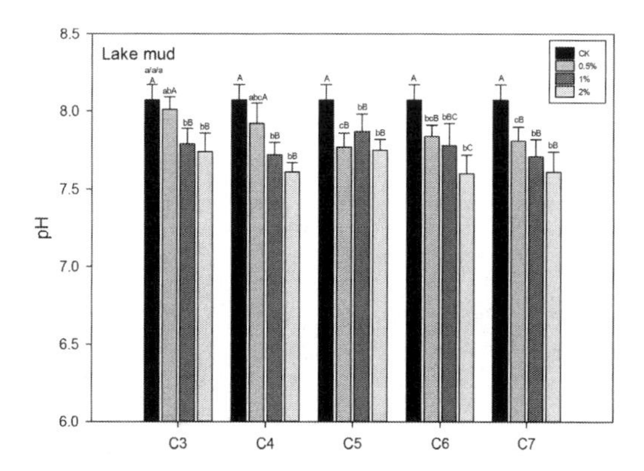

Figure 2. The effect of biochar on lake mud's pH.

the change of pH value of lake mud. It has been proved that biochar can increase the pH value of acid soil (Oguntunde et al., 2004), but the alkaline biochar in this study significantly lowered the pH value of alkaline soil.

3.2 *Effects of biochar on soil EC*

The EC value of biochar is also related to preparation conditions. Rehraha et al. (2014) suggested that EC value of biochar increased with the increase of thermal cracking temperature and thermal cracking time. In this experiment, biochar prepared from wheat straws at five different temperatures with different EC values of 820 us/cm, 1055 us/cm, 1530 us/cm, 1685 us/cm and 1595 us/cm. Figure 3 and Figure 4 show the effects of biochar on the EC values of sand and lake mud. Throughout the above analysis, it can be observed that the application of biochar can significantly improve the EC value of the test soils, mainly because the biochar EC value is much higher than that of test soil, and with the increase of carbon deposition, this effect is more and more obvious.

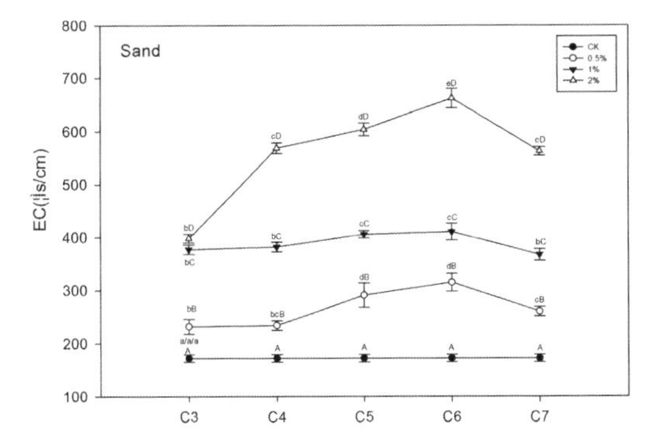

Figure 3. The effect of biochar on EC of sand.

3.3 *Effects of biochar on soil's color*

Soil color refers to the visible mixed shade that is reflected back by the soil (Wondafrash et al., 2005). This experiment used the Munsell soil colorimetric card to classify soil color, and the observation results are shown in Table 1.

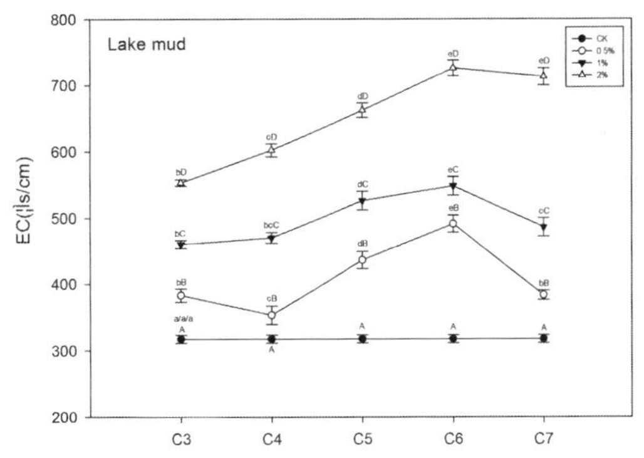

Figure 4. The effect of biochar on the EC value of lake mud.

Table 1. The observing result of the soil's color.

Sand						Lake mud					
Carbon content	Carbon spices	Tone	Brightness	Chroma	Color	Carbon content	Carbon spices	Tone	Brightness	Chroma	Color
SCK		10YR	7	4	very light brown	LCK		10YR	7	3	very light brown
0.50%	C3	10YR	6	4	light yellow-brown	0.50%	C3	10YR	5	4	Yellow-brown
	C4	10YR	6	3	light yellow-brown		C4	10YR	5	4	Yellow-brown
	C5	10YR	6	4	light yellow-brown		C5	10YR	6	4	light yellow-brown
	C6	10YR	6	4	light yellow-brown		C6	10YR	6	3	light brown
	C7	10YR	5	3	brown		C7	10YR	5	4	yellow-brown
1%	C3	10YR	6	4	light yellow-brown	1%	C3	10YR	5	4	yellow-brown
	C4	10YR	5	3	brown		C4	10YR	4	3	Brown
	C5	10YR	5	4	Yellow-brown		C5	10YR	5	4	yellow-brown
	C6	10YR	5	3	brown		C6	10YR	6	4	light yellow-brown
	C7	10YR	5	2	light grey-brown		C7	10YR	5	3	Brown
2%	C3	10YR	5	3	brown	2%	C3	10YR	4	4	Deep yellow-brown
	C4	10YR	4	4	Deep yellow-brown		C4	10YR	4	2	Deep grey-brown
	C5	10YR	4	3	brown		C5	10YR	5	3	Brown
	C6	10YR	4	2	Deep yellow-brown		C6	10YR	4	2	Deep grey-brown
	C7	10YR	4	2	Deep yellow-brown		C7	10YR	3	2	Very deep grey-brown

The results show that the higher the content of biochar in the soil, the stronger the effect on soil color will be. Biochar can absorb more heat, which is more likely to promote the rise of soil temperature.

3.4 *Effects of biological carbon on soil's aggregates*

Soil's aggregates constitute the smallest unit of soil, which plays an important role in determining soil's physical and chemical properties as an important indicator of soil's structure and stability (Brodowski et al., 2006; Sun & Lu, 2014).

This experiment was carried out mainly through Mean-Weight Diameter (MWD), geometgregates (WSA) and Percentage of Aggregate Disruption (PAD) to evaluate the composition and stability of soil aggregates. Figure 5 reflects the percentage of each particle size in the non-water stable aggregates of the sand.

It can be seen from Fig. 6 that the percentages of the non-water-stable aggregates of the lake mud are affected by the carbon content and the type of biochar.

The increase of soil aggregates (>0.25 mm) and its stability are important indexes of soil structural improvement. Therefore, related studies reported the non-water stability and water stability large aggregates of sand and lake mud, and calculated their agglomeration failure rate PAD, and the results are shown in Figure 7 and Figure 8.

The increase of soil aggregates (>0.25 mm) and its stability are important indexes of soil structural improvement. Therefore, related studies reported the non-water stability and water stability large aggregates of sand and lake mud, and calculated their agglomeration failure rate PAD, and the results are shown in Figure 7 and Figure 8.

The effects of biochar on non-water stable aggregates MWD and GMD are shown in Figure 9 and Figure 10.

Compared with MWD and GMD, the effect of biochar on promoting growth of MWD and GMD was not very strong and even negative. The results are shown in Figure 11 and Figure 12.

The effect of biochar on the stability of sandy soil and lake water stabilizer MWD and GMD is shown in Figure 13.

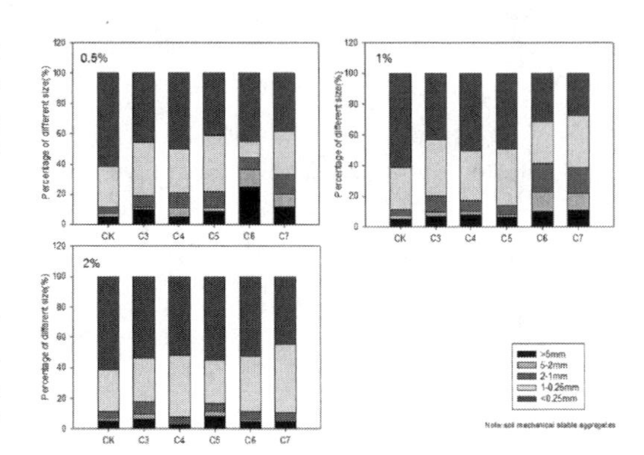

Figure 5. The percentage of different size (mechanical stable) in sand.

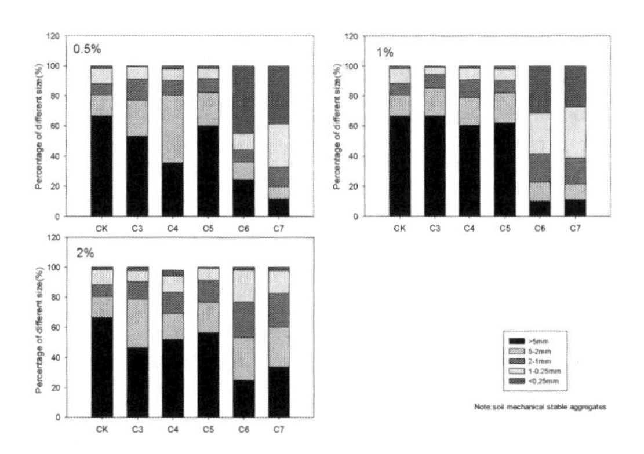

Figure 6. The percentage of different size (mechanical stable) in lake mud.

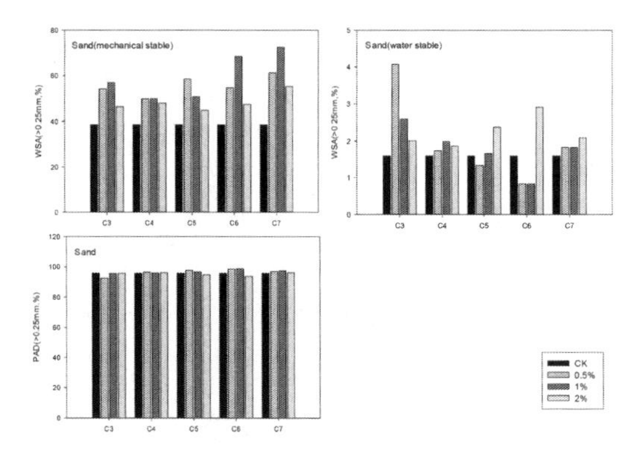

Figure 7. The WSA (>0.25 mm) and PAD (> 0.25 mm) of sand.

3.5 *Effects of biochar on fresh weight of ryegrass*

Fig. 14 and Fig. 15 respectively present the effects of biochar's fresh weight of ryegrass on sand and lime. The effects of carbon species and carbon content on the fresh weight of ryegrass were not significantly different between the two soils. As the experiment revealed, the initial application of biochar could inhibit the seed's germination (Denyes et al., 2012).

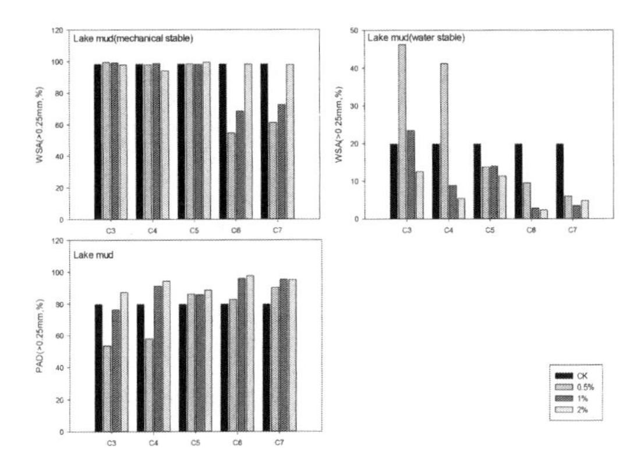

Figure 8. The WSA (> 0.25 mm) and PAD (> 0.25 mm) of lake mud.

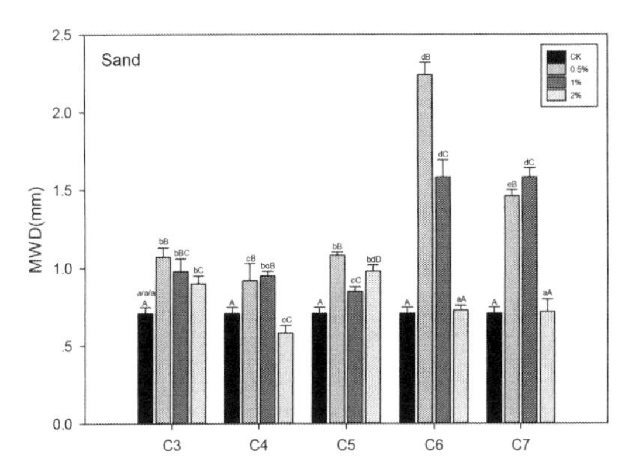

Figure 9. The MWD of sand mechanical stable aggregates.

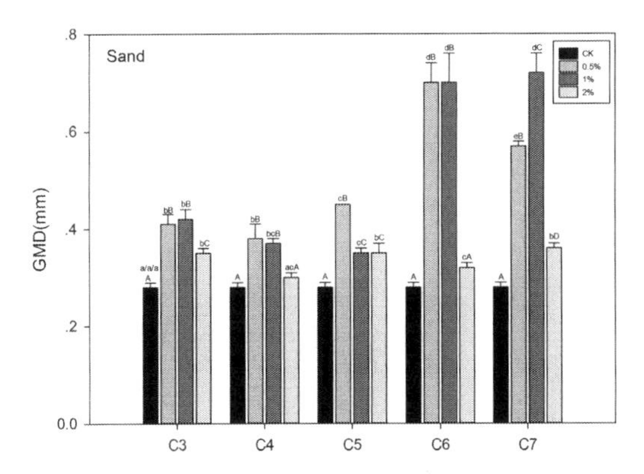

Figure 10. The GMD of sand mechanical stable aggregates.

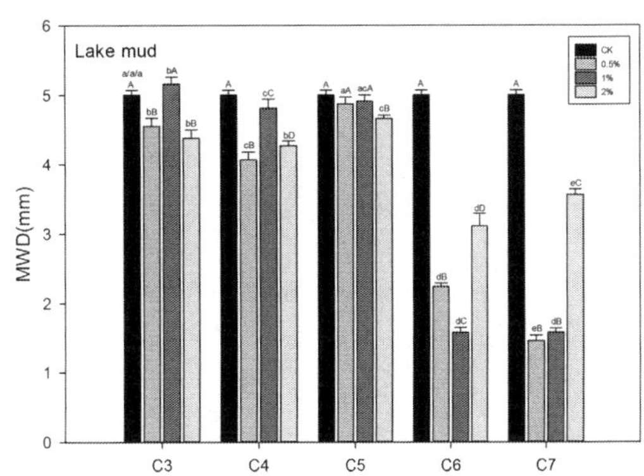

Figure 11. The MWD of lake mud mechanical stable aggregates.

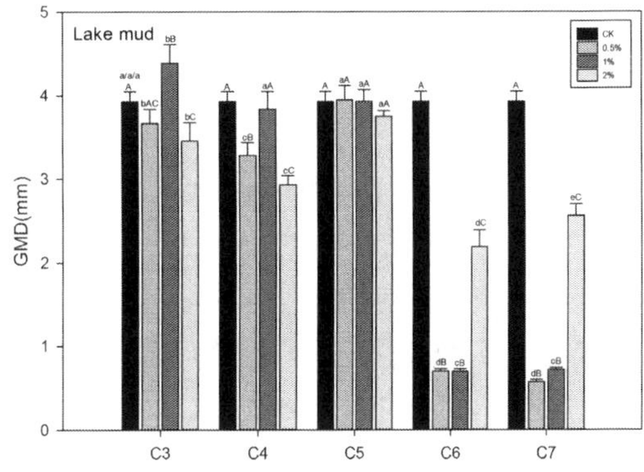

Figure 12. The GMD of lake mud mechanical stable aggregates.

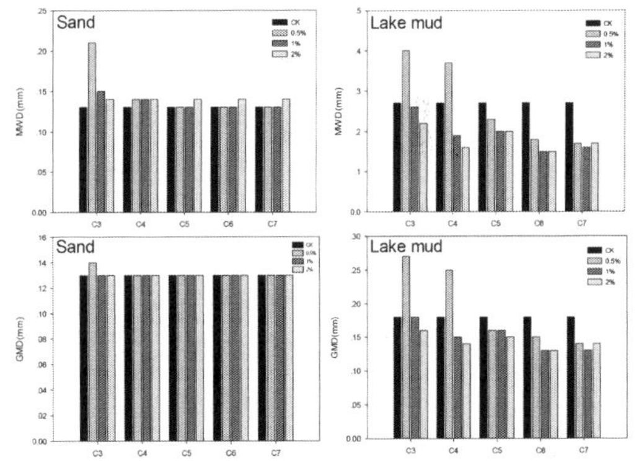

Figure 13. The MWD and GMD of soil water stable aggregates.

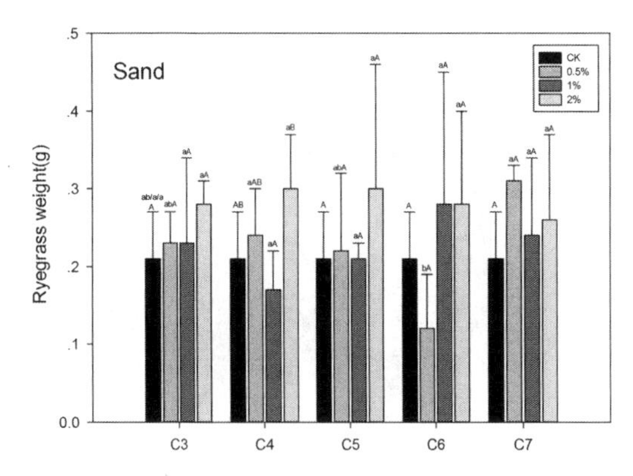

Figure 14. The effects of biochar's Ryegrass weight of sand.

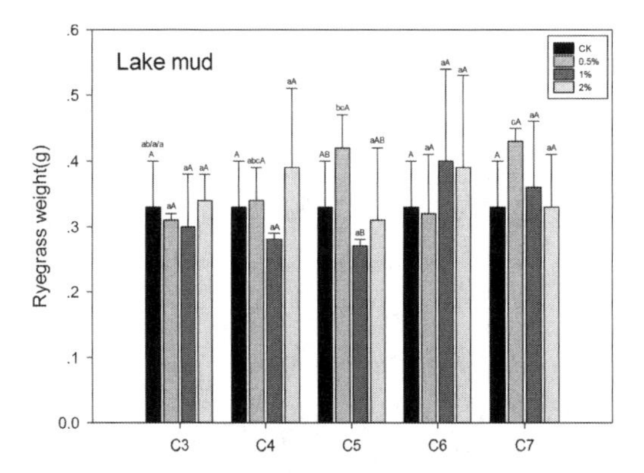

Figure 15. The effects of biochar's Ryegrass weight of lake mud.

4 CONCLUSION

From the experimental results, the application of biochar can effectively change soil's color, pH value, EC value, aggregate and improve plant growth, but the degree of influence and the direction of influence are different.

Because of its own color, the effect of biochar can significantly downgrade the soil's color, which reduces the sandy light by 1 to 3 units, and the lacustrine clarity by 1–4 units. And both the preparation temperature and the content are high (at high temperature (C6, C7) high carbon content (2%) is the most obvious). Biochar will reduce the pH of alkaline soils in the case of plant growth, but there is no explanation for this abnormal result. The only possible explanation is that the presence of biochar promotes the secretion of acidic substances in plant roots, and further tests are necessary. The significantly increased number of non-water stability aggregates the form of non-water stability aggregates mainly associated with clay in the soil, the sand itself contain little of clay, after the application of biochar, the clay in the sand is increased, so the number of non-water stability aggregates are increased, and presents no significant effect on green weight of ryegrass.

Biochar has a significant effect on soil's Electric Conductivity (EC) to some degree, which can be inferred from the results of the study:

1. Due to biochar itself, the effect of biochar on soil may not be possibly from other substances that are mixed with biochar. A large number of studies show that biochar has high pH value and EC value with the increase of temperature. In this regard, oxygen is also one important factor during the producing process.
2. Biochar's structure is stable, and it is difficult to be degraded. It can improve the soil's carbon content, but cannot effectively improve soil's organic content. It must be applied with organic matters, otherwise it cannot play a key role in forming soil's structure.
3. Pure biochar is carbon skeleton with a stable porous structure. The mechanism of improving soil's physical and chemical properties is via increasing its surface area and surface groups to cement, and furthermore to promote the formation of soil's structure, so as to further improve the biological environment and ultimately the soil quality.

ACKNOWLEDGEMENTS

This research was supported by the National Key Research Development Plan Program during the 13th Five-Year Plan Period (2016YF0501100) and the Key Technologies of Land Consolidation in Large-scale Coal Base (2016YF0501105). The authors are grateful for this financial support.

REFERENCES

Abujabhah, I.S., Bound, S.A, Doyle, R. & Bowman, J.P. (2016) Effects of biochar and compost amendments on soil physico-chemical properties and the total community within a temperate agricultural soil. *Applied Soil Ecology*, 98, 243–253.

Ahmad, M., Lee, S.S., Dou, X., Mohan, D., Sung, J.K., Yang, J.E. & Ok, Y.S. (2012) Effects of pyrolysis temperature on soybean stover-and peanut shell-derived biochar properties and TCE adsorption in water. *Bioresource Technology*, 118, 536–544.

Atkinson, C.J., Fitzgerald, J.D. & Hipps, N.A. (2010) Potential mechanisms for achieving agricultural benefits from biochar application to temperate soils: a review. *Plant and Soil*, 337(1–2), 1–18.

Brodowski, S., John, B., Flessa, H. & Amelung, W. (2006) Aggregate-occluded black carbon in soil. *European Journal of Soil Science*, 57, 539–546.
Case, S.D., McNamara, N.P., Reay, D.S. & Whitaker, J. (2012) The effect of biochar addition on N_2O and CO_2 emissions from a sandy loam soil—the role of soil aeration. *Soil Biology and Biochemistry*, 51, 125–134.
Deenik, J.L., McClellan, T., Uehara, G., Antal, M.J. & Campbell, S. (2010) Charcoal volatile matter content influences plant growth and soil nitrogen transformations. *Soil Science Society of America Journal*, 74(4),1259–1270.
Demirbas, A. (2004) Effects of temperature and particle size on bio-char yield from pyrolysis of agricultural residues. *Journal of Analytical and Applied Pyrolysis*, 72(2), 243–248.
Denyes, M.J., Langlois, V.S., Rutter, A. & Zeeb, B.A. (2012) The use of biochar to reduce soil PCB bioavailability to *Cucurbita pepo* and *Eisenia fetida*. *Science of the Total Environment*, 437, 76–82.
Domene, X., Enders, A., Hanley, K. & Lehmann, J. (2015) Ecotoxicological characterization of biochars: role of feedstock and pyrolysis temperature. *Science of the Total Environment*, 512: 552–561.
Dugan, E., Verhoef, A., Robinson, S., & Sohi, S. (2010, August). Bio-char from sawdust, maize stover and charcoal: Impact on Water Holding Capacities (WHC) of three soils from Ghana. In *19th World Congress of Soil Science, Symposium*, 4(2), pp. 9–12.
Grossman, J.M., O'Neill, B.E., Tsai, S.M., Liang, B., Neves, E., Lehmann, J. & Thies, J.E. (2010) Amazonian anthrosols support similar microbial communities that differ distinctly from those extant in adjacent, unmodified soils of the same mineralogy. *Microbial Ecology*, 60(1), 192–205.
Major, J., Rondon, M., Molina, D., Riha, S. J. & Lehmann, J. (2010) Maize yield and nutrition during 4 years after biochar application to a Colombian savanna oxisol. *Plant and Soil*, 333(1–2), 117–128.
Morales, M.M., Comerford, N., Guerrini, I.A., Falcão, N.P.S. & Reeves, J.B. (2013) Sorption and desorption of phosphate on biochar and biochar–soil mixtures. *Soil Use and Management*, 29(3), 306–314.
Oguntunde, P.G., Fosu, M., Ajayi, A.E. & Van De Giesen, N. (2004) Effects of charcoal production on maize yield, chemical properties and texture of soil. *Biology and Fertility of Soils*, 39(4), 295–299.
Rehrah, D., Reddy, M.R., Novak, J.M., Banslde R.R., Schimmel, K.A., Yu, J., Watts, D.W. & Ahmedna, M. (2014) Production and characterization of biochars from agricultural by-products for use in soil quality enhancement. *Journal of Analytical and Applied Pyrolysis*, 108, 301–309.
Suliman, W., Harsh, J.B., Abu-Lail, N.I., Fortuna, A.M., Dallmeyer, I. & Garcia-Perez, M. (2016) Influence of feedstock source and pyrolysis temperature on biochar bulk and surface properties. *Biomass and Bioenergy*, 84, 37–48.
Sun, F. & Lu, S. (2014) Biochars improve aggregate stability, water retention, and pore space properties of clayey soil. *Journal of Plant Nutrition and Soil Science*, 177(1), 26–33.
Troy, S.M., Lawlor, P.G., O'Flynn, C.J. & Healy, M.G. (2013) Impact of biochar addition to soil on greenhouse gas emissions following pig manure application. *Soil Biology and Biochemistry*, 60, 173–181.
Wondafrash, T.T., Sancho, I.M., Miguel, V.G. & Serrano, R.E. (2005) Relationship between soil color and temperature in the surface horizon of Mediterranean soils: A laboratory study. *Soil Science*, 170(7), 495–503.
Woolf, D., Amonette, J.E., Street-Perrott, F.A., Lehmann, J. & Joseph, S. (2010) Sustainable biochar to mitigate global climate change. *Nature communications*, 56.
Xie, Z.B., Liu, G., Bei, Q.C, Tang, H.Y., Liu, J.S., Sun, H.F., Xu, Y.P., Zhu, J.G. & Cadisch, G. (2010) CO_2 mitigation potential in farmland of China by altering current organic matter amendment pattern. *Science China Earth Sciences*, 53(9), 1351–1357.

Land Reclamation in Ecological Fragile Areas – Hu (Ed.)
ISBN 978-1-138-05103-4

Characteristics and adsorption properties of cotton-stalk derived biochars

L. Yang & X.Q. Zhou
College of Agriculture, Shihezi University, Shihezi, Xin Jiang, China

H. Deng
School of Chemistry and Chemical Engineering, Shihezi University, Shihezi, Xin Jiang, China

D.F. Ma
College of Agriculture, Shihezi University, Shihezi, Xin Jiang, China

ABSTRACT: A series of biochars were prepared by pyrolyzing cotton stalk at different carbonization temperature (300–700°C). The element composition and pore structure were characterized by elemental analyzer and BET surface area analyzer, and the adsorption capability of biochars for the removal of p-nitrotoluene from aqueous solution was evaluated. The results showed that the carbon content and aromaticity of the biochar produced from cotton stalk increased with the increase of pyrolysis temperature, whereas the polarity of biochar decreased. The micropore of biochar was suddenly developed and surface area was sharply enlarged when the carbonization temperature rised from 600°C to 700°C. The pseudo-second-order kinetic equation could be used to better describe the sorption process of p-nitrotoluene onto biochar, while the adsorption isotherms was better fitted by langmuir equation, and the five kinds of biochars followed an order of CB700 > CB600 > CB500 > CB400 > CB300.

1 INTRODUCTION

Water pollution by aromatic compounds has become an issue of major concern in recent years. These substances, due to high toxicity and long persistence, are widely present in waste water, which may lead to mutagenic and carcinogenic responses through the bioaccumulation (Lua and Jia, 2007; Williams, 1990). This may pose the risk to both human health and the aquatic environment. The p-nitrotoluene is one of the most important intermediate and material for the chemical, medication and pesticide industry, and is difficult to be biodegraded (Cheng et al., 2016). Hence, it is crucial to removal of p-nitrotoluene from wastewater before discharging into the environment.

According to previous studies, adsorption is the most suitable and effective method for the removal of persistent organic pollutants from wastewater, and is also the most attractive technique due to simple operation, broad availability and regeneration of adsorbent, low-energy demand (Ania et al., 2007). The commercial activated carbon has been the most extensively used as an adsorbent for pollutants' removal from aqueous solutions. However, as a consequence of high-cost materials and regeneration cost, there is a quest for relatively cheap and renewable materials as precursors to produce low cost adsorbent.

Xinjiang is the largest cotton-producing area, and it is inevitable to produce a large quantity of cotton stalk after the harvest. However, except for stalk returning to field, cotton stalk is disposed of by either open burning or discarding, which has result in many environmental problems (Deng et al., 2010). Hence, there is an urgent need to diversify the applications of cotton stalk and lower the risk of environmental pollution. The objective of this study was to prepare biochar from cotton stalk and investigate the effect of pyrolysis temperature on the element composition and pore structure of cotton-stalk derived biochar. Furthermore, the removal capacity of p-nitrotoluene from aqueous solution was also studied.

2 MATERIALS AND METHODS

2.1 *Biochar production*

The feedstock for the production of biochar, Cotton stalk, was provided from the farm in Shihezi, Xinjiang. After naturally air-dried, the feedstock was milled to less than 2.0 mm and stored in airtight bottle until ready for use. In order to avoid contact with air during the pyrolysis, 15 g of dried samples were filled into a sealed crucible. Then the sealed crucibles were placed in a furnace and were heated for 6 h at five different temperatures of 300°C, 400°C, 500°C, 600°C and 700°C, respectively. After pyrolysis, the carbonized samples were washed several times with distilled water. Finally, the dried samples under the pyrolysis temperature was denoted as CB300, CB400, CB500, CB600 and CB700, respectively.

2.2 *Characterization of biochar*

The elemental compositions in the biochar were measured by using a CHNS/O analyzer (VarioEL cube, Elementar), and the O content was obtained by weight difference. The pH of the samples was measured by a pH meter with a ratio 1:2.5 sample to water. The pore structure characteristics of the biochar were determined by N2 adsorption at 77 K by using sorption analyzer (ASAP 2010M + C, USA).

2.3 *Batch adsorption experiments*

In the batch adsorption experiments, the carbonized samples were used as the absorbents of p-nitrotoluene from aqueous solution. In the experiment, 0.1 g of the carbonized samples were put in serum bottles containing 100 mL of different p-nitrotoluene concentration solution (20, 40, 80, 120, 160 and 200 mg/l). Then the serum bottles were placed on a shaker and agitated at 150 rpm for a predetermined period at 25°C. After agitation, the solutions were filtered and the concentration of p-nitrotoluene was measured using UV-Vis spectrophotometer at the wavelength of 275 nm. The amount of p-nitrotoluene adsorbed by the carbonized samples was calculated using the following equation:

$$q_t = \frac{(C_0 - C_t)V}{m} \tag{1}$$

where C_0 and C_t are the initial and final concentrations of p-nitrotoluene at solutions (mg/L), respectively. V is the solution volume (L) and m is the mass of the carbonized sample (g).

2.4 *Kinetic adsorption models*

In order to study the kinetics of p-nitrotoluene adsorption onto the carbonized samples, three kinetic models were tested. The pseudo-first order model can represented by the linear form (Aydin et al., 2008):

$$\log(q_e - q_t) = \log(q_e) - \frac{k_1}{2.303}t \tag{2}$$

The pseudo-second order model can be written in the linear form (El-Ashtoukhy et al., 2008):

$$\frac{t}{q_t} = \frac{1}{k_2 q_e^{\,2}} + \frac{t}{q_e} \tag{3}$$

The linear form of the intraparticle diffusion model can be written as (El-Ashtoukhy et al., 2008):

$$q_t = k_{id} t^{1/2} + C \tag{4}$$

where q_e and q_t are the amounts of p-nitrotoluene adsorption at equilibrium and at time t, respectively (mg/g), respectively. The k_1, k_2 and k_{id} are the pseudo -first order, pseudo-second order and intraparticle diffusion adsorption rate constants (h^{-1}), respectively, t is the time and C is an intraparticle diffusion constant.

2.5 *Adsorption isotherms models*

In order to investigate the equilibrium behavior of p-nitrotoluene adsorption onto the carbonized samples, the Langmuir isotherm and the Freundlich isotherm were tested.

The Langmuir isotherm can represented by the linear form (Langmuir, 1918):

$$\frac{C_e}{q_e} = \frac{1}{q_m k_L} + \frac{C_e}{q_m} \tag{5}$$

The Freundlich isotherm can represented by the linear form (Freundlich, 1906):

$$\ln(q_e) = \ln(k_F) + \frac{1}{n}\ln(C_e) \tag{6}$$

where q_e is the amounts of p-nitrotoluene adsorption at equilibrium (mg/g), and and C_e is the p-nitrotoluene concentration at equilibrium (mg/L), k_L is Langmuir constant (L/mg), k_F is the Freundlich distribution coefficient (L/mg), 1/n is a Freundlich dimensionless parameter.

3 RESULTS AND DISCUSSION

3.1 *Characteristics of various biochar*

As shown in Table 1, the elemental compositions were also significantly affected by the pyrolysis temperatures. The C content significantly increased from 60.19% in CB300 to 68.19% in CB500, while above that, the C content of the biochars slightly increased from 69.39% to 71.58%. Conversely, due to emission of the volatile matter, the contents of H, O, and N dramatically decreased from 3.26%, 33.77%, 2.08% (CB300) to 1.22%, 25.30%, 1.33% (CB700), respectively. Furthmore, atomic ratios such as H/C, O/C and (O + N)/C were a decreasing trend with increasing temperature. This illustrated, during the pyrolysis, the aromaticity of the biochar produced from cotton stalk increased with

Table 1. Physico-chemical characteristics of samples.

Characteristics	Samples CB300	CB400	CB500	CB600	CB700
pH	7.78	8.26	9.48	10.15	10.35
C/%	60.19	63.66	68.19	69.39	71.58
N/%	2.08	2.02	1.82	1.68	1.33
H/%	3.26	2.88	2.25	2.06	1.22
(N+O)/C	0.596	0.515	0.424	0.403	0.372
O/C	0.561	0.483	0.398	0.379	0.353
H/C	0.054	0.045	0.033	0.030	0.017

Table 2. Adsorption kinetics parameters of Pseudo-first-order model for p-nitrotoluene.

Samples	$q_{e.exp}$ (mg/g)	Pseudo-first-order model $q_{e.cal}$ (mg/g)	k_1 (h^{-1})	R^2
CB300	22.58	4.21	0.4721	0.7168
CB400	28.24	7.39	0.3369	0.9033
CB500	34.88	8.92	0.4343	0.9238
CB600	37.01	8.76	0.4410	0.9191
CB700	92.86	18.62	0.4583	0.9104

Table 3. Adsorption kinetics parameters of Pseudo-second-order Model for p-nitrotoluene.

Samples	$q_{e.exp}$ (mg/g)	Pseudo-second-order model $q_{e.cal}$ (mg/g)	k_2 (h^{-1})	R^2
CB300	22.58	22.78	0.2439	0.9997
CB400	28.24	28.49	0.2053	0.9999
CB500	34.88	35.21	0.1792	0.9998
CB600	37.01	37.17	0.2335	0.9999
CB700	92.86	93.46	0.0954	0.9999

Table 4. Adsorption equilibrium parameters for p-nitrotoluene.

	Langmuir			Freundlich		
Samples	q_m (mg/g)	K_L (L/mg)	R^2	K_F (L/mg)	1/n	R^2
CB300	24.27	0.093	0.9997	6.589	0.2578	0.9445
CB400	30.96	0.073	0.9978	7.403	0.2773	0.9815
CB500	39.68	0.055	0.9916	6.833	0.3373	0.9873
CB600	42.92	0.051	0.9892	7.334	0.3385	0.9885
CB700	100	0.251	0.9898	41.22	0.1864	0.9904

the increase of pyrolysis temperature, whereas the polarity of biochar decreased.

The effect of pyrolysis temperature on the pore structure characteristics were analyzed by using the N_2 adsorption/desorption method. As the pyrolysis temperature increased, BET surface area and total pore volume of the samples gradually increased, but there were not remarkably change from 300°C to 600°C. However, as temperature increased to 700°C, the surface area of the sample dramatically increased from 9.489 m^2/g to 382.4 m^2/g, and average pore size decreased from 8.83 nm to 1.87 nm. In accordance to the classification of the International Union of Pure and Applied Chemistry (IUPAC, 1972), CB700 was dominant in micropore region.

3.2 *Adsorption kinetics*

The pseudo-first order and pseudo-second order model were fitted the experimental data and the results were shown in Table 2 and Table 3. The results showed that the pseudo-second order model best described adsorption kinetics for all the samples, with highest correlation coefficients values and the values of $q_{e,cal}$, which were very close to the experimental values.

3.3 *Adsorption equilibrium*

The equilibrium characteristic of p-nitrotoluene adsorption on the samples were described by Langmuir and Freudlich isotherms (Deng et al., 2010). As shown in Table 4, besides the CB700, the adsorption isotherms of p-nitrotoluene onto biochar were better fitted by langmuir equation, indicating that the sorption was dominated with monolayer sorption. In terms of p-nitrotoluene saturated sorption quantity, the five kinds of biochar followed an order of CB700 > CB600 > CB500 > CB400 > CB300.

4 CONCLUSIONS

The pyrolysis temperature could affect the characteristics and the adsorption capability of cotton-stalk biochar. With the increase of the pyrolysis temperature, the polarity of biochar decreased, whereas aromaticity dramatically increased. Increasing the temperature to 700°C promoted the formation of microspores. The maximum sorption capacity of CB700 was 100 mg/g, which could regard as an effective adsorbents to remove p-nitrotoluene.

ACKNOWLEDGMENTS

This work was supported by financial support from the National Natural Science Foundation of China (Grant No. 41401349, 51162024).

Yang Le and Deng Hui contributed equally to this work

REFERENCES

Ania, C.O., Cabal, B., Pevida, C., Arenillas, A., Parra, J.B., Rubiera, F. & Pis, J.J. (2007) Removal of naphthalene from aqueous solution on chemically modified activated carbons. *Water Res*, 41, 333–340.

Aydın, H., Bulut, Y., & Çiğdem Yerlikaya. (2008). Removal of copper (ii) from aqueous solution by adsorption onto low-cost adsorbents. *Journal of Environmental Management*, 87(1), 37–45.

Benaïssa, H. & Elouchdi, M.A. (2007). Removal of copper ions from aqueous solutions by dried sunflower leaves. *Chemical Engineering & Processing Process Intensification*, 46(7), 614–622.

Cheng, W.P., Gao, W., Cui, X., Ma, J.H. & Li, R.F.. (2016) Phenol adsorption equilibrium and kinetics on zeolite X/activated carbon composite. *Journal of the Taiwan Institute of Chemical Engineers*, 62:192–198.

Deng, H., Li, G.X., Yang, H.B., Tang, J.P., & Tang, J.Y. (2010). Preparation of activated carbons from cotton stalk by microwave assisted KOH and K_2CO_3 activation. Chemical Engineering Journal, 163(3), 373–381.

El-Ashtoukhy, E.S.Z., Amin, N.K., & Abdelwahab, O. (2008). Removal of lead (ii) and copper (ii) from aqueous solution using pomegranate peel as a new adsorbent. *Desalination*, 223(1–3), 162–173.

Freundlich HMF. (1906) Über die adsorption in lösungen. *Z Phys Chem*, 57:385–470.

Langmuir, I. (1918) The adsorption of gases on plane surfaces of glass silica and mica. *J. Am. Chem. Soc.* 40, 1361–1403.

Lua, A. & Jia, Q.P. (2016) Adsorption of phenol by oil-palm-shell activated carbons in a fixed bed. *Adsorption*, 13(2):129–137.

Williams, P.T. (1990) Sampling and analysis of polycyclic aromatic compounds from combustion systems-a review. *J. Inst. Energy*, 63, 22–30.

Land Reclamation in Ecological Fragile Areas – Hu (Ed.)
© 2017 Taylor & Francis Group, London, ISBN 978-1-138-05103-4

Spatial distribution and risk assessment of metals in overlapped areas of farmland and coal resources soils, Xuzhou, China

X.F. Sun & X. Xiao
School of Environment Science and Spatial Informatics, China University of Mining and Technology, Xuzhou, Jiang Su, China

Y.Y. Hu
School of Electric Power Engineering, China University of Mining and Technology, Xuzhou, Jiang Su, China

J. Chang
School of Architecture and Design, China University of Mining and Technology, Xuzhou, Jiang Su, China

H. Wang
School of Environment Science and Spatial Informatics, China University of Mining and Technology, Xuzhou, Jiang Su, China

ABSTRACT: In this case study, we studied pollution in overlapped areas of farmland and coal resources in the north of Xuzhou, Jiangsu Province, China. Totally 107 soil samples taken from the grid distributed locations were assayed for the trace elements, Cu, Cd, Pb, Zn, Cr, Hg and As. The spatial distribution of the pollutant concentrations was evaluated with respect to the locations of the potential contamination sources. The contamination level of this region was assessed based on the Soil Pollution Control Regulation of China. Results indicated that the contents of Cd, Hg, Cu, Pb, Zn, and Cr, especially the former two, are higher in the soil than the background. The ArcGIS data showed that pollutants are more likely to be detected in the vicinity of the contamination sources. Pb, Zn, Cu, and Cr are mainly spotted in the southwest and east of the region, particularly fly ash yards and new industry areas. Cd and As are more spotted in the straight north of the region, and Hg in the east and west. The results from the single factor pollution index showed that Cd pollutes 43.9% of sample locations heavily and 28.0% moderately, whereas the contents of the other six trace elements are safely below the detectable level of pollution. According to the pollution load index, twenty-six sample locations are heavily polluted. The potential ecological risk from the seven trace elements, in the order of decline, is Hg, Cd, Cu, As, Pb, Cr, and Zn. Five sample locations are of the severe ecological risk.

Keywords: overlapped areas of farmland and coal resources, metal spatial distribution, risk assessment

1 INTRODUCTION

Overlapped areas of Farmland and Coal resources (OFC) cover thirteen provinces and municipalities, and more than 26% of farmland in China (Hu, Z., et al., 2014). Coal Mining and utilization have a long history of negative impacts on environments due to emission of particulate matters, and inorganic and organic pollutants (Zarić, N.M., et al., 2016). Toxic metals contamination and vegetation damage have been found in coal mining regions (Bhuiyan, M.A., et al., 2010, Halim, M.A.,et al., 2015) and around coal-based thermal power plants (Singh, S., et al., 2015). Toxic metals in soil are indiscernible, permanent and irreversible. The excessive accumulation of toxic metals in agriculture soils degrades land and reduces yield (Zhao, C. and H. Wu, 2017).

Xuzhou is a typical OFC area as it has exploited coal mines for over 124 years, and it has produced 259.7 million tons of coal and generated 7.2 × 109 kWh of electric power in 2006 (Huang, S., et al., 2009) and Huaihe River Basin (HRB) is one of the most important food production areas in China. This work is, therefore, carried out to explore the degree and spatial distribution of Cu, Cd, Pb, Zn, Cr, Hg and As pollution of Overlapped areas of farmland and coal resources soil in north of Xuzhou.

2 STUDY AREA

The OFC area in this study is located 8 km northwest of the downtown of Xuzhou, 34°32′–34°36′ latitude and 117°07′–117°19′ longitude, The climate in this area is of a typical warm humid monsoon with an average annual temperature of 14°C and rainfall of 900 mm. The bedrock consists mainly of carboniferous grey limestone. Extensive diluvial plains and sporadic uplands form the local micro-land with fluvo-aquic soil layered upon the alluvium. There are five coal mines, three coal-fired electric plants, two large fly ash yards, and several coal yards on our concerned site. More than two-thirds of the area is cropland, surrounding almost all of the coal mines and electric plants. Wheat, paddy rice, maize, and soybean are the major crops grown in this area.

3 MATERIALS AND METHODS

3.1 *Samples collection*

The area concerned was divided into 100 × 100 m grids after a review of topographic and geologic maps. All together 107 samples were collected (Figure 1) and the sample locations were recorded using hand-held Global Positioning System (GPS). Samples were shoveled from the topsoil of 0–15 cm, and then placed into polyethylene bags before brought back to the laboratory. The soil was dried at room temperature, and sieved with an opening size of 2 mm after stones and roots were removed. The samples were stored in desiccators.

3.2 *Determination of heavy metals*

The total concentrations of the metals, Cu, Cd, Pb, Zn, Cr and As, in the bulk soil were analyzed by using ICP-AES (PE, optima8000) after digested in the solution of $HCl\text{-}HNO_3\text{-}HClO_4\text{-}HF$ (3:1:3:1) (GB/T17140). Hg was analyzed through Direct-Mercury Analyzer (Milestore, DMA 80). The accuracy of the metal concentration was determined by using two soil certified reference materials, GWB 07404 and GWB07427.

Figure 1. Location map of the study area.

Table 1. Statistical processing and comparison of the studied metals.

	Metal concentrations (mg/kg)						
	Cu	Cd	Pb	Zn	Cr	Hg	As
Mean	27.34	0.98	21.68	74.34	61.64	0.06	11.11
Maximum	76.72	2.09	45.78	176.15	84.87	0.44	28.88
Minimum	11.36	0.34	12.34	44.50	45.81	0.01	1.12
std.Dev	10.56	0.42	5.23	20.21	8.99	0.05	6.05
CV(%)	38.64	42.31	24.15	27.19	14.58	93.27	54.45
Background (Zhang, M.M., et al., 2013)	12.61	0.29	16.30	91.01	55.50	0.01	11.20
Guide value (GB15618–1995)	100.00	0.30	300.00	250.00	200.00	0.50	30.00

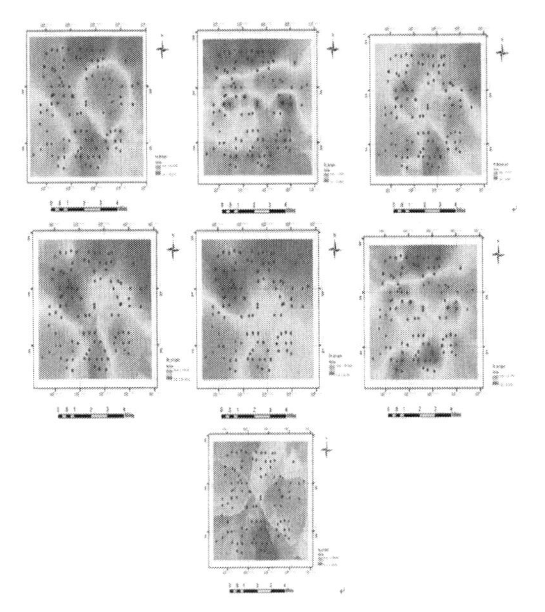

Figure 2. The spatial distributions of heavy metals and metalloids in surface soil.

3.3 *Quantification of soil pollution*

3.3.1 *Pollution Load Index(PLI)*

To assess the environmental quality of the soil, an integrated pollution load index of seven metals was calculated according to Suresh (Suresh, G., et al., 2012). The PLI was defined as the nth root of the multiplication of the contents (CFmetals),

$$PLI = (CF_1 \times CF_2 \times CF_3 \times CF_n)^{1/n} \quad (1)$$

where CF metals is the ratio of the content of each metal and the guide value from the Chinese Environmental Quality Standard for Soil.

3.3.2 *Potential Ecological Risk (PER)*

The equations for calculating the PER are as follows:

$$c_f^i = {c^i}/{c_n^i}, PER = \sum_{i=1}^{m} T_r^i \times c_{fi}^i \quad (2)$$

where C_f^i is the single element pollution factor, c^i is the content of the element in the samples, and c_n^i is the reference pollution degree of the environment, T_r^i is the biological toxic factor of the individual element, and is defined for Cr = 2, Cu = Pb = 5, Zn = 1, Cd = 30 and Hg = 40 (Guo, W., et al., 2010).

3.3.3 *Statistical analysis and kriging*

Analysis of variance (ANOVA) was performed using SPSS 13.0 and the soil quality maps of 7 heavy metals spatial distributing pattern were produced using a Kriging interpolation technique in ArcGIS 9.3.

4 RESULTS AND DISCUSSION

4.1 *Spatial distribution of heavy metals and comparison of contents with guide values*

The total heavy metal contents in the soil of the OFC area in the north of Xuzhou decrease in the order of Zn, Cr, Cu, Pb, As, Cd and Hg, as listed in Table 1. Mean levels of all the analyzed metals in the soil are higher than the background values except for Zn and As. The studied area lies in the Yellow River floodplain. Flooding soil results in reduction of sulphates and formation of volatile acids, and subsequently leads to the release and availability of heavy metals during a long period of time, which explains the reduce of Zn and As (Chen, H., et al.,2015,Marrugo-Negrete, J., et al., 2017). Anthropogenic activities are another reason of heavy metals accumulation in soil.

The spatial distributions of the heavy metals, Cu, Cd, Pb, Zn, Cr, Hg and As, are shown in Figure 2. The spatial distributions of Cu, Pb and Zn are similar in the studied area. Higher Cu, Cr, Pb and As are mainly found around the fly ash yards near Jiahe Coal Mine (east of the area) and the new

Table 2. Basic statistics of PLI and PER of soil.

	CF								Er							
	Cu	Cd	Pb	Zn	Cr	Hg	As	PLI	Cd	Cr	Cu	Hg	Pb	Zn	As	PER
Average	0.29	3.08	0.07	0.3	0.31	0.11	0.36	2.2	101.84	2.22	10.84	226.03	6.65	0.82	9.92	358.32
Max	0.77	6.95	0.15	0.7	0.51	0.87	0.86	5.00	215.69	3.03	30.42	1746.60	14.04	1.94	25.79	1860.55
Min	0.11	0.62	0.04	0.2	0.2	0.01	0.04	0.5	35.64	1.65	4.50	29.60	3.79	0.49	1.00	102.61

Table 3. Grade standards for Er and PER.

Potential ecological risk index of individual element	Ecological risk index	Potential ecological index	Grades of potential ecological risk
<40	Low	150	low-grade
40–80	Moderate	150–300	Moderate
80–160	Higher	300–600	Severe
160–320	Much higher	>600	Serious
>320	Serious		

industry area near Huamei Electric Plant (southwest). The distributions of Cd and As are similar to the higher heavy metals detected along the major roads in the area (north). The high content of Hg exists in the fly ash yards (east) and Zhangxiaoluo Coal Mine (west). These results might indicate that atmospheric deposition from coal mining and traffic are the primary sources of soil contamination.

4.2 *Assessment of soil contamination*

4.2.1 *Pollution Load Index*

The PLI values are ranged from 0.50–5.00 with an average of 2.20. PLI values equal to 0 indicates perfection; values of 1 indicate baseline levels of pollutants present and value above 1 indicate progressive deterioration (Suresh, G., et al., 2011), values above 3 indicate heavy pollution. 12.4% soil sample locations were below 1. 87.6% soils are polluted as the PLI values are higher than 1 and twenty-six sample locations are heavily polluted as PLI values above 3. Higher PLI value is observed at the area near Zhangxiaolou coal mine, which is similar as the CF_{Cd}. 43.9% sample locations are heavily polluted and 28.0% moderately polluted by Cd, whereas the contents of the other six trace elements are safely below the detectable level of pollution.

4.2.2 *Potential ecological risk*

The contamination degree of the seven metals is of the order, Hg > Cd > Cu > As > Pb > Cr > Zn. Combining the potential ecological risk index of the individual element (Er) (Table 2) with the grade classification (Table 3), Hg in the present soil shows much higher potential ecological risk, Cd higher ecological risk, and other metals low ecological risk. Coal burning is one of the major sources of Hg pollution. More than 95% Hg can be quickly absorbed or fixed by clay minerals and organic matters after depositing in the soil, which raises its accumulation in soil (Chen, X., et al., 2016).

The calculated PER values are ranged from 102.61–1860.55 with an average of 358.32 (Table 2). The highest PER vale is observed at the soil site of 45, which is in a cropland close to a construction material factory using fly ash. Only three samples are of low PER. More than half samples are of severe risk, and five of serious risk.

5 CONCLUSION

Contents and spatial distributions of the metals, Cu, Cd, Pb, Zn, Cr, Hg and As, are analyzed for the soil of the FCO areas in the north of Xuzhou. The mean levels of all the metals in the soil are higher than the background values except for Zn and As. Pb, Zn, Cu and Cr are mainly spotted in the southwest and east of the region, particularly around the fly ash yards near Jiahe Coal Mine and new industry areas near Huamei Electric Plant. Cd and As are more spotted along the major roads in the area, and Hg exists in the fly ash yards and Zhangxiaoluo Coal Mine. The results of PLI show that 12.4% soil sample locations are not polluted and twenty-six locations are heavily polluted. Higher PLI value is observed in the area near Zhangxiaolou Coal Mine, which is similar to the CFCd. The potential ecological risk index contamination degree of the seven metals is of the order, Hg > Cd > Cu > As > Pb > Cr > Zn. Hg in the present soil shows much higher potential ecological risk and Cd higher ecological risk. The calculated PER values are ranged from 102.61–1860.55 with an average of 358.32. More than half of sample locations are in severe risk, and five in serious risk.

ACKNOWLEDGEMENTS

This study was supported by Natural Science Foundation of China, 41671524.

REFERENCES

Bhuiyan, M.A., et al., (2010). Heavy metal pollution of coal mine-affected agricultural soils in the northern part of Bangladesh. *Journal of Hazardous Materials*, 173(1–3), 384–392.

Chen, H., et al., (2015). Contamination features and health risk of soil heavy metals in China. *Science of the Total Environments* 512–513, 143–153.

Chen, X., et al., (2016). Speciation and distribution of mercury in soils around gold mines located upstream of Miyun Reservoir, Beijing, China. *Journal of Geochemical Exploration*, 16, 1–9.

Guo, W., et al., (2010). Pollution and Potential Ecological Risk Evaluation of Heavy Metals in the Sediments around Dongjiang Harbor, Tianjin. *Procedia Environmental Sciences*, 2(1), 729–736.

Halim, M.A., R.K. Majumder, & M.N. Zaman (2015). Paddy soil heavy metal contamination and uptake in rice plants from the adjacent area of Barapukuria coal mine, northwest Bangladesh. *Arabian Journal of Geosciences*, 8(6), 3391–3401.

Hu, Z., et al., (2014). Farmland damage and its impact on the overlapped areas of cropland and coal resources in the eastern plains of China. *Resources Conservation & Recycling*, 86(3), 1–8.

Huang, S., et al., (2009). Assessment of selenium pollution in agricultural soils in the Xuzhou District, Northwest Jiangsu, China. *Acta Scientiae Circumstantiae*, 21(4), 481–487.

Marrugo-Negrete, J., J. Pinedo-Hernández & S. Díez (2017). Assessment of heavy metal pollution, spatial distribution and origin in agricultural soils along the Sinú River Basin, Colombia. *Environmental Research*, 154, 380–388.

Singh, S., N.J. Raju & S. Nazneen (2015). Environmental risk of heavy metal pollution and contamination sources using multivariate analysis in the soils of Varanasi environs, India. *Environmental Monitoring & Assessment*, 187(6), 4577.

Suresh, G., et al., (2012). Assessment of spatial distribution and potential ecological risk of the heavy metals in relation to granulometric contents of Veeranam lake sediments, India. *Ecotoxicology and Environmental Safety*, 84(10), 117–124.

Suresh, G., et al., (2011). Influence of mineralogical and heavy metal composition on natural radionuclide concentrations in the river sediments. *Applied Radiation & Isotopes Including Data Instrumentation & Methods for Use in Agriculture Industry & Medicine*, 69(10), 1466–1474.

Zarić, N.M., et al., (2016). Metal concentrations around thermal power plants, rural and urban areas using honeybees (Apis mellifera L.) as bioindicators. *International Journal of Environmental Science and Technology*, 13(2), 413–422.

Zhang, M.M., X. Xiao & M. Tan, (2013). Study on the Soil Heavy Metals Pollution Evaluation in the Northern Suburb of Xuzhou City. *Northern Horticulture*, 6, 177–179.

Zhao, C. & H. Wu (2017). Density functional investigation of mercury and arsenic adsorption on nitrogen doped graphene decorated with palladium clusters: a promising heavy metal sensing material in farmland. Applied Surface Science, 399, 55–66.

Land Reclamation in Ecological Fragile Areas – Hu (Ed.)
 ISBN 978-1-138-05103-4

The influence analysis of the heavy metals in the fly ash from the Da Tun power plant to the soil and plant around the fly ash yard

K. Zhu
Tian di Science and Technology Co. Ltd., Beijing, China

ABSTRACT: Through field sampling, laboratory tests and comparative analysis to relevant national standards, the research of the influence of the heavy metals in fly ash, which was stored for 1 year and 10 years, to soil and plant around the fly ash yard was done. The research result show that: the influence of Da Tun power plant fly ash to pH and heavy metals in soil was not obviously. The content of heavy metals in soil was even higher than fly ash; The content of heavy metals in root, stem and leaf, fruit of the plant which was grown on fly ash was decreasing, and the content was all lower than the limit value which was regulated in relevant national standards. The influence of the Da Tun power plant fly ash to soil and plant around the fly ash yard was low.

Keywords: fly ash; heavy metals; soil; plant

1 RESEARCH BACKGROUND AND SAMPLE INTRODUCTION

As an industrial waste, the fly ash was usually used as a filling material in reclamation of subsided area in Eastern China. However, because of many heavy metal element was contained in fly ash and some of them, such as As, Cd, Cr and Hg, could make bad effect on crops, and would impact on food safety. The study of the effect of the heavy metal element in fly ash to soil and plant was meaningful.

The fly ash, surrounding soil and plant sample in Da Tun power plant, NO.1 and NO.2 fly ash storage yard had been taken by project team. The fly ash sample was all produced by Da Tun power plant. The fly ash in Da Tun power plant was produced within one day, and the fly ash in NO.1 storage yard was produced within one year, and the fly ash in NO.2 storage yard was produced within 8 years. The soil sample was taken in 10 meters around the NO.1 and NO.2 storage yard. The plant sample, including beans, squash and sorghum, was growing on fly ash in NO.2 storage yard for over 2 years.

2 RESEARCH METHODS AND LABORATORY TESTING

2.1 *Research methods*

This research test the heavy metal element in fly ash sample to study the attenuation condition of heavy metal; This research test the heavy metal element in soil which was in 10 meters around the storage yard and the common soil to study the influence of fly ash to soil; This research test the heavy metal element in plants to study the influence of fly ash to plants. This research made a comparative analysis those heavy metal content to the national standard to study the pollution degree of heavy metal.

2.2 *Laboratory testing*

The methods of laboratory testing were referred to GB5085, GB/T15555 and CJ/T221-2005. There were 5 repeats in every test and the average value would be used. The analytical spectral line of every element was: As:188.98 nm, Cd:228.80 nm, Cr:267.71 nm, Cu:324.75 nm, Hg:194.16 nm, Ni:231.60 nm, Pb:182.14 nm, Zn:206.20 nm. Every element was set a minimum standard line and if the test value was lower than it, the test value would be "not detected". The minimum standard line of Hg was 0.001 mg/L(mg/kg) and the other elements were 0.01 mg/L(mg/kg).

3 DATA ANALYSIS

This research made comparison to the heavy metal element in three fly ash sample which was produced in different time, and made comparison to the heavy metal element in soil 10 meters around the storage yard. All test results were made comparison to the limit value in GB15618-2008 and HJ/T332-2006, and the effect of heavy metal element in

Table 1. The comparison of heavy metal in fly ash and soil.

Heavy metal	Fly ash within 1 day	Fly ash within 1 year	Fly ash within 8 day	Soil around NO.1 yard	Soil around NO.2 yard	Common soil	GB15618-2008 HJ/T332-2006
As	7.19	6.88	8.12	13.39	7	13.44	20
Cr	–	–				–	–
Cd	13.56	8.14	11.23	23.32	13	23.43	250
Cu	29.03	18.11	23.56	21.23	11	21.65	100
Hg	–	–				–	+0.44
Ni	11	21	9.12	21.36	13	20.96	60
Pb	11	4	5.33	9.120	6	9.190	50
Zn	37	23	19.26	51.36	23	38.12	300
pH	8.78	8.69	8.81	9.06	8.52	8.18	

Table 2. The comparison of heavy metal in fly ash and soil.

Heavy metal	Sorghum roots	Sorghum stems	Sorghum fruits	Soybean roots	Soybean stems	Soybean fruits	Pumpkin roots	Pumpkin fruits	GB2762-2012
As	–	–	–	–	–	–	–	–	0.5
Cr	–	–	–	–	–	–	–	–	0.1
Cd	–	–	–	–	–	–	–	–	0.5
Cu	3.12	3.11	6.23	4.02	4.31	1.11	3.19	1.01	–
Hg	–	–	–	–	–	–	–	–	0.01
Ni	–	–	–	–	–	–	–	–	1
Pb	–	–	–	–	–	–	–	–	0.1
Zn	24.35	12.11	16.45	6.99	18.02	9.01	8.94	5.02	–

fly ash on soil in open storage conditions could be reach. The detection result was list in Table 1.

Through the data in Table 1, the heavy metal element in fly ash in open storage condition did not significantly reduce. Only a few of heavy metal in the soil around the fly ash storage yard was lower than common soil. All test results were lower than the limit value in national standard.

This research text the heavy metal element in the roots, stems and fruits of the plants, and made comparison with the limit value in GB2762-2012. The detection result was list in Table 2.

Through the data in Table 2, the heavy metal element in the roots, stems and fruits of those plants which were growing on fly ash was low. The heavy metal in roots was morn than stems and morn than fruits.

4 RESULTS AND DISCUSSION

The background value of heavy metal in soil of DaTun in PeiXian was a bit high, and the heavy metal element in fly ash of Da Tun power plant was not easy to transfer, so the influence of the heavy metal in fly ash of Da Tun power plant to soil was low.

Heavy metals in fly ash were not easily absorbed by plants, so the influence of the heavy metal in fly ash of Da Tun power plant to plants was low.

From the angle of heavy metal pollution, the fly ash was used in land reclamation was feasible in Da Tun.

Land Reclamation in Ecological Fragile Areas – Hu (Ed.)

Research on the correlation between land use/land cover changes and Lyme disease in Maryland

Y.L. Gong, Y.J. Wu & Y.X. Du
College of Geoscience and Surveying Engineering, China University of Mining and Technology, Beijing, China

ABSTRACT: The purpose of this paper is to study the correlation between land use/land cover changes and the disease. To explore which kinds of land type variation can influence the disease transmission. We used ENVI and ArcGIS technology and combined with the data correlation and the method of geospatial overlay analysis to do data processing and results analysis that are the three period remote sensing images which about 2001, 2011, 2006 in Maryland. Results showed that: time scale, the relationship between Lyme disease and forest wetland and pasture land was negative correlation, but it positively correlated with bare land, grassland and waters. Spatial scale, Lyme disease mainly distributes in the western and southern areas in Maryland. There is spatial correlation between the farmland, wetland and the waters. We concluded that through the spatial-temporal distribution: waters and wet grasslands provide are so good habitat for ticks that the probability of Lyme disease spreading is becoming large. By increasing the contact frequency between human beings and animals makes ticks are more likely to spread disease, thus increasing the incidence of Lyme disease.

Keywords: Land Use/Land Cover Changes, Lyme disease, Spatial Distribution, Correlativity, Morbidity of Disease

1 INTRODUCTION

Land use/land cover changes are the core and hot topics in the global changes research. One thing, land use/land cover changes influence the natural basis of human existence and development, such as climate, soil, vegetation, water resources, bio-diversity and so on (Overmars, KP, 2005, 2006). One another thing, land use/land cover changes influence the structure and function of the biogeochemical cycle and energy and matter cycle of the earth system. What is more, it is closely related to global climate change, biodiversity reduction and ecological environment evolution (Turner BL, 1992).

Disease is an important human process and phenomenon in some regions. Land use/land cover changes bring about the change of regional elements and regional patterns and different habitats and geographical processes. And it causes a number of diseases. And it shows a certain spatial and temporal differences in the regional space (Geoghegan J, 2002). Therefore, the combination of disease and geography is also a popular trend. Micro scale, the researches on the disease mainly focus on the influence of the hydrological, climatic and soil factors; macroscopic scale, studies focus on the process of urbanization, population migration and socio-economic impact on the occurrence and spread of the disease.

2 SITE DESCRIPTION

Maryland is located in the northeastern United States and a total of 25 counties. It is rich in natural resources. Forests and open forests cover 40% of Maryland. The south region is the sand, The northeast coastal plain is with fertile land and rich agricultural planting (Sz MveczK, 2007; Coyle BS, 1996). In 2014, Lyme disease is a leptospira infection disease that is based on ticks. The distribution and occurrence have obvious regional and seasonal laws (Frank C, 2002). It has been found in 43 states in USA, but 96% of them come from 14 states, including northeastern Massachusetts, Maryland, west central Wisconsin, Minnesota, northwest California, Oregon and so on. The disease propagation time mainly occurred in June and July in 2014. The immature ticks mainly live in the dark wet soil, grass roots, etc. and parasite on rodents, insectivores, birds and other small animal. Lumberjacks or passers were often bitten by ticks that were on branches or leaves of grasses. So the infection rate of field workers and lumberjacks is high.

Maryland is rich in natural resources (SzMveczK, 2007). Around 40% of Maryland is made up of forests. The south is the sand, the northeast coastal plain is fertile land, agricultural development (Coyle BS, 1996). In 2014, there are 1373 cases of Lyme diseases in Maryland that accounts

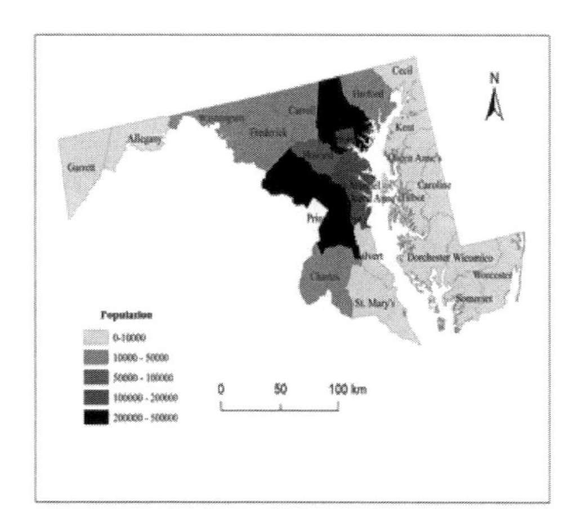

Figure 1. Land use cover map in Maryland.

for 4% of all the people. The average incidence of 23.78/10 million people, the incidence is high. So the study area is very typical. Figure 1 is distribution map of Maryland population.

3 MATERIAL AND METHODS

Based on ENVI and GIS technology, the remote-sensing images were classified. The land use cover map and the disease distribution map were obtained that were in 2001, 2006, 2011.

Land use/land cover changes are based on the NLCD classification system in 1992. Using ENVI5.1 to correct and handle the TM remote sensing images. Based on the method of unsupervised classification method and combined with the actual situation in Maryland to analysis of the outcome of the maps (Fig. 2 a, b, c). These maps including 15 land cover types. Study areas locate in the temperate plain region, so there are no permanent glaciers and snow cover types. Forest land and farmland were the main land use cover type in Maryland. The proportion of forest land and farmland were 31% and 30% respectively.

The incidence of Lyme disease in Maryland was obtained by the U.S. Centers for Disease Control and Prevention. Extraction of incidence data for three periods that about 2001, 2006, 2016 all three years in Maryland County. Importing administrative map by data overlay and corresponding with every county. According to the Lyme events happened every year to grade. We divided into five ranks for every one hundred thousand persons and they are 0–5, 5–50, 50–200, 200–500, 500-, and got the distribution map.

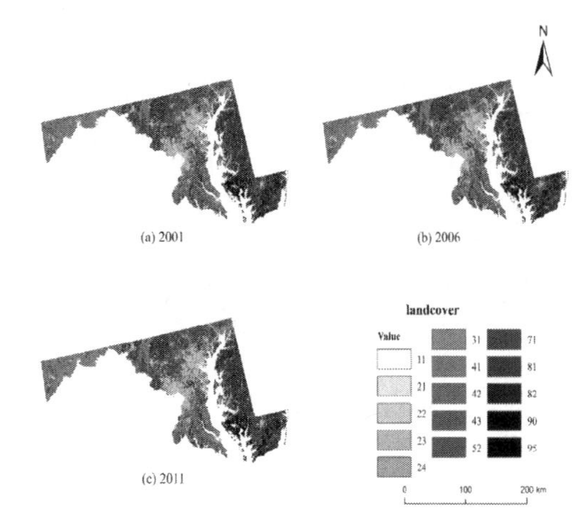

Figure 2. Land use cover map in Maryland.

4 DATA ANALYSIS

4.1 *Data correlation analysis*

Through Pearson correlation analysis found that wetland and grassland is negatively correlative with the occurrence of Lyme. However bare land, grassland and forages land is positively correlative with the occurrence of Lyme. Ticks are carriers of Lyme disease. And the larvae prefer to live in damp, dark soil or grass roots. So grassland, wetland, forest and farmland are benefit for the ticks living. Because of the change of forest land is not obvious, so it is difficult to form a strong correlation with Lyme disease. However others land types changed significantly, especially the increase of grassland and water area provides more suitable habitat for the survival of ticks. Affected by human and beast activity the area of forest, wetland and forages land is running down. With the increase of human or animal contacting with tick survival environments, which led to the occurrence probability of Lyme disease.

Therefore, there was obvious correlation between the land cover factors and Lyme disease there by means of correlation analysis. As for what kind of land cover types to produce more obvious effects on Lyme disease, we need to analyze the spatial distribution pattern of Lyme disease to do further analysis.

4.2 *Geospatial analysis*

The geographic distribution of Lyme disease in Maryland can be found (Table 2). The number of per one hundred thousand people that in Kent, Calvert, Somerset, Dorchester and several other counties was higher. These counties mainly distributed in the western and southern regions of

Table 1. The relationship of land use and land cover type changes and the incidence of Lyme disease.

Land-cover types	2001–2006	2006–2011	2001–2011	Pearson relevance	Significant (bilateral)
Open Water	0.42%	0.07%	0.49%	0.998	0.040
Open Space	0.69%	0.58%	1.27%	0.695	0.511
Low Intensity	1.95%	1.45%	3.43%	0.764	0.447
Medium Intensity	11.79%	5.95%	18.45%	0.896	0.293
High Intensity	10.50%	6.17%	17.32%	0.852	0.350
Barren Land	22.11%	5.44%	28.75%	0.984	0.114
Deciduous Forest	−1.05%	−0.81%	−1.85%	−0.749	0.461
Evergreen Forest	0.30%	−3.16%	−2.86%	0.479	0.682
Mixed Forest	−1.84%	−2.07%	−3.88%	−0.498	0.668
Shrub	−4.56%	8.73%	3.77%	−0.718	0.490
Grassland	65.75%	11.79%	85.28%	0.987	0.101
Pasture	−3.30%	−0.54%	−3.82%	−0.999	0.030
Cultivated Crops	0.94%	−0.69%	0.24%	0.857	0.345
Woody Wetland	−1.21%	0.09%	−1.13%	−0.987	0.101
Emergent Herbaceous Wetlands	0.56%	0.96%	1.52%	0.196	0.874
incid	93.01%	5.62%	103.86%	1.000	

Table 2. Geographic distribution of Lyme disease in Maryland.

ID	NAME	incid2001	incid2006	incid2011
1	Allegany	3	7	29
2	Anne Arundel	36	106	160
3	Baltimore	1	1	4
4	Calvert	487	561	655
5	Caroline	663	172	3132
6	Carroll	107	135	96
7	Cecil	115	75	83
8	Charles	141	405	322
9	Dorchester	435	879	1100
10	Frederick	90	50	89
11	Garrett	50	74	116
12	Harford	40	193	200
13	Howard	1	1	1
14	Kent	1719	2282	2277
15	Montgomery	35	47	117
16	Prince George's	16	9	13
17	Queen Anne's	719	980	2581
18	St. Mary's	328	300	264
19	Somerset	658	854	835
20	Talbot	170	183	281
21	Washington	26	23	22
22	Wicomico	144	109	129
23	Worcester	61	368	629
24	Baltimore City	4	15	15

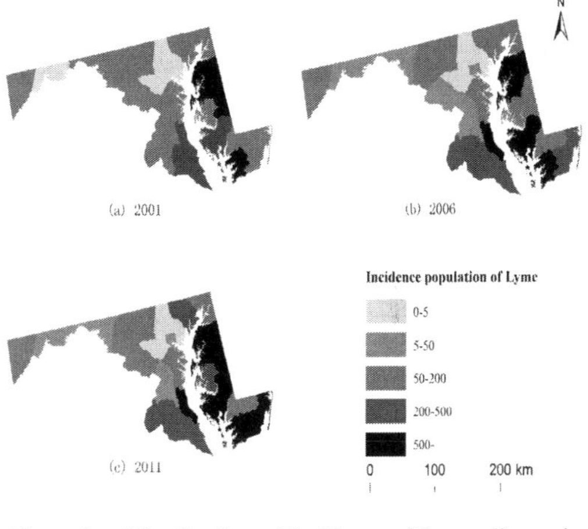

Figure 3. Distribution of incidence of Lyme disease in Maryland.

Maryland. The annual incidence of Lyme disease basically about 2000 on average in the Kent area from 2001 to 2011. It is a high incidence rate.

In order to analyze the relationship between Lyme disease occurrence and spatial distribution among all the counties, we combined all two classes to form water land, residential land, bare land, forest land, shrub, grassland, farmland and wetland. We put the land cover classification map and disease distribution map superposition analysis about 2011. We elided the few superposition effect of bare land and grassland. We formed geographic distribution map that was between water land, residential land, forest land, shrub, farmland, wetland and Lyme disease occurrence. We concluded from the Figure 4(a-f) and found water land, farmland and wetland distribution was more dense area that Lyme disease incidence is also higher. But Lyme disease incidence was lower in the residential land and forest land cluster area. The distribution of shrub land was more dispersed, and the occurrence of the disease could not be regular. It can

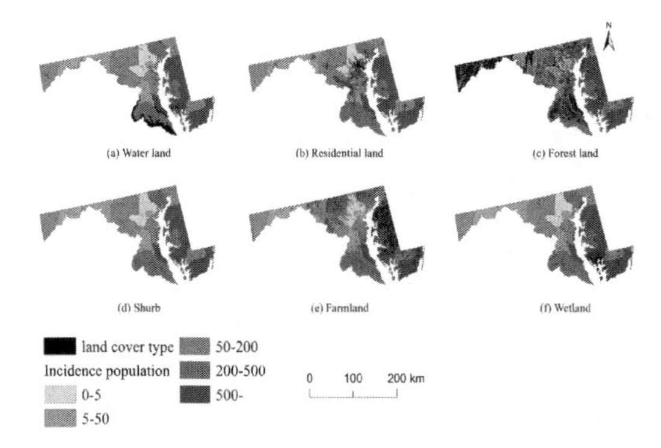

Figure 4. Distribution superimposed of land use/land cover types and Lyme disease.

be concluded that the high incidence of Lyme disease areas are mainly distributed in the water land, farmland and wetland areas.

The incidence of Lyme disease severity and various types of land use/land cover types distribution of 2011 finally concluded that there was a high correlation between water land, farmland and wetland with high intensity and severity of Lyme disease. The main reason is that these types of land cover can provide comfortable living environments for ticks. And in these areas are also human beings often contact areas especially arable land and wetland forest, due to long time contact the incidence of Lyme disease is much higher than that in the construction of intensive land area.

5 SUMMARY

Through spatial-temporal distribution we concluded that water land, forages land, farmland and wetland and their changes are correlative with Lyme disease in Maryland. And the incidence of Lyme was positively correlative with wetland, farmland and grassland. But Lyme disease was negatively correlative with water land. The phenomenon presented that humans and animals continued to contact the forages, grassland and wetland decreased the area, which increased the contact frequency of the ticks that was more easily spread disease. Finally the incidence of Lyme disease was higher.

REFERENCES

Coyle B S, Strickland G T, Liang Y Y. (1996) The public health impact of Lyme disease in Maryland. *Journal of Infectious Diseases*, [Online] 173(5), 1260–1262. Available from: https://oup.silverchair-cdn.com/ [Accessed 10 July 1995].

Frank C, Fix AD. (2002) Mapping Lyme Disease Incidence for Diagnostic and Preventive Decisions, Maryland. *Emerging Infectious Diseases*, [Online] 8(4):427–9. Available from: https://wwwnc.cdc.gov/

Geoghegan J. (2002) The value of open spaces in residential land use. *Land Use Policy*, 19(01):91–98. Available from: https://eurekamag.com/[16 August 2001].

Overmars K P, Verburg P H. (2006) Multilevel modelling of land use from field to village level in the Philippines. *Agricultural Systems*, 89(2–3):435–456. Available from: https://eurekamag.com/[14 October 2005].

Overmars K P, Verburg P H. (2005) Analysis of land use drivers at the watershed and household level: Linking two paradigms at the Philippine forest fringe. *International Journal of Geographical Information Science*, 19(2):125–152. Available from: http://www.tandfonline.com/Szlavecz K, Csuzdi C. (2007) Land use change affects earthworm communities in Eastern Maryland, USA. *European Journal of Soil Biology*, 43(s):79–85. Available from: http://ac.els-cdn.com/ [30 August 2007].

Turner B L, Skole D, Sanderson S. (1992) Land use and land cover change. *Ambio*, 21(23):308–324.

Land Reclamation in Ecological Fragile Areas – Hu (Ed.)
 ISBN 978-1-138-05103-4

Effects of land use change on soil organic carbon in a karst region

J.H. Tong, Y.C. Hu & Y.Y. Li
Department of Land Science and Technology, China University of Geosciences, Beijing, China

ABSTRACT: Soil Organic Carbon (SOC) which is regarded as the material basis for the soil fertility is a hot topic in the current study of the global carbon cycle and climate change problems. Land use change could make the surface coverage change directly, so it is the important driving force for terrestrial ecosystem carbon cycle. The complex natural environment makes the soil extremely sensitive to land use changes in the karst area. On the basis of a lot of reading literature, this paper outlines the influence of land use ways including Forestland, Grassland and Farmland change on SOC in typical karst area, and also analyzes the driving factors of SOC. On this basis, we put forward the research trend which is about the effect of land use change on SOC in the karst area of China.

1 INTRODUCTION

Soil carbon is not only an important carbon source of greenhouse gases, but also a material basis for soil quality. Land use change is an important means of human impact on carbon storage in terrestrial ecosystems and an important driver of carbon cycling. China's karst area is a typical ecologically fragile area, which is particularly sensitive to human activities, especially land use change.

2 EFFECTS OF LAND USE CHANGE ON SOC

The SOC is the result of dynamic equilibrium between input of organic matter and the amount of decomposition loss under action of soil microbes.

2.1 *Effects of land use reclamation on SOC*

Domestic and foreign scholars have done a lot of studies, because of the different land use patterns and research areas of these studies, which led to research results is inconsistent (Dawson, 2007; Davis *et al.*, 2007).

Forest land into grassland or farmland and grassland into farmland, SOC content generally showed a downward trend. Yuan et al., Wu et al. and Chen et al. have similar views (Chen *et al.*, 2015; Li, 2013; Lu *et al.*, 2012). However, most of these studies are based on specific research areas and ecological communities, which making the research results uncertain. China karst area still lacks a comprehensive study of regional scales.

2.2 *Effects of land use restoration on SOC*

Domestic and foreign scholars have also done a lot of research shows that vegetation restoration can be seen as an important carbon sink process (Shi *et al.*, 2010; Dawson, 2007). Huang (2012) and Wu (2016) *et al.* chose typical land use change studies in karst areas and found similar results. However, the problem is that the number of studies in this area is still too small to quantify the entire karst area.

3 INFLUENCING FACTORS OF SOC STORAGE

SOC can be considered as the result of carbon input and emission balance, the size of its reserves by natural factors and human factors such as the impact of many factors.

3.1 *Natural factors*

3.1.1 *Effects of climatic factors on SOC*

Climatic factors, especially the temperature and rainfall can affect the carbon input can also affect the carbon output. Reich *et al.* (2006) and Chen *et al.* (2004) have shown that warming and rainfall, CO_2 enrichment will accelerate the decomposition of soil organic matter and release the carbon stored in the soil into the atmosphere, further exacerbating global climate change. In addition to rainfall and temperature, soil water holding capacity and actual evapotranspiration also have a significant effect on SOC (Jeffery, 1997).

3.1.2 *Effects of vegetation types on SOC*

Usually, the Organic matter into the soil is determined by the vegetation type on the ground part. The most is the forest; into the soil is also the most. Followed by grassland, due to strong human interference, such as crop removal and frequent plowing, not only reduced the amount of litter also accelerated the decomposition of organic matter. On the other hand, vegetation types are different, soil organic matter into the soil in different ways. But there are studies that this law does not apply to all the circumstances (Ritter, 2007). This indicates that there are regional differences in the effects of vegetation factors on SOC.

3.1.3 *Effects of soil physical and chemical properties on SOC*

The effect of soil properties on soil carbon is the most direct, researchers are most concerned about the relationship between soil texture and SOC. Usually, SOC content and soil clay content showed a positive correlation. Also some studies have also shown that clay content cannot explain SOC variation (Percival, et al.). Soil pH, temperature and other factors will also affect the SOC content (Zhao, *et al.*).

3.2 *Human factors*

The impact of human activities includes changes in the use and management of the way.

3.2.1 *Effects of human activities on SOC in farmland*

Fertilization is the main driving factor, not only can change the organic carbon content, but also can affect the composition of SOC (Schmidt, 2011). In particular, the combination of organic fertilizers and inorganic fertilizers can improve SOC storage and stabilize the mineralization process of SOC (Zhao *et al.*, 2006). In addition, the promotion of straw, field tillage and no-tillage will increase soil carbon content in varying degrees. Straw removal or burning, no coverage and excessive use of pesticides can cause a large loss of SOC.

3.2.2 *Effects of human activities on SOC in grassland*

The most important pressure on grassland comes from grazing. Usually, SOC reserves are usually declining (Li *et al.*, 2008), and some studies show that short-term grazing will increase the retention rate of SOC through the experiment and monitoring of temperate grasslands in China, but in the long run, SOC reserves are reduced. Even studies have shown that rational grazing can promote grassland recovery and increase SOC levels (Li *et al.*, 2005); Grassland reclamation is another major factor contributing to the reduction of SOC. Grassland reclamation promotes soil respiration, accelerates the release of SOC, and reduces the amount of carbon supplied at the primary level, resulting in a decline in SOC reserves.

3.2.3 *Effects of human activities on SOC in forest*

In general, original balance was broken after forest was cut down, SOC reserves will be a clear downward trend. The appropriate intensity of logging, not only maintain the system balance, but also conducive to the transformation of internal factors within the system. The impact of fire on forests has received increasing attention. Different effects of fire are also different; low-intensity fire in a longer time scale will not lead to SOC reduction, high intensity fire will appear in varying degrees of decline (Johnson & Curtis, 2001).

In addition to the above factors, elevation, slope and other topographic factors and the research scale will have an impact on the results of the study (Bai *et al.*, 2015).

4 RESEARCH PROSPECTS

First of all, domestic and foreign scholars have studied the different scales of global, national, regional and ecological communities, but many studies are based on specific research fields and ecological communities, and the research results are not uniform. Therefore, a comprehensive study of regional scales should be strengthened.

Secondly, People are more concerned about the change in land reclamation than land use recovery. Therefore, we have to spend enough attention to it in the future.

At last, the impact of land use change on SOC is extremely complex. We have to establish a multi-factor impact model, to enhance the interaction of multi-factor research in the future.

ACKNOWLEDGMENT

Fund project: National Natural Science Foundation of China. "Study on the Interaction Effect and Regulation Mechanism of Land Use and Ecosystem in Karst Ecological Resettlement Area in Southwest China (41171440)."

REFERENCES

Chen, Q.S., Li, L.H., Han, X.G. et al. (2004) Adaptation of Soil Respiration to Temperature. Journal of Ecology, 24(11), 2649–2655.

Huang, Z.S. (2012) Characteristics of soil mineralizable carbon pool during natural restoration of karst for-

est vegetation. Chinese Journal of Applied Ecology, 23(8), 2165–2170.
Jeffery. (1997) Spatial patterns in SOC pool size in the Northwestern United States [J]. Boca Raton: CRC Press, 29–44.
Jiang, Y.J. (2005) Effects of Land Use Change on Soil Properties in Typical Karst Agricultural Region: A Case Study of Xiao Jiang River Basin. Journal of Geography, 60(5), 751–760.
Li, C.L., Zhao, M.L., Han, G.D. et al. (2008) Characteristics of SOC and Its Relationship with Vegetation in Desert Steppe under Different Grazing Pressure. Resources and Environment of Arid Land, 22(5), 134–138.
Lu, L.X., Song, T.Q., Peng, W.X. et al. (2012) Profile Distribution of SOC in Soil of Karst Peak—Cluster Depression. Journal of Applied Ecology, 23(5), 1167–1174.
Percival, Parfitt, Scott. (2000) Factors controlling soil carbon levels in New Zealand grasslands: Is clay content important? [J]. Soil Science Society of America Journal. 64(5), 1623–1630.
Reich, Russell, Kitayama. (2006) Temperature influences carbon accumulation in moist tropical forests[J]. Ecology, 87(1): 76–87.
Ritter. (2007) Carbon, nitrogen and phosphorus in volcanic soils following afforestation with native birch and introduced larch in Iceland. Plant and Soil, 295(1/2), 239–251.
Schmidt. (2011) Persistence of soil organic matter as an ecosystem property. Nature, 478(7367), 49–56.
Weng, B.Q. (2013) Advances in Effects of Vegetation Restoration on Soil Carbon and Nitrogen Cycle. Journal of Applied Ecology, 24(12), 3610–3616.
Wu, J.L., Liu, M.Y., Zhao, G.Q. et al. (2016) Land Use Effects on SOC Mineralization and Greenhouse Gas Emission in Loess Plateau. Journal of Agricultural Environmental Science, 15(5), 1029–1036.
Yuan, H.W., Su, Y.R., Zheng, H. et al. (2007) Characteristics of SOC and Nitrogen Distribution in Different Land Use Types in Karst Peak—Cluster. Chinese Journal of Ecology, 26(10), 1579–1584.
Zhao, M.S., Zhang, G.L., Li, D.C. et al. (2013) Variation of Soil Organic Matter and Its Main Influencing Factors in Jiangsu Province. Journal of Ecology, 33(16), 5058–5066.
Zhao, X, Yu, W.T., Li, J.D. et al. (2006) Research Progress of SOC and Its Components under Different Management and Management Conditions. Journal of Applied Ecology, 17(11), 2203–2209.

Land Reclamation in Ecological Fragile Areas – Hu (Ed.)

Adsorption performance investigation of heavy metals in acid mine drainage by the microbial method

Y.J. Zhu
Beijing General Research Institute of Mining and Metallurgy (BGRIMM), Beijing, China

ABSTRACT: Compared with traditional methods such as neutralization precipitation, and sulfide precipitation, microbial method has many advantages like low cost, high efficiency, wide pH range and small secondary pollution, recyclable etc. The study aimed at low pH and high concentration of heavy metals ions in the acidic mining wastewater, through the screening of strains, strengthen the resistance to acid, heavy metal irons, a plant under the condition of strong acid to remove heavy metal irons has higher ability of the strains come out, and analyzed the factors to affect the strain of cooper removal efficiency, primarily explore the mechanism of the bacteria to remove heavy metal irons, provide theoretical support for the microbial method.

1 INTRODUCTION

In the process of the ore mining, the sulfide mineral reacts with the air, water, and microbe, occurring a series of physical and chemical reactions, such as leaching, oxidation, hydrolysis, forming a low PH sulfuric acid-high iron sulfate solution, and a variety of metal ions were released, which mainly content Cu^{2+}, Zn^{2+}, Pb^{2+} and Cd^{2+} etc[1]. The toxicity of heavy metals in wastewater not only pollute the water seriously, effect the yield and quality of crop, but also enrich and expand in the food chain very easily, finally do harm to human health by accumulating in some organ cause chronic intoxication[2]. Therefore, seeking a economic and practical method to eliminate the harm based on the characters of the acidic mining wastewater, guaranteeing the sustainable development of mineral resources development has received much attention from the government and the society[3].

2 THE MATERIALS AND METHODS

2.1 *The materials*

The sludge used to screen strains taken from a certain copper mine waste water.

Enrichment medium: beef extract 5 $g{\cdot}L^{-1}$, peptone 10 $g{\cdot}L^{-1}$, NaCl 5 $g{\cdot}L^{-1}$, pH 7.2–7.4;

Selective medium: saccharose 5 $g{\cdot}L^{-1}$, $(NH_4)_2SO_4$ 1 $g{\cdot}L^{-1}$, K_2HPO_4 2 $g{\cdot}L^{-1}$, $MgSO_4{\cdot}7H_2O$ 0.5 $g{\cdot}L^{-1}$, NaCl 0.1 $g{\cdot}L^{-1}$, $CaCO_3$ 0.5 $g{\cdot}L^{-1}$, $CuSO_4{\cdot}5H_2O$ 0.39 $g{\cdot}L^{-1}$ (Cu^{2+} 100 $mg{\cdot}L^{-1}$), agar 20 $g{\cdot}L^{-1}$, pH = 5–6;

Domesticated culture medium: saccharose 5 $g{\cdot}L^{-1}$ $(NH_4)_2SO_4$ 1 $g{\cdot}L^{-1}$, K_2HPO_4 2 $g{\cdot}L^{-1}$, Mg $SO_4{\cdot}7H_2O$ 0.5 $g{\cdot}L^{-1}$, NaCl 0.1 $g{\cdot}L^{-1}$ yeast powder 0.5 $g{\cdot}L^{-1}$, agar 20 $g{\cdot}L^{-1}$ change the copper ion concentration in the medium by $CuSO_4{\cdot}5H_2O$, adjust medium PH by 0.1 $mol{\cdot}L^{-1}$ H_2SO_4 and 0.1 $mol{\cdot}L^{-1}$ NaOH.

2.2 *Test method*

According to the character of the mining acid wastewater, the study design simulation Cu^{2+}100 $mg{\cdot}L^{-1}$, pH = 3.

Sludge in sterile water is collected after shocks, take on the supernatant enriching, The Cu^{2+} concentration was 20 $mg{\cdot}L^{-1}$, 40 $mg{\cdot}L^{-1}$, 60 $mg{\cdot}L^{-1}$, 80 $mg{\cdot}L^{-1}$, 100 $mg{\cdot}L^{-1}$ medium plate coating, culturing 24 h, choose the train which can adapt to high concentration Cu^{2+}. Purify and separate the strains obtained after preliminary screening, decreased the pH value of medium 7, 6, 5, 4, 3. By measuring the residual concentration of the bacteria concentration and Cu^{2+} in liquid, the strain that toleranced to low pH and has high copper ion removal ability was selected out named strain M-1.

Through the single factor experiment, research on the type and dosage of the carbon source and nitrogen source of the strain and the other effect elements. Cultivation wastewater was 100 $mg{\cdot}L^{-1}$ Cu^{2+}, pH = 3, the certain conditions: carbon source, nitrogen source, temperature 30°C, shaker speed 140 r/min, inoculation amount 2 $g{\cdot}L^{-1}$, culture time 24 h.

2.3 *Analysis method*

Biomass: measured by spectrophotometer bacterium liquid OD_{600}, drawing the growth curve[6].

Heavy metal concentration of Cu^{2+} was determined by double cyclohexanone oxamide two hydResult and discussion.

3 RESULT AND DISCUSSION

3.1 *The efficiency on removal rate of Cu^{2+} of the carbon source of the strain*

Carbon provide body the energy needed to sustain life activities in the microbial growth process, and they were one of the essential nutrients of microbial[7]. Different kinds of microorganisms is selective for carbon source, the using ability is also diffeent. Therefore the choice of carbon, nitrogen source is not only conducive to microbial growth, so as to improve the copper ion removal rate, and suitable carbon dosage can effectively reduce the operation cost, and more beneficial to industrial application of microbial treatment in mine acid wastewater.

Choicing sodium acetate as the carbon source. And observe the effect of sodium acetate dosage on the growth of strain M-1 and removal rate of Cu^{2+}. Inoculate 10% M-1 strain into the simulated wastewater of sodium acetate with different concentrations, Under the condition of 30°C, 140 r/min and 24 h adsorption time, measure the concentration of the bacteria concentration and Cu^{2+} in the liquid, the results is shown as Fig. 1.

As you can see from Fig. 1, when the concentration of sodium acetate increases, the growth of the M-1 strain and the removal rate of Cu^{2+} rises as well. When the sodium acetate concentration is $3g{\cdot}L^{-1}$ and $5g{\cdot}L^{-1}$, removal rate Cu^{2+} is 46.8% and 49.2%. Considering the very small removal rate difference and the operation cost, choose the best concentration of sodium acetate is $3\ g{\cdot}L^{-1}$.

3.2 *Effect of pH on the strain ability of adsorbing Cu^{2+}*

For most biological adsorption processes, the pH value is the decisive factor to heavy metal adsorption. Due to the low pH of the mine acid wastewater s, mostly less than 7, so the test set the pH of the waste water at 1.0, 2.0, 3.0, 4.0, 5.0, 6.0, and 100 $mg{\cdot}L^{-1}$ Cu^{2+}, the inoculation amount of the strain is $2\ g{\cdot}L^{-1}$, under the condition of 30°C, 140 r/min and 24 h adsorption time, results are shown in Fig. 2.

Fig. 2 shows that the strain concentration increased with the increase of pH, the removal rate of Cu^{2+} was also increased. When $pH < 4$, the removal rate of Cu^{2+} increased obviously with the increase of pH. When $pH = 1$, the Cu^{2+} removal rate of strain M-1 was only 29.6%, when $pH = 4$, the removal rate increased to 51.2%, and when $pH = 6$, the removal rate could increase to 55.1%. The reason why this phenomenon appear can attribute to this: when pH is relatively low, there is competition of adsorption between hydronium ions and heavy metals Cu^{2+}, H^+ occupy plenty of active adsorption sites of the strain, so that the Cu^{2+} and active adsorption sites can not contact, and the adsorption rate decrease, low pH can affect the activity of enzyme, so the bacterial metabolism is limited and bacteria growth is affected at the same time. With the increase of pH and beyond the bacterial surface isoelectric point, the amount of negative charge on cell surface increased and the activity of the active group on bacteria surface enhanced, which is good at the adsorption of Cu^{2+}.

3.3 *Effect of adsorption time on the adsorption performance of Cu^{2+}*

In the wastewater of 100 mg/L Cu^{2+}, $pH = 3$, inoculation amount 2 g/L, 30°C, 140 r/min oscillating adsorption time are 1, 2, 4, 6, 8, 10 h, the test results are shown in Fig. 3.

It can be seen from the graph, in the 4 h of the start of the reaction, the removal rate of strains on heavy metal Cu^{2+} increased rapidly with time passed, and after 4 h, the removal rate increases slowly, tends to be stable, and it can reach 46.2%. The possible reason is that the adsorption of the M-1 strain to Cu^{2+} is mainly by surface, and it

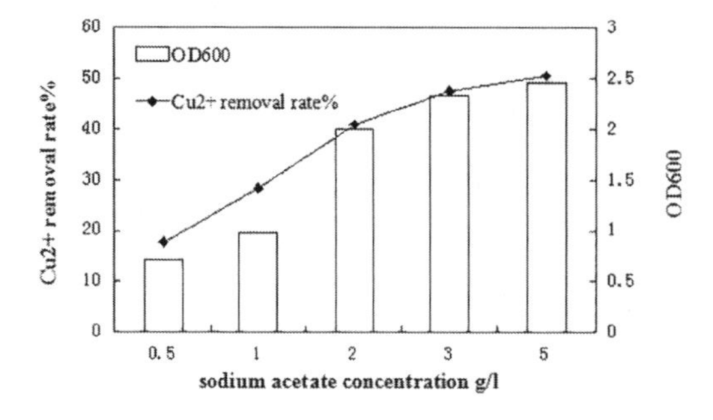

Figure 1. Effect of the mass concentration of sodium acetate on the growth of the strain M-1 and removal effeciency of Cu (II).

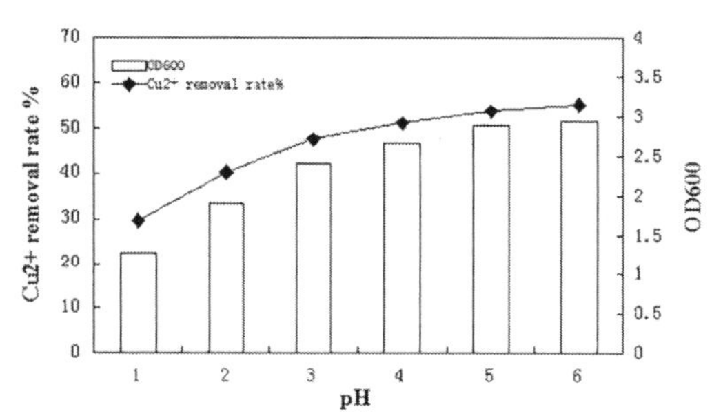

Figure 2. Effect of pH on the growth of the strain M-1 and removal effeciency of Cu (II).

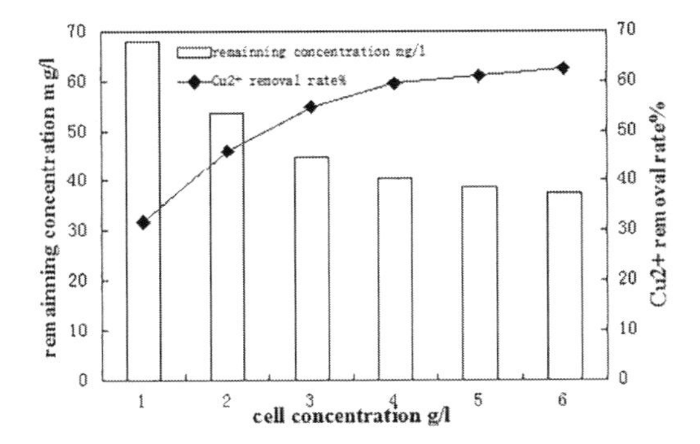

Figure 3. Effect of adsorption time on the biosorption of Cu (II) with the strain M-1.

doesn't rely on cell metabolism but rely on the combination of Cu^{2+} with amino, carboxyl, hydroxyl and other chemical groups on cell wall. This process was rapidly[4]. Therefore, the best adsorption time of this experiment is 4 h.

4 CONCLUSION

1. The test select ore dressing wastewater drainage from a copper ore to separate some microorganism which can resistant of copper. By selecting, separating and the concentration of Cu^{2+}, the pH of wastewater domesticated, select a strain M-1 which can absorb Cu^{2+} efficiently.
2. The test ensure the best carbon source of strains M-1 is sodium acetate, the best dosage is 3 $g \cdot L^{-1}$.
3. In condition that the microbial is 4 $g \cdot L^{-1}$, temperature is 30°C, the time of absorption is 4 h, The best absorption efficiency to waste water for strains M-1 is pH = 3. This time, the removal rate of Cu^{2+} in liquor at 59.7%.

REFERENCES

Chibber, P.K. & Majumdar, S.K. (1999). Foreign ownership and profitability: Property rights, control, and the performance of firms in Indian industry. *J. Law & Economics*, 42 (1), 209–238.

Donghee P, Yun Y S, Jong M P. (2010). The past, present and future trends of biosorption. *J. Biotechnology and Bioprocess Engineering*. 15:86–102.

Fang Di, Wang Fang (2010). Ecology and Environmental Science. *J.* 19(3):562–565.

Lasat M M, Baker A J M, Kochian L V. (1996). Physiological characterization of root Zn^{2+} absorption and translocation to shoots in Zn hyper accumulator and nonaccumulator species of thlaspi. *J. Plant Physiol*, 112:1715–1722.

Wang J L, Chen C. (2006). Biosorption of heavy metals by Saccha-romyces cerevisiae: A review. *J. Biotechnology Advances*, 24(5): 427–451.

Land Reclamation in Ecological Fragile Areas – Hu (Ed.)
 ISBN 978-1-138-05103-4

Comparative study of ecological restoration technology of a mining brownfield

Y.X. Feng
Institute of Mechanics and Engineering, China University of Mining and Technology, Beijing, China

ABSTRACT: There are soil remediation, water remediation and phytoremediation that included in ecological rehabilitation technologies of mining brown field. Soil remediation technologies contain physical methods, chemical methods, microbiologic methods and Plant methods. Water ecological remediation can be realized by natural filtration cycle, chemical treatment which combined with project and reclaimed water treatment and soon. Phytoremediation contain vegetation buffer, hard rock reclamation affore station and hyper accumulation plants on the contaminated land. Comparing with China and foreign countries to find the gap and the best technical solution, by designing ecological remediation technology, artificial nature can be reconstructed on mining brownfield, and renaissance of brownfield can be realized.

1 INTRODUCTION

Mining brownfield is the land that destroyed by mining activities, and which couldn't be used unless doing some remediation. USBN define the mining brownfield as: "the idle or waste mining and exploration area without modification." In China, it's always called mining wasteland, which means "the land is destroyed or polluted during the course of mining, washing and processing, and become to unserviceable land unless being governed."[1] Mining brownfield is a kind of post-industrial landscape. In this paper, by comparing research at home and abroad in the view of ecological landscape plan and design which depends on ecological security tectonic, he emphasis is exploring optimized strategies about ecological landscape restoration technology on the mining brownfield.

2 COMPARATIVE STUDY

2.1 *Ecological restoration technology*

2.1.1 *Soil restoration*

From the soil restoration cases, we can summarize the technologies of soil restoration include physical method, chemical method, microbiologic method and plant method. Physical method usually excavates and moves the polluted soil, and backfill the fresh soil. Chemical method makes the pollutant in soil have chemical reaction with chemical reagent to wipe off pollutant. Well, microbiologic method is using active biological enzyme to degrade the hazardous substance. And the plant method is pointedly planting on the polluted soil, by the plant growing to degrade the hazardous substance. According to the different type of soil pollution to choose corresponding solution, by comparing with the Seattle gasworks and the Huangshi mine reclamation, it's obviously to show that physical method could have effect in the shortest time, but, for deep pollution, backfill is not just costly, but also leaving the invisible pollutant. Well, chemical method is costly and easily causing secondary pollution. Microbiologic method and plant method are adaptive methods in mining brownfield restoration. The combine with microbiologic and plant is also costly, and the biology has life cycles, can't be used frequently and in short term, but it can guarantee the ecological security and avoid secondary pollution.

2.1.2 *Water restoration*

The ecological restoration on water have technologies such as natural filtration cycle, combine with chemical treatment and project and reclaimed water treatment. The less pollution level water in mining brownfield could be degraded by natural filtration cycle. The self-circulation sediment in water could interaction with aquatic flora and fauna and microbiologic to play the water self-cleaning effect adequately. The treatment that combine with chemical and project means putting chemical reagent into the water, and turning the pollutant into sediment by chemical conversion, then reaching purification standard by filtration cycle. Reclaimed water treatment means pumping polluted water into filtration cycle system, and the waste water which reach to government standard after filtration purification could be used as industrial water again.

Tanshan south lake park takes means such as return green to the hillock to prevent seepage from hillock to pollute water environment.[2] Removing the 3.5 million m^3 coal gangue and 8.0 million m^3 coal ash to improve the soil. At the same time, using the surface runoff adequately, using reclaimed water recharge to guarantee water saturation, and running the ecological system. The main function of this kind of ecological wetland is purifying water quality and soil, and it's suitable to be built in medium and mild subsidence area (the depth of water above 0.5 m). The researches show that, the depth of water about 0.5 m should plant emergent aquatic plant to make "the reed wetlands", by the plant's absorbing function to remove or reduce the heavy metalion, organic pollutant, bacteria and virus in water, and finishing the deep purification treatment for industrial waste water and sewage. While, the water area could plant some submerged plant to purify water when the depth of water is above 1.0 m.

Emscher river in Germany takes comprehensive governance methods, combining with flood control, pollution discharge and the landscape transformation on ecological riverways, bringing the site selection on canals and wetlands into completed plan, remoulding the existing facilities for reutilization in the progress of riverway naturalization and landscape design, in order to turn the unaesthetic facilities such as drainage pipelines into a part of new landscape at Emscher river. It's particularly worth mentioning that during the Emscher river ecological wetlands' construction process, implementing the water edges softening technology, dismantling the scleroid revetments and adding sinuosity of water, that is useful to increase ecological diversity of aquatic organism. This kind of winding water's edge is helpful to play the self-purification, and achieve the ecological recover aim. This kind of ecological restoration technology shows great ecological effect.

2.1.3 *Plant restoration*

The plant regeneration is mainly used in the site pollution in mining brownfield ecological landscape restoration. The main aim to prevent pollution after the mining brownfield halt production is to prevent the water pollution, especially the surface water flowing, the soil and water loss is the main carrier of brownfield pollutant. The source of pollution mainly are tailing heap, hillock, slagheap and so on, the only effective way for this kind of site to restore the water is setting plant buffer. According to the research, when the width of plant buffer is greater than 10 m, the totally interception rate could be above 80% for silt passing by, and adsorbing the dissolved pollutant more than 50%, and this should be regard as theoretical reference value of the buffer.[3]

The plant's protection and regeneration should be superior to protect the remaining species, then according to the current situation to find adaptive local species to plant. In the seriously polluted area where couldn't grow plants anymore, it's necessary to construct artificial nature to promote ecological recover quickly.

Contraposing the mining brownfield which contains exceeding heavy metal, selecting plants to restore ecology is the most economic and the most ecological method. Technology is selecting hyper accumulation plant, which can degrade the organics and enrichment itself, such as ciliate desert-grass, leucaena glauca and so on. There are more than 400 types plants that gathering heavy metal have been discovered, including 320 more types for nickel, 20 types for selenium, 30 types for cobalt, 14 types for plumbum and 34 types for copper. Nearly 200 types plants could grow and settle down on different types of tailing heap. Poaceae and solanaceae plants have strong tolerance for wicked environment such as lead-zinc tailings, some annual and biennial plants such as cogon, pennisetum, adiantum and so on could be chosen for pioneer plant. This kind of plant has strong adaptation, not only the strong viability, but also the ability to adsorb heavy metal and decompose organics, after several cycle, it has obvious effect to improve the heavy metal in soil.

2.1.4 *Reclamation afforestation on hard rock*

In the reclamation process of Huangshan mine, because of the hard diorite and marble and its' bad water binding capacity, it's very hard for reclamation. The reclamation is better to choose adaptable local plants to improve soil and to stabilize plant site. The local plants are growing on the original bare soil freely, generally speaking, it will become steady biocoenosis after life cycle. The local plants have features like fitting the local climate well and short cycle, it also could improve the soil, so it's better to keep the original plants in mining brownfield. The robinia pseudoacacia forest in Huangshi is a successful example.

3 SUMMARY

By analyzing and comparing with ecological restoration technology cases at home and abroad, we can know that there are many ecological restoration technology applications on mining brownfield landscape ecological design, however, there is definite technology gap on almost every project. Thus, based on mining brownfield, under the purport of ecological restoration in mining brownfield, it's indicating the direction of landscape planning design theory and practical exploration by artificial nature ecological landscape restoration technology, which by means of ecological restoration technology to rebuild artificial nature on mining brownfield.

REFERENCES

Gemmel. Plant settlement on industrial waste land [M] Ni pengnian Trans, Beijing, Beijing, Science Press, 1987:6.

Haoweiguo (2010). Reincarnation of the old mining Area of kailuan—The Planning and Construction of Tangshan Nanhu Ecological Park[J]. *Landscape Design*, (38): 96–98.

Uusi KJ, Braskerud B, Jansson H, et al. (2000) Buffer-zones and construcyed wetlands as filtes for agricul-turalphosphorus [J]. *J. Environ Qual*, (29): 151–158.

Land Reclamation in Ecological Fragile Areas – Hu (Ed.)
© 2017 Taylor & Francis Group, London, ISBN 978-1-138-05103-4

Author index